Autonomous Mobile Robots: Control, Planning, and Architecture

1951-1991
40 YEARS OF SERVICE

IEEE COMPUTER SOCIETY
A member society of the
Institute of Electrical and Electronics Engineers, Inc.

AUTONOMOUS MOBILE ROBOTS

S. Sitharama Iyengar
Alberto Elfes

1951-1991
IEEE Computer Society Press

The Institute of Electrical and
Electronics Engineers, Inc.

Autonomous Mobile Robots:
Control, Planning,
and Architecture

S.S. Iyengar and Alberto Elfes

1951-1991

IEEE Computer Society Press
Los Alamitos, California

Washington ● Brussels ● Tokyo

IEEE COMPUTER SOCIETY PRESS TUTORIAL

Library of Congress Cataloging-in-Publication Data

Autonomous mobile robots / edited by S.S. Iyengar and Alberto Elfes.
    p.   cm.
   Includes bibliographical references.
   Contents: v. 1. Perception, mapping, and navigation -- v.
2. Control, planning, and architecture.
   ISBN 0-8186-2018-8 (pbk. : v. 1). -- ISBN 0-8186-9018-6 (case : v. 1). -- ISBN
   0-8186-2116-8 (pbk. : v. 2). -- ISBN 0-8186-9116-6 (case : v. 2)
  1. Mobile robots. I. Iyengar, S. Sitharama. II. Elfes, Alberto. TJ211.415.A87 1991
629.8'92--dc20   CIP 91-13280

Published by the
IEEE Computer Society Press
10662 Los Vaqueros Circle
PO Box 3014
Los Alamitos, CA 90720-1264

IEEE Computer Society Press Order Number 2116
Library of Congress Number 91-13280
IEEE Catalog Number 91EH0342-6
ISBN 0-8186-6116-X (microfiche)
ISBN 0-8186-9116-6 (case)

Additional copies can be ordered from

IEEE Computer Society Press	IEEE Service Center	IEEE Computer Society	IEEE Computer Society
Customer Service Center	445 Hoes Lane	13, avenue de l'Aquilon	Ooshima Building
10662 Los Vaqueros Circle	PO Box 1331	B-1200 Brussels	2-19-1 Minami-Aoyama
PO Box 3014	Piscataway, NJ 08855-1331	BELGIUM	Minato-ku, Tokyo 107
Los Alamitos, CA 90720-1264			JAPAN

Technical Editor: Krishna Kavi
Production Editor: Robert Werner
Copy editing by Henry Ayling
Printed in the United States of America by McNaughton & Gunn, Inc.

Dedication

This book is dedicated to Manorama,
who is both my wife and my best friend,
and also to my children.

S.S. Iyengar

This book is dedicated to my beloved wife Noemia,
and to our children Cristiane and Albert,
precious companions on life's journey.

Alberto Elfes

TABLE OF CONTENTS

duction

Mapping,

mobile

ehicles,

istorical

le robot

otics and

nd sensor

onitoring

building

a context

ned above.

real-world

n than on

echnologies

by robotic

is, the major

investigated

mitantly, the

a result, the

s were made

n mobile robot

of the 1980's,

resurgence was

new round of

gies advanced to

these potential

cles in various

between distant

ine excavation,

es;

mployment of

These potential applications have further motivated advances in robotics-component technologies — particularly in areas of sensing, planning, and control.

A brief historical overview. Robots in general, and mobile robots in particular, have long been a fascinating subject. Developing a mechanism that has a certain degree of mobility and autonomy, and allowing it to interact with the real world, has spurred various efforts. The first mobile-robot projects can be traced back to the early 1970's, when experiments began in the coupling of processors to sensors and to mobile bases. Although a detailed historical review exceeds the scope of this introduction, we will highlight the following representative examples of research projects conducted in the mobile robotics domain:

• **Shakey** — One of the first mobile robots, Shakey was developed at the Stanford Research Institute, Stanford University, California, as part of a project lasting from 1966 through 1972. Shakey had a TV camera, an optical range finder, and several touch sensors. It operated in a highly constrained artificial environment populated with large blocks, and used a precompiled map of its surroundings. The project led to significant developments in development of logic-based planning and problem-solving techniques (resulting in the Strips and Abstrips systems) but did not focus on sensing or real-world modeling issues, thereby producing a system with extremely limited performance.

• **Jason** — Developed at the University of California, Berkeley, Jason had an on-board manipulator, as well as ultrasound range sensors and infrared proximity detectors. A substantial part of Jason's computation was done on board, but experimentation with it seems to have been largely confined to tele-operation work.

• **The JPL Rover** — Since the early 1970's, NASA's Jet Propulsion Laboratory, Pasadena, California, has maintained an ongoing research program to develop planetary explorers. Its first prototype was a semi-autonomous vehicle equipped with a laser range finder, a stereo pair of TV cameras, and a manipulator arm. This vehicle was intended to operate in rough terrain with partial human supervision. The cancellation in 1978 of the originally programmed NASA Mars mission temporarily slowed this research. More recently, new prospects for an unmanned NASA Mars sample-and-return mission in the late 1990's or the first decade of the next century have led to a vigorous research program and the development of a substantially improved semi-autonomous planetary rover.

• **The Stanford Cart** — Built at the Stanford Artificial Intelligence Laboratory, Stanford University, California, the Cart was a simple remotely controlled mobile platform equipped with a camera mounted on a slider mechanism, and a radio transmitter. It used sparse stereo vision for obstacle mapping by tracking sets of points over multiple steps. This milestone project emphasized low-level vision and operation in real-world environments.

• **Hilare** — The Hilare robot — built at the Laboratoire d'Automatique et d'Analyse des Systèmes, CNRS, Toulouse, France — is equipped with ultrasound sensors, a camera, and a triangulation-based laser range system. Research has concentrated on multisensory perception, planning, and navigation in maps represented as graphs.

• **Vesa** — The Vesa robot — developed at the Laboratoire d'Applications des Techniques Electroniques Avancées, of the Institut National des Sciences Appliquées, INRIA, Rennes, France — has a bumper skirt, ultrasound range finders, and an on-board manipulator. Issues in multisensing, mapping, and autonomous path planning and navigation have been investigated.

• **The Yamabiko robots** — The Yamabiko series of mobile robots was developed by Kanayama and Yuta at the Institute of Information Science and Electronics, University of Tsukuba, Japan. One of several Japanese research efforts in autonomous vehicles, the Yamabiko project has developed several mobile robots and explored issues in mobile robot programming, path planning, and navigation.

• **The MRL robots** — Research done at Carnegie Mellon University's Mobile Robot Lab, Pittsburgh, Pennsylvania, has addressed various problems in perception, navigation, and control for autonomous mobile robots. Specific projects have included the design, construction, and operation of several mobile vehicles; formulation of control strategies for wheeled mobile robots; research in sparse and dense stereo vision; range-based mapping, path planning, and navigation; and the investigation of high-level planning and control issues.

• **The Autonomous Land Vehicle Project** — Sponsored by the Defense Advanced Research Projects Agency of the Department of Defense, the autonomous land vehicle project is a multiyear effort that concentrated on road following and rough terrain traversal, using detailed stored maps and relying on sensor data for simple local navigation and obstacle avoidance. Our references provide a sample of specific research issues and the systems implemented.

• **The Hermies Robots** — The Hermies series of robots — developed at the Oak Ridge National Laboratory's Center for Engineering Systems Advanced Research, Oak Ridge, Tennessee — incorporate sonar and vision sensors, manipulators mounted on a mobile platform, and on-board processing, and have been used for research in visual perception, goal recognition, navigation in unknown dynamic environments, and the development of task-oriented manipulation strategies.

• **The VaMoRs Robot** — The VaMoRs autonomous vehicle project, developed at the Universität der Bundeswehr, Germany, has developed an integrated spatiotemporal approach to automatic visual guidance of autonomous vehicles.

• **The KAMRO Robot** — The KAMRO project, developed at the Institute for Real-Time Computer Systems and Robotics at the University of Karlsruhe, Germany, focused on the development of a mobile two-armed robot for experiments in autonomous navigation, docking, and assembly.

• **The FRC Robots** — The Field Robotics Center, at the Robotics Institute, Carnegie Mellon University, has concentrated on the development of tele-operated and semi-autonomous robots for large-scale applications, including inspection and decommissioning of nuclear facilities, operation in hazardous environments, mining, and road following.

• **Blanche** — The Blanche mobile robot — an experimental vehicle developed at the AT&T Bell Labs, Princeton, New Jersey, — is designed to operate autonomously within structured environments, including offices and factory areas. It uses an optical range finder, and has been used for experiments in robotic programming languages, sensor integration, and techniques for error detection and recovery.

• **The Stanford Mobi** — An omnidirectional platform equipped with a stereo camera system and sonar sensors, the Mobi vehicle has been used for indoor navigation. Issues in sensor integration, uncertainty handling, recovery of geometric world models, and their interaction with precompiled world maps have been investigated.

• **The MIT Mobots** — Several widely different vehicles have been developed at MIT's Artificial Intelligence Laboratory, Cambridge, Massachusetts, to investigate robust low-level sensing-and-control mechanisms, and their use in layered control architectures.

• **The LSU-RRL** — Research at the LSU Robotics Research Laboratory, Baton Rouge, Louisiana, involves the development of autonomous mobile robots. The laboratory concentrates on key areas, including intelligent sensing and navigation, exploring different methods of sensing and navigation by autonomous mobile robots in unknown environments. Related projects involve manipulator trajectory planning in "noisy" sensory environments (robots learn about the environment through sensors), and parallelization of vision systems.

• **The ARL Robots** — The Autonomous Robotics Laboratory at the IBM T.J. Watson Research Center, Yorktown Heights, New York, has focused on developing robots with agile and robust sensing, navigation, and control mechanisms, and on developing automatic robotic learning strategies.

We can observe certain global historical trends by examining issues addressed and approaches employed in the projects listed above. Research emphasis has slowly shifted from high-level AI symbolic-reasoning-based and problem-solving-based systems to more robust sensor and low-level control-based approaches. This can be construed as a natural consequence of the incremental realization that robotic perception-and-navigation issues are extremely difficult and have not been satisfactorily addressed within the AI framework.

Capabilities required for autonomous mobile robots. What capabilities does an autonomous mobile robot require? In general, using various sensors, a mobile robot needs to acquire and manipulate a substantial model of its operational environment by extracting and interpreting information from the real world. However, different range sensors are subject to intrinsic limitations that can be overcome only by utilizing multisensory systems. Additionally, mobile robots travel over extensive areas and must combine views obtained from many different locations into a single consistent world model reflecting information acquired and hypotheses proposed so far.

Consequently, sensor interpretation and world modeling processes must (1) incorporate information supplied by qualitatively different sensors, (2) cope gracefully with sensor errors and noise, and (3) address uncertainty arising from imprecise knowledge of the robot's location over time. The resulting world model should serve as the basis for crucial operations, including path planning and obstacle avoidance, position and motion estimation, navigation, and landmark identification.

When developing autonomous-mobile-robot systems to operate in such scenarios, a typical list of the sensory and perception capabilities required would include

• Development of adequate mechanisms for sensor interpretation and the modeling of sensor-specific characteristics, including uncertainty and noise;

• Generation of world models from sensory information, leading to robust and useful representations for robotic spatial-reasoning tasks;

• Explicit representation of uncertainty and unknown information in sensor-derived world models;

• Integration of multiple sensory-information sources, provided either by similar or by qualitatively different sensors, into a uniform and coherent description of the robot's environment;

• Composition of sensory information recovered from multiple viewpoints, obtained as the robot moves in its operational environment;

• Handling of robot and sensor position uncertainty caused by vehicular motion, and the incorporation of this uncertainty into the mapping process;

• Incorporation of information from precompiled maps of the robot's world, when such maps are available;

• Prediction and confirmation of possible sensor observations, using previously acquired maps of the robot's environment that may have been generated autonomously by the robot or precompiled by an external source (a human operator, for example);

• Efficient use of sensor-derived world models for spatial reasoning, path planning, obstacle avoidance, landmark recognition, motion solving, and other related navigational and planning problems; and

• Maintenance of maps covering wide geographic areas, as would be required for the representation of large indoor spaces and outdoor terrains.

Components of a mobile robotic architecture. We can identify numerous problem-solving activities as being integral parts of a conceptual autonomous-mobile-robotic architecture, and can classify these activities in three major conceptual areas: *perception,* which encompasses tasks including sensor interpretation, sensor integration, real-world modeling, and recognition; *planning and control,* which performs task-level planning, scheduling, and execution monitoring of overall robotic activity; and *actuation,* which comprises navigational activities, detailed motion and action planning, and actuator control. These various conceptual tasks clearly interact with and depend upon each other. In fact, experience with real systems reveals that complex interconnections and interdependencies exist between the various subsystems, with multiple flows of control and data.

Robotics and computer vision research has addressed the requirements and activities outlined above. Some areas have shown substantial progress, while others are still at a preliminary stage. Path planning, task-level planning, plan execution and monitoring, reactive planning, and manipulator and motion control have received much attention in recent years and have advanced to a relatively mature level when examined against the current overall state of the field. Other areas — most conspicuously, robotic perception and the recovery of robust world models — have only recently started to show more realistic performance.

Chapter 2: Mobile Robot Modeling and Control

The development of kinematic and dynamic models for analyzing and designing robot manipulators was one of the earliest contributions in robotics. These models are essential for robotic trajectory planning and for controlling manipulator motions, and have been fundamental in constructing and deploying reliable and successful industrial robot arms.

Although mobile robotic research dates back to the late 1960's and early 1970's, only recently have similar modeling issues been addressed in the context of mobile platforms. Muir and Neuman report on kinematic modeling of wheeled mobile robots and, in a second paper, discuss dynamic modeling of robotic mechanisms. Both papers present substantial research that parallels work done in robotic manipulators. Careful modeling of vehicle dynamics and vehicular interaction with the road are also an integral part of the road-following work discussed in Chapter 4 by Dickmanns and Graefe.

Our references provide other useful material on mobile robot modeling and control. [1,2]

References

1. P. Muir, *Kinematic and Dynamic Modeling and Control of Wheeled Mobile Robots,* PhD dissertation, Electrical and Computer Engineering Department, Carnegie Mellon University, Pittsburgh, Pa., 1988.
2. *Autonomous Robot Vehicles,* I.J. Cox and G.T. Wilfong, eds., Springer-Verlag, Berlin, 1990.

Kinematic Modeling of Wheeled Mobile Robots*

Patrick F. Muir and Charles P. Neuman
*Department of Electrical and Computer Engineering, The Robotics Institute,
Carnegie Mellon University, Pittsburgh, Pennsylvania 15213*
Received April 22, 1986; accepted November 25, 1986

We formulate the kinematic equations of motion of wheeled mobile robots incorporating conventional, omnidirectional, and ball wheels.[1] We extend the kinematic modeling of stationary manipulators to accommodate such special characteristics of wheeled mobile robots as multiple closed-link chains, higher-pair contact points between a wheel and a surface, and unactuated and unsensed wheel degrees of freedom. We apply the Sheth-Uicker convention to assign coordinate axes and develop a matrix coordinate transformation algebra to derive the equations of motion. We introduce a wheel Jacobian matrix to relate the motions of each wheel to the motions of the robot. We then combine the individual wheel equations to obtain the composite robot equation of motion. We interpret the properties of the composite robot equation to characterize the mobility of a wheeled mobile robot according to a mobility characterization tree. Similarly, we apply actuation and sensing characterization trees to delineate the robot motions producible by the wheel actuators and discernible by the wheel sensors, respectively. We calculate the sensed forward and actuated inverse solutions and interpret the physical conditions which guarantee their existence. To illustrate the development, we formulate and interpret the kinematic equations of motion of Uranus, a wheeled mobile robot being constructed in the CMU Mobile Robot Laboratory.

従来の車輪、全方向式の車輪およびボール式の車輪を含む車輪型移動ロボットの運動学的運動方程式を定式化する。多数の閉リンク機構、車輪と地面の間の多数の接触点、駆動されず計測されない車輪の自由度などの車輪型移動ロボットの特殊な性質を考慮するために静的なマニピュレータの運動学的モデリング法を拡張する。運動方程式の導出のための座標系の設定および行列による座標変換の代数的取扱はSheth-Uickerの慣習に従う。各車輪の運動とロボットの運動を関係ずけるために車輪ヤコビ行列を導入する。次いで、各々の車輪方程式を合わせてロボットの合成運動方程式を得る。車輪型移動ロボットの移動性を調べるために移動特性木によってロボットの合成運動方程式の性質を説明する。同様に、車輪のアクチュエータによって作り出すことのできるロボットの運動と車輪センサによって認識できるロボットの運動を明らかにするためにそれぞれ駆動特性木と計測特性木を適用する。計測の前向解法と駆動の逆解法を与え、それらの存在を保証する物理的条件を説明する。以上の結果を例証するために、CMU移動ロボット研究室で製作中の車輪型移動ロボットUranusの運動方程式を定式化し説明を行なう。

*This research has been supported by the Office of Naval Research under Contract No. N00014-81-K-0503 and the Department of Electrical and Computer Engineering, Carnegie Mellon University.

I. INTRODUCTION

Over the past 20 years, as robotics has matured into a scientific discipline, research and development have concentrated on stationary robotic manipulators,[2,3] primarily because of their industrial applications. Less effort has been directed to mobile robots. Although legged[4] and treaded[5] locomotion has been studied, the overwhelming majority of the mobile robots which have been built and evaluated utilize wheels for locomotion. Wheeled mobile robots (WMRs) are more energy efficient than legged or treaded robots on hard, smooth surfaces[6,7]; and WMRs will potentially be the first mobile robots to find widespread commercial application, because of the hard, smooth plant floors in existing industrial environments. Wheeled transport vehicles, which automatically follow paths marked by reflective tape, paint, or buried wire, have already found industrial applications.[8] WMRs find application in space and undersea exploration, nuclear and explosives handling, warehousing, security, agricultural machinery, military, education, mobility for the disabled, and personal robots.

The wheeled mobile robot literature documents investigations which have concentrated on the application of mobile platforms to perform intelligent tasks,[9,10] rather than on the development of methodologies for analyzing, designing, and controlling the mobility subsystem. Improved mechanical designs and mobility control systems will enable the application of WMRs to tasks in unstructured environments and to autonomous mobile robot operation. A kinematic methodology is the first step toward achieving these goals.

Even though the methodologies for modeling and controlling stationary manipulators are applicable to WMRs, there are inherent differences which cannot be addressed with these methodologies. Examples include:

(1) WMRs contain multiple closed-link chains[11]; whereas stationary manipulators form closed-link chains only when in contact with stationary objects.
(2) The contact between a wheel and a planar surface is a higher-pair joint; whereas stationary manipulators contain only lower-pair joints.[12-14]
(3) Only some of the degrees-of-freedom (DOFs) of a wheel on a WMR are actuated; whereas all of the DOFs of each joint of a stationary manipulator are actuated.
(4) Only some of the DOFs of a wheel on a WMR have position or velocity sensors; whereas all of the DOFs of each joint of a stationary manipulator have both position and velocity sensors.

Wheeled mobile robot control requires a methodology for modeling, analyzing, and design which parallels the technology of stationary manipulators.

Our objective is thus to model the kinematics of WMRs. Kinematics is the study of the geometry of motion.[13] In the context of WMRs, we determine the motion of the robot from the geometry of the constraints imposed by the motion of the wheels. Our kinematic analysis is based upon the assignment of coordinate axes within the robot and its environment, and the application of (4 × 4) matrices to transform between coordinate systems. Each step is defined precisely to lay a solid foundation for the dynamic modeling and feedback control of WMRs. Dynamic models may then be applied to design dynamics-based feedback controllers and simulators. A kinematic methodology may also be applied to design WMRs which satisfy such mobility characteristics as three DOFs (i.e., two translations and a rotation in the plane).

Our kinematic analysis of WMRs parallels the development of kinematics for stationary manipulators. An established methodology for modeling the kinematics of stationary robotic manipulators begins by applying the Denavit-Hartenberg convention[15] to assign coordinate axes to each of the robot joints. Successive coordinate systems on the robot are related by (4×4) homogeneous transformation A-matrices. The A-matrices are specified completely by four characteristic parameters (two displacements and two rotations) between consecutive coordinate systems. Each A-matrix describes both the shape and size of a robot link, and the translation (for a prismatic joint) or rotation (for a rotational joint) of the associated joint. We assign coordinate axes to the steering links and wheels of a WMR, and apply the Sheth-Uicker convention[16] to define transformation matrices. The Sheth-Uicker convention separates the constant shape and size parameters from the variable wheel joint parameters, and simplifies the matrix formulation. The Sheth-Uicker convention allows us to model the higher-pair relationship between each wheel on a WMR and the floor.

The position and orientation in base coordinates of the end-effector of a stationary manipulator is found by cascading the A-matrices from the basic link to the end effector.[17] Velocity and acceleration transformations are found by differentiating the coordinate transformations.[18] Velocities of the individual joints are related to the velocities of the end effector by the manipulator Jacobian matrix[19] in the forward solution. The inverse Jacobian matrix is applied in the inverse solution to calculate the velocities of the joint variables from the velocities of the end effector. We develop the wheel Jacobian matrix to relate the velocities of each wheel on a WMR to the robot body velocities. Since WMRs are multiple closed-link chains, the forward and inverse solutions are obtained by solving simultaneously the kinematic equations of motion of all of the wheels.

In this paper, we advance the kinematic modeling of WMRs, from the motivation of the kinematic methodology through its development and application.[1] In Section II, we describe the three wheels (conventional, omnidirectional, and ball wheels) utilized in all existing and foreseeable WMRs. Then, in Section III, we develop our approach for the kinematic modeling of WMRs. In Section IV, we adjoin the equations of motion of all of the wheels to formulate, solve, and interpret the solutions of the composite robot equation of motion. We apply, in Section V, our kinematic modeling methodology to Uranus, a WMR being constructed in the Mobile Robot Laboratory of Carnegie Mellon University. Finally, in Section VI, we provide concluding remarks and outline our plans for continuing research in dynamic modeling and feedback control of WMRs.

II. WHEEL TYPES

Three wheel types are used in WMR designs: conventional, omnidirectional, and ball wheels. In addition, conventional wheels are often mounted on a steering link to provide an additional DOF. Schematic views of the three wheels are shown in Figure

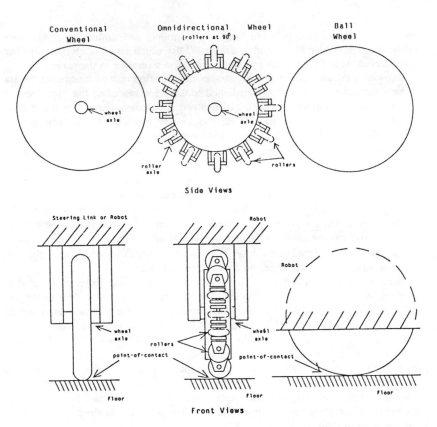

Figure 1. Conventional, omnidirectional, and ball wheels.

1 and the DOFs of each wheel are indicated by the arrows in Figure 2. The kinematic relationships between the angular velocity of the wheel and its linear velocity along the surface of travel are also compiled in Figure 2.

The conventional wheel having two DOFs is the simplest to construct. It allows travel along a surface in the direction of the wheel orientation, and rotation about the point of contact between the wheel and the floor. We note that the rotational DOF is slippage, since the point of contact is not stationary with respect to the floor surface.* Even though we define the rotational slip as a DOF, we do not consider slip transverse to the wheel orientation a DOF, because the magnitude of force required for the transverse motion is much larger than that for rotational slip. The conventional wheel is by far the most widely used wheel; automobiles, roller skates, and bicycles utilize this wheel.

---

*Two bodies are in rolling contact if the points of contact of the two bodies are stationary relative to each other.[14]

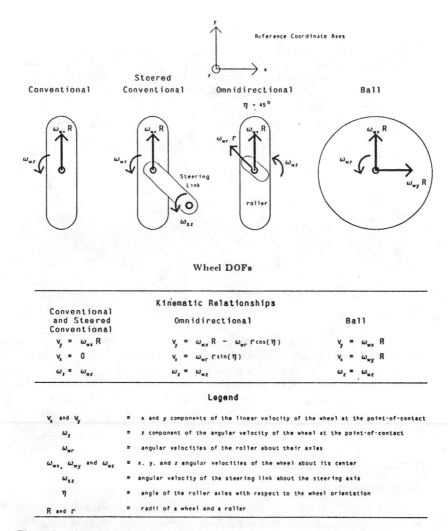

Figure 2. Wheel equations of motion.

The omnidirectional wheel has three DOFs. One DOF is in the direction of the wheel orientation. The second DOF is provided by motion of rollers mounted around the periphery of the main wheel. In principle, the roller axles can be mounted at any nonzero angle η with respect to the wheel orientation. The omnidirectional wheels in Figures 1 and 3 have roller axle angles of 90°[20-22] and 45°,[23] respectively. The third DOF is rotational slip about the point of contact. It is possible, but not common, to actuate the rollers of an omnidirectional wheel[24] with a complex driving arrangement. When sketching WMRs having omnidirectional wheels, the rollers on the underside

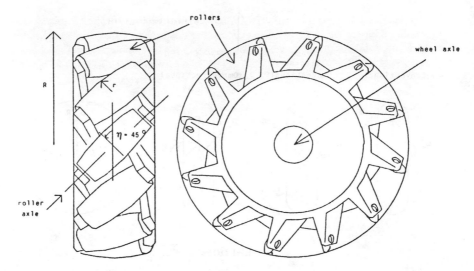

Figure 3. Omnidirectional wheel (rollers at 45°).

of the wheel (i.e., those touching the surface of travel), rather than the rollers which are actually visible from a top view, are drawn to facilitate kinematic analysis.

The most maneuverable wheel is a ball which possesses three DOFs without slip. Schemes have been devised for actuating and sensing ball wheels,[25] but we are unaware of any existing implementations. An omnidirectional wheel which is steered about its point of contact is kinematically equivalent to a ball wheel, and may be a practical design alternative.

III. KINEMATIC MODELING

A. Introduction

In this section, we apply and extend standard robotic nomenclature and methodology[19] to model the kinematics of WMRs. The novel aspects are our treatment of the higher-pair joint between each wheel and the floor, and the development of a transformation

We begin (in Section III B) by defining a WMR and enumerating our modeling assumptions to constrain the class of mobile robots to which our modeling methodology applies. To include all existing and foreseeable WMRs, we would have to generalize our methodology and thereby complicate the modeling of the overwhelming majority of WMRs. In Section III C, we assign coordinate systems to the robot body, wheels, and steering links to facilitate kinematic modeling. It is essential to define instantaneously coincident coordinate systems to model the higher-pair joints at the point of contact between each wheel and the floor. In Section III D, we assign homogeneous (4 × 4) transformation matrices to relate coordinate systems. We present (in Section III E) a matrix coordinate transformation algebra to formulate the equations of motion of a WMR. All kinematics are derived by straightforward application to the axioms and corollaries of the transformation algebra. Position kinematics are treated in Section III F. We demonstrate that transforming the coordinates of a point between coordinate systems is equivalent to finding a path in a transformation graph. Then, in Section III G, we formulate the velocity kinematics. Since the relationships between the wheel velocities and the robot velocities are linear, we develop a wheel Jacobian matrix to calculate the vector of robot velocities from the vector of wheel velocities. Finally, in Section III H, we apply our matrix coordinate transformation algebra to acceleration kinematics.

To summarize the development, we enumerate in Section III I our kinematic modeling procedure. In Section IV, we adjoin the equations of motion of all of the wheels to form the composite robot equation. We then proceed to solve the composite robot equation and interpet the solutions.

B. Definitions And Assumptions

The Robot Institute of America defines a robot as "a programmable, multifunction manipulator designed to move material, parts, tools, or specialized devices through variable programmed motions for the performance of a variety of tasks."[24] Kinematic models of WMRs are inherently different from those of stationary robotic manipulators and legged or treaded mobile robots. We thus introduce an operational definition of a WMR to specify the spectrum of robots to which the kinematic methodology presented in this paper applies.

Wheeled Mobile Robot. A robot capable of locomotion on a surface solely through the actuation of wheel assemblies mounted on the robot and in contact with the surface. A wheel assembly is a device which provides or allows relative motion between its mount and a surface on which it is intended to have a *single* point of *rolling* contact.

Each wheel (conventional, omnidirectional, or ball wheel) and all links between the robot body and the wheel constitute a wheel assembly.

We introduce the following six practical assumptions to make the modeling problem tractable.

Design Assumptions
(1) The WMR does not contain flexible parts.
(2) There is zero or one steering link per wheel.
(3) All steering axes are perpendicular to the surface.

Operational Assumptions
(4) The WMR moves on a planar surface.
(5) The translational friction at the point of contact between a wheel and the surface is large enough so that no translational slip may occur.
(6) The rotational friction at the point of contact between a wheel and the surface is small enough so that rotational slip may occur.

We discuss our assumptions in turn. Assumption 1 states that the dynamics of such WMR components as flexible suspension mechanisms and tires are negligible. We make this assumption to apply rigid body mechanics to kinematic modeling. We recognize that flexible structures may play a significant role in the kinematic analysis of WMRs. A dynamic analysis to determine the changes in kinematic structure due to forces/torques acting on flexible components is required to model these components. Such an analysis is appropriate for WMRs even though it has not conventionally been addressed for stationary open-link manipulators because WMRs are inherently closed-link mechanisms. Flexible components that allow compliance in the multiple closed-link chains of a WMR lead to a consistent kinematic model. Without compliant structures, there cannot be a consistent kinematic model for WMRs in the presence of surface irregularities, inexact component dimensions and inexact control actuation.[26] A simultaneous kinematic and dynamic analysis of WMRs is thus a natural continuation of our research.

We introduce Assumptions 2 and 3 to reduce the spectrum of WMRs that our methodology must address and thereby limit the complexity of our kinematic model. WMRs which have more than one link per wheel can be analyzed by our methodology if only one steering link is allowed to move. We require that all non-steering links are stationary, as if they are extensions of the robot body or wheel mounts. By constraining the steering links to be perpendicular to the surface of travel in Assumption 3, we reduce all motions to a plane. We thus constrain all component motions to a rotation about the normal to the surface, and two translations in a plane parallel to the surface.

Assumption 4 neglects irregularities in the actual surface upon which a WMR travels. Even though this assumption restricts the range of practical applications, environments which do not satisfy this assumption (e.g., rough, bumpy, or rock surfaces) do not lend themselves to energy-efficient wheeled vehicle travel.[7]

Assumption 5 ensures the applicability of the theoretical kinematic properties of a wheel in rolling contact[13,27] for the two translational degrees of freedom. This assumption is realistic for dry surfaces, as demonstrated by the success of braking mechanisms on automobiles. Automobiles also illustrate the practicality of Assumption 6. The wheels must rotate (i.e., slip) about their points of contact to navigate a turn. Since WMRs also rely on rotational wheel slip, we include Assumption 6.

C. Coordinate System Assignments

1. Sheth-Uicker Convention

Coordinate system assignment is the first step in the kinematic modeling of a stationary manipulator.[19] Lower-pair mechanisms* (such as revolute and prismatic joints) function with two surfaces in relative motion. In contrast, the wheels of a WMR are higher pairs which function ideally by point contact. Because the A matrices which model stationary manipulators depend upon the relative position and orientation of

---

*Lower-pair mechanisms are pairs of components whose relative motions are constrained by a common surface contact; whereas higher pairs are constrained by point or line contact.[27]

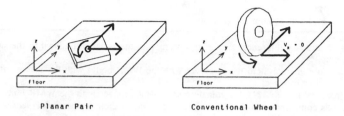

Figure 4. Planar pair model of a wheel.

two successive joints, the Denavit-Hartenberg convention[15] leads to ambiguous assignments of coordinate transformation matrices in multiple closed-link chains[16] which are inherent in WMRs. The ambiguity arises in deciding the joint ordering when there are more than two joints on a single link.

We apply the Sheth-Uicker convention[16] to assign coordinate systems and model each wheel as a planar pair at the point of contact. This convention allows the modeling of the higher-pair wheel motion and eliminates ambiguities in coordinate transformation matrices. The planar pair allows three DOFs as shown in Figure 4: X and Y translation, and rotation about the point of contact. The Sheth-Uicker convention is ideal for modeling ball wheels; the angular velocities of the wheel are converted directly into translational velocities along the surface. The planar pair motions must be constrained to include wheels which do not allow three DOFs. For example, the coordinate system assigned at the point of contact of a conventional wheel is aligned with the y axis parallel to the wheel. The wheel model is completed by constraining the x component of the wheel velocity to zero to satisfy Assumption 5 (in Section III B) and avoid translational slip.

2. WMR Coordinate Systems

We assign coordinate systems at both ends of each link of the WMR. The links of the closed-link chain of a WMR are the floor, the robot body, and the steering links. The joints are a revolute pair at each steering axis, a planar pair to model each wheel, and a planar pair to model the robot body. When the joint variables are zero, the coordinate systems of the two links which share the joint coincide. We summarize our approach to the modeling of a WMR having N wheels with the coordinate system assignments defined in Table I. Placement of the coordinate systems is illustrated in Figure 5 for the pictorial view of a WMR. For a WMR with N wheels, we assign $3N + 1$ coordinate systems to the robot and one stationary reference frame. There are also $N + 1$ instantaneously coincident coordinate systems (described in Section III 3) which need not be assigned explicitly.

The floor coordinate system F is stationary relative to the surface of travel and serves as the reference coordinate frame for robot motions. The robot coordinate system R is assigned to the robot body so that the position of the WMR is the displacement from the floor coordinate system to the robot coordinate system. The hip coordinate system H_i is assigned at the point on the robot body which intersects the steering axis

Table I. Coordinate system assignments.

F   *Floor:* Stationary reference coordinate system with the z-axis orthogonal to the surface of travel.

R   *Robot:* Coordinate system which moves with the WMR body, with the z-axis orthogonal to the surface of travel.

H_i *Hip* (for $i = 1, \ldots, N$): Coordinate system which moves with the WMR body, with the z-axis coincident with the axis of steering joint i if there is one; coincident with the contact point coordinate system C_i if there is no steering joint.

S_i *Steering* (for $i = 1, \ldots, N$): Coordinate system which moves with steering link i, with the z-axis coincident with the z-axis of H_i, and the origin coincident with the origin of H_i.

C_i *Contact Point* (for $i = 1, \ldots, N$): Coordinate system which moves with steering link i, with the origin at the point of contact between the wheel and the surface; the y axis is parallel to the wheel (if the wheel has a preferred orientation; if not, the y axis is assigned arbitrarily) and the x-y plane is tangent to the surface.

$\overline{R}$ *Instantaneously Coincident Robot:* Coordinate system coincident with the R coordinate system and stationary relative to the F coordinate system.

$\overline{C_i}$ *Instantaneously Coincident Contact Point* (for $i = 1, \ldots, N$): Coordinate system coincident with the C_i coordinate system and stationary relative to the F coordinate system.

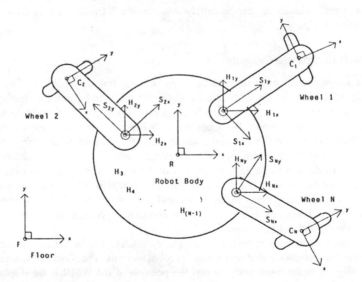

Figure 5. Placement of coordinate systems on a WMR.

of wheel i. The steering coordinate system S_i is assigned at the same point along the steering axis of wheel i, but is fixed relative to the steering link. We assign a contact point coordinate system C_i at the point of contact between each wheel and the floor.

Coordinate system assignments are not unique. There is freedom to assign the coordinate systems at positions and orientations which lead to convenient structures of the kinematic model. For example, all of the hip coordinate systems may be assigned parallel to the robot coordinate system resulting in sparse robot-hip transformation matrices and thus simplifying the model. Alternatively, for WMR control, the x axes of the hip coordinate systems can be aligned with the zero position of the steering joint position encoders so that the hip-steering transformation is expressed in terms of the actual steering angle.

3. Instantaneously Coincident Coordinate Systems

To introduce the concept of instantaneously coincident coordinate systems, we consider the one-dimensional example of a ball rolling in a straight line on a flat surface. The position of the ball is depicted by the point r in Figure 6.

The ball is moving right to left with velocity v_r and acceleration a_r. The stationary reference point $\bar{r}$ lies in the path of the moving ball. At the instant the ball (point r) and the reference (point $\bar{r}$) coincide in Figure 7, we observe that: (1) the position of the ball relative to the reference point $^{\bar{r}}p_r$ is zero; and (2) the velocity $^{\bar{r}}v_r$ and acceleration $^{\bar{r}}a_r$ of the ball relative to the reference point are non-zero. We call the point $\bar{r}$ an instantaneously coincident reference point for the moving ball at the instant shown in Figure 7.

We continuously assign coincident reference point $\bar{r}$ during the motion of the ball to generalize our observations for all time t. The position of the ball relative to its instantaneously coincident reference point is zero (i.e., $^{\bar{r}}p_r(t) = 0$), and the velocity and acceleration of the ball relative to its instantaneously coincident reference point are nonzero (i.e., $^{\bar{r}}v_r(t) \neq 0$ and $^{\bar{r}}a_r(t) \neq 0$). In the framework of instantaneously coincident reference points, we emphasize that we cannot differentiate the position (velocity) equation of motion to obtain the velocity (acceleration) equation of motion.

The stationary reference point f in Figure 7 is a conventional reference point whose position is fixed. Since both reference points f and $\bar{r}$ are stationary, the velocity (acceleration) of the ball relative to the point f is equal to the velocity (acceleration) of the ball relative to the point $\bar{r}$ in this one-dimensional example. Consequently, it

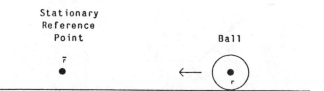

Figure 6. Ball in motion before instantaneous coincidence.

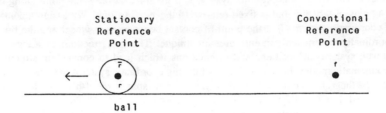

Figure 7. Ball in motion at instantaneous coincidence.

is not advantageous to introduce instantaneously coincident references in the one-dimensional example. The practical need for instantaneously coincident coordinate systems arises in the multi-dimensional example as depicted in Figure 8.

The coordinate system R is moving in three dimensions: X, Y, and θ. The coordinate systems $\bar{R}$ and F are stationary; $\bar{R}$ is an instantaneously coincident coordinate system and F is a conventional reference coordinate system. We make the analogous observations. The position of the moving coordinate system relative to its instantaneously coincident coordinate system is zero (i.e., $^{\bar{R}}p_R = 0$). The position of the moving coordinate system relative to the conventional reference coordinate system is non-zero (i.e., $^Fp_R \neq 0$). The non-zero velocity $^{\bar{R}}v_R$ (acceleration $^{\bar{R}}a_R$) of the moving coordinate system relative to the instantaneously coincident coordinate system is not equal to the velocity ^Fv_R (acceleration ^Fa_R of the moving coordinate system relative to the conventional reference coordinate system. The velocity (acceleration) of the moving coordinate system relative to the conventional reference coordinate system F depends upon the position and orientation of the moving coordinate system relative to the reference coordinate system. The motivation for assigning instantaneously coincident coordinate systems is that the velocities (accelerations) of a multi-dimensional moving coordinate system can be computed or specified independently of the position of the moving coordinate system. The instantaneously coincident coordinate system is a conceptual tool which enables us to calculate the velocities and accelerations of a moving coordinate system relative to its instantaneous current position and orientation.

For stationary serial link manipulators, all joints are one-dimensional lower pairs: prismatic joints allow Z motion and revolute joints allow θ motion. In contrast, WMRs have three-dimensional higher-pair wheel-to-floor and robot-to-floor joints allowing

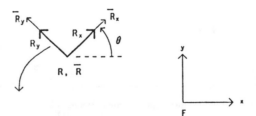

Figure 8. Coordinate system R in motion.

simultaneous X, Y, and θ motions. We assign an instantaneously coincident robot coordinate system $\bar{R}$ at the same position and orientation in space as the robot coordinate system R. In Table I, we define the instantaneously coincident robot coordinate system to be stationary relative to the floor coordinate system F. By design, the position and orientation of the robot coordinate system R and the instantaneously coincident robot coordinate system $\bar{R}$ are identical, but (in general) the relative velocities and accelerations between the two coordinate systems are non-zero. When the robot coordinate system moves relative to the floor coordinate system, we assign a different instantaneously coincident coordinate system for each time instant. The instantaneously coincident robot coordinate system facilitates the specification of robot velocities (accelerations) independently of the robot position. Similarly, the instantaneously coincident contact point coordinate system $\bar{C}_i$ (in Table I) coincides with the contact point coordinate system C_i and is stationary relative to the floor coordinate system. Since the position of the wheel contact point is not sensed, we require the instantaneously coincident contact point coordinate system to specify wheel velocities and accelerations.

D. Transformation Matrices

Homogeneous (4×4) transformation matrices express the relative positions and orientations of coordinate systems.[19] The homogeneous transformation matrix $^{A}\Pi_B$ transforms the coordinates of the point $^{B}\mathbf{r}$ in coordinate frame B to its corresponding coordinates $^{A}\mathbf{r}$ in the coordinate frame A:

$$^{A}\mathbf{r} = {}^{A}\Pi_B \, {}^{B}\mathbf{r}. \tag{1}$$

We adopt the following notation. Scalar quantities are denoted by lower case letters (e.g., w). Vectors are denoted by lower case boldface letters (e.g., $\mathbf{r}$). Matrices are denoted by upper case boldface letters (e.g., Π). Pre-superscripts denote reference coordinate systems. For example, $^{A}\mathbf{r}$ is the vector $\mathbf{r}$ in the A coordinate frame. The pre-superscript may be omitted if the coordinate frame is transparent from the context. Post-subscripts are used to denote coordinate systems or components of a vector or matrix. For example, the transformation matrix $^{A}\Pi_B$ defines the position and orientation of coordinate system B relative to coordinate frame A; and r_x is the x component of the vector $\mathbf{r}$.

Vectors denoting points in space, such as $^{A}\mathbf{r}$ in (1), consist of three Cartesian coordinates and a scale factor as the fourth element:

$$^{A}r = \begin{pmatrix} ^{A}r_x \\ ^{A}r_y \\ ^{A}r_z \\ 1 \end{pmatrix}. \tag{2}$$

We always use a scale factor of unity. Transformation matrices contain the (3×3) rotational matrix ($\mathbf{n} \ \mathbf{o} \ \mathbf{a}$), and the ($3 \times 1$) translational vector $\mathbf{p}$[19]

$$^A\Pi_B = \begin{pmatrix} n_x & o_x & a_x & p_x \\ n_y & o_y & a_y & p_y \\ n_z & o_z & a_z & p_z \\ 0 & 0 & 0 & 1 \end{pmatrix}. \tag{3}$$

The three vector components **n**, **o**, and **a** of the rotational matrix in (3) express the orientation of the x, y, and z axes, respectively, of the B coordinate system relative to the A coordinate system and are thus orthonormal. The three components p_x, p_y, and p_z of the translational vector **p** express the displacement of the origin of the B coordinate system relative to the origin of the A coordinate system along the x, y, and z axes of the A coordinate system, respectively.

The aforementioned properties of a transformation matrix guarantee that its inverse always has the special form:

$$^A\Pi_B^{-1} = \begin{pmatrix} n_x & n_y & n_z & -(\mathbf{p} \cdot \mathbf{n}) \\ o_x & o_y & o_z & -(\mathbf{p} \cdot \mathbf{o}) \\ a_x & a_y & a_z & -(\mathbf{p} \cdot \mathbf{a}) \\ 0 & 0 & 0 & 1 \end{pmatrix}. \tag{4}$$

Before we define the transformation matrices between the coordinate systems of our WMR model, we compile in Table II our nomenclature for rotational and translational displacements, velocities, and accelerations.

In general, any two coordinate systems A and B in our WMR model are located at non-zero x, y and z coordinates relative to each other. The transformation matrix must therefore contain the translations $^A d_{Bx}$, $^A d_{By}$, and $^A d_{Bz}$. We have assigned all coordinate systems with the z axes perpendicular to the surface of travel, so that all rotations between coordinate systems are about the z axis. A transformation matrix in our WMR model thus embodies a rotation $^A\theta_B$ about the z axis of coordinate system A and the translations $^A d_{Bx}$, $^A d_{By}$, and $^A d_{Bz}$ along the respective coordinate axes:

$$^A\Pi_B = \begin{pmatrix} \cos {}^A\theta_B & -\sin {}^A\theta_B & 0 & {}^A d_{Bx} \\ \sin {}^A\theta_B & \cos {}^A\theta_B & 0 & {}^A d_{By} \\ 0 & 0 & 1 & {}^A d_{Bz} \\ 0 & 0 & 0 & 1 \end{pmatrix}. \tag{5}$$

For zero rotational and translational displacements, the coordinate transformation matrix in (5) reduces to the identity matrix.

In Section III F, we apply the inverse of the transformation matrix in (5) to calculate position kinematics. By applying the inverse in (4) to the transformation matrix in (5), we obtain

Table II. Scalar variables.

Scalar Rotational and Translational Displacements

$^A\theta_B$: The rotational displacement about the z-axis of the A coordinate system between the x-axis of the A coordinate system and the x-axis of the B coordinate system (counterclockwise by convention).

$^Ad_{B_j}$: (for $j \in [x,y,z]$): The translational displacement along the j axis of the A coordinate system between the origin of the A coordinate system and the origin of the B coordinate system.

Scalar Rotational and Translational Velocities

$^A\omega_B$: The rotational velocity $^A\dot{\theta}_B$ about the z axis of the A coordinate system between the x axis of the A coordinate system and the x axis of the B coordinate system.

$^Av_{B_j}$: (for $j \in [x,y]$): The translational velocity $^A\dot{d}_{B_j}$ along the j axis of the A coordinate system between the origin of the A coordinate system and the origin of the B coordinate system. Since all motion is in the x-y plane, the z-component $^A\dot{d}_{B_z}$ of the translational velocity is zero.

Scalar Rotational and Translational Accelerations

$^A\alpha_B$: The rotational acceleration $^A\ddot{\theta}_B = {}^A\dot{\omega}_B$ about the z axis of the A coordinate system between the x axis of the A coordinate system and the x axis of the B coordinate system.

$^Aa_{B_j}$: (for $j \in [x,y]$): The translational acceleration $^A\ddot{d}_{B_j} = {}^B\dot{v}_A$ along the j axis of the A coordinate system between the origin of the A coordinate system and the origin of the B coordinate system. Since all motion is parallel to the x-y plane, the z-component $^A\ddot{d}_{B_z}$ of the translational acceleration is zero.

$$^A\Pi_B^{-1} = \begin{pmatrix} \cos\,{}^A\theta_B & \sin\,{}^A\theta_B & 0 & -{}^Ad_{Bx}\cos\,{}^A\theta_B - {}^Ad_{By}\sin\,{}^A\theta_B \\ -\sin\,{}^A\theta_B & \cos\,{}^A\theta_B & 0 & {}^Ad_{Bx}\sin\,{}^A\theta_B - {}^Ad_{By}\cos\,{}^A\theta_B \\ 0 & 0 & 1 & -{}^Ad_{Bz} \\ 0 & 0 & 0 & 1 \end{pmatrix}. \quad (6)$$

In Section III G, we differentiate the transformation matrix in (5) componentwise to calculate robot velocities:

$$^A\dot{\Pi}_B = \begin{pmatrix} -{}^A\omega_B\sin\,{}^A\theta_B & -{}^A\omega_B\cos\,{}^A\theta_B & 0 & {}^Av_{Bx} \\ {}^A\omega_B\cos\,{}^A\theta_B & -{}^A\omega_B\sin\,{}^A\theta_B & 0 & {}^Av_{By} \\ 0 & 0 & 0 & 0 \\ 0 & 0 & 0 & 0 \end{pmatrix}, \quad (7)$$

and in Section III H, we differentiate the transformation matrix in (7) componentwise to calculate robot accelerations:

$$(8)$$

$$^A\ddot{\Pi}_B = \begin{pmatrix} -^A\alpha_B \sin {^A\theta_B} - {^A\omega_B^2} \cos {^A\theta_B} & -^A\alpha_B \cos {^A\theta_B} + {^A\omega_B^2} \sin {^A\theta_B} & 0 & ^Aa_{Bx} \\ ^A\alpha_B \cos {^A\theta_B} - {^A\omega_B^2} \sin {^A\theta_B} & -^A\alpha_B \sin {^A\theta_B} - {^A\omega_B^2} \cos {^A\theta_B} & 0 & ^Aa_{By} \\ 0 & 0 & 0 & 0 \\ 0 & 0 & 0 & 0 \end{pmatrix}.$$

The assignment of coordinate systems results in two types of transformation matrices between coordinate systems: constant and variable. The transformation matrix between coordinate systems fixed at two different positions on the same link is constant. Transformation matrices relating the position and orientation of coordinate systems on different links include joint variables and thus are variable. Constant and variable transformation matrices are denoted by $^A\mathbf{T}_B$ and $^A\mathbf{\Phi}_B$, respectively.[16] In Table III, we compile the transformation matrices in our WMR model. The constant transformation matrices are the floor-robot transformation ($^F\mathbf{T}_{\overline{R}}$), the robot-hip transformation ($^R\mathbf{T}_{\overline{H}_i}$), the steering-contact transformation ($^{S_i}\mathbf{T}_{C_i}$), and the floor-contact transformation ($^F\mathbf{T}_{\overline{C}_i}$). Since the instantaneously coincident coordinate systems $\overline{R}$ and $\overline{C}_i$ are stationary relative to the floor coordinate system, all transformation matrices between the floor coordinate system and the instantaneously coincident coordinate systems are constant. The variable transformation matrices are the robot-robot transformation ($^{\overline{R}}\mathbf{\Phi}_R$), the hip-steering transformation ($^{H_i}\mathbf{\Phi}_{S_i}$), and the contact-contact transformation ($^{\overline{C}_i}\mathbf{\Phi}_{C_i}$). The transformation matrix from a coordinate system to its instantaneously coincident counterpart (or vice versa) is variable because there is relative motion. We compile the first and second time-derivatives of the variable transformation matrices in Tables III and IV, respectively. The matrix derivatives involving instantaneously coincident coordinate systems (i.e., $^{\overline{R}}\dot{\mathbf{\Phi}}_R$, $^{\overline{C}_i}\dot{\mathbf{\Phi}}_{C_i}$, $^{\overline{R}}\ddot{\mathbf{\Phi}}_R$, and $^{\overline{C}_i}\ddot{\mathbf{\Phi}}_{C_i}$) are formed by differentiating and simplifying the elements of the transformation matrices $^{\overline{R}}\mathbf{\Phi}_R$ and $^{\overline{C}_i}\mathbf{\Phi}_{C_i}$, respectively, by substituting $^{\overline{R}}\theta_R = 0$ and $^{\overline{C}_i}\theta_{C_i} = 0$. Because of the simplifying substitutions, the second time-derivative of a transformation matrix involving an instantaneously coincident coordinate system cannot be obtained by differentiating the first time-derivative. Time-derivatives of instantaneously coincident coordinate systems are calculated in Section III E by applying matrix coordinate transformation algebra. The time-derivatives of constant transformation matrices are zero.

For wheels which do not have steering links, the hip and steering coordinate systems are assigned to coincide with the contact point coordinate system, so that the hip-steering and steering-contact transformation matrices reduce to identity matrices and thereby simplify the ensuing kinematic modeling.

E. Matrix Coordinate Transformation Algebra

The kinematics of stationary manipulators are modeled by exploiting the properties of transformation matrices.[18] We formalize the manipulation of transformation matrices in the presence of instantaneously coincident coordinate systems by defining a matrix

Table III Transformation matrices of the WMR model.

Floor–$\bar{R}$obot Transformation: $\quad {}^{F}\mathbf{T}_{\bar{R}} \;=\; \begin{pmatrix} \cos{}^{F}\theta_{\bar{R}} & -\sin{}^{F}\theta_{\bar{R}} & 0 & {}^{F}d_{\bar{R}x} \\ \sin{}^{F}\theta_{\bar{R}} & \cos{}^{F}\theta_{\bar{R}} & 0 & {}^{F}d_{\bar{R}y} \\ 0 & 0 & 1 & {}^{F}d_{\bar{R}z} \\ 0 & 0 & 0 & 1 \end{pmatrix}$

$\bar{R}$obot–Robot Transformation: $\quad {}^{\bar{R}}\boldsymbol{\Phi}_{R} \;=\; \begin{pmatrix} \cos{}^{\bar{R}}\theta_{R} & -\sin{}^{\bar{R}}\theta_{R} & 0 & {}^{\bar{R}}d_{Rx} \\ \sin{}^{\bar{R}}\theta_{R} & \cos{}^{\bar{R}}\theta_{R} & 0 & {}^{\bar{R}}d_{Ry} \\ 0 & 0 & 1 & {}^{\bar{R}}d_{Rz} \\ 0 & 0 & 0 & 1 \end{pmatrix}$

Robot–Hip Transformation: $\quad {}^{R}\mathbf{T}_{H_i} \;=\; \begin{pmatrix} \cos{}^{R}\theta_{R} & -\sin{}^{R}\theta_{H_i} & 0 & {}^{R}d_{H_ix} \\ \sin{}^{R}\theta_{R} & \cos{}^{R}\theta_{H_i} & 0 & {}^{R}d_{H_iy} \\ 0 & 0 & 1 & {}^{R}d_{H_iz} \\ 0 & 0 & 0 & 1 \end{pmatrix}$

Hip–Steering Transformation: $\quad {}^{H_i}\boldsymbol{\Phi}_{S_i} \;=\; \begin{pmatrix} \cos{}^{H_i}\theta_{S_i} & -\sin{}^{H_i}\theta_{S_i} & 0 & 0 \\ \sin{}^{H_i}\theta_{S_i} & \cos{}^{H_i}\theta_{S_i} & 0 & 0 \\ 0 & 0 & 1 & 0 \\ 0 & 0 & 0 & 1 \end{pmatrix}$

Steering–Contact Transformation: $\quad {}^{S_i}\mathbf{T}_{C_i} \;=\; \begin{pmatrix} \cos{}^{S_i}\theta_{C_i} & -\sin{}^{S_i}\theta_{C_i} & 0 & {}^{S_i}d_{C_ix} \\ \sin{}^{S_i}\theta_{C_i} & \cos{}^{S_i}\theta_{C_i} & 0 & {}^{S_i}d_{C_iy} \\ 0 & 0 & 1 & {}^{S_i}d_{C_iz} \\ 0 & 0 & 0 & 1 \end{pmatrix}$

$\bar{C}$ontact–Contact Transformation: $\quad {}^{\bar{C}_i}\boldsymbol{\Phi}_{C_i} \;=\; \begin{pmatrix} \cos{}^{\bar{C}_i}\theta_{C_i} & -\sin{}^{\bar{C}_i}\theta_{C_i} & 0 & {}^{\bar{C}_i}d_{C_ix} \\ \sin{}^{\bar{C}_i}\theta_{C_i} & \cos{}^{\bar{C}_i}\theta_{C_i} & 0 & {}^{\bar{C}_i}d_{C_iy} \\ 0 & 0 & 1 & {}^{\bar{C}_i}d_{C_iz} \\ 0 & 0 & 0 & 1 \end{pmatrix}$

Floor–$\bar{C}$ontact Transformation: $\quad {}^{F}\mathbf{T}_{\bar{C}_i} \;=\; \begin{pmatrix} \cos{}^{F}\theta_{\bar{C}_i} & -\sin{}^{F}\theta_{\bar{C}_i} & 0 & {}^{F}d_{\bar{C}_ix} \\ \sin{}^{F}\theta_{\bar{C}_i} & \cos{}^{F}\theta_{\bar{C}_i} & 0 & {}^{F}d_{\bar{C}_iy} \\ 0 & 0 & 1 & {}^{F}d_{\bar{C}_iz} \\ 0 & 0 & 0 & 1 \end{pmatrix}$

Table IV. Transformation matrix time derivatives.

Robot–Robot:
$$^{\bar{R}}\dot{\Phi}_R = \begin{pmatrix} 0 & -^{\bar{R}}\omega_R & 0 & ^{\bar{R}}v_{Rx} \\ ^{\bar{R}}\omega_R & 0 & 0 & ^{\bar{R}}v_{Ry} \\ 0 & 0 & 0 & 0 \\ 0 & 0 & 0 & 0 \end{pmatrix}$$

Hip–Steering:
$$^{H_i}\dot{\Phi}_{S_i} = \begin{pmatrix} -^{H_i}\omega_{S_i}\sin{}^{H_i}\theta_{S_i} & -^{H_i}\omega_{S_i}\cos{}^{H_i}\theta_{S_i} & 0 & 0 \\ ^{H_i}\omega_{S_i}\cos{}^{H_i}\theta_{S_i} & -^{H_i}\omega_{S_i}\sin{}^{H_i}\theta_{S_i} & 0 & 0 \\ 0 & 0 & 0 & 0 \\ 0 & 0 & 0 & 0 \end{pmatrix}$$

Contact–Contact:
$$^{\bar{C}_i}\dot{\Phi}_{C_i} = \begin{pmatrix} 0 & -^{\bar{C}_i}\omega_{C_i} & 0 & ^{\bar{C}_i}v_{C_ix} \\ ^{\bar{C}_i}\omega_{C_i} & 0 & 0 & ^{\bar{C}_i}v_{C_iy} \\ 0 & 0 & 0 & 0 \\ 0 & 0 & 0 & 0 \end{pmatrix}$$

coordinate transformation algebra. The algebra consists of a set of operands and a set of operations which may be applied to the operands. The operands of matrix coordinate transformation algebra are transformation matrices and their first and second time derivatives (in Section III D). The operations are listed in Table VI as seven axioms. In Table VI, A, B, and X are coordinate systems and Π denotes either a constant T transformation matrix or a variable Φ transformation matrix. Matrix coordinate transformation algebra allows the direct calculation of the relative positions, velocities, and

Table V. Transformation matrix second time derivatives.

Robot–Robot:
$$^{\bar{R}}\ddot{\Phi}_R = \begin{pmatrix} -^{\bar{R}}\omega_R^2 & -^{\bar{R}}\alpha_R & 0 & ^{\bar{R}}\alpha_{Rx} \\ ^{\bar{R}}\alpha_R & -^{\bar{R}}\omega_R^2 & 0 & ^{\bar{R}}\alpha_{Ry} \\ 0 & 0 & 0 & 0 \\ 0 & 0 & 0 & 0 \end{pmatrix}$$

Hip–Steering:
$$^{H_i}\ddot{\Phi}_{S_i} = \begin{pmatrix} -^{H_i}\alpha_{S_i}\sin{}^{H_i}\theta_{S_i} - ^{H_i}\omega_{S_i}^2\cos{}^{H_i}\theta_{S_i} & -^{H_i}\alpha_{S_i}\cos{}^{H_i}\theta_{S_i} + ^{H_i}\omega_{S_i}^2\sin{}^{H_i}\theta_{S_i} & 0 & 0 \\ ^{H_i}\alpha_{S_i}\cos{}^{H_i}\theta_{S_i} - ^{H_i}\omega_{S_i}^2\sin{}^{H_i}\theta_{S_i} & -^{H_i}\alpha_{S_i}\sin{}^{H_i}\theta_{S_i} - ^{H_i}\omega_{S_i}^2\cos{}^{H_i}\theta_{S_i} & 0 & 0 \\ 0 & 0 & 0 & 0 \\ 0 & 0 & 0 & 0 \end{pmatrix}$$

Contact–Contact:
$$^{\bar{C}_i}\ddot{\Phi}_{C_i} = \begin{pmatrix} -^{\bar{C}_i}\omega_{C_i}^2 & -^{\bar{C}_i}\alpha_{C_i} & 0 & ^{\bar{C}_i}a_{C_ix} \\ ^{\bar{C}_i}\alpha_{C_i} & -^{\bar{C}_i}\omega_{C_i}^2 & 0 & ^{\bar{C}_i}a_{C_iy} \\ 0 & 0 & 0 & 0 \\ 0 & 0 & 0 & 0 \end{pmatrix}$$

Table VI. Matrix coordinate transformation algebra axioms.

Identity:	$^A\Pi_B = I \quad for\ B = A\ or\ B = \bar{A}$
Cascade:	$^A\Pi_B = {^A\Pi_X}\,{^X\Pi_B}$
Inversion:	$^A\Pi_B = {^B\Pi_A^{-1}}$
Zero—Velocity:	$^A\dot{\Pi}_B = 0 \quad for\ B = A\ or\ \Pi = T$
Velocity:	$^A\dot{\Pi}_B = {^A\dot{\Pi}_X}\,{^X\Pi_B} + {^A\Pi_X}\,{^X\dot{\Pi}_B}$
Zero—Acceleration:	$^A\ddot{\Pi}_B = 0 \quad for\ B = A\ or\ \Pi = T$
Acceleration:	$^A\ddot{\Pi}_B = {^A\ddot{\Pi}_X}\,{^X\Pi_B} + 2{^A\dot{\Pi}_X}\,{^X\dot{\Pi}_B} + {^A\Pi_X}\,{^X\ddot{\Pi}_B}$

accelerations of robot coordinate systems (including instantaneously coincident coordinate systems).

The identity axiom is self-evident since neither rotations nor translations are required to transform from a coordinate system to itself or to its instantaneously coincident coordinate system. The cascade axiom specifies the order in which transformation matrices are multiplied: the coordinate transformation matrix from the reference system to the destination is the cascade of two coordinate transformation matrices, the first from the reference system to an intermediate coordinate system, and the second from the intermediate coordinate system to the destination. The inversion axiom states that the coordinate transformation matrix from a reference coordinate system to a destination coordination system is the inverse of the coordinate transformation matrix from the destination coordinate system to the reference coordinate system.

Just as the multiplication of transformation matrices is specified by the cascade axiom, time differentiation of transformation matrices is specified by the four velocity and acceleration axioms. Specifically, we cannot differentiate both sides of a matrix transformation equation. For example, if we were to differentiate both sides of the equation $^A\Pi_{\bar{A}} = I$, we would obtain the incorrect result that $^A\dot{\Pi}_{\bar{A}} = 0$ since the velocities between a coordinate system and its instantaneously coincident counterpart are (in general) non-zero. The zero-velocity axiom states that the relative velocities between a coordinate system A and itself $(B = A)$ or another coordinate system assigned to the same link $(\Pi = T)$ are zero. This is because two coordinate systems assigned to the same link are stationary relative to the link and to each other. Similarly, the zero-acceleration axiom states that the relative accelerations between a coordinate system A and itself $(B = A)$ or another coordinate system assigned to the same link $(\Pi = T)$ are zero. The velocity axiom specifies how the time derivative of a transformation matrix may be expressed in terms of the two cascaded transformation matrices and their time derivatives. Finally, the acceleration axiom specifies how the second time-derivative of a transformation matrix may be expressed in terms of the two cascaded transformation matrices and their first and second time-derivatives.

The matrix coordinate transformation axioms in Table VI lead to the corollaries in Table VII, which we apply to the kinematic modeling of WMRs. We develop the instantaneous coincidence corollary by applying the identity and cascade axioms. The instantaneous coincidence corollary simplifies transformation matrix expressions by eliminating the instantaneously coincident coordinate systems. The cascade position corollary calculates the transformation matrix from a reference coordinate system to a destination coordinate system, which may be kinematically separated from the ref-

Table VII. Matrix coordinate transformation algebra corollaries.

Instantaneous Coincidence: $\quad {}^{\bar{A}}\Pi_B = {}^A\Pi_{\bar{B}} = {}^{\bar{A}}\Pi_{\bar{B}} = {}^A\Pi_B$

Cascade Position: $\quad {}^A\Pi_Z = {}^A\Pi_B \, {}^B\Pi_C \, {}^C\Pi_D \ldots {}^Y\Pi_Z$

Cascade Velocity: $\quad {}^A\dot{\Pi}_Z = {}^A\dot{\Pi}_B \, {}^B\Pi_Z + {}^A\Pi_B \, {}^B\dot{\Pi}_C \, {}^C\Pi_Z + \ldots + {}_A\Pi_Y \, {}^Y\dot{\Pi}_Z$

Cascade Acceleration:

$$
\begin{aligned}
{}^A\ddot{\Pi}_Z = \; & {}^A\ddot{\Pi}_B \, {}^B\Pi_Z + {}^A\Pi_B \, {}^B\ddot{\Pi}_C \, {}^C\Pi_Z + \ldots + {}^A\Pi_Y \, {}^Y\ddot{\Pi}_Z \\
& + 2\, {}^A\dot{\Pi}_B [{}^B\dot{\Pi}_C \, {}^C\Pi_Z + {}^B\Pi_C \, {}^C\dot{\Pi}_D \, {}^D\Pi_Z + \ldots + {}^B\Pi_Y \, {}^Y\dot{\Pi}_Z] \\
& + 2\, {}^A\Pi_B \, {}^B\dot{\Pi}_C [{}^C\dot{\Pi}_D \, {}^D\Pi_Z + \ldots + {}^C\Pi_Y \, {}^Y\dot{\Pi}_Z \\
& + \ldots + 2\, {}^A\Pi_X \, {}^X\dot{\Pi}_Y \, {}^Y\dot{\Pi}_Z
\end{aligned}
$$

erence system by a number of cascaded intermediate coordinate systems. The cascade position corollary, which is derived by repeated applications of the cascade axiom, is the foundation of position kinematics (in Section III F). The cascade velocity corollary is derived by repeated applications of the velocity axiom and the cascade axiom. The cascade acceleration corollary is derived by repeated applications of the cascade, velocity, and acceleration axioms. In Sections III G and III H; we apply the cascade velocity and cascade acceleration corollaries to relate linear and angular velocities and accelerations between coordinate systems. Throughout Section III G, we apply the axioms and corollaries of the matrix coordinate transformation algebra to derive the wheel Jacobian matrix.

F. Position Kinematics

We apply the transformation matrices (in Section III D) and the matrix coordinate transformation algebra (in Section III E) to calculate position kinematics. The practical position relations in WMR control require the calculation of the position of a point (e.g., $\mathbf{r}$) relative to one coordinate system (e.g., A) from the position of the point relative to another coordinate system (e.g., Z). For example, we calculate the position of the point mass relative to the floor coordinate system from the position of the point mass in a steering link relative to the steering coordinate system.

We transform position vectors by applying the transformation matrix in (1):

$$ {}^A\mathbf{r} = {}^A\Pi_Z \, {}^Z\mathbf{r}. \tag{9} $$

When the transformation matrix ${}^A\Pi_Z$ is not known directly, we apply the cascade position corollary to calculate ${}^A\Pi_Z$ from known transformation matrices:

$$ {}^A\Pi_Z = {}^A\Pi_B \, {}^B\Pi_C \, {}^C\Pi_D \ldots {}^Y\Pi_Z. \tag{10} $$

We apply transformation graphs to determine whether there is a complete set of known transformation matrices which can be cascaded to create the desired ${}^A\Pi_Z$. In Figure 9, we display a transformation graph of a WMR with one steering link per wheel.

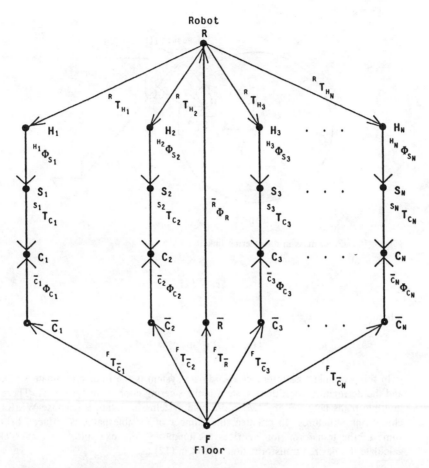

Figure 9. Transformation graph of a WMR.

The origin of each coordinate system is represented by a dot, and transformations between coordinate systems are depicted by directed arrows. The transformation in the direction opposing an arrow is calculated by applying the inversion axiom. Finding a cascade of transformations to calculate a desired transformation matrix (e.g., $^F\Pi_{S_1}$) is thus equivalent to finding a path from the reference coordinate system of the desired transformation (F) to the destination coordinate system (S_1). The matrices to be cascaded are listed by traversing the path in order. Each transformation in the path which is traversed from the tail to the head of an arrow is listed as the matrix itself, while transformations traversed from the head to the tail are listed as the inverse of the matrix.

For example, the point mass in Figure 10 located at position **r** relative to the steering coordinate system S_1 is transformed to its position relative to the floor coordinate system F according to:

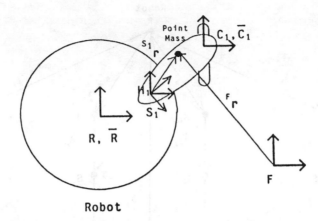

Figure 10. Point mass in the steering link.

$$^F\mathbf{r} = {}^F\Pi_{S_1} \, {}^{S_1}\mathbf{r}, \tag{11}$$

where

$$^F\Pi_{S_1} = {}^F\mathbf{T}_{\bar{R}} \, {}^R\Phi_R \, {}^R T_{H_1} \, {}^{H_1}\Phi_{S_1}. \tag{12}$$

In this example, the reference coordinate system is the floor coordinate system F and the destination coordinate system is the steering coordinate system S_1. There are multiple paths between any two coordinate systems in Figure 9 because WMRs are closed-link structures. In practice, the number of feasible paths is reduced because some of the transformation matrices are unknown. For example, we may seek to calculate the desired transformation matrix in (12) as:

$$^F\Pi_{S_1} = {}^F\Phi_{\bar{C}_1} \, {}^{\bar{C}_1}\Phi_{C_1} \, {}^{C_1}T_{S_1}, \tag{13}$$

but the transformation matrix from the floor to the wheel contact point ${}^F\mathbf{T}_{\bar{C}_1}$ is typically unknown.

G. Velocity Kinematics

1. Introduction

We relate the velocities of the WMR by applying the matrix coordinate transformation algebra axioms and the cascade velocity corollary. In Section III G2, we calculate the velocity of a point (e.g., $\mathbf{r}$) relative to a coordinate system (e.g., A) when the position of the point is fixed relative to another moving coordinate system (e.g., Z). This solution is directly applicable to the dynamic modeling of WMRs for computing the velocity of a differential mass element on the WMR relative to the floor coordinate system. Then, in Section III G3, we apply this methodology to calculate the velocities of the robot relative to the instantaneously coincident robot coordinate system when the velocities of a wheel* are sensed. We introduce the wheel Jacobian matrix to calculate the robot velocity vector from the wheel velocity vector. Finally, we calculate (in Section III G4) the robot velocity vector relative to the floor coordinate system when the robot velocity vector is sensed relative to the instantaneously coincident robot coordinate system.

2. Point Velocities

We differentiate the point transformation in (9) with respect to time to compute the velocity of the point $\mathbf{r}$ in the A coordinate system:

$$^A\dot{\mathbf{r}} = {}^A\dot{\Pi}_Z \,{}^Z\mathbf{r}. \tag{14}$$

When the matrix $^A\dot{\Pi}_Z$ is not known directly, we apply the cascade velocity corollary to calculate $^A\dot{\Pi}_Z$ from known transformation matrices and known transformation matrix time derivatives according to:

$$^A\dot{\Pi}_Z = {}^A\dot{\Pi}_B \,{}^B\Pi_Z + {}^A\Pi_B \,{}^B\dot{\Pi}_C \,{}^C\Pi_Z + \ldots + {}^A\Pi_Y \,{}^Y\dot{\Pi}_Z. \tag{15}$$

For example, (11) relates the position $\mathbf{r}$ of a point mass in the steering coordinate system S_1 to its position in the floor coordinate system F. We calculate the velocity of the point $\mathbf{r}$ relative to the floor coordinate system by differentiating (11):

$$^F\dot{\mathbf{r}} = {}^F\dot{\Pi}_{S_1} \,{}^{S_1}\mathbf{r}. \tag{16}$$

Since the vector $^{S_1}\mathbf{r}$ is constant, its time derivative is zero. We apply the cascade velocity corollary and the WMR transformation graph to obtain an expression for the unknown transformation matrix derivative in (16):

$$^F\dot{\Pi}_{S_1} = {}^F\dot{T}_{\bar{R}} \,{}^{\bar{R}}\Pi_{S_1} + {}^F T_{\bar{R}} \,{}^{\bar{R}}\dot{\Phi}_R \,{}^R\Pi_{S_1} + {}^F\Pi_R \,{}^R\dot{T}_{H_1} \,{}^{H_1}\Phi_{S_1} + {}^F\Pi_{H_1} \,{}^{H_1}\dot{\Phi}_{S_1}. \tag{17}$$

We simplify (17) to require only known transformation matrices and known transformation matrix derivatives.

$$
\begin{aligned}
^F\dot{\Pi}_{S_1} &= {}^F T_{\bar{R}} \,{}^{\bar{R}}\dot{\Phi}_R \,{}^R\Pi_{S_1} + {}^F\Pi_{H_1} \,{}^{H_1}\dot{\Phi}_{S_1} && \text{Zero-Velocity Axiom} \\
&= {}^F T_{\bar{R}} \,{}^{\bar{R}}\dot{\Phi}_R \,{}^R T_{H_1} \,{}^{H_1}\Phi_{S_1} + {}^F T_{\bar{R}} \,{}^{\bar{R}}\Phi_R \,{}^R T_{H_1} \,{}^{H_1}\dot{\Phi}_{S_1} && \text{Cascade Corollary} \\
&= {}^F T_{\bar{R}} \,{}^{\bar{R}}\dot{\Phi}_R \,{}^R T_{H_1} \,{}^{H_1}\Phi_{S_1} + {}^F T_{\bar{R}} \,{}^R T_{H_1} \,{}^{H_1}\dot{\Phi}_{S_1} && \text{Identity Axiom} \\
&= {}^F\Pi_R \,{}^{\bar{R}}\dot{\Phi}_R \,{}^R T_{H_1} \,{}^{H_1}\Phi_{S_1} + {}^F\Pi_R \,{}^R T_{H_1} \,{}^{H_1}\dot{\Phi}_{S_1} && \text{Instantaneous Coincidence.}
\end{aligned}
\tag{18}
$$

---

*The wheel velocities are the steering velocity ω_{sz}, the wheel velocity about its axle ω_{wx}, the rotational slip velocity ω_{wz}, the roller velocities ω_{wr} (for omnidirectional wheels), and the rotational velocity ω_{wy} (for ball wheels).

In (18), the robot velocity (in $^R\dot{\Phi}_R$) is calculated in the sensed forward solution (in Section IV G), the steering position (in $^{H_i}\Phi_{S_i}$) and velocity (in $^{H_i}\dot{\Phi}_{S_i}$) are sensed, the robot position (in $^F\Pi_R$) is calculated by dead reckoning,* and the robot-to-hip transformation ($^R\mathbf{T}_{H_i}$) is specified by WMR design. The right-hand side of (18) is thus known. We then substitute (18) into (16) to calculate the velocity of the point mass **r** relative to the floor coordinate system.

3. Wheel Jacobian Matrix

We formulate the equations of motion to calculate the velocities of the robot from the velocities of a wheel. We begin our development by applying the cascade velocity corollary to write the matrix equation in (19) with the unknown dependent variables (i.e., robot velocities, $^R\dot{\Phi}_R$) on the left-hand side, and the independent variables (i.e., the wheel i velocities, $^{H_i}\dot{\Phi}_{S_i}$ and $^{C_i}\dot{\Phi}_{C_i}$) on the right-hand side:

$$^R\dot{\Pi}_R = {}^F\mathbf{T}_{\bar{R}}^{-1}\, {}^F\mathbf{T}_{\bar{C}_i}\, {}^{\bar{C}_i}\hat{\dot{\Phi}}_{C_i}\, {}^{S_i}\mathbf{T}_{C_i}^{-1}\, {}^{H_i}\hat{\dot{\Phi}}_{S_i}^{-1}\, {}^R\mathbf{T}_{H_i}^{-1}$$
$$+ {}^F\mathbf{T}_{\bar{R}}^{-1}\, {}^F\mathbf{T}_{\bar{C}_i}\, {}^{\bar{C}_i}\hat{\dot{\Phi}}_{C_i}\, {}^{S_i}\mathbf{T}_{\bar{C}_i}^{-1}\, {}^{H_i}\dot{\Phi}_{S_i}^{-1}\, {}^R\mathbf{T}_{H_i}^{-1}. \quad (19)$$

The transformation graph in Figure 9 is utilized to determine the order in which to cascade the transformation matrices; the inversion axiom is applied when an arrow in the transformation graph is traversed from head-to-tail and the zero-velocity axiom is applied to eliminate the matrices which multiply the derivatives of constant **T** matrices. Since the position of the wheel contact point relative to the floor is typically unknown, we apply the cascade position corollary to write an alternative expression to the floor-contact transformation matrix:

$$^F\mathbf{T}_{\bar{C}_i} = {}^F\mathbf{T}_{\bar{R}}\, {}^R\Phi_R\, {}^R\mathbf{T}_{H_i}\, {}^{H_i}\Phi_{S_i}\, {}^{S_i}\mathbf{T}_{C_i}\, {}^{\bar{C}_i}\Phi_{C_i}^{-1}. \quad (20)$$

We substitute (20) into (19) to obtain:

$$^R\dot{\Pi}_R = {}^R\Phi_R\, {}^R\mathbf{T}_{H_i}\, {}^{H_i}\Phi_{S_i}\, {}^{S_i}\mathbf{T}_{C_i}\, {}^{\bar{C}_i}\Phi_{C_i}^{-1}\, {}^{\bar{C}_i}\dot{\Phi}_{C_i}\, {}^{S_i}\mathbf{T}_{C_i}^{-1}\, {}^{H_i}\Phi_{S_i}^{-1}\, {}^R\mathbf{T}_{H_i}^{-1}$$
$$+ {}^R\Phi_R\, {}^R\mathbf{T}_{H_i}\, {}^{H_i}\Phi_{S_i}\, {}^{H_i}\dot{\Phi}_{S_i}^{-1}\, {}^R\mathbf{T}_{H_i}^{-1}. \quad (21)$$

We apply the identity axiom to simplify (21):

$$^R\dot{\Pi}_R = {}^R\mathbf{T}_{H_i}\, {}^{H_i}\Phi_{S_i}\, {}^{S_i}\mathbf{T}_{C_i}\, {}^{\bar{C}_i}\dot{\Phi}_{C_i}\, {}^{S_i}\mathbf{T}_{C_i}^{-1}\, {}^{H_i}\dot{\Phi}_{S_i}^{-1}\, {}^R\mathbf{T}_{H_i}^{-1}$$
$$+ {}^R\mathbf{T}_{H_i}\, {}^{H_i}\Phi_{S_i}\, {}^{H_i}\dot{\Phi}_{S_i}^{-1}\, {}^R\mathbf{T}_{H_i}^{-1}. \quad (22)$$

We next apply Tables III and IV to write the transformation matrices and the transformation matrix derivatives and multiply the result to obtain:

*Dead reckoning[28] is the real-time calculation of the WMR position in floor coordinates from the wheel sensor measurements.[1]

$$\begin{pmatrix} 0 & -^{\bar{R}}\omega_R & 0 & ^{\bar{R}}v_{Rx} \\ ^{\bar{R}}\omega_R & 0 & 0 & ^{\bar{R}}v_{Ry} \\ 0 & 0 & 0 & 0 \\ 0 & 0 & 0 & 0 \end{pmatrix} \tag{23}$$

$$= \begin{pmatrix} 0 & -^{\bar{C_i}}\omega_{C_i} & 0 & ^{\bar{C_i}}\omega_{C_i}{}^{R}d_{C_iy} + ^{\bar{C_i}}v_{C_ix}\cos{}^{R}\theta_{C_i} - ^{\bar{C_i}}v_{C_iy}\sin{}^{R}\theta_{C_i} \\ ^{\bar{C_i}}\omega_{C_i} & 0 & 0 & -^{\bar{C_i}}\omega_{C_i}{}^{R}d_{C_ix} + ^{\bar{C_i}}v_{C_ix}\sin{}^{R}\theta_{C_i} + ^{\bar{C_i}}v_{C_iy}\cos{}^{R}\theta_{C_i} \\ 0 & 0 & 0 & 0 \\ 0 & 0 & 0 & 0 \end{pmatrix}$$

$$+ \begin{pmatrix} 0 & ^{H_i}\omega_{S_i} & 0 & -^{H_i}\omega_{S_i}{}^{R}d_{H_iy} \\ -^{H_i}\omega_{S_i} & 0 & 0 & ^{H_i}\omega_{S_i}{}^{R}d_{H_ix} \\ 0 & 0 & 0 & 0 \\ 0 & 0 & 0 & 0 \end{pmatrix}.$$

To simplify the notation in (23), we have made the following substitutions:

$$^{R}\theta_{H_i} + ^{H_i}\theta_{S_i} + ^{S_i}\theta_{C_i} = ^{R}\theta_{C_i}$$

$$^{S_i}d_{C_ix}\cos({}^{R}\theta_{H_i} + ^{H_i}\theta_{S_i}) - ^{S_i}d_{C_iy}\sin({}^{R}\theta_{H_i} + ^{H_i}\theta_{S_i}) + ^{R}d_{H_ix} = ^{R}d_{C_ix} \tag{24}$$

$$^{S_i}d_{C_ix}\sin({}^{R}\theta_{H_i} + ^{H_i}\theta_{S_i}) + ^{S_i}d_{C_iy}\cos({}^{R}\theta_{H_i} + ^{H_i}\theta_{S_i}) + ^{R}d_{H_iy} = ^{R}d_{C_iy}.$$

Upon equating the elements in (23), we obtain the robot velocities:

$$^{\bar{R}}\dot{\mathbf{p}}_R = \begin{pmatrix} ^{\bar{R}}v_{Rx} \\ ^{\bar{R}}v_{Ry} \\ ^{\bar{R}}\omega_R \end{pmatrix} \tag{25}$$

$$= \begin{pmatrix} \cos{}^{R}\theta_{C_i} & -\sin{}^{R}\theta_{C_i} & ^{R}d_{C_iy} & -^{R}d_{H_iy} \\ \sin{}^{R}\theta_{C_i} & \cos{}^{R}\theta_{C_i} & -^{R}d_{C_ix} & ^{R}d_{H_ix} \\ 0 & 0 & 1 & 1 \end{pmatrix} \begin{pmatrix} ^{\bar{C_i}}v_{C_ix} \\ ^{\bar{C_i}}v_{C_iy} \\ ^{\bar{C_i}}\omega_{C_i} \\ ^{H_i}\omega_{S_i} \end{pmatrix}$$

$$= \hat{\mathbf{J}}_i\,\dot{\mathbf{q}}_i,$$

where $i = 1 \ldots N$ is the wheel index, $^{\bar{R}}\dot{\mathbf{p}}_R$ is the vector of robot velocities in the robot frame, $\hat{\mathbf{J}}_i$ is the pseudo-Jacobian matrix for wheel i, and $\dot{\mathbf{q}}_i$ is the pseudo-velocity vector for wheel i. We define the number of wheel variables of wheel i to be w_i. Since typical wheels possess fewer than four wheel variables, the physical velocity vector $\dot{\mathbf{q}}_i$ of typical wheels contains fewer than the four component velocities in (25). Furthermore, since all physical wheel motions are rotations about physical wheel axes, the wheel velocity vector $\dot{\mathbf{q}}_i$ contains the angular velocities of the wheels rather than the linear velocities of the point of contact along the surface of travel. We relate the

(4×1) pseudo-velocity to the $(w_i \times 1)$ physical velocity vector $\dot{\mathbf{q}}_i$ by the $(4 \times w_i)$ wheel matrix $\mathbf{W}_i$:

$$\dot{\mathbf{q}} = \mathbf{W}_i \, \dot{\mathbf{q}}_i. \tag{26}$$

We substitute (26) into (25) to calculate the robot velocities from the wheel velocity vector:

$$^{\bar{R}}\dot{\mathbf{p}}_R = \hat{\mathbf{J}}_i \, \mathbf{W}_i \, \dot{\mathbf{q}}_i = \mathbf{J}_i \, \dot{\mathbf{q}}_i. \tag{27}$$

The product $\mathbf{J}_i = \hat{\mathbf{J}}_i \mathbf{W}_i$ is the $(3 \times w_i)$ wheel Jacobian matrix of wheel i. The rank of the wheel Jacobian matrix indicates the number of DOFs of the wheel. A wheel having fewer DOFs than wheel variables is redundant. The Jacobian matrix of a redundant wheel has dependent columns. We thus formulate the following computational test to determine whether a wheel is non-redundant:

Non-Redundant Wheel Criterion

$$det[\mathbf{J}_i^T \, \mathbf{J}_i] \neq 0. \tag{28}$$

The wheel Jacobian matrices for non-steered conventional wheels, steered conventional wheels, omnidirectional wheels, and ball wheels are detailed in Appendix 1. In Section IV, we utilize (27) to develop the inverse and forward solutions. In Section V, we apply the matrices in Appendix 1 to calculate the inverse and forward solutions of Uranus.

4. Transforming Robot Velocities

We equate the components in (15) to compute the translational $^A v_{Zx}$, and $^A v_{Zy}$ and rotational velocities $^A \omega_Z$ velocities* of the coordinate system Z relative to coordinate system A. We apply this methodology to the practical problem of transforming velocities of the robot from $\bar{r}$obot coordinates $\bar{R}$ to floor coordinates F. We assume that the floor-robot transformation matrix $^F\mathbf{T}_R$ (i.e., the position and orientation of the robot relative to the floor) and the matrix $^{\bar{R}}\dot{\mathbf{\Phi}}_R$ (i.e., the velocities of the robot relative to its current position and orientation) are known. The velocities to be calculated (i.e., the velocities of the robot relative to the floor) are the components of the matrix $^F\dot{\mathbf{\Pi}}_R$. We apply the cascade velocity corollary (in Section III E) and the WMR transformation graph (in Section III F) to write the matrix equation

$$^F\dot{\mathbf{\Pi}}_R = {}^F\dot{\mathbf{T}}_{\bar{R}} \, {}^{\bar{R}}\mathbf{\Phi}_R + {}^F\mathbf{T}_{\bar{R}} \, {}^{\bar{R}}\dot{\mathbf{\Phi}}_R \tag{29}$$

*There are no translational velocities along the z axis or angular velocities about the x and y axes because of our coordinate assignments.

in terms of known matrices. To simplify (29), we apply the zero-velocity axiom and the instantaneous coincidence corollary:

$$^F\dot{\Pi}_R = {}^F T_R \, {}^{\bar{R}}\dot{\Phi}_R. \tag{30}$$

We expand each matrix into scalar components: the matrix derivative $^F\dot{\Pi}_R$ according to (7), the transformation matrix $^F T_R$ according to (5), and the transformation matrix derivative $^{\bar{R}}\dot{\Phi}_R$ according to Table IV. Upon multiplying, we obtain:

$$\begin{pmatrix} -{}^F\omega_R \sin {}^F\theta_R & -{}^F\omega_R \cos {}^F\theta_R & 0 & {}^F v_{Rx} \\ {}^F\omega_R \cos {}^F\theta_R & -{}^F\omega_R \sin {}^F\theta_R & 0 & {}^F v_{Ry} \\ 0 & 0 & 0 & 0 \\ 0 & 0 & 0 & 0 \end{pmatrix} .$$

$$= \begin{pmatrix} -{}^{\bar{R}}\omega_R \sin {}^F\theta_R & -{}^{\bar{R}}\omega_R \cos {}^F\theta_R & 0 & {}^{\bar{R}} v_{Rx} \cos {}^F\theta_R - {}^{\bar{R}} v_{Ry} \sin {}^F\theta_R \\ {}^{\bar{R}}\omega_R \cos {}^F\theta_R & -{}^{\bar{R}}\omega_R \sin {}^F\theta_R & 0 & {}^{\bar{R}} v_{Rx} \sin {}^F\theta_R + {}^{\bar{R}} v_{Ry} \cos {}^F\theta_R \\ 0 & 0 & 0 & 0 \\ 0 & 0 & 0 & 0 \end{pmatrix} . \tag{31}$$

We obtain the angular velocity of the robot $^F\omega_R$ from elements $(1,1)$ and $(2,1)$ and read the translational velocities $^F v_{Rx}$ and $^F v_{Ry}$ directly from elements $(1,4)$ and $(2,4)$ of (31), respectively. We find that:

$$^F\dot{\mathbf{p}}_R = \begin{pmatrix} {}^F v_{Rx} \\ {}^F v_{Ry} \\ {}^F\omega_R \end{pmatrix} = \begin{pmatrix} \cos {}^F\theta_R & -\sin {}^F\theta_R & 0 \\ \sin {}^F\theta_R & \cos {}^F\theta_R & 0 \\ 0 & 0 & 1 \end{pmatrix} \begin{pmatrix} {}^{\bar{R}} v_{Rx} \\ {}^{\bar{R}} v_{Ry} \\ {}^{\bar{R}}\omega_R \end{pmatrix} = \mathbf{V} \, {}^{\bar{R}}\dot{\mathbf{p}}_R. \tag{32}$$

In (32), we observe that the angular velocity of the robot is equal in both coordinate frames; whereas the translational velocities in the floor coordinate frame and the (3×3) motion matrix $\mathbf{V}$ depend explicitly upon the robot orientation.

H. Acceleration Kinematics

We calculate the accelerations of the WMR by applying the cascade acceleration corollary. Since the development parallels that of the velocity kinematics in Section III G, we omit the computational details and concentrate on interpreting the results. We cannot formulate the acceleration equations of motion by differentiating the results of Section III G, because differentiation of both sides of a transformation matrix equation is not an allowable operation in our matrix coordinate transformation algebra. This is in contrast to the acceleration kinematics of mechanisms containing only lower pairs (e.g., stationary manipulators) which are formulated by differentiating velocity kinematics.

The acceleration of the point $\mathbf{r}$ fixed relative to the moving coordinate system Z is transformed to the A coordinate frame according to:

$$^A\ddot{\mathbf{r}} = {}^A\ddot{\mathbf{\Pi}}_Z {}^Z\mathbf{r}.\qquad(33)$$

We apply the cascade acceleration corollary to calculate the second time derivative of the transformation matrix $^A\ddot{\mathbf{\Pi}}_Z$ and find that the component accelerations of the robot ($^{\bar{R}}a_{Rx}$, $^{\bar{R}}a_{Ry,}$ and $^{\bar{R}}\alpha_R$) are related to the wheel accelerations ($^{H_i}a_{S_i}$, $^{\bar{C}_i}a_{C_i x}$, $^{\bar{C}_i}a_{C_i y}$, and $^{\bar{C}_i}a_{C_i}$) as the cascade velocity corollary, in Section III G3, relates the robot velocities to the wheel velocities. In the notation of (24), the robot accelerations are:

$$
\begin{pmatrix} ^{\bar{R}}a_{Rx} \\ ^{\bar{R}}a_{Ry} \\ ^{\bar{R}}\alpha_R \end{pmatrix} =
\begin{pmatrix}
\cos {}^R\theta_{C_i} & -\sin {}^R\theta_{C_i} & {}^Rd_{C_i y} & -{}^Rd_{H_i y} \\
\sin {}^R\theta_{C_i} & \cos {}^R\theta_{C_i} & -{}^Rd_{C_i x} & {}^Rd_{H_i x} \\
0 & 0 & 1 & -1
\end{pmatrix}
\begin{pmatrix} ^{\bar{C}_i}a_{C_i x} \\ ^{\bar{C}_i}a_{C_i y} \\ ^{\bar{C}_i}\alpha_{C_i} \\ ^{H_i}\alpha_{S_i} \end{pmatrix}.\qquad(34)
$$

$$
+ \begin{pmatrix}
^Rd_{C_i x} & ^Rd_{H_i x} & ^Rd_{H_i x} \\
^Rd_{C_i y} & ^Rd_{H_i y} & ^Rd_{H_i y} \\
0 & 0 & 0
\end{pmatrix}
\begin{pmatrix} ^{\bar{C}_i}\omega_{C_i}^2 \\ -2^{\bar{C}_i}\omega_{C_i}{}^{H_i}\omega_{S_i} \\ ^{H_i}\omega_{S_i}^2 \end{pmatrix}
$$

The robot accelerations in (34) have three components: the self-accelerations ($^{\bar{C}_i}a_{C_i x}$, $^{\bar{C}_i}a_{C_i y}$, $^{\bar{C}_i}\alpha_{C_i}$, and $^{H_i}\alpha_{S_i}$); the centripetal accelerations ($^{\bar{C}_i}\omega_{C_i}^2$ and $^{H_i}\omega_{S_i}^2$) having squared velocities; and the Coriolis accelerations ($^{\bar{C}_i}\omega_{C_i}{}^{H_i}\omega_{S_i}$) having products of different velocities.

Transforming robot accelerations from $\bar{r}$obot coordinates to floor coordinates is analogous to transforming robot velocities (in Section III G4). We find that the robot accelerations are transformed from the $\bar{r}$obot to the floor coordinate frame by the motion matrix $\mathbf{V}$ that transforms the velocities in (32):

$$
^F\ddot{\mathbf{p}}_R = \begin{pmatrix} ^Fa_{Rx} \\ ^Fa_{Ry} \\ ^F\alpha_R \end{pmatrix} =
\begin{pmatrix}
\cos {}^F\theta_R & -\sin {}^F\theta_R & 0 \\
\sin {}^F\theta_R & \cos {}^F\theta_R & 0 \\
0 & 0 & 1
\end{pmatrix}
\begin{pmatrix} ^{\bar{R}}a_{Rx} \\ ^{\bar{R}}a_{Ry} \\ ^{\bar{R}}\alpha_R \end{pmatrix} = \mathbf{V}\,^{\bar{R}}\ddot{\mathbf{p}}_R.\qquad(35)
$$

The equations of motion are not solved in practice because accurate acceleration measurements are difficult to obtain.

I. Summary

We have formulated a systematic procedure for modeling the position, velocity, and acceleration kinematics of a WMR. In this section, we outline a step-by-step enumeration of the methodology to facilitate engineering applications.

---

1. Make a sketch of the WMR. Show the relative positioning of the wheels and the steering links. The sketch need not be to scale. A top and side view are typically sufficient.

2. Assign the coordinate systems. The robot, hip, steering, contact point and floor coordinate systems are assigned according to the conventions introduced in Table I.
3. Develop the (4 × 4) coordinate transformation matrices. The robot-hip, hip-steering, and steering-contact transformation matrices are written according to Table III.
4. Formulate the position equations of motion. The relative positions and orientations of two coordinate systems are determined by applying the cascade position corollary. The transformation graph of Figure 9 is utilized to determine the order in which to cascade the matrices.
5. Formulate the velocity equations of motion. The equations relating velocities are formulated by applying the cascade velocity corollary. The wheel Jacobian matrix, which relates wheel velocities to robot velocities, may be written directly by substituting components of the transformation matrices into the symbolic wheel Jacobian matrices compiled in Appendix 1.
6. Formulate the acceleration equations of motion. The equations relating accelerations are formulated by applying the cascade acceleration corollary.

The non-redundant wheel criterion in (28) tests the Jacobian matrix in (27) to determine whether a wheel has as many DOFs as wheel variables. In our companion technical report,[1] we apply (28) to illustrate disadvantages of redundant wheels. A kinematic model; i.e., the position, velocity and acceleration equations-of-motion, may be applied to the dynamic modeling, design, and control of a WMR. In these applications, the equations of motion are solved to compute unknown variables from constant and sensed variables. In Section IV, we compute the inverse and forward solutions by utilizing the wheel Jacobian matrix (introduced in Section III G3) as the foundation.

IV. THE COMPOSITE ROBOT EQUATION

A. Introduction

We combine the kinematic equations of motion of all of the wheels on a WMR to form the composite robot equation. We then investigate solutions of the composite robot equation and their properties and implications for WMR locomotion. Our investigation illuminates WMR mobility (in Section IV D), actuation (in Sections IV E and IV F) and sensing (in Sections IV G and IV H).

In Section IV B, we formulate the composite robot equation and in Section IV C we discuss the conditions for its solution. We apply the results of Section IV C to develop a mobility characterization tree in Section IV D which enables us to interpret the solubility conditions in terms of the mobility characteristics of the WMR. The mobility characterization tree indicates whether the mobility structure is determined, overdetermined, or undetermined, and associates specific mobility characteristics with each possibility. For example, we may apply the mobility characterization tree to determine whether a WMR allows three DOF motion; and if it does not, the tree indicates the motion constraints.

We proceed to solve the composite robot equation by addressing two classical kinematic modeling problems: the actuated inverse solution (in Section IV E) and the sensed forward solution (in Section IV G). The actuated inverse solution computes the actuated wheel velocities from the robot velocities. For WMR control, we solve only for the velocities of the actuated wheel variables. The solution for all of the wheel velocities is a special case which may be obtained by assuming that all of the wheel variables are actuated.

The actuated inverse solution does not guarantee that the specified robot velocities will be attained when the actuated wheel variables are driven to the calculated velocities. We investigate the possible robot motions when the actuated wheel variables attain the velocities computed by the actuated inverse solution in Section IV F. We develop an actuation characterization tree, analogous to the mobility characterization tree, which enables us to determine the actuation structure (determined, overdetermined, or undetermined) of a WMR. The actuation characterization tree is applicable for WMR design to avoid overdetermined actuation (which may cause actuator conflict) and undetermined actuation (which allows the WMR uncontrollable DOFs). From our analysis, we are able to determine whether the actuated wheel variables are sufficient for producing all of the motions allowed by the mobility structure.

The sensed forward solution in Section IV G computes the robot velocities from the sensed wheel velocities and positions. Since a WMR consists of closed kinematic chains, it is not required to sense all of the wheel positions and velocities, and in practice, it is difficult to do so.

In Section IV H, we develop a sensing characterization tree which enables us to determine the character (undetermined, determined, or overdetermined) of the WMR sensing. We thus are able to determine whether the sensed wheel variables are sufficient for discerning all of the motions allowed by the mobility structure. Finally, in Section IV I, we summarize our development.

B. Formulation of the Composite Robot Equation

In Section III G3, we developed the wheel Jacobian matrix $\mathbf{J}_i$ by applying velocity kinematics to compute the robot velocity vector $\dot{\mathbf{p}}$ from the wheel velocity vector $\dot{\mathbf{q}}_i$:

$$\dot{\mathbf{p}} = \mathbf{J}_i\dot{\mathbf{q}}_i \qquad \text{for } i = 1, \ldots, N, \tag{36}$$

where i is the wheel index, N is the total number of wheels, $\dot{\mathbf{p}}$ is the vector of robot velocities, $\mathbf{J}_i$ is the $(3 \times w_i)$ Jacobian matrix for wheel i, w_i is the number of variables for wheel i, and $\dot{\mathbf{q}}_i$ is the $(w_i \times 1)$ vector of wheel velocities.

The $3N$ wheel equations in (36) must be solved simultaneously to characterize the WMR motion. We combine the wheel equations to form the composite robot equation:

$$\begin{pmatrix} \mathbf{I}_1 \\ \mathbf{I}_2 \\ \vdots \\ \mathbf{I}_N \end{pmatrix} \dot{\mathbf{p}} = \begin{pmatrix} \mathbf{J}_1 & 0 & \ldots & 0 \\ 0 & \mathbf{J}_2 & \ddots & \vdots \\ \vdots & \ddots & \ddots & 0 \\ 0 & \ldots & 0 & \mathbf{J}_N \end{pmatrix} \begin{pmatrix} \dot{\mathbf{q}}_1 \\ \dot{\mathbf{q}}_2 \\ \vdots \\ \dot{\mathbf{q}}_N \end{pmatrix} \tag{37}$$

or

$$A_0 \dot{\mathbf{p}} = B_0 \dot{\mathbf{q}} \tag{38}$$

where the I_i, for $i = 1, \ldots, N$, are (3×3) identity matrices, A_0 is a $(3N \times 3)$ matrix, B_0 is a $(3N \times w)$ block diagonal matrix, $w = w_1 + w_2 + \ldots w_N$ is the total number of wheel variables, and $\dot{\mathbf{q}}$ is the composite wheel velocity vector.

Having formulated the matrix equation in (38) to model the robot motion, we proceed to investigate the solution for the robot velocity vector $\dot{\mathbf{p}}$ in Section IV C and its implications for WMR locomotion in Section IV D.

C. Solution of Ax = By

We characterize WMR mobility (in Section IV D), actuation (in Section IV F), and sensing (in Section IV H) by examining the properties of the solutions of the composite robot equation in (38). We extend the standard criteria[29] for the system of linear algebraic equations $\mathbf{Ax} = \mathbf{b}$, where $\mathbf{A}$ is an $(m \times n)$ matrix, $\mathbf{x}$ is a $(n \times 1)$ vector, and $\mathbf{b}$ is a $(m \times 1)$ vector, to the solution of the system of linear algebraic equations

$$\mathbf{Ax} = \mathbf{By}, \tag{39}$$

where $\mathbf{B}$ is an $(m \times p)$ matrix and $\mathbf{y}$ is a $(p \times 1)$ vector. Since the composite robot equation (38) has the form of (39), solutions of (39) are directly applicable to the solution of the composite robot equation.

We apply the method of least squares[29] to compute the vector $\mathbf{x}$ for overdetermined (i.e., having fewer variables than independent equations) and determined (i.e., having the same number of variables as independent equations) systems of linear algebraic equations:

$$\mathbf{x} = (\mathbf{A}^T \mathbf{A})^{-1} \mathbf{A}^T \mathbf{By}. \tag{40}$$

The necessary condition for applying the least-squares solution in (40) is that rank $(\mathbf{A}) = n$. There is no unique solution for undetermined systems (i.e., systems having fewer independent equations than independent variables).

The residual error of the least-squares solution is:

$$\mathbf{Ax} - \mathbf{By} = [\mathbf{A}(\mathbf{A}^T \mathbf{A})^{-1} \mathbf{A}^T - \mathbf{I}] \, \mathbf{By} = \Delta(\mathbf{A}) \, \mathbf{By}. \tag{41}$$

For expository convenience, we define the Delta matrix function $\Delta(\bullet)$ as:

$$\Delta(\mathbf{U}) = \begin{cases} -\mathbf{I} & \text{for } \mathbf{U} = \text{null} \\ \mathbf{U}(\mathbf{U}^T \mathbf{U})^{-1} \mathbf{U}^T - \mathbf{I} & \text{Otherwise} \end{cases} \tag{42}$$

where the argument $\mathbf{U}$ is a $(c \times d)$ matrix of rank d.

To characterize WMR motion, we must determine whether the least-squares error in (41) is zero for all $\mathbf{y}$. To do so, we may apply either of the following equivalent tests:

$$\Delta(\mathbf{A}) \, \mathbf{B} = 0 \qquad\qquad (43)$$

or

$$\text{rank}[\mathbf{A}; \, \mathbf{B}] = \text{rank}[\mathbf{A}]. \qquad\qquad (44)$$

If either test (43) or (44) is satisfied, the least-squares error is zero for all $\mathbf{y}$. The first test in (43) is apparent from the expression for the least-squares error in (41). The second test in (44) states that if the columns of the matrix $\mathbf{B}$ lie in the vector space spanned by the columns of the matrix $\mathbf{A}$, then the vector $\mathbf{By}$ must also lie in the vector space spanned by the columns of $\mathbf{A}$ for all $\mathbf{y}$. The vector $\mathbf{By}$ can then be expressed as a linear combination of the columns of $\mathbf{A}$ by proper choice (via the least-squares solution) of $\mathbf{x}$. Similarly, we may determine whether the least-squares error is zero for a specific $\mathbf{y}$ by applying either of the following two equivalent tests:

$$\Delta(\mathbf{A}) \, \mathbf{By} = 0, \qquad\qquad (45)$$

or

$$\text{rank}[\mathbf{A}; \, \mathbf{By}] = \text{rank}[\mathbf{A}]. \qquad\qquad (46)$$

We depict in Figure 11 a tree illustrating the nature of all possible solutions for the vector $\mathbf{x}$ of the system of linear algebraic equations in (39). The tree branches (directed arrows) indicate tests on the matrices $\mathbf{A}, \mathbf{B},$ and $\mathbf{y}$ and are numbered for future reference. The leaves (boxes) indicate the corresponding properties of the solution.

As depicted in Figure 11, the system of linear algebraic equations in (39) may be determined, overdetermined, or underdetermined. The top branches, (0) and (1), determine whether the least-squares solution is applicable by testing the rank of the matrix $\mathbf{A}$. If the rank of $\mathbf{A}$ is n (branch (0)), the least-squares solution is applicable and there is a unique solution for some $\mathbf{y}$. If the rank of $\mathbf{A}$ is less than n (branch (1)), the least-squares solution is not applicable—indicating that the system is undetermined and there is no unique solution for any $\mathbf{y}$. An undetermined system has more unknowns than independent equations.

A determined system is one in which the number of independent equations (less than or equal to m) equals the number of unknowns (n). The least-squares error is zero for all $\mathbf{y}$ and thus tests (43) and (44) apply at branch (00).

An overdetermined system is one in which the number of independent equations is greater than the number of unknowns. The least-squares error of an overdetermined system is thus non-zero for some $\mathbf{y}$ (branch (01)). Tests (45) and (46) are applied at branch (010) to determine whether the least-squares is zero for a specific $\mathbf{y}$. If so, the system is consistent and there is a unique solution. If the least-squares error is non-zero for a specific $\mathbf{y}$ (branch (011)), the system is inconsistent and there is no exact solution.

In Section IV D, we apply the solution tree in Figure 11 to the composite robot equation in (38) and discuss the implications for WMR mobility characterization.

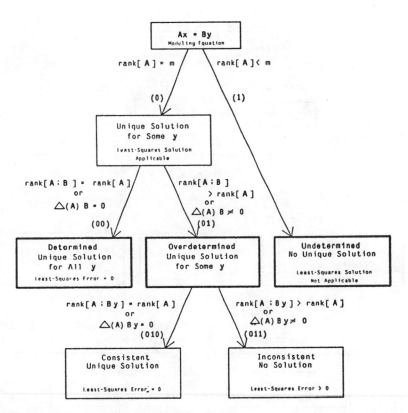

Figure 11. The solution tree for the vector x in (39).

D. Robot Mobility Characteristics

The composite robot equation in (38) has the form of the system of linear algebraic equations depicted in Figure 11, in which A_0, B_0, $\dot{p}$, and $\dot{q}$ play the roles of A, B, x, and y, respectively. Since the robot velocity vector $\dot{p}$ plays the role of the dependent variable, we investigate the conditions under which the forward solution may be computed. In Figure 12, we apply the solution tree in Figure 11 to the composite robot equation in (38).

By inspection of (37), we observe that the rank of the $(3N \times 3)$ matrix A_0 is 3 and thus branch (0) always applies. Since branch (1) does not apply, the solution cannot be undetermined; and hence the robot motion is completely specified by the motion of the wheels. From the structure of the matrices A_0 and B_0 in (37), we observe that the rank of the augmented matrix $[A_0, B_0]$ is greater than 3 when there is more than one wheel. A WMR with one wheel is determined [branch (00)], and a WMR with more than one wheel is overdetermined [branch (01)]. The overdetermined nature of WMRs having more than one wheel is a consequence of the closed-link kinematic

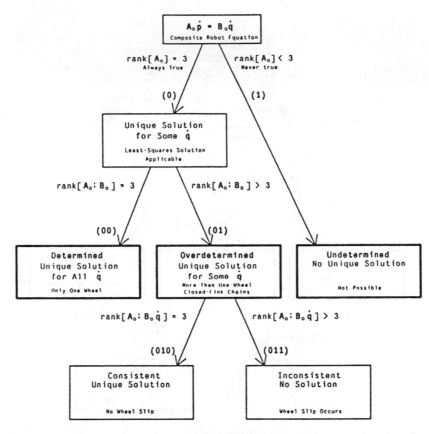

Figure 12. The solution tree for the robot velocity vector $\dot{\mathbf{p}}$.

structure of a WMR. As indicated in Figure 12, the composite robot equation in (38) will be consistent [and have a solution at branch (010)] or inconsistent [and have no solution at branch (011)] depending upon the wheel velocity vector $\dot{\mathbf{q}}$. Our no-slip assumption (in Section III B) ensures that the motions of the wheels and the robot are consistent and that there is thus an exact solution.

We depict in Figure 12 the solution of the robot velocity vectory $\dot{\mathbf{p}}$ from the complete wheel velocity vector $\dot{\mathbf{q}}$. In practice, the wheel velocity vector must be measured by sensors. It is difficult to sense some of the wheel velocities, such as the rotational wheel slip. Since a WMR with more than one wheel has closed-link chains, it is not necessary to sense all of the wheel velocities to calculate the robot velocity because many of the sensor motions are dependent. In Sections IV G and IV H, we investigate the solution of the robot velocity vector from the sensed wheel velocities.

Although the nature of the forward solution of the composite robot equation provides us with little physical insight, we gain significant understanding of WMR motion by investigating the nature of the inverse solution. For WMR control it is not necessary to compute all of the wheel variables in the inverse solution since they are not all actuated. Because of the closed-link chains, moreover not all of the wheel variables must be actuated. In Section IV E, we compute the actuated inverse solution for the actuated wheel variables. In the remainder of this section, we focus on the complete inverse solution to gain physical insight into WMR mobility characteristics.

We investigate the inverse solution by interchanging the roles of the right- and left-hand sides of the composite robot equation in (38) and applying the solution tree in Figure 11. Thereby, $\mathbf{B}_0$, $\mathbf{A}_0$, $\dot{\mathbf{q}}$, and $\dot{\mathbf{p}}$ in (38) play the roles of $\mathbf{A}$, $\mathbf{B}$, $\mathbf{x}$, and $\mathbf{y}$ in (39), respectively. The solution tree for the inverse solution, subsequently referred to as the mobility characterization tree, is depicted in Figure 13. The branch tests indicated within curly brackets "{ }" are simplified tests which apply if there are no couplings between wheels.

The inverse solution can be determined, undetermined, or overdetermined depending upon the kinematics (i.e., $\mathbf{B}_0$ and $\mathbf{A}_0$). The top branches test the rank of the $(3N \times w)$

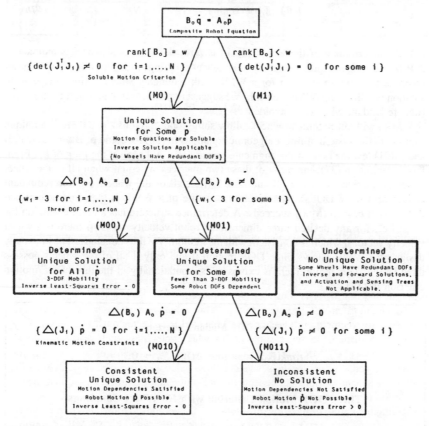

Figure 13. The mobility characterization tree.

matrix $\mathbf{B}_0$ against the total number of wheel variables w. Since the rank $\mathbf{B}_0$ is the sum of the ranks of all of the wheel Jacobian matrices when there are no wheel couplings, we test the rank of each wheel Jacobian matrix $\mathbf{J}_i$ against the number of wheel variables w_i for all wheels $i = 1, \ldots, N$. The rank of the $(3 \times w_i)$ wheel Jacobian matrix $\mathbf{J}_i$ is w_i if the determinant of the matrix $[\mathbf{J}_i^T \mathbf{J}_i]$ is non-zero as indicated by the non-redundant wheel criterion in (28). We refer to branch test (M0) as the soluble motion criterion because it determines whether the composite robot equation can be solved.

Soluble Motion Criterion

$$rank[\mathbf{B}_0] = w, \tag{47a}$$

Soluble Motion Criterion with No Wheel Couplings

$$det[\mathbf{J}_i^T \mathbf{J}_i] \neq 0 \; for \; i = 1, \ldots, N. \tag{47b}$$

If the determinant of the matrix $[\mathbf{J}_i^T \mathbf{J}_i]$ is zero, the associated wheel is redundant. A WMR having redundant wheels and no wheel couplings is undetermined. We cannot compute the inverse solution for a WMR with redundant wheels. Since the inverse solution is utilized for WMR control,[1] we suggest that undetermined mobility structures (i.e., redundant wheels) be avoided.

WMRs without redundant wheels allow some robot motions since there is a unique solution to the system of linear algebraic equations in (38) for some $\dot{\mathbf{p}}$. Branches (M00) and (M01) test the rank of the augmented matrix $[\mathbf{B}_0; \mathbf{A}_0]$ against the rank of $\mathbf{B}_0$. From their structure in (37), the ranks of these two matrices are equal when all of the wheel Jacobian matrices are (3×3) and rank 3 (i.e., all of the wheels are non-redundant and possess three DOFs). The mobility structure of a WMR is therefore determined if the test at branch (M00) succeeds. A determined structure has a unique solution for all $\dot{\mathbf{p}}$; i.e., for any desired three-dimensional robot velocity vector $\dot{\mathbf{p}}$ there is a wheel velocity vector $\dot{\mathbf{q}}$ which is consistent with the motion. We thus conclude: The kinematic design of a WMR allows three DOF motion if and only if all of the wheels possess three DOFs. This requirement is expressed computationally in the three DOF motion criterion in (48).

Three DOF Motion Criterion

$$rank[\mathbf{B}_0] = w \; and \; \Delta(\mathbf{B}_0) \, \mathbf{A}_0 = 0, \tag{48a}$$

Three DOF Motion Criterion with No Wheel Couplings

$$det[\mathbf{J}_i^T \mathbf{J}_i] \neq 0 \; and \; w_i = 3 \; for \; i = 1, \ldots, N. \tag{48b}$$

If branch (M0) succeeds and the WMR does not possess three DOFs, the solution is overdetermined [branch (M01)]. The robot does not allow some motions because some of the robot DOFs are dependent. For example, a WMR with a non-steered conventional wheel which satisfies branch (M0) must have an overdetermined mobility structure because no motions perpendicular to the wheel orientation may occur without slip. Branches (M010) and (M011) indicate the possible robot motions $\dot{\mathbf{p}}$ without slip. If the least-squares error is zero, the solution is consistent, and the motion may occur. We thus determine the kinematic constraints on the robot motion by equating the least-squares error to zero in (49). By examining the structure of the error in (49a), we find an equivalent computationally simpler test in (49b) when there are no couplings between wheels.

Kinematic Motion Constraints

$$\Delta(\mathbf{B}_0)\,\mathbf{A}_0\dot{\mathbf{p}} = 0, \qquad (49a)$$

Kinematic Motion Constraints with No Wheel Couplings

$$\Delta(\mathbf{J}_i)\,\dot{\mathbf{p}} = 0 \; for \; i = 1, \ldots, N. \qquad (49b)$$

We may thus determine the kinematic motion constraints for a WMR without redundant wheels or wheel couplings by considering each wheel independently.

The augmented matrix $[\Delta(\mathbf{B}_0)\,\mathbf{A}_0]$ indicates whether the WMR possesses three DOFs at branch (M00) or fewer than three DOFs at branch (M01). When there are fewer than three DOFs, the number of independent columns of the matrix $[\Delta(\mathbf{B}_0)\,\mathbf{A}_0]$ specifies the number of dependent robot DOFs. The number of DOFs of a WMR having no redundant wheels is:

Number of WMR DOFs

$$DOFs = 3 - rank[\Delta(\mathbf{B}_0)\,\mathbf{A}_0]. \qquad (50)$$

The test at branch (M0) determines whether the complete inverse solution for all of the wheel variables can be calculated by the least-squares solution. In Section IV E, we apply the least-squares solution to calculate the actuated inverse solution for the actuated wheel variables. Although the actuated inverse solution may exist for some robot velocities $\dot{\mathbf{p}}$ for which the complete inverse solution does not, it is not practical to apply such an actuated inverse solution because the desired robot velocities are constrained by the unactuated wheel variables. We thus utilize the soluble motion criteria in (47) to indicate when the actuated inverse solution in Section IV E is practically applicable.

E. Actuated Inverse Solution

We calculate the actuated inverse solution by solving for the actuated wheel velocities in (38). Because of the closed-link chains in WMRs, we need not actuate all of the wheel variables. To separate the actuated and unactuated wheel variables, we partition the wheel equation in (36) into two components:

$$\dot{\mathbf{p}} = \mathbf{J}_{ia}\,\dot{\mathbf{q}}_{ia} + \mathbf{J}_{iu}\,\dot{\mathbf{q}}_{iu}. \tag{51}$$

The "a" subscripts denote the actuated components and the "u" subscripts denote the unactuated components. We let a_i denote the number of actuated variables, and u_i denote the number of unactuated variables for wheel i (i.e., $a_i + u_i = w_i$). We define the total number of actuated wheel variables to be $a = a_1 + a_2 + \ldots + a_N$ and the total number of unactuated wheel variables to be $u = u_1 + u_2 + \ldots + u_N$. We combine the partitioned wheel equations in (51) to rewrite the composite robot equation in (37) as

$$
\begin{pmatrix} \mathbf{I}_1 \\ \mathbf{I}_2 \\ \vdots \\ \mathbf{I}_N \end{pmatrix} \dot{\mathbf{p}} =
\begin{pmatrix}
\mathbf{J}_{1a} & 0 & \cdots & 0 & \mathbf{J}_{1u} & 0 & \cdots & 0 \\
0 & \mathbf{J}_{2a} & \ddots & \vdots & 0 & \mathbf{J}_{2u} & \ddots & \vdots \\
\vdots & \ddots & \ddots & 0 & \vdots & \ddots & \ddots & 0 \\
0 & \cdots & 0 & \mathbf{J}_{Na} & 0 & \cdots & 0 & \mathbf{J}_{Nu}
\end{pmatrix}
\begin{pmatrix} \dot{\mathbf{q}}_{1a} \\ \dot{\mathbf{q}}_{2a} \\ \vdots \\ \dot{\mathbf{q}}_{Na} \\ \dot{\mathbf{q}}_{1u} \\ \dot{\mathbf{q}}_{2u} \\ \vdots \\ \dot{\mathbf{q}}_{Nu} \end{pmatrix}
$$

$$\tag{52}$$

or

$$\mathbf{A}_0\dot{\mathbf{p}} = \mathbf{B}_{0p}\dot{\mathbf{q}}_p. \tag{53}$$

The ($3N \times w$) matrix $\mathbf{B}_{0p}$ and the ($w \times 1$) vector $\dot{\mathbf{q}}_p$ are the partitioned counterparts of the matrix $\mathbf{B}_0$ and the vector $\dot{\mathbf{q}}$ in (37). The soluble motion criterion in (47) indicates under what conditions the least-squares solution may be practically applied to compute the inverse solution (i.e., rank$[\mathbf{B}_0] = w$). We henceforth assume that the least-squares solution is applicable and that all matrix inverses encountered in its application are computable. We apply the least-squares solution in (40) to calculate the vector of the wheel variables from the robot velocity vector:

$$\dot{\mathbf{q}}_p = (\mathbf{B}_{0p}^T\,\mathbf{B}_{0p})^{-1}\,\mathbf{B}_{0p}^T\,\mathbf{A}_0\,\dot{\mathbf{p}}. \tag{54}$$

In Appendix 2, we compute the vector of actuated wheel velocities $\dot{\mathbf{q}}_a = [\dot{\mathbf{q}}_{ia}^T \ldots \dot{\mathbf{q}}_{Na}^T]^T$ in (52) as

$$\boxed{\begin{array}{c} \textbf{Actuated Inverse Solution} \\[2mm] \dot{\mathbf{q}}_a = \begin{pmatrix} [\mathbf{J}_{1a}^T \, \Delta \, (\mathbf{J}_{1u})\mathbf{J}_{1a}]^{-1} \, \mathbf{J}_{1a}^T \, \Delta \, (\mathbf{J}_{1u}) \\ [\mathbf{J}_{2a}^T \, \Delta (\mathbf{J}_{2u})\mathbf{J}_{2a}]^{-1} \, \mathbf{J}_{2a}^T \, \Delta(\mathbf{J}_{2u}) \\ \vdots \\ [\mathbf{J}_{Na}^T \, \Delta \, (\mathbf{J}_{Nu})\mathbf{J}_{Na}]^{-1} \, \mathbf{J}_{Na}^T \, \Delta \, (\mathbf{J}_{Nu}) \end{pmatrix} \dot{\mathbf{p}} = \mathbf{J}_a\dot{\mathbf{p}}. \end{array}} \qquad (55)$$

Each $(a_i \times 3)$ block row of the matrix on the right-hand side of (55), corresponding to the actuated velocities $\dot{\mathbf{q}}_{ia}$, involves only the Jacobian matrix of wheel i. The inverse solution for each wheel is thus independent of the kinematic equations of all of the $(N-1)$ other wheels. When wheel i is non-redundant with three DOFs and all three wheel variables are actuated, each block row of (55) simplifies to

$$\dot{\mathbf{q}}_{ia} = (\mathbf{J}_i^{-1})\dot{\mathbf{p}}. \qquad (56)$$

We may therefore assume that all of the wheel variables of all of the non-redundant wheels having three DOFs are actuated, apply the inverse Jacobian matrix in (56) to calculate the wheel velocities, and extract the actuated velocities for robot control. This approach requires approximately one-tenth of the arithmetic operations required for the direct application of (55).

F. Robot Actuation Characteristics

A WMR control engineering application of the actuated inverse solution (in Section IV E) is to command the velocities of the actuated wheel variables to their calculated values. We investigate the characteristics of the robot motion when the actuated wheel velocities attain the values computed by the actuated inverse solution. We relate the robot velocity vector to the actuated wheel velocities by eliminating the unactuated wheel velocities from the composite robot equation in (37). Under the no-slip assumption, the unactuated wheel velocities will be consistent and comply to the robot motion. We compute in (57) the unactuated wheel velocities from the robot velocities in the actuated inverse solution by interchanging the roles of the actuated ("a" subscripts) and unactuated ("u" subscripts) variables:

$$\dot{\mathbf{q}}_u = \begin{pmatrix} [\mathbf{J}_{1u}^T \, \Delta \, (\mathbf{J}_{1a}) \, \mathbf{J}_{1u}]^{-1} \, \mathbf{J}_{1u}^T \, \Delta \, (\mathbf{J}_{1a}) \\ [\mathbf{J}_{2u}^T \, \Delta \, (\mathbf{J}_{2a}) \, \mathbf{J}_{2u}]^{-1} \, \mathbf{J}_{2u}^T \, \Delta \, (\mathbf{J}_{2a}) \\ \vdots \\ [\mathbf{J}_{Nu}^T \, \Delta \, (\mathbf{J}_{Na}) \, \mathbf{J}_{Nu}]^{-1} \, \mathbf{J}_{Nu}^T \, \Delta \, (\mathbf{J}_{Na}) \end{pmatrix} \dot{\mathbf{p}}. \qquad (57)$$

The conditions guaranteeing the computability of the unactuated and actuated inverse solutions are identical and are indicated in the soluble motion criterion in (47). We substitute (57) into the partitioned composite robot equation in (52) to obtain:

$$\begin{pmatrix} \mathbf{I} - \mathbf{J}_{1u} \left[\mathbf{J}_{1u}^T \Delta \left(\mathbf{J}_{1a} \right) \mathbf{J}_{1u} \right]^{-1} \mathbf{J}_{1u}^T \Delta \left(\mathbf{J}_{1a} \right) \\ \mathbf{I} - \mathbf{J}_{2u} \left[\mathbf{J}_{2u}^T \Delta \left(\mathbf{J}_{2a} \right) \mathbf{J}_{2u} \right]^{-1} \mathbf{J}_{2u}^T \Delta \left(\mathbf{J}_{2a} \right) \\ \vdots \\ \mathbf{I} - \mathbf{J}_{Nu} \left[\mathbf{J}_{Nu}^T \Delta \left(\mathbf{J}_{Na} \right) \mathbf{J}_{Nu} \right]^{-1} \mathbf{J}_{Nu}^T \Delta \left(\mathbf{J}_{Na} \right) \end{pmatrix} \dot{\mathbf{p}}$$

$$= \begin{pmatrix} \mathbf{J}_{1a} & 0 & \dots & 0 \\ 0 & \mathbf{J}_{2a} & \ddots & 0 \\ \vdots & \ddots & \ddots & 0 \\ 0 & \dots & 0 & \mathbf{J}_{Na} \end{pmatrix} \dot{\mathbf{q}}_a, \qquad (58)$$

or

$$\mathbf{A}_a \, \dot{\mathbf{p}} = \mathbf{B}_a \, \dot{\mathbf{q}}_a. \qquad (59)$$

The robot actuation equation in (58) has the form of (39) with $\mathbf{A}_a$, $\mathbf{B}_a$, $\dot{\mathbf{p}}$, and $\dot{\mathbf{q}}_a$ playing the roles of $\mathbf{A}$, $\mathbf{B}$, $\mathbf{x}$ and $\mathbf{y}$, respectively. We apply the solution tree in Figure 11 to (59) and obtain the actuation characterization tree in Figure 14.

The actuation characterization tree, in analogy with the mobility characterization tree, indicates the properties of the actuation structure of a WMR. The branch tests are developed from the solution tree in Figure 11. We concentrate on the implications of the solutions.

The system of linear algebraic equations in (59) representing the actuation structure of the WMR may be determined, undetermined, or overdetermined. If branch (A1) succeeds, the actuation structure is undetermined and there is no unique solution for the robot motion $\dot{\mathbf{p}}$. Since we cannot calculate the robot motion, it is unpredictable, and some robot DOFs are uncontrollable. We suggest that undetermined actuation structures be avoided.

If branch (A0) succeeds, we are assured that all robot DOFs are actuated. Specifically, all robot motions allowed by the mobility structure can be produced by the actuators. Consequently, we refer to branch test (A0) as the adequate actuation criterion:

Adequate Actuation Criterion

$$det(\mathbf{A}_a^T \, \mathbf{A}_a) \neq \mathbf{0}. \qquad (60)$$

If the actuation structure is overdetermined [branch (A01)], some of the actuator motions are dependent. If the dependent actuator motions are consistent [at branch (A010)] robot motion is produced, otherwise (at branch (A011)) wheel slip occurs. Any mechanical couplings between actuated wheel variables must satisfy the actuator dependencies to allow robot motion; we therefore refer to branch test (A010) as the actuator coupling criterion:

Actuator Coupling Criterion

$$\Delta(A_a)\, B_a\, \dot{q}_a = 0. \tag{61}$$

If the dependent actuator motions are not consistent [branch (A011)], wheel slip must occur because the least-squares error is non-zero. Since a control system cannot guarantee zero actuator tracking errors, the actuated wheel velocities may deviate from the values computed by the actuated inverse solution. In the presence of these tracking errors, the actuator coupling criterion is not satisfied and the system of linear algebraic equations in (59) becomes inconsistent with no solution. We refer to this situation as

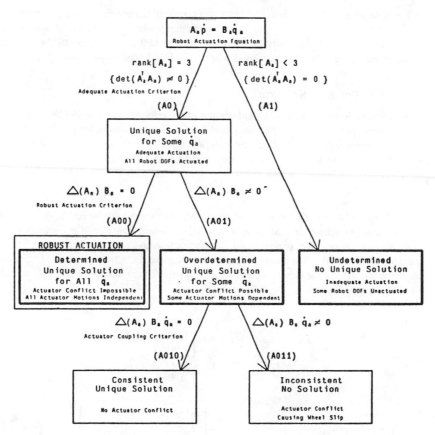

Figure 14. The actuation characterization tree.

actuator conflict because the forces and torques produced by the inconsistent actuator motions generate stress forces and torques within the WMR structure causing wheel slip instead of generating robot motion. A determined actuation structure (when branch (A00) succeeds) is robust in the sense that actuator conflict cannot occur in the presence of actuator tracking errors. The actuator motions are independent and all possible actuated wheel velocity vectors map into unique robot velocity vectors. Branch test (A00) is thus referred to as the robust actuation criterion:

Robust Actuation Criterion

$$\Delta(\mathbf{A}_a) \, \mathbf{B}_a = \mathbf{0}. \tag{62}$$

Because of actuator conflict, we suggest that overdetermined actuation structures be avoided. We recommend actuator arrangements leading to a robust (determined) actuation structure. In Sections IV G and IV H, we turn our attention to the sensed forward solution and relate the sensed wheel variables to the robot motion.

G. Sensed Forward Solution

The sensed forward solution calculates the robot velocity vector $\dot{\mathbf{p}}$ in (38) from the sensed wheel positions and velocities $\mathbf{q}_s$ and $\dot{\mathbf{q}}_s$. The development of the sensed forward solution parallels the actuated inverse solution in Section IV D. The first step is to separate the sensed and not-sensed wheel velocities and write (36) as:

$$\dot{\mathbf{p}} = \mathbf{J}_{is} \, \dot{\mathbf{q}}_{is} + \mathbf{J}_{in} \, \dot{\mathbf{q}}_{in}. \tag{63}$$

The subscripts "s" and "n" denote the sensed and not-sensed quantities, respectively. The numbers of sensed and not-sensed variables of wheel i are s_i and n_i, respectively (i.e., $s_i + n_i = w_i$). We assume that both the position and velocity of a sensed wheel variable are available. We combine the wheel equations in (63) for $i = 1, \ldots, N$ to form the partitioned robot sensing equation, with all of the unknown robot and wheel positions and velocities on the left-hand side:

$$\begin{pmatrix} \mathbf{I}_1 & -\mathbf{J}_{1n} & 0 & \cdots & 0 \\ \mathbf{I}_2 & 0 & -\mathbf{J}_{2n} & \cdots & \vdots \\ \vdots & \vdots & \ddots & \ddots & 0 \\ \mathbf{I}_N & 0 & \cdots & 0 & -\mathbf{J}_{Nn} \end{pmatrix} \begin{pmatrix} \dot{\mathbf{p}} \\ \dot{\mathbf{q}}_{1n} \\ \dot{\mathbf{q}}_{2n} \\ \vdots \\ \dot{\mathbf{q}}_{Nn} \end{pmatrix}$$

$$= \begin{pmatrix} \mathbf{J}_{1s} & 0 & \cdots & 0 \\ 0 & \mathbf{J}_{2s} & \ddots & \vdots \\ \vdots & \ddots & \ddots & 0 \\ 0 & \cdots & 0 & \mathbf{J}_{Ns} \end{pmatrix} \begin{pmatrix} \dot{\mathbf{q}}_{1s} \\ \dot{\mathbf{q}}_{2s} \\ \vdots \\ \dot{\mathbf{q}}_{Ns} \end{pmatrix} \tag{64}$$

or

$$A_n \dot{\mathbf{p}}_n = B_s \dot{\mathbf{q}}_s. \tag{65}$$

We define the total number of sensed wheel variables to be $s = s_1 + \ldots s_N$ and the total number of not-sensed wheel variables to be $n = n_1 + \ldots + n_N$. Thereby, A_n is $(3N \times [3 + n])$, $\dot{\mathbf{p}}_n$ is $([3 + n] \times 1)$, B_s is $(3N \times s)$, and $\dot{\mathbf{q}}_s$ is $(s \times 1)$. We apply the least-squares solution in (40) to calculate the vector of robot and not-sensed wheel velocities $\dot{\mathbf{p}}_n$ from the sensed wheel velocity vector $\dot{\mathbf{q}}_s$:

$$\dot{\mathbf{p}}_n = (A_n^T A_n)^{-1} A_n^T B_s \dot{\mathbf{q}}_s. \tag{66}$$

In contrast to the actuated inverse solution, the least-squares forward solution need not produce a zero error because of sensor noise and wheel slippage. In the presence of these error sources, we cannot calculate the exact velocity of the robot. Our least-squares solution does provide an optimal solution by minimizing the sum of the squared errors in the velocity components. We have applied our least-squares forward solution practically to dead reckoning for a WMR in the presence of sensor noise and wheel slippage.[1]

In Appendix 3, we assume that the matrix inverse $(A_n^T A_n)^{-1}$ is computable, and solve (66) for the robot velocities $\dot{\mathbf{p}}$. We find that

---

Sensed Forward Solution

$$\dot{\mathbf{p}} = [\Delta(J_{1n}) + \Delta(J_{2n}) + \ldots$$
$$+ \Delta(J_{Nn})]^{-1}[\Delta(J_{1n})J_{1s} \ \Delta(J_{2n})J_{2s} \ldots \Delta(J_{Nn})J_{Ns}]\dot{\mathbf{q}}_s$$

or

$$\dot{\mathbf{p}} = J_s \dot{\mathbf{q}}_s. \tag{67}$$

In Section IV H, we develop an adequate sensing criterion to indicate the conditions under which the sensed forward solution in (67) is applicable. A wheel without sensed variables does not contribute any columns $\Delta(J_{in})J_{is}$ to (67). Furthermore, if three independent wheel variables are not sensed, the matrix $\Delta(J_{in})$ is zero. We may thus eliminate the kinematic equations of motion of any wheel which has three not-sensed DOFs in the calculation of the sensed forward solution. We note that the Jacobian matrix of a steered wheel depends upon the steering angle. Therefore, if any wheel variables of a steered wheel are sensed, the steering angle must also be sensed so that J_{in} and J_{is} are computable. Since the matrix $[\Delta(J_{1n}) + \Delta(J_{2n}) + \ldots + \Delta(J_{Nn})]$ is (3×3), solving the system of linear algebraic equations in (67) for the robot velocities $\dot{\mathbf{p}}$ is not a computational burden.

H. Robot Sensing Characteristics

The relationship between the sensed wheel variables and the robot motion is the dual of the relationship between the actuated wheel variables and the robot motion. Our development thus parallels the discussion in Section IV F on actuation characteristics. We begin by rewriting the composite robot equation in (37) to relate the robot velocity vector to the sensed wheel velocity vector. We express the not-sensed wheel velocities in terms of the robot velocities by applying the actuated inverse solution in (55) with the not-sensed ("n" subscripts) and sensed ("s" subscripts) wheel velocities playing the roles of the actuated ("a" subscripts) and unactuated ("u" subscripts) wheel velocities, respectively:

$$
\dot{\mathbf{q}}_n = \begin{pmatrix} [\mathbf{J}_{1n}^T \Delta(\mathbf{J}_{1s}) \mathbf{J}_{1n}]^{-1} \mathbf{J}_{1n}^T \Delta(\mathbf{J}_{1s}) \\ [\mathbf{J}_{2n}^T \Delta(\mathbf{J}_{2s}) \mathbf{J}_{2n}]^{-1} \mathbf{J}_{2n}^T \Delta(\mathbf{J}_{2s}) \\ \vdots \\ [\mathbf{J}_{Nn}^T \Delta(\mathbf{J}_{Nn}) \mathbf{J}_{Nn}]^{-1} \mathbf{J}_{Nn}^T \Delta(\mathbf{J}_{Nn}) \end{pmatrix} \dot{\mathbf{q}}. \tag{68}
$$

The inverse solution is applicable for any WMR satisfying the soluble motion criterion in (47). We partition the sensed and not-sensed wheel velocities in the composite robot equation in (37) and substitute (68) for the not-sensed wheel velocities to obtain:

$$
\begin{pmatrix} \mathbf{I} - \mathbf{J}_{1n}[\mathbf{J}_{1n}^T \Delta(\mathbf{J}_{1s}) \mathbf{J}_{1n}]^{-1} \mathbf{J}_{1n}^T \Delta(\mathbf{J}_{1s}) \\ \mathbf{I} - \mathbf{J}_{2n}[\mathbf{J}_{2n}^T \Delta(\mathbf{J}_{2s}) \mathbf{J}_{2n}]^{-1} \mathbf{J}_{2n}^T \Delta(\mathbf{J}_{2s}) \\ \vdots \\ \mathbf{I} - \mathbf{J}_{Nn}[\mathbf{J}_{Nn}^T \Delta(\mathbf{J}_{Nn}) \mathbf{J}_{Nn}]^{-1} \mathbf{J}_{Nn}^T \Delta(\mathbf{J}_{Nn}) \end{pmatrix} \dot{\mathbf{p}} = \begin{pmatrix} \mathbf{J}_{1s} & 0 & \cdots & 0 \\ 0 & \mathbf{J}_{2s} & \cdots & 0 \\ \vdots & \ddots & \ddots & 0 \\ 0 & \cdots & 0 & \mathbf{J}_{Ns} \end{pmatrix} \dot{\mathbf{q}}_s, \tag{69}
$$

or

$$
\mathbf{A}_s \dot{\mathbf{p}} = \mathbf{B}_s \dot{\mathbf{q}}_s. \tag{70}
$$

The robot sensing equation in (70) has the form of (39) with $\mathbf{A}_s$, $\mathbf{B}_s$, $\dot{\mathbf{p}}$, and $\dot{\mathbf{q}}_s$ playing the roles of $\mathbf{A}$, $\mathbf{B}$, $\mathbf{x}$, and $\mathbf{y}$, respectively. We apply the solution tree of Figure 11 to the robot sensing equation in (70) to obtain the sensing characterization tree in Figure 15.

The solution of the robot velocity $\dot{\mathbf{p}}$ from the sensed wheel velocities $\dot{\mathbf{q}}_s$ may be determined, undetermined or overdetermined, depending on the matrices $\mathbf{A}_s$ and $\mathbf{B}_s$. In parallel with WMR actuation, undetermined systems are undesirable because one or more DOFs of the robot motion cannot be discerned from the sensed wheel velocities. Both determined and overdetermined sensing structures allow a unique solution for consistent sensor motions $\dot{\mathbf{q}}_s$. Branch (S0) thus provides the adequate sensing criteria in (71) which specifies whether all WMR motions allowed by the mobility structure are discernable through sensor measurements:

Adequate Sensing Criterion

$$det(\mathbf{A}_s^T \mathbf{A}_s) \neq 0. \qquad\qquad (71)$$

The adequate sensing criterion also specifies the conditions under which the sensed forward solution in (67) is applicable.

Determined sensing structures provide sufficient information for discerning the robot motion. Overdetermined sensing structures become inconsistent in the presence of sensor noise, which is analogous to the impact of actuator tracking errors on overdetermined actuation structures. Our forward solution in (67) anticipates the overde-

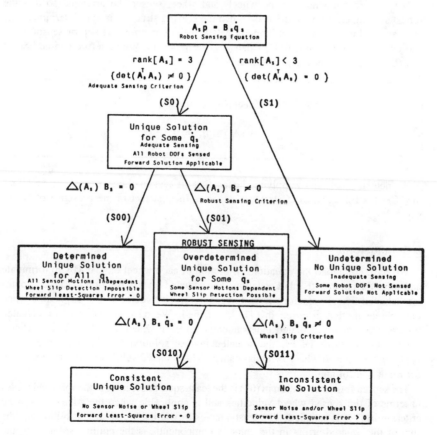

Figure 15. The sensing characterization tree.

termined nature of the sensor measurements and provides the least-squares solution. In the case of actuation, an overdetermined actuator structure causes undesirable actuator conflict. In contrast, redundant (and even inconsistent) information is desirable for the least-squares solution of the robot velocity from sensed wheel velocities. Redundant information in the least-squares solution reduces the effects of sensor noise on the solution of the robot velocity. Overdetermined sensing structures are thereby robust and branch test (S01) is referred to as the robust sensing criterion:

Robot Sensing Criterion

$$\Delta \, (\mathbf{A}_s) \, \mathbf{B}_s \neq \mathbf{0}. \qquad (72)$$

We thus recommend that the wheels and wheel sensors be arranged so that the robust sensing criterion is satisfied. When the sensing structure is overdetermined, the least-squared error is zero [at branch (S010)] if there is no wheel slip or sensor noise and non-zero [at branch (S011)] when wheel slip occurs. We therefore denote branch test (S011) as the wheel slip criterion:

Wheel Slip Criterion

$$\Delta \, (\mathbf{A}_s) \, \mathbf{B}_s \dot{\mathbf{q}}_s \neq \mathbf{0}. \qquad (73)$$

We detect wheel slip by applying the fact that the system of linear algebraic equations in (70) of a robust sensing structure becomes inconsistent in the presence of wheel slip.[1]

I. Conclusions

We have combined the equations of motion of each wheel on a WMR to formulate and solve the composite robot equation. The actuated inverse solution in (55) computes the actuated wheel velocities from the robot velocity vector and is applicable when the soluble motion criterion in (47) is satisfied. We have shown that the actuated inverse solution is calculated independently for each wheel on a WMR. For wheels which possess three DOFs, the actuated inverse solution is calculated directly by applying the inverse wheel Jacobian matrix. The actuated velocities are then extracted for robot control applications.

The sensed forward solution in (67) is the least-squares solution of the robot velocities in terms of the sensed wheel velocities and is applicable when the adequate sensing criterion in (71) is satisfied. The least-squares forward solution, which minimizes the sum of the squared errors in the velocity components, is the optimal solution of the robot velocities in the presence of sensor noise and wheel slippage. We have found that the sensed forward solution may be simplified by eliminating the equations of motion of wheels having three not-sensed DOFs because they do not affect the solution. If any variables of a steered wheel are sensed, the steering angle must also be sensed.

We have discussed the nature of solutions of the composite robot equation and their implications for robot mobility (in Section IV D), actuation (in Section IV F), and sensing (in Section IV H). We have developed the mobility characterization tree in Figure 13 to characterize the motion properties of a WMR. The implications of the mobility characterization tree are summarized by the following insights. If the soluble motion criterion in (47) is satisfied, the actuated inverse solution, actuation and sensing trees, and the WMR DOF calculation in (50) are applicable. The three DOF motion criterion in (48) indicates whether the WMR kinematic structure allows three DOF motion. If the kinematic structure does not allow three DOF motion, the kinematic motion constraints are computed according to (49). The number of WMR DOFs are calculated from (50).

The implications of the actuation characterization tree in Figure 14 are summarized by three criteria. The adequate actuation criterion in (60) indicates whether the number and placement of the actuators is adequate for producing all motions allowed by the mobility structure. If the adequate actuation criterion is not satisfied, some robot DOFs are uncontrollable. The robust actuation criterion in (62) determined whether the actuation structure is robust; i.e., actuator conflict cannot occur in the presence of actuator tracking errors. If the actuation structure is adequate but not robust, some actuator motions are dependent. The actuator coupling criterion in (61) calculates these actuator dependencies which must be satisfied to avoid actuator conflict and forced wheel slip.

The sensing characterization tree in Figure 15 indicates properties of the sensing structure of a WMR. The adequate sensing criterion in (71) indicates whether the number and placement of the wheel sensors is adequate for discerning all robot motions allowed by the mobility structure. The robust sensing criterion in (72) indicates whether the sensing structure is such that the calculation of the robot position from wheel sensor measurements is minimally sensitive to wheel slip and sensor noise. The wheel slip criterion in (73) provides a computational algorithm for detecting wheel slip in robust sensing structures.

V. URANUS

To illustrate the kinematic modeling of Uranus, we provide two kinematic descriptions in Figure 16: a written description, and a top and side view sketch. We assign the coordinate systems to create the coordinate transformation matrices. We then form the wheel Jacobian matrices by substituting elements of the coordinate transformation matrices into the symbolic wheel Jacobian matrices in Appendix 1. We determine the nature of the mobility, actuation, and sensing structures to gain insight into the mobility characteristics of Uranus. We compute the wheel motor velocities from the robot velocity vector (i.e., actuated inverse solution) and the least-squares robot velocity vector from the sensed wheel velocities and positions (i.e., sensed forward solution). We conclude with remarks on its kinematic structure and suitability for particular tasks.

Uranus[30] has the kinematic structure of the Wheelon wheelchair[31]: four omnidirec-

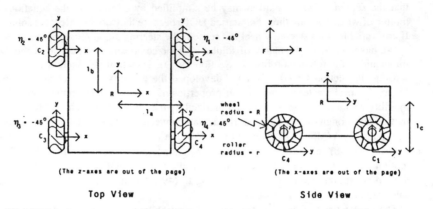

Figure 16. Coordinate system assignments for Uranus.

tional wheels with rollers at 45° angles to the wheels. The coordinate system assignments and robot dimensions are shown in Figure 16.

Since there are no steering links, the coordinate transformation matrices for Uranus are:

$$^R\mathbf{T}_{H_1} = \begin{pmatrix} 1 & 0 & 0 & l_a \\ 0 & 1 & 0 & l_b \\ 0 & 0 & 1 & -l_c \\ 0 & 0 & 0 & 1 \end{pmatrix} \qquad {}^{H_1}\mathbf{\Phi}_{S_1} = {}^{S_1}\mathbf{T}_{C_1} = \mathbf{I}$$

$$^R\mathbf{T}_{H_2} = \begin{pmatrix} 1 & 0 & 0 & -l_a \\ 0 & 1 & 0 & l_b \\ 0 & 0 & 1 & -l_c \\ 0 & 0 & 0 & 1 \end{pmatrix} \qquad {}^{H_2}\mathbf{\Phi}_{S_2} = {}^{S_2}\mathbf{T}_{C_2} = \mathbf{I}$$

$$^R\mathbf{T}_{H_3} = \begin{pmatrix} 1 & 0 & 0 & -l_a \\ 0 & 1 & 0 & -l_b \\ 0 & 0 & 1 & -l_c \\ 0 & 0 & 0 & 1 \end{pmatrix} \qquad {}^{H_3}\mathbf{\Phi}_{S_3} = {}^{S_3}\mathbf{T}_{C_3} = \mathbf{I}$$

$$^R\mathbf{T}_{H_4} = \begin{pmatrix} 1 & 0 & 0 & l_a \\ 0 & 1 & 0 & -l_b \\ 0 & 0 & 1 & -l_c \\ 0 & 0 & 0 & 1 \end{pmatrix} \qquad {}^{H_4}\mathbf{\Phi}_{S_4} = {}^{S_4}\mathbf{T}_{C_4} = \mathbf{I}.$$

The radius assignments are $R_1 = R_2 = R_3 = R_4 = R$, and $r_1 = r_2 = r_3 = r_4 = r$, and the roller angles are $\eta_1 = \eta_3 = -45°$, and $\eta_2 = \eta_4 = 45°$. The Jacobian matrix for omnidirectional wheels in (A1.9) enables us to write the equation of motion for each wheel:

$$\dot{\mathbf{p}} = \begin{pmatrix} v_{Rx} \\ v_{Ry} \\ \omega_R \end{pmatrix} = \begin{pmatrix} 0 & -r\sqrt{2}/2 & l_b \\ R & -r\sqrt{2}/2 & -l_a \\ 0 & 0 & 1 \end{pmatrix} \begin{pmatrix} \omega_{w_1 x} \\ \omega_{w_1 r} \\ \omega_{w_1 z} \end{pmatrix} = \mathbf{J}_1 \dot{\mathbf{q}}_1 \tag{74}$$

$$\dot{\mathbf{p}} = \begin{pmatrix} v_{Rx} \\ v_{Ry} \\ \omega_R \end{pmatrix} = \begin{pmatrix} 0 & r\sqrt{2}/2 & l_b \\ R & -r\sqrt{2}/2 & l_a \\ 0 & 0 & 1 \end{pmatrix} \begin{pmatrix} \omega_{w_2 x} \\ \omega_{w_2 r} \\ \omega_{w_2 z} \end{pmatrix} = \mathbf{J}_2 \dot{\mathbf{p}}_2 \tag{75}$$

$$\dot{\mathbf{p}} = \begin{pmatrix} v_{Rx} \\ v_{Ry} \\ \omega_R \end{pmatrix} = \begin{pmatrix} 0 & -r\sqrt{2}/2 & -l_b \\ R & -r\sqrt{2}/2 & l_a \\ 0 & 0 & 1 \end{pmatrix} \begin{pmatrix} \omega_{w_3 x} \\ \omega_{w_3 r} \\ \omega_{w_3 z} \end{pmatrix} = \mathbf{J}_3 \dot{\mathbf{q}}_3, \tag{76}$$

$$\dot{\mathbf{p}} = \begin{pmatrix} v_{Rx} \\ v_{Ry} \\ \omega_R \end{pmatrix} = \begin{pmatrix} 0 & r\sqrt{2}/2 & -l_b \\ R & -r\sqrt{2}/2 & -l_a \\ 0 & 0 & 1 \end{pmatrix} \begin{pmatrix} \omega_{w_4 x} \\ \omega_{w_4 r} \\ \omega_{w_4 z} \end{pmatrix} = \mathbf{J}_4 \dot{\mathbf{p}}_4. \tag{77}$$

Since the soluble motion criterion is satisfied, the actuated inverse solution is applicable and none of the wheels has redundant DOFs. Furthermore, the three DOF criterion is satisfied and the motion structure is capable of three DOF motion.

The adequate actuation criterion yields: $\det[\mathbf{A}_a^T \mathbf{A}_a] = 64(l_a + l_b)^2$. The actuators are thus able to provide motion in all three DOFs. We find that the robust actuation criterion is not satisfied. The actuation structure is thus not robust and actuator conflict may occur. The sensed and actuated wheel variables are identical so that the sensing structure is robust which allows the detection of wheel slip.[1] The sensed forward solution is therefore applicable.

Since the mobility structure of Uranus allows three DOFs, the actuated inverse solution in (55) is exact for all robot motions. The actuated inverse solution is:

$$\begin{pmatrix} \omega_{w_1 x} \\ \omega_{w_2 x} \\ \omega_{w_3 x} \\ \omega_{w_4 x} \end{pmatrix} = \frac{1}{R} \begin{pmatrix} -1 & 1 & l_a + l_b \\ 1 & 1 & -l_a - l_b \\ -1 & 1 & -l_a - l_b \\ 1 & 1 & l_a + l_b \end{pmatrix} \begin{pmatrix} v_{Rx} \\ v_{Ry} \\ \omega_R \end{pmatrix}. \tag{78}$$

The actuated inverse solution in (78) may be obtained by assuming that all wheel variables are actuated, applying the inverse solution in (56) and extracting only the actuated wheel variables. This alternate approach is less computationally intensive because the inverse solution for each wheel simplifies to inverting each of the Jacobian matrices.

We apply the least-squares sensed forward solution in (67) to obtain:

$$\begin{pmatrix} v_{Rx} \\ v_{Ry} \\ \omega_R \end{pmatrix} = \frac{R}{4(l_a + l_b)} \begin{pmatrix} -(l_a + l_b) & (l_a + l_b) & -(l_a + l_b) & (l_a + l_b) \\ (l_a + l_b) & (l_a + l_b) & (l_a + l_b) & (l_a + l_b) \\ 1 & -1 & -1 & 1 \end{pmatrix} \begin{pmatrix} \omega_{w_1x} \\ \omega_{w_2x} \\ \omega_{w_3x} \\ \omega_{w_4x} \end{pmatrix}.$$

$$(79)$$

Uranus is a general-purpose three DOF WMR, with the kinematic capabilities of the Unimation robot.[32] The actuation structure is adequate and the sensing structure is robust in comparison with Unimation's robust actuation and adequate sensing. Uranus has more ground clearance because of the arrangement of the wheels. Also, the wheel profiles are exact circles because the rollers are at 45° angles avoiding the discontinuity of wheels with 90° rollers. To utilize practically the three DOF capabilities of this robot, we envision the simultaneous operation of an on-board manipulator.

We have illustrated that our kinematic modeling methodology in Section III and the solutions in Section IV establish the foundation for developing and solving the kinematic equations of motion of a WMR. Uranus shows that formulating the equations of motion for an actual WMR is a straightforward procedure which does not require insight into the operation of the robot.

VI. CONCLUDING REMARKS

We have developed and illustrated a methodology for the kinematic modeling of WMRs which is summarized in Section III I. Our kinematic model enables the determination of the salient WMR mobility characteristics summarized in Section IV I. In Section V, we have demonstrated the practical application of our methodology to the kinematic modeling of Uranus.

In our companion technical report,[1] we build upon our kinematic modeling methodology for WMRs by: surveying existing WMRs; developing naming and diagramming conventions for WMR kinematic structures; applying the kinematic model to design, kinematics-based feedback control, dead-reckoning, and slip detection; and formulating the kinematic models of the Unimation robot,[32] Newt,[33] Neptune,[34], Rover,[35] and the Stanford Cart.[10]

We summarize our results from our companion technical report.[1] Our nomenclature provides a convenient literal and verbal representation of the essential kinematic information of a WMR; and our symbolic representation displays pictorially the essential kinematic relationships between the robot body, wheels, and steering links using mnemonic symbols. Both representations are convenient tools for the design and mobility analysis of WMRs. Just as the mobility characterization tree in Figure 13 enables us to determine the motion characteristics of an existing WMR, we may utilize the tree to design WMRs to possess such desired characteristics as two or three DOFs. We suggest that three DOF WMRs are applicable for use with an on-board manipulator. The mobility of the base extends the workspace of the manipulator. The majority of practical applications (i.e., parts, tools, and materials transport) require only two DOFs. We conclude that a WMR having two diametrically opposed driven wheels (such as

Newt[33]) is ideal for this application because of the simplicity of its mechanical design and kinematic model. The actuation characterization tree in Figure 14 may be applied to design a WMR to have a robust actuation structure, thus avoiding actuator conflict.[1] Similarly, the sensing characterization tree in Figure 15 may be applied to design a WMR with a robust sensing structure to minimize the adverse effects of wheel slip on the calculation of the WMR position. We have found that the set of actuated wheel variables and sensed wheel variables cannot coincide if both robust actuation and robust sensing are desired. This finding is contrary to the designs of existing WMRs in which only the actuated wheel variables are sensed. We have detailed a kinematics-based robot level control system for WMRs which utilizes the sensed forward and actuated inverse solutions. We have developed a dead-reckoning update calculation by integrating the robot velocity computed by the sensed forward solution. We have also developed an algorithm for detecting wheel slip in robust sensing structures by calculating the least-squares error in the sensed forward solution.

Even though our research is motivated by, and tailored to, WMRs, our methodology may be applied to the kinematic modeling of other mechanisms, such as legged or treaded vehicles. The analysis of mechanisms having higher-pair joints, multiple closed-link chains, or unactuated and unsensed joint variables may benefit from our methodology. In particular, our matrix coordinate transformation algebra (in Section III E) may be applied to the transformation matrices expressing the relationships between lower and high-pair joints.

We are extending our WMR research to include the dynamic modeling and dynamics-based feedback control of WMRs. The few WMR control systems which have been documented are wheel level control systems[36,37] which do not incorporate a dynamic model of the WMR. The documented designs are tailored to the specific WMR being controlled. A dynamic WMR model is required to apply such manipulator control algorithms as robust computed torque[38] to the feedback control of WMRs.

APPENDIX I: WHEEL JACOBIAN MATRICES

A. Introduction

In this appendix, we develop the wheel Jacobian matrices for conventional wheels, steered conventional wheels, omnidirectional wheels, and ball wheels. The wheel Jacobian matrix (introduced in Section III G 3) relates the velocities of the WMR to the velocities of the wheel. The wheel Jacobian matrix is the product of the pseudo-Jacobian matrix $\hat{J}_i$ and the wheel matrix W_i:

$$J_1 = \hat{J}_i \, W_1. \tag{A1.1}$$

The pseudo-Jacobian matrix relates the wheel pseudo-velocities to the robot velocities, as described in Section III G 3:

$$\hat{J}_i = \begin{pmatrix} \cos {}^R\theta_{C_i} & -\sin {}^R\theta_{C_i} & {}^Rd_{C_iy} & -{}^Rd_{H_iy} \\ \sin {}^R\theta_{C_i} & \cos {}^R\theta_{C_i} & -{}^Rd_{C_ix} & {}^Rd_{H_ix} \\ 0 & 0 & 1 & -1 \end{pmatrix}. \tag{A1.2}$$

The wheel matrix in (A1.1) relates the pseudo-velocities to the actual wheel velocities. The wheel equations of motion in Figure 2 are applied to construct the wheel matrices. The pseudo-velocities $^{C_i}v_{C_i x}$, $^{C_i}v_{C_i y}$, and $^{C_i}\omega_{C_i}$ are the velocities v_x, v_y, and ω_z in Figure 2. The actual wheel velocities are the angular velocities of the wheel and rollers $\omega_{w_i x}$, $\omega_{w_i y}$, $\omega_{w_i z}$, and $\omega_{w_i r}$ about their respective axes. With these observations, the wheel matrix for each wheel is written directly from the wheel equations of motion in Figure 2. The wheel Jacobian matrix is then formed by multiplying the pseudo-Jacobian matrix in (A1.2) by the wheel matrix. We consider each of the aforementioned wheels in turn.

B. Conventional Non-Steered Wheel

The conventional non-steered wheel has two DOFs: motion in the direction of the wheel orientation, and rotational slip about the point of contact, corresponding to the two wheel pseudo-velocities $^{C_i}v_{C_i y}$, and $^{C_i}\omega_{C_i}$, respectively. The actual wheel velocities are the angular velocity of the wheel about its axle $\omega_{w_i x}$ and the angular velocity of the rotational slip $\omega_{w_i z}$. These velocities are related by the (4×2) wheel matrix $\mathbf{W}_i$ in (A1.3).

$$\dot{\mathbf{q}}_i = \begin{pmatrix} 0 & 0 \\ R & 0 \\ 0 & 1 \\ 0 & 0 \end{pmatrix} \begin{pmatrix} \omega_{w_i x} \\ \omega_{w_i z} \end{pmatrix} = \mathbf{W}_i \dot{\mathbf{q}}_i \qquad (A1.3)$$

The wheel matrix is multiplied by the pseudo-Jacobian matrix in (A1.2) to form the (3×2) Jacobian matrix:

Conventional Non-Steered Wheel Jacobian Matrix

$$\mathbf{J}_1 = \begin{pmatrix} -R_i \sin {}^R\theta_{C_i} & {}^R d_{C_i y} \\ R_i \cos {}^R\theta_{C_i} & -{}^R d_{C_i x} \\ 0 & 1 \end{pmatrix}. \qquad (A1.4)$$

This wheel is termed degenerate because the Jacobian is non-square and thus non-invertible. Even though a robot velocity vector can be calculated from a wheel velocity vector, it is not always possible to compute a wheel velocity vector from a robot velocity vector. The degenerate nature of the kinematic equations-of-motion of the non-steered conventional wheel precludes its application to three DOF WMRs.

C. Conventional Steered Wheel

The conventional steered wheel has an additional DOF provided by the steering joint corresponding to the pseudo-velocity $^{H_i}\omega_{S_i}$. The actual steering velocity $\omega_{s_i z}$ (in

Fig. 2) is equal to the steering pseudo-velocity. The (4×3) wheel matrix and the (3×3) wheel Jacobian matrix are, respectively:

$$\dot{\hat{q}}_i = \begin{pmatrix} 0 & 0 & 0 \\ R & 0 & 0 \\ 0 & 1 & 0 \\ 0 & 0 & 1 \end{pmatrix} \begin{pmatrix} \omega_{w_ix} \\ \omega_{w_iz} \\ \omega_{s_iz} \end{pmatrix} = \mathbf{W}_i \, \dot{q}_i \tag{A1.5}$$

and

Conventional Steered Wheel Jacobian Matrix

$$\mathbf{J}_1 = \begin{pmatrix} -R_i \sin {}^R\theta_{C_i} & {}^Rd_{C_iy} & -{}^Rd_{H_iy} \\ R_i \cos {}^R\theta_{C_i} & -{}^Rd_{C_ix} & {}^Rd_{H_ix} \\ 0 & 1 & -1 \end{pmatrix}. \tag{A1.6}$$

The Jacobian matrix is invertible if its determinant is nonzero; i.e., if

$$det(\mathbf{J}_i) = R_i \, ({}^{S_i}d_{C_iy} \cos {}^{S_i}\theta_{C_i} - {}^{S_i}d_{C_ix} \sin {}^{S_i}\theta_{C_i}) \neq 0. \tag{A1.7}$$

The determinant is zero and the conventional steered wheel is redundant if the steering axis intercepts the wheel point of contact (i.e., if ${}^{S_i}d_{C_ix} = {}^{S_i}d_{C_iy} = 0$) or if the wheel is oriented perpendicular to the steering link (i.e., if ${}^{C_i}d_{S_iy} = {}^{S_i}d_{C_ix} \sin {}^{S_i}\theta_{C_i} - {}^{S_i}d_{C_iy} \cos {}^{S_i}\theta_{C_i} = 0$).

D. Omnidirectional Wheel

The omnidirectional wheel possesses three DOFs without a steering joint. The DOFs are motion in the direction of the wheel orientation, motion in the direction of the roller orientation, and rotational slip, which correspond respectively to the actual wheel velocities ω_{w_ix}, ω_{w_ir}, and ω_{w_iz}. The pseudo-velocities $\dot{\hat{q}}_i$ are linear combinations of the actual velocities $\mathbf{q}_i$:

$$\dot{\hat{q}}_i = \begin{pmatrix} 0 & r \sin \eta & 0 \\ R & -r \cos \eta & 0 \\ 0 & 0 & 1 \\ 0 & 0 & 0 \end{pmatrix} \begin{pmatrix} \omega_{w_ix} \\ \omega_{w_ir} \\ \omega_{w_iz} \end{pmatrix} = \mathbf{W}_i \, \dot{q}_i. \tag{A1.8}$$

The wheel Jacobian matrix is:

Omnidirectional Wheel Jacobian Matrix

$$\mathbf{J_i} = \begin{pmatrix} -R_i \sin {}^R\theta_{C_i} & r_i \sin ({}^R\theta_{C_i} + \eta_i) & {}^Rd_{C_iy} \\ R_i \cos {}^R\theta_{C_i} & -r_i \cos ({}^R\theta_{C_i} + \eta_i) & -{}^Rd_{C_ix} \\ 0 & 0 & 1 \end{pmatrix} \qquad (A1.9)$$

The determinant of the omnidirectional wheel Jacobian matrix is $-R_i r_i \sin \eta_i$, and consequently, the Jacobian matrix is invertible whenever the rollers are not aligned with the wheel (i.e., whenever $\eta_i \neq 0$).

E. Ball Wheel

The ball wheel possesses three DOFs of rotation about the three normal axes positioned at the wheel center. The wheel matrix relating the actual wheel velocities ω_{w_ix}, ω_{w_iy}, and ω_{w_iz} to the pseudo-velocities is:

$$\dot{\hat{q}}_i = \begin{pmatrix} R & 0 & 0 \\ 0 & R & 0 \\ 0 & 0 & 1 \\ 0 & 0 & 0 \end{pmatrix} \begin{pmatrix} \omega_{w_ix} \\ \omega_{w_iy} \\ \omega_{w_iz} \end{pmatrix} = \mathbf{W}_i \, \dot{q}_i. \qquad (A1.10)$$

The wheel Jacobian matrix is:

Ball Wheel Jacobian Matrix

$$\mathbf{J_1} = \begin{pmatrix} R_i \cos {}^R\theta_{C_i} & -R_i \sin {}^R\theta_{C_i} & {}^Rd_{C_iy} \\ R_i \sin {}^R\theta_{C_i} & R_i \cos {}^R\theta_{C_i} & -{}^Rd_{C_ix} \\ 0 & 0 & 1 \end{pmatrix}. \qquad (A1.11)$$

Since the determinant of the ball wheel Jacobian matrix is R_i^2, it is invertible for all non-zero wheel radii.

In Section V, the wheel Jacobian matrices developed in this appendix are applied to obtain the kinematic equations of motions of Uranus.

APPENDIX II: ACTUATED INVERSE SOLUTION MATRIX CALCULATIONS

In this appendix, we detail the matrix manipulations leading to the actuated inverse solution in Section IV E. We solve the composite partitioned robot equation in (52)

$$\mathbf{A}_0 \dot{\mathbf{p}} = \begin{pmatrix} \mathbf{I}_1 \\ \mathbf{I}_2 \\ \vdots \\ \mathbf{I}_N \end{pmatrix} \dot{\mathbf{p}}$$

$$= \begin{pmatrix} \mathbf{J}_{1a} & 0 & \ldots & 0 & \mathbf{J}_{1u} & 0 & \ldots & 0 \\ 0 & \mathbf{J}_{2a} & \ddots & \vdots & 0 & \mathbf{J}_{2u} & \ddots & \vdots \\ \vdots & \ddots & \ddots & 0 & \vdots & \ddots & \ddots & 0 \\ 0 & \ldots & 0 & \mathbf{J}_{Na} & 0 & \ldots & 0 & \mathbf{J}_{Nu} \end{pmatrix} \begin{pmatrix} \dot{\mathbf{q}}_{1a} \\ \dot{\mathbf{q}}_{2a} \\ \vdots \\ \dot{\mathbf{q}}_{Na} \\ \dot{\mathbf{q}}_{1u} \\ \dot{\mathbf{q}}_{2u} \\ \vdots \\ \dot{\mathbf{q}}_{Nu} \end{pmatrix} \quad \text{(A2.1)}$$

$$= \mathbf{B}_{0p} \begin{pmatrix} \dot{\mathbf{q}}_a \\ \dot{\mathbf{q}}_u \end{pmatrix}$$

to calculate the actuated wheel velocities $\dot{\mathbf{q}}_a$ in the least-squares solution in (54):

$$\begin{pmatrix} \dot{\mathbf{q}}_a \\ \dot{\mathbf{q}}_u \end{pmatrix} = (B_{0p}^T B_{0p})^{-1} \mathbf{B}_{0p}^T \mathbf{A}_0 \dot{\mathbf{p}}. \quad \text{(A2.2)}$$

We begin by forming the matrix product:

$$(\mathbf{B}_{0p}^T \mathbf{B}_{0p}) = \begin{pmatrix} \mathbf{J}_{1a}^T \mathbf{J}_{1a} & 0 & \ldots & 0 & \mathbf{J}_{1a}^T \mathbf{J}_{1u} & 0 & \ldots & 0 \\ 0 & \mathbf{J}_{2a}^T \mathbf{J}_{2a} & \ddots & \vdots & 0 & \mathbf{J}_{2a}^T \mathbf{J}_{2u} & \ddots & \vdots \\ \vdots & \ddots & \ddots & 0 & \vdots & \ddots & \ddots & 0 \\ 0 & \ldots & c & \mathbf{J}_{Na}^T \mathbf{J}_{Na} & 0 & \ldots & 0 & \mathbf{J}_{Na}^T \mathbf{J}_{Nu} \\ \mathbf{J}_{1u}^T \mathbf{J}_{1a} & 0 & \ldots & 0 & \mathbf{J}_{1u}^T \mathbf{J}_{1u} & 0 & \ldots & 0 \\ 0 & \mathbf{J}_{2u}^T \mathbf{J}_{2a} & \ddots & \vdots & 0 & \mathbf{J}_{2u}^T \mathbf{J}_{2u} & \ddots & \vdots \\ \vdots & \ddots & \ddots & 0 & \vdots & \ddots & \ddots & 0 \\ 0 & \ldots & 0 & \mathbf{J}_{Nu}^T \mathbf{J}_{Na} & 0 & \ldots & 0 & \mathbf{J}_{Nu}^T \mathbf{J}_{Nu} \end{pmatrix}$$

$$= \begin{pmatrix} \mathbf{D}_{aa} & \mathbf{D}_{au} \\ \mathbf{D}_{au}^T & \mathbf{D}_{uu} \end{pmatrix}. \quad \text{(A2.3)}$$

To invert $(\mathbf{B}_{0p}^T \mathbf{B}_{0p})$, we have written the matrix in block form with four components, each one a block diagonal matrix. We let the block matrix $\mathbf{X}$ be the inverse of the matrix in (A2.3). To compute the block components of the matrix inverse in terms of the block components of the matrix in (A2.3), we apply the fact that the inverse of a matrix times the matrix itself is the identity matrix; i.e.,

$$\begin{pmatrix} \mathbf{X}_{11} & \mathbf{X}_{12} \\ \mathbf{X}_{21} & \mathbf{X}_{22} \end{pmatrix} \begin{pmatrix} \mathbf{D}_{aa} & \mathbf{D}_{au} \\ \mathbf{D}_{au}^T & \mathbf{D}_{uu} \end{pmatrix} = \begin{pmatrix} \mathbf{I} & \mathbf{0} \\ \mathbf{0} & \mathbf{I} \end{pmatrix}. \quad \text{(A2.4)}$$

Since we seek only the upper (actuated) components of the wheel velocity vector $\dot{\mathbf{q}}_a$ in (A2.1), we calculate only the two components in the top row of the block matrix inverse. We thus separate the solution of the actuated wheel velocities

$$\dot{\mathbf{q}}_a = (\mathbf{X}_{11} \ \mathbf{X}_{12}) \ \mathbf{B}_{0p}^T \ \mathbf{A}_0 \dot{\mathbf{p}} \tag{A2.5}$$

from the solution of the unactuated ones. We expand (A2.4) to obtain

$$\mathbf{X}_{11} \ \mathbf{D}_{aa} + \mathbf{X}_{12} \mathbf{D}_{au}^T = \mathbf{I} \tag{A2.6}$$

and

$$\mathbf{X}_{11} \mathbf{D}_{au} + \mathbf{X}_{12} \mathbf{D}_{uu} = \mathbf{0}. \tag{A2.7}$$

From (A2.6) and (A2.7), we find

$$\mathbf{X}_{12} = -\mathbf{X}_{11} \mathbf{D}_{au} \mathbf{D}_{uu}^{-1} \tag{A2.8}$$

and

$$\mathbf{X}_{11} = (\mathbf{D}_{aa} - \mathbf{D}_{au} \mathbf{D}_{uu}^{-1} \mathbf{D}_{au}^T)^{-1}, \tag{A2.9}$$

where

$$\mathbf{D}_{uu}^{-1} = \begin{pmatrix} (\mathbf{J}_{1u}^T \mathbf{J}_{1u})^{-1} & 0 & \cdots & 0 \\ 0 & (\mathbf{J}_{2u}^T \mathbf{J}_{2u})^{-1} & \ddots & \vdots \\ \vdots & \ddots & \ddots & 0 \\ 0 & \cdots & 0 & (\mathbf{J}_{Nu}^T \mathbf{J}_{Nu})^{-1} \end{pmatrix}. \tag{A2.10}$$

The matrix $\mathbf{X}_{11}$ in (A2.9) is

$$\mathbf{X}_{11} = \begin{pmatrix} -[\mathbf{J}_{1a}^T \Delta(\mathbf{J}_{1u}) \mathbf{J}_{1a}]^{-1} & 0 & \cdots & 0 \\ 0 & -[\mathbf{J}_{2a}^T \Delta(\mathbf{J}_{2u}) \mathbf{J}_{2a}]^{-1} & \ddots & \vdots \\ \vdots & \ddots & \ddots & 0 \\ 0 & \cdots & 0 & -[\mathbf{J}_{Na}^T \Delta(\mathbf{J}_{Nu}) \mathbf{J}_{Na}]^{-1} \end{pmatrix} \tag{A2.11}$$

The matrix $\mathbf{X}_{12}$ in (A2.8) is

$$\mathbf{X}_{12} = \begin{pmatrix} \Lambda_1 & 0 & \cdots & 0 \\ 0 & \Lambda_2 & \ddots & \vdots \\ \vdots & \ddots & \ddots & 0 \\ 0 & \cdots & 0 & \Lambda_N \end{pmatrix} \tag{A2.12}$$

where,

$$\Lambda_i = [\mathbf{J}_{ia}^T \Delta(\mathbf{J}_{iu}) \mathbf{J}_{ia}]^{-1} \mathbf{J}_{ia}^T \mathbf{J}_{iu} (\mathbf{J}_{iu}^T \mathbf{J}_{iu})^{-1}. \tag{A2.13}$$

We substitute (A2.11) and (A2.12) into (A2.5) to obtain the actuated wheel velocity vector

$$\dot{\mathbf{q}}_a = \begin{pmatrix} [\mathbf{J}_{1a}^T \Delta(\mathbf{J}_{1u}) \mathbf{J}_{1a}]^{-1} \mathbf{J}_{1a}^T \Delta(\mathbf{J}_{1u}) \\ [\mathbf{J}_{2a}^T \Delta(\mathbf{J}_{2u}) \mathbf{J}_{2a}]^{-1} \mathbf{J}_{2a}^T \Delta(\mathbf{J}_{2u}) \\ \vdots \\ [\mathbf{J}_{Na}^T \Delta(\mathbf{J}_{Nu}) \mathbf{J}_{Na}]^{-1} \mathbf{J}_{Na}^T \Delta(\mathbf{J}_{Nu}) \end{pmatrix} \dot{\mathbf{p}}. \tag{A2.14}$$

Equation (A2.14) is the least-squares solution for the actuated wheel velocity vector. We note that this solution is applicable only when the matrix in (A2.3) is invertible. The conditions under which this solution is applicable are specified by the soluble motion criterion in (47).

APPENDIX III: SENSED FORWARD SOLUTION MATRIX CALCULATIONS

In this appendix, we detail the matrix manipulations leading to the least-squares sensed forward solution. We solve the partitioned robot sensing equation in (64)

$$
\begin{aligned}
\mathbf{A}_n \begin{pmatrix} \dot{\mathbf{p}} \\ \dot{\mathbf{q}}_n \end{pmatrix} &= \begin{pmatrix} \mathbf{I}_1 & -\mathbf{J}_{1n} & 0 & \ldots & 0 \\ \mathbf{I}_2 & 0 & -\mathbf{J}_{2n} & \cdots & \vdots \\ \vdots & \vdots & \cdots & \cdots & 0 \\ \mathbf{I}_N & 0 & \ldots & 0 & \mathbf{J}_{Nn} \end{pmatrix} \begin{pmatrix} \dot{\mathbf{p}} \\ \dot{\mathbf{q}}_{1n} \\ \dot{\mathbf{q}}_{2n} \\ \vdots \\ \dot{\mathbf{q}}_{Nn} \end{pmatrix} \\
&= \begin{pmatrix} \mathbf{J}_{1s} & 0 & \ldots & 0 \\ 0 & \mathbf{J}_{2s} & \ldots & \vdots \\ \vdots & \cdots & \ldots & 0 \\ 0 & \ldots & 0 & \mathbf{J}_{Ns} \end{pmatrix} \begin{pmatrix} \dot{\mathbf{q}}_{1s} \\ \dot{\mathbf{q}}_{2s} \\ \vdots \\ \dot{\mathbf{q}}_{Ns} \end{pmatrix} = \mathbf{B}_s \dot{\mathbf{q}}_s
\end{aligned} \tag{A3.1}
$$

to calculate the robot velocities $\dot{\mathbf{p}}$ in the least-squares solution in (66):

$$\begin{pmatrix} \dot{\mathbf{p}} \\ \dot{\mathbf{q}}_n \end{pmatrix} = (\mathbf{A}_n^T \mathbf{A}_n)^{-1} \mathbf{A}_n^T \mathbf{B}_s \dot{\mathbf{q}}_s. \tag{A3.2}$$

We begin by forming the matrix product

$$
\begin{aligned}
(\mathbf{A}_n^T \mathbf{A}_n) &= \begin{pmatrix} N\mathbf{I} & -\mathbf{J}_{1n} & -\mathbf{J}_{2n} & \ldots & -\mathbf{J}_{Nn} \\ -\mathbf{J}_{1n}^T & \mathbf{J}_{1n}^T \mathbf{J}_{1n} & 0 & \ldots & 0 \\ -\mathbf{J}_{2n}^T & 0 & \mathbf{J}_{2n}^T \mathbf{J}_{2n} & \cdots & \vdots \\ \vdots & \vdots & \cdots & \cdots & 0 \\ -\mathbf{J}_{Nn}^T & 0 & \ldots & 0 & \mathbf{J}_{Nn}^T \mathbf{J}_{Nn} \end{pmatrix} \\
&= \begin{pmatrix} N\mathbf{I} & \mathbf{T} \\ \mathbf{T}^T & \mathbf{D} \end{pmatrix},
\end{aligned} \tag{A3.3}
$$

where N is the number of wheels and $\mathbf{I}$ is the (3×3) identity matrix. We let the block matrix $\mathbf{X}$ be the inverse of the symmetric matrix $(\mathbf{A}_n^T\mathbf{A}_n)$ in (A3.3). Since the inverse of a matrix times the matrix is the identity matrix,

$$\begin{pmatrix} \mathbf{X}_{11} & \mathbf{X}_{12} \\ \mathbf{X}_{21} & \mathbf{X}_{22} \end{pmatrix} \begin{pmatrix} N\mathbf{I} & \mathbf{T} \\ \mathbf{T}^T & \mathbf{D} \end{pmatrix} = \begin{pmatrix} \mathbf{I} & 0 \\ 0 & \mathbf{I} \end{pmatrix} \qquad (A3.4)$$

We use the top block row of the matrix inverse to separate the robot velocity vector $\dot{\mathbf{p}}$ from the non-sensed wheel velocity vector $\dot{\mathbf{q}}_n$:

$$\dot{\mathbf{p}} = (\mathbf{X}_{11} \ \mathbf{X}_{12}) \ \mathbf{A}_n^T\mathbf{B}_s\dot{\mathbf{q}}_s. \qquad (A3.5)$$

From (A3.4), we obtain

$$\mathbf{X}_{11} N\mathbf{I} + \mathbf{X}_{12}\mathbf{T}^T = \mathbf{I} \qquad (A3.6)$$

and

$$\mathbf{X}_{11}\mathbf{T} + \mathbf{X}_{12}\mathbf{D} = 0, \qquad (A3.7)$$

from which

$$\mathbf{X}_{12} = -\mathbf{X}_{11}\mathbf{T}\mathbf{D}^{-1} \qquad (A3.8)$$

and

$$\mathbf{X}_{11} = (N\mathbf{I} - \mathbf{T}\mathbf{D}^{-1}\mathbf{T}^T)^{-1} \qquad (A3.9)$$

The inverse of the block diagonal matrix $\mathbf{D}$ is:

$$\mathbf{D}^{-1} = \begin{pmatrix} (\mathbf{J}_{1n}^T\mathbf{J}_{1n})^{-1} & 0 & \cdots & 0 \\ 0 & (\mathbf{J}_{2n}^T\mathbf{J}_{2n})^{-1} & \cdots & \vdots \\ \vdots & \ddots & \ddots & 0 \\ 0 & \cdots & 0 & (\mathbf{J}_{Nn}^T\mathbf{J}_{Nn})^{-1} \end{pmatrix}. \qquad (A3.10)$$

We expand the block elements in (A3.8) and (A3.9) to obtain

$$\mathbf{X}_{12} = -\mathbf{X}_{11} \ [-\mathbf{J}_{1n}(\mathbf{J}_{1n}^T\mathbf{J}_{1n})^{-1}$$
$$-\mathbf{J}_{2n}(\mathbf{J}_{2n}^T\mathbf{J}_{2n})^{-1} \ \cdots \ -\mathbf{J}_{Nn}(\mathbf{J}_{Nn}^T\mathbf{J}_{Nn})^{-1}] \quad (A3.11)$$

where

$$\mathbf{X}_{11} = [N\mathbf{I} - \mathbf{J}_{1n}(\mathbf{J}_{1n}^T\mathbf{J}_{1n})^{-1}\mathbf{J}_{1n}^T - \mathbf{J}_{2n}(\mathbf{J}_{2n}^T\mathbf{J}_{2n})^{-1}\mathbf{J}_{2n}^T - \ldots - \mathbf{J}_{Nn}(\mathbf{J}_{Nn}^T\mathbf{J}_{Nn})^{-1}\mathbf{J}_{Nn}^T]^{-1}.$$
$$= -[\Delta(\mathbf{J}_{1n}) + \Delta(\mathbf{J}_{2n}) + \ldots + \Delta(\mathbf{J}_{Nn})]^{-1} \qquad (A3.12)$$

Finally, we substitute (A3.11) and (A3.12) into (A3.5) to obtain the least-squares solution for the robot velocity vector:

$$\dot{\mathbf{p}} = [\Delta(\mathbf{J}_{1n}) + \Delta(\mathbf{J}_{2n}) + \ldots$$
$$+ \Delta(\mathbf{J}_{Nn})]^{-1} [\Delta(\mathbf{J}_{1n})\mathbf{J}_{1s} \quad \Delta(\mathbf{J}_{2n})\mathbf{J}_{2s} \ldots \Delta(\mathbf{J}_{Nn})\mathbf{J}_{Ns}]\dot{\mathbf{q}}_s. \quad \text{(A3.13)}$$

In Section IV H, we develop the adequate sensing criterion, which ensures the invertibility of the matrix $(\mathbf{A}_n^T\mathbf{A}_n)$ in (A3.3) and thereby the applicability of the least-squares sensed forward solution in (A3.13).

References

1. P. F. Muir and C. P. Neuman, "Kinematic Modeling of Wheeled Mobile Robots," Technical Report No. CMU-RI-TR-86-12, The Robotics Institute, Carnegie Mellon University, Pittsburgh, PA, 15213, July, 1986.
2. M. Brady et al., Eds., *Robot Motion, Planning and Control,* MIT Press, Cambridge, MA, 1982.
3. J. Y. S. Luh, "An Anatomy of Industrial Robots and Their Controls," *IEEE Transactions on Automatic Control,* **AC-28**(2), 133–153 (1983).
4. M. Raibert et al., "Dynamically Stable Legged Locomotion," Robotics Institute Technical Report No. CMU-RI-TR-83-20, Carnegie Mellon University, Pittsburgh, PA, December, 1983.
5. T. Iwamoto, H. Yamamoto, and K. Honma, "Transformable Crawler Mechanism with Adaptability to Terrain Variations," International Conference on Advanced Robotics, Tokyo, Japan, September, 1983.
6. M. G. Bekker, *Off-The-Road Locomotion,* University of Michigan Press, Ann Arbor, MI, 1960.
7. M. G. Bekker, *Introduction to Terrain-Vehicle Systems,* University of Michigan Press, Ann Arbor, MI, 1969.
8. J. F. Derry, "Roving Robots," *Robotics Age,* **4**(5), 18–23 (1982).
9. N. J. Nilsson, "Shakey the Robot," Technical Note 323, Artificial Intelligence Center, Computer Science and Technology Division, SRI International, Menlo Park, CA, April, 1984.
10. H. P. Moravec, "Obstacle Avoidance and Navigation in the Real World by a Seeing Robot Rover," PhD Thesis, Department of Computer Science, Stanford University, 1980.
11. D. E. Orin and S. Y. Oh, "Control of Force Distribution in Robotic Mechanisms Containing Closed Kinematic Chains," *Journal of Dynamic Systems, Measurement and Control,* **103**(2), 134–141 (1981).
12. J. Angeles, *Spatial Kinematic Chains, Analysis, Synthesis, Optimization,* Springer-Verlag, Berlin, 1982.
13. J. E. Shigley, *Kinematic Analysis of Mechanisms,* McGraw-Hill, New York, 1969.
14. J. E. Shigley and J. J. Uicker, Jr., *Theory of Machines and Mechanisms,* McGraw-Hill, New York, 1980.
15. J. Denavit and R. S. Hartenberg, "A Kinematic Notation for Lower-Pair Mechanisms Based on Matrices," *Journal of Applied Mechanics,* **77**(2), 215–221 (1955).
16. P. N. Sheth and J. J. Uicker, Jr., "A Generalized Symbolic Notation for Mechanisms," *Journal of Engineering for Industry, Series B,* **93**, 70-Mech-19, 102–112 (1971).
17. D. L. Pieper, "The Kinematics of Manipulators under Computer Control," Stanford Artificial Intelligence Report, Memo No. AI-72, Computer Science Department, Stanford University, Stanford, CA, October 1968.
18. J. Denavit, R. S. Hartenberg, R. Razi, and J. J. Uicker, Jr., "Velocity, Acceleration, and Static-Force Analysis of Spatial Linkages," *Journal of Applied Mechanics,* **87**(4), 903–910 (1965).

19. R. P. Paul, *Robot Manipulators: Mathematics, Programming and Control*, MIT Press, Cambridge, MA, 1981.
20. J. F. Blumrich, "Omnidirectional Wheel," U.S. Patent No. 3,789,947, 1974.
21. H. M. Bradbury, "Omni-Directional Transport Device," U.S. Patent No. 4,223,753, 1980.
22. J. Grabowiecki, "Vehicle-Wheel," U.S. Patent No. 1,305,535, 1919.
23. B. E. Ilon, "Wheels for a Course Stable Selfpropelling Vehicle Movable in any Desired Direction on the Ground or Some Other Base," U.S. Patent No. 3,876,255, 1975.
24. J. M. Holland, *Basic Robotics Concepts*, Howard W. Sams & Co., Indianapolis, IN, 1983, pp. 107–170.
25. H. P. Moravec, Personal Communication, Carnegie Mellon University, Pittsburgh, PA, January, 1982.
26. P. F. Muir, "Digital Servo Controller Design For Brushless DC Motors," Master's Project Report, Department of Electrical and Computer Engineering, Carnegie Mellon University, Pittsburgh, PA, 15213, April 1984.
27. J. S. Beggs, *Advanced Mechanism*, Macmillan, New York, 1966.
28. *McGraw-Hill Encyclopedia of Science & Technology*, 5th ed., McGraw-Hill, New York, 1982, Vol. 4, p. 36.
29. P. M. Cohn, *Linear Equations*, Library of Mathematics, Routledge and Kegan Paul, London, 1958.
30. H. P. Moravec, Ed., "Autonomous Mobile Robots Annual Report—1985," Robotics Institute Technical Report No. CMU-RI-MRL-86-1, Carnegie Mellon University, Pittsburgh, PA, January 1986.
31. Alvema Rehab, "Wheelon—the New Movement," (advertisement), P.O. Box 17017,S-16117 Bromma, Sweden.
32. B. Carlisle, "Omni-Directional Mobile Robot," in *Developments in Robotics*, IFS Publishing Ltd., Kempston, Bedfordshire, England, 1983.
33. R. Hollis, "Newt: A Mobile, Cognitive Robot," *Byte*, 2(6), 30–45 (1977).
34. G. Podnar, K. Dowling, and M. Blackwell, "A Functional Vehicle for Autonomous Mobile Robot Research," Robotics Institute Technical Report No. CMU-RI-TR-84-28, Carnegie Mellon University, Pittsburgh, PA, April, 1984.
35. H. P. Moravec, "The Stanford Cart and the CMU Rover," *Proceedings of the IEEE*, 71(7), 872–884 (1983).
36. D. J. Daniel, "Analysis, Design, and Implementation of Microprocessor Control for a Mobile Platform," Master's Project Report, Department of Electrical and Computer Engineering, Carnegie Mellon University, Pittsburgh, PA, 15213, August 1984.
37. T. Hongo et al., "An Automatic Guidance System of a Self-Controlled Vehicle: The Command System and the Control Algorithm," in *IEEE Proceedings of the IECON*, San Francisco, CA, November 1985, pp. 18–22.
38. V. D. Tourassis and C. P. Neuman, "Robust Nonlinear Feedback Control for Robotic Manipulators," *IEE Proceedings—D: Control Theory and Applications*, Special Issue on Robotics, 132(4), 134–143 (1985).

The authors acknowledge Hans Moravec, Head of the Mobile Robot Laboratory, for his encouragement and support of this research.

DYNAMIC MODELING OF MULTIBODY ROBOTIC MECHANISMS:
Incorporating Closed-Chains, Friction, Higher-Pair Joints, and Unactuated and Unsensed Joints*

Patrick F. Muir and Charles P. Neuman

Department of Electrical and Computer Engineering, The Robotics Institute
Carnegie Mellon University, Pittsburgh, PA 15213

ABSTRACT

We introduce a novel formulation for the dynamic modeling of multibody robotic mechanisms to incorporate closed-chains, higher-pair joints, friction, (including stiction, Coulomb, rolling and viscous friction) and unactuated and unsensed joints. Although we have developed this formulation for the dynamic modeling of *Uranus*, an *omnidirectional* wheeled mobile robot (WMR) designed and constructed in the Robotics Institute of Carnegie Mellon University, our methodology is directly applicable to a spectrum of multibody robotic mechanisms. Our methodology is based upon Newtonian dynamics, our kinematic methodology [10], and the concepts of *force/torque propagation* and *frictional coupling* at a joint which we introduce in this paper. Our extensible *matrix-vector* dynamics formulation allows the application of classical methodologies for the solution of systems of linear algebraic equations (e.g., inverse and forward dynamic solutions [13]). To illustrate the procedure, we apply our dynamics formulation to a planar double pendulum and a biped in the frontal plane.

1. Introduction

The Mobile Robot Laboratory of the Robotics Institute of Carnegie Mellon University has designed and built *Uranus*, an omnidirectional wheeled mobile robot, as a testbed to study servo-control, vision, navigation and sensing [9]. A goal of our servo-control research is to determine whether existing manipulator servo-controller designs (e.g., resolved motion rate control [22], computed torque control [7], and robust computed torque control [21]) are applicable to wheeled vehicles. We thus require a dynamic model of Uranus for the design of *dynamics-based* servo-controllers. We have identified the following *five* salient characteristics of WMRs which require special consideration in the dynamic modeling process: (1) Closed-chains; (2) Friction; (3) Higher-pair joints; (4) Unactuated joints; and (5) Unsensed joints.

Conventional stationary manipulators are open-chain mechanisms; whereas the wheels of a WMR form a *closed-chain* when in contact with the surface-of-travel. WMR mobility stems from the translational *friction* between the wheels and the surface-of-travel; whereas friction is oftentimes neglected in comparison with the inertial and gravitational forces/torques of stationary manipulators. Lack of sufficient friction at the wheel point-of-contact leads to wheel slippage, a problem not encountered in stationary manipulator operation. Friction at the wheel bearings also has a significant effect. For a WMR with the kinematic structure of

Uranus, the roller bearing frictions in the omnidirectional wheels can dissipate as much as 80% of the total available energy [1], in direct contrast to manipulator bearing friction which is typically small in comparison with the inertial and gravitational forces/torques [18].

All stationary manipulator joints, prismatic and revolute, are lower-pairs. The WMR joint between each wheel and the surface-of-travel is a *higher-pair*. A lower-pair allows a common surface contact between adjacent links providing holonomic (positional) constraints; whereas a higher-pair allows point or line contact providing nonholonomic (velocity) constraints. To control the motion of an open-chain, all of the joints must be actuated and sensed. In contrast, a closed-chain mechanism may be adequately controlled with some joints *unactuated*, and its motions may be adequately discerned with some joints *unsensed* [10]. Moreover, the WMR higher-pair wheel joints do not allow the actuation and sensing of the rotational θ degree-of-freedom of each wheel about the point-of-contact with the surface-of-travel.

Our dynamics methodology, unlike existing dynamics methodologies, is applicable to the modeling of multibody robotic mechanisms exhibiting these five characteristics. We highlight in Section 2 existing dynamics formulations and subsequently introduce in Section 3 our dynamic modeling procedure. To illustrate our dynamic modeling procedure, we apply our dynamics formulation to a planar double pendulum (in Section 4.1) and a biped in the frontal plane (in Section 4.2). The complete development of our dynamics formulation (discussed in Section 3) and the dynamic model of Uranus, including solutions of the dynamic equations-of-motion and extensions of our dynamics methodology, are documented in our companion technical report [11].

2. Existing Dynamics Formulations

The Lagrange [18] and Newton-Euler [6] formulations are the two dynamics methodologies most widely applied to stationary manipulator modeling [14]. Neither the Lagrange nor the Newton-Euler formulations are adequate for WMR modeling. Both model open-chains containing lower-pair joints. Because WMRs are closed-chains the kinematic and dynamic equations-of-motion must be computed in *parallel*, thus disabling the direct application of the recursive Newton-Euler dynamics algorithm. When dry friction is incorporated [18], the frictional force/torque is added to the actuator force/torque for each joint. This dry friction modeling procedure does not generalize to chains containing *unactuated* joints since dry friction at the unactuated joints does not affect the computed actuator forces/torques. The dependence of the dry frictional forces/torques on the normal force is also neglected in such a model. Viscous friction has been

---

* This research has been supported by the Office of Naval Research under Contract N00014-81-K-0503 and the Department of Electrical and Computer Engineering, Carnegie Mellon University.

EH0342-6/91/0000/0067$01.00 © 1988 IEEE

incorporated for the actuators, but not for the robot links [19]. Finally, application of these dynamics formulations to robot servo-control requires that *all* of the joint positions (angles) and velocities be sensed.

The Lagrange and Newton-Euler formulations and their extensions to closed-chains and nonholonomic systems (e.g., Draganoiu [2], Kane [4], Luh and Zheng [8], Orin and Oh [17] and Wittenburg [23]) are inadequate to achieve our WMR dynamic modeling goals. Although existing formulations model nonholonomic contraints and closed-chains, none address unactuated and unsensed joints, and none are amenable for incorporating dry friction (i.e., stiction, and Coulomb and rolling friction at the wheel point-of-contact and at the bearings) and viscous friction [11].

3. Our Dynamics Formulation

In our companion technical report [11], we develop a dynamics methodology which satisfies the WMR dynamic modeling requirements outlined in Section 1. Our approach is to construct the conceptually complex dynamic robot model from conceptually simple force/torque models by a conceptually simple force/torque manipulation method. Newtonian dynamics form the basis for modeling inertial and gravitational force/torques, and computing the dynamic equations-of-motion. We introduce the method of *force/torque propagation* to compute the effect at coordinate system B (within the robot mechanism) of forces/torques which originate at coordinate system A. We introduce the viewpoint that dry friction (i.e., stiction, Coulomb friction and rolling friction) is a force/torque *coupling* phenomenon in contrast with the conventional view of dry friction as a force/torque source originating at a joint. Newtonian dynamics, force/torque propagation, frictional couplings and our kinematic modeling methodology [10] thus provide the foundation for our dynamic modeling methodology. We formulate the kinematic and dynamic models independently. In this section we expound upon the concepts of force/torque propagation and frictional coupling at a joint and their roles in our dynamic modeling formulation.

Our dynamics formulation is designed for the *simple closed-chain* mechanical system of rigid bodies depicted in Figure 1. Orin and Oh [17] describe a simple closed-chain mechanism as "*one in which the removal of a particular member of the system breaks all closed-chains*".

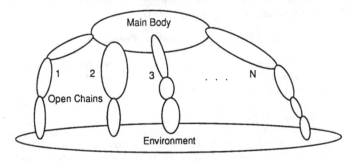

Figure 1: A Simple Closed-Chain Mechanical System of Rigid Bodies

The mechanical system in Figure 1 consists of a main body in contact with N open-chains of rigid bodies. Each pair of adjoining bodies contact at a joint, and the distal rigid body of each open-chain contacts the environment (i.e., a body external to the system). The mechanical configuration in

Figure 1 applies to a spectrum of conventional robotic mechanisms [11], including m-DOF robotic manipulators, multi-manipulator systems, WMRs, legged robots, and robotic hands. For each of these robotic mechanisms, *our goal is to formulate the dynamic equations-of-motion of the main body as a function of the motion (i.e., the positions, velocities and accelerations) of all of the bodies in the system and the actuator and environmental forces/torques.

We utilize the following notation throughout our development: lower case letters denote scalars (e.g., m), lower case bold letters denote vectors (e.g., f), upper case letters denote coordinate systems (e.g., A), upper case italics letters denote bodies (e.g., *A*), and upper case bold letters denote matrices (e.g., **M**). Pre-superscripts denote reference coordinate systems. For example, ^{A}f is the vector f in the A coordinate system. The pre-superscript may be omitted if the coordinte system is transparent from the context. Post-subscripts denote coordinate systems, bodies, or components of a vector or matrix, as indicated in each application. We place the three force components f_x, f_y, and f_z and the three torque components τ_x, τ_y, and τ_z in the force/torque *six*-vector $f = \begin{pmatrix} f_x & f_y & f_z & \tau_x & \tau_y & \tau_z \end{pmatrix}^T$. Linear and angular positions, velocities and accelerations are similarly placed in six-vectors with the x, y, and z rotations according to the the roll-pitch-yaw convention [18].

Coordinate systems play a key role in dynamic modeling. We *fix* each coordinate system with a body within the system so that the motion of the coordinate system is exactly that of the body with which it is fixed. Positions $^A\mathbf{p}_B$, velocities $^A\mathbf{v}_B$, and accelerations $^A\mathbf{a}_B$ always denote the motion of the coordinate system B relative to the coordinate system A. To specify the position, velocity or acceleration of body C, we thus specify the position $^{A_1}\mathbf{p}_{C_1}$, velocity $^{A_1}\mathbf{v}_{C_1}$, or acceleration $^{A_1}\mathbf{a}_{C_1}$ of coordinate system C_1 which is fixed with body C relative to coordinate system A_1 which is fixed with body A. An *instantaneously coincident coordinate system* $\overline{X}$ coincides with the coordinate system X but is fixed with the absolute (i.e., stationary) coordinate system at the instant of interest [10]. We assign coordinate systems to joints according to the Sheth-Uicker convention [20] which is applicable to both lower and higher-pairs.

The forces/torques acting on each rigid body within the system originate from inertial, gravitational, actuation, viscous friction, and environmental contact. These forces/torques may be applied at a point (as with actuation and environmental forces/torques) or may be distributed over the mass (as with inertial and gravitational forces/torques) or surface (as with viscous friction forces/torques) of the body; however, these forces/torques are all conventionally modeled as originating at a point. We assign a coordinate system which is fixed with the body and located at that point-of-application as a reference for labeling the forces/torques. The dynamic model of each of these forces/torques is simplified conceptually in this particular *natural* coordinate system. For example, the inertial forces acting on a rigid body are conceptually simple to model in a natural coordinate system located at the center-of-mass of the body and aligned with the principal axes [5]. We utilize the conceptually simple dynamic models of all forces/torques as the modular building blocks for the systematic formulation of the dynamic model of the complex mechanical system in Figure 1.

The dynamic model of the system is then formulated by *propagating* all forces/torques acting on all of the bodies within the system to a common coordinate system. Even though we may model each force/torque at any coordinate system fixed with the body on which the force/torque is acting, the components of the force/torque vector depend upon the location of the coordinate system. We may therefore model a force/torque acting on body A at two distinct coordinate systems A_1 and A_2 both fixed with body A. The force/torque vector at coordinate systems A_1 and A_2 are then $^{A_1}f_A$ and $^{A_2}f_A$, respectively. The two force/torque vectors describe the same force/torque at different coordinate systems. The force/torque $^{A_1}f_A$ applied at coordinate system A_1 thus has the identical effect on body A as the force/torque $^{A_2}f_A$ applied at coordinate system A_2. The force/torque $^{A_2}f_A$, which is a *linear* function of the force/torque $^{A_1}f_A$, is computed according to $^{A_2}f_A = {^{A_1}L_{A_2}^T}\,{^{A_1}f_A}$, where the *link Jacobian matrix* $^{A_1}L_{A_2}$ is computed directly from the position six-vector $^{A_2}p_{A_1}$ [11,18]. We refer to computing the force/torque $^{A_2}f_A$ at coordinate system A_2 from the force/torque $^{A_1}f_A$ at coordinate system A_1 as *force/torque propagation*.

Forces/torques propagate through joints according to the *coupling* characteristics of the joint. Forces/torques in directions which do not correspond to joint degrees-of-freedom propagate across the joint as if the joint were a rigid link because no relative motion is possible. Forces/torques aligned with the degrees-of-freedom of the joint propagate across the joint according to the frictional characteristics. For example, Coulomb friction couples a normal force exerted by one body to a force opposing the motion of the contacting body. Since the force/torque due to Coulomb friction would not exist without the normal force, we consider Coulomb friction a coupling phenomenon rather than a force/torque source. Forces/torques $^{A_2}f_A$ on body A at joint coordinate system A_2 are coupled through the joint to adjoining body B according to $^{A_2}f_B = {^{A_2}C_{BA}}\,{^{A_2}f_A}$, where the *coupling matrix* $^{A_2}C_{BA}$ is formulated according to the degrees-of-freedom and the frictional characteristics of the joint [11]. We incorporate friction as an integral component of our dynamics formulation to unify the static and dynamic modeling of forces/torques.

We *cascade* transposed link Jacobian and coupling matrices to propagate force/torques from their *natural* coordinate system A_1 fixed to body A to coordinate system Z_2 fixed to body Z separated from body A by intermediate links and joints. For example, $^{Z_2}f_Z = {^{Z_2}p_{A_1}}\,{^{A_1}f_A}$ where the *propagation matrix* $^{Z_2}p_{A_1} = {^{Z_1}L_{Z_2}^T}\,{^{Y_2}C_{ZY}}\,{^{Y_1}L_{Y_2}^T}\,{^{X_2}C_{YX}}\ldots{^{A_1}L_{A_2}^T}$. In analogy with the propagation of light and sound waves through a medium, forces/torques originating at a point within a mechanical system propagate their effects throughout the system.

We have implemented the aforementioned force/torque propagation and frictional coupling concepts to formulate a *step-by-step* dynamic modeling procedure for the simple closed-chain multibody mechanical system in Figure 1 [11]. The forces/torques acting on the system are the inertial (i),

gravitational (g), actuation (a), viscous friction (v) and environmental (e) forces/torques originating from all of the bodies within the system. We propagate all of these forces/torques to the center-of-mass coordinate system M(M) of the main body M. We then equate the sum of these propagated forces/torques to zero (according to Newton's equilibrium law) to formulate the six *primary* dynamic equations-of-motion:

$$\sum_X \sum_s \left({^{M(M)}P_{N(s,X)}}\,{^{N(s,X)}f_{sX}} \right) = 0 . \qquad (1)$$

In (1), the inner summation is over all force/torque sources (i.e., s = i, g, a, v, and e) and the outer summation is over all bodies X within the system, and N(s,X) is the natural coordinate system for source s acting on body X. The force/torque acting on body X from source s is $^{N(s,X)}f_{sX}$ and the corresponding propagation matrix is $^{M(M)}P_{N(s,X)}$.

The constraints between forces/torques along the degrees-of-freedom of the joint axes are not included in the formulation of the six primary dynamic equations-of-motion in (1). We must, therefore, include these force/torque constraints to complete our dynamic model of the system. We thus formulate N_s *secondary* dynamic equations-of-motion where N_s is the number of joint degrees-of-freedom in the system. The secondary dynamic equations-of-motion for the joint between bodies A and B at joint coordinate system B_1 are formulated according to:

$$[\mathbf{I} - {^{B_1}C_{BA}}\,{^{B_1}C_{AB}}]\sum_Y \sum_s \left({^{B_1}P_{N(s,Y)}}\,{^{N(s,Y)}f_{sY}} \right) = 0. \qquad (2)$$

In (2), the outer summation is over all bodies Y within the system between the joint and the environment, and the coupling matrix $^{B_1}C_{AB}$ is computed by negating the nondiagonal elements of the coupling matrix $^{B_1}C_{BA}$. From the six equations in (2), we obtain $(6-d_{AB})$ null equations and d_{AB} secondary force/torque equations-of-motion where d_{AB} is the number of DOFs of the joint between bodies A and B. We formulate (2) for all of the joints in the system.

The complete dynamic model consisting of the six *primary* and the N_s *secondary* dynamic equations-of-motion in (1) and (2) are *linear* in the actuator and environmental forces/torques and the accelerations of the center-of-mass of all of the system bodies. The velocities and accelerations in the model are referenced to natural instantaneously coincident coordinate systems. We thus apply our kinematic methodology [10] to compute these instantaneously coincident velocities and accelerations from the velocities and accelerations of the joints and the main body. We then substitute *componentwise* these separate kinematic computations into the dynamic equations-of-motion to formulate the closed-form dynamic model [11].

4. Examples

4.1 Planar Double Pendulum

We apply our dynamics formulation to the planar double pendulum (an *open-chain* mechanism) sketched in Figure 2.

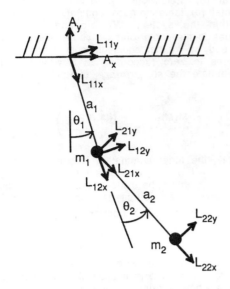

Figure 2: A Planar Double Pendulum

The four position six vectors of the planar double pendulum are:

$${}^A p_{L_{11}} = \begin{pmatrix} 0 & 0 & 0 & 0 & 0 & (\theta_1 - 90°) \end{pmatrix}^T \quad {}^{L_{11}} p_{L_{12}} = \begin{pmatrix} a_1 & 0 & 0 & 0 & 0 & 0 \end{pmatrix}^T$$

$${}^{L_{12}} p_{L_{21}} = \begin{pmatrix} 0 & 0 & 0 & 0 & 0 & \theta_2 \end{pmatrix}^T \text{ and } {}^{L_{21}} p_{L_{22}} = \begin{pmatrix} a_2 & 0 & 0 & 0 & 0 & 0 \end{pmatrix}^T;$$

the coupling matrices are

$${}^A C_{EL_1} = {}^{L_{12}} C_{L_1 L_2} = \begin{pmatrix} 1 & 0 & 0 & 0 & 0 & 0 \\ 0 & 1 & 0 & 0 & 0 & 0 \\ 0 & 0 & 1 & 0 & 0 & 0 \\ 0 & 0 & 0 & 1 & 0 & 0 \\ 0 & 0 & 0 & 0 & 1 & 0 \\ 0 & 0 & 0 & 0 & 0 & 0 \end{pmatrix};$$

and the nine force/torque vectors are

$${}^{L_{12}} f_{iL_1} = \begin{pmatrix} -m_1 {}^{\overline{L}_{12}} a_{L_{12}x} & -m_1 {}^{\overline{L}_{12}} a_{L_{12}y} & 0 & 0 & 0 & 0 \end{pmatrix}^T$$

$${}^{L_{22}} f_{iL_2} = \begin{pmatrix} -m_1 {}^{\overline{L}_{22}} a_{L_{22}x} & -m_1 {}^{\overline{L}_{22}} a_{L_{22}y} & 0 & 0 & 0 & 0 \end{pmatrix}^T$$

$${}^{L_{12}} f_{gL_1} = \begin{pmatrix} 0 & -m_1 g & 0 & 0 & 0 & 0 \end{pmatrix}^T \quad {}^{L_{22}} f_{gL_2} = \begin{pmatrix} 0 & -m_2 g & 0 & 0 & 0 & 0 \end{pmatrix}^T$$

$${}^A f_{eL_1} = \begin{pmatrix} f_x & f_y & 0 & 0 & 0 & \tau_z \end{pmatrix}^T \quad {}^{L_{11}} f_{aL_1} = -{}^A f_{aL_1} = \begin{pmatrix} 0 & 0 & 0 & 0 & 0 & \tau_1 \end{pmatrix}^T$$

and $\quad {}^{L_{21}} f_{aL_2} = -{}^{L_{21}} f_{aL_2} = \begin{pmatrix} 0 & 0 & 0 & 0 & 0 & \tau_2 \end{pmatrix}^T.$

The constants m_1 and m_2 are the masses at the ends of links L_1 and L_2, respectively, g is the gravitational constant, and τ_1 and τ_2 are the actuator torques applied at the bases of links L_1 and L_2, respectively. Each actuator produces a torque on the link it is driving and an equal and opposite reactional torque on the link to which it is mounted. We obtain the six primary dynamic equations-of-motion by propagating all of the forces/torques to the end of the pendulum (coordinate system L_{22}), and one secondary equation-of-motion at each joint; i.e., the coordinate systems L_{12} and A. The six primary dynamic equations-of-motion lead to the three non-trivial scalar dynamic equations-of-motion:

$$-c_2 m_1 {}^{\overline{L}_{12}} a_{L_{12}x} - s_2 m_1 {}^{\overline{L}_{12}} a_{L_{12}y} - m_2 {}^{\overline{L}_{22}} a_{L_{22}x}$$
$$+ c_{12} m_1 g + c_{12} m_2 g + s_{12} f_x - c_{12} f_y = 0 \quad (3)$$

$$s_2 m_1 {}^{\overline{L}_{12}} a_{L_{12}x} - c_2 m_1 {}^{\overline{L}_{12}} a_{L_{12}y} - m_2 {}^{\overline{L}_{22}} a_{L_{22}y}$$
$$- s_{12} m_1 g - s_{12} m_2 g + c_{12} f_x + s_{12} f_y = 0 \quad (4)$$

$$-a_2 s_2 m_1 {}^{\overline{L}_{12}} a_{L_{12}x} + a_2 c_2 m_1 {}^{\overline{L}_{12}} a_{L_{12}y}$$
$$+ a_2 s_{12} m_1 g + \tau_2 - a_2 c_{12} f_x - a_2 s_{12} f_y = 0 \quad (5)$$

where $s_i = \sin(\theta_i)$, $c_i = \cos(\theta_i)$, $s_{ij} = \sin(\theta_i + \theta_j)$, and $c_{ij} = \cos(\theta_i + \theta_j)$. The two secondary dynamic equations are:

$$\tau_1 - \tau_2 - a_1 c_1 f_x - a_1 s_1 f_y = 0 \quad (6)$$

$$-\tau_1 + \tau_z = 0 \ . \quad (7)$$

The four acceleration equations required to complete (3)-(7) are:

$${}^{\overline{L}_{12}} a_{L_{12}x} = -a_1 \omega_1^2 \quad (8)$$

$${}^{\overline{L}_{12}} a_{L_{12}y} = a_1 \alpha_1 \quad (9)$$

$${}^{\overline{L}_{22}} a_{L_{22}x} = -a_1 c_2 \omega_1^2 + a_1 s_2 \alpha_1 - a_2 (\omega_1 + \omega_2)^2 \quad (10)$$

$${}^{\overline{L}_{22}} a_{L_{22}y} = a_1 s_2 \omega_1^2 + a_1 c_2 \alpha_1 + a_2 (\alpha_1 + \alpha_2) \ . \quad (11)$$

In (8)-(11), ω_1 and ω_2 are the angular velocities and α_1 and α_2 are the angular accelerations of the joints. We solve (3)-(11) for the two joint torques τ_1 and τ_2 and obtain the classical inverse dynamic model of the planar double pendulum [12].

4.2 Biped in the Frontal Plane

We next apply our dynamics formulation to the biped in the frontal plane (a *closed-chain* mechanism) depicted in Figure 3 [3]. The twelve position six-vectors of the biped are:

$${}^A p_{A_1} = \begin{pmatrix} -d_1/2 & 0 & 0 & 0 & 0 & 0 \end{pmatrix}^T \quad {}^{A_1} p_{L_{11}} = \begin{pmatrix} 0 & 0 & 0 & 0 & 0 & -\theta_1 \end{pmatrix}^T$$

$${}^{L_{11}} p_{C_1} = \begin{pmatrix} 0 & k_1 & 0 & 0 & 0 & 0 \end{pmatrix}^T \quad {}^{C_1} p_{L_{12}} = \begin{pmatrix} 0 & (l_1 - k_1) & 0 & 0 & 0 & 0 \end{pmatrix}^T$$

$${}^{L_{12}} p_{L_{21}} = \begin{pmatrix} 0 & 0 & 0 & 0 & 0 & (\theta_1 - \theta_2) \end{pmatrix}^T \quad {}^{L_{21}} p_{C_2} = \begin{pmatrix} d_2 & k_2 & 0 & 0 & 0 & 0 \end{pmatrix}^T$$

$${}^A p_{A_3} = \begin{pmatrix} d_1/2 & 0 & 0 & 0 & 0 & 0 \end{pmatrix}^T \quad {}^{A_3} p_{L_{32}} = \begin{pmatrix} 0 & 0 & 0 & 0 & 0 & (180° - \theta_3) \end{pmatrix}^T$$

$${}^{L_{32}} p_{C_3} = \begin{pmatrix} 0 & k_3 & 0 & 0 & 0 & 0 \end{pmatrix}^T \quad {}^{C_3} p_{L_{31}} = \begin{pmatrix} 0 & (l_3 - k_3) & 0 & 0 & 0 & 0 \end{pmatrix}^T$$

$${}^{L_{31}} p_{L_{32}} = \begin{pmatrix} 0 & 0 & 0 & 0 & 0 & (\theta_3 - \theta_2 - 180°) \end{pmatrix}^T$$

$${}^{L_{22}} p_{C_2} = \begin{pmatrix} -d_2 & k_2 & 0 & 0 & 0 & 0 \end{pmatrix}^T.$$

The coupling matrices are:

$${}^{A_1} C_{EL_1} = {}^{L_{12}} C_{L_1 L_2} = {}^{L_{31}} C_{L_2 L_3} = {}^{A_3} C_{EL_3} = \begin{pmatrix} 1 & 0 & 0 & 0 & 0 & 0 \\ 0 & 1 & 0 & 0 & 0 & 0 \\ 0 & 0 & 1 & 0 & 0 & 0 \\ 0 & 0 & 0 & 1 & 0 & 0 \\ 0 & 0 & 0 & 0 & 1 & 0 \\ 0 & 0 & 0 & 0 & 0 & 0 \end{pmatrix}$$

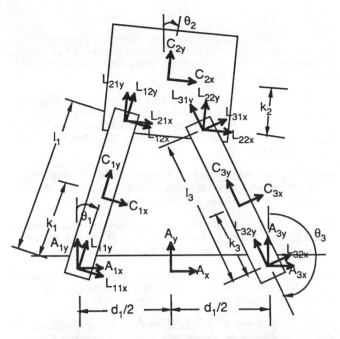

Figure 3: Biped in the Frontal Plane

And the sixteen force/torque vectors are:

$$^{C_1}f_{iL_1} = \begin{pmatrix} -m_1\,^{\bar{C}_1}a_{C_1x} \\ -m_1\,^{\bar{C}_1}a_{C_1y} \\ -m_1\,^{\bar{C}_1}a_{C_1z} \\ 0 \\ 0 \\ -I_1\,^{\bar{C}_1}\alpha_{C_1z} \end{pmatrix} \qquad ^{C_2}f_{iL_2} = \begin{pmatrix} -m_2\,^{\bar{C}_2}a_{C_2x} \\ -m_2\,^{\bar{C}_2}a_{C_2y} \\ -m_2\,^{\bar{C}_2}a_{C_2z} \\ 0 \\ 0 \\ -I_2\,^{\bar{C}_2}\alpha_{C_2z} \end{pmatrix}$$

$$^{C_3}f_{iL_3} = \begin{pmatrix} -m_3\,^{\bar{C}_3}a_{C_3x} \\ -m_3\,^{\bar{C}_3}a_{C_3y} \\ -m_3\,^{\bar{C}_3}a_{C_3z} \\ 0 \\ 0 \\ -I_3\,^{\bar{C}_3}\alpha_{C_3z} \end{pmatrix}$$

$$^{G(L_1)}f_{gL_1} = \begin{pmatrix} 0 & -m_1g & 0 & 0 & 0 & 0 \end{pmatrix}^T$$

$$^{G(L_2)}f_{gL_2} = \begin{pmatrix} 0 & -m_2g & 0 & 0 & 0 & 0 \end{pmatrix}^T$$

$$^{G(L_3)}f_{gL_3} = \begin{pmatrix} 0 & -m_3g & 0 & 0 & 0 & 0 \end{pmatrix}^T$$

$$^{A_1}f_{eL_1} = \begin{pmatrix} F_1 & G_1 & 0 & 0 & 0 & H_1 \end{pmatrix}^T$$

$$^{A_3}f_{eL_3} = \begin{pmatrix} -F_4 & -G_4 & 0 & 0 & 0 & H_4 \end{pmatrix}^T$$

$$^{A_1}f_{aE} = -^{L_{11}}f_{aL_1} = \begin{pmatrix} 0 & 0 & 0 & 0 & 0 & u_1 \end{pmatrix}^T$$

$$^{L_{12}}f_{aL_1} = -^{L_{21}}f_{aL_2} = \begin{pmatrix} 0 & 0 & 0 & 0 & 0 & u_2 \end{pmatrix}^T$$

$$^{L_{22}}f_{aL_2} = -^{L_{31}}f_{aL_3} = \begin{pmatrix} 0 & 0 & 0 & 0 & 0 & u_3 \end{pmatrix}^T$$

$$^{L_{32}}f_{aL_3} = -^{A_3}f_{aE} = \begin{pmatrix} 0 & 0 & 0 & 0 & 0 & u_4 \end{pmatrix}^T$$

The principal moment of inertia of body i about the z-axis is I_i, the joint actuator torques are u_i, the environmental forces in the x and y directions are F_i and G_i, respectively, and the environmental torque about the z-axis is H_i. The coordinate systems $G(L_i)$ for i=1,2,3 which are not explicitly drawn in Figure 3 are gravitational coordinate systems located at the center-of-mass of body L_i and aligned with the gravitational field.

We obtain the six primary dynamic equations-of-motion by propagating the sixteen forces/torques to the center-of-mass of the biped body (coordinate system C_2), and one secondary equation-of-motion at each of the four joints;i.e., at the four coordinate systems X = A_1, L_{12}, L_{31}, and A_3. We then substitute the velocities and accelerations relative to the instantaneously coincident coordinate systems (computed from the joint velocities and accelerations) into the dynamic equations-of-motion. Finally, we apply the two positional geometries

$$l_1s_1 + 2d_2c_2 + l_3s_3 = d_1 \quad \text{and} \quad l_1c_1 - 2d_2s_2 + l_3c_3 = 0$$

and their derivatives to obtain the Hemami and Wyman dynamic equations-of-motion for the biped in the frontal plane [3].

5. Concluding Remarks and Further Research

We have designed a dynamics formulation to incorporate the special characteristics of WMRs [11]. We model the dynamics of simple *closed chains* by propagating all forces/torques within the system to a common coordinate system. We incorporate dry *frictional* coupling at a joint by introducing coupling matrices. The coupling matrix for a joint contains ones along the diagonal corresponding to joint degrees-of-freedom, and frictional coefficients off-of-the diagonal corresponding to dry friction couplings. *Higher pair joints* are modeled by applying the Sheth-Uicker convention [20] and instantaneously coincident coordinate systems [10]. Actuator forces/torques are incorporated in the dynamic model along with the inertial, gravitational, viscous frictional and environmental contact forces/torques. We thereby model an *unactuated joint* by the absence of an actuator force/torque. We apply our kinematic modeling methodology [10] to compute the *unsensed joint* velocities and accelerations from the sensed joint velocities and accelerations, and therby formulate the dynamic model from sensed joint velocities and accelerations.

We have introduced a dynamics formulation which is conceptually simple to apply. Our matrix-vector dynamics formulation permeated by sparce matrices provides a conceptual framework for the design of computationally efficient algorithms. For example, servo-control algorithms emanating from our dynamics formulation may be designed for efficient computation by eliminating all scalar additions and multiplications by zero and multiplications by plus and minus one [15-16]. Because the dynamic equations are based upon matrix-vector products, the servo-control algorithms may be most amenable to direct implementation on parallel and vector (array) processors.

We are continuing our study of WMR modeling and servo-control. We are applying our kinematic [10] and dynamic WMR models to evaluate servo-control algorithms for real-time WMR trajectory tracking. We are applying the forward dynamic solution to simulate the WMR motion and control the simulation by implementing the inverse kinematic and dynamic solutions with resolved motion rate [22], computed torque [7], and robust computed torque [21] servo-control algorithms. Extensions to our dynamics formulation include: the modeling of flexible links; the modeling of such material effects as stress, strain, fracture, and wear; and the modeling of complex closed-chains.

6. References

[1] D. J. Daniel, "Analysis, Design, and Implementation of Microprocessor Control for a Mobile Platform," Master's Project Report, Department of Electrical and Computer Engineering, Carnegie Mellon University, Pittsburgh, PA 15213, August 1984.

[2] G. Draganoiu, "Computer Method for Setting Dynamical Model of an Industrial Robot with Closed Kinematic Chains," *Proceedings of the Twelfth International Symposium on Industrial Robots*, Paris, France, June 1982, pp. 371-379.

[3] H. Hemami and B. F. Wyman, "Modeling and Control of Constrained Dynamic Systems with Application to Biped Locomotion in the Frontal Plane," *IEEE Transactions on Automatic Control*, Vol. AC-24, No. 4, August 1979, pp. 526-535.

[4] T. R. Kane, "Dynamics of Nonholonomic Systems," *Journal of Applied Mechanics*, Vol. 83, No. 4, December 1961, pp. 574-578.

[5] L. D. Landau and E. M. Lifshitz, *Mechanics*, Third Edition, Pergamon Press, New York, 1976.

[6] J. Y. S. Luh, M. W. Walker, and R. P. C. Paul, "On-Line Computational Scheme for Mechanical Manipulators," *Journal of Dynamic Systems, Measurement, and Control*, Vol. 102, No. 2, June 1980, pp. 69-76.

[7] J. Y. S. Luh, M. W. Walker, and R. P. C. Paul, "Resolved Acceleration Control of Mechanical Manipulators," *IEEE Transactions on Automatic Control*, Vol. AC-25, No. 3, June 1980, pp. 468-474.

[8] J. Y. S. Luh and Y. Zheng, "Computation of Input Generalized Forces for Robots with Closed Kinematic Chain Mechanism," *IEEE Journal of Robotics and Automation*, Vol. RA-1, No. 2, June 1985, pp. 95-103.

[9] H. P. Moravec (editor), "Autonomous Mobile Robots Annual Report - 1985," Robotics Institute Technical Report No. CMU-RI-MRL-86-1, Carnegie Mellon University, Pittsburgh, PA 15213, January 1986.

[10] P. F. Muir and C. P. Neuman, "Kinematic Modeling of Wheeled Mobile Robots," *Journal of Robotic Systems*, Vol. 4, No. 2, April 1987, pp. 281-340.

[11] P. F. Muir and C. P. Neuman, "Dynamic Modeling of Wheeled Mobile Robots," Robotics Institute Technical Report No. CMU-RI-TR-87-10, Carnegie Mellon University, Pittsburgh, PA 15213, February 1988.

[12] J. J. Murray, *Computational Robot Dynamics*, PhD Dissertation, Electrical and Computer Engineering Department, Carnegie Mellon University, Pittsburgh, PA 15213, September 1986.

[13] C. P. Neuman and P. F. Muir, "From Linear Algebraic Equations to Wheeled Mobile Robot Modeling," in K. S. Narendra (editor), *Proceedings of the Fifth Yale Workshop on Applications of Adaptive Systems Theory*, Yale University, New Haven, CT, May 1987.

[14] C. P. Neuman and J. J. Murray, "Computational Robot Dynamics: Foundations and Applications," *Journal of Robotic Systems*, Vol. 2, No. 4, Winter 1985, pp. 425-452.

[15] C. P. Neuman and J. J. Murray, "Customized Computational Robot Dynamics," *Journal of Robotic Systems*, Vol. 4, No. 4, August 1987, pp. 503-526.

[16] C.P. Neuman and J. J. Murray, "Symbolically Efficient Formulations for Computational Robot Dynamics," *Journal of Robotic Systems*, Vol. 4, No. 6, December 1987.

[17] D. E. Orin and S. Y. Oh, "Control of Force Distribution in Robotic Mechanisms Containing Closed Kinematic Chains," *Journal of Dynamic Systems, Measurement, and Control*, Vol. 102, No. 2, June 1981, pp. 134-141.

[18] R. P. Paul, *Robot Manipulators: Mathematics, Programming and Control*, The MIT Press, Cambridge MA, 1981.

[19] M. S. Pfeifer and C. P. Neuman, "An Adaptable Simulator for Robot Arm Dynamics," *Computers in Mechanical Engineering*, Vol. 3, No. 3, November 1984, pp. 57-64.

[20] P. N. Sheth and J. J. Uicker, Jr., "A Generalized Symbolic Notation for Mechanisms," *Journal of Engineering for Industry*, Series B, Vol. 93, No. 70-Mech-19, 1971, pp. 102-112.

[21] V.D. Tourassis and C. P. Neuman, "Robust Nonlinear Feedback Control for Robotic Manipulators," *IEE Proceedings - D: Control Theory and Applications*, Special Issue on Robotics, Vol. 132, No. 4, July 1985, pp. 134-143.

[22] D. E. Whitney, "Resolved Motion Rate Control of Manipulators and Human Protheses," *IEEE Transactions on Man-Machine Systems*, Vol. MMS-10, No. 2, June 1969, pp. 47-53.

[23] J. Wittenburg, *Dynamics of Systems of Rigid Bodies*, B. G. Teuber Stuttgart, 1977.

Chapter 3: Task-Level Planning, Decision Making, and Control

All autonomous robots need an intelligent control system. This chapter highlights basic elements of task-level planning and outlines relevant research. More specifically, the chapter explores autonomous control structures for mobile robots; this exploration includes task-level planning, system-level decision making and control, and some architectures for planning and control. Within task-level planning, we will examine precompiled plans, reactive planning, dynamic planning, and the integration of prior plans and local information. System-level decision making and control consists of AI approaches and control theory approaches, including optimal decision making under uncertainty, and integrated approaches. Architectures for planning and control include conventional architectures (sequential control), blackboard-based distributed architectures, plus subsumption and massively parallel architectures. Georgeff provides a tutorial introduction to planning.

Dean et al. argue that task planning and execution belong fundamentally to the same class. Specifically, the authors discuss these issues as they apply to robotic control, and provide sketches of analytical-approach perspectives in designing and assessing the performance of complex control systems.

In his 1986 journal article, Elfes presents a distributed control architecture for an autonomous mobile robot that provides scheduling and coordination of multiple concurrent activities. An expert system module attached to each task communicates with other modules through messages and a blackboard.

Noreils and Chatila offer a control system for mobile robots. Their system consists of three entities — modules, processes, and functional units. Modules accomplish basic computations on data from sensors. Processes establish dynamic links between perception and action. Functional units provide specific functionalities.

Brooks provides a layered-control-structure architecture for mobile robots that gives increasing competence levels. Each layer consists of asynchronous modules, providing fairly simple computational communication over low-bandwidth channels with other modules. Arkin, Payton et al., and Maes and Brooks explore reactive decision making.

In his 1991 proceedings paper, Elfes stresses the use of stochastic and information-theoretic models for the active control of robotic perception. Doyle et al., Gat et al., and Dean et al. (in Dean's second contribution to this chapter) discuss the connection between planning and perception.

Finally, Premvuti and Yuta address the communication and cooperation of multiple robots.

PLANNING

Michael P. Georgeff [1]

Artificial Intelligence Center, SRI International, Menlo Park, California
94025

1. INTRODUCTION

The ability to act appropriately in dynamic environments is critical to the survival of all living creatures. For lower life forms, it seems that sufficient capability is provided by stimulus-response and feedback mechanisms. Higher life forms, however, must be able to anticipate the future and form plans of action to achieve their goals. Reasoning about actions and plans can thus be seen as fundamental to the development of intelligent machines that are capable of dealing effectively with real-world problems.

Researchers in artificial intelligence (AI) have long been concerned with this area of investigation (McCarthy 1968). But, as with most of AI, it is often difficult to relate the different streams of research and to understand how one technique compares with others. Much of this difficulty derives from the varied (and sometimes confused) terminology and the great diversity of problems that arise in real-world planning. Indeed, there are few practical planning systems for which the class of appropriate applications can be clearly delineated.

This article attempts to clarify some of the issues that are important in reasoning about actions and plans. As the field is still young, it would be premature to expect us to have a stable foundation on which to build a discipline of planning. Nevertheless, I hope that the following discussion contributes toward that objective and that it will help the reader to evaluate the pertinent literature.

[1] Also affiliated with the Center for the Study of Language and Information, Stanford University, Stanford, California.

2. THE REPRESENTATION OF ACTIONS AND EVENTS

Humans spend a great deal of time deciding and reasoning about actions, some with much deliberation and some without any forethought. They may have numerous desires that they wish fulfilled, some more strongly than others. It is often necessary to accommodate conflicting desires, to choose among them, and to reason about how best to accomplish those that are chosen. This choice, and the means chosen to realize these ends, will depend upon currently held beliefs about present and future situations, and upon any commitments or intentions that may have been earlier decided upon. Often it will be necessary to obtain more information about the tasks to be performed, either prior to choosing a plan of action or during its execution. Furthermore, our knowledge of the world itself is frequently incomplete, making it necessary for us to have some means of forming reasonable assumptions about the possible occurrence of other events or the behaviors of other agents.

All this has to be accomplished in a complex and dynamic world populated with many other agents. The agent planning or deciding upon possible courses of action can choose from an enormous repertoire of actions, and these in turn can influence the world in exceedingly complicated ways. Moreover, because of the presence of other agents and processes, the environment is subject to continuous change—even as the planner deliberates on how best to achieve its goals.

2.1 *Models of Action*

To tackle the kind of problems mentioned above, we first have to understand clearly what entities we are to reason about. The traditional approach has been to consider that, at any given moment, the world is in one of a potentially infinite number of *states* or *situations*. A world state may be viewed as a snapshot of the world at a given instant of time. A sequence of world states is usually called a *behavior*; one that stretches back to the beginning of time or forward to its end is called a *world history* or a *chronicle*. Such a world history, for example, may represent the past history of the actual world, or some potential future behavior.

The world can change its state only by the occurrence of an *event* or *action*. An *event type* is usually modeled as a set of behaviors, representing all possible occurrences of the event in all possible world histories. Thus, the event type "John running around a track three times" corresponds to all possible behaviors in which John does exactly that—namely, runs around *some* track during *some* interval at *some* location exactly three times. An *event instance* is a particular occurrence of an event type in a particular world history. However, where there is no ambiguity, we shall call event types simply events.

An *action* is a special kind of event—namely, one that is *performed* by some agent, usually in some intentional way. For example, a tree's shedding of its leaves is an event but not an action; John's running around a track is an action (in which John is the agent). Philosophers make much of this distinction between actions and events, primarily because they are interested in activities that an agent decides upon, rather than those events that are not caused by the agent (such as leaves falling from a tree) or that involve the agent in some unintentional way (such as tripping over a rug) (Davis 1979). For our purposes, however, we can treat these terms synonymously.

We shall begin be restricting our attention to domains in which there is no concurrent activity, as could be used to represent a single agent acting in a static environment. In these domains, it is only necessary to consider the initial and final states of any given event, as nothing can happen during the event to change its outcome. Consequently, an event can be modeled as a set of pairs of initial and final states, rather than as a set of complete behaviors. If, in addition, we limit ourselves to deterministic events, this relation between initial and final states will be functional; that is, the initial state in which an event occurs will uniquely determine the resulting final state. Of course, there may be certain states in which an event cannot be initiated; that set of states in which it *can* is usually called the *domain* of the event.

Events may be composed from other events in a number of ways. As they are just relations (or, in the more general case, sets of behaviors), the two simplest means of composition are set union and intersection. For example, the intersection of the event in which Mary hugs John and the event in which Mary kisses John is the event in which Mary both hugs *and* kisses John. We can also compose events sequentially; for example, to yield the event in which Mary first hugs John and then kisses him.

We also want to be able to say that certain *properties* hold of world states. For example, in some given state, it might be that a specified block is on top of some other block, or that its color is red. But what kind of entities are such properties? For example, consider the property of redness. In a static world, we might model this property as a set of individuals (or objects)—namely, those that are red. However, in dynamic worlds, the individuals that are red can vary from state to state; we therefore cannot model redness in this way.

One approach is to view the properties of the domain as relating to particular states. Thus, instead of representing redness as a set of individuals, it is instead represented by a relation on individuals and states; a pair $[A, s]$ would be a member of this relation just in case A were red in state s. Such entities are commonly called *situational relations*.

Another, more elegant way to handle this problem is to introduce the notion of a *fluent* (McCarthy & Hayes 1969), which is a function defined on world states. Essentially, a given fluent corresponds to some property of world states, and its value in a given state is the value of that property in that state. For example, the property of redness could be represented by a fluent whose *value* in a given state is the set of individuals that are red in that state.

Fluents come in a variety of types. A fluent whose value in a given state is either *true* or *false* is usually called a *propositional* fluent. For example, the property of it being raining could be represented by a propositional fluent that has the value *true* in those states in which it is raining and the value *false* when it is not raining. An equivalent view of propositional fluents identifies them with the set of states for which the property is true.

Fluents can also represent individuals (such as agents and objects); their value in a given state will be some specific individual that exists in that state. For example, one may have a fluent representing Caesar; the value of this fluent in any state will be whoever happens to be the emperor of Rome at that point in time. Similarly, one can introduce fluents whose value in a given state is a relation over individuals, or fluents whose values are functions from individuals to individuals (Montague 1974). The former are typically called *relational fluents*, the latter *functional fluents*.[2]

2.2 *The Situation Calculus*

Of course, in any interesting domain, it is infeasible to specify explicitly the functions and relations representing events and fluents. We therefore need some formal language for describing and reasoning about them. [Those unfamiliar with logical formalism and its use in AI should refer to the article by Levesque on knowledge representation in Volume 1 of this series (Levesque 1986) and Hayes' beautifully clear exposition of naive physics (Hayes 1985).]

McCarthy (McCarthy & Hayes 1969) proposed a formal calculus of situations (states) which has become the classical approach to this problem. In the variant we describe here, the logical terms of the calculus are used to denote the states, events, and fluents of the problem domain. For example, the event term *puton*(A, B) could be used to denote the action in

---

[2] A good introduction to many of these concepts, including some of the logics mentioned in Section 2.2, can be found in the first six chapters of Dowty et al (1981). Note, however, that fluents are therein called *intensions*.

which block A is placed on top of block B. Similarly, the fluent term $on(A, B)$ could designate the fluent representing the proposition that A is on top of B. We could also introduce other event terms to denote composite events and other fluent terms to denote the different kinds of fluents and various individuals (such as A and B) in the domain.

The predicates in this situation calculus are used primarily to make statements about the values of fluents in particular states. For propositional fluents, we shall use the expression $holds(f, s)$ to mean that the fluent f has value *true* in state s. For example, $holds(on(A, B), s)$ will be true if the fluent denoted by $on(A, B)$ has value *true* in state s; that is, if block A is on top of B in s. We can use other predicates and function symbols to describe the properties of other kinds of fluents (Manna & Waldinger 1987; McCarthy & Hayes 1969; Montague 1974).

We must also be able to specify the state transitions associated with any particular event in the problem domain. The usual way to do this is to introduce the term $result(e, s)$ to designate the state resulting from the performance of event e in state s. For example, $result(puton(A, B), s)$ denotes the state that results when the action $puton(A, B)$ is initiated in state s. We can also use the *result* function to characterize those states that are *reachable* by the agent from some given state. That is, for any state s and any performable action e, the state denoted by $result(e, s)$ will be reachable from s and, in turn, from any other state from which s is itself reachable.

The well-formed formulas of this situation calculus may also contain the usual logical connectives and quantifiers. With this machinery, we can now express general assertions about the effects of actions and events when carried out in particular situations. For example, we can express the result of putting block A on top of block B as follows:

$\forall s \cdot holds(clear(A), s) \land holds(clear(B), s)$

$$\supset holds(on(A, B), result(puton(A, B), s)).$$

This statement is intended to mean that if blocks A and B are initially clear, then after the action $puton(A, B)$ has been performed, block A will be on top of B.

One problem with the above approach is the apparently large number of axioms needed to describe what properties are *unaffected* by events. For example, if block B were known to be red prior to our placing block A upon it, we would not be able to conclude, on the basis of the previous axiom alone, that block B would still be red afterwards. To do so, we require an additional axiom stating that the movement of block A does not change the color of block B:

$$\forall s \cdot holds(color(B, red), s) \supset holds(color(B, red), result(puton(A, B), s)).$$

In fact, we would have to provide similar axioms for every property of the domain left unaffected by the action. These are called *frame axioms*; being forced to specify them is commonly known as the *frame problem* (Hayes 1973).

Various other logical formalisms have been developed for representing and reasoning about dynamic domains. The most common are the *modal logics*, which avoid the explicit use of terms representing world state. One type of modal logic, called *temporal logic*, introduces various *temporal operators* for describing properties of world histories (Prior 1967). For example, the fact that it will rain sometime in the future can be represented by the formula $\Diamond raining$. Here, the temporal operator $\Diamond$ represents the temporal modality "at some time in the future"; the formula $\Diamond \phi$ means that there exists some future state for which the formula ϕ is true. The use of temporal operators corresponds closely to the way *tense* is used in natural languages; thus, it is claimed, these logics provide a natural and convenient means for describing the temporal properties of given domains.

Process logics are another kind of modal logic in which explicit mention of state is avoided (Nishimura 1980). These logics are based on the same model of the world as described above, but introduce programs (or plans) as additional entities in the domain (see Section 3.1). *Dynamic logics* can be viewed as a special class of process logics that are concerned solely with the input-output behavior of programs (Harel 1979). Hence, these logics are concerned with binary relations on world states rather than with entire behaviors. Harel (1984) provides a good review of dynamic and related logics.

2.3 *The STRIPS Representation*

The STRIPS representation of actions, originally proposed by Fikes & Nilsson (1971), is one of the most widely used alternatives to the situation calculus. It was introduced to overcome what were seen primarily as computational difficulties in using the situation calculus to construct plans. The major problem was to avoid (*a*) the specification of a potentially large number of frame axioms, and (*b*) the necessity of having the planner consider these axioms in determining the properties that hold at each point in the plan.

In the STRIPS representation, a world state is represented by a set of logical formulas, the conjunction of which is intended to describe the given state. Actions or events are represented by so-called *operators*. An operator consists of a *precondition*, an *add list*, and a *delete list*. Given a description of a world state *s*, the precondition of an operator is a logical formula that specifies whether or not the corresponding action can be performed in s, and the add and delete lists specify how to obtain a representation of the world state resulting from the performance of the action in s. In particular, the add list specifies the set of formulas that are true in the resulting state and must therefore be added to the set of formulas representing s, while the delete list specifies the set of formulas that may no longer be true and must therefore be deleted from the description of s. This scheme for determining the descriptions of successive states is called the *STRIPS rule*.

For example, the following STRIPS operator can be taken to represent the action that moves block A from location 0 to location 1.

Precondition: $loc(A, 0) \wedge clear(A)$
Add list: $\{loc(A, 1)\}$
Delete list: $\{loc(A, 0)\}$.

Let's say that some world is described by the formulas $\{loc(A, 0), clear(A), red(A)\}$. Given this set of formulas, it is possible (trivially in this case) to prove that the precondition holds, so that the operator is then considered applicable to this world description. The description of the world resulting from application of this operator is $\{loc(A, 1), clear(A), red(A)\}$.

It is important to note that the formulas appearing in the delete list of an operator are not necessarily *false* in the resulting state; rather, the truth value of each of these formulas is considered unknown (unless it can be deduced from other information about the resulting state). Operators can also be parameterized and thus can represent a whole class of actions.

Although the operators in STRIPS are intended to describe actions that transform world states into other world states, they actually define *syntactic* transformations on *descriptions* of world states. STRIPS should thus be viewed as a form of logic and the STRIPS rule as a *rule of inference* within this logic. Given this perspective, it is necessary to specify the conditions under which the STRIPS rule is *sound*. That is, for each operator and its associated action, the formulas generated by application of the operator should indeed be true in the state resulting from the performance of the action. Surprisingly, only very recently has anyone attempted to provide such a semantics, though the importance of doing so has long been recognized.

The problem is that soundness is not possible to achieve if the STRIPS rule is allowed to apply to arbitrary formulas. One way around this difficulty is to specify a set of *allowable* formulas and require that only such formulas occur in world descriptions, operator add lists, and operator delete lists (although the preconditions of an operator can involve arbitrary formulas). Lifschitz (1987b) shows that such a system is sound if, for every operator and its associated action, (a) every allowable formula that appears in the operator's add list is satisfied in the state resulting from the performance of the action, and (b) every allowable formula that is satisfied in the state in which the action is initiated and does not belong to the operator's delete list is satisfied in the resulting state. The latter condition is commonly known as the *STRIPS assumption*.

As described, the STRIPS representation avoids the specification of frame axioms that state what properties are left unchanged by the occurrence of actions. Furthermore, the lack of frame axioms allows a planner to better focus its search effort. On the other hand, STRIPS is not nearly as expressive as the situation calculus (Waldinger 1977). In particular, the STRIPS representation compels us to include in an operator's delete list all allowable formulas that could possibly be affected by the action, even if the truth value of some of these could be deduced from other axioms. For example, even if we were given an axiom stating that when Fred dies he stops breathing, an operator representing the fatal shooting of Fred would nonetheless have to include in its delete list *both* effects of the shooting.

To overcome this difficulty, it is tempting to modify the STRIPS rule so that formulas that can be *proved* false in the resulting state need not be included in an operator's delete list. This leads to the *extended STRIPS assumption*, which states that any formula that is satisfied in the initiating state and does not belong to the delete list will be satisfied in the resulting state, *unless* it is inconsistent to assume so. Unfortunately, no one has yet provided an adequate semantics for such an approach (Reiter 1980).

Another alternative is to allow any kind of formula to appear in state and operator descriptions and to modify the STRIPS rule so that only a certain class of *basic* formulas is passed from state to state. The idea is that the truth value of a nonbasic formula in the state resulting from application of an operator cannot be determined using the STRIPS rule; it must instead be derived from other formulas that are true in that state. In this way, we can often simplify considerably the add lists of operators. [Some early attempts to implement this idea (Fahlman 1974; Fikes 1975) were flawed (Waldinger 1977), and later implementations (Wilkins 1984) appear to work only under certain restrictions.]

Yet another variant representation is described by Pednault (1986). Each action is represented by an operator that describes how performance of the action affects the relations, functions, and constants of the problem domain. As with the STRIPS representation, the state variable is suppressed and frame axioms need not be supplied. For a restricted but commonly occurring class of actions, the representation appears as expressive as the situation calculus.

3. PLAN SYNTHESIS

Plan synthesis concerns the construction of some plan of action for one or more agents to achieve some specified goal or goals, given the constraints of the world in which these agents are operating. In its most general form, it is necessary to take into account the various degrees to which the agents desire that their goals be fulfilled, the various risks involved, and the limitations to further reasoning arising from the real-time constraints of the environment. However, we shall begin with the simpler problem in which an agent's goals are consistent and all of the same utility; we shall disregard reasoning about the consequences of plan failure; and we shall not concern ourselves with real-time issues. [In philosophy, this kind of planning is commonly called *means-ends reasoning*, and is considered to be just one of the many components comprising rational activity (Bratman 1987; Davidson 1980).]

3.1 *Plans*

The essential component of a plan is that, when given to an agent or machine to perform or *execute*, it will produce some behavior. For example, a program for a computer is a particular kind of plan. Exactly what behavior occurs in a given situation will depend on the agent (or machine) that attempts to execute the plan, as well as on the environment in which that agent is embedded. In domains populated by more than one agent, plans may be assigned to a number of agents to execute cooperatively; such plans are often called *multiagent* plans.

The execution of a plan, of course, need not always be successful. Thus, there are at least two types of behavior associated with any plan: those that can be considered successful, in that the agent manages to execute each part of the plan without failure, and those that are generated by the plan but that, for some reason or another, turn out to be unsuccessful. As a rule, an unsuccessful behavior is one in which the agent has executed part of the plan successfully but then failed to execute some subsequent step. In many applications, both kinds of behavior must be taken into consideration—one's choice of plans often depends on the likelihood of plan failure and its consequences (Georgeff & Lansky 1986a).

Plans usually have a definite structure that depends on how the plan has been composed from more primitive components. The standard ways of composing plans include *sequencing* (resulting in sequential plans), *choice* (conditional plans), *iteration* (iterative plans), and *recursion* (recursive plans). One can also define *nondeterministic* operators that allow an arbitrary choice of which component to execute next and *parallel* operators to allow for concurrent activity.

The components of such composite plans are usually called *subplans*; the basic plan elements admitting of no further decomposition are called *atomic* or *primitive* plans (or, somewhat misleadingly, *atomic actions*). Plans can have other properties as well; which ones are considered important will often depend on the domain of application. For example, a plan may be of a certain type, or have an associated risk or likelihood of success.

It is not necessary that a plan be composed solely of atomic elements. For example, a plan for the evening might simply be to get dressed and then go to the theater, without it being specified how either of these activities should be accomplished. Similarly, we may fully detail each of the steps in a plan but leave unspecified the order in which they should be performed. Such plans are often called *partial* plans.

To reason about plans, we have to introduce additional *plan terms* into our formal language of events and actions. A plan term simply denotes a plan, in the same way that state terms, event terms, and fluent terms denote respectively states, events, and fluents. Furthermore, we need to introduce various function symbols and predicates to describe the allowed plan composition operators and any other properties we choose to ascribe to plans.

As mentioned above, one of the most important properties of a plan is the set of successful behaviors it generates. In the case that there is no concurrent activity, this can be reduced to considering simply the transformation from initial to final states. Let's represent the fact that state s_2 results from the execution of plan **p** by agent M when initiated in state s_1 by the expression $generate(M, \mathbf{p}, s_1, s_2)$. With this, we can now describe the effects of plan execution and the properties of the various plan composition operators.

For example, for some particular agent M and plan **p**, we might have

$$\forall s_1, s_2 \cdot holds(\phi, s_1) \supset (generate(M, \mathbf{p}, s_1, s_2) \supset holds(\psi, s_2)).$$

That is, if **p** is executed by agent M in a state in which ϕ holds, at the completion of execution ψ will hold. Assuming a fixed agent, we shall write the above formula simply as $exec(\mathbf{p}, \phi, \psi)$.

Strictly speaking, plans are not actions. We might have some predicates that apply to plans (such as whether or not they are partial, conditional, or unreliable) but that clearly do not apply to actions. However, predicates such as *generate* (and *exec*) allow us to specify the relation between plans and the actions or behaviors they generate. Process logic and dynamic logic may be viewed as variant notations for describing the same relation. However, some authors (Green 1969; Pelavin & Allen 1986) *equate* plans with the state transformations or events they generate. Used carefully, and in well-circumscribed problem domains, this can be the most frugal way to do things; in more general settings, however, it is restrictive and can lead to unnecessary confusion.

Finally, we need to consider what a *goal* is. Goals are important because they are the things that agents try to accomplish in executing their plans of action. In most of the early AI literature, a goal was considered to be simply some specified set of world states; an agent would be said to accomplish or achieve its goal if it managed to attain one of these states. For this reason, these goals are often called *end goals* or *goals of attainment*. Other researchers take goals to be some desired *transformation* from some set of possible initial states to some set of final states. This is typically the case in the area of automatic programming, in which one attempts to synthesize programs that meet certain input-output requirements.

However, real-world agents have goals that are more complex than these. For example, an agent may have a goal to *maintain* some condition over an interval of time, such as to remain in a position of power. Some goals of maintenance correspond to the *prevention* of some condition. For example, one might have the goal of preventing Congress from discovering how certain illegally obtained funds are distributed. There may also be goals in which the properties of some final state are not particularly important, but where the intervening *sequence of activities* is—for example, the goal to call in at a Swiss bank prior to returning home. Thus, in general, we can consider a goal to be some set of state sequences or behaviors; an agent succeeds in achieving such a goal if the actual behavior of the world turns out to be an element of this set.

Because in the general case goals are just particular sets of behaviors, they may be composed in the same way that events are. For example, the goal to place the books on the bookshelf and the goal to place the cups on the table can be combined to form the single composite goal to place both the books on the bookshelf *and* the cups on the table. Goals can also be composed sequentially; in this case, the composite goal may be to *first* place the books on the bookshelf and *subsequently* place the cups on the table. The behaviors corresponding to the former composite goal, of course, may be different from those corresponding to the latter one. In the first case, the goal is to achieve a state in which both the books are on the bookshelf *and* the cups are on the table, but there is no requirement regarding the order in which these tasks should be performed. In the second case, the ordering of tasks is specified, but there is no requirement that the books remain on the bookshelf while the cups are placed on the table.

Goal descriptions are often viewed as *specifications* for a plan: They describe the successful behaviors that execution of the plan should produce. Goals of attainment can be specified by stating simply the conditions that should hold after execution of the plan. Thus, they can be adequately described using the language we introduced in Section 2 for describing the properties of world states. Goals of transformation can be similarly specified. For describing more general kinds of goal, however, we must be able to express properties of sequences of states or events. To do this, we need the more expressive formalisms developed for multiagent domains (see Section 4).

3.2 *General Deductive Approaches*

Given a formulation of actions and world states as described in Section 2, the simplest approach to planning is to prove—by means of some automatic or interactive theorem-proving system—the existence of a sequence of actions that will achieve the goal condition. More precisely, suppose that we have some end goal ψ and that the initial state satisfies some condition ϕ. Then the theorem to be proved is

$$\forall s \cdot holds(\phi, s) \supset \exists z \cdot holds(\psi, z) \wedge reachable(z, s).$$

That is, we are required to prove that there exists a state z, reachable from s, in which the goal ψ holds, given that ϕ holds in the initial state s.

For example, a plan to clear a block A, given an initial state in which B is on top of A and C is on B, could be constructed by proving the theorem

$$\forall s \cdot holds(on(C, B), s) \wedge holds(on(B, A), s)$$

$$\supset \exists z \cdot holds(clear(A), z) \wedge reachable(z, s)$$

If done carefully, the proof could lead to a solution of the form

$$\forall s \cdot holds(on(C, B), s) \wedge holds(on(B, A), s) \supset$$

$$holds(clear(A), result(puton(B, table), result(puton(C, table), s))).$$

That is, if C is initially on B, which in turn is on A, then A will be clear in the state resulting from depositing C and then B on the table.

Green (1969) was the first to implement this idea. As he observed, however, it is essential to have the theorem prover provide the right kind of constructive proof. For example, consider being faced with a choice of two doors, behind one of which is a ferocious lion and the other a young maiden. In trying to maximize your lifespan, a theorem prover may well suggest that you simply open the door behind which lies the young maiden. Unfortunately, you may only be able to ascertain the maiden's location after opening the door—too late for you but of little concern to the planning system. This difficulty arises because the sequence of actions constructed by the planner can be conditional on properties of *future* states; that is, on properties that the agent executing the plan is not in a position to determine.

Manna & Waldinger (1987) show how many of these problems can be solved by reasoning about plans rather than actions. The planning technique they develop is based on the following scheme. For a goal condition ψ and initial state satisfying ϕ, we attempt to prove the theorem

$$\exists p \cdot exec(p, \phi, \psi) \wedge executable(p).$$

The aim is to find a plan p that satisfies this theorem. For example, a solution to the problem given above would be

$$exec((puton(B, table) ; (puton(C, table)), on(B, A) \wedge on(C, B), clear(A)),$$

where the symbol ";" represents the sequential composition of plans. The requirement regarding executability is included to prevent the planner from returning trivial or nonexecutable plans; this requirement is usually left implicit.

Conditional plans are not difficult to construct in this framework. For example, consider that we can construct two plans p_1 and p_2 that satisfy respectively $exec(p_1, \gamma \wedge \phi, \psi)$ and $exec(p_2, \neg\gamma \wedge \phi, \psi)$. Then it is straightforward to show that the conditional plan $p = $ **if** γ **then** p_1 **else** p_2 satisfies $exec(p, \phi, \psi)$. However, one must be careful in introducing conditionals, as they can expand the search space of potential solutions considerably.

The construction of plans involving recursion is difficult. To handle recursion, we have to provide the theorem prover with an induction axiom. There are various kinds of induction axioms that one can use. Manna & Waldinger (1987) use the principle of well-founded induction — that is, induction over a well-founded relation. This is a general rule that applies to many subject domains, but there are two difficulties. First, each domain will have its own well-founded relations, which must be specified explicitly. Second, it is often necessary to strengthen the goal constraint so as to have the benefit of a strong enough induction hypothesis to make things work out properly. With human intuition, it may not be difficult to formulate such strengthened goals, but even in simple cases the requisite strengthening seems to be beyond the capability of current theorem provers.

Iteration is often as problematic as recursion. It is not difficult to insert a *fixed* iterative subplan into a plan, provided that we have appropriate axioms describing its behavior. However, it is not a simple matter to *synthesize* an iterative subplan. One approach would be to first form a recursive plan (with all its attendant problems) and then transform this, if possible, into an iterative plan. Strong (1971), for example, shows how to convert certain classes of recursive plans (programs) into iterative ones.

This part is not difficult; the truly hard task is constructing the recursive plan to begin with.

However, in some cases, the synthesis of iterative plans is straightforward. For example, if a certain goal condition has to be satisfied for some arbitrary number of objects, it is often possible to construct a plan that accomplishes the goal for one of the objects and then iterate that plan over all the remaining objects.

Instead of the situation calculus, any of the modal logics discussed in Section 2.2 could alternatively be used to represent knowledge of the problem domain (Lansky 1987a; S. Rosenschein 1981; Stuart 1985). Unfortunately, for most domains of interest it appears that we require the expressive power of first-order modal logics, for which suitable theorem provers are currently unavailable. An exception to this is the work of Stuart (1985). He is concerned with the synchronization properties of plans, which can be adequately described using propositional modal logic.

3.3 *Planning as Search*

The basis of planning is to find, out of all the possible actions that an agent can perform, which, if any, will result in the world's behaving as specified by the goal conditions, and in what order these actions should occur. It can thus be viewed as a straightforward search problem: Find some or all possible orderings of the agent's actions that would result in achieving the specified goal, given the constraints of the world in which the agent is embedded. The number of possible action orderings is equal to the factorial of the number of actions performable by the agent. This makes the *general* problem computationally intractable (or what is usually called NP-hard).

Thus, a considerable part of the research effort in planning has been directed to finding effective ways of reducing this search, either by formulating the problem in some appropriate way, restricting the class of problems that can be represented, or by careful choice in the manner in which potential plans are examined [see Levesque (1986) for a discussion of similar issues].

There are two common ways of viewing plan search techniques. One is to perceive the process as searching through a space of world states, with the transitions between states corresponding to the actions performable by the agent. Another view is that the search takes place through a space of partial plans, in which each node in the search space corresponds to a partially completed plan. The latter view is the more general, as the first can be seen as a special case in which the partial plan is extended by adding a primitive plan element to either end of the current partial plan.

Thus, we can characterize most approaches to the planning problem as follows. Each node in the search space corresponds to some possibly partial plan of action to achieve the given goal. The search space is expanded by further elaborating some component of plan formed so far. We shall call the ways in which plans can be elaborated *plan specialization operators*.[3] The plan space can be searched with a variety of techniques, both classical and heuristic (Nilsson 1980; Tate 1984).

At each stage, a very general plan (which admits of a potentially large number of behaviors) is increasingly refined until it becomes very specific. As it becomes more specific, the set of behaviors that can potentially satisfy the plan becomes smaller and smaller. The process continues until the plan is specific enough to be considered a solution to the problem; usually this is taken to mean that the plan can be executed by some specified agent. Alternatively, we can view the whole process as continually refining the *specifications* for a plan; that is, by reducing progressively the original goal to some composition of subgoals.

A different approach to planning involves plan *modification* rather than plan specialization. Thus, beginning with a plan that only *approximates* the goal specifications, various *plan modification operators* are applied repeatedly in an attempt to improve the approximation. Unlike plan specialization, the problem with plan modification is that it is often difficult to determine whether or not a particular modification yields a plan that is any closer to a solution. In many cases, a combination of plan specialization and modification can be used effectively. Furthermore, many techniques that are framed in terms of plan modification can be recast as equivalent plan-specialization techniques and vice versa.

In the next few sections, we shall examine the more important of these specialized planning techniques. For many of these, there exists a corresponding deductive method whereby certain constraints are imposed on the order in which inferences are drawn (Genesereth & Nilsson 1987; Kowalski 1979; Manna & Waldinger 1987).

3.4 *Progression and Regression*

Before we consider specific planning techniques, let us introduce some new terminology. Consider a primitive plan a that, for some conditions ϕ and ψ, satisfies $exec(a, \phi, \psi)$. That is, we are guaranteed that, after initiation of a in a state in which ϕ holds, ψ will hold at the completion of execution. If ψ is the strongest condition for which we can prove that this holds, we shall call ψ the *strongest provable postcondition* of a with respect to ϕ. We

---

[3] Note that plan specialization operators are not the same as plan composition operators, although often there is a correspondence between the two. The former operators map partial plans into more specific plans; the latter map some tuple of plans into a composite plan.

can similarly define the *weakest provable precondition* of **a** with respect to ψ to be the weakest condition ϕ that guarantees that ψ will hold if **a** is initiated in a state in which ϕ holds. Analogously, we can apply these terms to actions as well as plans.

Let's now consider how we could form a plan to achieve a goal ψ, starting from an initial world in which ϕ holds. That is, we are required to find a plan **p** that satisfies $exec(\mathbf{p}, \phi, \psi)$. For any primitive plan element **a**, this condition will be satisfied for **p** if:

1. $\mathbf{p} = \text{NIL}$ and $\forall s \cdot holds(\phi, s) \supset holds(\psi, s)$.
2. $\mathbf{p} = \mathbf{a} ; \mathbf{q}$, where **q** satisfies $exec(\mathbf{q}, \gamma, \psi)$ and γ is the strongest provable postcondition of **a**.
3. $\mathbf{p} = \mathbf{q} ; \mathbf{a}$, where **q** satisfies $exec(\mathbf{q}, \phi, \gamma)$ and γ is the weakest provable precondition of **a** and ψ.
4. $\mathbf{p} = \mathbf{q}_1 ; \mathbf{a} ; \mathbf{q}_2$, where, for some γ_1 and γ_2, **a** satisfies $exec(\mathbf{a}, \gamma_1, \gamma_2)$, $\mathbf{q}_1$ satisfies $exec(\mathbf{q}_1, \phi, \gamma_1)$, and $\mathbf{q}_2$ satisfies $exec(\mathbf{q}_2, \gamma_2, \psi)$.

Case 1 simply says that if the goal condition is already satisfied, we need not plan anymore—i.e. the empty plan (NIL) will do. Now consider case 2. Let's say that we are guaranteed that if we execute some primitive plan **a** in a state in which ϕ holds, γ will be true in the resulting state. Thus, if the plan begins with the element **a**, the rest of the plan must take us from a state in which γ is true to one in which ψ is true. We can take γ to be any condition that is guaranteed to hold after the execution of **a** but, to spare ourselves from planning for situations that cannot possibly occur, it is best to take γ to be the strongest of these conditions. Thus, case 2 amounts simply to forward-chaining from the initial state and is usually called *progression*. Case 3 is similar to case 2, except that we chain backward from the goal. It is usually called *regression*; the condition γ is often called the *regressed goal*. Case 4 is tantamount to choosing a primitive plan element somewhere in the middle of the plan, then trying to patch the plan at either end. In fact, case 4 is a generalization of cases 2 and 3.

It is straightforward to construct a simple planner that uses these rules recursively to build a plan. Initially the planner starts with the fully unelaborated plan **p**, then specializes this plan recursively, applying rules 2, 3, or 4 until, finally, rule 1 can be applied. Clearly, whether or not a solution is obtained will depend on the choice of rules and the choice of primitive plan elements at each step. The algorithm works for any plan or action representation, requiring only that we be able to determine plan (or action) postconditions and preconditions, as described above. For example, GPS (Newell & Simon 1963) and STRIPS (Fikes & Nilsson 1971) use STRIPS-like action representations and rules 1 and 4, whereas S. Rosenschein

(1981) employs dynamic logic to describe the effects of actions and uses rules 1, 2, and 3.

Rather than specify the execution properties of primitive plans alone (using STRIPS or some other variant), it can also be useful to provide information on the execution of partial plans. For example. it may be that going out to dinner and then to the theater results in one being happy, irrespective of the way in which this plan is eventually realized. I have elsewhere (Georgeff & Lansky 1986a) called this *procedural knowledge*, on the grounds that such facts describe properties regarding the execution of certain procedures or plans. Lansky and I show how such procedural knowledge can be used effectively for handling quite complex tasks, such as fault diagnosis on the space shuttle (Georgeff & Lansky 1986b) and the control of autonomous robots (Georgeff et al 1987). The plan operators used in NOAH (Sacerdoti 1977), DEVISER (Vere 1983), and SIPE (Wilkins 1984) (see Section 3.6) also allow the representation of procedural knowledge, although in more restrictive forms.

3.5 *Exploiting Commutativity*

Unfortunately, simple progression and regression techniques are too inefficient to be useful in most interesting planning problems. But is there anything better we can do? In the worst case, the answer is no—we simply have to explore all possible action orderings. However, the real world is not always so unkind.

For example, consider that one has a goal of stacking cups on a table *and* putting books on a bookshelf. It is often possible to construct plans that satisfy each of these component goals without regard to the other, then to combine these plans in some way to achieve the composite goal. (This might not be the case, however, if the cups were initially on the bookshelf or the books strewn randomly atop the table.)

For any two goals, if a plan for achieving one goal does not interfere with a plan to achieve the other goal, we say that the two plans are *interference free* (Georgeff 1984) or *commutative* (Nilsson 1980; Pednault 1986) with respect to these goals.[4] If this condition holds, any interleaving of these plans will satisfy both goals (Georgeff 1984; Lassez & Maher 1983). Partitioning the problem-solving space in this way can lead to a substantial reduction in the complexity of the problem, as it reduces the number of action orderings that need be considered. The assumption that two given goals can be solved independently (i.e. that the plans constructed

---

[4] In fact, these two notions are not identical. Plans that are interference free with respect to particular goals need not be commutative, and vice versa.

for each goal will be interference free) is known as the *strong linearity* assumption (Sussman 1973).

Unfortunately, in many cases of interest, the plans that are produced for each subgoal turn out not to be interference free. It may nevertheless be possible to construct plans under the strong linearity assumption and, should they interfere with one another, patch them together in some way so that the desired goals are still achieved. Some early planners [notably HACKER (Sussman 1973) and INTERPLAN (Tate 1974)] adopted this approach: They would construct plans that are flawed by interference with one another and then try to fix them by reordering the operations in the plans. These systems backtrack when they find interference; they reorder a couple of subgoals and then start planning to achieve them in the new order. However, this can be inefficient and is somewhat restricted in the class of problems that can be solved (Waldinger 1977).

Waldinger (1977) developed a more general planning method that, in forming a plan for a particular subgoal, took account of the constraints imposed by the subplans for previously solved subgoals. That is, he first constructs a plan to achieve one of the subgoals, without regard to any others. He then tries to achieve the next subgoal while maintaining the constraints imposed by the first plan. In particular, the first subgoal and its regressed conditions, as obtained from the initial plan, become goals of maintenance that must be satisfied by the new plan. These goals of maintenance are usually called *protected* conditions.

Kowalski (1979) and Warren (1974) follow essentially the same approach as Waldinger. Kowalski also examines ways of improving the efficiency of planning by combining regression and progression in various ways. However, none of these methods is complete (that is, none guarantees to find a solution if one exists), primarily because they match the regressed goals with the initial state prematurely. This deficiency was corrected by Pednault, who presents perhaps the most advanced version of this technique (Pednault 1986).

The foregoing approaches can be used with any sufficiently expressive action representation. Warren restricts himself to a simple STRIPS-like representation. An extended form of the STRIPS assumption is embedded in both the Waldinger and Kowalski systems. This, however, is not dealt with adequately by either author, so that the semantics of their action representations is not made clear. For example, Waldinger simply states that he uses "a default rule stating that, if no other regression rule applies, a given relation is assumed to be left unchanged by [the] action." This leads to serious semantic difficulties (Reiter 1980). Kowalski (1979) introduces a predicate *Diff*, the truth value of which is determined according to the *syntactic* structure of the terms that appear as arguments. Again the meaning of this is unclear. Pednault (1986), on the other hand, uses an extended form of the STRIPS representation that does have a well-defined semantics.

For techniques that aim to exploit commutativity, it is clearly desirable that any composite goals be decomposed into maximally independent subgoals. However, unless one introduces some notion of independence (see Section 4.2), such decompositions are not possible to determine. In fact, all existing planners decompose composite goals on the basis of their *syntactic* structure alone, rather than on any properties of the domain itself. The reason that this often appears to work is probably a result of the way the predicates chosen to represent the world reflect some kind of underlying independence that has not been made explicit. For example, given the composite goal of stacking cups on the table and books on the bookshelf, one possible decomposition is into two subgoals, one of which is to stack half the cups and half the books and the other to stack the remainder of the cups and books. Clearly, in most situations, this turns out to be a poor decomposition of the composite goal; that most planners don't make this decomposition is simply a result of fortuitous choice of domain predicates. However, there are some cases in which this decomposition actually *is* the best one; for example, when half the cups are stored together with half the books in one container and the rest of the cups and books in another container.

3.6 *Improving Search*

The techniques outlined above provide only a limited number of plan specialization and modification operators and apply only to those cases in which the goal descriptions (plan specifications) consist of a *conjunction* of simpler goal conditions. Furthermore, these operators are such that their application always results in a linear ordering of primitive plan elements. They are thus often called *linear* planners.

However, the search can often be improved by deferring decisions on the ordering of plan elements until such decisions are forced; by that time, we could well have acquired the information we need to make a wise choice. This is the technique adopted by the so-called *nonlinear planners*. These planners allow the partial plans formed during the search to be arbitrary partial orders over plan elements. Thus, instead of arbitrarily choosing an ordering of plan elements, they are left unordered until some conflict is detected among them, at which time some ordering constraint is imposed to remove the conflict. Nonlinear planners therefore do not have to commit themselves prematurely to a specific linear order of actions and can get by with less backtracking than otherwise. Examples of such planners include NOAH (Sacerdoti 1977), NONLIN (Tate 1977), DEVISER (Vere 1983), and SIPE (Wilkins 1984).

Another way in which we can defer making decisions that, at some later stage, may have to be retracted, is to allow the individuals (objects) that appear in a plan to be partially specified. This can be achieved by accumulating *constraints* on the properties that these individuals must satisfy, and by deferring the selection of any particular individual as long as possible (Stefik 1981a; Wilkins 1984). For example, we may know that to perform a certain block-moving action the block being moved must weigh under five pounds. Instead of selecting a particular block that meets this constraint, we instead just post that constraint on the value of the logical variable denoting the block. Later, if we discover that the block should be red as well, we can then attempt to select a block that meets both requirements. This technique can be easily implemented by associating a list of constraints with each such variable and periodically checking these constraints for consistency. Mutual constraints among variables may be similarly handled.

Efficiency can often be further improved by introducing additional plan specialization and plan modification operators. Most nonlinear planners provide a great variety of such operators. For a restricted class of problems, Chapman (1985) and Pednault (1986) furnish a complete set of plan specialization and modification operators, and prove soundness and completeness for their systems. They also provide a good review of the techniques and failings of other nonlinear planners. Chapman also analyzes the complexity of various classes of these planners. Of course, these planners are no more efficient than their linear counterparts in the worst case.

3.7 *Hierarchical Planners*

The major disadvantage of the planners discussed so far is that they do not distinguish between activities that are critical to the success of the plan and those that are merely details. As a result, these planners can get bogged down in a mass of minutiae. For example, in planning a trip to Europe, it is usually a waste of time to consider, prior to sketching an itinerary, the purchase of tickets or the manner of travelling to the airport.

The method of *hierarchical planning* is first to construct an abstract plan in which the details are left unspecified, and then to refine these components into more detailed subplans until enough of the plan has been elaborated to ensure its success. The advantage of this approach is that the plan is first developed at a level at which the details are not computationally overwhelming.

For an approach to be truly hierarchical, we require that the abstract planning space be a *homomorphic image* of the original (ground) problem. Let us assume we have some function g that maps partial plans in the ground problem space into partial plans in the abstract space. Also, for any plan composition operator C in the ground space, let C' be the corresponding composition operator in the abstract space. Then, if p_1 and p_2 are partial plans in the ground space, we require that

$$g(C(p_1, p_2)) = C'(g(p_1), g(p_2)).$$

This simply captures the fact that the abstraction should preserve the structure of the original problem.

These requirements ensure that if we find a solution at the ground level there will exist a corresponding solution at the abstract level. Furthermore, they ensure that any plan constructed at the abstract level will be composed of plan elements that can be solved *independently* of one another at the lower levels of abstraction. Indeed, it is in this manner that the complexity of the problem is factored. The method can be extended to multiple levels of abstraction in a straightforward way.

There have been a number of attempts to devise hierarchical planning systems. The system ABSTRIPS (Sacerdoti 1973) induced appropriate homomorphisms by neglecting the predicates specifying details of the domain and only retaining the important ones. Thus, for example, in determining how to get from Palo Alto to London, the predicates describing the possession of plane tickets or the relative location of, say, Palo Alto to San Francisco airport may be initially neglected. This simplifies planning the global itinerary, leaving the details of how to collect the plane tickets and how to travel from home to the airport to be determined at lower levels of abstraction. S. Rosenschein (1981) describes another hierarchical planning method in which the homomorphism between levels is given by the relationship between primitive plan elements, at the one level, and elaborated plans at the next lower level—what Rosenschein describes as the relation of being "correctly implemented."

Unfortunately, the notion of hierarchical planning is confused in much of the AI literature. In particular, there is considerable misunderstanding about the requirement that the problems at lower levels of abstraction be independently solvable. This does not mean that the plans formed at the lower levels of abstraction must be independent of the features of other plans formed at these levels. The important point is that any features upon which these plans may mutually depend should be reflected at the higher abstraction levels (i.e. be retained by the homomorphic mapping g). This ensures that any interactions are taken into account in the abstract space and the lower levels can pursue their planning independently of one another. Of course, this approach only works if the number of features that have to be reflected at the higher levels is considerably smaller than the number of features that the lower levels must deal with in their planning.

For example, it may be that a certain stove can only accommodate two saucepans at a time. In planning the overall preparation of a meal, it is important that this information be retained. In this way, the plan formed at the abstract level can account for the potential interference between the cookings of various dishes, and the cookings themselves can then be separately and independently planned. The resource mechanism used in SIPE (Wilkins 1984) serves exactly this purpose.

Most hierarchical planners [e.g. ABSTRIPS (Sacerdoti 1973) and Rosenschein's hierarchical planner (S. Rosenschein 1981)] first form an abstract plan and only then consider construction of the lower-level plans. Stefik (1981), on the other hand, describes an approach in which the planning at the lower levels of abstraction may proceed before the abstract plan is fully elaborated. This is achieved by allowing the abstract plan to be partial with respect to the values of certain critical variables and letting the lower-level plans post global constraints on these values (although he describes this process somewhat differently). In this way, planning need not proceed top-down—from the abstract space to the lower-level spaces—but rather can move back and forth between levels of abstraction.

The plan specialization operators used in some of the nonlinear planners [such as NOAH (Sacerdoti 1977), NONLIN (Tate 1977), SIPE (Wilkins 1984), and DEVISER (Vere 1983)] also allow planning at higher levels of abstraction prior to concentrating on the finer details of a plan. However, these planners are *not* strictly hierarchical. The difficulty is that even if all the subplans produced at lower levels of abstraction are successful (i.e. satisfy the plan constraints imposed by the higher levels of abstraction) it may still not be possible to combine them into a solution of the original problem (S. Rosenschein 1981). Furthermore, the potential for global interactions is dependent on arbitrarily fine details of the lower-level solutions, and the advantages of hierarchy are thus largely lost. There are also difficulties in the way that the STRIPS assumption manifests itself in planning at the higher levels of abstraction; Wilkins (1985a) discusses this problem in more detail and offers a solution.

3.8 *Other Planning Techniques*

The planning techniques we have discussed so far are all domain independent, in the sense that the representation schemes and reasoning mechanisms are general and can be applied to a variety of problem domains. However, in many cases, techniques specific to the application may be preferable. Many such specialized planning systems have been developed, mostly concerned with robotic control, navigation, and manipulation. For example, Gouzenes (1984) has investigated techniques for solving collision-avoidance problems, Brooks (1983) considers planning collision-free motions for pick-and-place operations, and Myers (1985) describes a collision-avoidance algorithm suited to multiple robot arms operating concurrently. Krogh & Thorpe (1986) have combined algorithms for path planning and dynamic steering control into a scheme for real-time control of autonomous vehicles in uncertain environments, and Kuan (1984) describes a hierarchical method for spatial planning. Lozano-Perez (1980) was one of the first researchers to demonstrate the utility of changes of representation in spatial planning. McDermott & Davis (1984) describe an approach to path planning in uncertain surroundings, and Brooks (1985b) has devised a symbolic map representation whose primitives are suited to the task of navigation and that is explicitly grounded on the assumption that observations of the world are approximate and control is inaccurate.

In many applications, it is necessary to plan not only to achieve certain conditions in the world but also to *acquire* information about the world. Relatively little work has been done in this area. Both Moore (1980) and Morgenstern (1986) propose to treat this problem by explicitly reasoning about the knowledge state of the planning agent. In such a framework, a test is viewed as an action that increases the knowledge of the agent, and can be reasoned about in the same way as other actions. In many cases, however, we can get by with a much simpler notion of a test. For example, Manna & Waldinger (1987) allow a set of primitive tests to be specified directly and utilize these in forming conditional plans. Lansky and I (Georgeff & Lansky 1986a) represent tests simply as plans that are guaranteed to fail if the condition being tested is false at the moment of plan initiation. In this manner, plans involving complex tests can be constructed without having to reason about the knowledge state of the planner.

Sometimes, one must be able to reason explicitly about units of time and the expected duration of events. For example, in planning the purchase of a used car, it may be necessary to know that the bank closes at 3 PM, and that the time taken to travel to the bank is about 15 minutes, depending on traffic density. Furthermore, it is often necessary to reason about the probabilities of event occurrences and the likelihood of future situations. Some initial work in these areas can be found in Bell & Tate (1985), Dean (1984, 1985), Fox (1984), Vere (1983), and Wesley et al (1979).

3.9 *Planning and Scheduling*

The problem of scheduling can be considered a special case of the planning problem. Essentially, one has a set of activities to carry out while satisfying some set of goal and domain constraints. Unlike the aforementioned planning problems, however, these activities are usually completely detailed, requiring no means-ends analysis. The question is how to put these activities together while maintaining whatever constraints are imposed by the problem domain. These constraints usually concern the availability of resources (such as a certain machine being available to perform one activity at a time) and timing requirements (such that a certain plan must be completed by a specified time). More often than not, the problem involves more than one agent and therefore is really a special case of the multiagent planning problem discussed in Section 4, below.

The class of problems considered by scheduling systems, moreover, often goes beyond those handled by traditional planning systems. For example, the domain constraints are often much richer than those considered by traditional planning systems. In addition, one usually wants an optimal or nearly optimal solution. Furthermore, many of the goals and constraints are *soft*; that is, while it is desirable that they be met, they *can* be relaxed if necessary (at a certain cost). Thus, scheduling requires some kind of cost-benefit analysis to determine the optimal solution.

There are various standard mathematical techniques for solving scheduling problems. However, the combinatorics encountered makes them unsuitable for most real-world applications, especially when rescheduling is involved. On the other hand, there have been few attempts in AI to tackle this challenge. Fox (1984) has developed a number of systems for handling a broad range of scheduling constraints, and has explored the use of special heuristics to constrain the search. As yet, however, the general principles underlying this work remain to be examined (Fox 1986).

3.10 *Operator Choice and Metaplanning*

As more and more plan specialization operators are introduced, one is faced with the problem of when (i.e. in which order) to apply them. There are various ways of handling this problem. The simplest, and most common, is to specify some particular order and embed this directly in the planning algorithm. Another approach is to specify, for each operator, the conditions under which it should be used. The operators are then invoked (triggered) nondeterministically, depending on the current state of the planner. These techniques are usually called *opportunistic planners* (Hayes-Roth 1985).

Alternatively, one can consider the problem of how to go about constructing a plan (which operators should be expanded next, which goal to work on, how to decompose a goal, when to backtrack, how to make choices of means to a given end, etc) as itself a problem in planning. One would thus provide rules describing possible planning methodologies and let the system determine automatically, for each particular case, the best way to go about constructing a plan. This kind of approach is often referred to as *metaplanning* (Dawson & Siklossy 1977; Genesereth 1983; Georgeff & Lansky 1986a; Stefik 1981b; Wilensky 1981). The obvious problem with metaplanning is to ensure that the time saved in better constraining the search is not lost in reasoning about how to achieve this focus of effort.

Metaplanning can be viewed as describing or axiomatizing the process of planning. In its fullest generality, it could involve the synthesis of metalevel plans based on the properties of basic metalevel actions (Stefik 1981b). However, metaplanning is rarely used in this general way. Rather, the metalevel axiomatization usually serves as a convenient way to describe various planning strategies to some generic planning system, without hard-wiring them into the interpreter. Thus, for example, to have the system construct plans in some manner one would simply provide the system with an axiomatization of that particular planning technique. In addition, we can in this way describe plan specialization and modification operators that are intended specifically for certain domains.

4. MULTIAGENT DOMAINS

Most real worlds involve dynamic processes beyond the control of an agent. Furthermore, they may be populated with other agents—some cooperative, some adversarial, and others who are simply disinterested. The single-agent planners we have been considering are not applicable in such domains. These planners cannot reason about actions that the agent has no control over and that, moreover, may or may not occur concurrently with what the agent is doing. There is no way to express non-performance of an action, let alone to reason about it.

We therefore need to develop models of actions and plans that are different from those we have previously considered. We need theories of what it means for one action to interfere with another. Many interactions are harmful, leading to unforseen consequences or deadlock. Some are beneficial, even essential (such as lifting an object by simultaneously applying pressure from both sides). We should be able to state the result of the concurrence of two events or actions. We need to consider cooperative planning. planning in the presence of adversaries. and how to form contingency plans. In addition, we shall require systems capable of reasoning about the beliefs and intentions of other agents and how to communicate effectively both to exchange information and to coordinate plans of action. Furthermore, these systems will sometimes need to infer the beliefs, goals, and intentions of other agents from observation of their behaviors.

4.1 Action Representations

Multiagent domains are those having the potential for concurrent activity among multiple agents or other dynamic processes. The entities introduced in earlier sections—world states, histories, fluents, actions, events, and plans—can also form the basis for reasoning in these domains. However, most of the simplifying assumptions we made for handling single-agent domains cannot be usefully employed here. In particular, it is not possible to consider every action as a relation on states, as the effects of performing actions concurrently depends on what happens *during* the actions (Georgeff 1983; Pelavin & Allen 1986). For example, in a production line making various industrial components, it is important to know what machines are used during each activity so that potential resource conflicts can be identified.

In addition, we need more powerful and expressive formalisms for representing and reasoning about world histories. For example, we should be able to express environmental conditions such as "The bank will stay open until 3pm" and "If it rains overnight, it will be icy next morning." Similarly, we have to be able to reason about a great variety of goals, including goals of maintenance and goals satisfying various ordering constraints (Pelavin & Allen 1986).

As before, we can take an event type to be a set of state sequences, representing all possible occurrences of the event *in all possible situations* (Allen 1984; Georgeff 1987; McDermott 1982, 1985; Pelavin & Allen 1986). Unlike single-agent domains, however, the set of behaviors associated with a given event must include those in which other events occur *concurrently* or *simultaneously* with the given event. For example, the event type corresponding to "John running around a track three times" must include behaviors in which other events (such as the launch of the space shuttle, it being raining, or John being accompanied by other runners) are occurring concurrently.

One possible approach is to approximate concurrent activity by using an interleaving approximation (Georgeff 1983, 1984; Pednault 1987). This renders the problem amenable to the planning techniques that are used for single-agent domains. However, it is not possible to model simultaneous events using such an approach, which limits its usefulness in some domains.

Another approach to reasoning about multiagent domains is to extend the situation calculus to allow reasoning about world histories and simultaneous events. I show elsewhere (Georgeff 1987) how this can be done by introducing the notion of an atomic event, which can be viewed as an instantaneous transition from one world state to another. Atomic events cannot be modeled as functions on world state, as it would then be impossible for two such events to occur simultaneously (unless they had exactly the same effect on the world). Given this perspective, the transition relation of an atomic event places restrictions on those world relations that are directly affected by the event but leaves most others to vary freely (depending upon what else is happening in the world). This is in contrast to the classical approach, which views an event as changing some world relations but leaving most of them unaltered.

I also introduce a notion of *independence* to describe the region of influence of given events. This turns out to be critical for reasoning about the persistence of world properties and other issues that arise in multiagent domains. Indeed, what makes planning useful for survival is the fact that we can structure the world in a way that keeps most properties and events independent of one another, thus allowing us to reason about the future without complete knowledge of all the events that could possibly be occurring.

McDermott (1982) provides a somewhat different formalism for describing multiagent domains, although the underlying model of actions and events is essentially as described above. Perhaps the most important difference is that world histories are taken to be dense intervals of states, rather than sequences; that is, for any two states in any given world history, there always exists a distinct state that occurs between them. The aim of this extension is to allow reasoning about continuous processes and may also facilitate reasoning about hierarchical systems.

Allen and Pelavin (Allen 1984; Pelavin & Allen 1986) introduce yet another formalism based on a variation of this model of actions and events. The major difference is that fluents are viewed as functions on *intervals* of states, rather than as functions on states. Thus, in this formalism, *holds(raining, i)* would mean that it is raining over the interval of time *i*, which might be, for example, some particular period on some specific day. The aim is that, by using intervals rather than states, we obtain a more natural and possibly more tractable language for describing and reasoning about multiagent domains. A similar approach has been developed by Kowalski & Sergot (1986).

Note that in these interval-based calculi, world states per se need not be included in the underlying model; indeed, intervals become the basic entities that appear in world histories. The ways intervals relate to one another are more complex than for states, however. Whereas in a given world history, states can only either precede or succeed one another. intervals can also meet, contain, and overlap one another in a variety of ways. Some work on formalizing the interval calculus and its underlying models can be found in Allen (1982), Kowalski & Sergot (1986), Ladkin (1987), Ladkin & Maddux (1987), Pelavin & Allen (1986), Sadri (1986), and van Bentham (1984).

Yet another approach is suggested by Lansky (1987a), who considers events as primitive and defines state derivatively in terms of event sequences. Properties that hold of world states are then restricted to being temporal properties of event sequences. For example, one might identify the property "waiting for service" with the condition that an event of type "request" has occurred and has not been followed by an event of type "serve." Lansky uses a temporal logic for expressing general facts about world histories and, in part, for reasoning about them also.

4.2 *Causality and Process*

One problem I have not yet addressed is the apparent complexity of the axioms that describe the effects of actions. For example, while it might seem reasonable to state that the location of block B is independent of the movement of block A, this is simply untrue, as everyone knows, in most interesting worlds. Whether or not the location of B is independent of the movement of A will depend on a host of conditions, such as whether B is in front of A, on top of A, atop A but tied to a door, and so on.

One way to solve this problem is by introducing a notion of *causality*.[5] Two kinds of causality suggest themselves: one in which an event causes the simultaneous occurrence of another event; the other in which an event causes the occurrence of a subsequent event. These two kinds of causality suffice to describe the behavior of any procedure, process, or device that is based on discrete (rather than continuous) events.

For example, we might have a causal law to express the fact that, whenever a block is moved, any block atop it and not somehow restrained (e.g. by a string tied to a door) will also move. According to this view, causation is simply a relation between atomic events that is conditional on the state of the world at the time of the events. Causation must also be related to the temporal ordering of events; for example, one would want to assume that an event cannot cause another event that precedes it. Various treatments of causality can be found in several sources (Allen 1984; Lansky 1987a; McDermott 1982; Shoham 1986). Most of these view causality as a simple relation between events; Shoham (1987), however, takes a radical view and defines causality in terms of the *knowledge state* of the agent. Wilkins (1987) indicates how causal laws could be used effectively in traditional planning systems.

It is often convenient to be able to reason about groups of causally interrelated events as single entities. Such groupings of events, together with the causal laws that relate them to one another, are usually called

---

[5] Although it might appear that the notion of causality also arises in single-agent domains, there seems little point in developing a theory of causality suited to such worlds. The power of causal reasoning lies in describing the properties of dynamic environments in which the actions of an agent may affect or initiate actions by other agents or processes.

processes [although Allen (1984) uses the term quite differently]. For example, we might want to amalgamate the actions and events that constitute the internal workings of a robot, or those that pertain to each component in a complex physical system. A machine (or agent) together with a plan of action may also be viewed as a special kind of process.

Charniak & McDermott (1985) examine how everyday processes may be reasoned about. For example, they consider the problem of reasoning about the filling of a bathtub and how we can infer that it will eventually fill up if the tap is turned on (and the plug not pulled). However, they do not indicate how to provide an adequate formalism for describing such processes, particularly with regard to their interaction with other, possibly concurrent processes.

Indeed, the strength of the concept of process derives from the way the interaction among events in different processes is strictly limited. Surprisingly, there has been little work in AI in this direction, although similar notions have been around for a long time. For example, Hayes (1985) describes the situation in which two people agree to meet in a week. They then part, one going to London and the other remaining in San Francisco. They both lead eventful weeks, each independently of the other, and duly meet as arranged. To describe this using world states, we have to say what each of them is doing just before and just after each noteworthy event involving the other. As Hayes remarks, this is clearly silly.

One approach to this problem is to specify a set of processes and classify various events and fluents as being either internal or external with respect to these processes (Georgeff 1987a; Lansky 1987b). If we then require that there be no *direct* causal relationship between internal and external events, the only way the internal events of a given process can influence external events (or vice versa) is through indirect causation by an event that belongs to neither category. Within the framework of concurrency theory, these intermediary events (more accurately, event types) are often called *ports*. Processes thus impose causal boundaries and independence properties on a problem domain, and can thereby substantially reduce cominatorial complexity (Georgeff 1987a; Lansky 1987b).

The identifiability of processes depends strongly on the problem domain. In standard programming systems (at least those that are well structured), processes can be used to represent scope rules and are fairly easy to specify. In complex physical systems, it is often the case that many of the properties of one subsystem will be independent of the majority of actions performed by other subsystems; thus these subsystems naturally correspond to processes as defined here. Lansky (1987a,b) gives other examples in which processes are readily specified. In other situations, such specification might be more complicated. Moreover, in many real-world situations, dependence will vary as the spheres of influence and the potential for interaction change over time (Hayes 1985).

If we are to exploit the notion of process effectively, it is also important to define various composition operators and to show how properties of the behaviors of the composite processes can be determined from the behaviors of their individual components. For example, we should be able to write down descriptions of the behaviors of individual agents and, from these descriptions, deduce properties of groups of agents acting concurrently. We should *not* have to consider the internal behaviors of each of these agents to determine how the group as a whole behaves. The existing literature on concurrency theory (Hoare 1985; Milner 1980) provides a number of useful composition operators, though this area remains to be explored (Georgeff 1987a).

4.3 *Multiagent Planning*

Despite the variety of formalisms developed for reasoning about multiagent domains, relatively few planning systems have been fully implemented. Allen & Koomen (1983) describe a simple planner, based on a restricted form of interval logic (Allen 1984). While this technique is effective for relatively simple problems, it is not obvious that the approach would be useful in more complex domains. Furthermore, the semantics of their action representation is unclear, particularly with respect to the meaning of concurrent activity. Kowalski & Sergot (1986) also describe systems for reasoning about events based on interval calculi. Again, it appears that the class of problems that can be considered is limited; however, these techniques appear adequate for a large range of dynamic database applications.

Another issue concerns how separate plans can be combined in a way that avoids interference among the agents executing the plans. In such a setting, one could imagine a number of agents each forming their own plans and then, after communicating their intentions (plans) to one another or a centralized scheduler, modifying these to avoid interference. To solve this problem, it is necessary to ascertain, from descriptions of the actions occurring in the individual plans, which actions could interfere with one another and in what manner (Georgeff 1984). After this has been determined, a coordinated plan that precludes such interference must then be constructed. This plan can be formed by inserting appropriate synchronization actions (interagent communications) into the original plans to ensure that only interference-free orderings will be allowed (Georgeff 1983). Stuart (1985) formalized this approach and implemented a synchronizer based on techniques developed by Manna & Wolper (1981).

As the above work shows, the notion of intending or committing to a course of behavior is an important cooperative principal. Some researchers have investigated how multiple agents can cooperate and resolve conflicts by reasoning locally about potential payoffs and risks (J. Rosenschein 1986; J. Rosenschein & Genesereth 1984). Other research has focused on the use of global organizational strategies for coordinating the behavior of agents operating asynchronously (Corkill & Lesser 1983). Another interesting development is the work of Durfee et al (1985) on distributed analysis of sensory information. They show that, by reasoning about the local plans of others, individual tracking agents can form partial global plans that lead to satisfactory performance even in rapidly changing environments. Reasoning about communication is discussed by J. Rosenschein (1982) and Cohen & Levesque (1985).

Lansky (1987b) has developed a multiagent planner that exploits causal independencies. Unlike the approaches described above, constraints between events have to be specified explicitly. However, the system accommodates a wide class of plan-synchronization constraints. Also, the process of plan synchronization is not limited to a strategy of planning to separately achieve each component task and then combining the results. Instead, a more general, adaptable strategy is used that can bounce back and forth between local (i.e. single-agent) and global (multiagent) contexts, adding events where necessary for purposes of synchronization. Planning loci can be composed hierarchically or even overlap.

5. THE FRAME PROBLEM

Although the so-called frame problem has been regarded as presenting a major difficulty for reasoning about actions and plans, there is still considerable disagreement over what it actually is. Some researchers, for example, see the problem as largely a matter of combinatorics (McCarthy & Hayes 1969; Reiter 1980); others view it as a problem of reasoning with incomplete information (McDermott 1982); and yet others believe it relates to the difficulty of enabling systems to notice salient properties of the world (Haugeland 1985). I shall take the problem to be simply that of constructing a formulation in which it is possible to readily specify and reason about the properties of events and situations. This gives rise to at least five related subproblems, which I discuss below.

The first of these is what I shall call the *combinatorial problem*. While it does not appear too difficult to give axioms that describe the *changes* wrought by some given action, it seems unreasonable to have to write down axioms describing all the properties *unaffected* by the action. Axioms of the latter kind are usually called *frame axioms* (McCarthy & Hayes 1969) and, in general, they need to be given (or, at least, be deducible) for all property-action pairs. (Note that I use "unaffected" rather than "unchanging." This is an important distinction if we want to allow for concurrent events, as in most real-world domains.) The real problem here is to avoid *explicitly* writing down (or having to reason with) all the frame axioms for every property-action pair. Most solutions to this problem attempt to formalize some closed-world assumption regarding the specification of these dependencies.

As McCarthy was the first to observe (McCarthy 1980), two further problems arise as a result of the fact that specifying the effects of actions is usually subject to qualification. The first sort of qualification has to do with the conditions under which the action effects certain *changes* in the world. This is what I call the *precondition qualification* problem [also variously called the intraframe problem (Shoham 1986) and, simply, the qualification problem (McCarthy 1980)]. The second sort of qualification concerns the extent of influence of the action (or what remains *unaffected* by the action). I call it the *frame qualification* problem [also called the interframe problem (Shoham 1986) and the persistence problem (McDermott 1982)]. Let me give examples of these two kinds of qualification.

First, consider that we are trying to determine what happens if Mary fires a loaded gun (at point-blank range) at Fred (Hanks & McDermott 1986). Given such a scenario, we should be able to derive, without having to state a host of qualifications, that Fred dies as a result of the shooting. However, if we then discover that (or are given an extra axiom to the effect that) the gun was loaded with a blank round, the conclusion (that Fred dies) should be defeasible—i.e. we should be able to accommodate the notion of Fred's possibly being alive after the firing. This is the precondition qualification problem. Most solutions to this problem aim to formalize the rule: "These are the only *preconditions* that matter as far as the performance of the action is concerned, *unless* it can be shown otherwise."

As an example of the frame qualification problem, consider the point at which Mary loads the gun prior to firing it at Fred. All things being equal, it should be possible to derive that Fred's state of being is unaffected by loading the gun. However, if we discover that Fred actually died while the gun was being loaded, it should be possible (without changing the theory, except for the additional axiom about the time of Fred's death) to accommodate this without inconsistency and, if desired, to derive other theorems about the resulting situation. Most solutions to this problem attempt to formalize the rule thus: "These are the only *effects* of the action (given the preconditions), *unless* it can be shown otherwise." Solutions to this last problem often provide a solution to the combinatorial problem; the two problems, however, are clearly distinct.

The fourth problem concerns the ability to write down certain axioms of invariance regarding world states. Following Finger (1986) I shall call this the *ramification problem*. For example, I should be able to formulate an axiom stating that everyone stops breathing when they die. Then I should be able to state that the effect of shooting a loaded gun at Fred results in his death without having to specify that it also results in cessation of his breathing. This latter effect of the shooting action should be inferable from the first axiom describing the consequences of dying. This seems straightforward, but a problem arises from the fact that such axioms can complicate the solution of the previous problems.

The fifth problem, which I call the *independence problem*, arises primarily in multiagent domains. In a dynamic world, or one populated with many agents, we want out solutions to allow for the independent activities of other agents. Most importantly, we do not want to have to specify explicitly all the external events that might occur. But if we leave the occurrence of such events unspecified, we do not want a solution to the previously mentioned problems to overcommit us to a world in which these events are thereby assumed *not* to have occurred.

There are a great variety of approaches to these problems, including the use of default logics (Reiter 1980); nonmonotonic logics (McDermott 1982); consistency arguments (Dean 1984); minimization of the effects of actions (Lifschitz 1987b), abnormalities (McCarthy 1984), event occurrences (Georgeff 1987b), and ignorance (Shoham 1986); and some ad hoc devices (Hayes 1973). All can be viewed as metatheories regarding the making of appropriate *assumptions* about the given problem domain (Poole et al 1986). This kind of reasoning will necessarily be *nonmonotonic*: In the light of additional evidence, some assumptions may need to be withdrawn, together with any conclusions based on those assumptions.

There are, however, serious difficulties in providing an acceptable semantics for many of these approaches (Hayes 1973; Reiter 1980), and many don't yield the intended results (Hanks & McDermott 1986). Furthermore, it is not clear how to implement most of these schemes efficiently. Brown (1987) has collected the most recent papers regarding this issue.

6. EMBEDDED SYSTEMS

Of course, the ability to plan and reason about actions and plans is not much help unless the agent doing the planning can survive in the world in which it is embedded. This brings us to perhaps the most important and also most neglected area of planning research—the design of systems that are actually *situated* in the world and that must operate effectively given the real-time constraints of their environment.

6.1 Execution Monitoring Systems

Most existing architectures for embedded planning systems consist of a plan constructor and a plan executor. As a rule, the plan constructor plans an entire course of action before commencing execution of the plan (Fikes & Nilsson 1971; Vere 1983; Wilkins 1985b). The plan itself is usually composed of primitive actions—that is, actions that are directly performable by the system. The rationale for this approach, of course, is to ensure that the planned sequence of actions will actually achieve the prescribed goal. As the plan is executed, the system performs the primitive actions in the plan by calling various low-level routines. Usually, execution is monitored to ensure that these routines achieve the desired effects; if they do not, the system may return control to the plan constructor so that it can modify the existing plan appropriately.

Various techniques have been developed for monitoring the execution of plans and replanning upon noticing potential plan failure (Fikes & Nilsson 1971; Wilkins 1985b). The basis for most of these approaches is to retain with the plan an explicit description of the conditions that are required to hold for correct plan execution. Throughout execution, these conditions are periodically checked. If any condition turns out to be unexpectedly false, a replanning module is invoked. This module uses various plan-modification operators to change the plan, or returns to some earlier stage in the plan formation process and attempts to reconstruct the plan given the changed conditions.

However, in real-world domains, much of the information about how best to achieve a given goal is acquired during plan execution. For example, in planning to get from home to the airport, the particular sequence of actions performed depends on information acquired on the way—such as which turnoff to take, which lane to get into, when to slow down and speed up, and so on. Traditional planners can only cope with this uncertainty in two ways: (a) by building highly conditional plans, most of whose branches will never be used, or (b) by leaving low-level tasks to be accomplished by fixed primitive operators that are themselves highly conditional [e.g. the intermediate level actions (ILAs) used by SHAKEY (Nilsson 1984)]. The first approach only works in limited domains—the environment is usually too dynamic to anticipate all possible contingencies. The second approach simply relegates the problem to the primitive operators themselves, and does not provide any mechanism by which the higher-level planner can control their behavior.

To overcome this problem, at least in part, there has been some work on developing planning systems that interleave plan formation and execution (Davis & Chien 1977; Durfee & Lesser 1986). Such systems are better-suited to real worlds than the kind of systems described above, as decisions can be deferred until they *have* to be made. The reason for deferring decisions is that an agent can only acquire *more* information as time passes; thus, the quality of its decisions can only be expected to improve. Of course, there are limitations resulting from the need to coordinate activities in advance and the difficulty of manipulating large amounts of information, but some degree of deferred decision-making is clearly desirable.

6.2 *Reactive Systems*

Real-time constraints pose yet further problems for traditionally structured systems. First, the planning techniques typically used by these systems are very time-consuming. While this may be acceptable in some situations, it is not suited to domains where replanning is frequently necessary and where system viability depends on readiness to act. In real-world domains, unanticipated events are the norm rather than the exception, necessitating frequent replanning. Furthermore, the real-time constraints of the domain often require almost immediate reaction to changed circumstances, allowing insufficient time for this type of planning.

A second drawback of traditional planning systems is that they usually provide no mechanisms for responding to new situations or goals during plan execution, let alone during plan formation. Indeed, the very survival of an autonomous system may depend on its ability to react quickly to new situations and to modify its goals and intentions accordingly. These systems should be able to reason about their current intentions, changing and modifying these in the light of their possibly changing beliefs and goals. While many existing planners have replanning capabilities, none have yet accommodated modifications to the system's underlying set of goal priorities.

Finally, traditional planners are overcommitted to the planning strategy itself. No matter what the situation, or how urgent the need for action, these systems *always* spend as much time as necessary to plan and reason about achieving a given goal before performing any external actions whatsoever. They do not have the ability to decide when to stop planning, nor to reason about the trade-offs between further planning and longer available execution time. Furthermore, these planners are committed to a single planning technique and cannot opt for different methods in different situations. This clearly mitigates against survival in the real world.

Even systems that interleave planning and execution are still strongly committed to achieving the goals that were initially set them. They have no mechanisms for changing focus, adopting different goals, or reacting to sudden and unexpected changes in their environment.

A number of systems developed for the control of robots have a high degree of reactivity (Albus 1981; Albus et al 1981). Even SHAKEY (Nilsson 1984) utilized reactive procedures (ILAs) to realize the primitive actions of the high-level planner (STRIPS), and this idea is pursued further in some recent work by Nilsson (1985). Another approach is advocated by Brooks (1985a), who proposes decomposition of the problem into *task-achieving* units in which distinct behaviors of the robot are realized separately, each making use of the robot's sensors, effectors, and reasoning capabilities as needed. This is in contrast to the traditional approach in which the system is structured according to *functional* capabilities, resulting in separate, self-contained modules for performing such tasks as perception, planning, and task execution. Kaelbling (1987) proposes an interesting hybrid architecture based on similar ideas.

Such architectures could lead to more viable and robust systems than the traditionally structured systems. Yet most of this work has not addressed the issues of general problem-solving and commonsense reasoning; the work is instead almost exclusively devoted to problems of navigation and execution of low-level actions. It remains to extend or integrate these techniques with systems that have the ability to completely change goal priorities, to modify, defer, or abandon current plans, and to reason about what is best to do in light of the current situation.

6.3 *Rational Agents*

Another promising approach to providing the kind of high-level goal-directed reasoning capabilities, together with the reactivity required for survival in the real world, is to consider planning systems as rational agents that are endowed with the psychological attitudes of belief, desire, and intention. The problem that then arises is specifying the properties we expect of these attitudes, the ways they interrelate, and the ways they determine rational behavior in a situated agent.

The role of beliefs and desires in reasoning about action has a long history in the philosophical literature (Davidson 1980). However, only relatively recently has the role of intentions been carefully examined (Bratman 1987). Moreover, there remain some major difficulties in formalizing these ideas. One serious problem is simply to choose an appropriate semantics for these notions. In particular, it is important to take into account the fact that the beliefs, desires, and intentions of an agent are *intensional* objects rather than *extensional* ones. For example, someone who does not know that the President of the United States is Ronald Reagan may well desire to meet one but not the other. Halpern (1986) provides an excellent collection of papers on reasoning about knowledge and belief. The most serious attempt to formalize some basic principles governing the rational balance among an agent's beliefs, intentions, and consequent actions can be found in the work of Cohen & Levesque (1987).

Lansky and I have been largely concerned with means-ends reasoning in dynamic environments, and with the way partial plans affect practical reasoning and govern future behavior (Georgeff & Lansky 1986b). We have developed a highly reactive system, called PRS, to which is attributed attitudes of belief, desire, and intention. Because these attitudes are explicitly represented, they can be manipulated and reasoned about, resulting in complex goal-directed and reflective behaviors. The system consists of a *data base* containing current *beliefs* or facts about the world, a set of current *goals* or *desires* to be realized, a set of *procedures* or *plans* describing how certain sequences of actions and tests may be performed to achieve given goals or to react to particular situations, and an *interpreter* or *reasoning mechanism* for manipulating these components. At any moment, the system also has a *process stack*, containing all currently active plans, which can be viewed as the system's current *intentions* for achieving its goals or reacting to some observed situation.

The set of plans includes not only procedural knowledge about a specific domain, but also *metalevel* plans—that is, information about the manipulation of the beliefs, desires, and intentions of the system itself. For example, a typical metalevel plan would supply a method for choosing among multiple relevant plans, for achieving a conjunction of goals, or for deciding how much more planning or reasoning can be undertaken, given the real-time constraints of the problem domain.

The system operates by first forming a partial overall plan, then figuring out near-term means, executing any actions that are immediately applicable, further expanding the near-term plan, executing further, and so on. At any time, the plans the system intends to execute (i.e. the selected plans) are structurally partial—that is, while certain general goals have been decided upon, specific questions about the means to attain these ends are left open for future reasoning.

Furthermore, not all options that are considered by the system arise as a result of means-end reasoning. Changes in the environment may lead to changes in the system's beliefs, which in turn may result in the consideration of new plans that are not means to any already intended end. For example, the system may decide to drop its current goals and intentions completely, adopting new ones in their stead. This ability is vital in worlds in which emergencies of various degrees of severity can occur during the performance of other, less important tasks.

While the above work attempts to show how means-ends reasoning may be accomplished by systems situated in real-world environments, little research has been done in providing theories of *decision-making* that are appropriate to resource-bounded agents. Researchers in philosophy, as well as decision theory, have long been concerned with the question of how a rational agent weighs alternative courses of action (Jeffrey 1983). This work has largely assumed, either explicitly or implicitly, idealized agents with unbounded computational resources. In reality, however, agents do not have arbitrarily long to decide how to act, for the world is changing around them while they deliberate. If deliberation continues for too long, they very beliefs and desires upon which deliberation is based, as well as the real circumstances of the action, may change. These and related issues are explored by Bratman (1987) and Thomason (1987). Dean (1987) discusses some methods whereby a planning system can recognize the difficulty of the problems it is attempting to solve and, depending on the time it has to consider the matter and what it stands to gain or lose, produce solutions that are reasonable given the circumstances.

Systems that are situated in worlds populated with other agents also have to be able to reason about the behaviors and capabilities of these other systems. This requires complex reasoning about interprocess communication (Appelt 1985; Cohen & Levesque 1985) and the ability to infer the beliefs, goals, and intentions of agents from observations of their behavior (Pollack 1986, 1987). The challenge remains, however, to design situated planning systems capable of even the simplest kinds of rational behavior.

ACKNOWLEDGMENTS

I wish to thank particularly Amy Lansky, Nils Nilsson, Martha Pollack, and Dave Wilkins for their critical readings of earlier drafts of this paper. I am also indebted to Margaret Olender for her patient preparation of the bibliography and to Savel Kliachko for his erudite editorial advice.

The preparation of this paper has been made possible by a gift from the System Development Foundation, by the Office of Naval Research under Contract No. N00014-85-C-0251, and by the National Aeronautics and Space Administration, Ames Research Center, under Contract No. NAS2-12521.

Literature Cited

Albus, J. S., Anthony, A. J., Nagel, R. N. 1981. Theory and practice of hierarchical control. In *Proc. Twenty-Third IEEE Comput. Soc. Int. Conf.*

Albus, J. S. 1981. *Brains, Behavior, and Robotics.* Peterborough, NH: McGraw-Hill

Allen, J. F. 1982. Maintaining knowledge about temporal intervals. *Commun. ACM* 26: 832–43

Allen, J. F. 1984. Towards a general theory of action and time. *Artif. Intell.* 23: 123–54

Allen, J. F., Koomen, J. A. 1983. Planning using a temporal world model. In *Proc. Eighth Int. Joint Conf. Artif. Intell.*, Karlsruhe, West Germany, pp. 741–47

Appelt, D. E. 1985. Planning English referring expressions. *Artif. Intell.* 26: 1–34

Bell, C. E., Tate, A. 1985. Using temporal constraints to restrict search in a planner. In *Proc. Third Workshop Alvey IKBS Programme Planning Spec. Interest Group*, Sunningdale, Oxfordshire, UK

Bratman, M. 1987. *Intention, Plans, and Practical Reason.* Cambridge, Mass: Harvard Univ. Press. Forthcoming

Brooks, R. A. 1983. Planning collision-free motions for pick-and-place operations. *Int. J. Robot. Res.* 2(4): 19–40

Brooks, R. A. 1985a. *A Robust Layered Control System for a Mobile Robot. Tech. Rep. 864, Artif. Intell. Lab.* Cambridge, Mass: MIT

Brooks, R. A. 1985b. Visual map making for a mobile robot. In *Proc. IEEE Conf. Robot. Automat.* St. Louis, Missouri, pp. 824–29

Brown, F. 1987. *The Frame Problem in Artificial Intelligence: Proceedings of the 1987 Workshop.* Los Altos, Calif: Morgan Kaufmann

Chapman, D. 1985. *Planning for conjunctive goals.* Master's thesis MIT-AI-802, MIT

Charniak, E., McDermott, D. 1985. *Introduction to Artificial Intelligence.* Reading, Mass: Addison-Wesley

Cohen, P. R., Levesque, H. J. 1985. Speech acts and the recognition of shared plans. In *Proc. Twenty-Third Conf. Assoc. Comput. Linguist.*, Stanford, California, pp. 49–59

Cohen, P. R., Levesque, H. J. 1987. Persistence, intention, and commitment. In *Reasoning about Actions and Plans: Proceedings of the 1986 Workshop*, pp. 297–340. Los Altos, Calif: Morgan Kaufmann

Corkill, D. D., Lesser, V. R. 1983. The use of meta-level control for coordination in a distributed problem solving network. In *Proc. Eighth Int. Joint Conf. Artif. Intell.*, pp. 748–56

Davidson, D. 1980. *Actions and Events.* Oxford: Clarendon Press

Davis, L. H. 1979. *Theory and Action, Foundations of Philosophy Series.* Engelwood Cliffs, NJ: Prentice-Hall

Davis, P. R., Chien, R. T. 1977. Using and reusing partial plans. In *Proc. Fifth Int. Joint Conf. Artif. Intell.*, Cambridge, Massachussets, p. 494

Dawson, C., Siklossy, L. 1977. The role of preprocessing in problem solving systems. In *Proc. Fifth Int. Joint Conf. Artif. Intell.*, Cambridge, Massachussets, pp. 465–71

Dean, T. 1984. Planning and temporal reasoning under uncertainty. In *Proc. IEEE Workshop Knowledge-Based Syst.*, Denver, Colorado

Dean, T. 1985. *An Approach to Reasoning about Time for Planning and Problem Solving. Tech. Rep. 433.* New Haven, Conn: Comput. Sci. Dept., Yale Univ.

Dean, T. 1987. Intractability and time-dependent planning. In *Reasoning about Actions and Plans: Proceedings of the 1986 Workshop*, pp. 245–66. Los Altos, Calif: Morgan Kaufmann

Dowty, D. R., Wall, R. E., Peters, S. 1981. *Introduction to Montague Semantics. Synthese Language Library.* Boston, Mass: Reidel

Durfee, E. H., Lesser, V. R. 1986. Incremental planning to control a blackboard-based problem solver. In *Proc. Fifth Natl. Conf. Artif. Intell.* Philadelphia, Penn., pp. 58–64

Durfee, E. H., Lesser, V. R., Corkill, D. D. 1985. Increasing coherence in a distributed problem solving network. In *Proc. Eighth Int. Joint Conf. Artif. Intell.*, Los Angeles, California, pp. 1025–30

Fahlman, S. E. 1974. A planning system for robot construction tasks. *Artif. Intell.* 5: 1–49

Fikes, R. E. 1975. Deductive retrieval mechanisms for state description models. In *Proc. Fourth Int. Joint Conf. Artif. Intell.*, Tbilisi, USSR, pp. 99–106

Fikes, R. E., Nilsson, N. J. 1971. STRIPS: a new approach to the application of theorem proving to problem solving. *Artif. Intell.* 2: 189–208

Finger, J. J. 1986. *Exploiting Constraints in Design Synthesis.* PhD thesis, Stanford Univ., Stanford, California

Fox, M. S. 1984. ISIS—a knowledge-based system for factory scheduling. *Expert Syst.* 1(1): 24–49

Fox, M. S. 1986. Observations on the role of constraints in problem solving. In *Proc. Sixth Can. Conf. Artif. Intell.*, Montreal, pp. 172–87

Genesereth, M. R. 1983. An overview of meta-level architecture. In *Proc. Third Natl. Conf. Artif. Intell.*, pp. 119–24

Genesereth, M. R., Nilsson, N. J. 1987. *Logical Foundations of Artificial Intelligence.* Los Altos, Calif: Morgan Kaufmann

Georgeff, M. P., Lansky, A. L., Schoppers, M. 1987. *Reasoning and Planning in Dynamic Domains: An Experiment with a Mobile Robot. Tech. Note 380.* Menlo Park, Calif: Artif. Intell. Cent., SRI Int.

Georgeff, M. P. 1983. Communication and interaction in multiagent planning. In *Proc. Third Natl. Conf. Artif. Intell.*, Washington, DC, pp.125–29

Georgeff, M. P. 1984. A theory of action for multiagent planning. In *Proc. Fourth Natl. Conf. Artif. Intell.*, Austin, Texas, pp. 121–25

Georgeff, M. P. 1987a. Actions, processes,

and causality. In *Reasoning about Actions and Plans: Proceedings of the 1986 Workshop*, pp. 99–122. Los Altos, Calif: Morgan Kaufmann

Georgeff, M. P. 1987b. Many agents are better than one. In *The Frame Problem in Artificial Intelligence: Proceedings of the 1987 Workshop*. Los Altos, Calif: Morgan Kaufmann

Georgeff, M. P., Lansky, A. L. 1986a. Procedural knowledge. In *Proc. IEEE. Spec. Iss. Knowledge Representation*, 74: 1383–98

Georgeff, M. P., Lansky, A. L. 1986b. *A System for Reasoning in Dynamic Domains: Fault Diagnosis on the Space Shuttle. Tech. Note 375*. Menlo Park, Calif: Artif. Intell. Cent., SRI Int.

Gouzenes, L. 1984. Strategies for solving collision-free trajectories problems for mobile and manipulator robots. *Int. J. Robot. Res.* 3(4): 51–65

Green, C. C. 1969. Application of theorem proving to problem solving. In *Proc. First Int. Joint Conf. Artif. Intell.*, Washington, DC, pp. 219–39

Halpern, J. 1986. *Theoretical Aspects of Reasoning about Knowledge: Proceedings of the 1986 Conference*. Los Altos, Calif: Morgan Kaufmann

Hanks, S., McDermott, D. 1986. Default reasoning, nonmonotonic logics, and the frame problem. In *Proc. Fifth Natl. Conf. Artif. Intell.*, Philadelphia, Pennsylvania, pp. 328–33

Harel, D. 1979. *First Order Dynamic Logic. Springer Lect. Notes in Comput. Sci. 68*

Harel, D. 1984. *Dynamic Logic. Handbook of Philosophical Logic, Vol. II*. NY: Reidel

Haugeland, J. 1985. *Artificial Intelligence: The Very Idea*. Cambridge, Mass: MIT Press

Hayes, P. J. 1973. The frame problem and related problems in artificial intelligence. In *Artificial and Human Thinking*, ed. A. Elithorn, D.Jones, pp. 45–59. San Francisco: Jossey-Bass

Hayes, P. J. 1985. The second naive physics manifesto. In *Readings in Knowledge Representation*, pp. 467–85. Los Altos, Calif: Morgan Kaufmann

Hayes-Roth, B. 1985. A blackboard architecture for control. *Artif. Intell.* 26(3): 251–321

Hoare, C. A. R. 1985. *Communicating Sequential Processes. Series in Computer Science*. Englewood Cliffs, NJ: Prentice Hall

Jeffrey, R. 1983. *The Logic of Decision*. Chicago: Univ. Chicago Press

Kaelbling, L. P. 1987. An architecture for intelligent reactive systems. In *Reasoning about Actions and Plans: Proceedings of the 1986 Workshop*, pp. 395–410. Los Altos, Calif: Morgan Kaufmann

Kowalski, R. 1979. *Logic for Problem Solving*. NY: North Holland

Kowalski, R. A., Sergot, M. J. 1986. A logic-based calculus of events. *New Generation Comput.*

Krogh, B. H., Thorpe, C. E. 1986. Integrated path planning and dynamic steering control for autonomous vehicles. In *Proc. 1986 IEEE Int. Conf. Robot. Automat.*, San Francisco, California, pp. 1664–69

Kuan, D. T. 1984. Terrain map knowledge representation for spatial planning. In *Proc. IEEE First Natl. Conf. Artif. Intell. Appl.*, Denver, Colorado, December, pp. 578–84

Ladkin, P. B. 1987. The completeness of a natural system for reasoning with time intervals. In *Proc. Tenth Int. Joint Conf. Artif. Intell.*, Milan, Italy

Ladkin, P. B., Maddux, R. D. 1987. *The algebra of convex time intervals. Kestrel Inst. Tech. Rep. KES.U.87.2*

Lansky, A. L. 1987a. A representation of parallel activity based on events, structure, and causality. In *Reasoning about Actions and Plans: Proceedings of the 1986 Workshop*, pp. 123–59. Los Altos, Calif: Morgan Kaufmann

Lansky, A. L. 1987b. *Localized Representation and Planning Methods for Parallel Domains*. In *Proc. Natl. Conf. Artif. Intell.*, Seattle, Washington

Lassez, J. L., Maher, M. 1983. The denotational semantics of horn clauses as a production system. In *Proc. Natl. Conf. Artif. Intell.*, Washington, DC, pp. 229–31

Levesque, H. J. 1986. Knowledge representation and reasoning. *Ann. Rev. Comput. Sci.* 1: 255–87

Lifschitz, V. 1987a. Formal theories of action. In *The Frame Problem in Artificial Intelligence: Proceedings of the 1987 Workshop*. Los Altos, Calif: Morgan Kaufmann

Lifschitz, V. 1987b. On the semantics of STRIPS. In *Reasoning about Actions and Plans: Proceedings of the 1986 Workshop*, Timberline, Oregon

Lozano-Perez, T. 1980. *Spatial Planning: A Configuration Space Approach. AI Memo 605*. Cambridge, Mass: AI Lab, MIT

Manna, Z., Waldinger, R. J. 1987. A theory of plans. In *Reasoning about Actions and Plans: Proceedings of the 1986 Workshop*, pp. 11–45. Los Altos, Calif: Morgan Kaufmann

Manna, Z., Wolper, P. 1981. *Synthesis of Communicating Processes from Temporal Logic Specifications. Tech. Rep. STAN-CS-81-872*. Stanford, Calif: Comput. Sci. Dept., Stanford Univ.

McCarthy, J. 1968. Programs with common sense. In *Semantic Information Processing*, ed. M. Minsky. Cambridge. Mass: MIT Press

McCarthy, J. 1980. Circumscription—a form of nonmonotonic reasoning. *Artif. Intell.* 13: 27-39

McCarthy, J. 1984. Applications of circumscription to formalizing common sense knowledge. In *Proc. AAAI Non-Monotonic Reasoning Workshop*, pp. 295-324

McCarthy, J., Hayes, P. J. 1969. Some philosophical problems from the standpoint of artificial intelligence. *Mach. Intell.* 4: 463-502

McDermott, D. 1982. A temporal logic for reasoning about processes and plans. *Cognit. Sci.* 6: 101-55

McDermott, D. 1985. Reasoning about plans. In *Formal Theories of the Commonsense World*, ed. J. R. Hobbs, R. C. Moore, pp. 269-317. Norwood. NJ: Ablex

McDermott, D., Davis, E. 1984. Planning routes through uncertain territory. *Artif. Intell.* 22(2): 107-56

Milner, R. 1980. *A Calculus of Communicating Systems. Springer Lect. Notes in Comput. Sci. 92*

Montague, R. 1974. Deterministic theories. In *Formal Philosophy: Selected Papers of Richard Montague*, ed. R. H. Thomason. New Haven: Yale Univ. Press

Moore, R. C. 1980. *Reasoning about Knowledge and Action. Tech. Note 191.* Menlo Park, Calif: Artif. Intell. Cent., SRI Int.

Morgenstern, L. 1986. A first order theory of planning, knowledge, and action. In *Theoretical Aspects of Reasoning about Knowledge: Proceedings of the 1986 Conference*, ed. J. Halpern, pp. 99-114. Los Altos, Calif: Morgan Kaufmann

Myers, J. K. 1985. Multiarm collision avoidance using the potential-field approach. *Soc. Photo-Optic. Instru. Eng.* 580: 78-87

Newell, A., Simon, H. A. 1963. GPS, a program that simulates human thought. In *Computers and Thought*, ed. E. A. Feigenbaum, J. Feldman, pp. 279-93. NY: McGraw-Hill

Nilsson, N. J. 1980. *Principles of Artificial Intelligence*. Palo Alto, Calif: Tioga Publ.

Nilsson, N. J. 1984. *Shakey the Robot. Tech. Note 323.* Menlo Park, Calif: Artif. Intell. Cent., SRI Int.

Nilsson, N. J. 1985. *Triangle Tables: A Proposal for a Robot Programming Language. Tech. Note 347.* Menlo Park, Calif: Artif. Intell. Cent., SRI Int.

Nishimura, H. 1980. Descriptively complete process logic. *Acta Inform.* 14: 359-69

Pednault, E. P. D. 1986. *Toward a mathematical theory of plan synthesis. PhD thesis.* Stanford Univ., Stanford, Calif

Pednault, E. P. D. 1987. Solving multiagent dynamic world problems in the classical planning framework. In *Reasoning about Actions and Plans: Proceedings of the 1986 Workshop*, pp. 42-82. Los Altos. Calif: Morgan Kaufmann

Pelavin, R., Allen, J. F. 1986. A formal logic of plans in a temporally rich domain. In *Proc. IEEE, Spec. Iss. Knowledge Representation*, 74: 1364-82

Pollack, M. E. 1986. *Inferring domain plans in question answering. PhD thesis.* Univ. Pennsylvania

Pollack, M. E. 1987. A model of plan inference that distinguishes between the beliefs of actors and observers. In *Reasoning about Actions and Plans: Proceedings of the 1986 Workshop*, pp. 279-95. Los Altos, Calif: Morgan Kaufmann

Poole, D. L., Goebel, R. G., Aleliunas, R. 1986. *Theorist: a Logical Reasoning System for Defaults and Diagnosis.* NY: Springer-Verlag

Prior, A. N. 1967. *Past, Present and Future.* Oxford: Clarendon Press

Reiter, R. 1980. A logic for default reasoning. *Artif. Intell.* 13: 81-132

Rosenschein, J. S. 1982. Synchronization of multiagent plans. In *Proc. Conf. Artif. Intell.*, Stanford, California, pp. 115-19

Rosenschein, J. S., Genesereth, M. R. 1984. *Communication and Cooperation. Tech. Rep. 84-5, Heur. Program. Proj.* Stanford, Calif: Comput. Sci. Dept., Stanford Univ.

Rosenschein, J. S. 1986. *Rational interaction: cooperation among intelligent agents. PhD thesis.* Stanford Univ.

Rosenschein, S. J. 1981. Plan synthesis: a logical perspective. In *Proc. Seventh Int. Joint Conf. Artif. Intell.*, pp. 331-37. Vancouver, British Columbia

Sacerdoti, E. D. 1973. Planning in a hierarchy of abstraction spaces. In *Proc. Third Int. Joint Conf. Artif. Intell.*, Stanford, California, pp. 412-30

Sacerdoti, E. D. 1977. *A Structure for Plans and Behavior.* NY: Elsevier, North Holland

Sadri, F. 1986. Representing and reasoning about time and events: three recent approaches. Department of Computing. Imperial College, London

Shoham, Y. 1986. Chronological ignorance: time, nonmonotonicity, necessity and causal theories. In *Proc. Fifth Natl. Conf. Artif. Intell.*, Philadelphia, Pennsylvania, pp. 389-93

Shoham, Y. 1987. What is the frame problem. In *Reasoning about Actions and Plans: Proceedings of the 1986 Workshop*, pp. 83-98. Los Altos, Calif: Morgan Kaufmann

Stefik, M. 1981a. Planning with constraints (MOLGEN: Part 1). *Artif. Intell.* 16(2): 111–40

Stefik, M. 1981b. Planning with constraints (MOLGEN: Part 2). *Artif. Intell.* 16(2): 141–70

Strong, H. R. 1971. Translating recursive equations into flowcharts. *J. Comput. Syst. Sci.* 5(3): 254–85

Stuart, C. J. 1985. *Synchronization of Multiagent Plans Using A Temporal Logic Theorem Prover. Tech. Note 350.* Menlo Park, Calif: Artif. Intell. Cent., SRI Int.

Sussman, G. J. 1973. *A Computational Model of Skill Aquisition. Tech. Rep. AI TR-297.* Cambridge, Mass: Artif. Intell. Lab., MIT

Tate, A. 1974. *INTERPLAN: A plan generation system which can deal with interactions between goals. Memo MIP-R-109.* Edinburgh: Mach. Intell. Res. Unit, Univ. Edinburgh

Tate, A. 1977. Generating project networks. In *Proc. Fifth Int. Joint Conf. Artif. Intell.*, Cambridge, Massachusetts, pp. 888–93

Tate, A. 1984. Planning in expert systems. In *Alvey IKBS Expert Systems Theme — First Workshop*, Oxford, March. Also *D.A.I. Res. Pap. 221*, Univ. Edinburgh

Thomason, R. H. 1987. The context-sensitivity of belief and desire. In *Reasoning about Actions and Plans: Proceedings of the 1986 Workshop*, pp. 341–60. Los Altos, Calif: Morgan Kaufmann

van Betham, J. 1984. Tense logic and time. *Notre Dame J. Formal Logic* 25(1): 1–16

Vere, S. 1983. Planning in time: windows and durations for activities and goals. *IEEE Trans. Pattern Analysis Mach. Intell.* 5(3): 246–67

Waldinger, R. 1977. Achieving several goals simultaneously. *Mach. Intell.* 8: 94–136

Warren, D. H. D. 1974. *WARPLAN: A system for generating plans. Tech. Rep.* Edinburgh: Univ. Edinburgh

Wesley, L. P., Lowrance, J. D., Garvey, T. D. 1979. *Reasoning about Control: An Evidential Approach. Tech. Note 324.* Menlo Park, Calif: Artif. Intell. Cent., SRI Int.

Wilensky, R. 1981. Meta-planning: representing and using knowledge about planning in problem solving and natural language understanding. *Cognit. Sci.* 5: 197–233

Wilkins, D. E. 1984. Domain independent planning: representation and plan generation. *Artif. Intell.* 22: 269–301

Wilkins, D. E. 1985a. *Hierarchical Planning: Definition and Implementation. Tech. Note 370.* Menlo Park, Calif: Artif. Intell. Cent., SRI Int.

Wilkins, D. E. 1985b. Recovering from execution errors in SIPE. *Comput. Intell.* 1: 33–45

Wilkins, D. E. 1987. *Using causal rules in planning. Tech. Note 410.* Menlo Park, Calif: Artif. Intell. Cent., SRI Int.

Hierarchical planning involving deadlines, travel time, and resources

THOMAS DEAN[1]

Department of Computer Science. Brown University. Box 1910. Providence. RI 02912. U.S.A.

R. JAMES FIRBY

Department of Computer Science. Yale University. Box 2158. Yale Station. New Haven. CT 06520. U.S.A.

AND

DAVID MILLER

The Artificial Intelligence Group. Jet Propulsion Laboratory. California Institute of Technology. 4800 Oak Grove Drive. Pasadena. CA 91109. U.S.A.

Received July 30. 1987

Revision accepted September 19, 1988

This paper describes a planning architecture that supports a form of hierarchical planning well suited to applications involving deadlines. travel time. and resource considerations. The architecture is based upon a temporal database, a heuristic evaluator, and a decision procedure for refining partial plans. A partial plan consists of a set of tasks and constraints on their order, duration, and potential resource requirements. The temporal database records the partial plan that the planner is currently working on and computes certain consequences of that information to be used in proposing methods to further refine the plan. The heuristic evaluator examines the space of linearized extensions of a given partial plan in order to reject plans that fail to satisfy basic requirements (e.g., hard deadlines and resource limitations) and to estimate the utility of plans that meet these requirements. The information provided by the temporal database and the heuristic evaluator is combined using a decision procedure that determines how best to refine the current partial plan. Neither the temporal database nor the heuristic evaluator is complete and. without reasonably accurate information concerning the possible resource requirements of the tasks in a partial plan. there is a significant risk of missing solutions. A specification language that serves to encode expectations concerning the duration and resource requirements of tasks greatly reduces this risk, enabling useful evaluations of partial plans. Details of the specification language and examples illustrating how such expectations are exploited in decision making are provided.

Cet article décrit une architecture de planification qui soutient une forme de planification hiérarchique adaptée aux applications mettant en cause des limites. des temps de déplacement et des considérations au niveau des ressources. Cette architecture est fonction d'une base de données temporelle. d'un évaluateur heuristique et d'un processus décisionnel permettant de perfectionner des plans partiels. Un plan partiel consiste en une série de tâches ainsi que leurs contraintes au niveau de l'ordre. de la durée et des exigences possibles en matière de ressources. La base de données temporelle enregistre le plan partiel sur lequel le planificateur travaille et évalue certaines conséquences de ces données qui serviront à proposer des méthodes de perfectionnement du plan. L'évaluation heuristique examine l'espace des extensions linéarisées d'un plan partiel donné afin de rejeter les plans qui ne satisfont pas aux exigences de base (par ex. : dates limites serrées et limitations des ressources) et d'évaluer l'utilité des plans qui respectent ces exigences. Les renseignements fournis par la base de données temporelle et l'évaluation heuristique sont combinés à l'aide d'un processus décisionnel qui détermine la meilleure façon de perfectionner le plan partiel actuel. Ni la base de données temporelle. ni l'évaluateur heuristique ne sont complets. Sans des renseignements relativement précis concernant les exigences possibles au niveau des ressources des tâches d'un plan partiel. il y a un important risque que des solutions nous échappent. Un langage de spécification qui sert à encoder les attentes concernant les exigences au niveau de la durée et des ressources des tâches réduit grandement ce risque et permet des évaluations utiles de plans partiels. Des détails au sujet du langage de spécification et des exemples illustrant comment de telles attentes sont exploitées dans les processus décisionnel sont fournis.

[Traduit par la revue]

Comput. Intell. 4. 381-398 (1988)

1. Introduction

Planning is generally viewed as a process of incremental construction. At any point in the process, a planner will have before it a partially constructed plan that it is working to finish; the actions in the plan or the order in which they are to be executed may not be completely specified. For instance. while constructing a plan to manufacture a particular part, a planner may commit to using a specific manufacturing process without committing to a specific machine with which to carry out that process. Planning proceeds as a series of decisions concerning how to refine (or provide more detail to) the partially constructed plan. These decisions determine a search through a space in which each state is a partially constructed plan and each search operator serves to refine a partially constructed plan.

To direct the search, most planners employ some means of evaluating a partially constructed plan. Evaluation typically involves predicting the effects of actions already in the plan to determine whether or not they are likely to bring about some desired state of affairs. With appropriate knowledge, a planner can reason about actions and their effects at many different levels of detail and evaluate a partially constructed plan by predicting whether it brings about the desired state of affairs. given the current level of detail.

To support plan evaluation, the planner needs domain knowl-

---

[1] Much of the work described in this paper was carried out while all three of the authors were at Yale University. The order of names is alphabetical and has no other significance.

edge of two types: a database of rules that establish the cause-and-effect relations between actions and their results, and a database of rules that define the conditions for using certain methods for refining partially constructed plans. In addition to this static knowledge, a planner must maintain a database holding its partially constructed plan and the observed and predicted events corresponding to the context in which that plan is to be executed. The planner uses this temporal database and its rules for refining partially constructed plans to decide how best to add to the plan it is building.

In this paper, we build upon an approach to planning referred to as *hierarchical planning* (Sacerdoti 1977; Tate 1977; Vere 1983; Wilkins 1984). The main intuition that we borrow from hierarchical planning is that the search can often be directed so that decisions made early in planning are independent of decisions made later. To provide the necessary direction, the planning system described in these pages makes use of a powerful representation language that allows one to encode domain-specific information about the possible consequences of partially constructed plans. In addition, the system employs two programs that can make use of this information to predict further consequences. The first program, the temporal database manager, keeps track of consequences that are true of every completion of the currently partially constructed plan. The second program, the heuristic task scheduler, ensures that at all times there exists some completion of the current partially constructed plan that satisfies the planner's top level goals. The way these two programs are used in deciding what refinements to make to a partially constructed plan is central to this paper. Neither program is guaranteed to be complete; hence, there are situations in which the system will claim that there is no solution when there actually is one. However, if the system proposes a solution, then, barring inaccuracies in the system's knowledge, the solution is guaranteed correct.

The rest of this paper describes the FORBIN planner: a system for generating plans in domains involving mobile robots, deadlines, and limited resources. Throughout the paper we will use examples drawn from an automated factory domain. Most of the examples do not require detailed knowledge of the domain; Appendix A provides the details for those interested.

2. Planning as search in the FORBIN system

Each state in the FORBIN search space is a partially constructed plan. A partially constructed plan consists of a set of actions and a set of constraints on the expected duration and order of execution of those actions. An action is said to be *primitive* if breaking it down into simpler actions will reveal no additional useful information. Examples of primitive actions in the FORBIN domain are "Pick up the widget" and "Push the lathe start button."

A *task* is simply an action in the planner's current partially constructed plan. A task corresponding to a primitive action is said to be primitive; otherwise it is said to be *problematic*. Most planners deal with tasks that involve making certain facts true of the world. For example, if the desired fact is that block A is on block B (i.e., (ON A B)), then the associated task is (ACHIEVE (ON A B)). Carrying out such a task corresponds to executing primitive actions that bring about a state of the world in which the desired fact is true. Using this notion of task, planning is defined as the search for a sequence of primitive actions that will carry out all of the tasks that the planner has been assigned.

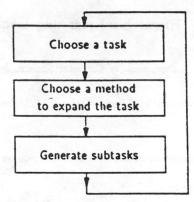

FIG. 1. Basic algorithm for task expansion.

In some approaches, each primitive action is associated with a unique search operator;[2] a partially constructed plan is a sequence of primitive actions, and applying an operator consists of appending the action associated with the operator to the end of the current partially constructed plan. In FORBIN, as in most hierarchical planners, the notion of a search operator is somewhat more complex. Transforming the current partially constructed plan may involve splicing in new tasks and imposing additional constraints on existing tasks. Planning proceeds by selecting a problematic task and attempting to replace or supplement its current specification with a set of other, hopefully less problematic, tasks. This process of transformation is referred to as *expansion*. For each nonprimitive action, the planner has a set of *methods* for carrying out these expansions.

Each method specifies a set of actions and constraints on their order and duration. In addition, each method is annotated with information that describes how good it is in a given set of circumstances. Applying a search operator in the FORBIN system consists of combining the actions and constraints in a particular method with those of the current partially constructed plan. Simplifying somewhat, the process of planning consists of choosing a problematic task that has yet to be expanded, choosing a method to expand that task, and transforming the current partially constructed plan using the chosen method. If there are no problematic tasks requiring expansion, then the plan is complete and the process stops. The basic cycle of activity is depicted in Fig. 1.

In FORBIN, the initial set of tasks describes a partially constructed plan, albeit a plan at a fairly high level of abstraction. Each task is associated with an interval of time during which it is to be carried out, and generally there are constraints on the order and completion time of the tasks. This partially constructed plan is represented in a *task network* (McDermott 1977). Figure 2a illustrates a simple task network corresponding to two initially supplied tasks. Expansion results in adding subtask and precedence links to the network. Figure 2b shows the result of expanding each of the two tasks in Fig. 2a. The task network encodes the basic decisions of the planner. As we will see, there is a great deal more that has to be represented to support hierarchical planning. In particular, the effects of the planner's proposed actions as well as the effects of events outside the planner's control all have to be taken into account in evaluating a partially constructed plan.

---

[2] In these approaches, operators and actions are in one-to-one correspondence, so the two terms are often used interchangeably.

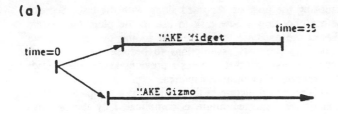

(a)

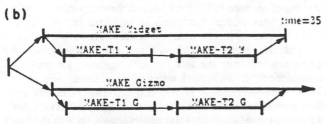

(b)

FIG. 2. Simple task networks.

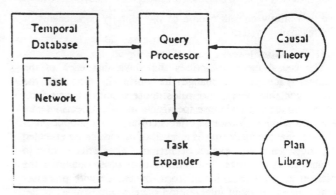

FIG. 3. Basic architecture for hierarchical planning.

Many hierarchical planners can be viewed as having the basic architecture of Fig. 3. Arrows indicate the flow of information between modules. circles represent static knowledge sources, and the box labeled "temporal database" indicates the dynamic component of the planner's knowledge. The task expander implements the algorithm sketched in Fig. 1. and the query processor assists in making choices by answering questions concerning the contents of the temporal database. Planners can be distinguished by the expressiveness of the languages used for representing plans and causal rules and by the techniques used for reasoning about such plans. The temporal database contains a set of cached deductions concerning the state of the partially constructed plan and (some of) the effects of the proposed actions. In the NASL planner (McDermott 1977). the temporal database and the task network are identical. In other planners. the temporal database may consist of a task network annotated with information concerning the truth of propositions corresponding to the effects of each task i.e., the *table of multiple effects* of NOAH (Sacerdoti 1977).

In general. the effects of one task are contingent upon the effects of preceding tasks. An *interaction* corresponds to a situation in which the effect of one task serves to undermine the intended effect of another task. The main difficulty with hierarchical planning is sorting out unexpected interactions involving the expansion method chosen for different tasks. The actions and the constraints on their order and duration within a single method will have been chosen so that they do not interfere with each other. However, when a plan is built for several tasks. subtasks from the different expansions may become interspersed and cause interactions that compromise the success of the plan under construction.

To notice and deal with such interactions, the planner must keep track of the expected state of the world resulting from each task in the plan it has built so far. By carefully choosing expansion methods and orderings that preserve the conditions required by actions already specified, interactions can often be avoided. The hard part is anticipating such interactions. One way to accomplish this is annotate methods with expectations concerning what conditions they will require. These expectations are then used to predict certain consequences of the current partially constructed plan and guide decision making so that any interactions that are encountered can easily be

resolved. To make good decisions concerning which expansion method to use. the FORBIN planner must continually predict the expected future at each point in its plan. This process of prediction is referred to as *projection* and plays an integral role in decision making.

3. An overview of the planner and its algorithm

Broadly speaking, the FORBIN planning algorithm consists of two operations — expansion and projection — that are repeated until a completely refined plan is reached. The contribution of the FORBIN system is the way each operation is accomplished without becoming computationally intractable. As mentioned. the knowledge that the planner brings to each problem can be thought of as residing in two separate databases: one containing cause-and-effect relationships for use in projecting the expected results of the plan being constructed and one containing expansion methods for refining that plan to more detail. The plan under constructing is represented in a temporal database that keeps track of the tasks already in the plan. any events the planner does not control but knows will occur. and the expected effects of both the planned tasks and uncontrolled events. Each time that the planner refines one or the tasks in the plan. temporal projection is used to update the temporal database to include any new effects that can be derived from the new detail added to the plan.

To refine a partially constructed plan. the planner chooses a task to work on and consults the database of expansions to find available refinement methods. The temporal database is then consulted to determine which of those methods is appropriate under the circumstances expected at execution time. This consultation generally produces a small number of candidate methods suitable for the expansion. To choose between these candidates. a detailed projection is performed to determine which method is best. This projection step involves a search through the possible total orderings of the steps in the unfinished plan. and the only backtracking done by the FORBIN system is that required for this search. Once the best refinement method is found. it is integrated into the current plan and the temporal database is updated accordingly. Should FORBIN ever find itself with no appropriate refinement method for a particular task. it simply stops and signals failure. The FORBIN system does employ very sophisticated routines for tracking the reasons behind its expansion decisions. but currently there is no theory for recovering when those decisions turn out to be incorrect. As we will see. however. FORBIN goes to great lengths to make the right decisions the first time around.

Before we describe the FORBIN algorithm in more detail,

we will briefly consider some of the primary components of the reasoning system.

3.1. The causal theory

The causal theory describes the ways that each of the planner's actions will change the world. In particular, the theory contains two types of rules: those specifying the effects of executing tasks and those specifying the way effects interact with one another. Most task effects correspond to atomic propositions that are made true as a result of actions being executed or, more generally, events occurring. Other effects refer to quantities that change continuously over time: quantities the planner can add to, subtract from, or change with processes that continue without intervention once begun. For example, the action of starting an automatic milling machine has the simple effect of changing its state from off to on and the more complex effects of using up raw materials to make the part, causing wear on the cutting bit, and at some point in the future causing a finished part to exist where there was none before.

The causal theory must also concisely capture the interactions that occur between effects. Many of the results of a task come about indirectly and it is inappropriate to encode them in the task's effects. Turning on a light might cause a solar-powered sculpture to begin turning, but there should be no mention of solar-powered sculptures in the rules describing the effects of toggling the light switch. Rather, the connection between the switch and the sculpture should be encoded in a separate set of rules that refer to much more general physical principles.

3.2. The temporal database

The heart of the FORBIN system is its temporal datbase. The database maintains a picture of the expected future, given the plan so far, the causal theory, and any known external events. This representation of the future is used for two different purposes: (1) it gives the expected situation that each task in the plan will encounter (to use when choosing an expansion for the task) and (2) it prevents the plan under construction from becoming invalid by ensuring that task expansion preconditions do not get changed.

When planning begins, the temporal database contains the initial world state and the time and nature of any events under external control. As the plan is constructed, each new subtask the system commits to is added to the database along with its expected duration and starting time. The database then applies rules from the causal theory to determine the effects of that task and to propagate those effects into the future to compute the expected world state at all times.[3]

The choice of any particular action to add to the plan will usually depend on the situation expected when that action is to be executed. For example, the planner might decide that the best way to get a new widget by 2:00 p.m. is to make it on the lathe and add such a subtask to the database. During later planning, however, the planner might decide that to fill an important order, it would like to use the lathe to make gizmos all day. Simply adding this use of the lathe to the database would lead to an inconsistent future and to the planner mistakenly thinking it could make the widget and gizmos at the same time. To avoid such situations two precautions are taken. First, when a task is added to the database, facts assumed true when

---

[3]In general, projection need not terminate. FORBIN gets around this by imposing a lower limit on the duration of events and an upper limit on the interval over which projected events will be considered.

choosing the task are recorded along with the task. Second, before choosing a new task to use in the plan, the system checks the database to ensure that adding the task will not cause any previous assumptions to become invalid. Never adding a new task that violates a previous assumption avoids the creation of inconsistent futures.

This nice neat picture of the database as a complete description of the expected future is complicated by the fact that planned tasks may be only partially ordered. Projection of the sort required by FORBIN involving partially ordered events is *NP-hard* (Chapman 1987) even without the temporally varying quantities that FORBIN allows. To maintain computational tractability, an important heuristic is used: the temporal database only represents (or caches) those atomic assertions that can be derived from the task network, ignoring the possible consequences of additional ordering constraints (see Dean 1985). Order-dependent effects and quantity variations are taken into account through a separate detailed projection that explores a subset of all total task orders consistent with the partial order. This separate projection ensures the temporal database has at least one consistent ordering without having to constrain the task network to keep that ordering for the rest of the planning process.

3.3. The task expander

Refining a partially constructed plan proceeds by expanding a task into a set of simpler subtasks until only primitive actions requiring no further refinement remain. The routines responsible for performing the expansions are called the *task expander*. The database of refinement methods used by the task expander consists of "recipes" for expanding various task types into subtasks. Refinement in FORBIN proceeds hierarchically through several levels of methods culminating in expansions involving only primitive actions. Hierarchical refinement is used so that the temporal database always contains all planning goals, and hence the whole expected future, at some level of detail. Representing the entire future in the database provides a solid basis for each level of expansion and allows FORBIN to take account of important interactions early in planning.

3.4. The basic FORBIN algorithm

The basic planning algorithm is easy to understand in terms of the task expander, the temporal database, and the causal theory. The task expander selects a nonprimitive task from the database and looks up all appropriate refinement methods in the expansion library. It then asks the database which methods can be used in the context of the current task network (i.e., some methods will be based on assumptions that are not expected to be true, and some methods will have effects that invalidate previous assumptions). If no methods will work, then the task cannot be carried out in the current plan and FORBIN fails. If one or more methods will work, then the task expander scores them and chooses the "best." The best expansion is then added to the temporal database, and the database is updated in accordance with the causal theory. This cycle of choosing a task to work on, picking the best expansion based on the expected future, and then updating the temporal database continues until either some task cannot be expanded and failure occurs or all tasks have been reduced to primitive actions and the plan is complete. This basic cycle of activity is sketched in Fig. 4.

To support the selection of potential expansion methods, the temporal database handles a class of queries known as *abduc-*

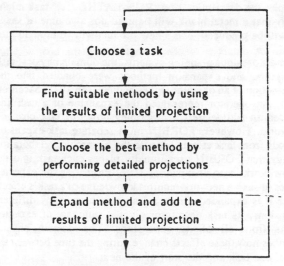

FIG. 4. FORBIN algorithm for task expansion.

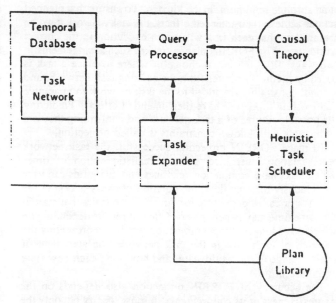

FIG. 5. FORBIN architecture.

tive queries. An abductive query is made to see whether the assumptions for a method are satisfied in the current temporal database. The answer returned from such a query includes a set of additional ordering constraints that may be added to the database to satisfy the assumptions. If a particular query can be satisfied in some totally ordered extension of the current partial order, then the temporal database is guaranteed to respond positively. However, because the temporal database only caches certain projection results and not all, it is possible that the database will respond positively even in cases where the query is not satisfied in any totally ordered extension.

To verify abductive queries, FORBIN employs an additional set of procedures for projecting the detailed consequences of the partially ordered task network. If a query is false in all totally ordered extensions of the current partial order, then these routines would detect this. The query routines of the temporal database act as an initial filter: the detailed projection routines are the final arbiter. The detailed projection routines are also used to score expansion methods by installing each expansion into the plan in a temporary database and then simulating the result considering various consistent total orders. Scoring is handled using domain-specific utility functions encoded with the various expansion methods. By actually looking at total orders, the simulation procedures are able to make very accurate predictions concerning the potential value of a given expansion. To avoid the possibility of doing an intractable amount of work in performing its evaluations, the simulation routines employ a strong heuristic component (for details see Miller 1985) which limits the number of total orders examined to only the few most promising ones. This method of evaluation is incomplete, since possible good total orderings may be missed and the corresponding methods rejected, but it is always correct. Within the limits of the causal theory, if a method is determined to work by the evaluation routines, it is correct.

Our initial hope was that the search done by the evaluation routines could be limited by passing them only the "relevant" parts of the current plan. Unfortunately, we never developed a theory of "relevance" and FORBIN was forced to apply the heuristic task scheduler, as the simulation routines were collectively called, to the entire set of tasks. It is our opinion that the knowledge necessary for restricting the application of the heu-

ristic task scheduler would be largely domain specific, as our attempts to derive a purely domain-independent theory met with little success.

3.5. The basic FORBIN architecture

Figure 5 shows the architecture of the FORBIN planner simplified to emphasize its interesting aspects. The temporal database manager is responsible for selectively caching deductions (limited projection) concerning the state of the partially constructed plan. The task expander implements the algorithm sketched in Fig. 4. The heuristic task scheduler handles the detailed predictions required for selecting among methods. In addition, FORBIN employs a *task queue manager* (not shown in Fig. 5) that keeps track of the tasks in the database that still need further refinement and decides on the best order to make those refinements.

An important consideration when designing the FORBIN planning system was that planning and execution might occur together. In particular, refinement should concentrate on those tasks to be executed earliest so that their constituent primitive actions can be "peeled off the front" of the plan and executed even while tasks later in the plan are unexpanded. Also, new tasks should be allowed to be introduced by the user at any time and should be incorporated into the growing and executing plan with as little disruption as possible. The task queue manager handles both of these duties in a uniform manner by giving priority for refinement to those tasks whose deadlines are approaching and by inserting user-added tasks into the queue just like subtasks spawned through expansion. The FORBIN goal of concurrent planning and execution has been only partially fulfilled so far, but the need for a system with the right properties has been a driving force behind its structure.

4. The FORBIN plan language: causal theory

The FORBIN system gives the calculation of expected states of the world a central role in plan construction. Each new task added to the plan is designed to have a particular effect on the world but, typically, it will have that effect only if the world is in an appropriate state to begin with. For example, putting down a cup will not leave it on the kitchen table unless we are,

121

at an absolute minimum. in the kitchen. To ensure that planned actions achieve their intended effects. the task network must be arranged so that each task will begin execution in the correct world state. The FORBIN planner makes such arrangements in two ways: first. the initial choice of where to place a task is based on predicted futures from the existing task network. and second. constraints are added to the task network to guarantee that existing tasks will have their intended effects. Predicting the possible futures of a task network and confirming they are compatible with known constraints is called projection.

To make FORBIN projection successful. the task network representation must incorporate a realistic notion of time. When tasks have completion deadlines and quantities can vary continuously through time. the duration of every action in the plan is critical in predicting the future. Two tasks that start at the same time may generate very different results depending on which takes longer. Proper projection requires representing the duration of each task in the task network. the start time of every change to the world state. and how long each new state continues in time.

The accuracy of FORBIN projection also depends on the expressiveness of its *causal theory*. It must specify not only the effect each task has on the world. but also the way those effects evolve through time after the task is complete. In most previous planning systems. causal theories are composed of rules specifying a list of assertions to add or delete in order to model each action being executed. The world state at any one time is seen as a set of assertions like (ON *A B*) or (COLOR *X* RED). By applying the add and delete lists from each action in the plan in turn. the expected world state at all times can be determined. Projection rules employing simple add and delete lists are quite powerful even in complex worlds. but they are inadequate for representing actions that make relative changes to quantities or actions that begin processes which change continuously on their own. For example. the action (DEPOSIT $10) might have the effect of increasing one's bank account by $10. The resulting balance. however. must be calculated from the previous balance and may depend on a complex function of previous changes that includes interest and instant-teller charges. Similarly. the action (TURN-ON FAUCET) might start filling the sink with water. but no set of add and delete lists alone can predict what time it will be filled. FORBIN must represent both *static effects*. previously handled with add and delete lists. and *dynamic effects* which have results that change with time and context.

4.1. Specifying the causal theory

The FORBIN causal theory comes in two parts: the *domain physics* and the *task descriptors*. The domain physics contains the system's knowledge about every aspect of the world relevant to projection. For example. the FORBIN factory domain physics declares that there are such things as lathes, that they can be used to make only one thing at a time, that if one is turned off it will stay off unless acted upon explicitly, and so on. The domain physics mentions all fact types to be used. how these facts evolve through time once they become true. and how facts interact with one another if they become true at the same time.

A task descriptor encodes the way a task is expected to change the world when executed. Task descriptors are analogous to the add and delete lists of past planners with extensions to include statements about dynamic effects as well. For example. the description of a (RUN-LATHE . . .)[4] task might specify that a metal blank will be used up. a volume of shavings will be produced. and a new product will be brought into being.

In most previous planning systems. the domain physics. task descriptors. and expansion methods were bundled into the single notion of an operator (Fikes and Nilsson 1971; Sacerdoti 1977). An operator represented the expansion of a task and included an add/delete list describing how that task changed the world. However. FORBIN must separate task-expansion methods from task effects because the hierarchical elaboration strategy that FORBIN uses initially places each task into the task network in an unexpanded state. Projecting the expected futures of such a network requires knowledge of a task's effects before it is expanded. Since a task may have many different expansions. its task descriptor gives a summary of expected effects from all of its expansions. The domain physics describes how these effects change during the time *between* the end of one task and the start of the next.

4.2. Domain physics

There are two subsections to the domain physics: one to deal with the static-fact types and the other to deal with dynamic-fact types. Static facts are used to represent facets of the world that become true and remain unchanged until specifically acted upon. while dynamic facts are used to represent quantities that are influenced by actions and may change over time without any intervention. For example. the task descriptor for (MOVE-TO WIDGET IN-BOX) might specify that the static fact (LOCATION WIDGET IN-BOX) becomes true and the dynamic fact representing the weight of the box is increased: (BOX-WEIGHT +(WEIGHT-OF WIDGET)).

The distinction between static and dynamic facts is not logically required. since static effects are a subset of dynamic ones. However. the FORBIN system separates the two for pragmatic efficiency reasons that allow different processing for each type during projection. Assertions that result from projecting static facts alone are cached to make up the temporal database used in choosing candidate task expansions. In contrast. the detailed results from projecting dynamic facts are not cached in the task network. After they are used to choose between candidate expansions and ensure the choice is compatible with previous commitments. they are dropped so as not to overconstrain future choices.

4.2.1. Static facts

Static facts represent aspects of the world state that are made true and remain true until explicitly changed. They are modeled in the FORBIN system as predicate calculus assertions and defined with the form:

(DEFINE-FACT (⟨name⟩ ⟨var1⟩ ⟨var2⟩ . . .))

For example. many of the properties of a lathe are static in nature and might be defined as:[5]

---

[4] In the FORBIN plan language. all tasks. facts. and states are referred to in terms of *patterns*. A pattern is a list with the first element representing the task, fact, or quantity type. Remaining elements in the pattern are arguments that modify the type. Arguments will often appear as variables in the language but during planning all variables will be instantiated.

[5] Variables are specified by prefacing their name with a "?". This notation is drawn from the DUCK programming language (McDermott 1985) used to implement much of the FORBIN system.

(DEFINE-FACT (SETUP-LATHE ?lathe ?type))
(DEFINE-FACT (BIT ?lathe ?type))
(DEFINE-FACT (STATUS ?machine ?status))

All static features of the world referred to by task descriptors must be defined in this way. Static facts are exactly facts previously represented in add and delete lists.

Once a static fact has been defined, the projection machinery knows that whenever such a fact becomes true, it will remain true until specifically altered. The traditional way of altering a static fact is to delete it from the current state and add a new fact reflecting the change. However, the explicit deletion of a fact is subsumed in the FORBIN system by the clipping rule. Clipping rules declare that the beginning of one static fact causes the end of another. For example, asserting (STATUS LATHE-1 RUNNING) should cause the fact (STATUS LATHE-1 IDLE) to stop being true. There is no need to explicitly delete (STATUS LATHE-1 IDLE) as long as that rule is known. Clipping rules are specified with the form:

(DEFINE-CLIP ⟨fact 1⟩ ⟨fact 2⟩ ⟨condition fact⟩)

which states that whenever ⟨fact 1⟩ becomes true while the ⟨condition fact⟩ is true, it causes ⟨fact 2⟩ to end (and vice versa). For the example above one might use

(DEFINE-CLIP (STATUS ?machine ?status 1)
             (STATUS ?machine ?status2)
             (NOT (= ?status1 ?status2)))

Clipping rules allow task descriptors to assert only static facts they make true because facts they clip are taken care of automatically. Clipping rules are more general than delete lists because they work with facts generated by temporal forward chaining rules as well as those from task descriptors.

The FORBIN causal theory can represent more complex interactions between static facts than just clipping between static facts through the use of temporal forward chaining rules. The two most important rule types are DEFINE-OVERLAP and DEFINE-CAUSE which declare that whenever certain static facts overlap in time, new facts should be asserted.[6] With DEFINE-OVERLAP the new fact extends over the interval of time the initiating facts overlap, whereas a new fact spawned by DEFINE-CAUSE persists indefinitely (unless clipped). The basic temporal forward chaining rule syntax is

(DEFINE-[OVERLAP | CAUSE]
             (AND ⟨fact 1⟩ ⟨fact 2⟩ ...) ⟨result fact⟩)

For example, to encode that the lathe and milling machine running together are very noisy, one might write

(DEFINE-OVERLAP (AND
     (STATUS LATHE RUNNING)
     (STATUS MILLING-MACHINE RUNNING))
          (NOISE-LEVEL HIGH))

while stating they will crack the window requires

(DEFINE-CAUSE (AND (STATUS LATHE RUNNING)
             (STATUS MILLING-
                  MACHINE RUNNING))
          (HEALTH WINDOW CRACKED))

All three types of temporal inference (i.e., clipping, overlap, and cause) might best be thought of as if-then rules with different temporal scoping on the initiating and resulting facts.

A final static-fact type is the *pool*. A pool is a collection of objects that are essentially interchangeable and are referred to primarily by type rather than name. By referring to the pool, an object of the appropriate type can be identified. Pools are managed using an interconnected set of facts and clipping rules set up with a special syntax because of their frequent use. A pool is initialized with the form:

(DEFINE-POOL ⟨type⟩ (⟨member 1⟩ ⟨member 2⟩...))

For example, a factory containing two identical lathes might represent them as a pool from which either can be selected for a particular job:

(DEFINE-POOL LATHE (LATHE-1 LATHE-2))

The task-expansion language contains forms for reserving an object from a pool and freeing it when finished. Pools are used extensively to manage discrete resources in the FORBIN example problems.

Static facts, clipping rules, and temporal forward chaining rules give the FORBIN planner a powerful model for many aspects of the world (see Dean (1985) for more background on reasoning about static facts). However, to handle task effects that vary with detailed context, dynamic facts must be used as well.

4.2.2. Dynamic facts

Dynamic facts represent those facets of the world that are best modeled as continuous quantities.[7] Static facts change value instantaneously and are always specified in terms of their new value. Dynamic facts, on the other hand, are often specified in terms of a change to their value (hence the new value depends on the old) and once changed, a dynamic fact may continue to change without further intervention. In the FORBIN plan language, dynamic facts are called *quantities*.

The schema for a quantity is shown below and consists of four parts: a name and three functions called MOVE, DELAY, and UPDATE.

(DEFINE-QUANTITY ⟨name⟩
   (MOVE   (lambda (⟨current-value⟩ ⟨change⟩
                  ⟨time⟩)...))
    DELAY (lambda (⟨current-value⟩ ⟨change⟩
                  ⟨time⟩)...))
    UPDATE (lambda (⟨current-value⟩ ⟨change⟩
                  ⟨time⟩)...))

---

[6]The temporal database manager (T.D.M.) which manages the interaction of static facts actually supports several more types of temporal chaining rules. However, rules other than these two were used infrequently in our trials. For further discussion of temporal reasoning and the abilities of the TDM, see Dean (1985).

[7]The syntax given here for quantities is a simplified version of what is possible. A quantity can be generalized from a scalar quantity to an arbitrary LISP object and can carry a great deal of internal information around. The three quantity functions can then refer to this information in an arbitrary manner. For such examples, the patterns used to refer to the quantity in expansions and descriptors will have more arguments. See Miller (1985) for more background on the heuristic task scheduler used to reason about quantity representations.

(a) Example A

```
(DEFINE-QUANTITY PLASTIC-RESIN
  (MOVE   (lambda (old change time) (positive?
            (+ change old))))
  (DELAY  (lambda (old change time) 0))
  (UPDATE (lambda (old change time) (+ change old))))
```

(b) Example B

```
(DEFINE-QUANTITY POSITION
  (MOVE   (lambda (begin end time)
            (and (member? begin *valid-locations*)
                 (member? end *valid-locations*)
                 (know-route-between? begin end))))
  (DELAY  (lambda (begin end time)
            (* (distance-between begin end) *nominal-speed*)))
  (UPDATE (lambda (begin end time)
            (ACTION (move-primitive begin end))
            end)))
```

FIG. 6. Examples of quantity definitions.

Each of the functions takes three arguments: the current value of the quantity, the change desired in the value, and the current time. During projection, these functions are called by the system to find out whether, and then how, a quantity can be changed as required by a task. The current value and time are filled in by the system for the temporal interval of interest and the change is taken from the quantity change statement that appears in the task's descriptor. The MOVE function returns a boolean value saying whether the desired change is possible, and the DELAY function returns an estimation of the time it will take to carry out that change. The UPDATE function returns the new value of the quantity after the change has been made. For example, the quantity representing a robot's location might be called POSITION with a current value of AT-LATHE-CHUCK. If a task has the effect of changing the robot's location to AT-WORKBENCH, then the MOVE function applied to AT-LATHE-CHUCK, AT WORKBENCH and the current time would return whether or not the move is possible. The DELAY function would return how long the move will take, and the UPDATE function would return the new value of POSITION (i.e., AT-WORKBENCH).

The three quantity functions are written by the causal-theory builder to reflect the way different dynamic facts are used. In the robot position example above, task descriptors specify changes to POSITION as new, absolute robot locations. It is just as easy to write quantity functions that take changes specified as relative to the current value. For example, the robot might have a store of plastic resin that is often taken from and occasionally replenished. Such a quantity could be represented as shown in Fig. 6a. Notice that in the PLASTIC-RESIN quantity functions, the change is treated in a relative way. In the plan language such changes would be represented as (PLASTIC-RESIN +10) or (PLASTIC-RESIN −10).

An important feature of quantity definitions is that the UPDATE function may generate action requests of the form: (ACTION ⟨pattern⟩). An action-request pattern corresponds to a primitive task to be incorporated as part of the final plan that FORBIN builds. During the planning process itself, however, action requests are ignored and play no part in decisions that influence the plan. Action requests allow the task network to grow without regard to certain necessary tasks and yet have those tasks added into the network at the finish. The most common example of such actions are robot travel tasks. In the

```
(TASK-DESCRIPTOR ⟨id⟩
  (TASK ⟨pattern⟩)
  (EXPECTED-DURATION ⟨low⟩ ⟨high⟩)
  (GENERATED-FACTS ...)
  (QUANTITY-CHANGES ...))
```

FIG. 7. The basic form of a task descriptor.

FORBIN factory, the robot must travel from work station to work station and the final plan must contain a primitive task for each change of locale. However, the start and end position of these travel tasks cannot be determined until all other parts of the plan have been elaborated and the planner knows where the robot needs to be for every task. Thus, travel tasks are generated using action requests from the position quantity, shown in Fig. 6b, once the final task network has been determined. This ability to add actions to the final plan as required is used extensively in FORBIN example problems to deal flexibly, but effectively, with travel time.

4.3. Task descriptors

Each type of task in the system has a task descriptor that specifies the way it changes facts in the world. The task descriptor also gives the expected duration for its task. FORBIN task descriptors can specify only the *expected* behavior of a task because there may be several different ways for the task to be expanded. For example, to make a widget it may be possible to use either the drill press or the milling machine. Each of these expansions will tie up a different machine and will alter different facets of the world. However, the (MAKE WIDGET) task must be incorporated into the task network, with its expected behavior, before an expansion is chosen. Thus, the task descriptor for the (MAKE WIDGET) task must make some compromise between the facts that might be changed by the drill press expansion and the facts that might be changed by the milling machine expansion.

The schema for a task descriptor is given in Fig. 7 and consists of four sections: the TASK section of the descriptor identifies the task type to which this descriptor applies, the EXPECTED-DURATION gives an estimation of how long the task will take, and the QUANTITY-CHANGES and GENERATED-FACTS sections together specify the effects that the task is expected to have on the world.

4.3.1. Task specifier

Whenever a task is added to the FORBIN task network, a task descriptor with matching specification is taken from the causal theory and used to describe the effects of that task to the projection machinery. Task descriptor specification may contain variables and hence it is possible for a task in the network to match more than one descriptor. For this reason, task descriptors are ordered (like prolog clauses) and the first one to match a task is chosen. Tasks in the task network will never have variables in their specifications and, after matching, all of the variables in a task descriptor will be instantiated. Some example task specifications are

```
(TASK (MAKE ?thing))
(TASK (SETUP-LATHE ?lathe WIDGET))
(TASK (SETUP-LATHE ?lathe ?type))
```

4.3.2. Expected duration

The expected duration of each task is crucial when temporal considerations are important to the success of a plan. Planning for deadlines requires knowledge of how long execution of all tasks in the plan is expected to take. The expected duration section of the task descriptor gives this information. A duration is specified as an estimated interval with a lower and upper bound. Some sample duration specifications are

```
(EXPECTED-DURATION 5.0 6.0)
(EXPECTED-DURATION (* 3 (SIZE-OF ?thing))
                   (* 10 (SIZE-OF ?thing)))
```

Once an expansion is chosen for a task, its expected duration is replaced with the composite durations of the subtasks in the expansion. Thus, as tasks are refined, so are the estimates of how long they will take.

4.3.3. Generated facts

The generated facts section of the task descriptor specifies those static effects that are expected to be made true by the task. The specification is a simple list of facts like a traditional add list. It is assumed during projection that each of these facts will be made true at some time during the execution of the task. No delete list is required because of the clipping rules specified in the domain physics. The following generated facts specification declares that executing the task will cause the lathe to be ready, the lathe bit to be of the correct type, and for the robot to be positioned at the lathe's chuck:

```
(GENERATED-FACTS
  (READY ?lathe ?type)
  (BIT ?lathe (BIT ?type)
  (LOCATION ROBOT (LOC-OF CHUCK ?lathe)))
```

4.3.4. Quantity changes

The final section of a task descriptor specifies a list of quantity changes or dynamic effects. Each change is augmented with an indication of when during the task the quantity will take on (or be changed by) the value specified. A quantity may be specified to change at the start, at the end, or sometime during the execution of the task. A quantity change has the form (⟨when⟩ ⟨pattern⟩), where ⟨when⟩ can be of START, END, or DURING. It is most enlightening to think of quantity changes as specifying "island states" or places in time when particular changes to dynamic facts will occur. For example, the following specification for the quantity changes of a (MOVE ?thing ?loc1 ?loc2) task will put the robot's position at ?loc1 to start and ?loc2 at the end:

```
(GENERATED-STATES
  (START (ROBOT-POSITION ?loc1))
  (END (ROBOT-POSITION ?loc2)))
```

As with the expected duration, after a task is expanded, its generated facts and quantity changes are replaced with those of its composite subtasks.

4.3.5. An example

To pull everything together, consider the complete task descriptor in Fig. 8. This descriptor specifies the causal information required for projection of the effects of a lathe setup task. It states that each setup task is expected to take from 5.0 to 6.0 time units and will result in the lathe being set up for the correct type of operation with the correct bit installed. Also, at the end of the task the robot will be positioned at the lathe's chuck. Whenever a task of this type is added to the task network, it will be annotated with all of this information for use in projection.

This task descriptor also illustrates an interesting point with

```
(TASK-DESCRIPTOR SETUP-T1
  (TASK (SETUP-LATHE ?lathe ?type))
  (EXPECTED-DURATION 5.0 6.0)
  (GENERATED-FACTS
    (SETUP-FOR ?lathe ?type)
    (BIT ?lathe (BIT ?type))
    (LOCATION ROBOT (LOC-OF CHUCK ?lathe)))
  (QUANTITY-CHANGES
    (END (POSITION (LOC-OF CHUCK ?lathe)))) )
```

FIG. 8. The task descriptor for a lathe setup task.

respect to the different processing of facts and quantities. Notice that there is both a generated fact entry for the robot's LOCATION and a quantity change entry for the robot's POSITION: obviously these two entries represent the same facet of the world and will take on the same values. However, the LOCATION models the robot's location as a static fact that can be changed at will, while the POSITION quantity is continuous and requires action requests to change it. The results of projecting the two together are the same as projecting either one, but the POSITION quantity will have more temporal accuracy. The reason for using both is to let LOCATION facts be cached for later use in temporal queries while having the POSITION quantity to calculate detailed travel times and preserve overall plan consistency. Also, in the final plan, the POSITION quantity will add the action requests necessary to generate actual robot movement tasks. Thus, the two complement each other with the LOCATION used as a basis for future plan choices and the POSITION used to add more detail.

5. The FORBIN plan language: expansion library

The task-expansion library contains the methods that FORBIN has available for accomplishing its tasks. The system cannot build a plan for a task that does not have an expansion method in the library. Each entry in the library describes one way of carrying out an abstract task. When there is more than one way to carry out a task, there is more than one entry in the library. For example, it might be possible to build a new gizmo with either the lathe or the milling machine. Since each one requires a different set of bits and setup procedures, it would make sense to enter each one into the library as a separate expansion. Each expansion entry is known as a *method* for expanding its type of task.

This section of the paper describes the notation used to specify task-expansion methods.

5.1. The expansion method

A FORBIN task-expansion method presents one possible way to carry out a task. Each method consists of four sections as shown in Fig. 9. The TASK section of the method is used as an index and specifies the type of task that the method can be used to accomplish. The ASSUMPTIONS section describes characteristics that must be true of the world state before and during execution of this method to ensure correct performance of the desired task. The actual subtasks, and their order, are detailed in the SUBTASKS section of the method, and the UTILITY section gives a way of comparing this method with other methods that perform the same task. All other things being equal, the highest utility method with satisfied assumptions is the best one to choose.

The ⟨identifier⟩ of the plan descriptor is a unique symbol used to differentiate one expansion method from another. It plays no role in the FORBIN algorithm. The other elements are described in some detail below.

5.1.1. Task type

The (TASK ⟨pattern⟩) portion of the plan descriptor is used to identify what task this method is an expansion for. The ⟨pattern⟩ portion may contain parameters or constants depending on whether the method may be used as an expansion for several tasks or just one. The TASK field serves the identical function in expansion methods as it does in the task descriptors described in the previous section.

5.1.2. Utility

The (UTILITY ⟨number⟩) portion of the plan descriptor is used to give relative "goodness" ratings to the various methods of carrying out the task. When a task has more than one plan descriptor, the one with the highest utility is given some small degree of preference over the others by the heuristic task scheduler (HTS) when it is choosing the best plan to use.

The utility value can either be a constant or a lambda expression. Thus the utility for a particular expansion can vary depending on the exact task for which the expansion is to be used. For example, consider

(UTILITY (lambda () (cond ((= ?type widget) 4)
                          ((= ?type gizmo) 2))))

The method containing this statement has a higher utility if it is operating on widgets than if it is being used for working on a gizmo.

5.1.3. Assumptions

The (ASSUMPTIONS ...) section in the plan descriptor details those things that must be true before the plan can be used as a method for completing the task. Each assumption is a predicate that must be true in the temporal database. All of the assumptions together make up a conjunctive query to the database. A valid assumption is any predicate that might be true in the temporal database, but for the most part, the only predicates necessary are

- (TT ⟨begin⟩ ⟨end⟩ ⟨effect⟩) stating that the formula ⟨effect⟩ must be true throughout the time interval ⟨begin⟩ to ⟨end⟩.
- (RESERVE ⟨begin⟩ ⟨end⟩ ⟨pool⟩ ⟨thing⟩ ⟨tag⟩) stating that the object ⟨thing⟩ from the ⟨pool⟩ must be available to be reserved by the task ⟨tag⟩ throughout the time interval ⟨begin⟩ to ⟨end⟩.

Throughout an expansion method, the special symbol *SELF* stands for the task that is being expanded. For example, the assumptions necessary for the GENERATE task in the FORBIN domain are

(ASSUMPTIONS
  (RESERVE (BEGIN *SELF*) (END *SELF*)
    LATHE ?LATHE (TAG *SELF*)))

The assumption states that this method will work only if a LATHE can be reserved for this task over the entire duration of the task. LATHE refers to a pool of lathes—several machines, any of which would be suitable for performing the GENERATE task. ?LATHE will be bound to a particular machine in the lathe pool. For a more detailed discussion of the RESERVE predicate, see Dean (1985).

(DEFINE-METHOD ⟨identifier⟩
  (TASK ⟨pattern⟩)
  (UTILITY ⟨number⟩)
  (ASSUMPTIONS ⟨assumption1⟩
               ⟨assumption2⟩...)
  (SUBTASKS (⟨tag⟩ ⟨pattern⟩)
             ⟨reason1⟩ FOR ⟨tag⟩
             ⟨reason2⟩ FOR ⟨tag⟩...)
           (⟨tag⟩ ⟨pattern⟩
             ⟨reason1⟩ FOR ⟨tag⟩
             ⟨reason2⟩ FOR ⟨tag⟩...))

Fig. 9. The basic form of a task-expansion method.

The MOVE method of Appendix B requires the TT assumption:

ASSUMPTIONS
  (TT (BEGIN *SELF*) (BEGIN *SELF*)
    (LOCATION ?THING ?START)))

This assumption demands that the fact (LOCATION ?THING ?START) be true during the interval defined by the start of that MOVE expansion method. In other words, that fact must be true at the start of the move task if this method of doing the move is to be successful.

For most cases in which there are alternative methods for accomplishing a task, the assumptions will play an important role in choosing a particular method. The assumptions for a particular method constrain the time and the order in which that particular method may be used in the overall plan. The HTS, using the assumptions, picks the method that fits in "best" with the overall plan, as already developed. Thus, different assumptions will cause different methods to be picked—causing an overall different plan to be created.

5.1.4. Subtasks

The (SUBTASK ...) portion of the expansion method is used to define the piece of the task network into which the task should expand. This portion of the method includes some number of subtasks, each identified within the method by a ⟨tag⟩. Following the tag is some ⟨pattern⟩ which should match the TASK pattern of one or more expansion methods in the task library. After each pattern is a list of the ⟨reason⟩s for that subtask to be done. Reasons often will reference the tags, thereby defining some partial order over the subtasks.

There are three general types of reasons for a subtask:

- (ACHIEVE ⟨effect⟩): the purpose of the subtask is to create this effect. The effect will be entered into the temporal database.
- (CREATE ⟨pool⟩ ⟨object⟩): this subtask creates new object ⟨object⟩ that can be added to the group of objects specified by ⟨pool⟩.
- (CONSUME ⟨pool⟩ ⟨object⟩): the opposite of CREATE, this reason specifies an object to be taken out of a pool.

The reasons for a subtask are connected to other subtasks with the FOR operator. FOR indicates that the reason given is to satisfy a precondition for the execution of the named subtask (referenced by its tag). The subtasks and their reasons are then loaded into the temporal database where the projection machinery forms the appropriate partial order. Normally, all of the subtasks are constrained to fall within the temporal bounds of the task being expanded (i.e., *SELF*). However, the FOR

```
(SUBTASKS
  (T1 (REMOVE-BIT ?OTHER-BIT ?LATHE)
      (ACHIEVE (BIT ?LATHE NONE)) < FOR T2)
  (T2 (PUT ?BIT (LOC-OF CHUCK ?LATHE))
      (ACHIEVE (BIT ?LATHE ?BIT)))))
```

Fig. 10. The subtasks for the INSTALL-1 method.

```
(SUBTASKS
  (T1 (GET (BIT ?TYPE) (LOC-OF CHUCK ?LATHE))
      (ACHIEVE (HAVE *ME* ?BIT)) < FOR T2)
  (T2 (PUT (BIT ?TYPE) (LOC-OF BIT-RACK ?TYPE))
      FOR >)))
```

Fig. 11. The subtasks for the REMOVE-BIT method.

operator has three other forms: <FOR. FOR>. and <FOR>: which indicate that the indicated subtask may be positioned before. after. or on either side of the main task for which it is in service.[*] For example. the INSTALL-1 method has the two subtasks shown in Fig. 10. The first subtask takes the old bit out of the lathe chuck so that the lathe chuck will be empty. By using the <FOR. the fact that the lathe chuck is empty as a precondition for doing the second subtask is added into the temporal database. The < indicates that the first subtask may be accomplished any time prior to the second subtask. as long as the fact that the lathe chuck is empty remains true until the second subtask is started.

The REMOVE-BIT method has a slightly different subtask scheme (see Fig. 11). The first subtask must be performed before the second and during the scope of the REMOVE-BIT task (as shown by the vanilla FOR). The second subtask may be done any time after the first. The FOR> indicates it is not bound by the end of the REMOVE-BIT task.

5.2. An example method

Now it is time to put everything together and describe a full expansion method. Figure 12 shows the expansion methods for the MAKE. GENERATE. and MOVE tasks. From the preceeding discussion it should be fairly obvious how these methods work. What may not be obvious is why they were written this way rather than some other seemingly more simple way. The next section goes through an expansion in detail. keeping track of what is being added into the temporal database. The reasons for the detailed syntax should then become more apparent.

6. The FORBIN planning algorithm

Section 3 of this paper describes the FORBIN system as a planner that builds its task network using a combination of hierarchical expansion, which adds detail. and temporal projection. which ensures correctness. Section 4 describes the causal reasoning methods used to project the expected effects of a partially built task network and Sect. 5 describes the language used to write expansion strategies for specific task types that might appear in the network. This section of the paper gives a detailed description of the hierarchical search algorithm at the heart of the FORBIN planner.

---

[*]The FORs may have associated with them metric intervals that specify exactly how close or how far two subtasks may be separated in time. These intervals have been left out of this paper to help maintain clarity.

```
(DEFINE-METHOD MAKE
  (TASK (MAKE ?TYPE))
  (UTILITY 1.0)
  (ASSUMPTIONS nil)
  (SUBTASKS
    (T1 (GENERATE ?TYPE $NEW-THING)
        (CREATE ?TYPE $NEW-THING) FOR T2)
    (T2 (MOVE $NEW-THING (LOC-OF SHELF ?TYPE))
        (ACHIEVE (LOCATION $NEW-THING
                  (LOC-OF SHELF ?TYPE)))))))
(DEFINE-METHOD GENERATE
  (TASK (GENERATE ?TYPE ?THING))
  (UTILITY 1.0)
  (ASSUMPTIONS
    (RESERVE (BEGIN *SELF*) (END *SELF*)
    LATHE ?LATHE (NAME *SELF*)))
  (SUBTASKS
    (T1 (SETUP-LATHE ?LATHE ?TYPE)
        (ACHIEVE (READY ?LATHE ?TYPE)) FOR T2)
    (T2 (RUN-LATHE ?LATHE ?TYPE)
        (CREATE ?TYPE ?THING)
        (ACHIEVE (LOCATION ?THING
                  (LOC-OF HOPPER ?LATHE)))))))
(DEFINE-METHOD MOVE
  (TASK (MOVE ?THING ?FINISH))
  (UTILITY 1.0)
  (ASSUMPTIONS
    (TT (BEGIN *SELF*) (BEGIN *SELF*)
    (LOCATION ?THING ?START)))
  (SUBTASKS
    (T1 (GET ?THING ?START)
        (ACHIEVE (HAVE *ME* ?THING)) FOR T2)
    (T2 (PUT ?THING ?FINISH)
        (ACHIEVE (LOCATION ?THING ?FINISH))))))
```

Fig. 12. The methods for MAKE.

The FORBIN search algorithm is conceptually quite simple. At any point in the search there is a partially elaborated plan represented by the current task network. The object of the search is to expand each task in the network into more and more detail until only primitive tasks remain. The final plan produced by the system is the network of primitive tasks.

A naive backtracking implementation of this search algorithm is computationally much too expensive: therefore. the FORBIN system makes use of many heuristics to limit the number of partial task networks considered and the number of projected futures calculated. The detailed discussion below attempts to make these heuristics explicit and thus clarify which planning problems FORBIN can solve and which it cannot.

6.1. Introducing a new task into the task network

Throughout the remainder of this section we will illustrate the FORBIN algorithm on a simple example in the FORBIN domain. The causal theory and task-expansion methods are contained in appendices B and C. Initially FORBIN is given the task (or more precisely, tasks) of constructing two items in its factory: a gizmo and a widget. The system is given the further constraint that the widget must be completed (i.e., constructed and properly shelved) by time 35. An implicit constraint is that both projects be finished as quickly as is mechanically possible.

Figure 13 shows the initial task network (the state of the tem-

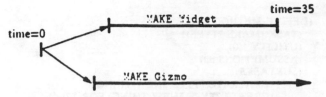

FIG. 13. The initial task network for constructing a widget and gizmo.

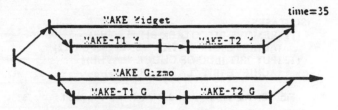

FIG. 14. The task network after one round of expansion.

poral database) after FORBIN has been given its task. The vertical line segments indicate the points in time actually dealt with by the planner. Time flows from left to right. The thin lines connect pieces of the network and represent ordering constraints, but not definite periods of time. The labeled horizontal segments indicate time to be spent doing whatever the label indicates; several activities may be going on in parallel. The network illustrated indicates that at some time after time zero the (MAKE WIDGET) task will be started; it will continue for a while, but will end by time 35. Meanwhile, some time after time zero (but not necessarily the same "some time" as in the previous sentence) the (MAKE GIZMO) task will be started. This task will also continue on for a while, but its end point is not presently constrained.

The network in Fig. 13 includes some information from the task descriptors for each task. In particular, the descriptors give the initial estimates of how long each task will take to execute.

Task-expansion choices are based on more than the task network shown in the diagram: they are also based on the temporal database derived from the task network and the task descriptors. There are many other things actually in the temporal database than are shown in the picture: the position of the robot, the bit currently in the lathe, what the robot is holding, and all other tasks known to the system, both past and future. However, for simplicity's sake, we will concentrate directly on the tasks at hand and show only the network itself.

6.2. Selecting a task to expand

Whenever FORBIN has unexpanded nonprimitive tasks in the task network, it attempts to expand them. Task expansions are done one at a time in an order based on several criteria:

- Since planning takes time, and some tasks have deadlines, tasks with approaching deadlines are given some priority in expansion.
- Tasks very high in the abstraction hierarchy should be expanded to sufficient detail that they may be placed in proper perspective in the task network.
- Heavily constrained tasks should be expanded so that none of their constraints are accidentally violated.

FORBIN will not retreat on an expansion decision, so the order that tasks are expanded may be critical to the plan that eventually evolves. Decisions about the way a task should be expanded (i.e., what method to use) are based on the facts in the task network and depend on the level of expansion for the tasks in the network. Since FORBIN does not retract expansion decisions, the algorithm may fail to find a successful plan where one exists. However, the search algorithm as a whole makes expansion decisions flexibly, by considering many alternatives before committing, so that the system is usually steered in an appropriate direction.

6.3. Queries and projections of possible expansions

Figure 14 shows the task network after both the widget and gizmo tasks have gone through one expansion. These expansions are easy to make because there is only one way to expand a MAKE task and it cannot interfere with other tasks in the network. Thus there are really no expansion decisions to worry about. FORBIN's decisions are not always this easy.

The MAKE plan is very high level. The only details that come out from its expansion are that two subtasks are required, the first generates the item, the second involves moving the item to its proper storage place. Both of these subtasks, as defined in the method, are constrained within the original bounds of their supertask. The unlabeled arrows between the subtasks indicate that the second subtask must follow the first by some amount of time that is only indirectly constrained by the starting constraint on the first subtask and the completion constraint on the second.

After another expansion cycle, the task network looks like that shown in Fig. 15. The expansion centered on the widget task because it is constrained by a deadline. The GENERATE subtask has been expanded and an assumption that the lathe is reserved was added to the temporal database. No conflicts with the assumption were detected, so the expansion was adopted. The left and right angle brackets linking items in the network (e.g., MAKE-T2 Widget and (MOVE WIDGET)) indicate the passing on of exact constraints (i.e., (MOVE WIDGET) has exactly the same constraints as the subtask from which it was expanded).

The choice of task to expand now switches to the gizmo task. The reason for this switch is that the (MAKE GIZMO) is now at too high an abstraction level with respect to the expanded (GENERATE WIDGET).

Exactly as with (MAKE WIDGET), the first subtask of (MAKE GIZMO) is expanded and a query is made using the temporal database machinery. This query asks when the assumption about lathe reservation that goes with the GENERATE method will be true. The query returns a disjunctive ordering constraint: the (GENERATE GIZMO) can occur either before or after the (GENERATE WIDGET) and have the lathe RESERVE come out true. Adopting this expansion and constraint leaves the task network with two distinct possible futures: one with the gizmo made on the lathe before the widget, and another with the widget produced first (shown in Fig. 16).

It should be noted that both of the orderings may not actually be feasible. The tasks have not really been expanded to sufficient detail to know for certain whether either will actually lead to a feasible plan, and the FORBIN algorithm, during each cycle, runs the task network through more detailed local projection where adverse interactions of dynamic effects may be spotted. As long as one feasible ordering for the task network can be found, however, it is left as is.

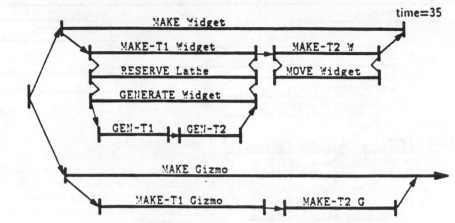

FIG. 15. The further expansion of the (MAKE WIDGET) task.

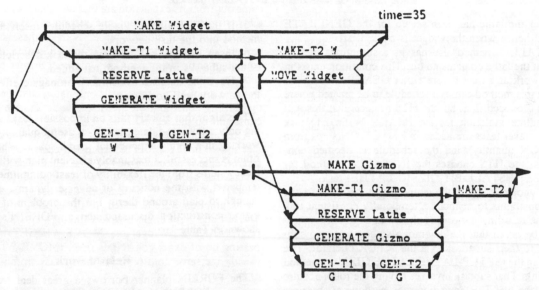

FIG. 16. The further expansion of the (MAKE GIZMO) task.

6.4. Comparing possible expansions

So far. in this discussion of the FORBIN algorithm. the relevant projection of the task network has been accomplished almost exclusively by the temporal database manager. This machinery has allowed the planner to identify possible trouble spots due to conflicts arising from separate tasks (e.g.. the need for a lathe by both MAKE tasks) and it has automatically inserted disjunctive constraints to keep these potential conflicts from becoming actual plan bugs.

For some tasks FORBIN has several different methods for expansion. The first subtask of the GENERATE method is to set up the lathe. There is one method for expanding this task. It involves the FORBIN robot acquiring the correct lathe bit. and then installing that bit into the lathe. There are multiple methods for installing the bit. Which to use depends on whether there is a bit already in the lathe. and what type of bit is needed. Committing to a specific method for installing the bit into the lathe is committing the world to be in a particular state (or at least to having certain facts be true) at the time when that method is to be expanded. A partial order may not be sufficient to guarantee this, and further constraints may have to be added. The temporal database can determine what

constraints are necessary to keep a method's assumptions valid when those assumptions are queried. but it cannot decide which method and (or) constraint set is the "best." Thus. when FORBIN gets to the point of choosing between method and (or) constraint sets. all options are loaded into the heuristic task scheduler for more detailed projection.

The HTS efficiently explores the space of total orderings of the task network. propagating dynamic effects that occur in each ordering. As it explores these orderings. it may place any of the possible expansion methods into the schedule it is building, pursuing those with the greatest utility (and implicitly those with the shortest overall duration) first. When a feasible schedule is derived, the expansion method used is returned. This is the method that is inserted into the task network.

In the case of expanding the (INSTALL-BIT GIZMO LATHE) task, the assumptions can be met for either of two expansion methods. The assumptions for the INSTALL-1 method demand that the lathe already be set up. but for the wrong type of product. Thus this method includes steps for removing the old bit and inserting the new one. These assumptions are met in the task network if the INSTALL takes place after the lathe has been reserved for the GENERATE Gizmo

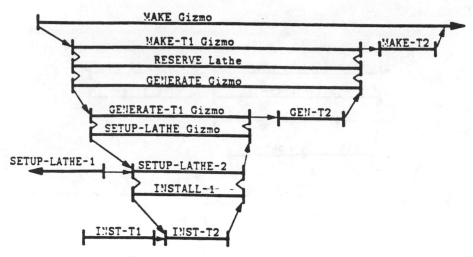

FIG. 17. Choosing the INSTALL-1 method.

task and after the lathe has been set up for the GENERATE Widget task (hence when the widget is made first).

The INSTALL-2 method also has its assumptions met. It demands that the lathe contains no bit. This condition exists in the task network at a time before either GENERATE task is done. Using this method causes a schedule to be created where the gizmo is produced first, but the HTS projects that this will cause a deadline violation for the (MAKE WIDGET) task (after all the travel tasks are added in by ACTION steps from the POSITION quantity) and the schedule is deemed non-viable. Thus, the HTS chooses the INSTALL-1 method for expanding the (INSTALL-BIT GIZMO LATHE) task.

Once this choice is made, the temporal database inserts the necessary constraints to maintain a temporally consistent task network. The resulting network is shown in Fig. 17.

It should be noted that the diagram also contains two subtasks linked to their supertasks via the < FOR operator. The first subtask in the (SETUP-LATHE LATHE GIZMO) method has such a link. That subtask involves having the robot acquire the proper lathe bit. This aquisition may be done at any time prior to the installation of the bit into the lathe. The first subtask in the INSTALL-1 method is similarly linked, but that task (which involves removing the old bit from the lathe) is constrained by having the lathe reserved.

6.5. Summary of FORBIN algorithm and solution

The FORBIN algorithm involves a cycle of operations that are performed on the task network. These are

1. Select a non primitive task descriptor from the network to expand. This selection is based on several heuristics involving the constraints on the tasks and the state of the network. Here the network acts as task queue.

2. From the library of expansion methods, extract all those whose pattern fields match the task being expanded.

3. Using the temporal database projection machinery, query the task network with the assumptions from each of the selected expansion methods. Eliminate those methods whose assumptions cannot be met. If all methods are eliminated then FORBIN has failed to successfully find a plan for that set of tasks.

4. Have the HTS try to create a viable schedule from the task network and its choice of expansion method. If the HTS is unable to find a viable schedule, FORBIN fails in planning for that set of tasks.

5. If the HTS finds a viable schedule, insert its choice of method into the task network.

6. Insert into the task network the task descriptors for each of the subtasks in the method just added.

7. Have the database machinery propagate effects through the network, and go to step 1.

This algorithm quickly fails on impossible tasks after exploring only a few candidate plans — a useful quality in a planner working in a highly exponential search space. When the algorithm is successful, a reasonably efficient plan will result. Efficiency stems from the flexibility of least-commitment planning tempered with the noticing of adverse dynamic effects early enough to plan around them. For the problem of widget and gizmo construction discussed above, FORBIN's solution is shown in Table 1.

7. Related work

The FORBIN planner borrows a great deal from existing systems. FORBIN builds on previous work in hierarchical planning (Sacerdoti 1977; Tate 1977; Wilkins 1984) by providing a specification language that allows one to encode expectations concerning deadlines, resources, and continuously changing quantities. The FORBIN specification language makes use of a temporal notation based loosely on McDermott's temporal logic (McDermott 1982). While FORBIN lacks even the full expressive power of propositional temporal logic, it is sufficiently expressive to encode many complex planning problems.

The FORBIN specification language is similar in many respects to the language used in the SIPE planner (Wilkins 1984). There are some problems that SIPE can handle and FORBIN cannot (e.g., certain restricted forms of quantification), but FORBIN excels in its treatment of metric time and continuously changing quantities.

FORBIN can also be viewed as a constraint-posting planner (Stefik 1981; Wilkins 1984; Chapman 1987) it attempts to avoid costly mistakes by making the constraints between tasks explicit. It is the manner in which FORBIN treats temporal constraints that sets it apart from other planners; FORBIN employs special routines (specifically, the heuristic scheduler and the temporal database query routines) that exploit the structure of time to efficiently detect possible inconsistencies and hence anticipate interactions involving resources and dead-

TABLE 1. FORBIN's solution to the example problem

Task	Place	Time		
		Travel	Task	Elapsed
(get (bit widget) hand1)	bit-rack	0	1	1
(put (bit widget) hand1)	lathe-chuck	3	1	5
(push (button widget) hand1)	lathe-control	1	1	7
(wait widget)	—	0	11	—
(get (bit gizmo) hand1)	bit-rack	3	1	11
(get (bit widget) hand2)	lathe-chuck	3	1	19ʷ
(put (bit gizmo) hand1	lathe-chuck	0	1	20
(push (button gizmo) hand1)	lathe-control	1	1	22
(wait gizmo)	—	0	14	—
(get widget hand1)	lathe-hopper	3	1	26
(put widget hand1)	(shelf widget)	5	1	32ᵐ
(get gizmo hand1)	lathe-hopper	5	1	38ᵍ
(put (bit widget) hand2)	bit-rack	3	1	42
(put gizmo hand1)	(shelf gizmo)	3	1	46ᵍ

ʷEnd (wait widget) at 18.
ᵐEnd (make widget).
ᵍEnd (wait gizmo) at 36.
ᵍEnd (make gizmo).

lines. Other planning systems have taken time into account (Allen and Koomen 1983; Cheeseman 1984), and some have even employed specification languages that allow metric constraints (Vere 1983; Hendrix 1973; Smith 1983), but FORBIN combines an expressive language for describing continuous processes with a sophisticated strategy for coping with the complexity that comes with the increased expressive power.

The FORBIN architecture was borne out of a realization that the combinatorics of temporal reasoning had to be dealt with directly. FORBIN attempts to solve a class of problems similar to that which the ISIS program (Fox and Smith 1985) and its various extensions were meant to handle. FORBIN attempts to capture much of the knowledge for reasoning about time and resources that is scattered throughout the ISIS knowledge base in a set of general-purpose routines and a strategy for using them.

8. Summary

Any useful approach to solving planning problems will involve some means of encoding knowledge concerning the domain of application. In addition to simply encoding this knowledge, there must be some means of drawing appropriate conclusions from this domain-specific knowledge and particular problem instances. In the case of reasoning about metric time and continuously changing quantities involving partially ordered events, computing all consequences that one might need to make the best choice is computationally prohibitive. The FORBIN architecture embodies one strategy for coping with the complexity of reasoning about time. Specifically, the FORBIN specification language allows one to encode descriptions of plans at several levels of detail, including information concerning expected resource use. This information is used during planning to eliminate plan expansions that cannot possibly lead to a solution, and rank those that can. This process of eliminating and ranking expansions is itself a potentially exponential process, and FORBIN employs a two-stage strategy for making these determinations. First, a set of candidates is generated using a temporal database query program that ignores interactions between partially ordered events. Second, this set of candidates is trimmed using a heuristic evaluation

program that eliminates candidates for which it cannot find an interaction-free total order. Using this two-stage evaluation strategy, if FORBIN terminates claiming success, then the solution it provides is correct. Moreover, FORBIN always terminates in time polynomial in the size of its knowledge base, and, supplied with reasonable expectations concerning the resource requirements of abstract actions, FORBIN will quite often terminate signalling success. The temporal database and heuristic evaluation routines exploit the structure of time and causation to realize high performance. FORBIN represents a practical and theoretically motivated concession to the complexity of real-world planning.

Acknowledgements

This paper was a long time in the writing and would not have been possible at all without the patience and support of Drew McDermott. The clarity of the paper, such as it is, also owes much to long discussions with Yoav Shoham and Steve Hanks. We would like to thank these three people and the many others whom we have harassed over the last two years.

This report describes work done by all three authors at Yale University supported in part by the Advanced Research and Projects Agency of the Department of Defense under Office of Naval Research contract N00014-83-K-0281. Subsequent discussion and writing was continued after two authors moved on and was further supported in part by DARPA Office of Naval Research contract N00014-85-K-0301 at Yale University, NSWC contract N60921-83-G-A165 at Virginia Polytechnic, and NSF grant IRI-8612644 at Brown University. In addition, Thomas Dean was supported at Brown University through an IBM Faculty Development Award and Jim Firby was supported at Yale University through a Canadian NSERC 1967 Science and Engineering Scholarship.

ALLEN, J., and KOOMEN, J. A. 1983. Planning using a temporal world model. Proceedings of the International Joint Conference on Artificial Intelligence, Karlsruhe, West Germany, pp. 741–747.
CHAPMAN, D. 1987. Planning for conjunctive goals. Artificial Intelligence, 32: 333–377.
CHEESEMAN, P. 1984. A representation of time for automatic plan-

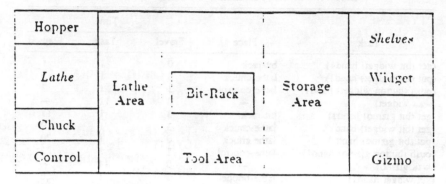

FIG. 18. Layout of the factory.

ning. Proceedings of the IEEE International Conference on Robotics.

DEAN. T. 1985. Temporal imagery: an approach to reasoning about time for planning and problem solving. Technical Report 433. Computer Science Department. Yale University. New Haven. CT.

FIKES. R. and NILSSON. N. J. 1971. STRIPS: a new approach to the application of theorem proving to problem solving. Artificial Intelligence. 2: 189–208.

FOX. M. S.. and SMITH. S. 1985. ISIS: a knowledge-based system for factory scheduling. Expert Systems. 1: 25–49.

HENDRIX. G. 1973. Modeling simultaneous actions and continuous processes. Artificial Intelligence. 4: 145–180.

McDERMOTT. D. V. 1977. Flexibility and efficiency in a computer program for designing circuits. Technical Report 402. Artificial Intelligence Laboratory. Massachusetts Institute of Technology. Cambridge. MA.

———1982. A temporal logic for reasoning about processes and plans. Cognitive Science. 6: 101–155.

———1985. The DUCK manual. Technical Report 399. Computer Science Department. Yale University. New Haven. CT.

MILLER. D. P. 1985. Planning by search through simulations. Technical Report 423. Computer Science Department. Yale University. New Haven. CT.

SACERDOTI. E. 1977. A structure for plans and behavior. American Elsevier Publishing Company. Inc.. New York. NY.

SMITH. S. F. 1983. Exploiting temporal knowledge to organize constraints. Technical Report CMU-RI-TR-83-12. Intelligent Systems Laboratory. Carnegie-Mellon University. Pittsburgh. PA.

STEFIK. M. J. 1981. Planning with constraints. Artificial Intelligence. 16: 111–140.

TATE. A. 1977. Generating project networks. Proceedings of the International Joint Conference on Artificial Intelligence. Cambridge. MA.

VERE. S. 1983. Planning in time: windows and durations for activities and goals. IEEE Transactions on Pattern Analysis and Machine Intelligence. 5: 246–267.

WILKINS. D. 1984. Domain independent planning: representation and plan generation. Artificial Intelligence. 22: 269–302.

Appendix A. The FORBIN factory domain

The FORBIN planner is a general-purpose system. but for illustration we will present examples from a small automated factory domain. The factory consists of an automatic lathe. a storage area. and a robot operator that builds *widgets* and *gizmos* and places them in the storage area (see Fig. 18). Making either a widget or gizmo requires the robot to carry out the following steps:

1. Make sure the correct bit is in the lathe.
2. Start the lathe and let it run until finished (11 units of time for a gizmo and 14 units for a widget).

TABLE 2. Travel times in the factory

	Hopper	Chuck	Control	Rack	Widget	Gizmo
Hopper	0	2	3	3	5	6
Chuck	2	0	1	3	5	5
Control	3	1	0	3	6	5
Rack	3	3	3	0	3	3
Widget	5	5	6	3	0	2
Gizmo	6	5	5	3	2	0

3. Remove the finished item from the lathe hopper.
4. Place item on the correct shelf.

There are three primitive actions. get. put. and push. each taking one unit of time to complete and requiring the use of one of the robot's hands. The robot has two hands. each of which can carry one item or be used to operate a control. The robot can travel throughout the factory at a fixed velocity and the travel times between the various work stations are given in Table 2. A typical problem is to construct a number of widgets and gizmos where some of each must be done within specific deadlines.

This factory differs from typical job-shop factories like those handled by the ISIS program (Fox and Smith 1985) in two important respects. First. a complete "job" does not travel from work station to work station as a single entity (e.g.. not all of the widgets in an order need to be turned on the lathe before any can be moved to the storage shelves) and second. the travel time of the robot from one place in the factory to another can be a significant part of the overall factory production time.

This domain illustrates the importance of temporal representation and plan efficiency issues that have not been adequately handled by previous planning systems.

Appendix B. The FORBIN factory task expansion library

This appendix contains all of the task expansion methods referenced in Sect. 6 of the text. These methods are supplied for the sake of completeness. The world physics and task descriptors for this domain are supplied in Appendix C.

```
; – (MAKE ?TYPE) –

(DEFINE-METHOD MAKE
  (TASK (MAKE ?TYPE))
  (UTILITY 1.0)
  (ASSUMPTIONS nil)
  (SUBTASKS
```

```
(T1 (GENERATE ?TYPE $NEW-THING)
    (CREATE ?TYPE $NEW-THING) FOR T2)
(T2 (MOVE $NEW-THING (LOC-OF SHELF ?TYPE))
    (ACHIEVE (LOCATION $NEW-THING
             (LOC-OF SHELF ?TYPE))))))

(DEFINE-METHOD GENERATE
 (TASK (GENERATE ?TYPE ?THING))
 (UTILITY 1.0)
 (ASSUMPTIONS
  (RESERVE (BEGIN *SELF*) (END *SELF*)
   LATHE ?LATHE (NAME *SELF*)))
 (SUBTASKS
  (T1 (SETUP-LATHE ?LATHE ?TYPE)
      (ACHIEVE (READY ?LATHE ?TYPE)) FOR T2)
  (T2 (RUN-LATHE ?LATHE ?TYPE)
      (CREATE ?TYPE ?THING)
      (ACHIEVE (LOCATION ?THING
               (LOC-OF HOPPER ?LATHE)))))))

: -- (SETUP-LATHE ?LATHE ?TYPE) --

(DEFINE-METHOD SETUP-LATHE
 (TASK (SETUP-LATHE ?LATHE ?TYPE))
 (UTILITY 1.0)
 (ASSUMPTIONS
  (TT (BEGIN *SELF*) (BEGIN *SELF*)
      (LOCATION (BIT-FOR ?TYPE) ?LOC1)))
 (SUBTASKS
  (T1 (GET (BIT-FOR ?TYPE) ?LOC1)
      (ACHIEVE (HAVE *ME* (BIT-FOR ?TYPE)))
      <FOR T2)
  (T2 (INSTALL-BIT (BIT-FOR ?TYPE) ?LATHE)
      (ACHIEVE (READY ?LATHE ?TYPE)))))

-- (RUN-LATHE ?LATHE ?TYPE) --

(DEFINE-METHOD RUN-LATHE
 (TASK (RUN-LATHE ?LATHE ?TYPE))
 (UTILITY 1.0)
 (ASSUMPTIONS nil)
 (SUBTASKS
  (T1 (PUSH BUTTON ?TYPE)
      (LOC-OF CONTROL ?LATHE)) FOR T2)
  (T2 (WAIT (TIME ?TYPE)))))

-- (INSTALL-BIT ?BIT ?LATHE) --

(DEFINE-METHOD INSTALL-1
 (TASK (INSTALL-BIT ?BIT ?LATHE))
 (UTILITY 2.0)
 (ASSUMPTIONS
  (TT (BEGIN *SELF*) (BEGIN *SELF*)
      (BIT ?LATHE ?OTHER-BIT))
  (NOT ?OTHER-BIT BIT)
  (NOT ?OTHER-BIT NONE))
 (SUBTASKS
  (T1 (REMOVE-BIT ?OTHER-BIT ?LATHE)
      (ACHIEVE (BIT ?LATHE NONE)) <FOR T2)
  (T2 (PUT ?BIT (LOC-OF CHUCK ?LATHE))
      (ACHIEVE (BIT ?LATHE ?BIT)))))

(DEFINE-METHOD INSTALL-2
 (TASK (INSTALL-BIT ?BIT ?LATHE))
 (UTILITY 1.0)
 (ASSUMPTIONS
```

```
(TT (BEGIN *SELF*) (BEGIN *SELF*)
    (BIT ?LATHE NONE)))
 (SUBTASKS
  (T1 (PUT ?BIT (LOC-OF CHUCK ?LATHE))
      (ACHIEVE (BIT ?LATHE ?BIT)))))

-- (REMOVE-BIT ?BIT ?LATHE) --

(DEFINE-METHOD REMOVE-BIT
 (TASK (REMOVE-BIT (BIT-FOR ?TYPE) ?LATHE))
 (UTILITY 1.0)
 (ASSUMPTIONS nil)
 (SUBTASKS
  (T1 (GET (BIT-FOR ?TYPE)
      LOC-OF CHUCK ?LATHE))
      (ACHIEVE (HAVE *ME* ?BIT)) FOR T2)
  (T2 (PUT (BIT-FOR ?TYPE)
      (LOC-OF BIT-RACK ?TYPE))
      (ACHIEVE (LOCATION (BIT-FOR ?TYPE)
      (LOC OF BIT-RACK ?TYPE))) FOR>)))

-- (MOVE ?THING ?FINISH)

(DEFINE-METHOD MOVE
 (TASK (MOVE ?THING ?FINISH))
 (UTILITY 1.0)
 (ASSUMPTIONS
  (TT (BEGIN *SELF*) (BEGIN *SELF*)
      (LOCATION ?THING ?START)))
 (SUBTASKS
  (T1 (GET ?THING ?START)
      (ACHIEVE (HAVE *ME* ?THING)) FOR T2)
  (T2 (PUT ?THING ?FINISH)
      (ACHIEVE (LOCATION ?THING ?FINISH)))))
```

Appendix C. The FORBIN factory causal theory

This appendix contains the causal theory definitions necessary to complete the domain-specific knowledge FORBIN uses in the factory domain.

Fact declarations:

```
(DEFINE-FACT (BIT ?lathe ?bit))
(DEFINE-FACT (READY ?lathe ?type))
(DEFINE-FACT (LOCATION ?thing ?place))
(DEFINE-FACT (HAVE ?robot ?thing))
```

Clipping rules:

```
(DEFINE-CLIP (BIT ?lathe ?bit) (BIT ?lathe ?other)
             (NOT (= ?bit ?other)))
(DEFINE-CLIP (READY ?lather ?type1)
             (READY ?lathe ?type2)
             (NOT (= ?type1 ?type2)))
(DEFINE—CLIP (LOCATION ?thing ?place1)
             (LOCATION ?thing ?place2)
             (NOT (= ?place1 ?place2)))
(DEFINE CLIP (HAVE ?robot ?thing)
             (LOCATION ?thing ?place)
             (NOT (= ?robot ?place)))
```

; State declarations:

```
(DEFINE-QUANTITY POSITION
 (MOVE   (LAMBDA (ORIG DEST TIME) T))
 (DELAY  (LAMBDA (ORIG DEST TIME)
```

```
              (LOOKUP-TRAVEL-TIME ORIG DEST)))
    (UPDATE(LAMBDA (ORIG DEST TIME)
              (ACTION (ROBOT-MOVE-TO DEST))
              DEST)))
```

: Pools of objects to be referenced:

```
(DEFINE-POOL LATHE   (LATHE-A LATHE-B))
(DEFINE-POOL WIDGET ())
(DEFINE-POOL GIZMO   ())
```

: Task Descriptors:

```
(TASK-DESCRIPTOR
  (TASK (MAKE ?TYPE))
  (ESTIMATED-DURATION 21.0 37.0)
  (QUANTITY-CHANGES
    ())
  (GENERATED-FACTS
    ())))

(TASK-DESCRIPTOR
  (TASK (GENERATE ?TYPE ?NAME))
  (ESTIMATED-DURATION 17.0 31.0)
  (QUANTITY-CHANGES
    ())
  (GENERATED-FACTS
    ())))

(TASK-DESCRIPTOR
  (TASK (SETUP-LATHE ?TYPE))
  (ESTIMATED-DURATION 5.0 6.0)
  (QUANTITY-CHANGES
    (END (POSITION (CHUCK ?LATHE))))
  (GENERATED-FACTS
    (READY ?LATHE ?TYPE)
    (BIT ?LATHE (BIT-FOR ?TYPE))
    (LOCATION *ME* (LOC-OF CHUCK ?LATHE))))

(TASK-DESCRIPTOR
  (TASK (INSTALL-BIT ?BIT ?LATHE))
  (ESTIMATED-DURATION 1.0 2.0)
  (QUANTITY-CHANGES
    ())
  (GENERATED-FACTS
    (BIT ?LATHE ?BIT)))

(TASK-DESCRIPTOR
  (TASK (REMOVE-BIT ?BIT ?LATHE))
  (ESTIMATED-DURATION 1.0 2.0)
  (QUANTITY-CHANGES
    ())
  (GENERATED-FACTS
    ())))
```

```
(TASK-DESCRIPTOR
  (TASK (RUN-LATHE ?LATHE ?TYPE))
  (ESTIMATED-DURATION 12.0 14.0)
  (QUANTITY-CHANGES
    (START (POSITION (CONTROL ?LATHE))))
  (GENERATED-FACTS
    ()))

(TASK-DESCRIPTOR
  (TASK (MOVE ?THING ?FINISH))
  (ESTIMATED-DURATION 7.0 8.0)
  (QUANTITY-CHANGES
    (START (POSITION (HOPPER ?LATHE)))
    (END (POSITION (SHELF ?TYPE))))
  (GENERATED-FACTS
    (LOCATION ?THING ?FINISH)
    (LOCATION *ME* ?FINISH)))
  -- PRIMITIVES --

(TASK-DESCRIPTOR
  (TASK-PRIMITIVE (GET ?THING ?PLACE))
  (ESTIMATED-DURATION 1.0 1.0)
  (QUANTITY-CHANGES
    (START (POSITION ?PLACE)))
  (GENERATED-FACTS
    (HAVE *ME* ?THING)
    (LOCATION *ME* ?PLACE)))

(TASK-DESCRIPTOR
  (TASK-PRIMITIVE (PUT ?THING ?PLACE))
  (ESTIMATED-DURATION 1.0 1.0)
  (QUANTITY-CHANGES
    (START (POSITION ?PLACE)))
  (GENERATED-FACTS
    (LOCATION ?THING ?PLACE)
    (LOCATION *ME* ?PLACE)))

(TASK-DESCRIPTOR
  (TASK-PRIMITIVE (PUSH ?BUTTON ?PLACE))
  (ESTIMATED-DURATION 1.0 1.0)
  (QUANTITY-CHANGES
    (START (POSITION ?PLACE)))
  (GENERATED-FACTS
    (LOCATION *ME* ?PLACE)))

(TASK-DESCRIPTOR
  (TASK-PRIMITIVE (WAIT ?TIME))
  (ESTIMATED-DURATION ?TIME ?TIME)
  (QUANTITY-CHANGES
    ())
  (GENERATED-FACTS
    ())))
```

A distributed control architecture for an autonomous mobile robot

Alberto Elfes

Mobile Robot Lab – The Robotics Institute and Engineering Design Research Center, Carnegie-Mellon University, Pittsburgh, PA 15213, USA

This paper describes a Distributed Control Architecture for an autonomous mobile robot. We start by characterizing the Conceptual Levels into which the various problem-solving activities of a mobile robot can be classified. In sequence, we discuss a Distributed Control System that provides scheduling and coordination of multiple concurrent activities on a mobile robot. Multiple Expert Modules are responsible for the various tasks and communicate through messages and over a Blackboard. As a testbed, the architecture of a specific system for Sonar-Based Mapping and Navigation is presented, and a distributed implementation is described.

Key Words: mobile robot software architecture, sonar mapping, autonomous navigation, distributed problem-solving, distributed control.

1. INTRODUCTION

Research in autonomous mobile robots provides a very rich environment for the development and test of advanced concepts in a variety of areas, such as Artifical Intelligence, Robotics, Sensor Understanding and Integration, Real-World Modelling, Planning and High-Level Control.

Much mobile robot research in the past has concentrated only on very specific problem areas, such as stereo-based mapping or path-planning. While the development of these techniques is crucial, it is also essential to investigate how these subsystems are integrated into an overall software architecture for an autonomous vehicle. This kind of effort can show what areas are already well understood and suitable for real-world, real-time applications, while also helping to identify weaknesses or prioritize research topics.

The work described here is part of an investigation of some of the issues involved in the development of a software architecture for an autonomous mobile robot[30]. The topics addressed in this paper include:

- A characterization of the Conceptual Processing Levels into which the various problem-solving activities needed for a mobile robot can be classified.
- A description of a High-Level Distributed Control Architecture for Mobile Robots. A Distributed Control System responsible for scheduling and coordinating multiple concurrent activities on an autonomous mobile robot is discussed. In this System, Expert Modules communicate through messages and maintain globally relevant information in a Blackboard. As an actual test of this framework, we are currently working on a distributed implementation of a previously developed Sonar-Based Mapping and Navigation System.

1.1. Overview

We start by characterizing the conceptual levels into which the various processing activities required by a mobile robot can be classified. This classification identifies multiple interacting levels of representation and problem-solving, such as robot control, sensor processing, real-world modelling or planning. It reflects a conceptual partitioning of the problem-solving activities needed for an autonomous vehicle. This framework has served as a paradigm for the implementation of the Dolphin autonomous navigation system, discussed below.

The multiplicity of complex cognitive and physical tasks necessary to achieve autonomous control and navigation for a mobile robot, as well as the parallelism and interaction characteristics of these tasks, led us to develop a Distributed Control System where the necessary cooperation can be expressed naturally. The grain level where this framework offers parallelism is at the process level. Conceptually, the problem-solving activity is seen as a computing environment where independent modules communicate through messages and are able to post on or retrieve relevant information from multiple Blackboards. A set of primitives is used that provide message-based communication, process control, blackboard creation and access, and event handling.

This research has been supported in part by the Office of Naval Research under Contract N00014-81-K-0503, and in part by the Western Pennsylvania Advanced Technology Center. The author is supported in part by the Conselho Nacional de Desenvolvimento Científico e Tecnológico – CNPq, Brazil, under Grant 200.986-80, in part by the Instituto Tecnológico de Aeronáutica – ITA, Brazil, in part by The Robotics Institute, Carnegie-Mellon University, and in part by the Design Research Center – CMU.

The views and conclusions contained in this document are those of the author and should not be interpreted as representing the official policies, either expressed or implied, of the funding agencies.

Much work in mobile robots has either concentrated on very specific subareas, such as path-planning, or has been conceptual work that never made it to a real application, such as several of the planning systems (e.g., Refs. 17 and 33). This is partially due to the complexity of the overall task, the unavailability of adequate testbeds and the difficulty of doing actual experiments in something close to real-time. As part of the research at the Mobile Robot Lab, we have recently developed a Sonar-Based Mapping and Navigation System called Dolphin. This system has proven successful and can be used as a testbed for higher-level experiments. It provides a sufficiently rich description of the robot's environment so as to call for more complex tasks; additionally, Dolphin operates in a sufficiently real-time situation so as to allow actual experiments to be done within a reasonable time frame. We are currently finishing a distributed implementation of the Dolphin system as part of an actual test of the Distributed Control System.

2. CONCEPTUAL PROCESSING LEVELS FOR AN AUTONOMOUS MOBILE ROBOT

2.1. Introduction

To widen the range of application of robotic devices, both in industrial and scientific applications, it is necessary to develop systems with high levels of autonomy and able to operate in unstructured environments with little *a priori* information. To achieve this degree of independence, the robot system must have an understanding of its surroundings, by acquiring and manipulating a rich model of its environment of operation. For that, it needs a variety of sensors to be able to interact with the real world, and mechanisms to extract meaningful information from the data being provided. Systems with little or no sensing capability are usually limited to fixed sequence operations in highly structured working areas, and cannot provide any substantial degree of autonomy or adaptability.

A central need, both for manipulators and for mobile robots, is the ability to acquire and handle information about the existence and localization of objects and empty spaces in the environment of operation of the device. This is essential for such fundamental operations as path-planning, obstacle avoidance, and spatial and geometric reasoning. Typically, due to limitations intrinsic to any kind of sensor, it is important to integrate information coming from multiple sensors, and to build a coherent world-model that reflects the information acquired and the hypotheses proposed so far. This world model can then serve as a basis for several of the operations mentioned previously: path-planning, landmark identification, position estimation, etc.

Any system with the scope outlined above becomes bewilderingly complex. Deciding what sensor, actuator or processing action to perform next — in other words, doing scheduling of the various activities — calls for task-level planning to provide the appropriate sequence. A problem-solving environment is needed with multiple agents that run semi-independently and cooperate in handling these various activities in a coherent manner. Finally, to interpret the plan and oversee the global behaviour of the system, a supervisory module becomes essential.

The different activities mentioned above can be conceptually classified into major groups or levels of processing. These Levels are described in the following Section.

2.2. Conceptual processing levels

In a software architecture for an autonomous mobile robot, we characterize several conceptual processing levels (Fig. 1): the *Robot Control*, *Sensor Interpretation*, *Sensor Integration*, *Real-World Modelling*, *Navigation*, *Control*, *Global Planning* and *Supervisor* Levels. These Levels are briefly discussed below:

- *Robot Control:* This level takes care of the physical control of the different sensors and actuators available to the robot. It provides a set of primitives for locomotion, actuator and sensor control, data acquisition, etc., that serve as the robot interface, freeing the higher levels of the system from low-level details. It includes activities such as vehicle-based motion estimation and monitoring of internal sensors. *Internal Sensors* provide information on the status of the different physical subsystems of the robot, while *External Sensors* are used to acquire data about the outside world.

- *Sensor Interpretation:* On this level the acquisition of sensor data and its interpretation by Sensor Modules is done. Each Sensor Module is specialized in one type of sensor or even in extracting a specific kind of information from the sensor data. The modules provide information to the higher levels using a common representation and compatible frames of reference.

- *Sensor Integration:* Due to the intrinsic limitations of any sensory device, it is essential to integrate information coming from qualitatively different sensors, such as stereo vision systems, sonar devices, laser range sensors, etc. Specific assertions provided by the Sensor Modules are correlated to each other on this level. For example, the geometric boundaries extracted from an obstacle detected by sonar can be used to provide connectivity information to a set of scattered 3D points generated by the stereo vision subsystem. On this level, information is aggregated and assertions about specific portions of the environment can be made.

- *Real-World Modelling:* To achieve any substantial degree of autonomy, a robot system must have an understanding of its surroundings, by acquiring and manipulating a rich model of its environment of operation. This model is based on assertions integrated from the various sensors, and reflects the data obtained and the hypotheses proposed so far. On this level, local pieces of information are used in the incremental construction of a coherent global Real-World Model; this Model can then be used for several other activities, such as landmark recognition, matching of newly acquired information against previously stored maps, and generation of expectations and goals.

- *Navigation:* For autonomous locomotion, a variety of problem-solving activities are necessary, such as short-term and long-term path-planning, obstacle-avoidance, detection of emergencies, etc. These different activities are performed by modules that provide specific services.

VIII. Supervisor

- Global Supervision of System Behaviour
- User Interface

VII. Global Planning

- Task-Level Planning to provide sequences of sensory, actuator and processing (software) actions
- Simulation
- Error-Recovery and Replanning in case of failure or unexpected events

VI. Control

- Scheduling of Activities
- Integration of Plan-Driven with Data-Driven Activities

V. Navigation

- Navigation Modules provide services such as Path-Planning and Obstacle Avoidance

IV. Real-World Modelling

- Integration of local pieces of correlated information into a Global Real-World Model that describes the robot's environment of operation
- Matching acquired information against stored maps
- Object Identification
- Landmark Recognition

III. Sensor Integration

- Information provided by different Sensor Modules is correlated and abstracted
- Common representations and compatible frames of reference are used

II. Sensor Interpretation

- Acquisition of Sensor Data (Vision, Sonar, Rangefinder, etc.)
- Interpretation of Sensor Data

I. Robot Control

- Set of Primitives for Robot Operation
- Actuator Control (e.g., locomotion)
- Sensor Control
- Internal Sensor Monitoring

Fig. I. Conceptual Processing Levels in a Mobile Robot Software Architecture

- *Control:* This level is responsible for the scheduling of the different activities and for combining Plan-Driven and Data-Driven activites in an integrated manner so as to achieve coherent behaviour. In other words, this level tries to execute the task-level plan that was handed to it, while adapting to changing real-world conditions as detected by the sensors.
- *Global Planning:* To achieve a global goal proposed to the robot, this level provides task-level planning for autonomous generation of sequences of actuator, sensor and processing actions. Other activities needed include simulation, error detection, diagnosis and recovery, and replanning in the case of unexpected situations or failures.
- *Supervisor:* Finally, on this level a Supervisory Module oversees the various activities and provides an interface to a human user.

By identifying these levels, we are not implying that communication between processing modules is only possible between adjacent levels. On the contrary, experience with real systems shows that usually there are very complex interconnections and interdependencies between the various subsystems, with multiple flows of control and data. Additionally, a specific Module (such as Stereo Vision or Sonar Interpretation) may be a very complex system in itself, with sophisticated control, planning and problem-solving activities.

Clearly, none of the presently existing mobile robot systems cover all of the levels described above. This conceptual structure provides, however, a context within which several of our research efforts situate themselves[8,30]. The Dolphin Sonar-Based Mapping and Navigation System, in particular, embodies several of the elements of this framework, as mentioned in Section 4.

3. A DISTRIBUTED CONTROL SYSTEM

3.1. Introduction

Mobile robots pose a number of fascinating problems from the point of view of overall software system design. A large number of semi-independent activities are necessary to achieve autonomous mobility. These tasks include:

- Control of different kinds of actuators and sensors.
- Monitoring and interpretation of data provided by several qualitatively different sensors.
- Representation of sensory information along multiple levels of interpretation and abstraction and integration of this information into a coherent framework.
- Planning and problem-solving activities that have to be performed in a multiplicity of areas and on various levels of abstraction.

Other factors that have to be taken into consideration when designing an overall software system structure for a mobile robot are:

- A mobile entity is likely to encounter unforeseen situations that may render previous plans and interpretations useless.
- Many of the tasks to be performed are quasi-independent and should logically proceed in parallel.
- An asynchronous mode of interaction between the various subsystems is required to provide adequate response.

- Multiple levels of control and decision-making have to coexist in the system to allow abstraction and distribution of responsibilities.
- The system has to integrate Data-Driven (responding to sensory information) and Plan-Driven (following a plan to execute a given task) modes of behaviour.

These problems are aggravated by the perspective that, to achieve real-time response, large amounts of processing power are necessary. One way of achieving this is to apply several processors to the problem. This, however, brings the need to develop new and adequate distributed control and problem-solving mechanisms.

Some of the advantages of Distributed Problem-Solving systems include the fact that many problems can be naturally represented in a distributed computational framework, allowing *decomposition* of the problem and *cooperation* among the problem-solvers.

In this Section, we present a Distributed Control System initially developed for the CMU Rover[8], the first robot developed at the Mobile Robot Lab.

In the Distributed Control System, Expert Modules run as independent process and exchange globally relevant information over a Blackboard. The Modules are distributed over a processor network and communicate through messages. They are used to control the operation of the sensors and acuators, interpret sensory and feedback data, build an internal model of the robot's environment, devise strategies to accomplish proposed tasks and execute these strategies. Each Expert Module is composed of a Master process and a Slave process, where the Master controls the scheduling and operation of the Slave. Communication among Expert Modules occurs asynchronously over a Blackboard structure. Information specific to the execution of a given task is provided through a Control Plan. The system is distributed over a network of processors. An Executive local to each processor and an interprocess message communication mechanism ensure transparency of the underlying network structure.

3.2. Distributed problem-solving

Several sophisticated control mechanisms have been developed in the context of specific languages and applications. The goal in these efforts was to develop new paradigms to cope with increasingly complex tasks and to develop more 'natural' ways to express the problem-solving activities involved. These include co-routining[18], automatic backtracking mechanisms[23], priority queues[1,2], pattern-directed systems[19,20,22,38] and agenda-based systems[21].

Two problem-solving architectures that have had a major impact and have been applied in complex domains are Blackboard Systems and Message-Based Systems. These are briefly discussed below:

- **Blackboard Systems:** a blackboard is a globally accessible data base where different processes that are cooperating towards the solution of a certain problem can share data and results, suggest hypotheses to be examined, communicate the state of their own computations, etc. This paradigm subsumes several other kinds of problem-solving and control organizations including priority queues, pattern-directed systems and agenda-based systems. It was initially developed within the context of the

Hearsay-II speech understanding system[13], and has been tried in several other complex applications, such as Image Understanding[21] and Vehicle Tracking[6,28,29]. The knowledge which is brought to bear on the problem is encoded in *Knowledge Sources* (*KSs*); when the *Condition* part of a KS is satisfied (depending on the state of the Blackboard), the *Action* part is scheduled for subsequent execution. The blackboard is subdivided so as to topologically correspond to the different subareas of problem-solving and levels of knowledge representation.

- **Message-Based Systems**: in this paradigm, the problem-solving work is performed by several independent program modules that communicate with each other through *Messages*. These messages are used to distribute work and responsibilities, determine the status of a module, provide synchronization, exchange information and coordinate the problem-solving behaviour of the system as a whole. Work in this area includes the Actors system[24], the PLITS language[16], and the Sprite system[26].

A recent area of research called Distributed AI[15] investigates distributed problem-solving frameworks. As part of this effort, some of the advanced control mechanisms mentioned above are being explored in a Distributed Processing context[5,15], and new distributed problem-solving mechanisms are being developed. Some of the more promising paradigms are mentioned below:

- **Parallel Production Systems**: This work concentrates on applying the pattern-directed invocation approach in a distributed processing environment (see, for example, Ref. 27).
- **Message-Module Systems**: The message-based communication mechanism of Actor-like systems can be naturally expanded to a distributed processing environment, with completely independent modules that interact and work on a common goal[24,26] by interchanging messages. More complex ways of establishing responsibilities within a set of cooperating modules by using 'contract nets' are explored in Ref. 34. Recent work that uses the Message/Module approach in a distributed environment is discussed in Refs 5, 9, 35, 36, 37, 39.
- **Distributed Blackboard Systems**: The blackboard-based problem-solving paradigm is one of the most powerful and general ones, and several extensions have been suggested to apply it in a distributed environment. These include improving the scheduling strategies by using separate 'Data' and 'Goal' blackboards[3,6,14]; the use of multiple blackboards, each one dedicated to one specific sub-problem[28,29,35]; and the generalization of the Hearsay-II Condition/Action Knowledge Sources to a more general Master/Slave approach, the implementation of Expert Knowledge Sources as independent modules and the use of messages to provide communication between them in a distributed environment[8,35,37]. A truly distributed Blackboard system is presented in Ref. 4.
- **Team Systems**: Within this framework, several of the mechanisms mentioned above are combined. The general problem-solving paradigm to be followed is to have a number of Cooperating Experts responsible

for specific tasks, communicating with each other through messages. Several Knowledge Sources are brought to bear on a specific problem or parts of a problem. For numerical applications, for example, one frequently encounters situations where several algorithms are available, but no one algorithm is successful over the entire range of situations. The work described in Refs 35 and 37 uses the approach of combining several algorithms into a team and allowing them to attack the problem in parallel. A blackboard is used to store global relevant information; a global manager oversees the team members' progress and coordinates the problem-solving effort. The advantages of using asynchronous algorithms instead of synchronous ones are discussed in Ref. 35, and the use of homogeneous and heterogeneous teams in Ref. 37. Other researchers have examined hierarchical and heterarchical organizations[39].

3.3. The distributed control system

The complexity, inherent parallelism and patterns of interaction characteristic of a mobile robot software organization require a framework that allows the problem-solving activities to be expressed in a natural way. The Distributed Control System (DCS) was developed to address these needs.

3.3.1. Global Structure

The general architecture of the Distributed Control System (see Fig. 2) consists of an expandable community of *Expert Modules*, which communicate asynchronously among themselves through *Messages*. Information needed for the common good is posted on and retrieved from a *Control Blackboard*. A given task is performed under the direction of a *Control Plan*. The Expert Modules are specialized subsystems used to monitor the

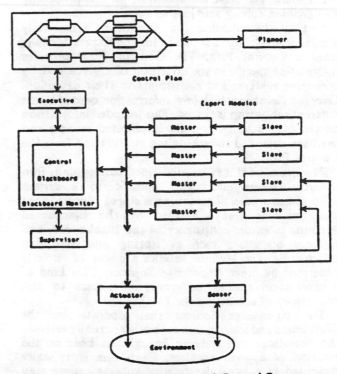

Fig. 2. *Structure of the Distributed Control System*

139

```
Navigate:[BootstrapRobot;
        SetGoal(Final-Position);
        Go:<Scan;
            ON (Present-Position ≠ Final-Position)
            DO [GetScan;
                BuildSonarMap;
                <DisplayMap;
                CalculatePath(New-Position);
                >
                ]
            ON (Present-Position ≠ New-Position)
            DO <Move(New-Position);
                ON (Nearing-Object)
                DO [Stop(Move);
                    Abort(Go);
                    Run(Go);
                    ]
                >
            ON (Present-Position = Final-Position)
            DO [StopRobot]
            >
        ].
```

Fig. 3. Example of a Simple Plan: Moving to a Goal and Avoiding Obstacles

sensors and actuators, interpret sensory and feedback data, build an internal model of the robot's environment, plan strategies to accomplish proposed tasks, supervise the execution of the plan and do global system management. Each Expert Module is composed of two closely coupled components: a *Master* process and a *Slave* process. The Master process controls the scheduling and the activities of the Slave process and provides an interface to other modules. It retrieves needed data from the Blackboard, changes the status (*run/suspend/terminate/continue*) of the Slave, and posts relevant information generated by the Slave process on the Blackboard. The Slave is responsible for the processing and problem-solving activities as such.

One of the modules, the *Executive*, dynamically abstracts scheduling information for the Expert Modules from a Control Plan. The Control Plan provides information specific to the execution of a given task by specifying subtasks and constraints in their execution. Over the Blackboard high-level information needed by the different subsystems is shared. This includes information on the robot's status, relevant interpreted sensory and feedback data, and the scheduling information from the Control Plan.

The Blackboard may be subdivided conceptually into subareas, where information specific to a certain subproblem or level of processing is stored. Access to the Blackboard is always handled by the Blackboard Monitor to ensure consistency of the Blackboard, and includes operations such as storing and retrieving information. The Monitor handles a queue of requests generated by other Knowledge Sources. This kind of encapsulation ensures uniformity of access to and consistency of updates on the Blackboard.

The Knowledge Sources that operate on the Blackboard and interact with each other actually embody the knowledge that is being brought to bear on the solution of a given problem. Each one is typically dedicated to a particular type of subtask; some may embody procedural knowledge and formal algorithms;

others may use heuristic knowledge or inexact reasoning and pattern-directed problem-solving. Some of these modules may be responsible for an evaluation of the progress towards a solution being made by several Knowledge Sources that are working on a given problem; others may plan and proposed intermediary goals to be achieved, suggest ways of decomposing a problem into simpler subproblems or favourable directions to explore in the search for a solution, or organize the sequencing of relevant tasks.

The Expert Modules are distributed over the processor network. An *Executive* local to each processor is responsible for process scheduling, and an interprocess message communication mechanism ensures transparency of the underlying network structure. Besides using the Blackboard, processes also exchange data of more specific interest directly among themselves. Interprocess messages have a *header* with routing information (including source, destination, subject and priority) and a *body*, which essentially stores a frame with (⟨name⟩, ⟨value⟩) slots. The system is built on top of a set of primitives that provide process handling, message-based interprocesses communication and access to the blackboard[4].

Integrated into the overall control structure of the robot is the *Control Plan*. A simplified example is shown in Fig. 3. The syntax is partially based on Ref. 7. Processes can be specified to execute in parallel (those within ⟨ ⟩ brackets) or sequentially (those within [ ] brackets). Responses to events are defined by the use of 'ON ⟨event⟩ DO ⟨action⟩' rules. From the Plan, information about parallel and sequential execution of processes, as well as reactions to events, is abstracted, and posted dynamically on the blackboard as the Plan is executed.

3.3.2. Discussion

Taking into account the multiprocessing network and the need to dynamically respond to the changing conditions of the real-world environment, the role of the 'condition part' of the Hearsay-II Knowledge Sources[13] was expanded in the context of a more general *master/slave* relation. In this way, both Master and Slave can be run in parallel and the Master can evaluate the global situation, activating or stopping the Slave when appropriate. With this organization, irrelevant or dangerous activities can be discontinued. Additionally, a separate control plan, integrated into the overall structure and absent in Hearsay-II, was considered essential in our case. Other researchers have also felt the need to use separate focusing and goal-proposing mechanisms[9,21].

The system presented reflects the structure of a community of *cooperating experts*. These experts communicate asynchronously over the processor network, generating and absorbing streams of data. Concomitantly, the system embodies a *hierarchical model* of distributed computation. This reflects the decision-making model: control decisions are, whenever possible, made locally in the module which is confronted with a problem. Otherwise, the problem or data is broadcast recursively to the next higher level of decision-making. Another result from this model is that commands and data can exist within the system at several levels of abstraction. At each level only the necessary degree of detail is present. Higher levels are able to deal with the

140

same information in a more abstract form and do not become cluttered with unnecessary details.

The system described is *loosely coupled*, since the rate of communication between modules, especially beyond the motor and sensor subsystem levels, is relatively small. This results from the use of asynchronous processes and the blackboard. Such an approach can lead to higher performance and better adaptability to dynamically changing conditions.

4. ROBOT SYSTEM ARCHITECTURE

4.1. Introduction

As an actual test of the Distributed Control System described above, we are implementing a distributed version of a system for autonomous mobile robot navigation. To provide a context for this discussion, we present in this Section the overall architecture of the Dolphin Sonar-Based Mapping and Navigation System. The functions of the major modules and their interaction with the various sonar map representations are discussed.

The Dolphin system is intended to provide sonar-based mapping and navigation for an autonomous mobile robot operating in unknown and unstructured environments. The system is completely autonomous in the sense that it has no *a priori* model or knowledge of its surroundings and also carries no user-provided map. It acquires data from the real world through a set of sonar sensors and uses the interpreted data to build a multi-leveled and multi-faceted description of the robot's operating environment. This description is used to plan safe paths and navigate the vehicle toward a given goal.

4.2. Sonar-based mapping and navigation

The Dolphin Sonar-Based Mapping and Navigation System[11,31] uses wide-angle sonar range sensors to obtain information about the robot's environment. Multiple range measurements are used to build a probabilistic sonar map. Each sonar range reading is interpreted as providing information about *empty* and *occupied* volumes in the space subtended by the sonar beam (in our case, a 30° cone in front of the sensor). This occupancy information is modelled by probability profiles that are projected onto a rasterized two-dimensional map, where OCCUPIED, EMPTY and UNKNOWN areas are represented. Sets of range measurements taken from multiple sensors on the robot and from different positions occupied by the robot over time provide multiple views that are systematically integrated into the sonar map. In this way, the accuracy of the sonar map is progressively improved and the uncertainty in the positions of objects is reduced. Overlapping empty areas reinforce each other, and serve to sharpen the boundaries of the occupied areas, so that the map definition improves as more readings are added. The final map shows regions probably OCCUPIED, probably EMPTY, and UNKNOWN. The method deals effectively with noisy data and with the limitations intrinsic to the sensor.

The resulting sonar maps are very useful for navigation and landmark recognition. They are much denser that the ones produced from current stereo vision programs, and computationally about an order of magnitude faster to produce. We have demonstrated an autonomous navigation system[11] that uses an A*-based path-planner to obtain routes in these maps. A first version of this system was tested in cluttered indoor environments using the *Neptune*[32] mobile robot, and outdoors in open spaces, operating among tress, using the *Terregator* robot. The system operated successfully in both kinds of environments, navigating the robot towards a given destination.

Conceptually, two modes of operation are possible: in *Cruising Mode*, the system acquires data, builds maps, plans paths and navigates toward a given goal. In *Exploration Mode*, it can wander around and collect enough information so as to able to build a good description of its environment. The system is intended for indoor as well as outdoor use; outdoors, it may be coupled to other systems, such as vision[25], to locate landmarks that would serve as intermediate or final destinations.

4.3. Multiple axis of representation of sonar mapping information

The probabilistic sonar map serves as the basis for a multi-levelled and multi-faceted description of the robot's operating environment. Sonar maps are represented in the system along several dimensions: the *Abstraction axis*, the *Geographical* axis and the *Resolution* axis[12]:

- THE ABSTRACTION AXIS: Along this axis we move from a sensor-based, low-level, data-intensive representation to increasingly higher levels of interpretation and abstraction. Three levels are defined: the *Sensor Level*, the *Geometric Level* and the *Symbolic Level*.
- THE GEOGRAPHICAL AXIS: Along this axis we define *Views*, *Local Maps* and *Global Maps*, depending on the extent and characteristics of the area covered.
- THE RESOLUTION AXIS: Sonar Maps are generated at different values of grid resolution for different applications. Some computations can be performed satisfactorily at low levels of detail, while others need high or even multiple degrees of resolution.

4.4. System architecture

The overall architecture of the Sonar Mapping and Navigation System is shown in Fig. 4. The functions of the major modules and their interaction with the different sonar map representations are described below[10]:

Sonar Control:	Interfaces to and runs the Sonar Sensor Array, providing range readings.
Scanner:	Preprocesses and filters the sonar data. Annotates it with the position and orientation of the corresponding sensor, based on the robot's motion estimate.
Mapper:	Using the information provided by the Scanner, generates a View obtained from the current position of the robot. This View is then integrated into a Local Map.
Cartographer:	Aggregates sets of Local Maps into Global Maps. Provides map handling and bookkeeping functions.
Matcher:	Matches a newly acquired Local Map against portions of Global Maps for operations such as

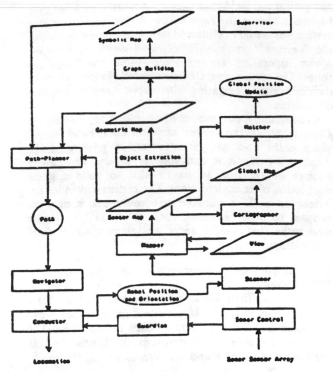

Fig. 4. *Architecture of the Dolphin Sonar-Based Mapping and Navigation System*

Navigator: Takes care of the overall navigation issues for the vehicle. This includes examining already planned paths to determine whether they are still usable, invoking the path-planner to provide new paths, setting intermediary goals, overseeing the actual locomotion, etc.

Conductor: Controls the physical locomotion of the robot along the planned path. The latter is smoothed and approximated by sequences of line segments, using a line-fitting approach. This module also returns an estimate of the new position and orientation of the robot.

Guardian: During actual locomotion, this module continuously checks the incoming sonar readings and signals a stop if the robot is coming too close to a (possibly moving) obstacle not detected previously. It serves as a 'sonar bumper'.

Supervisor: Oversees the operation of the various modules and takes care of the overall control of the system. It also provides a user interface.

Comparing this architecture with the conceptual framework outlined in Section 2, we can identify an immediate correspondence between the subsystems of the Dolphin system and some of the processing Levels described previously: the Sonar Control and Conductor modules belong to Level I; Scanning and Mapping provide functions on Level II; the Object Extraction, Graph Building, Cartographer and Matcher operate on Level IV; Path-Planning, Navigation and the Guardian are situated in Level V; and the Supervisor is on Level VIII.

landmark identification or update of the robot's absolute position estimate.

Object Extraction: Provides geometric information about obstacles. Objects are extracted by merging regions of OCCUPIED cells and determining the corresponding polygonal boundaries. A region-colouring approach is used for unique labelling.

Graph Building: Searches for larger regions that are either empty or else have complex patterns of obstacles, labelling them as 'free' or 'interesting' spaces.

Path-Planning: Path-Planning can occur on three different levels: *Symbolic Path-Planning* is done over wider areas (Global Maps) and at a higher level of abstraction (Symbolic Maps); *Geometric Path-Planning* can be used as an intermediary stage, when the uncertainty in Local Map is low, and has the advantage of being faster than finding routes in the Sensor Map; finally, *Sensor Map Path-Planning* generates detailed safe paths. The latter performs an A* search over the map cells, with the cost function taking into account the OCCUPIED and EMPTY certainty factors, as well as the UNKNOWN areas and the distance to the goal. The path found by this module is provided to the Navigator.

5. A DISTRIBUTED IMPLEMENTATION OF THE MAPPING AND NAVIGATION SYSTEM

5.1. Introduction

To provide a real-life testbed for the Distributed Control System described above, we are currently applying it to the Sonar-Based Mobile Robot Navigation system discussed in Section 4. The original version of the system works by passing control to each module in sequence. In the distributed version of the system, the Expert Modules are activated by their masters, which watch the Blackboard for conditions that warrant a change in the status of their slaves. Information concerning the availability of data or results, the status of the robot, the activities of the Expert Modules and other relevant high-level data and control information is shared over the Blackboard. The Executive provides additional scheduling information to achieve an overall integrated and coherent behaviour.

5.2. Primitives

As a basis for implementing the distributed version of the Sonar System, we are using a set of primitives that provide message-based communication, process control, blackboard creation and access, and event handling[4].

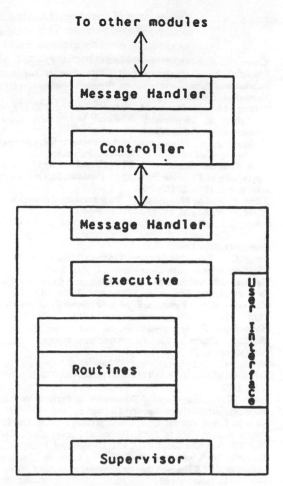

To other modules

Message Handler

Controller

Message Handler

Executive

Routines

Supervisor

User Interface

Fig. 5. Module Architecture

These primitives include send and receive of messages; run, suspend, continue, terminate and inquire the status of a process; post and retrieve information from a Blackboard; and assert and wait for events.

5.3. Module structure

In the following discussion, a *Module* refers to a ⟨master, slave⟩ pair of processes. The architecture of a typical Module in the distributed version of the Dolphin system is shown in Fig. 5. Each Module has the following subsystems:

Message Handler: Handles incoming and outgoing traffic of information, either to other modules or to the Control Blackboard.

Controller: This is a knowledge-based subsystem that knows about the capabilities of its associated Slave. It invokes the Slave when appropriate, gathers the necessary data to provide to it, oversees the execution and changes the scheduling status of the Slave when necessary, and handles the results provided by it.

Module Executive: The Module Executive is responsible for the activities of the Slave as such. On the basis of the incoming requests and the data

provided, it invokes the internal Routines and performs all the processing required to execute the Slave's task. The results are returned to the Master.

Module Supervisor: This subsystem allows the Slave Module to run in stand-alone mode. It permits independent testing of the module, to facilitate debugging and integration.

User Interface: Through this subsystem, the user can inspect and alter internal parameters of the Module.

5.4. The distributed sonar mapping system

In Fig. 6, a simplified view of the architecture of the distributed version of the Dolphin system is shown, with the major subsystems and the main flow of data. Instead of a sequential, cyclical activation sequence, we now have the various Modules responding in an asynchronous fashion to the various problem-solving needs as they arise. The bulk of the data is still passed directly between the modules themselves, since it consists of information relevant only to specific routines. While high-level status and control information is posted on the Blackboard, other data structures such as the Sonar Maps are maintained separately.

6. CONCLUSIONS

In this paper, a distributed control architecture for an autonomous mobile robot was presented. We briefly described the various levels of processing needed for a mobile robot, presented a Distributed Control System

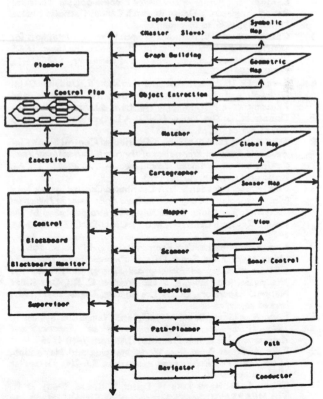

Fig. 6. A Distributed Implementation of the Dolphin Sonar-Based Mapping and Navigation System

where independent processes communicate through messages and store relevant information in a Blackboard, and showed a distributed implementation of a sonar-based mapping and navigation system.

We are currently finishing the implementation of the distributed version of the system, and hope to gain further insights from actual tests of the Distributed Control System.

ACKNOWLEDGEMENTS

I would like to thank Professors Hans P. Moravec and Sarosh N. Talukdar for their interest and support. Eleri Cardozo for providing access to his Distributed Problem-Solving Kernel and for many useful discussions, and Gregg W. Podnar for assistance with the Neptune robot.

REFERENCES

1 Aho, A. V., Hopcroft, J. E. and Ullman, J. D. *The Design and Analysis of Computer Algorithms*, Addison-Wesley, Reading, MA, 1974

2 Ballard, D. H. Model-Directed Detection of Ribs in Chest Radiographs in *Proceedings of the Fourth IJCPR*, Kyoto, Japan, 1978

3 Balzer, R., Erman, L. D., London, P. and Williams, C. Hearsay-III: A Domain-Independent Framework for Expert Systems in *Proceedings of the First National Conference on Artificial Intelligence*, AAAI, Stanford, CA, August 1980

4 Cardozo, E. *A Kernel for Distributed Problem-Solving*, Technical Report, Engineering Design Research Center, Carnegie-Mellon University, 1986, to be published

5 Chandrasekaran, B. Natural and Social System Metaphors for Distributed Problem Solving: Introduction to the Issue, *IEEE Transactions on Systems, Man and Cybernetics*, SMC-11(1), January 1981

6 Corkill, D. D., Lesser, V. R. and Hudlicka, E. Unifying Data-Directed and Goal-Directed Control: An Example and Experiments in *Proceedings of the AAAI-82*, August 1982

7 Donner, M. D. The Design of OWL: A Language for Walking in *SIGPLAN 83*, ACM, July 1983

8 Elfes, A. and Talukdar, S. N. A Distributed Control System for the CMU Rover in *Proceedings of the Ninth International Joint Conference on Artificial Intelligence – IJCAI-83*, IJCAI, Karlsruhe, German, August 1983

9 Elfes, A. and Talukdar, S. N. A Distributed Control System for a Mobile Robot in *I. Congresso Nacional de Automação Industrial – CONAI*, São Paulo, Brazil, July 1983, also published by the Design Research Center – CMU as CMU-DRC-TR-18-65-83, 1983

10 Elfes, A. Multiple Levels of Representation and Problem-Solving Using Maps From Sonar Data in *Proceedings of the DOE/CESAR Workshop on Planning and Sensing for Autonomous Navigation*, (eds Howe, E. S. and Weisbin, C. R.), Oak Ridge National Laboratory, UCLA, Los Angeles, August 18–19 1985, invited paper

11 Elfes, A. A Sonar-Based Mapping and Navigation System in *1986 IEEE International Conference on Robotics and Automation*, IEEE, San Francisco, CA, April 7–10 1986

12 Elfes, A. Sonar-Based Real World Mapping and Navigation, *IEEE Journal of Robotics and Automation*, RA-2(4), December 1986, to appear

13 Erman, L. D., Hayes-Roth, F., Lesser, V. R. and Reddy, D. R. The HEARSAY-II Speech Understanding System: Integrating Knowledge to Resolve Uncertainty, *Computing Surveys*, June 1980, 12(2), 213–253

14 Erman, L., London, P. and Fickas, S. The Design and an Example Use of Hearsay-III in *Proceedings of the Seventh International Joint Conference on Artificial Intelligence*, IJCAI, Vancouver, Canada, August 1981

15 Fehling, M. and Erman, L. Report on the Third Annual Workshop on Distributed Artificial Intelligence, *SIGART Newsletter* (84), April 1984

16 Feldman, J. A. High-Level Programming for Distributed Computing, *CACM*, July 1979, 22(6), 363–368

17 Fikes, R. E. and Nilsson, N. J. STRIPS: A New Approach to the Application of Theorem Proving to Problem Solving, *Artificial Intelligence*, 1971, 2, 189–208

18 Floyd, R. W. The Paradigms of Programming, *Communications of the ACM*, August 1979, 22(8), 455–460

19 Forgy, C. L. *OPS5 User's Manual*, Technical Report CMU-CS-81-135, Computer Science Department, Carnegie-Mellon University, July 1981

20 Forgy, C. L. *The OPS83 Report*, Technical Report CMU-CS-84-133, Computer Science Department, CMU, May 1984

21 Hanson, A. R. and Riseman, E. M. Visions: A Computer System for Interpreting Scenes in *Computer Vision Systems*, (eds Hanson, A. R. and Riseman, E. M.), Academic Press, New York, 1978

22 Hayes-Roth, F., Waterman, D. A. and Lenat, D. B. (eds) *Building Expert Systems*, Addison-Wesley, Reading, Massachusetts, 1983

23 Hewitt, C. *Description and Theoretical Analysis (Using Schemata) of PLANNER*, Technical Report AI-TR-258, AI Lab, MIT, 1972

24 Hewitt, C. Viewing Control Structures as Patterns of Passing Messages, *Artificial Intelligence Journal* 8, June 1977

25 Kanade, T. and Thorpe, C. E. *CMU Strategic Computing Vision Project Report: 1984 to 1985*, Technical Report CMU-RI-TR-86-2, The Robotics Institute, CMU, November 1985

26 Kornfeld, W. A. and Hewitt, C. E. The Scientific Community Metaphor, *IEEE Transactions on Systems, Man and Cybernetics*, SMC-11(1), January 1981

27 Leão, L. A. V. *COPS – A Concurrent Production System*, Technical Report, The Design Research Center, CMU, May 1986

28 Lesser, V. R. and Corkill, D. D. Functionally Accurate, Cooperative Distributed Systems, *IEEE Transactions on Systems, Man and Cybernetics*, SMC-11(1):81–96, January 1981

29 Lesser, V. R. and Corkill, D. D. The Distributed Vehicle Monitoring Testbed, *The AI Magazine*, Fall, 1983

30 Moravec, H. P. et al. Towards Autonomous Vehicles, *1985 Robotics Research Review*, The Robotics Institute, Carnegie-Mellon University, Pittsburgh, PA, 1985

31 Moravec, H. P. and Elfes, A. High Resolution Maps from Wide Angle Sonar in *International Conference on Robotics and Automation*, IEEE, March 1985

32 Podnar, G. W., Blackwell, M. K. and Dowling, K. *A Functional Vehicle for Autonomous Mobile Robot Research*, Technical Report, CMU Robotics Institute, April 1984

33 Sacerdoti, E. D. *A Structure for Plans and Behavior*, Elsevier, New York, NY, 1977

34 Smith, R. G. and Davis, R. Frameworks for Cooperation in Distributed Problem-Solving, *IEEE Transactions on Systems, Man and Cybernetics*, SMC-11(1), January 1981

35 Talukdar, S. N., Pyo, S. S. and Elfes, A. *Distributed Processing for CAD – Some Algorithmic Issues*, Technical Report, The Design Research Center – CMU, 1983

36 Talukdar, S. N., Pyo, S. S. and Giras, T. G. Asynchronous Procedures for Parallel Processing in *Power Industry Computer Applications Conference*, 1983, to appear in IEEE Transactions on PAS

37 Talukdar, S. N., Elfes, A., Tyle, N. and Vidovic, N. Blackboards – A Means for Integrating Expert Systems with Other Computer Aids in *Proceedings of the 1984 Power System Society Conference*, Power System Society, IEEE, 1984

38 Waterman, D. A. and Hayes-Roth, F. (eds) *Pattern-Directed Inference Systems*, Academic Press, New York, 1978

39 Wesson, R., Hayes-Roth, F., Burge, J. W., Stasz, C. and Sunshine, C. A. Network Structures for Distributed Situation Assessment, *IEEE Transactions on Systems, Man and Cybernetics*, SMC-11(1), January 1981

Control of Mobile Robot Actions

Fabrice R. Noreils and Raja G. Chatila
L.A.A.S- CNRS
Robotics And Artificial Intelligence Group
7, Avenue du Colonel Roche
31077 Toulouse-Cedex FRANCE

Abstract

This paper deals with the control architecture of mobile robots. It is based on three types of entities: *Modules* that accomplish basic computations on data from sensors (resp. commands to effectors); *Processes*, in charge to establish *dynamic links* between perception and action to achieve closed-loop behaviors, and *Functionnal Units* providing specific functionalities. These modules are organized hierarchically, and are considered to be the robot's resources. We introduce and discuss a generic control system structure, based on a decomposition of its functions mainly into an *executive* managing robot resources, and a *surveillance manager* for detecting and reacting to asynchronous events. The control system enables the robot to execute missions (plans) expressed in a command language. Several running examples are given.

1 Introduction

In order to achieve a high degree of robustness and autonomy, a mobile robot should be able to plan and execute actions, and adapt its behavior to environment changes. The variety of environment conditions emphasizes the importance of multisensory perception on the one hand, and of robot reactivity at the various representation levels of its internal knowledge on the other hand.

We present in this paper an approach to the problem of action control for mobile robots, and the underlying software architecture. The hardware architecture on which this system is implemented is a general purpose multiprocessor system.

The control system manages the robot's resources (modules, sensors, effectors, processes and functional units) to enable it to execute a task given by a planner or a user, while providing reactivity to external stimuli. This includes the detection of, and reaction to, asynchronous events by adapting the robot's behavior so that the final goals are reached, and if not, the known reasons of failure reported. The software architecture should be robust and enable dynamic interactions between the robot's modules. It should also be flexible in order to integrate new functions, and be easy to implement and debbug [5].

Our approach to build a generic control structure meeting these requirements is to consider that the robot possesses functionalities achieved by related modules that can be defined and extended as the robot system grows more complex (section 3). The control structure does not merely integrate them "statically", but also provides for a mechanism for achieving programmable dynamic links between some of them to produce given behaviors. This structure (section 4) is composed of a *supervisor* that handles the global task sending each action to an *executive* that is in a way similar to an operating system, a *surveillance manager* for detecting some given events or situations and adapting the robot's behavior accordingly, and a *diagnostic* module for assessing the situations when an error occurs, and mending the on-going plan. Examples (section 6) showing the behavior of this system will be detailed. A command language for describing tasks (section 5), and corresponding to the robot's possible actions, is also introduced. Before detailing this architecture, we briefly overview some existing systems in the following section.

2 Related Approaches

For the A.L.V project [26], an architecture called the "Driving Pipeline" [14] that integrates multiperception and provides continuous motion was developed. The implementation is based on a "WhiteBoard" [23], derived from the BlackBoard paradigm, but wherein modules can asynchronously access the database by a token exchange mechanism. There are dedicated modules (Pilot, Helm, Obstacle Avoidance and Color Vision) with intrinsic synchronization through messages exchange. Elfes [11] also presents an architecture based on a Blackboard paradigm and proposes a plan formalism (which handles parallelism). These are centralized systems (information contained in a central database); modules must have low granularity due to the centralized communication, and low communication bandwidth. Other authors propose a distributed approach. Brooks [5,13] claims for a behavioral decomposition and builds an architecture in layers of control corresponding to levels of behaviors to ensure reactivity. A given layer includes a number of modules with low bandwidth communication. A low level can be "subsumed" by a higher one. This is done by inhibition of the layer's inputs/outputs and timeout. The control is entirely wired, so it is irrevocable. An important issue is conflict resolution between behaviors [3,8]. This is a research area which has not been sufficiently studied and a necessary step in the development of the behavioral theory. Payton [22] introduces *Virtual sensors* to recognize features in the environment and send regularly their data to *reflexive behaviors* which in turn send commands (currently these are commands to motors) on a blackboard. A control module, using a priority arbitration, selects one message which is fed into vehicle actuators. Arkin [2] develops *motor and perceptual schemas*. A schema can be viewed as a computing agent. For each motor schema activated, perceptual schemas are created to detect events. If such an event occurs, a new motor schema is instanciated and enters in concurrency with the others. To ensure a high degree of concurrency, schemas communicate through a blackboard. Kaelbling [15] also uses behaviors but there are no connections between them. Behavior management is done by *mediators* which may inhibit behavior outputs

according to internal conditions. Firby [12] also builds a reactive system but he uses a more traditional Artificial Intelligence approach to do it. He introduces *reactive action packages* (R.A.P) which are programs running until either the goal is reached or the R.A.P fails. Activation and conflict resolution are done by the *R.A.P interpreter*. Meystel [19] develops a strongly hierarchical architecture including rule-based systems. Kanayama [16] uses a decomposition of tasks in processes and develops an environment named RCS based on a multitasking paradigm.

3 Robot System Architecture

We describe now the robot architecture that the control system as such will manage. The atomic component of the robot system is called *module*. Figure 1 shows the modules integrated to date. There exist several kinds of modules.

EH0342-6/91/0000/0145$01.00 © 1989 IEEE

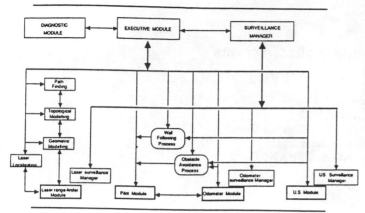

Figure 1: General Organization of The System

a) *Sensor* and *effector* modules that process sensor outputs or effector commands. These modules are layered according to a hierarchical internal representation of knowledge corresponding to levels of interpretation and abstraction, and therefore to levels of action of the robot.

b) Some modules, called *functional units*, provide a given specific function (*e.g.,* robot localization selecting and/or combining the available means to do it: vision, odometry, laser ranging, *etc.*)

c) A third kind of module implements the notion of *process* that provides, in general, for a dynamic closed-loop between perception and action, possibly using functional units.

3.1 Sensor and Effector Modules

The data acquired by a given sensor is processed by a sensor module and made available to the following higher-level ones for their own processings and interpretations. Some modules in this chain may have their input from different sensors. A similar hierarchy applies for the effectors, but here, the commands are interpreted downward. A sensor module achieves two main functions:

- Data processing: A series of processings are performed on incoming data to obtain different levels of interpretation that will then be used by the higher levels of our architecture.

- Reactivity to events: a level of interpretation corresponds to the extraction of a numerical or symbolic feature defined by a collection of parameters. The control system can instanciate a set of conditions which are applied to these parameters. Whenever the conditions are verified the control system is informed. These conditions are the left-hand-side member of what will be referred to as *surveillance monitors* discussed later in the paper (section 4.1.1).

As an example, let us consider a laser range finder on a pan scanning platform. First, each measurement ρ is associated with the sensor's orientation θ. Depth information is then extracted (from sensor calibration) providing a local position (x, y) in the frame related to the robot. A local map is built by linear approximations of depth points. This constitutes a chain of laser sensor modules. Successive local models can be integrated in a global map suited for navigation [6].
Other sensor modules are the Odometer and Ultrasonic Modules. The Pilot Module is an example of effector module. It provides a variety of trajectories that the robot can follow (such as lines, clothoids ...) or a tracking mode. In this mode the robot tracks a target (in its local frame) supplied by another module.

3.2 Functional Units

The information provided by a given level of interpretation can be used for a specific purpose by a dedicated module called functional unit. It is activated upon an explicit call and execute some computations, possibly making use of other modules. The algorithms included in a functional unit may be very complex. For example, robot localization is a functional unit. This unit may use different resources to provide robot position, for example a laser range-finder, or vision, by matching a local perception with a global model, or odometry, or a combination thereof. It also may compute robot position by deducing it from other knowledge (*e.g.,* the robot is docked to a known workstation). Robot localization can be given as a set of coordinates (and associated uncertainties), as well as a symbolic expression (*e.g.,* (IN_ROOM ROBOT R3)).

3.3 Processes

When we want to realize a *dynamic link* between perception and action, an entity called *process* is activated.

A process can be represented by a finite state automaton. Its inputs are data from sensor modules (regardless of the interpretation level), and its outputs are commands toward an effector. It has a control line required for:

- activating or halting the process by the control system. In our definition, a process is never generated dynamically but is continuously present in the system. Along with its activation, the control system can send some information (parameters) needed by the process for local decisions;

- Reporting process failures. If the input data requested from a sensor module is not sufficient to enable the process to accomplish its task, and if this situation occurs repeatedly, the process reports to the control system. Besides this case, a process does not stop by itself.

Instances of processes are Docking (to a workstation for example), and Local Obstacle Avoidance. An application of such processes with a detailed description of sensor/effector modules is presented in [21].

4 The Control System

We presented the elements that the robot control system will make use of for actually operating the robot. We shall now describe in detail the control system (or controller) itself. In our approach, its structure has both centralized and distributed components.
The controller has four components: **Supervisor (SV), Executive Module (EM), Surveillance Manager (SM), and Diagnostic Module (DM)** (figure 1). Currently, the Supervisor and Diagnostic Module are not implemented, and their part is assumed by the Executive or by default decisions, as we shall see later. We will thus focus an the SM and EM, and firstly, we present the surveillance system that the SM controls. In former papers [6,25], this structure was presented in a first implementation wherein the functions of all these modules were achieved by a single one, called *decisional kernel*, that sets distributed surveillance monitors and is informed when they are triggered to decide for further actions after the first reflex reaction. We have refined this structure considerably.

4.1 The Surveillance Monitoring System

4.1.1 Surveillance monitors

Surveillance monitors play a key role in mission execution, and in robot operation. Their purpose is to set conditions on sensed data or computed results, and to trigger some immediate (reflex) reaction when these conditions are met. Formally, a surveillance monitor is equivalent to a classical rule [25]:

$$\mathbf{MNTR} \ \text{<conditions>} \Rightarrow \text{<actions>}$$

Where <conditions>contains conditions on sensors or robot state variables, and <actions>contains the set of reactive actions to carry when the left-hand side conditions are true. For example:

$$(\text{S\_zone IN} <\text{prmts}>) \Rightarrow (\text{set-speed } v)$$

is a surveillance monitor of type zone; if the robot enters in an area defined by a sequence of points (prmts), its linear speed is changed to v. In section 5, a more complete definition of surveillance monitors is given.

We define two main kinds of surveillance monitors: P-surveillances related to the on-going plan or mission, and provided along with it to detect some events or situation and cope with them, and O-surveillances related to the operation of the robot, even if no task is being executed (*e.g.*, battery power level).

All monitors are set by the Surveillance Manager that is informed when they are fired. However, for each sensor module, there is also a local surveillance manager.

4.1.2 Sensors and Surveillance Monitors

Conditions can be set at any level of data processing in a chain of sensor modules on their results in order to guarantee an immediate reaction when the events that produced them occur. On the same sensor module, several surveillance monitors can be active at the same time and are controlled by a *Sensor Surveillance Manager* (SSM). Its role is shown in the following example.

Several conditions may be monitored for instance by the ultrasound module (Figure 2) that performs a number of computation at different levels of representation. Conditions that can be monitored are, for example, related to the detection of a given pattern in the local environment model (such as <corner>or <edge>), specified with the appropriate parameters. The pattern may be defined by a logical condition on several ultrasonic sensors such as $(\text{AND} (\text{sensor}_i \ S_i)(\text{sensor}_j \ S_j))$, where S_i and S_j are values of sensor measurements. This actually corresponds to as many surveillance monitors. The SSM detects the simultaneous triggering of all the monitors in order to validate the global one, and then sends the corresponding message to the SM.

4.1.3 The Surveillance Manager

At the top level of the Surveillance system is the *Surveillance Manager*. Its function is twofold:

- It receives surveillance monitors from the Executive and decomposes the left-hand side member, sending the various components to the adequate SSMs.

- It receives the messages from the SSMs indicating that a surveillance has been triggered. If the left-hand side of a given monitor included a logical condition on different sensors, the SM verifies the logical formula. Then the reflex actions corresponding to the triggered monitors are executed through the Executive.

  If different surveillance monitors are triggered (almost) simultaneously, a *priority order* is used to decide what action to perform first. This solution is rather limited, and we are developing a new approach to this problem based on a precomputed decision tree.

4.2 The Executive Module

This module is in fact similar to an *Operating System*. The Executive Module (EM) receives the commands (requests, missions) through the user interface. Missions may include macro-commands that correspond to predefined procedures, and are interpreted by the executive by decoding a stored script. This first function will be in further implementations carried out by a separate module (the Supervisor) that sends the tasks to the executive each step at a time, leaving to the executive itself the task of managing robot resources (Sensor/effector modules, Functional Units, processes) to achieve the mission.

If an action is executed correctly, the executive launches the next one. Otherwise, in case of failure, the system is in one of the following situations:

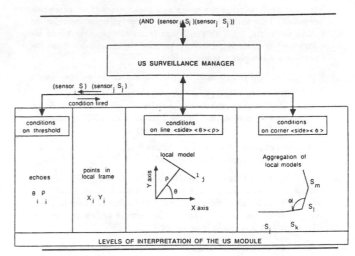

Figure 2: Levels of Interpretation and conditions of the U.S module

a) Process failure: A running process does not receive the data necessary to its computations because the sensor does not detect anything (*e.g.*, an obstacle avoidance has terminated, or the obstacle moved away), or because it is not adapted to changing external (environment) conditions (*e.g.*, insufficient lighting for vision).

b) Sensor/effector module failure: a module is unable to answer to its request, mainly because of physical malfunction (*e.g.*, battery power level, blocked wheel).

c) P-Surveillance triggering: In this case, an asynchronous event occured, and the right-hand-side of the surveillance monitor redirects the executive's action as programmed in the mission itself. The surveillance might be related to a running process, and in this case the process is halted.

Cases a) and b) might result in triggering related surveillance monitors (O-surveillances in general). All cases are detected because appropriate P-surveillances or O-surveillances were triggered. For instance, in case a), luminosity level is monitored, and if insufficient while a vision-based tracking process is active, the process is halted and a ultrasound-based process is activated instead [21]. After the execution of the reflex action, the Executive calls the Diagnostic module, providing it with the information from the failing module for further possible recovery action. The role of the diagnostic module is to assess a failure situation if necessary, after the first reflex reactions, and to provide a "local" plan to recover from it to resume the planned mission if possible. In order to be able to carry out its task, this module may require some data from sensor modules for example, through the executive. After diagnosis, error recovery may consist in executing a stored procedure. This module is not implemented yet as such, and will be composed of a compiled rule-based system. Currently, the executive system itself includes this function.

5 Mission Formalism

A mobile robot executes plans or missions provided by a planner or a human user. It is important to define the formalism in which the plan is expressed, and the information that it includes to enable the control system to execute it while complying with varying conditions. Our longer term goal is to define a command language.

The basic element in this formalism is a *command*. A command has two fields:

- The first field specifies the name and parameters of an *execution operator*. This is an order sent to sensor/effector modules or functional units, such as read-pos (read the position given by odometer) or smooth-move() (motion command to Pilot for excuting a path given as argument).

- The second field is optional. It specifies one or more surveillance monitors.

Before giving the formal command syntax, let us give some details on the role of surveillance monitors during plan execution. The monitors can fall in three different categories to this respect:

Terminal Surveillance Monitors : When triggered, the current action (execution operator) is stopped.

Non Terminal Surveillance Monitors : When triggered, they do not interfere with the current action, but start another one concurrently, provided it does not share the same resources. However, these monitors are linked to the current action and are inhibited or killed when it is achieved.

Static Surveillance Monitors : When a plan fails, the Executive cancels the current action and the related surveillance monitors, and calls the diagnostic module. However, it might be necessary that some monitors remain active, or be reactivated. We call them *static* monitors.

A command's syntax is then the following:

$$< cmd > \quad ::= \quad (< operator > \{< surv >\}^*)$$
$$< surv > \quad ::= \quad \text{MNTR [STATIC]} < condition >$$
$$\Rightarrow < \text{reflex-action}>\{<\text{control-action}>\}$$

The reflex-action field is the immediate reaction to the event, launched by the Surveillance Manager. The last field {<control-action>} specifies the reaction through the Executive module after a surveillance is triggered and the reflex-action executed. The control-action may include a decision concerning the continuation or not of the mission -see examples below-).

A mission is then defined as follows:

$$<mission> \quad ::= \quad \text{(MISSION <mission-name>(}$$
$$\text{[(PRE: <precondition-list>)]}$$
$$\text{[(ENV: <environment-list>)]}$$
$$\text{[(LSURV: <surveillance-list >)]}$$
$$\text{(MAIN: <body >)))}$$
$$<body > \quad ::= \quad <mission>$$
$$| \quad <cmd>$$

where:

- <precondition-list>is the set of conditions that must be true for executing the mission.

- <environment-list>sets the execution environment or model by specifying some indications to the controller for error recovery and local decisions.

- <surveillance-list>is the set of surveillance monitors that have to be active during mission execution.

- <body>is the sequence of commands to be executed. Sub-plans (or macros, or elementary tasks) can be included. Control structures (*e.g.*, loops) may also be used.

The control system will execute the mission step by step, but should be able to stop before the end of an action in some cases. For this purpose, *key-words* appear in the *control-action* field of the surveillance monitors. Currently these key-words may have two possible values:

NEXT : means that, the current action being stopped, its related monitors are killed and the next action executed.

FAIL : the whole mission is stopped and the diagnostic module is called.

A great flexibility is thus given to the control system to modify the programmed course of a plan according to new perceived environment or robot conditions. **The control system not only has the capacity of executing plans, but also provides with a programmable reactivity.** This is actually the central issue in our approach.

Other systems in the litterature include sensor perceptions in plans. Doyle [10] presents a system (GRIPE) which inserts perception requests as well as the expected results in order to check that an action has been accuratly executed. For navigation in a known environment (defined as a set of labelled areas) Miller [20] develops a language which includes closed-loops on sensor data as well as actions on effectors. Chochon and Alami [7] describe an on-line system (NNS) for executing assembly tasks that include an execution model specifying alternatives based on sensor conditions. Gini and Smith [24] propose a system which expands a program also inserting conditions to check before executing the next action.

6   Application And Examples

Before presenting several examples to illustrate the control system operation, we describe the current implementation status on our mobile robot HILARE.

6.1   Implementation

HILARE has two independant driving wheels and a free wheel. Currently, its on-board sensors are:

- A ultrasonic sensor belt (16 sensors) [4],

- A Laser range-finder on a pan platform.

- An odometry system

- A trinocular stereovision system [9] (computations made on a ground workstation).

The Pilot Module [17] is the only effector module. Figure 1 shows the system's architecture, without detailing the sensor/effector modules (detailed in [21]).

The implementation is partly on-board, and partly on a SUN workstation (figure 3). The robot is radio-linked to the SUN.

6.1.1   On-Board Implementation

Sensor modules for the ultrasonics, odometer, and laser range-finder, as well as the effector module for motor control (Pilot module) are on-board. Two processes have also been implemented associating the ultrasonics and pilot modules: Docking (wall following) and Obstacle Avoidance.

The SSMs are not implemented yet at the sensor level, and the SM actually assumes their function. Therefore, the surveillance conditions can only concern the interpretation levels.

All programs are written in PASCAL or PLM86 and implemented on five INTEL 80286 and 8086s on MULTIBUS I (this configuration is soon to be changed to 68020s on VME).

6.1.2   Off-Board Implementation

The man/machine and graphics interface and part of the control system, which is currently in its development phase, are implemented on a SUN workstation. Eventually, the controller will be completely on-board.

The off-board system is constituted by:

- A parser (written with YACC/LEX) which parses the commands and generates a new structure used by the Executive Module.

- The Executive Module.

- The User and graphics interface.

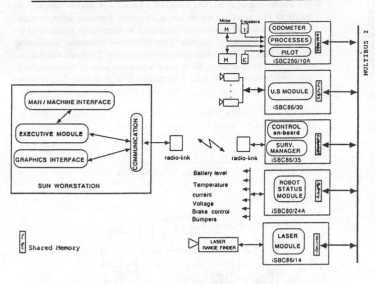

Figure 3: On-Board/ Off-Board Implementation

The basic function of the user interface is to send commands or requests to the Executive system and keap a history record of all previous commands. The graphics interface is used to display robot movements and sensor readings. These tasks communicate using a system developed at LAAS based on sockets, also enabling remote inter-process communication [1].

6.2 Examples of Mission Execution

We show on a first simple example the mission formalism and the system's behavior, and then an application of the approach in a more complex situation (example 2).

6.2.1 Example 1: Wall Following

Let us consider a robot moving in a room. We would like to reach a given location, near a wall, by navigating in the room to the wall and then change navigation mode automatically to wall following until the goal is reached. Goal coordinates are known (figure 4). This task is expressed as follows:

(MISSION mission1
(PRE: (eq motors ON))
(MAIN:
    (smooth-move(...) MNTR(S_us-perception line (ρ, θ))
                $\Rightarrow$ ((stop)){(NEXT)})
    (RUN (wall-following (right-side 25))
        MNTR(S_straight-line (x, y, ϕ)) $\Rightarrow$ ((stop)){(NEXT)})
)
)

Let us examin how the control system handles this mission. Figure 5 shows the exchanges between modules for the first command; messages are numbered as we reference them here for the sake of clarity. The executive module sends the first command (smooth-move) to the pilot module (1). Smooth-move is a command whose argument is a sequence of points (a broken line) along which the robot has to move, smoothing orientation changes by clothoid curves. When SM receives the surveillance monitor associated to the operator (2) it returns on the one hand an identifier to the executive (<n1>in (3)) and on the other hand sends the left member of the surveillance labelled by the identifier to the Ultrasonic SSM (4). The control-action field {(NEXT)} is memorized by the executive and referenced by the identifier sent by SM. During the movement, if the condition of the surveillance is triggered

(5) (detection of a line (wall) with the given parameters -see figure 2), US-SSM emits a message to SM indicating that the condition identified by <n1>is true. SM executes the right member of the surveillance (6-1) and informs the executive (6-2). The Pilot module which receives the stop order, informs the executive that the movement has been interrupted (7). Then, the executive performs the control-action which is NEXT, meaning that the next command should be carried out. After destroying the previous monitor, this step is executed. It consists in running the *process* wall-following on the right side and at 25 cm distance from the wall, associated to a surveillance monitor for ending it when the robot crosses a straight-line defined by the parameters (x, y, ϕ) defining the goal. When this happens, the robot stops, and the NEXT control-action exits from the mission.

Figure 4 shows the robot executing this mission on the graphics interface. The command to provide this display is (show-robot on 50), specifying the refreshing period to the pilot for sending robot position. The ultrasonic echoes, along with the sensor identifier (between 1 and 16) are also shown.

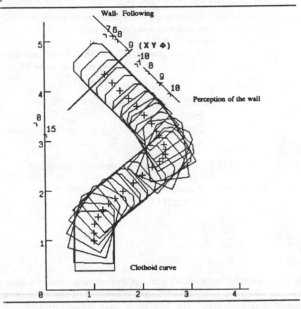

Figure 4: Wall Detection and Following

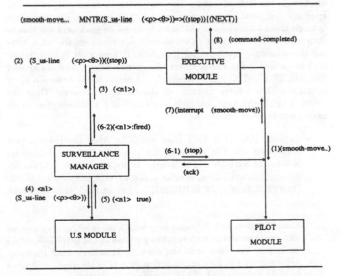

Figure 5: Flow of messages between modules

6.2.2 Exemple 2

In this example, the robot reports its ultrasonic sensor readings while passing by an object expected to be along its computed trajectory inside a specified zone. Furthermore, if an unexpected obstacle lies on the path in this zone, the mission terminates with a failure.

```
(define Tclist[4] ZONE109 ((150 500)(150 700)(350 700)(350 500)))
(define Tclist[4] sensors-list ((1 40)(2 50)(3 50)(4 40)))
(define Tpath PATH ((180 520)(350 520)(360 320 2700)))
...
...
(MISSION mission2
(PRE: (eq motor ON)(not (eq robot-in-zone $ZONE109)))
(ENV: (set LOA T))
(MAIN:
  (smooth-move(PATH)
            MNTR(S_zone IN ZONE109) ⇒ ((set-speed 20)){
              (set robot-in-zone ZONE109)
              (set LOA NIL)
              (MNTR(S_zone OUT ZONE109) ⇒ ((set-speed 50)){
                              (set robot-in-zone NIL)
                              (set LOA T)
              }
              (WHILE (robot-in-zone)
                (read-all-us-data)
                (draw %G_pos%G_echoes r-position r-us-data)
              )
            }
            MNTR STATIC (S_us-threshold sensors-list) ⇒ ((stop)) {
              (IF (LOA)
                ((avoid-obstacle)(resume-order))
                ((FAIL))
              )
            }
  )
)
```

In this example, the mission includes nested structures, variable definitions with type (e.g., define Tclist ZONE109), a control loop, and a conditional statement. The first three lines define variables that will be used later in the mission. The precondition field specifies that the mission can only be executed if the motors are on, and the robot not already in the defined zone. The execution environment, or model, specifies that local obstacle avoidance (LOA) is active, i.e., the robot is to go around unexpected obstacles lying on its path, and not just to stop when detecting them. Two surveillance monitors are active during the mission. The first one (S1: MNTR(S_zone IN ZONE109) ⇒ ...) is verified if the robot enters the specified zone, and the second (S2: MNTR STATIC (S_us-threshold ...) reacts to the detection of an obstacle within a given distance of the ultrasonic sensors. Once the robot is inside the zone, obstacle avoidance is prohibited. Two cases may occur in these conditions:

- Suppose that S1 is fired (and not S2). After the reflex action set-speed the EM executes the corresponding control-action: it sets the variable robot-in-zone to its new value, inhibits local obstacle avoidance, creates a new surveillance monitor (S1-2: (MNTR(S_zone OUT ZONE109) ...) to be fired when the robot

  exits the zone, and defines a loop which reads ultrasonics sensors and sends results as well as robot position to the graphic interface as long as the robot is in the zone. If S2 is then fired, the executive carries out its control-action (the IF statement), and since LOA is desactivated, FAIL is executed and the overall mission aborts.

- If S2 is fired before S1, the elementary-task avoid-obstacle, which includes the avoidance process, is executed because LOA is activated. During its execution, if the robot enters the zone, (S1) is fired, and LOA is set to NIL. The executive stops the avoidance task because its execution environment (LOA T) is not valid anymore. At this moment, S2 is fired again because it is *static*, and its conditions are still true (the obstacle is still there), but now LOA is NIL and the FAIL is executed thus aborting the mission.

Figure 6 shows the robot avoiding an obstacle and returning echoes to the graphics interface while in the specified zone.

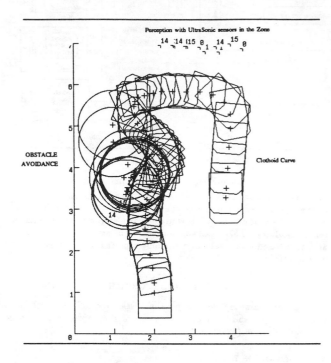

Figure 6: Obstacle Avoidance and U.S. data transmission in the specified zone

7 Future Work

We presented a control system architecture for mobile robots able to execute missions specified by a user, and having reactivity and error recovery capacities. This system uses the concept of process to achieve a controlled reactivity through the use of surveillance monitors providing a flexible control of these processes.

Future developements will concern mainly:

Integration of vision: This approach enables to easily integrate new sensors or modules to a mobile robot, the control system being able of using them according to the conditions specified in surveillance monitors for example. Vision for localization [18], object tracking and motion control in a structured environment (corridor) is currently being integrated.

Flexibility : Integration of vision will intensively make use of processes. Currently, their inputs and outputs are fixed. In order to reconfiger the various processes according to the context of use, their inputs/outputs will be typed, and a formalism to manipulate them developed.

Mission Formalism: Including control structures in the mission formalism will lead eventually to a language for programming mobile robots.

This approach provides the possibility for integrating new sensors and modules easily through the definition of their conditions of use. It will eventually make use of a language, of which elements were presented, useful to define "(re)programmable mobile robots" that could be reconfigured according to applications.

Acknowledgements: This work would not have been possible without the help of G. Bauzil, G. Vialaret and C. Lemaire for HILARE's hardware, and A. Khoumsi who implemented the Pilot Module.

References

[1] Rachid Alami. CPU Manuel d'Utilisation. Technical Note, L.A.A.S., Robotics Group, 1988.

[2] Ronald C. Arkin. Motor shema based navigation for a mobile robot: an approach to programming by behavior. In *IEEE, International Conference on Robotics and Automation, Raleigh*, pages 264 – 271, 1987.

[3] J. C. Aylett. Some speculations on the behavioural paradigm. Discussion Paper No 56, 1988. Department of Artificial Intelligence, University of Edinburgh.

[4] G. Bauzil, M. Briot, and P. Ribes. A navigation sub-system using ultrasonic sensors for the mobile robot HILARE. In *1st International Conference on Robot Vision and Sensory Control, Stratford-Upon-Avon*, April 1981.

[5] Rodney A. Brooks. A robust layered control system for a mobile robot. *IEEE journal of Robotics and Automation*, RA-2(1):14–23, 1986.

[6] Raja G. Chatila. Mobile robot navigation : space modeling and decisional processes. In *3rd ISRR, Gouvieux, France*, 1985.

[7] H. Chochon and R. Alami. NNS, a knowledge-based on-line system for an assembly workcell. In *IEEE, International Conference on Robotics and Automation, San Francisco*, pages 603 – 609, 1986.

[8] Peter W. Cudhea and Rodney A. Brooks. Coordinating multiple goals for a mobile robot. In *Intelligent Autonomous Systems*, pages 168 – 174, 1986.

[9] A. Robert de St Vincent. A 3d perception system for the mobile robot HILARE. In *IEEE, International Conference on Robotics and Automation, San Francisco*, 1986.

[10] Richard J. Doyle, David J. Atkinson, and Rajkumar S. Doshi. Generating perception requests and expectations to verify the execution of plans. In *5th national conference on artificial intelligence*, pages 81 – 88, 1986.

[11] Alberto Elfes and Sarosh N. Talukdar. A distributed control system for the C.M.U rover. In *8th IJCAI, Karlsruhe (FRG)*, 1983.

[12] R. J. Firby. An investigation into reactive planning in complex domains. In *6th national conference on artificial intelligence*, pages 202–206, 1987.

[13] Anita M. Flynn and Rodney A. Brooks. MIT mobile robots: what's next? In *IEEE, International Conference on Robotics and Automation, Philadelphie*, pages 611 – 617, 1988.

[14] Yoshimasa Goto, Steven A. Shafer, and Anthony Stentz. *The Driving Pipeline: A Driving Control Scheme For Mobile Robots*. Technical Report CMU-RI-TR-88-8, Robotics Institute, Carnegie Mellon University, 1988.

[15] Leslie P. Kaelbling. *An Architecture for Intelligent Reactive Systems*. Technical Report, Artificial Intelligence Center SRI International, April 1986.

[16] Y. Kanayama. Concurrent programming of intelligent robots. In *8th IJCAI, Karlsruhe (FRG)*, pages 834 – 838, 1983.

[17] Ahmed Khoumsi. Pilotage, asservissement sensoriel et localisation d'un robot mobile autonome. Doctorat de l' Université Paul Sabatier, June 1988.

[18] Eric P. Krotkov. *Mobile Robot Localization using a Single Image*. Technical Report 88125, L.A.A.S., May 1988.

[19] A. Meystel. Planning in a hierarchical nested autonomous control system. In *SPIE vol. 727 Mobile Robots*, pages 42 – 76, 1986.

[20] David P. Miller. *Planning by Search Through Simulations*. PhD thesis, Yale University, Department of Computer Science, October 1985.

[21] Fabrice R. Noreils, A. Khoumsi, G. Bauzil, and R. Chatila. Reactive processes for mobile robot control. To appear in International Conference on Advanced Robotics (ICAR), 1989.

[22] David W. Payton. An architecture for reflexive autonomous vehicle control. In *IEEE, International Conference on Robotics and Automation, San Francisco*, pages 1838 – 1845, 1986.

[23] S. A. Shafer, A. Stentz, and C. E. Thorpe. *An Architecture for Sensor Fusion in a Mobile Robot*. Technical Report CMU-RI-TR-86-9, Robotics Institute, 1986.

[24] Richard E. Smith and Maria Gini. Reliable real-time robot operation employing intelligent forward recovery. *Journal of Robotic Systems*, 3(3):281– 300, 1986.

[25] Ralph P. Sobek and Raja G. Chatila. Integrated planning and execution control for an autonomous mobile robot. *The Int. Journal for Artificial Intelligence in Engineering*, 3, April 1988.

[26] Charles Thorpe and Takeo Kanade. *1987 Year End Report for Road Following at Carnegie Mellon*. Technical Report CMU-RI-TI-88-4, Robotics Institute, Carnegie Mellon University, 1988.

A Robust Layered Control System
For A Mobile Robot

RODNEY A. BROOKS, MEMBER, IEEE

Abstract—A new architecture for controlling mobile robots is described. Layers of control system are built to let the robot operate at increasing levels of competence. Layers are made up of asynchronous modules that communicate over low-bandwidth channels. Each module is an instance of a fairly simple computational machine. Higher-level layers can subsume the roles of lower levels by suppressing their outputs. However, lower levels continue to function as higher levels are added. The result is a robust and flexible robot control system. The system has been used to control a mobile robot wandering around unconstrained laboratory areas and computer machine rooms. Eventually it is intended to control a robot that wanders the office areas of our laboratory, building maps of its surroundings using an onboard arm to perform simple tasks.

I. INTRODUCTION

A CONTROL SYSTEM for a completely autonomous mobile robot must perform many complex information processing tasks in real time. It operates in an environment where the boundary conditions (viewing the instantaneous control problem in a classical control theory formulation) are changing rapidly. In fact the determination of those boundary conditions is done over very noisy channels since there is no straightforward mapping between sensors (e.g. TV cameras) and the form required of the boundary conditions.

The usual approach to building control systems for such robots is to decompose the problem into a series (roughly) of functional units as illustrated by a series of vertical slices in Fig. 1. After analyzing the computational requirements for a mobile robot we have decided to use *task-achieving behaviors* as our primary decomposition of the problem. This is illustrated by a series of horizontal slices in Fig. 2. As with a functional decomposition, we implement each slice explicitly then tie them all together to form a robot control system. Our new decomposition leads to a radically different architecture for mobile robot control systems, with radically different implementation strategies plausible at the hardware level, and with a large number of advantages concerning robustness, buildability and testability.

Manuscript revised February 3, 1986. This work was supported in part by an IBM Faculty Development Award, in part by a grant from the Systems Development Foundation, in part by an equipment grant from Motorola, and in part by the Advanced Research Projects Agency under Office of Naval Research contracts N00014-80-C-0505 and N00014-82-K-0334.

The author is with the Artificial Intelligence Laboratory, Massachusetts Institute of Technology, 545 Technology Square, Cambridge, MA 02139, USA.

IEEE Log Number 8608069.

Fig. 1. Traditional decomposition of a mobile robot control system into functional modules.

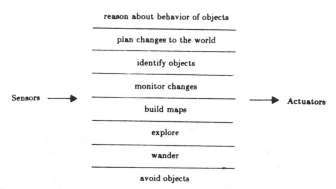

Fig. 2. Decomposition of a mobile robot control system based on task-achieving behaviors.

A. Requirements

We can identify a number of requirements of a control system for an intelligent autonomous mobile robot. They each put constraints on possible control systems that we may employ. They are identified as follows.

Multiple Goals: Often the robot will have multiple goals, some conflicting, which it is trying to achieve. It may be trying to reach a certain point ahead of it while avoiding local obstacles. It may be trying to reach a certain place in minimal time while conserving power reserves. Often the relative importance of goals will be context-dependent. Getting off the railroad tracks when a train is heard becomes much more important than inspecting the last ten track ties of the current track section. The control system must be responsive to high priority goals, while still servicing necessary "low-level" goals (e.g., in getting off the railroad tracks, it is still important that the robot maintains its balance so it doesn't fall down).

Multiple Sensors: The robot will most likely have multiple sensors (e.g., TV cameras, encoders on steering and drive mechanisms, infrared beacon detectors, an inertial navigation system, acoustic rangefinders, infrared rangefinders, access to a global positioning satellite system, etc.). All sensors have an error component in their readings. Furthermore, often there is no direct analytic mapping from sensor values to desired physical quantities. Some of the sensors will overlap in the physical quantities they measure. They will often give inconsistent readings—sometimes due to normal sensor error and sometimes due to the measurement conditions being such that the sensor (and subsequent processing) is used outside its domain of applicability. Often there will be no analytic characterization of the domain of applicability (e.g. under what precise conditions does the Sobel operator return valid edges?). The robot must make decisions under these conditions.

Robustness: The robot ought to be robust. When some sensors fail it should be able to adapt and cope by relying on those still functional. When the environment changes drastically it should be able to still achieve some modicum of sensible behavior, rather then sit in shock or wander aimlessly and irrationally around. Ideally it should also continue to function well when there are faults in parts of its processor(s).

Extensibility: As more sensors and capabilities are added to a robot it needs more processing power; otherwise, the original capabilities of the robot will be impaired relative to the flow of time.

B. Other Approaches

Multiple Goals: Elfes and Talukdar [4] designed a control language for Moravec's robot [11], which tried to accommodate multiple goals. It mainly achieved this by letting the user explicitly code for parallelism and to code an exception path to a special handler for each plausible case of unexpected conditions.

Multiple Sensors: Flynn [5] explicitly investigated the use of multiple sensors, with complementary characteristics (sonar is wide angle but reasonably accurate in depth, while infrared is very accurate in angular resolution but terrible in depth measurement). Her system has the virtue that if one sensor fails the other still delivers readings that are useful to the higher level processing. Giralt *et al.* [6] use a laser range finder for map making, sonar sensors for local obstacle detection, and infrared beacons for map calibration. The robot operates in a mode in which one particular sensor type is used at a time and the others are completely ignored, even though they may be functional. In the natural world multiple redundant sensors are abundant. For instance [10] reports that pigeons have more than four independent orientation sensing systems (e.g., sun position compared to internal biological clock). It is interesting that the sensors do not seem to be combined but rather, depending on the environmental conditions and operational level of sensor subsystems, the data from one sensor tends to dominate.

Robustness: The above work tries to make systems robust in terms of sensor availability, but little has been done with making either the behavior or the processor of a robot robust.

Extensibility: There are three ways this can be achieved without completely rebuilding the physical control system. 1) Excess processor power that was previously being wasted can be utilized. Clearly this is a bounded resource. 2) The processor(s) can be upgraded to an architecturally compatible but faster system. The original software can continue to run, but now excess capacity will be available and we can proceed as in the first case. 3) More processors can be added to carry the new load. Typically systems builders then get enmeshed in details of how to make all memory uniformly accessible to all processors. Usually the cost of the memory to processor routing system soon comes to dominate the cost (the measure of cost is not important—it can be monetary, silicon area, access time delays, or something else) of the system. As a result there is usually a fairly small upper bound (on the order of hundreds for traditional style processing units; on the order to tens to hundreds of thousands for extremely simple processors) on the number of processors which can be added.

C. Starting Assumptions

Our design decisions for our mobile robot are based on the following nine dogmatic principles (six of these principles were presented more fully in [2]).

1) Complex (and useful) behavior need not necessarily be a product of an extremely complex control system. Rather, complex behavior may simply be the reflection of a complex environment [13]. It may be an observer who ascribes complexity to an organism—not necessarily its designer.

2) Things should be simple. This has two applications. a) When building a system of many parts one must pay attention to the interfaces. If you notice that a particular interface is starting to rival in complexity the components it connects, then either the interface needs to be rethought or the decomposition of the system needs redoing. b) If a particular component or collection of components solves an unstable or ill-conditioned problem, or, more radically, if its design involved the solution of an unstable or ill-conditioned problem, then it is probably not a good solution from the standpoint of robustness of the system.

3) We want to build cheap robots that can wander around human-inhabited space with no human intervention, advice, or control and at the same time do useful work. Map making is therefore of crucial importance even when idealized blueprints of an environment are available.

4) The human world is three-dimensional; it is not just a two-dimensional surface map. The robot must model the world as three-dimensional if it is to be allowed to continue cohabitation with humans.

5) Absolute coordinate systems for a robot are the source of large cumulative errors. Relational maps are more useful to a mobile robot. This alters the design space for perception systems.

6) The worlds where mobile robots will do useful work are not constructed of exact simple polyhedra. While polyhedra may be useful models of a realistic world, it is a mistake to build a special world such that the models can be exact. For

this reason we will build no artificial environment for our robot.

7) Sonar data, while easy to collect, does not by itself lead to rich descriptions of the world useful for truly intelligent interactions. Visual data is much better for that purpose. Sonar data may be useful for low-level interactions such as real-time obstacle avoidance.

8) For robustness sake the robot must be able to perform when one or more of its sensors fails or starts giving erroneous readings. Recovery should be quick. This implies that built-in self calibration must be occurring at all times. If it is good enough to achieve our goals then it will necessarily be good enough to eliminate the need for external calibration steps. To force the issue we do not incorporate any explicit calibration steps for our robot. Rather we try to make all processing steps self calibrating.

9) We are interested in building *artificial beings*—robots that can survive for days, weeks and months, without human assistance, in a dynamic complex environment. Such robots must be self-sustaining.

II. Levels and Layers

There are many possible approaches to building an autonomous intelligent mobile robot. As with most engineering problems, they all start by decomposing the problem into pieces, solving the subproblems for each piece, and then composing the solutions. We think we have done the first of these three steps differently to other groups. The second and third steps also differ as a consequence.

A. Levels of Competence

Typically, mobile robot builders (e.g., [3], [6], [8], [11], [12], [14], [Tsuji 84], [Crowley 85]) have sliced the problem into some subset of

- sensing
- mapping sensor data into a world representation
- planning
- task execution
- motor control.

This decomposition can be regarded as a horizontal decomposition of the problem into vertical slices. The slices form a chain through which information flows from the robot's environment, via sensing, through the robot and back to the environment, via action, closing the feedback loop (of course most implementations of the above subproblems include internal feedback loops also). An instance of each piece must be built in order to run the robot at all. Later changes to a particular piece (to improve it or extend its functionality) must either be done in such a way that the interfaces to adjacent pieces do not change, or the effects of the change must be propagated to neighboring pieces, changing their functionality, too.

We have chosen instead to decompose the problem vertically as our primary way of slicing up the problem. Rather than slice the problem on the basis of internal workings of the

solution, we slice the problem on the basis of desired external manifestations of the robot control system.

To this end we have defined a number of *levels of competence* for an autonomous mobile robot. A level of competence is an informal specification of a desired class of behaviors for a robot over all environments it will encounter. A higher level of competence implies a more specific desired class of behaviors.

We have used the following levels of competence (an earlier version of these was reported in [1]) as a guide in our work.

0) Avoid contact with objects (whether the objects move or are stationary).
1) Wander aimlessly around without hitting things.
2) "Explore" the world by seeing places in the distance that look reachable and heading for them.
3) Build a map of the environment and plan routes from one place to another.
4) Notice changes in the "static" environment.
5) Reason about the world in terms of identifiable objects and perform tasks related to certain objects.
6) Formulate and execute plans that involve changing the state of the world in some desirable way.
7) Reason about the behavior of objects in the world and modify plans accordingly.

Notice that each level of competence includes as a subset each earlier level of competence. Since a level of competence defines a class of valid behaviors it can be seen that higher levels of competence provide additional constraints on that class.

B. Layers of Control

The key idea of levels of competence is that we can build layers of a control system corresponding to each level of competence and simply add a new layer to an existing set to move to the next higher level of overall competence.

We start by building a complete robot control system that achieves level 0 competence. It is debugged thoroughly. We never alter that system. We call it the zeroth-level control system. Next we build a another control layer, which we call the first-level control system. It is able to examine data from the level 0 system and is also permitted to inject data into the internal interfaces of level 0 suppressing the normal data flow. This layer, with the aid of the zeroth, achieves level 1 competence. The zeroth layer continues to run unaware of the layer above it which sometimes interferes with its data paths.

The same process is repeated to achieve higher levels of competence (Fig. 3). We call this architecture a *subsumption architecture*.

In such a scheme we have a working control system for the robot very early in the piece—as soon as we have built the first layer. Additional layers can be added later, and the initial working system need never be changed.

We claim that this architecture naturally lends itself to solving the problems for mobile robots delineated in Section I-A.

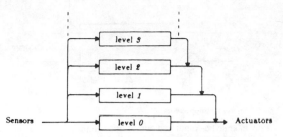

Fig. 3. Control is layered with higher level layers subsuming the roles of lower level layers when they wish to take control. The system can be partitioned at any level, and the layers below form a complete operational control system.

Multiple Goals: Individual layers can be working on individual goals concurrently. The suppression mechanism then mediates the actions that are taken. The advantage here is that there is no need to make an early decision on which goal should be pursued. The results of pursuing all of them to some level of conclusion can be used for the ultimate decision.

Multiple Sensors: In part we can ignore the sensor fusion problem as stated earlier using a subsumption architecture. Not all sensors need to feed into a central representation. Indeed, certain readings of all sensors need not feed into central representations—only those which perception processing identifies as extremely reliable might be eligible to enter such a central representation. At the same time however the sensor values may still be being used by the robot. Other layers may be processing them in some fashion and using the results to achieve their own goals, independent of how other layers may be scrutinizing them.

Robustness: Multiple sensors clearly add to the robustness of a system when their results can be used intelligently. There is another source of robustness in a subsumption architecture. Lower levels that have been well debugged continue to run when higher levels are added. Since a higher level can only suppress the outputs of lower levels by actively interfering with replacement data, in the cases that it can not produce results in a timely fashion the lower levels will still produce sensible results—albeit at a lower level of competence.

Extensibility: An obvious way to handle extensibility is to make each new layer run on its own processor. We will see below that this is practical as there are in general fairly low bandwidth requirements on communication channels between layers. In addition we will see that the individual layers can easily be spread over many loosely coupled processors.

C. Structure of Layers

But what about building each individual layer? Don't we need to decompose a single layer in the traditional manner? This is true to some extent, but the key difference is that we don't need to account for all desired perceptions and processing and generated behaviors in a single decomposition. We are free to use different decompositions for different sensor-set task-set pairs.

We have chosen to build layers from a set of small processors that send messages to each other. Each processor is a finite state machine with the ability to hold some data structures. Processors send messages over connecting "wires." There is no handshaking or acknowledgement of messages. The processors run completely asynchronously, monitoring their input wires, and sending messages on their output wires. It is possible for messages to get lost—it actually happens quite often. There is no other form of communication between processors, in particular there is no shared global memory.

All processors (which we refer to as modules) are created equal in the sense that within a layer there is no central control. Each module merely does its thing as best it can.

Inputs to modules can be suppressed and outputs can be inhibited by wires terminating from other modules. This is the mechanism by which higher level layers subsume the role of lower levels.

III. A Robot Control System Specification Language

There are two aspects to the components of our layered control architecture. One is the internal structure of the modules, and the second is the way in which they communicate. In this section we flesh out the details of the semantics of our modules and explain a description language for them.

A. Finite State Machines

Each module is a finite state machine, augmented with some instance variables, which can actually hold Lisp data structures.

Each module has a number of input lines and a number of output lines. Input lines have single-element buffers. The most recently arrived message is always available for inspection. Messages can be lost if a new one arrives on an input line before the last was inspected. There is a distinguished input to each module called *reset*. Each state is named. When the system first starts up all modules start in the distinguished state named NIL. When a signal is received on the reset line the module switches to state NIL. A state can be specified as one of four types.

Output	An output message, computed as a function of the module's input buffers and instance variables, is sent to an output line. A new specified state is then entered.
Side effect	One of the module's instance variables is set to a new value computed as a function of its input buffers and variables. A new specified state is then entered.
Conditional dispatch	A predicate on the module's instance variables and input buffers is computed and depending on the outcome one of two subsequent states is entered.

Event
dispatch   A sequence of pairs of conditions and states
            to branch to are monitored until one of the
            events is true. The events are in combina-
            tions of arrivals of messages on input lines
            and the expiration of time delays. [1]

An example of a module defined in our specification
language is the Avoid module in Listing 1.

Listing 1.   Avoid module in Lisp.

```
(defmodule avoid 1
  :inputs (force heading)
  :outputs (command)
  :instance-vars (resultforce)
  :states
    ((nil (event-dispatch (and force heading) plan))
     (plan (setf resultforce (select-direction force heading))
           go)
     (go (conditional-dispatch (significant-force-p resultforce 1.0)
                               start
                               nil))
     (start (output command (follow-force resultforce))
            nil)))
```

Here, select-direction, significant-force-p, and follow-force
are all Lisp functions, while setf is the modern Lisp assignment special form.

The force input line inputs a force with magnitude and
direction found by treating each point found by the sonars as
the site of a repulsive force decaying as the square of distance.
Function select-direction takes this and combines it with the
input on the heading line considered as a motive force. It
selects the instantaneous direction of travel by summing the
forces acting on the robot. (This simple technique computes
the tangent to the minimum energy path computed by [9].)

The function significant-force-p checks whether the resulting force is above some threshold—in this case it determines
whether the resulting motion would take less than a second.
The dispatch logic then ignores such motions. The function
follow-force converts the desired direction and force magnitude into motor velocity commands.

This particular module is part of the level 1 control system
(as indicated by the argument "1" following avoid, the name
of the module), which is described in Section IV-B. It
essentially does local navigation, making sure obstacles are
avoided by diverting a desired heading away from obstacles. It
does not deliver the robot to a desired location—that is the task
of level 2 competence.

B. Communication

Fig. 4 shows the best way to think about these finite state
modules for the purposes of communications. They have some
input lines and some output lines. An output line from one
module is connected to input lines of one or more other

The exact semantics are as follows. After an event dispatch is executed all
input lines are monitored for message arrivals. When the next event dispatch
is executed it has access to latches which indicate whether new messages
arrived on each input line. Each condition is evaluated in turn. If it is true then
the dispatch to the new state happens. Each condition is an and/or expression
on the input line latches. In addition, condition expressions can include delay
terms, which become true a specified amount of time after the beginning of the
execution of the event dispatch. An event dispatch waits until one of its
condition expressions is true.

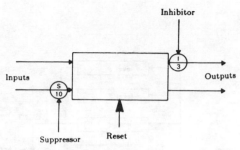

Fig. 4.   A module has input and output lines. Input signals can be suppressed
and replaced with the suppressing signal. Output signals can be inhibited.
A module can also be reset to state NIL.

modules. One can think of these lines as wires, each with
sources and a destination. Additionally, outputs may be
inhibited, and inputs may be suppressed.

An extra wire can terminate (i.e. have its destination) at an
output site of a module. If any signal travels along this wire it
inhibits any output message from the module along that line
for some predetermined time. Any messages sent by the
module to that output during that time period is lost.

Similarly, an extra wire can terminate at an input site of a
module. Its action is very similar to that of inhibition, but
additionally, the signal on this wire, besides inhibiting signals
along the usual path, actually gets fed through as the input to
the module. Thus it suppresses the usual input and provides a
replacement. If more than one suppressing wire is present they
are essentially OR-ed together. For both suppression and
inhibition we write the time constants inside the circle.

In our specification language we write wires as a source
(i.e. an output line) followed by a number of destinations (i.e.
input lines). For instance the connection to the force input of
the Avoid module might be the wire defined as

(defwire 1 (feelforce force) (avoid force)).

This links the force output of the Feelforce module to the
input of the Avoid module in the level one control system.

Suppression and inhibition can also be described with a
small extension to the syntax above. Below we see the
suppression of the command input of the Turn module, a level
0 module by a signal from the level 1 module Avoid.

(defwire 1 (avoid command) ((suppress (turn command) 20.0))).

In a similar manner a signal can be connected to the reset
input of a module.

IV. A ROBOT CONTROL SYSTEM INSTANCE

We have implemented a mobile robot control system to
achieve levels 0 and 1 competence as defined above, and have
started implementation of level 2 bringing it to a stage which
exercises the fundamental subsumption idea effectively. We
need more work on an early vision algorithm to complete level
2.

A. Zeroth Level

The lowest level layer of control makes sure that the robot
does not come into contact with other objects. It thus achieves
level 0 competence (Fig. 5). If something approaches the robot
it will move away. If in the course of moving itself it is about

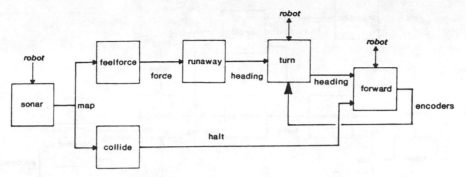

Fig. 5. Level 0 control system.

to collide with an object it will halt. Together these two tactics are sufficient for the robot to flee from moving obstacles, perhaps requiring many motions, without colliding with stationary obstacles. The combination of the tactics allows the robot to operate with very coarsely calibrated sonars and a wide range of repulsive force functions. Theoretically, the robot is not invincible of course, and a sufficiently fast-moving object or a very cluttered environment might result in a collision. Over the course of a number of hours of autonomous operation, our physical robot (see Section V-B) has not collided with either a moving or fixed obstacle. The moving obstacles have, however, been careful to move slowly.

The Turn and Forward modules communicate with the actual robot. They have extra communication mechanisms, allowing them to send and receive commands to and from the physical robot directly. The Turn module receives a heading specifying an in-place turn angle followed by a forward motion of a specified magnitude. It commands the robot to turn (and at the same time sends a busy message on an additional output channel described in Fig. 7) and on completion passes on the heading to the Forward module (and also reports the shaft encoder readings on another output line shown in Fig. 7). The Turn module then goes into a wait state ignoring all incoming messages. The Forward module commands the robot to move forward but halts the robot if it receives a message on its halt input line during the motion. As soon as the robot is idle, it sends out the shaft encoder readings. The message acts as a reset for the Turn module, which is then once again ready to accept a new motion command. Notice the any heading commands sent to the Turn module during transit are lost.

The Sonar module takes a vector of sonar readings, filters them for invalid readings, and effectively produces a robot centered map of obstacles in polar coordinates.

The Collide module monitors the sonar map and if it detects objects dead ahead, it sends a signal on the halt line to the Motor module. The Collide module does not know or care whether the robot is moving. Halt messages sent while the robot is stationary are essentially lost.

The Feelforce module sums the results of considering each detected object as a repulsive force, generating a single resultant force.

The Runaway module monitors the 'force' produced by the sonar detected obstacles and sends commands to the turn module if it ever becomes significant.

Fig. 5 gives a complete description of how the modules are connected together.

B. First Level

The first level layer of control, when combined with the zeroth, imbues the robot with the ability to wander around aimlessly without hitting obstacles. This was defined earlier as level 1 competence. This control level relies in a large degree on the zeroth level's aversion to hitting obstacles. In addition it uses a simple heuristic to plan ahead a little in order to avoid potential collisions which would need to be handled by the zeroth level.

The Wander module generates a new heading for the robot every ten seconds or so.

The Avoid module, described in more detail in Section III, takes the result of the force computation from the zeroth level and combines it with the desired heading to produce a modified heading, which usually points in roughly the right direction, but is perturbed to avoid any obvious obstacles. This computation implicitly subsumes the computations of the Runaway module, in the case that there is also a heading to consider. In fact the output of the Avoid module suppresses the output from the Runaway module as it enters the Motor module.

Fig. 6 gives a complete description of how the modules are connected together. Note that it is simply Fig. 5 with some more modules and wires added.

C. Second Level

Level 2 is meant to add an exploratory mode of behavior to the robot, using visual observations to select interesting places to visit. A vision module finds corridors of free space. Additional modules provide a means of position servoing the robot to along the corridor despite the presence of local obstacles on its path (as detected with the sonar sensing system). The wiring diagram is shown in Fig. 7. Note that it is simply Fig. 6 with some more modules and wires added.

The Status module monitors the Turn and Forward modules. It maintains one status output which sends either hi or lo messages to indicate whether the robot is busy. In addition, at

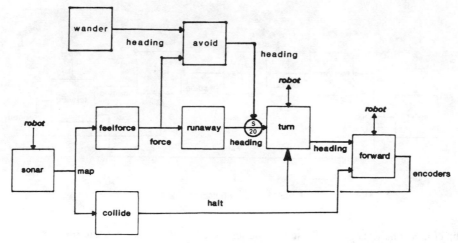

Fig. 6. Level 0 control system augmented with the level 1 system.

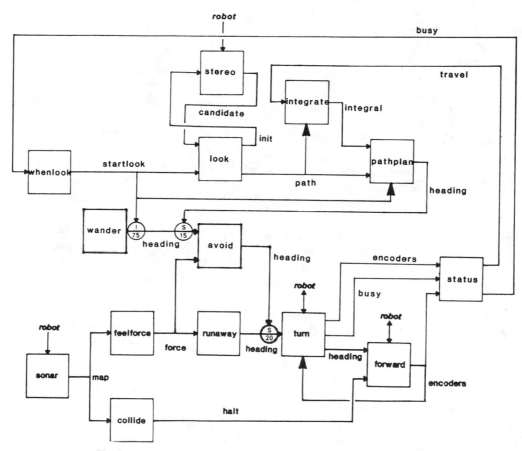

Fig. 7. Level 0 and 1 control systems augmented with the level 2 system.

the completion of every turn and roll forward combination it sends out a combined set of shaft encoder readings.

The Whenlook module monitors the busy line from the Status module, and whenever the robot has been sitting idle for a few seconds it decides its time to look for a corridor to traverse. It inhibits wandering so it can take some pictures and process them without wandering away from its current location, and resets the Pathplan and Integrate modules. This latter action ensures that the robot will know how far it has

moved from its observation point should any Runaway impulses perturb it.

The Look module initiates the vision processing, and waits for a candidate freeway. It filters out poor candidates and passes any acceptable one to the Pathplan module.

The Stereo module is supposed to use stereo TV images [7], which are obtained by the robot, to find a corridor of free space. At the time of writing final version of this module had not been implemented. Instead, both in simulation and on the physical robot, we have replaced it with a sonar-base corridor finder.

The Integrate module accumulates reports of motions from the status module and always sends its most recent result out on its integral line. It gets restarted by application of a signal to its reset input.

The Pathplan module takes a goal specification (in terms of an angle to turn, a distance to travel) and attempts to reach that goal. To do this, it sends headings to the Avoid module, which may perturb them to avoid local obstacles, and monitors its integral input which is an integration of actual motions. The messages to the Avoid module suppress random wanderings of the robot, so long as the higher level planner remains active. When the position of the robot is close to the desired position (the robot is unaware of control errors due to wheel slippage etc., so this is a dead-reckoning decision) it terminates.

The current wiring of the second level of control is shown in Fig. 7, which augments the two lower level control systems. The zeroth and first layers still play an active roll during normal operation of the second layer.

V. PERFORMANCE

The control system described here has been used extensively to control both a simulated robot and an actual physical robot wandering around a cluttered laboratory and a machine room.

A. A Simulated Robot

The simulation tries to simulate all the errors and uncertainties that exist in the world of the real robot. When commanded to turn through angle α and travel distance d the simulated robot actually turns through angle $\alpha + \delta\alpha$ and travels distance $d + \delta d$. Its sonars can bounce off walls multiple times, and even when they do return they have a noise component in the readings that model thermal and humidity effects. We feel it is important to have such a realistic simulation. Anything less leads to incorrect control algorithms.

The simulator runs off a clock and runs at the same rate as would the actual robot. It actually runs on the same processor that is simulating the subsumption architecture. Together they are nevertheless able to perform a real-time simulation of the robot and its control and also drive graphics displays of robot state and module performance monitors. Fig. 8 shows the robot (which itself is not drawn) receiving sonar reflections at some of its 12 sensors. Other beams did not return within the time allocated for data collection. The beams are being reflected by various walls. There is a small bar in front of the robot perpendicular to the direction the robot is pointing.

Fig. 9 shows an example world in two dimensional projection. The simulated robot with a first level control system connected was allowed to wander from an initial position. The squiggly line traces out its path. Note that it was wandering aimlessly and that it hit no obstacles.

Fig. 10 shows two examples of the same scene and the motion of the robot with the second level control system connected. In these cases the Stereo module was supplanted with a situation-specific module, which gave out two precise

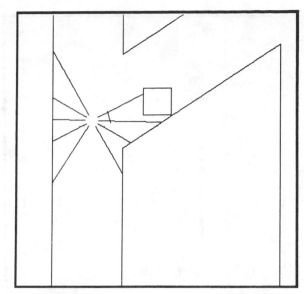

Fig. 8. Simulated robot receives 12 sonar readings. Some sonar beams glance off walls and do not return within a certain time.

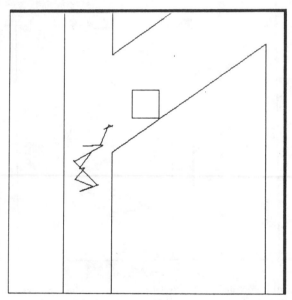

Fig. 9. Under levels 0 and 1 control the robot wanders around aimlessly. It does not hit obstacles.

corridor descriptions. While achieving the goals of following these corridors the lower level wandering behavior was suppressed. However the obstacle avoiding behavior of the lower levels continued to function—in both cases the robot avoided the square obstacle. The goals were not reached exactly. The simulator models a uniformly distributed error of ± 5 percent in both turn and forward motion. As soon as the goals had been achieved satisfactorily the robot reverted to its wandering behavior.

B. A Physical Robot

We have constructed a mobile robot shown in Fig. 11. It is about 17 inches in diameter and about 30 inches from the ground to the top platform. Most of the processing occurs offboard on a Lisp machine.

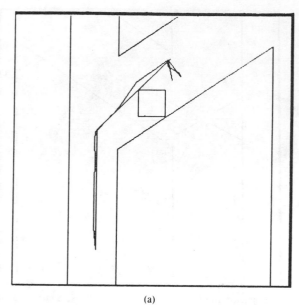

(a)

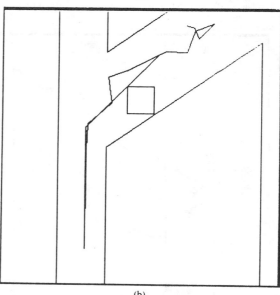

(b)

Fig. 10. (a) With level 2 control the robot tries to achieve commanded goals. The nominal goals are the two straight lines. (b) After reaching the second goal, since there are no new goals forthcoming, the robot reverts to aimless level 1 behavior.

The drive mechanism was purchased from Real World Interface of Sudbury, MA. Three parallel drive wheels are steered together. The two motors are servoed by a single microprocessor. The robot body is attached to the steering mechanism and always points in the same direction as the wheels. It can turn in place (actually it inscribes a circle about 1 cm in diameter).

Currently installed sensors are a ring of twelve Polaroid sonar time-of-flight range sensors and two Sony CCD cameras. The sonars are arranged symmetrically around the rotating body of the robot. The cameras are on a tilt head (pan is provided by the steering motors). We plan to install feelers that can sense objects at ground level about six inches from the base extremities.

Fig. 11. The M.I.T. AI Lab mobile robot.

A central cardcage contains the main on-board processor, an Intel 8031. It communicates with off-board processors via a 12 Kbit/s duplex radio link. The radios are modified Motorola digital voice encryption units. Error correction cuts the effective bit rate to less than half the nominal rating. The 8031 passes commands down to the motor controller processor and returns encoder readings. It controls the sonars and the tilt head, and it switches the cameras through a single channel video transmitter mounted on top of the robot. The latter transmits a standard TV signal to a Lisp machine equipped with a demodulator and frame grabber.

The robot has spent a few hours wandering around a laboratory and a machine room.

Under level 0 control, the robot finds a large empty space and then sits there contented until a moving obstacle approaches. Two people together can successfully herd the robot just about anywhere—through doors or between rows of disk drives, for instance.

When level 1 control is added the robot is no longer content to sit in an open space. After a few seconds it heads off in a random direction. Our uncalibrated sonars and obstacle repulsion functions make it overshoot a little to locations where the Runaway module reacts. It would be interesting to make this the basis of adaption of certain parameters.

Under level 2 a sonar-based corridor finder usually finds the most distant point in the room. The robot heads of in the direction. People walking in front of the robot cause it to detour, but the robot still gets to the initially desired goal, even when it involves squeezing between closely spaced obstacles. If the sonars are in error and a goal is selected beyond a wall, the robot usually ends up in a position where the attractive force of the goal is within a threshold used by Avoid of the repulsive forces of the wall. At this point Avoid does not issue any heading, as it would be for some trivial motion of the robot. The robot sits still defeated by the obstacle. The Whenlook module, however, notices that the robot is idle and initiates a new scan for another corridor of free space to follow.

C. Implementation Issues

While we have been able to simulate sufficient processors on a single Lisp machine up until now, that capability will soon pass as we bring on line our vision work (the algorithms have been debugged as traditional serial algorithms, but we plan on re-implementing them within the subsumption architecture). Building the architecture in custom chips is a long-term goal.

One of the motivations for developing the layered control system was extensibility of processing power. The fact that it is decomposed into asynchronous processors with low-bandwidth communication and no shared memory should certainly assist in achieving that goal. New processors can simply be added to the network by connecting their inputs and outputs at appropriate places—there are no bandwidth or synchronization considerations in such connections.

The finite state processors need not be large. Sixteen states is more than sufficient for all modules we have written so far. (Actually, eight states are sufficient under the model of the processors we have presented here and used in our simulations. However we have refined the design somewhat towards gate-level implementation, and there we use simpler more numerous states.) Many such processors could easily be packed on a single chip.

The Lisp programs that are called by the finite state machines are all rather simple. We believe it is possible to implement each of them with a simple network of comparators, selectors, polar coordinate vector adders, and monotonic function generators. The silicon area overhead for each module would probably not be larger than that required for the finite state machine itself.

VI. Conclusion

The key ideas in this paper are the following.

1) The mobile robot control problem can be decomposed in terms of behaviors rather than in terms of functional modules.
2) It provides a way to incrementally build and test a complex mobile robot control system.
3) Useful parallel computation can be performed on a low bandwidth loosely coupled network of asynchronous simple processors. The topology of that network is relatively fixed.
4) There is no need for a central control module of a mobile robot. The control system can be viewed as a system of agents each busy with their own solipsist world.

Besides leading to a different implementation strategy it is also interesting to note the way the decomposition affected the capabilities of the robot control system we have built. In particular, our control system deals with moving objects in the environment at the very lowest level, and it has a specific module (Runaway) for that purpose. Traditionally mobile robot projects have delayed handling moving objects in the environment beyond the scientific life of the project.

Note: A drawback of the presentation in this paper was merging the algorithms for control of the robot with the implementation medium. We felt this was necessary to convince the reader of the utility of both. It is unlikely that the subsumption architecture would appear to be useful without a clear demonstration of how a respectable and useful algorithm can run on it. Mixing the two descriptions as we have done demonstrates the proposition.

Acknowledgment

Tomás Lozano-Pérez, Eric Grimson, Jon Connell, and Anita Flynn have all provided helpful comments on earlier drafts of this paper.

References

[1] R. A. Brooks, "Aspects of mobile robot visual map making," in *Robotics Research 2,* Hanafusa and Inoue, Eds. Cambridge, MA: M.I.T., 1984, pp. 369–375.
[2] ——, "Visual map making for a mobile robot," in *Proc. 1985 IEEE Conf. Robotics and Automat.,* pp. 824–829.
[3] James L. Crowley, "Navigation for an intelligent mobile robot," *IEEE J. Robotics Automat.,* vol. RA-1, no. 1, Mar. 1985, pp. 31–41.
[4] A. Elfes and S. N. Talukdar, "A distributed control system for the CMU rover," in *Proc. IJCAI,* 1983, pp. 830–833.
[5] A. Flynn, "Redundant sensors for mobile robot navigation," M.S. Thesis, Department of Electrical Engineering and Computer Science, M.I.T., Cambridge, MA, July 1985.
[6] G. Giralt, R. Chatila, and M. Vaisset, "An integrated navigation and motion control system for autonomous multisensory mobile robots," in *Robotics Research 1,* Brady and Paul, Eds. Cambridge, MA: M.I.T. 1983, 191–214.
[7] W. L. Grimson, "Computational experiments with a feature based stereo algorithm," *IEEE Trans. Patt. Anal. Mach. Intell.,* vol. PAMI-7, pp. 17–34, Jan. 1985.
[8] Y. Kanayama, "Concurrent programming of intelligent robots," in *Proc. IJCAI,* 1983, pp. 834–838.
[9] O. Khatib, "Dynamic control of manipulators in operational space," *Sixth IFTOMM Cong. Theory of Machines and Mechanisms,* Dec. 1983.
[10] M. L. Kreithen, "Orientational strategies in birds: a tribute to W. T. Keeton," in *Behavioral Energetics: The Cost of Survival in Vertebrates.* Columbus, OH: Ohio State University, 1983, pp. 3–28.
[11] H. P. Moravec, "The stanford cart and the CMU rover," *Proc. IEEE,* vol. 71, pp. 872–884, July 1983.
[12] N. J. Nilsson, "Shakey the robot," SRI AI Center, tech. note 323, Apr. 1984.
[13] H. A. Simon, *Sciences of the Artificial.* Cambridge, MA: M.I.T., 1969.
[14] S. Tsuji, "Monitoring of a building environment by a mobile robot," in *Robotics Research 2,* Hanafusa and Inoue, Eds. Cambridge, MA: M.I.T., 1985, pp. 349–356.

Rodney A. Brooks was born in Adelaide, South Australia, on December 30, 1954. He received the B.Sc. and M.Sc. degrees in mathematics from the Flinders University of South Australia and the Ph.D. degree in computer science from Stanford University, Stanford, CA, in 1981.

Since 1981 he has held research positions at Carnegie-Mellon University, Pittsburgh, PA, and the Massachusetts Institute of Technology (M.I.T.), Cambridge, and faculty positions at Stanford and M.I.T. He is currently an Assistant Professor in the Department of Electrical Engineering and Computer Science at M.I.T., and a member of the Artificial Intelligence Laboratory. His research interests include computer vision, geometric reasoning, robot planning, and Lisp systems and compilers. His books include *Model-Based Computer Vision* from UMI Research Press and *Programming in Common Lisp* from John Wiley and Sons.

MOTOR SCHEMA BASED NAVIGATION FOR A MOBILE ROBOT:
An Approach to Programming by Behavior

Ronald C. Arkin

VISIONS / Laboratory for Perceptual Robotics
Computer and Information Science Department, University of Massachusetts,
Graduate Research Center, Amherst, Massachusetts, 01003

Abstract

Motor schemas are proposed as a basic unit of behavior specification for the navigation of a mobile robot. These are multiple concurrent processes which operate in conjunction with associated perceptual schemas and contribute independently to the overall concerted action of the vehicle. The motivation behind the use of schemas for this domain is drawn from neuroscientific, psychological and robotic sources. A variant of the potential field method is used to produce the appropriate velocity and steering commands for the robot. An implementation strategy based on available tools at UMASS is described. Simulation results show the feasibility of this approach.

1. Introduction

Path planning and navigation, at the execution level, can most easily be described as a collection of behaviors. *Don't run into things! Go to the end of the sidewalk then turn right! Stay to the right side of the sidewalk except when passing! Watch out for the library - the turn is just beyond it! Follow that man!* This collection of commands constitutes some of the possible behaviors for an entity trying to move from one location to another. Traditional programming – using an inflexible, rigid, hard-coded approach – does not provide the essential adaptability necessary for coping with unexpected events. These events might include unanticipated obstacles, moving objects, or the recognition of a landmark in a seemingly inappropriate location. These unexpected occurrences should influence, in an appropriate manner, the course which a vehicle (or person) takes in moving from start to goal.

A potential solution can be drawn from models that have been developed in the domains of brain theory and robotics. Schemas, a model used to describe the interaction between perception and action, can be adapted to yield a mobile robot system that is highly sensitive to the currently perceived world. Motor schemas operating in a concurrent and independent, yet communicating, manner can produce paths that reflect the uncertainties in the detection of objects. Additionally they can cope with conflicting data arising from diverse sensor modalities and strategies.

The purpose of this paper is to provide insights into the design of a motor-schema-based control system for mobile robots. Section 2 will describe the motivations for the use of schema theory in this domain – drawing from work in both brain theory and robotics. Section 3 will discuss the tack being taken for a motor-schema-based control system in the UMASS autonomous robot architecture (AuRA), utilizing a mobile robot equipped with ultrasonic and video sensors; specifically the role of the pilot and the motor schema manager. Section 4 will present the results of simulations using schemas that specify different behaviors and draw on simulated sensor input. A summary and projection of future work will conclude this report.

2. Motivation

The concept of schemas originated in psychology [1,2,3] and neurology [4,5]. Webster [6] defines a schema as "a mental codification of experience that includes a particular organized way of perceiving cognitively and responding to a complex situation or set of stimuli". The model used for this paper draws on more recent sources: the applications of schema theory to brain modeling and robotics. As brain theory can unequivocally be called a sound basis for the study of intelligent behavior, the first part of this section will present the contributions of brain science that influenced the design of the schema control system described below. Roboticists for some time have drawn on schema theory, not always in the form envisioned by brain theoreticians. The previous work in robotics that relates to the schema-based approach to navigation will be described in the final part of this section.

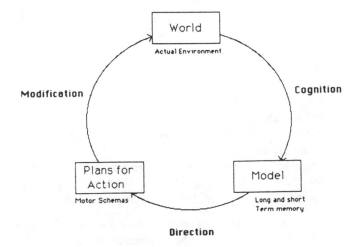

Figure 1. Action – perception cycle

This research was supported in part by the NSF under grant DCR-8318776 and the U.S. Army under ETL grant DACA76-85-C-008.

2.1 Brain Theory and Psychology

The action-perception cycle (fig. 1) provides a principal motivation for the application of schema theory [7]. Sensor-driven expectations provide the plans (schemas) for appropriate motor action, which when undertaken provide new sensory data that is fed back into the system to provide new expectations. This cycle of cognition (the altering of the internal world model), direction (selection of appropriate motor behaviors), and action (the production of environmental changes and resultant availability of new sensory data) is central to the way in which schemas must interact with the world.

Most significantly, perception should be viewed as action-oriented. There is no need to process all available sensor data, only that data which is pertinent to the task at hand. The question for the roboticist would be: how do we select from the wealth of sensor data available that which is relevant? By specifying schemas, each individual component of the overall task can make its demands known to the sensory subsystem, and thus guide the focus of attention mechanisms and limited sensory processing that is available.

Guided by Arbib's work [8,9] in the study of the frog and its machine analog *Rana Computatrix*, the frog prey selection mechanism serves as a basis for analysis. In particular, Arbib and House [10] have developed a model for worm acquisition by the frog in an obstacle-cluttered environment (a spaced fence - fig. 2). Although Arbib and House describe two models to account for the behavior of the frog, the second is the most readily applicable to the mobile robot's domain (the first model is based on visual orientation). In their work, they describe primitive vector fields (fig. 3): a prey-attractant field, a barrier-repellent field, and a field for the animal itself. These fields, when combined, yield a model of behavior (fig. 4) that is consistent with experimental observations of the frog.

In the mobile robot system described below, analogs of these fields will be used (prey-attractant ⇒ **move-to-goal**, barrier-repellent ⇒ **avoid-static-obstacle**). Additionally, new fields will be added to describe additional motor tasks (**stay-on-path**, **avoid-moving-obstacle**, etc.)

This model, in conjunction with expectation-driven sensing, provides a basic correlate with the functioning of the brain (albeit the frog brain). Although the brain has been handling visually guided detours since time immemorial, the benefits of using a neuroscience model would wane if it proved impractical for

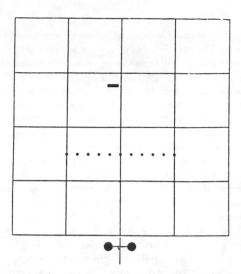

Figure 2. A depiction of a frog prey-selection scenario. The two large blackened circles at the bottom of the figure denote the frog's eyes, the smaller circles are fence-posts, and the darkened rectangle a supply of worms. (fig. 2,3,4 reprinted from [10] with permission).

a mobile robot. In the sections following, the practicality of this approach will be demonstrated, especially regarding the decomposition of the task to a form which is readily adaptable to distributed processing. This is essential if the real-time demands of mobile robot environmental interaction are to be met.

2.2 Robotics

Schema theory as applied to robotics has almost as many different definitions as there are developers. In the realm of robotic manipulators, Lyons' schemas [14] and Geschke's servo processes [12], (a schema analog), are used as approaches to task level control. Overton [15] has described the use of motor schemas in the assembly domain. The UMASS VISIONS group, guided by Hanson and Riseman, has applied perceptual schemas to the interpretation of natural scenes; Weymouth's thesis is the prime example of this work [13]. Although AuRA will, in the future, include perceptual schemas running in the context of the VISIONS system, perceptual schemas as they appear in the VISIONS system are not a principal concern of this paper.

One of the simplest and most straightforward definitions for a schema is "a generic specification of a computing agent" [14]. This definition fits well with the concept of a behavior (an individual's response to its environment) – each schema represents a generic behavior. Schema-based control systems are signifi-

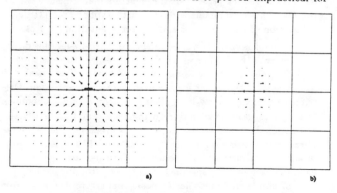

Figure 3. Primitive vector fields associated with figure 2.
a) Prey-attractant field
b) Barrier repellent field

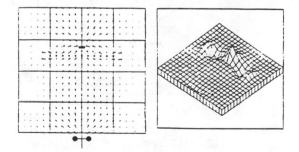

Figure 4. Resultant frog-prey selection field.

cantly more than a collection of frames or templates for behavior, however. The way in which they are set into action and interact immediately distinguish them from simpler representational forms. The instantiations of these generic schemas (SI – schema instantiation) provide the potential actions for the control of the robot. A schema instantiation is created when a copy of a generic schema is parameterized and activated as a computing agent.

Lyons further defines a motor schema as a control system or motor program which describes a task. Overton [15] describes a motor schema as "a control system which continually monitors feedback from the system it controls to determine the appropriate pattern of action for achieving the motor schema's goals, (these will, in general, be subgoals within some higher-level coordinated control program)". This more constrained definition is also in accord with the system described below. Sensory perception provides the feedback to affect individual instantiations of motor schemas, each SI thus providing an appropriate behavior which collectively determine the overall system's behavior. Some other definitions for motor schema include an "interaction plan" [25] or "unit of motor behavior" [16].

Other work in the path planning domain, although not schema based, bears a resemblance to the schema control system. Brooks [17] uses a planning system with a "horizontal decomposition" which effectively emulates multiple behaviors. Although related, there is still a rigid layering present which distinguishes it from a schema-based approach. Payton [23] describes a multi-behavior approach for reflexive control of an autonomous vehicle. The association of virtual sensors with a selected set of reflexive behaviors bears a similarity to the schema-based approach. An arbitrary choice of behavior, however, based on a priority system, is made during execution, without provision for a mechanism to combine the results of concurrent behaviors. Kadonoff et al [18] also incorporate multiple behaviors for the control of a mobile robot and similarly arbitrate between these behaviors, proposing a production system for arbitrating competitive strategies and the use of an optimal filter for the treatment of complementary strategies.

The schema system described below is strongly influenced by Krogh's [19] generalized potential fields approach and to a lesser degree by Lyons' [11] tagged potential fields. It bears a superficial resemblance to the integrated path planning and dynamic steering control system described by Krogh and Thorpe [20]. Potential fields are used, in each case, to produce the steering commands for a mobile robot. A major distinction between their system and our schema model lies in the tracking of the individual obstacles (individual SIs for each obstacle - important for the treatment of uncertainty) and the incorporation of additional behaviors such as road following and treatment of moving obstacles. The state of the each obstacle's SI is dynamically altered by newly acquired sensory information. The potential functions for each SI reflect the measured uncertainty associated with the perception of each object. The schema approach is not limited to obstacle avoidance, but is versatile enough for road following, object tracking and other behavioral patterns.

3.   Approach

Motor schemas, when instantiated, must drive the robot to interact with its environment. On the highest level, this will be to satisfy a goal developed within the planning system; on the lowest level, to produce specific translations and rotations of the robot vehicle. The schema system enables the software designer to deal with conceptual structures that are easy to comprehend and handle. The task of robot programming is fundamentally simplified through the use of a divide and conquer strategy.

This section will first describe the overall UMASS autonomous robot architecture's planning subsystem; particularly the roles of the pilot and motor schema manager. Implementation strategies will then be described.

3.1   Path Planning and Navigation System

The AuRA high-level path planner (fig. 5) is hierarchical in design; consisting of a mission planner, navigator and pilot. The mission planner is delegated the responsibility for interpreting high level commands, determining the nature of the mission, setting criteria for mission, navigator and pilot failure, and setting appropriate navigator and pilot parameters. The mission planner, although part of the overall design, is not yet fully implemented, and has a relatively low priority.

The navigator accepts a start and goal point from the mission planner and using a "meadow map", a hybrid vertex-graph free-space representation, determines a path to achieve that goal. The navigator produces a piecewise linear path that avoids all modeled obstacles present in the *a priori* map constructed by the map-builder component of the cartographer. See [21] for a description of the navigator and the representations it uses.

The pilot is charged with implementing leg-by-leg this piecewise linear path. To do so, the pilot chooses from a repertoire of available sensing strategies and motor behaviors (schemas) and passes them to the motor schema manager for instantiation. Distributed control and low-level planning occur within the confines of the motor schema manager during its attempt to satisfy the navigational requirements. As the robot proceeds, the cartographer, using sensor data, builds up a model of the perceived world in short-term memory. If the actual path deviates too greatly from the path initially specified by the navigator due to the presence of unmodeled obstacles or positional errors, the navigator will be reinvoked and a new global path computed. If the deviations are within acceptable limits, (as determined by higher levels in the planning hierarchy), the pilot and motor schema

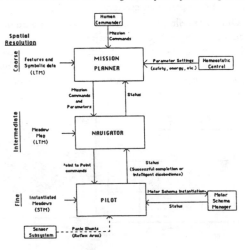

Figure 5. Hierarchical planner for AuRA.

manager will, in a coordinated effort, attempt to bypass the obstacle, follow the path, or cope with other problems as they arise. Additionally, the problem of robot localization is constantly addressed through the monitoring of short-term memory and appropriate **find-landmark** schemas. Multiple concurrent behaviors (schemas) may be present during any leg, for example:

- **Stay-on-path** (a sidewalk or a hall)
- **Avoid-static-obstacles** (parked cars etc.)
- **Avoid-moving-obstacles** (people etc.)
- **Find-intersection** (to determine end of path)
- **Find-landmark**(building) (for localization)

The first three are examples of motor schemas, the last two perceptual schemas. To provide the correct behavior, perceptual schemas must be associated with each motor schema. For example, in order to stay on the sidewalk, a **find-terrain**(sidewalk) perceptual schema must be instantiated to provide the necessary data for the **stay-on-path** motor schema to operate. If the uncertainty in the actual location of the sidewalk can be determined, the SI's associated velocity field, applying pressure to remain on the sidewalk, will reflect this uncertainty measure. The same holds for obstacle avoidance: if a perceptual schema for obstacle detection returns the position of a suspected obstacle and the relative certainty of its existence, the actual avoidance maneuvering will depend not only on whether an obstacle is detected but also on how certain we are that it exists. A more concrete example follows.

The robot is moving across a field in a particular direction (**move-ahead** schema). The **find-obstacle** schema is constantly on the look-out for possible obstacles within a subwindow of the video image (windowed by the direction and velocity of

the robot). When an event occurs, (e.g. a region segmentation algorithm detects an area that is distinct from the surrounding backdrop or an interest operator locates a high-interest point in the direction of the robot's motion), the **find-obstacle** schema spawns off an associated perceptual schema (**static-obstacle** SI) for that portion of the image. It is now the static-obstacle SI's responsibility to continuously monitor that region. Any other events that occur elsewhere in the image spawn off separate **static-obstacle** SIs. Additionally an **avoid-static-obstacle** SI motor schema is created for each detected potential obstacle.

The motor schema SI hibernates waiting for notification that the perceptual schema is sufficiently confident in the obstacle's existence to warrant motor action. If the perceptual schema proves to be a phantom (e.g. shadow) and not an obstacle at all, both the perceptual and related motor SIs are deinstantiated before producing any motor action. On the other hand, if the perceptual SI's confidence (activation level) exceeds the motor SI's threshold for action, the motor schema starts producing a repulsive field surrounding the obstacle.[1] The sphere of influence (spatial extent of repulsive forces) and the intensity of repulsion of the obstacle are affected by the distance from the robot and the obstacle's perceptual certainty. Eventually, when the robot moves beyond the range for perception of the obstacle, both the motor and perceptual SIs are deinstantiated. In summary, when obstacles are detected with sufficient certainty, the motor schema associated with a particular obstacle (its SI) starts to produce a force moving away from the object. Fig. 6a shows a typical repulsive field for an **avoid-static-obstacle** SI. The control of the priorities of the behaviors, (e.g. when is it more important to follow the sidewalk than to avoid uncertain but possible obstacles) is partially dependent on the uncertainty associated with the obstacle's representation. Other isolated motor schema velocity fields are shown in fig. 6b-d. Various combinations of motor schemas are illustrated in fig. 7.

If each schema functions independently of each other, how can any semblance of realistic and consistent behavior be achieved? Two components are required to satisfactorily answer this question. First a combination mechanism must be applied to all the SI-produced vector fields. The result is then used to provide the necessary velocity changes to the robot. The simplest approach is vector addition. By having each motor SI create a normalized velocity field, a single **move-robot** schema monitors the posted data for each SI, adds them together, makes certain it is within acceptable bounds and then transmits it to the low-level robot control system. In essence, the specific velocity and direction for the robot can be determined at any point in time by summing the output vectors of the individual SIs. As each motor SI is a distributed computing agent, preferably operating on separate processors on a parallel machine, and needs only to compute the velocity at the point the robot is currently located (and not the entire velocity field), real-time operation is within reach.

The second component of the response to the question posed in the previous paragraph is communication. Potential fields can have problems with dead spots or plateaus where the robot can become stranded. By allowing communication mechanisms between the SIs, the forces of conflicting actions can be reconciled. Lyons [14] proposes message passing between ports on one SI and connected ports on another SI as a schema communication mechanism. Alternatively, a blackboard mechanism is used in the Schema Shell system (discussed below). In either case, communication mechanisms can solve problems that might otherwise prove intractable. An example to illustrate this point follows.

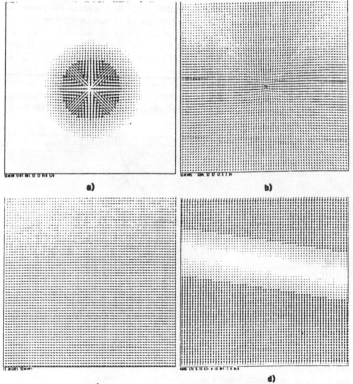

Figure 6. Isolated motor schema SI vector fields.
a) **Avoid-static-obstacle**   b) **Move-to-goal**
c) **Move-ahead**   d) **Stay-on-path**

---

[1] The obstacle is first grown in a configuration space manner [27] to enable the robot to be treated henceforth as a point for path planning purposes.

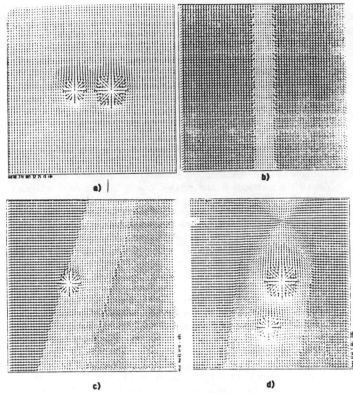

Figure 7. Several combined motor schemas.
a) **Move-ahead SI + 2 Avoid-static-obstacle SIs.**
b) **Move-ahead SI + Stay-on-path SI.**
c) **Move-ahead SI + Stay-on-path SI + 1 Avoid-static-obstacle SI.**
d) **Move-to-goal SI + Stay-on-path SI + 2 Avoid-static-obstacle SIs.**

The robot is instructed to move in a particular direction, stay on the sidewalk and avoid static obstacles. Suppose that the sidewalk is completely blocked by an obstacle; eventually the velocity would drop to 0 and the robot stop (fig. 8a). The stoppage of the robot is detected by the **stay-on-path** SI through inter-schema communication with the **move-robot** SI (the **move-robot** SI combines the individual motor SIs and communicates

the results to the low-level control system). The **stay-on-path** SI, when created, was instructed to yield if an obstacle blocks the path. The **stay-on-path** motor schema reduces its field (fig. 8b) and allows the robot to wander off the sidewalk thus circumnavigating the obstacle. As soon as the direction of the force produced by the offending obstacle indicates it has been successfully passed, the **stay-on-path** field returns to its original state forcing the robot back on the path (fig. 8c).

Suppose, however, the **stay-on-path** SI was instantiated for a hall. Then, under no circumstances, would the force field associated with the **stay-on-path** SI be reduced or else the robot would crash into the wall. The robot would instead stop, and signal for the navigator (higher level component of the planner) to be reinvoked and produce an alternate global path that avoids the newly discovered blocked passageway. These communication pathways are specified within the schema structures themselves.

Another approach to be explored is the addition of a background stochastic **noise** schema. This SI would produce a low-magnitude random direction velocity vector that would change at random time intervals, but persist sufficiently long to produce a change in the robot's position if the robot's velocity was otherwise zero. The behavior produced by this schema would correspond to the "wander" layer in Brook's horizontally layered architecture [17]. This schema would serve to remove the robot from any potential field plateaus or ridges upon which the robot becomes perched (e.g. from a direct approach to an obstacle - fig. 9). Other traps common to potential field approaches (e.g. box canyons) can be handled by establishing hard time deadlines for goal attainment. If these deadlines are violated, the pilot would be reinvoked to establish an alternate route using STM data gathered by the cartographer during the route traversal.

It is worth noting that a single sensory event may have two or more SIs associated with it. For example: if the robot is looking for a mailbox to get its bearings for localization purposes, a perceptual schema for localization (**find-landmark**) would process portions of the image that are likely to be mailboxes. If the mailbox happens to be in the path of the vehicle, a concurrent **avoid-static-obstacle** SI would view that object not as a mailbox but rather as an obstacle, only concerned with avoiding a collision with it. This "divide and conquer" approach based on action-oriented perception simplifies programming and overall system design. A more complex scenario appropriate for a mobile robot appears in fig. 10.

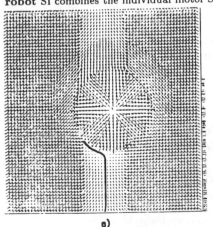

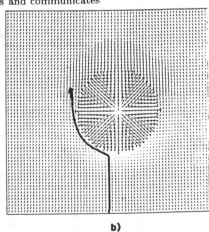

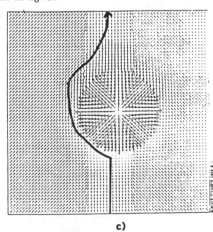

Figure 8. Blocked sidewalk scenario.
a) Robot stops in dead spot due to pressure to both remain on sidewalk and avoid the obstacle.
b) Gain lowered on **stay-on-path** SI allows robot to bypass obstacle.
c) Once obstacle is passed **stay-on-path** SI returns to normal, forcing robot back onto the sidewalk.

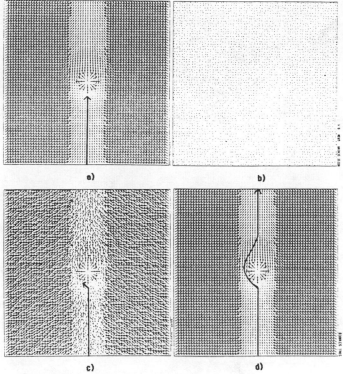

Figure 9. Stall scenario.

a) If the robot approachs an obstacle exactly head-on, it is possible for it to become stalled.

b) **Noise SI** provides small magnitude random direction vector to push robot off of the tiny plateau.

c) **Noise schema** added to a).

d) The robot can now succssfully bypass the obstacle. The **noise SI** is then deinstantiated.

4. Implementation Strategy

The implementation tool chosen for the motor schema system is the Schema Shell [22]: a system developed by the VISIONS group at UMASS for use in the perceptual schema analysis of natural scenes. It currently runs on a Texas Instruments Explorer workstation and is tied to the Computer Science Department's VAXen over Chaosnet. The schema communication mechanism is blackboard-based. The Schema Shell system is expected to be moved to the department's newly acquired Sequent parallel processor. Although the Explorer only simulates distributed processing, everything points towards the availability of a truly distributed environment in the not too distant future.

The schemas themselves (in the Schema Shell) consist of a schema template and multiple strategies. Associated with each instantiated schema is an object hypothesis maintenance (OHM) strategy. This part of the SI monitors the blackboard for new events (e.g. sensory data) that would produce changes in the SI's posted output. Other strategy components for each SI handle conflict resolution, cooperative enhancement, initialization and other relevant factors. Not all strategies are necessary or desirable for all schemas.

Multiple instantiations of a single schema are frequently the case. Each generic "skeleton" is parameterized when instantiated. Consequently, it is entirely possible that two different instantiations of the same generic schema produce significantly different fields under similar sensory conditions. The parameters set at instantiation may depend on the sensory events that triggered the instantiation. Fig. 11 shows a typical generic motor schema cast in the Schema Shell format.

Pilot issues instructions to follow sidewalk while avoiding obstacles Continue approximately 200 ft on sidewalk then turn right at lamppost onto intersection (first encountered) Watch for landmarks (mailbox on left, building edge on right) for localization

Motor Schemas instantiated by pilot

→ **Complete_pilot_maneuver**(schema_list,schema_base_priority_vector) creates:

- **Stay_on**(identify_terrain(ahead,60%))   (assumes sidewalk is ahead to start)
- **Move_ahead**(200,10,start_heading)   (nominal distance, then add some slop)
- **Avoid_static_obstacles**(15,identify_obstacle(robot_heading,nil,70%))
   Start maneuvering around when within 10 feet
   nil denotes static obstacle
- **Avoid_dynamic_obstacles**(20,identify_obstacle(robot_heading,- robot_heading,40%))
   Start evasive action when head on approach within 20 feet
- **Follow_dynamic_obstacle**(8,start_heading,identify_obstacle(True,start_heading95%))
   when an obstacle is moving in the right direction within 8 feet of the robot, follow it (regardless of robot's current heading)
- **Avoid_dynamic_obstacles**(5,identify_obstacle(robot_heading,True,40%))
   Start evasive action when within 5 feet for any dynamic obstacle
   (includes crossing dynamic obstacles)
- **Turn_when**(recognize_landmark(lamppost_1+5ft,90,90%)) - right 90 degrees
- **Turn_when**(recognize_landmark(intersection_3a,90,90%)) - right 90 degrees
- **Localize**(recognize_landmark(mailbox_7,90%))
- **Localize**(recognize_landmark(building_2a,edge3, 85%))
   prune spatial error map on landmark recognition
- **Stop_when**(not (sidewalk_1 = identify_terrain(ahead,90%)))   (missed turn)

Perceptual Schemas instantiated by pilot:

- **Identify_obstacle**(robot_heading,obstacle_heading,certainty)
   Only detects obstacles in the way of the robot (distinct from landmarks)
   robot_heading and obstacle_heading are filters
   certainty is threshold for identification
   returns obstacle position and type
   1 identify_obstacle spawned for each strategy type above
   * obstacle - generic - many spawned for each identify_obstacle
   returns certainty
   tracks motion over time
   types:
   static_obstacle - predicate
   dynamic_obstacle - predicate
   obstacle_heading (nil if static)
   speed   (0 if static)
- **Recognize_landmark**(LTM_model,certainty) - 1 spawned per landmark
   Assumes robot's current position for observation is available in global coordinates (spatial error map)
   certainty is threshold for recognition
   returns landmark location
   (not necessarily in direct track of robot, could be anywhere)
   * landmark(LTM_model) - many/landmark spawned off
   returns certainty
- **Identify_terrain**(position,certainty)
   returns terrain type

At end of maneuver, deinstantiate all obstacle schemas.

Figure 10. Example mobile robot schema scenario.

```
;-------------------------------------------------
;           MOVE-AHEAD Motor Schema
;    ->
;    ->      x-axis is 0; frame of reference is robot's initial heading afterturn;
;    ->
;    ->
(make-motor-schema
    :name "MOVE-AHEAD"
    :default-argument-list (list '("heading" 0))
    :body
    (progn
       (de-schema  move-ahead (original-heading current-heading move-ahead-force-table))

          (de-strategy move-ahead ohm ()
             (progn
                (call-strategy 'move-ahead 'init)
   (loop
                (setf #!current-heading
                   (read-or-wait '(current-orientation-message robot-position)))

                (write-to-blackboard
                   (list "move-ahead" (look-up-move-ahead-force #!current-heading) (time-stamp))
                   'vector-section)  ; write resultant force to vector section
                )
    ))

       strategies for move-ahead

       (de-strategy move-ahead init()
          (build-move-ahead-force-table)               ; initialize lookup table
          (setf #!original-heading (read-or-wait '(orientation-message robot-position)))
          (write-to-window "move-ahead init done")
          )

    ; conflict in case of reverse direction- send message-to-mover-to-terminate
    ; contains list of contradictory motions or evidence

    ; support in case of confirmed direction
    ; support contains list of related schemas use to confirm hypothesis
    ; e.g. to stop-when's
    )); End make-motor-schema
;-------------------------------------------------
```

Figure 11.
Prototype **move-ahead** schema as implemented in the Schema Shell.

5. Simulation

Simulations were run on a VAX 750 using the following motor schemas: **stay-on-path**, **move-ahead**, **move-to-goal**, **avoid-static-obstacle**. Each simulation run (fig. 12–13) shows the sequence of resultant overall force fields based on perceived entities. These entities include path borders and obstacles. The grid size is 64 units by 64 units and sensory sampling update time (once per second) is based on a nominal velocity of 1 unit/second. The maximum vector length for display purposes has been set to 2.0 normal velocity units. The actual vector magnitude within the obstacles is set to infinity (a discrete approximation). All obstacles are currently modeled as circles (as in Moravec's tangent space [24]).

The field equations for both the **avoid-static-obstacle** and **stay-on-path** schemas are linear. An example showing the velocity produced by an obstacle (O) is below:

$$O_{magnitude} = \begin{cases} 0 & for \ d > S \\ \frac{S-d}{S-R} & for \ R < d \leq S \\ \infty & for \ d \leq R \end{cases}$$

where:

S = Sphere of Influence (radial extent of force from the center of the obstacle)
R = Radius of obstacle
d = Distance of robot to center of obstacle
$O_{direction}$ = along a line from robot to center of obstacle moving away from obstacle

More complex equations could be used (e.g. cubic as in [20]) but were deemed unnecessary in these early stages of the research.

Figure 12 illustrates the robot's course on a sidewalk moving towards a goal. The course is studded with 8 obstacles, only 7 of which are perceptible to the robot during its journey (fig. 12a). Note how the vector fields change as the robot encounters more obstacles along the way (fig. 12b-d). When it has successfully navigated obstacles and they have moved out of range, their representation is dropped from short-term memory and the associated motor schema is deinstantiated (fig. 12d). The robot stays on the path for the complete course successfully achieving its goal while avoiding each obstacle. An expanded version could update long-term memory as a result of experience, thus incorporating learning.

Figure 13 shows the robot's path to a specified goal through a field of 9 obstacles. This simulation prevents perceived objects that have too great an uncertainty from producing a repulsive field. In this case, the uncertainty increases with the distance from the obstacle. The simulation in figure 13 uses a **move-to-goal** SI. Actually the robot would operate under the control of a **move-ahead** SI until the goal is perceived (assuming dead-reckoning or inertial guidance is not used). At the moment of goal perception, the **move-ahead** SI would be deinstantiated and a **move-to-goal** SI created in its stead.

6. Summary and Future Work

Motor schemas are proposed as a means for navigation of a mobile robot. This schema-based methodology affords many advantages. These include the use of distributed processing, which facilitates real-time performance, and the modular construction of schemas for ease in the development, testing and debugging of new behavioral and navigational patterns. Complex behavioral patterns can be emulated by the concurrent execution of individual primitive SIs.

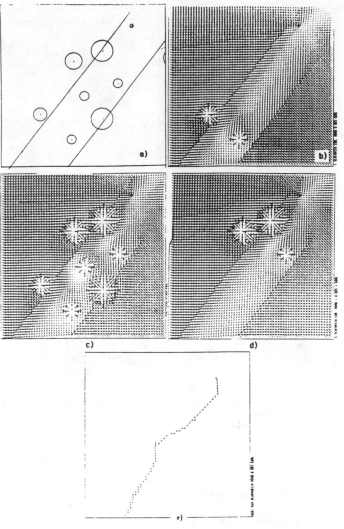

Figure 12. Simulation run.

This simulation shows 7 avoid-static-obstacle SI and a stay-on-road SI.
a) Shows the layout of the obstacle ridden course.
b-d) With the robot starting at the lower left, the robot's progress through the course can be observed. Note that the obstacles are added as they are perceived by the sensory system. No *a priori* knowledge of their whereabouts is assumed.
e) The robot's path through the course.

The next logical step is to complete the implementation of the system on the Explorer within the framework of the Schema Shell and to interface it with the high-level planning component of AuRA.

Work is underway for the acquisition of road edges using a new fast line-finding algorithm that can serve as the perceptual schema for the **stay-on-path** motor schema. Obstacle location using a multiple frame depth-from-motion algorithm [26,28] is being explored as a perceptual schema for the associated **avoid-static-obstacle** SI. Additionally, the use of ultrasonic data as input for the **avoid-obstacle** SIs is anticipated.

Motor schemas for following a moving object (tracking) and avoiding moving obstacles are being developed. These will enable the vehicle to emulate both follow-the-leader and dodging behaviors.

Long term goals include tying in the VISIONS system as the

means for providing sensor-independent object-based input to the motor schemas.

Acknowledgments

The author would like to thank Prof. Ed Riseman, Prof. Michael Arbib, Dr. Damian Lyons and the members of the Laboratory for Perceptual Robotics and VISIONS groups for their assistance in the preparation of this paper.

References

1. Bartlett, F.C.,*Remembering: A Study in Experimental and Social Psychology*, London, Cambridge at the Univ. Press, 1932.
2. Oldfield, R.C. and O.L. Zangwill, "Head's Concept of the Schema and its Application in Contemporary British Psychology – Parts I and II", *Br.J. Psychol.* 32:267-286 1942, 33: 58-64,113-129,143-149, 1943.
3. Piaget, J., *Biology and Knowledge*, Univ. of Chicago Press, 1971.
4. Head, H. and Holmes, G., "Sensory disturbances from cerebral lesions", *Brain* 34:102-254, 1911.
5. Frederies, J.A.M., "Disorders of the Body Schema", *Handbook of Clinical Neurology, Disorders of Speech Perception and Symbolic behavior*, Ed. Vinken and Bruyn, North Holland, 1969, Vol. 4, pp. 207-240.
6. Webster's *Ninth New Collegiate Dictionary*, Merriam-Webster, 1984.
7. Neisser, U., *Cognition and Reality: Principles and Implications of Cognitive Psychology*, Freeman, 1976.
8. Arbib, M., "Perceptual Structures and Distributed Motor Control", *Handbook of Physiology – The Nervous System II*, ed. Brooks, pp. 1449-1465, 1981.
9. Arbib, M., *The Metaphorical Brain*, Wiley, 1972.
10. Arbib, M. and House, D., "Depth and Detours: An Essay on Visually Guided Behavior", *COINS TR* 85-20, Univ. of Mass, 1985.
11. Lyons D., "Tagged Potential Fields: An Approach to Specification of Complex Manipulator Configurations", *IEEE Conf. on Robotics and Auto.*, pp. 1749-1754, 1986.
12. Geschke, C., "A System for Programming and Controlling Sensor-based Robot Manipulators", *IEEE Trans. on PAMI*, PAMI-5, No. 1, 1983, pp. 1-7.
13. Weymouth, T., "Using Object Descriptions in a Schema Network for Machine Vision", *Ph.D. Dissertation, COINS Tech. Rep.* 86-24, Univ. of Massachusetts, Amherst, May 1986.
14. Lyons, D., "RS: A Formal Model of Distributed Computation for Sensory-Based Robot Control", *Ph.D. Dissertation, COINS Tech. Rep.* 86-43, , Univ. of Massachusetts, Amherst, 1986.
15. Overton, K., "The Acquisition, Processing, and Use of Tactile Sensor Data in Robot Control", *Ph.D. Dissertation, COINS Tech. Rep.* 84-08, Chapter 6, Univ. of Massachusetts, Amherst, 1984.
16. Lyons, D. and Arbib, M., "A Task-Level Model of Distributed Computation for Sensory-Based Control of Complex Robot Systems", *COINS Tech. Rep.* 85-30 (also *IFAC Symposium on Robot Control*, Barcelona), 1985.
17. Brooks, R., "A Robust Layered Control System for a Mobile Robot", *IEEE Jour. of Robotics and Auto.*, Vol. RA-2, No. 1, 1986 pp. 14-23.
18. Kadonoff, M., Benayad-Cherif, F., Franklin, A., Maddox, J., Muller, L., Sert, B. and Moravec, H., "Arbitration of Multiple Control Strategies for Mobile Robots", to appear in MOBILE ROBOTS - *SPIE proc. Vol. 727*, 1986.
19. Krogh, B., "A Generalized Potential Field Approach to Obstacle Avoidance Control", *SME - RI Technical Paper* MS84-484, 1984.
20. Krogh, B. and Thorpe, C., "Integrated Path Planning and Dynamic Steering Control for Autonomous Vehicles", *IEEE Conf. on Robotics and Auto.*, pp. 1664-1669, 1986.
21. Arkin, R., "Path Planning for a Vision-based Mobile Robot", to appear in MOBILE ROBOTS - *SPIE Proc. Vol. 727*, 1986.
22. Draper, B., "Introduction to the Schema Shell", and "Programmer's Reference for using the Schema Shell", *VISIONS working papers*, Univ. of Massachusetts, Amherst, 1986.
23. Payton, D., "An Architecture for Reflexive Autonomous Vehicle Control", *IEEE Conf. on Robotics and Auto.*, pp. 1838-1845, 1986.
24. Moravec, H., *Robot Rover Visual Navigation*, UMI Press, 1981.

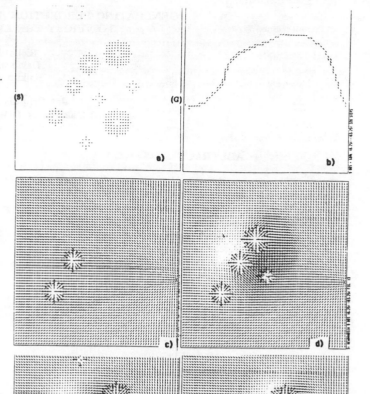

Figure 13. Simulation run.

This simulation include 9 avoid-static-obstacle SIs and 1 **move-to-goal** SI.

a) Location of the 9 obstacles.

b) Path of robot as it crosses from left to right around obstacles to the goal.

c-f) Velocity fields based on robot's perceptions as it moves from left to right as in b).

This simulation includes an uncertainty measure for obstacles which increases with the distance of the obstacle from the robot. If the obstacle is relatively uncertain, its position is shown but it produces no field (e.g. the two rightmost obstacles in fig. c). As the robot approaches, it becomes more certain of the obstacle and starts to produce a repulsive field surrounding the obstacle.

25. Arbib, M., Iberall, T. and Lyons, D., "Coordinated Control Programs for Movements of the Hand", *COINS TR* 83-25, Univ. of Mass., 1983 (also *Exp. Brain Res. Suppl.*, 10, pp. 111-129, 1985.
26. Bharwani, S., Riseman, E. and Hanson, A., "Refinement of Environmental Depth Maps over Multiple Frames", *Proc. IEEE Workshop on Motion Representation and Analysis*, pp. 73-80, May 1986.
27. Lozano-Perez, T., "Automatic Planning of Manipulator Transfer Movements", in *Robot Motion: Planning and Control*, ed. Brady et al, pp. 499-535,1982.
28. Arkin, R., Riseman, E. and Hanson, A., "Visual Strategies for Mobile Robot Navigation", *COINS Tech. Rep.*, in preparation, Univ. of Massachusetts, Amherst, 1987.

GENERATING PERCEPTION REQUESTS AND EXPECTATIONS TO VERIFY THE EXECUTION OF PLANS

Richard J. Doyle
David J. Atkinson
Rajkumar S. Doshi

Jet Propulsion Laboratory
4800 Oak Grove Drive, Pasadena, CA 91109

ABSTRACT

This paper addresses the problem of verifying plan execution. An implemented computer program which is part of the execution monitoring process for an experimental robot system is described. The program analyzes a plan and automatically inserts appropriate perception requests into the plan and generates anticipated sensor values. Real-time confirmation of these expectations implies successful plan execution. The implemented plan verification strategy and knowledge representation are described. Several issues and extensions of the method are discussed, including a language for plan verification, heuristics for constraining plan verification, and methods for analyzing plans at multiple levels of abstraction to determine context-dependent verification strategies.

1. THE PROBLEM

In a partially-modelled real world, an agent executing a plan may not actually achieve desired goals. Failures in the execution of plans are always possible because of the difficulty in eliminating uncertainty in world models and of a priori determining all possible interventions. Given these potential failures, the expected effects of actions in a plan must be verified at execution time.

In this paper, we address the problem of providing an execution monitoring system with the information it needs to verify the execution of a plan in real time. Our solution is to identify acquirable **perceptions** which serve as more reliable verifications of the successful execution of actions in a plan than do the inferences directly derivable from the plan itself. Assertions which appear as preconditions and postconditions in plan actions are mapped to appropriate sensor requests and expectations describing a set of values. Observing a value from the expectation set on the indicated sensor at the appropriate time during execution of the action implies that the assertion holds. The strategies for verifying the execution of actions are derived from the intentions behind their use. The knowledge of which perceptions and expectations are appropriate for which actions is represented by **verification operators**.

These ideas have been implemented in a working computer program called **GRIPE** (Generator of Requests Involving Perceptions, and Expectations). After describing this program, we propose several generalizations of its results by examining the issues involved in **Selection**, or determining which actions in a plan to monitor, and **Generation**, or how to verify the successful execution of particular actions.

1.1. CONTEXT OF THE PROBLEM

Generating perception requests and expectations to verify the execution of actions in a plan is only one aspect of a robust control system for an intelligent agent. Research on such a control system is underway at the Jet Propulsion Laboratory. In our system, known as **PEER** (Planning, Execution monitoring, Error interpretation and Recovery) [Atkinson, 1986] [Friedman, 1983] [Porta, 1986] several cooperating knowledge-based modules communicate through a blackboard [James, 1985] in order to generate plans, monitor those plans, interpret errors in their execution, and attempt to recover from those errors. In this paper we concentrate on part of the execution monitoring task.

The primary application for PEER is the proposed NASA/JPL Telerobot, intended for satellite servicing at the U.S. space station. The current testbed scenarios for the Telerobot include a subset of the Solar Max satellite repairs recently accomplished by shuttle astronauts.

Broadly speaking, the tasks which must be accomplished by execution monitoring are **Selection**, **Generation**, **Detection/Comparison**, and **Interpretation**. The **Selection** task must determine which effects of actions in the plan require monitoring. The **Generation** task involves determining the appropriate sensors to employ to verify assertions, and the nominal sensors values to expect. This is the task accomplished by the GRIPE system, which we discuss in detail below. The **Detection/Comparison** monitoring task handles the job of recognizing significant events on sensors and then comparing these events with the corresponding expectations. Finally, the **Interpretation** task involves explicating the effects of failed expectations on subsequent plan actions. We will discuss all of these in more detail.

1.2. OTHER WORK

Monitoring task execution and feedback have been the topic of research in AI for quite some time. Attention has been focused on monitoring at the task level, the geometric and physical levels, and also at the servo level. Early work which exposed the role of uncertainty in planning and other problems in error recovery includes [Fikes, 1972], [Munson, 1972], and [Chien, 1975]. Sacerdoti discussed the issues of monitoring and verification in NOAH in detail [Sacerdoti, 1974] [Sacerdoti, 1977]. This work illustrated the role which planning at multiple levels of abstraction could play in monitoring. NOAH used the plan hierarchy as a guide for asking a human to verify plan assertions.

Some recent research in planning and execution monitoring has focused on handling uncertainty. A planner implemented by Brooks reasons about the propagation and accumulation of errors [Brooks, 1982]. It modifies a plan by inserting sensing operations and constraints which ensure that the plan does not become untenable. Erdmann's method for planning trajectory motions utilizes a backprojection algorithm that geometrically captures the uncertainty in motion [Erdmann, 1985]. Donald also addressed the the problem of motion planning with uncertainty in sensing, control and the geometric models of the robot and its environment [Donald, 1986]. He proposed a formal framework for error detection and recovery.

[Wilkins, 1982] and [Wilkins, 1985] deals extensively with planning actions to achieve goals. Wilkins also deals with Error Recovery which is the problem of recovering from errors that could occur at execution time. To be precise, Wilkins does not deal with the problem of planning to monitor the plan generated by the planner.

[Tate, 1984] discusses the usefulness of the intent and the rich represtation of plans. He also mentions the issues in goal ordering, goal interaction, planning with time, cost and resource limitations, and interfacing the planner with other subsystems. He also throws some light on some solutions to the problem of Error Recovery. He mentions the problem of Execution Monitoring but has not discussed the problems or issues or any related solutions.

Other recent research has also addressed the problem of using sensors to verify plan execution. Van Baalen [VanBaalen, 1984] implemented a planner that inserts sensory action requests into a plan if an assertion of an operator is manually tagged "MAYBE".

Miller [Miller 1985] includes continuous monitoring and monitoring functions in a route navigation planner. His focus is on the problem of coordination of multiple time dependent tasks in a well-known environment, including sensor and effector tasks involving feedback.

Gini [Gini et al., 1985] have developed a method which uses the intent of a robot plan to determine what sensor conditions to check for at various points in the plan. In the final executable plan, the system inserts instructions after each operator to check approriate sensors for all possible execution errors.

Fox and Smith [Fox et al., 1984] have also acknowledged the need to detect and react to unexpected events in the domain of job shop scheduling.

2. IMPLEMENTATION

The algorithms presented in this paper have been implemented in a working computer program called GRIPE. In addition to the knowledge sources supplied to the program, the basic input is a plan specification as described below. GRIPE's output consists of a modified plan which includes sensing operations, expectations about sensor values to be used by a sensor monitoring program, and subgoals for the planner to plan required sensor operations or establish preconditions for sensing. GRIPE has been tested on a segment of the JPL TeleRobot demonstration scenario and generates a plan of 67 steps modified to include perception requests and additional output, as described above. The examples shown below are drawn from this test case. GRIPE has been tested on

examples from the Solar Max satellite repair domain. The system generates a modified plan which include perception requests, as well as expectations about those perceptions and subgoals for acquiring those perceptions. The following sections describe this process in more detail.

2.1. VERIFICATION STRATEGY

The basic input to the GRIPE system is a plan specification. GRIPE prepares a plan for execution monitoring by examining the preconditions and postconditions of each action in the plan. For each of these assertions GRIPE generates an appropriate perception request and an expectation which, if verified, implies that the assertion holds. During execution, an action is commanded when all of its preconditions are verified and its successful execution is signalled when all of its postconditions are verified.

GRIPE prepares a plan for execution monitoring by examining the preconditions and postconditions of each action in the plan. For each of these assertions GRIPE generates an appropriate perception request and an expectation which, if verified, implies that the assertion holds. An action is commanded when all of its preconditions are verified and its successful execution is signalled when all of its postconditions are verified. GRIPE uses dependency information between conditions established as postconditions in one action and required as preconditions in another. If the establishment and use of a condition occurs in consecutive actions, the condition is verified only once. Otherwise, the condition is verified when it is established, and re-verified when it is needed.

The method of verifying the execution of an action is derived by examining the intention behind its use. The knowledge of which perceptions and expectations are appropriate for which actions is encoded in **verification operators**, described below. In the prototype GRIPE implementation, we assume that actions have a single intention. In general, however, the intent of an action may vary according to the context in which it appears.

As an example, consider moving a robot's arm as a precondition to a grasp. This operation may require high accuracy. A combination of sensors such as position encoders in the arm, proximity sensors at the end effector and vision could be used to ensure that the end-effector is properly placed to grasp this object. On the other hand, moving the arm away from the object after the release operation may require very little verification. The available latitude in the final position of the arm may be large. In this case, a cursory check on the position encoders may suffice. In the prototype GRIPE implementation, we assume that actions have a single intention.

2.2. REPRESENTATION

Before examining in detail how GRIPE generates perception requests and expectations, we describe our representation for plans and our models for actions and sensors in the JPL TeleRobot domain.

A plan is a totally ordered sequence of actions representing a schedule of commands to an agent. The dependencies among actions are maintained explicitly.

Actions are modelled in the situation calculus style [Fikes, 1971], with specified preconditions and postconditions. In addition, an explicit duration for each action is determined and represented by a start and stop time. Currently, we model all actions as having the same duration. As an example, the action operator for GRASP is shown in Figure #1.

```
(create-action-operator
:type GRASP
:action (GRASP End-Effector Object before after)
:preconditions
   ((MODEL= (POSITION-OF-MODELLED Object before)
       (POSITION-OF-MODELLED End-Effector before)
    (MODEL= (FORCE-OF-MODELLED End-Effector before)
       0)
    (MODEL= (WIDTH-OF-MODELLED End-Effector before)
       'Open))
:postconditions
   ((MODEL= (FORCE-OF-MODELLED End-Effector after)
       (COMPLIANT-GRASP-FORCE-FOR Object after))
    (MODEL= (WIDTH-OF-MODELLED End-Effector after)
       (GRASP-WIDTH-FOR Object after)))))
```

Figure 1:   GRASP Action Operator

There are four types of sensors currently modelled for the TeleRobot: position encoders for the arms, force sensors at the end-effectors, configuration encoders for the end-effectors which tell how wide the grippers are held, and vision system cameras. The table shown in Figure #2 associates with each sensor type the actual perception request which GRIPE grafts into plans. We assume that all the sensors except vision can be read passively at any time. The TeleRobot vision system uses CAD/CAM-type models and requires an expected position and orientation to effectively acquire objects.

Sensor	Perception Request
Arm-Kinesthetic	(WHERE End-Effector when)
Hand-Kinesthetic	(CONFIGURATION End-Effector when)
Force	(FEEL End-Effector when)
Vision	(SEE Object Position when)

Figure 2:   Perception Requests for Sensors

Note that there are two equivalence predictes, **MODEL=** and **SENSE=**. The **MODEL=** predicate appears in the preconditions and postconditions of action operators and represents a comparison between two assertions in the world model. All reasoning done during planning occurs within the world model. Assertions in the world model are identified by the suffix -**MODELLED**.

The **SENSE=** predicate, appearing in expectations generated by GRIPE, represents comparisons between a perception and an assertion in the world model. These comparisons are the essence of verification. Perceptions are identified by the suffix -**SENSED**.

2.3. VERIFICATION OPERATORS

The knowledge of how to verify the assertions which appear as preconditions and postconditions in actions is captured by **verification operators**. Verification operators map assertions to appropriate perception requests, expectations, and possibly subgoals. The definition of verification operators is shown in Figure #3. As an example, the verification operator for determining that an object is at a particular location is shown in Figure #4.

```
(define-verification-operator
   ;Assertion to be verified.
assertion
   ;Actions partially verified by this operator.
actions
   ;Constraints on assertion.
constraints
   ;The sensor to be used.
sensor
   ;Perception request which can verify assertion.
perception
   ;Preconditions for obtaining the perception.
preconditions
   ;Sensor value which verifies assertion.
expectation)
```

Figure 3:   Verification Operator Definition

Each verification operator is relevant to a single assertion which may appear as a precondition or postcondition in several different actions. Verification operators are indexed under the actions which they help to verify. In our example, there are three steps involved in determining the relevance of the verification operator shown in Figure #4 to preconditions of the GRASP action shown in Figure #1.

First, the relevant verification operators for the GRASP action are retrieved. Next, an attempt is made to unify the precondition against the assertion pattern specified in each retrieved verification operator. Finally, any constraints specified in the verification operator are checked. These constraints constitute a weak context mechanism; in our example, the specified constraint distinguishes the use of position encoders to verify the location of an end-effector from the use of the vision system to verify the location of an external object.

Once the relevant verification operator has been identified, a perception request and expectation for verifying that the precondition holds at execution time are generated from the appropriate fields of the verification operator. This information is then passed to the real-time execution monitor.

Perception requests are themselves actions to acquire perceptions via various sensors. The use of sensors may also be subject to the establishment of preconditions. In our example, the simulated vision system can acquire an object only if there are unobstructed views from the cameras to the object. Currently, the other three sensors we simulate are passive and do not have preconditions on their use. In this case, GRIPE generates and submits subgoals generated by particular perception

```
    (create-verification-operator
  :sensor
     VISION
  :actions
     (GRASP RELEASE)
  :assertion
     (MODEL= (POSITION-OF-MODELLED Object Moment)
             Position)
  :constraints
     ((NOT (MEMQ Object
            ' (Left-End-Effector Right-End-Effector))))
  :perception
     (SEE Object Position Moment)
  :preconditions
     ((UNOBSTRUCTED-PATH
        (POSITION-OF-MODELLED
             'Left-Camera Moment)
        Position
        Moment)
      (UNOBSTRUCTED-PATH
        (POSITION-OF-MODELLED
             'Right-Camera Moment)
        Position
        Moment))
  :expectation
     (SENSE= (POSITION-OF-SENSED Object Moment)
             Position))
```

Figure 4: Example Verification Operator

requests to the planner. The task of the planner is to
further modify the plan, which now includes perception
requests, so that preconditions on the use of sensors are
properly established. This process details the extent of the
interaction of monitoring and planning and suggests the
issue of how closely the two processes should be
interleaved, a problem which has not yet received much
close attention.

3. AN EXAMPLE

The following example is drawn from a satellite
repair scenario for the JPL Telerobot, described above.
Part of the servicing sequence in the previous example
involves grasping the handle of a hinged panel on the
satellite. A segment of this plan is shown in Figure #5
When GRIPE processes this plan segment for execution
monitoring, it inserts appropriate perception requests into
the plan and generates expectations about nominal sensor
values. The plans given to GRIPE have been
hand-generated.

```
Perform the action
   (MOVE right-end-effector handle
        (NEAR (POSITION-OF-MODELLED handle 2))
        (POSITION-OF-MODELLED handle 3)
        2 3).
Perform the action
   (GRASP right-end-effector handle 3 4).
```

Figure 5: Example plan input to GRIPE

Verify and do the action (MOVE ... 2 3)
using the **ARM-KINESTHETIC** *sensor*
```
   (WHERE right-end-effector 2).
   (SENSE= (POSITION-OF-SENSED right-end-effector 2)
           (NEAR (POSITION-OF-MODELLED handle 2)))
   (MOVE ... 2 3)
   (WHERE right-end-effector 3)
   (SENSE= (POSITION-OF-SENSED  right-end-effector 3)
           (POSITION-OF-MODELLED handle 3))
```

Verify and do the action (GRASP ... 3 4)
using **VISION** *sensor, the* **FORCE** *sensor,*
and the **HAND-KINESTHETIC** *sensor*
```
   (SEE handle
        (POSITION-OF-MODELLED right-end-effector 3) 3)
   (SENSE= (POSITION-OF-SENSED handle 3)
           (POSITION-OF-MODELLED
                right-end-effector 3))
   (FEEL right-end-effector 3)
   (SENSE= (FORCE-OF-SENSED right-end-effector 3) 0)
   (CONFIGURATION right-end-effector 3)
   (SENSE= (WIDTH-OF-SENSED right-end-effector 3)
           open)
   (GRASP ... 3 4)
   (CONFIGURATION right-end-effector 4)
   (SENSE= (WIDTH-OF-SENSED right-end-effector 4)
           (GRASP-WIDTH-FOR handle 4))
   (FEEL right-end-effector 4)
   (SENSE= (FORCE-OF-SENSED right-end-effector 4)
           (COMPLIANT-GRASP-FORCE-FOR handle 4))
```

Figure 6: Example plan output from GRIPE

GRIPE's strategy for verifying the successful
execution of these two actions is: Use the position
encoders of the arm to verify that the end-effector is in
the correct position before and after the MOVE-TO.
Before the GRASP, use the vision system to verify that
the handle is in the expected location, use the force
sensor to verify that the end-effector is not holding
anything, and use the configuration encoder of the
end-effector to verify that it is open. After the GRASP,
read the force sensor and configuration encoder of the
end-effector and verify that the values on these sensors
are appropriate for gripping the handle. The modified
plan is shown in Figure #6.

4. ISSUES

A number of issues have been raised during our
development of the GRIPE system, some of which were
handled in the initial implementation by making certain
assumptions. In this section we examine these issues in
detail and propose some preliminary solutions.

4.1. PERCEPTION VERSUS INFERENCE

The essence of verification is gathering a
perception which implies that an assertion holds. A
motivating assumption of our work is that inferences
which have a basis in perception are more reliable as
verifications than inferences (such as the specification of
postconditions in an action) which are not so based. Thus
our basic strategy of verification is to substitute relevant
perceptions for the assertions that appear in plans.

Verification operators embody essentially one-step inferences between perceptions and assertions. There is no reason why such inferences could not be more indirect. An example appears implicitly in the GRASP action in our example above.

One of the preconditions for the GRASP action is that there must be no forces at the end-effector. Implicit in this assertion is the inference that the gripper is not holding any object when the forces at the end-effector are zero. This reasoning can be made explicit by making the assertion that the gripper is empty appear as the precondition in the GRASP action. An additional inference rule relating no forces at the gripper to the gripper being empty allows the same perception request involving the force sensor to be generated. However, now it is possible to define other strategies (e.g. using the vision system) to verify this restatement of the precondition for the GRASP action.

We intend to develop our verification knowledge base so that GRIPE can construct more complicated chains of inferences to determine how to verify the assertions appearing in plans. Under this extended verification knowledge base, there should often be several ways to verify a particular assertion. The considerations involved in choosing a verification strategy are discussed in the remainder of this section.

4.2. WHEN SHOULD THE EFFECTS OF PLAN ACTIONS BE MONITORED?

As others have pointed out [VanBaalen, 1984] [Gini et al., 1985], it is too expensive to check all the assertions in a plan. In many domains, it may be impossible. Sensors should be viewed as a resource of the agent which must be planned and scheduled just like other resources [Miller, 1985]. However, the process is aided by the observation that exhaustive monitoring may not be necessary and that selection criteria exist which can effectively limit the scope of monitoring. Some of these criteria are listed below. How they may best be combined in an assertion selection process is an open research topic.

- **Uncertainty Criteria.** Uncertainty in a number of forms may exist which requires that actions be closely monitored. This area has been the most extensively investigated [Brooks, 1982], [Donald, 1986], [Erdmann, 1985] and [Gini et al. 1985]. Uncertainty may exist in the world model which is used for planning. Uncertainty may exist about the effects of actions themselves; multiple outcomes may be possible. The effects of actions may be "fragile" and easily become undone (e.g., balancing operations). Actions may have known failure modes which should be explicitly checked. If the effects of actions have a duration, there may be uncertainty about their persistence.

- **Dependency Criteria.** There is a class of assertions which do not need to be verified at all. These are assertions which appear as postconditions of an action but are not required as preconditions of later actions, i.e., side effects. The assertions not on the this *critical path* of explicit dependencies between actions in a plan can be ignored in the verification process. The dependency information in the plan can be used to prune out these irrelevant effects of actions.

- **Importance Criteria.** If we have explicit representation of the dependencies among effects and actions in a plan, we can prioritize assertions for monitoring based on their criticality. The simplest metric is the number of subsequent actions which depend directly or indirectly on an assertion. More complicated metrics might take into account the importance of the dependent actions as well. The failure to achieve highly critical effects could have profound implications for subsequent error recovery.

- **Recovery Ease Criteria.** These criteria interact with the importance criteria. If an effect may be trivially re-achieved after a failure, the effects of a failure to verify even highly critical assertions is somewhat mitigated. Consequently, the need to monitor the assertion closely is not so severe.

4.3. WHICH PERCEPTION(S) CAN BEST VERIFY AN ASSERTION?

The current set of verification operators for GRIPE provide only a single, context-independent perception request for verifying individual assertions. In previous sections, we discussed how a more extensive verification knowledge base could support reasoning about multiple ways to verify assertions. These options often will be necessary.

For example, consider the difference between an arm movement which sets up a GRASP and a movement after a RELEASE. The location of the end-effector is critical to the success of a GRASP. In this case a battery of sensors such as position encoders, proximity sensors, force sensors, and vision might be indicated to verify that the end-effector is properly in place. On the other hand, a movement of an arm after a RELEASE may be performed relatively sloppily, particularly if this movement terminates a task sequence. A simple check on a position encoder (or even no check at all) may be sufficient.

4.4. HOW ACCURATELY SHOULD ASSERTIONS BE VERIFIED?

Using the same example, the latitude in the position of the end-effector for a GRASP is small; this position must be verified with a great deal of precision. On the other hand, the latitude in the position of the arm after movement away from a RELEASE is presumably quite large.

4.5. SHOULD AN ASSERTION BE VERIFIED INSTANTANEOUSLY OR CONTINUOUSLY?

In the current version of GRIPE, we assume that the successful execution of actions can be verified by instantaneously verifying the actions's preconditions before its execution, and instantaneously verifying its postconditions after its execution. This approach proves inadequate for some actions.

For example, consider a MOVE-OBJECT action which transports an object gripped by the end-effector of an arm by moving the arm. The force sensors in the end-effector should be checked continuously because the object might be dropped at any point along the trajectory.

Instantaneous monitoring is also insufficient for those actions which involve looping, for example, when a running hose is being used to fill a bucket with water. In this case, monitoring must not only be continuous but conditionalize the performance of the filling action itself.

4.6. SHOULD ASSERTIONS WITH PERSISTENCE BE RE-VERIFIED?

GRIPE's strategy for verifying assertions which are established as postconditions in one action and required as preconditions in a later, non-consecutive action is to verify the assertion twice -- both at the time of its establishment and at the time of its use.

Like the issue above, this issue concerns assertions across actions rather than assertions during actions. An error interpretation and recovery system should know as soon as possible if a condition which is needed later during the execution of a plan becomes unsatisfied. For example, suppose a part is to be heated and used in a delayed, subsequent action. If there is uncertainty about how quickly the part will cool (i.e., the duration or persistence of the "heated" assertion), then the temperature of the part should be frequently checked to verify it stays within the desired parameters. If it cools too quickly, additional heat may need to be applied before the subsequent action can be executed.

5. THE VERIFICATION LANGUAGE

The verification operators described earlier capture a restricted style of verification. In this section, we develop an extended language for verification which makes explicit a set of issues relevant to determining how to verify assertions in plans (the language is not yet implemented in GRIPE).

The need to perform both **instantaneous** and **continuous** monitoring functions suggests two fundamental types of perception requests, called **Brief** and **Prolonged.** Any particular perception request is exclusively one of these types. Brief-type perception requests handle instantaneous monitoring tasks which involve simple pattern matches against sensor or data-base values. Prolonged-type perception requests handle continuous or repetitive monitoring tasks which may involve extended modification of the plan. However, both types of perception requests may require preconditions to the use of sensors to be established by the planner. In addition, the planner ensures that any sensor resources specified are appropriate and available at the desired time. If sensor resources are not explicitly specified, the planner must choose appropriately from those available. The current GRIPE implementation does yet interact with a planner and therefore its sensor resource managment is not this facile.

Since monitoring an assertion may itself involve planning and the generation of additional plan actions, the process may recursively involve monitoring of the plan generated to achieve the original monitoring request. In the current GRIPE implementation, we allow a maximum recursive depth of two. However, to be satisfactory, we need to first, relax the depth restriction, and second, to use heuristics to constrain the recursion depth. The second requires a priori assumptions about the success of some plan operations.

```
<Perception-Request>      == <Brief-Type> |
                             <Prolonged-Type>
<Brief-Type>              == IF <Quick-Condition>
                             THEN <Action>
<Quick-Condition>         == data-base-query |
                             NOT data-base-query |
                             <Cond-Operator>
                                 <Sensor>
                                 <Value-Spec>
<Cond-Operator>           == IN-RANGE |
                             <Relational-Op>
<Relational-Op>           == < | = | > | <= | =>
<Sensor>                  == any available
                             and appropriate sensor
<Value-Spec>              == [<integer> ... <integer>] |
                             <integer>
<Prolonged-Type>          == CHECK <assertion>
                                 <Time-Spec>
                                 <Stopping-Cond>
<Time-Spec>               == <Time-Relationship> |
                             <Time-Designation>
<Time-Relationship>       == <Timing> <Action-Spec>
<Timing>                  == BEFORE | AFTER | DURING
<Action-Spec>             == an instance
                             of a plan-action-node
<Time-Designation>        == FOR <Time-Designation-Spec>
                             WITH FREQUENCY <integer>
<Time-Designation-Spec>   == TIME <Relational-Op> <integer> |
                             <integer> NUMBER-OF-TIMES |
                             NEXT <integer> ACTION-NODES
<Stopping-Spec>           == STOP MONITORING <Brief-Type>
```

Figure 7: Verification Planning Language

Figure #7 gives a grammar for a **Verification Planning Language** which addresses these considerations. Requests of the syntax defined in Figure #7 are generated by an expectation generator module such as GRIPE and recursively input to the planner. Eventually, this iteration flattens Prolonged-type perception requests. The final executable perception request in the plan is always of the Brief-type. For example, if a Prolonged-type perception request stated that an assertion should be monitored 5 times then the final plan would state **IF** *predicate* **THEN** *action 5 times.* Miller [Miller, 1985] has discussed similar ideas.

6. DETERMINING CONTEXT AND THE INTENTS OF ACTIONS

Our overall approach to verifying the execution of plans is driven by the following observation: The appropriate means of verifying the execution of an action is constrained by the intent of the action. In general, the intent of an action may vary according to context. Our results so far are restricted by the assumption that actions have a single intention. In this section, we describe our approaches to determining the intent of actions from context. They are similar to those described in [Gini et al. 1985].

One approach is top-down and assumes the existence of a hierarchical planner, as in [Sacerdoti, 1974]. Recall the example concerning two movements of an arm, one to set up a GRASP operation, and one after a RELEASE operation. The movement in the context of the

GRASP operation needs to be verified quite accurately; the movement in the context of the RELEASE operation requires only cursory verification. For example, the expanded movement operator before the GRASP might be a MOVE-GUARDED which indicates the need for careful verification; the expanded movement operator after the RELEASE might be a MOVE-FREE which requires less exacting verification.

Even when actions are not distinguished during expansion, the context provided by higher-level actions of which they are a part may be sufficient to distinguish them for the purpose of verification. In our example, the GRASP might have been expanded from a GET-OBJECT task while the RELEASE might have been expanded from a LEAVE-OBJECT task. The knowledge that the MOVE within a GET-OBJECT task is critical while the MOVE within a LEAVE-OBJECT task is not can be placed in the verification knowledge base.

The context of an action also may be determinable through a more local, bottom-up strategy. In this same example, the two MOVE actions at the lower level might be distinguished by noting that one occurs before a GRASP and the other occurs after a RELEASE. These contexts then can be used in the same way to retrieve appropriate verification strategies from the verification knowledge base.

7. INTERFACING WITH GEOMETRIC AND PHYSICAL LEVEL REASONING SYSTEMS

GRIPE reasons at what is commonly referred to as task level. We envision GRIPE and the other knowledge-based modules of our proposed PEER system interfacing with systems that can reason directly about the geometry and physics of task situations. Examples of systems are described in [Erdmann, 1985] and [Donald, 1986].

Erdmann has refined a method for computing the accuracy required in the execution of motions to guarantee that constraints propagated backward from goals are satisfiable. His approach could be incorporated into our system to generate expectations for verifying the execution of motions (only). An expectation would be a volume; a perception which indicates that a motion has reached any point within the volume would verify the successful execution of the motion.

Donald has developed a complementary technique for planning motions in the presence of uncertainities in the world model (as opposed to uncertainities in the execution of motions). He also proposes a theoretical framework for constructing strategies for detecting and recovering from errors in the execution of motion planning tasks.

8. CONCLUSIONS

The problem addressed in this paper is that of verifying the execution of plans. We have implemented a system which analyzes a plan and generates appropriate perception requests and expectations about those perceptions which, when confirmed, imply successful execution of the actions in the plan.

Typically, not all the assertions in a plan can or should be verified. We have proposed a number of heuristic criteria which are relevant to the selection of assertions for verification.

In general, verification strategies must be context-dependent; this need can be supported by an ability to analyze plans at multiple levels of abstraction.

Finally, we are developing a language for verification which makes explicit the relevant considerations for determining verification strategies: the appropriate perceptions, the degree of accuracy needed, discrete vs. continuous verification, and the need for re-verification.

9. ACKNOWLEDGEMENTS

The work described in this paper was carried out by the Jet Propulsion Laboratory, California Institute of Technology, under contract with the National Aeronautics and Space Administration. We would like to thank Leonard Friedman of the USC Information Sciences Institute for instigating the PEER project at JPL and stimulating our research. Rajkumar Doshi would like to thank his advisor Professor Maria Gini, University of Minnesota, for providing inspiration and ideas.

REFERENCES

[1] Atkinson D., James M., Porta H., Doyle R.
Autonomous Task Level Control of a Robot.
In Proceedings of Robotics and Expert Systems, 2nd
Workshop. Instrument Society of America
June, 1986.

[2] Brooks, Rodney.
Symbolic Error Analysis and Robot Planning.
A.I.Memo 685, Massachusetts Institute of Technology,
September, 1982.

[3] Chien, R.T., Weismann, S.
*Planning and Execution in Incompletely Specified
Environments.*
In International Joint Conference on Artificial
Intelligence. 1975.

[4] Donald, Bruce.
*Robot Motion Planning with Uncertainty in the
Geometric Models of the Robot and Environment:
A Formal Framework for Error Detection
and Recovery.*
In IEEE International Conference on
Robotics & Automation
San Francisco, CA, 1986.

[5] Erdmann, Michael.
*Using Backprojections for Fine Motion Planning
with Uncertainty.*
In IEEE International Conference on
Robotics & Automation
St. Loius, MO, 1985.

[6] Fikes, R.E., Nilsson, N.J.
*STRIPS: A new approach to the application
of Theorem Proving to Problem Solving.*
Artficial Intelligence Journal 2(3-4), 1971.

[7] Fikes, R.E., Hart, P.E., Nilsson, N.J.
New Directions in Robot Problem Solving.
Machine Intelligence 7, 1972.

[8] Fox, Mark S., Smith, Stephen.
*The Role of Intelligent Reactive Processing in
Production Management.*
In CAM-I, 13th Annual Meeting &
Technical Conference
November, 1984.

[9] Friedman, Leonard.
Diagnosis Combining Empirical and Design Knowledge.
Technical Report JPL-D-1328, Jet Propulsion
Laboratory, December, 1983.

[10] Gini, Maria., Doshi, Rajkumar S., Garber, Sharon.,
Gluch, Marc., Smith, Richard., Zualkernain, Imran.
Symbolic Reasoning as a basis for Automatic Error
Recovery in Robots.
Technical Report 85-24, University of Minnesota
July 1985.

[11] James, Mark.
The Blackboard Message System.
Technical Memorandum
Jet Propulsion Laboratory, 1985.
Write to author, stating reason.

[12] Miller, David P.
Planning by Search through Simulations.
PhD thesis, Yale University, October, 1985.

[13] Munson, John.
*Robot Planning, Execution and Monitoring in an
Uncertain Environment.*
In International Joint Conference on Artificial
Intelligence. 1972.

[14] Porta Harry.
Dynamic Replanning.
In Proceedings of Robotics and Expert Systems, 2nd
Workshop. Instrument Society of America, June, 1986.

[15] Sacerdoti, Earl.
Planning in a Hierarchy of Abstraction Spaces.
Artficial Intelligence Journal 5(2), 1974.

[16] Sacerdoti, Earl.
A Structure for Plans and Behaviour.
Elsevier North-Holland Inc., 1977.

[17] Tate, Austin.
Planning and Condition Monitoring in a FMS
University of Edingburgh,
Artificial Intelligence Applications Institute,
AIAI TR #2, July 1984.

[18] Van Baalen, Jefffrey.
Exception Handling in a Robot Planning System.
IEEE Workshop on Principles of Knowledge-Based
Systems, Denver, CO, December, 1984.
Not Published due to late submission.

[19] Wilkins, David.
*Domain Independent Planning:
Representation & Plan Generation*
SRI, Technical Note #266, August 1982.

[20] Wilkins, David.
Recovering From Execution Errors in SIPE
SRI Technical Note #346, January 1985.

Path Planning and Execution Monitoring for a Planetary Rover

Erann Gat   Marc G. Slack   David P. Miller   R.James Firby

Jet Propulsion Laboratory/California Institute of Technology
4800 Oak Grove Drive
Pasadena, California 91109

ABSTRACT

In order to navigate through natural terrain an autonomous planetary rover must be able to sense its environment, plan and traverse a course through that environment, and react appropriately to unexpected situations as they appear. All this must be done while guiding the vehicle towards the goals that have been given to it by its operators on the Earth. This paper describes research at the Jet Propulsion Laboratory which concentrates on the planning and execution monitoring that must be carried out by the rover to ensure that a safe and efficient path is found and traversed correctly.

1. Introduction

In the next two decades, NASA is planning several missions involving an autonomous rover moving across the surface of a planet and collecting samples. Due to communication delays of up to tens of minutes the rover must be able to navigate autonomously to sites of scientific interest and back to the return vehicle. To do this the rover must be able to plan and execute paths over dozens of kilometers through natural terrain.

The major problem in traversing natural terrain autonomously is that the information available to the vehicle about its environment is necessarily noisy and incomplete. Therefore, the path planning algorithm must be as robust as possible in the face of incomplete and inaccurate data. Furthermore, the vehicle must react appropriately and in real time to unexpected situations which will inevitably arise.

Because of the variety of possible missions (e.g., Lunar site survey, Mars sample return, Mars survey, etc.) the actual rover hardware will probably be different from mission to mission. Our approach tries to standardize the interactions between the planning systems, the rover's sensors, and the rover's mobility system and other subsystems. We are using a general architecture for autonomous rovers consisting of a sensing and perception system, a path planner, an execution monitoring planner, and a plan execution and monitoring system. The system is being tested on two robot testbeds with very different mobility characteristics.

This paper concentrates on the path planner and execution monitoring planner that will enable the rover to safely and correctly navigate to its various destinations while detecting and avoiding hazards. The remainder of this paper is divided into five sections. Section 2 gives an overview of the complete architecture. Section 3 describes the path planner. Section 4 describes the execution monitoring system. Implementation and testbeds are described in section 5. In the final section we summarize the unique aspects of this research.

2. Architecture Overview

In order to support a wide variety of mission scenarios we have developed a general architecture for a semiautonomous navigation (SAN) system. In the SAN approach the vehicle navigates autonomously to destinations which are given to it by human operators. The rover uses a coarse-resolution global terrain map together with local sensor information to plan and execute paths.

The system consists of four major components: a sensing and perception system, a path planner, an execution-monitoring planner, and a vehicle control system. (See figure 1.) In addition there is a system executive which coordinates the operations of the various subsystems, and a vehicle simulator which is used by the path planner and the execution-monitoring planner.

The system operates in a modified sense-plan-execute cycle. At the start of a cycle, the system executive instructs the vehicle's sensing and perception system to construct a model of the terrain surrounding the vehicle. The sensing and perception system takes raw data from the vehicle's sensors and constructs a representation of the local environment [Wilcox87]. It also correlates this local data with a global terrain map obtained from orbital data to more accurately determine the rover's current position. Information from the local and global maps is integrated to construct the final local terrain model which is passed to the path planner.

The local terrain model consists of arrays of equally-spaced data points at various resolutions. Each data point contains the average height and slope of the

Reprinted from the *Proceedings of the IEEE International Conference on Robotics and Automation*, 1990, pages 20-25.

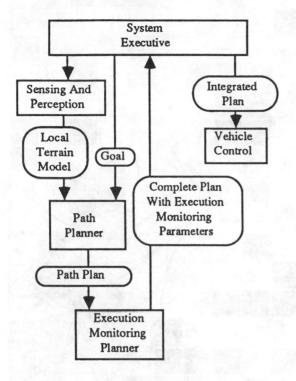

Figure 1. A block diagram of the semi-autonomous navigation system.

corresponding terrain area, as well as a 3x3 covariance matrix which indicates the uncertainties in the height and the slope and provides a measure of how rough the terrain is. The system currently produces three separate maps: a 1 meter resolution map covering a 64x64 meter square, a 0.25 meter map covering a 16x16 meter square, and a 0.125 meter map covering an 8x8 meter square. The arrays are centered at the rover's current position and are aligned to the coordinate frame of the global map.

The local terrain model is then passed to the path planner which plans a local path between five and ten meters long. This path is passed to a vehicle simulator which performs a detailed kinematic simulation of the vehicle as it traverses the planned path, and computes expected values for the rover's sensors over the path. These expected values are passed to the execution monitoring planner which produces execution monitoring profiles. These profiles tell the run-time execution monitoring system which sensors to monitor and when to monitor them. They also include recovery procedures to be executed in the event that something unexpected should occur.

The planned path and the execution monitoring parameters and recovery procedures are integrated by the execution monitoring planner and passed back to the system executive. The system executive checks that the plan conforms to any global constraints that the rover has. If the plan is acceptable, the system executive passes that plan to the vehicle control system which moves the vehicle along the planned path. During the traverse, if an execution monitoring parameter is violated, the vehicle immediately executes the recovery procedure associated with that violation.

When one path segment is complete, the system executive then begins a new cycle with the construction of a fresh local terrain model. The system architecture allows the new terrain model to be constructed during the traverse of a previous segment to allow interleaved operation of the various subsystems and continuous movement of the vehicle, though interleaved operation is not currently implemented.

3. Path Planning

Much work on path planning has been done for mobile robots [Linden87, Brooks82, Chatila85, Chubb87, Crowley85, Elfes87, McDermott84, Miller87, Nguyen84, Slack87, and Thorpe84]. However, most of this work has been done for vehicles designed to travel in structured environments where an area is either traversable or not. For a planetary rover in rough natural terrain, traversability is a continuous quantity. A path plan must take into account what kind of terrain the rover is currently on, what the transitions are like to the adjoining types of terrain, the size of the rocks, and other factors. Additionally the route planner must model non-geometric hazards that cannot be directly detected by the sensing and perception system, but can only be inferred by the geologic context, or the behavior of the rover as it makes its traverse.

Our path planner operates in two phases. In the first phase the rover is given a global terrain map and a goal. The goal is a description not only of the rover's desired location, but also its orientation and articulation state, together with acceptable error bounds. This information collectively is called an objective posture.

The planner begins by generating a global path gradient similar to [Payton88] from the global terrain map, using a spreading activation algorithm like that in [Slack87] and [Thorpe84] which propagates information from the goal to all other locations of the global map. Our algorithm takes into account the kinematic constraints of the vehicle as well as the traversability of the terrain. This is achieved by computing the gradient in two steps. In the first step, a gradient called the "quality gradient" is computed which takes into account only the locations of obstacles and terrain which is difficult to traverse. This gradient is then used as a cost function for computing the actual path gradient by spreading activation from the goal. The propagation of the gradient is constrained according to the vehicle kinematics. The result is a global gradient which conforms to the vehicle kinematic constraints as well as environmental constraints.

The global gradient is used to guide the second phase of planning which is a repeated incremental heuristic search through the local map that will move the robot closer to its goal. The approach exploits the hierarchical structure of the local terrain maps in a way that allows the robot to find local paths very quickly where the terrain is relatively free of obstacles, while retaining the ability to navigate cluttered areas through the use of progressively more refined planning techniques.

The first step in the local path planning process is to find a set of locations along the edge of the low resolution local map that will bring the robot closer to its goal as measured by the magnitude of the global gradient. These locations are called exit zones. A local gradient is then computed for the local terrain map as it was for the global map. The exit zones are then ranked according to the magnitude of the local gradient. A path from the current vehicle position to the top-rated exit zone is then determined by a simple spline-fitting algorithm.

Figure 2: A segment search.

This "quick-path" is sent to the vehicle simulator which determines if this is an acceptable path to traverse. The exit zones are checked until an acceptable path is found, or until some fixed number have been checked without finding an acceptable path.

This procedure is very efficient, and will usually yield a path when the terrain is relatively free of obstacles. However, if the terrain is cluttered then this procedure will fail. In this case, a detailed search through a space of short path segment sequences is performed. An example of a segment search is shown in figure 2. The grey areas are obstacles of varying heights. The short line segments indicate the direction of the local gradient. Note how the gradient tends to keep the rover moving in the white clear areas and away from the grey cluttered areas, especially along the top third of the map. The concentric circles are previously encountered non-geometric hazards (i.e. obstacles which were not detected by the remote sensors). The circles with the wedges along the right border are exit zones. The search has just finished finding a path to the lower exit zone, shown in light grey. The dark grey curves show the rest of the search tree.

If the segment search fails, the entire procedure is repeated at the medium-resolution map. In order to set up the exit-zones for the medium-resolution layer the low-resolution layer generates a path gradient (similar to the global path gradient) using the low-resolution exit-zones as its goals. The exit-zones of the medium resolution layer are then computed and the process of finding a path is repeated. If planning is not successful in the medium-resolution layer then the process is again repeated at the high resolution layer. If the search fails at the highest resolution layer then a skid-turn to align the robot to the local path gradient is returned.

If the region is extremely cluttered or full of non-geometric hazards, then the strategy of last resort is to back up some distance along the previously traversed path, recompute the global gradient, and start again from the beginning. This situation is extremely rare, resulting only when the global map is very inaccurate (such as when there are many non-geometric hazards), or when no traversable paths to the goal exist.

Our approach has two interesting features. The first is that the robot's kinematic constraints are taken into account implicitly through the path gradient and explicitly by the segment search. Second, the hierarchical search algorithms result in extremely fast path planning through open terrain. More expensive search techniques must be employed only in areas where there are many geographic features to help constrain the search. Thus the system is able to plan efficiently in almost any terrain.

4. Execution Monitoring and Planning

The execution monitoring system [Gat89] consists of two main components, the execution monitoring planner and the execution monitoring run-time system. The execution monitoring planner uses information generated by the traverse simulator as well as the local terrain model to produce a set of execution monitoring profiles. An execution monitoring profile defines an envelope of acceptable values for one sensor as a function of another sensor. Each envelope consists of a set of minimum and maximum values for a given sensor. These limits may be functions of the value of another sensor.

An execution monitoring profile is shown in figure 3. The expected value for the sensor is the dark line near the center, and the envelope is shown by the lighter lines above and below. This particular profile is for a path segment whose first part traverses terrain with high uncertainty in the local map. Therefore, the range of the expected values is higher for this part of the path.

Associated with each minimum and maximum value is a reflex action which is to be performed in the event that the value of the dependent sensor should violate one of these limits. The reflex action is simply an index into a table of precomputed reflex actions. Thus, the invocation of a reflex action once a parameter violation is detected is virtually instantaneous. The most common reflex is to stop the vehicle and back up far enough that the remote sensors can see the area where the expectation violation occurred. A new local terrain map is then generated and the traverse is attempted anew. If the expectation violation recurs, then the vehicle backs up again, and the area is considered a non-geometric hazard and marked untraversable in the global map.

There are three types of sensors that the execution monitoring system must deal with. First, there are physical sensors which do not require resource scheduling, such as wheel encoders and inclinometers. Their values are available continuously to any subsystem which needs them. Second, there are physical sensors which require resource scheduling such as cameras which must be aimed in the right direction at the right time and which require significant processing before useful information is available from them. Finally, there are virtual sensors which are mathematical functions defined over the values of the physical sensors. For example, there are virtual sensors for the vehicle's absolute spatial location in Cartesian coordinates. These values do not correspond to any physical sensor, but are computed using the values of many different sensors. A virtual sensor may require resource scheduling.

The execution monitoring planner uses expected values from the vehicle traverse simulator in order to generate execution monitoring profiles. The traverse simulator takes into account the uncertainty in the local terrain data in order to directly produce expected value ranges for all of the vehicle's physical non-scheduled sensors for every few centimeters along the path. These values are analyzed by the execution monitoring planner in order to construct a set of execution monitoring parameters. The planner selects segments of the path where the expected sensor values are more or less constant and sets the limits on that sensor to a value close to the expected deviations predicted by the simulator. The planner attempts to achieve maximum sensor coverage with a minimum of execution monitoring parameters since the performance of the runtime system can be impaired as the number of parameters grows large.

The execution monitoring run-time system is responsible for the actual monitoring of the vehicle sensors during a traversal. From the point of view of the execution monitoring runtime system, no distinction is made between a physical sensor and a virtual sensor. Thus, very complex interactions among physical sensors may be monitored by simply setting bounds on a single virtual sensor, which allows the runtime system to be very simple and efficient. This is essential in order to achieve the necessary real-time performance.

5. Implementation and Testing

The navigation system described in the previous sections is being implemented and tested on two testbed rovers at the Jet Propulsion Laboratory. For ease of prototyping and testing, parts of the navigation system have been implemented on a modified Cybermation K2A

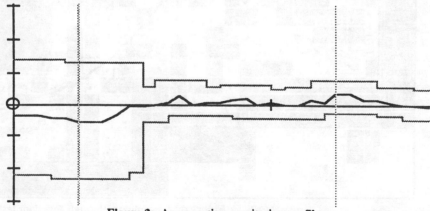

Figure 3: An execution monitoring profile.

industrial mobile robot. This robot is designed for indoor use, and uses a simulated sensing and perception system. The sensing and perception simulator is given an accurate map of the robot's surroundings. However, the planner has no access to this map. Instead, the sensing and perception system simulator uses this map to construct local terrain maps which are passed to the planner. The simulator adds gaussian noise to the data to make it more realistic. The resulting simulated terrain maps thus have quite similar characteristics to those generated by the actual sensing and perception system.

The indoor testbed has been successfully run using the path planning and execution planning/monitoring software over "rough indoor terrain." This terrain consists of a large open tiled area which is cluttered with obstacles. Some of these obstacles are traversable, e.g. carpeted areas or small books, while other obstacles are impassable, e.g. display cases, walls, and stairways. The testbed has motor encoders, contact sensors, and inclinometers so that the rover can know its position accurately from dead reckoning, its attitude from direct sensing, and whether it has run into anything from its contact sensors. Because of the absence of a vision system, physical objects not modeled in the vision simulator may be used to simulate non-geometric hazards,

since the rover cannot sense the obstacle until it can be detected by the inclinometers or the contact sensors.

For experiments in actual rough outdoor terrain an outdoor full-scale navigation testbed has been built that has mobility characteristics similar to that of an actual planetary rover [Miller89]. The outdoor testbed uses a stereo vision system [Gennery77] [Gennery80] for generating actual terrain maps from natural terrain. The outdoor testbed is currently in the final stages of system integration. The computational hardware on both testbeds is compatible.

A trace of an actual run of the indoor testbed is shown in figure 4. The path traversed by the rover starts in the upper left and ends at the lower right. Each small circle is the endpoint of a local path. The irregular section approximately two-thirds of the way along the path is where a non-geometric hazard was encountered. The grey areas are traversable obstacles of various heights, most of which are the result of sensor noise. The black areas are non-traversable obstacles. (In real life these are a number of display cases and a staircase.) The two small hollow rectangles just below the center of the picture are two obstacles which were not included in the simulated terrain maps which serve as non-geometric hazards. The short line segments indicate the direction and magnitude of the

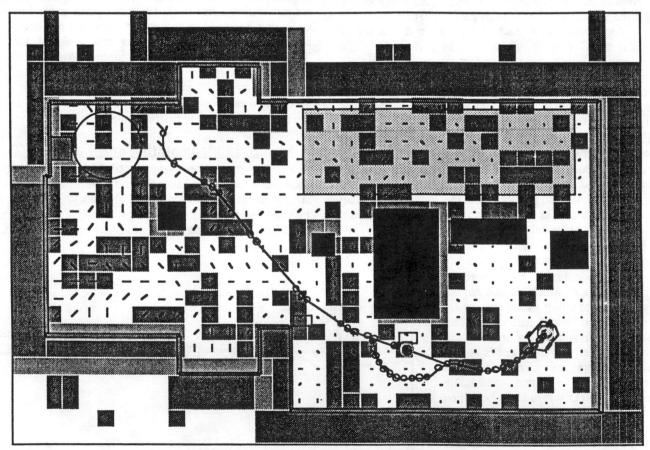

Figure 4: A sample run through an indoor test course.

global gradient. The objective posture (covered by the robot icon) is in the lower right hand corner facing northeast.

6. Conclusions and Future Research

The path planning and execution monitoring software has been successfully demonstrated on the indoor testbed. The robot can plan and execute a path through complex terrain using sensor and actuator interfaces which are essentially identical to those on the outdoor testbed. The robot can detect unforseen obstacles and take appropriate action. This includes having the rover back away from the hazard and mark the area as untraversable in the rover's internal map. Our experiments have consisted of paths roughly twenty meters in length.

Our architecture provides a number of capabilities not demonstrated by other mobile vehicles. It will work with a large variety of rover configurations with different kinematic constraints. The planner finds acceptable paths through natural terrain very efficiently, and the execution monitoring planner allows monitoring limits to be dynamically assigned, providing tight execution monitoring.

Over the next year we hope to add several capabilities to the navigation system. The path planner should be expanded to take into consideration such factors as the vehicle's power budget and the anticipated load bearing strength of the terrain. The execution monitoring system should be expanded to allow it to schedule pointable sensors in order to do landmark-based navigation [Miller86].

We also wish to explore the use of the execution monitoring system for doing accurate sensor-based positioning and reacting. For example, to position the rover a given distance from a rock in order to collect a sample, the execution monitoring planner could point a range sensor at the rock and set up a reflex action to stop the rover when the rock was within the proper range.

Acknowledgements: The research described in this paper was carried out by the Jet Propulsion Laboratory — California Institute of Technology under a contract with the National Aeronautics and Space Administration. The authors wish to thank John Loch, Brian Wilcox, Brian Cooper and Andy Mishkin for help in porting the system to the testbeds.

References

[Brooks82] R. A. Brooks, "Solving the find path problem by a good representation of free space," *Proceedings of AAAI 82*, pp. 381-386, 1982.

[Chatila85] R. Chatila, "Position referencing and consistent world modeling for mobile robots," *Proceedings of the International Conference on Robotics and Automation*, pp. 138-145, 1985.

[Chubb87] D. W. Chubb, "An introduction and analysis of a straight line path algorithm for use in binary domains," *Proceedings of the Workshop on Spatial Representation and Multi-Sensor Fusion*, St. Charles, Illinois, pp. 220-229, October 1987.[Crowley85] J. L. Crowley, "Dynamic world modeling for an intelligent mobile robot using rotating ultrasonic ranging device," *Proceedings of the International Conference on Robotics and Automation*, pp 128-135, 1985.

[Elfes87] A. Elfes, "Sonar based real world mapping and navigation," *IEEE Journal of Robotics & Automation*, vol. RA-3, pp. 249-265, 1987.

[Gat89] E. Gat, *et al.*,, "Planning for Execution Monitoring on a Planetary Rover", *Proceedings of the NASA Conference on Spacecraft Operations Autonomy and Roboics*, Houston, Texas., July 1989.

[Gennery80] D. B. Gennery, "Modelling the Environment of an Exploring Vehicle by Means of Stereo Vision," AIM-339 (Computer Science Dept. Report STAN-CS-80-805), Stanford University, (Ph.D. dissertation), 1980.

[Gennery77] D. B. Gennery, "A Stereo Vision System for an Autonomous Vehicle," *Proceedings of the Fifth International Joint Conference on Artificial Intelligence*, pp. 576-582, 1977.

[Linden87] T. Linden, J. Glicksman, "Contingency planning for an autonomous land vehicle," *Proceeding of the 10th International Joint Conference on Artificial Intelligence*, pp 1047-1054, 1987.

[Khatib85] O. Khatib, "Real-time obsticle avoidance for manipulators and mobile robots," *International Journal of Robotics Research*, vol. 5, pp. 90-98, 1986.

[McDermott84] D. V. McDermott and E. Davis, "Planning routes through uncertain territory," *Artificial intelligence*, vol. 22, pp. 107-156, 1984.

[Miller86] D. P. Miller, "Scheduling robot sensors for multisensory tasks," *Proceedings of the 1986 Robots West Conference*, Long Beach, California, September 1986.

[Miller87] D. P. Miller and M. G. Slack, "Efficient navigation through dynamic domains," *Proceedings of Workshop on Spatial Representation and Multi-Sensor Fusion*, St. Charles, Illinois, October 1987.

[Miller89] D. P. Miller, *et al.*, "Autonomous navigation and mobility for a planetary rover, " 27th Aerospace Sciences Meeting, AIAA, paper #89-0859, Reno, NV, January 1989.

[Nguyen84] V. Nguyen, "The find-path problem in the plane," MIT AI Technical Report 760, 1984.

[Payton88] D. W. Payton, "Internalized plans: a representation for action resources," *Proceedings of the Workshop on Representation and Learning in an Autonomous Agent*, Lagos, Portugal, November 1988.

[Slack87] M. G. Slack and D. P. Miller, "Path planning through time and space in dynamic domains," *Proceedings of the Tenth International Joint Conference on Artificial Intelligence*, pp 1067-1070, 1987.

[Thorpe84] C. E. Thorpe, "Path relaxation: path planning for a mobile robot," *Proceedings of AAAI*, pp. 318-321, 1984.

[Wilcox87] W. H. Wilcox, et al., "A vision system for a mars rover," *Procedings of SPIE Mobile Robots II*, vol. 852, November 1987.

Plan Guided Reaction

DAVID W. PAYTON, J. KENNETH ROSENBLATT, AND DAVID M. KEIRSEY

Reprinted from *IEEE Transactions on Systems, Man, and Cybernetics*, Vol. 20, No. 6, November/December 1990, pages 1370-1382. Copyright © 1990 by The Institute of Electrical and Electronics Engineers, Inc. All rights reserved.

Abstract —A set of architectural concepts that address the needs for integrating high-level planning activities with lower-level reactive or participatory behaviors is presented. Based on lessons learned from our experience with a hierarchical architecture for autonomous cross-country navigation, various pitfalls that may arise from the misuse of abstraction have been recognized. Consequently, a new approach that emphasizes the minimization of information loss both within and between system layers has been adopted. This change in perspective has allowed us to greatly enhance the overall capabilities and performance of our system.

I. INTRODUCTION

AN AUTONOMOUS mobile robot must be constantly involved in the processing of large amounts of sensory data in order to produce meaningful actions. The ability of a control architecture to support this immense processing task in a timely manner is significantly affected by the organization of information pathways within the architecture. Some architectures have been explicitly designed to maximize the amount of parallel information flow from sensing to action so as to provide minimal delay in responding to a constantly changing environment. Our own experience with such a system has led us to a better understanding of how action may be derived from the results of multiple independent decisionmaking processes and how plans may be used to guide these processes.

Recent efforts to develop intelligent autonomous robotic agents capable of interacting with a dynamic environment have led to a variety of alternatives to traditional planning methods that better meet the demands of real-time performance. Work by Brooks, for example, is aimed at obtaining purposeful action without any use of plans [6]. Instead, intelligent action is obtained as a manifestation of many simple processes interacting with and coordinated by a complex environment. While plans are not represented explicitly in such a system, they are in some sense implicitly designed into the system through the pre-established interactions between behaviors. Similarly,

Manuscript received April 1, 1989; revised January 1, 1990. This work was supported in part by Defense Advanced Research Projects Agency contract DACA76-85-C-0017. The material in this paper was partially presented at the Workshop for Representation and Learning in an Autonomous Agent, Lagos, Portugal, November 1988, and at the International Joint Conference on Neural Networks, Washington DC, June 1989.

The authors are with the Artificial Intelligence Center of Hughes Research Laboratories, 3011 Malibu Canyon Road, Malibu, CA, 90265.

IEEE Log Number 9037598.

Agre and Chapman have shown how a system that determines its actions through the constant evaluation of its current situation can perform complex tasks that might otherwise have been thought to require planning [1]. Rosenschein and Kaelbling take a similar approach, except they automatically compile their decision networks from world knowledge which is expressed in terms of a specialized formal logic [20]. In all these systems, emphasis is placed on the theme that action is obtained by always knowing what to do at any instant. Activities such as look-ahead and anticipation of future events are still seen as desirable yet these activities are not applied in the manner of traditional planning methods.

Our own efforts to address the problems of real-time robot control have led to an architecture which provides a very tight coupling between sensing and action. Although the overall system was hierarchical rather than layered, the lowest level of control had many similarities to the subsumption architecture developed by Brooks [5]. In this level, all action was determined by the output of multiple concurrent, independent decision-making processes called behaviors [18]. Each behavior received sensory input which was processed to suit its particular decision-making needs. In this way, we were able to shift our attention away from the computationally expensive problem of generating a single centralized model of the perceived world and focus instead on the production of coherent action. By appropriately fusing behavior commands through arbitration, the system was capable of responding to its environment without the delays imposed by sensor fusion. In effect, sensor fusion was supplanted by command fusion.

In a series of experiments performed by members of the Hughes Artificial Intelligence Center in August and December of 1987, this architecture was used to perform a number of successful tests of autonomous cross-county navigation with the DARPA autonomous land vehicle (ALV) [8], [12]. Under the control of a set of sensor-based behaviors, the vehicle was capable of avoiding locally sensed obstacles such as gullies, rocks, and trees. However, top–down control was also necessary to enable the vehicle to execute long-range traversals from one specified location to another. To this end, behaviors were specialized such that they could interpret a route plan and issue commands that would cause the vehicle to follow that route as closely as possible. The route plan was generated by a map-based planner that used digital map data to determine the traversability of various regions of terrain. The resulting integration of map and sensor-based control produced a system that could traverse complex natural terrain.

While the above approach has yielded many successful tests of autonomous navigation, we stand to gain most from a close examination of how it failed. Some of the complications encountered in these experiments have indicated certain consequences of the inappropriate use of abstraction both within and between system modules. Although abstraction serves as a useful tool to help designers conceptualize and understand the interactions between system components, we find that it can also severely limit ·critical information pathways. In this paper, we highlight some experiments that illustrate how the information barriers created by abstraction can lead to undesirable action. We then show how the elimination of these barriers can yield more flexible and opportunistic system performance. Finally, we illustrate how these concepts can be embodied within a system architecture for cross-country navigation.

II. Lessons Learned

Experience with the ALV has led to some valuable insights into the limitations of our control architecture. In particular, difficulties both with the interpretation of high-level map plans, and the implementation of low-level behaviors have vividly shown the consequences of inappropriate use of abstraction. In the case of low-level behaviors, we find that the abstraction which results from the modularization of behaviors and the arbitration of behavior commands causes difficulty in making certain types of decisions. In the case of map-based plans, we find that abstracting the results of search into a specific path description leads to the loss of information which otherwise could be quite useful as a resource for action.

A. Command Arbitration Problems

Our technique for combining road-following with obstacle avoidance provides a good illustration of how coherent vehicle action may be produced from the fusion of commands from multiple concurrent behaviors. In this particular instance, however, we found that the appropriate choices could not always be made. We first describe our road-following technique, and then explain how the addition of obstacle avoidance behaviors occasionally resulted in undesirable actions.

Our work on road-following relied on a set of perceptual algorithms developed by Turk *et al*. [25]. These algorithms processed video imagery and extracted a labeled set of points defining the left and right road edges. These sets of points, called scene models, were updated once every two seconds. Fig. 1 illustrates a typical scene model.

The basic task of road-following was performed by the concurrent operation of two independent behaviors. One behavior, called *maintain road speed*, was concerned only with controlling vehicle speed, simply making the speed

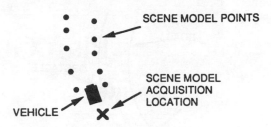

Fig. 1.   Example of a typical scene model used for road-following.

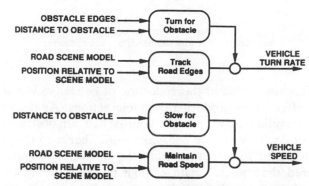

Fig. 2.   Behaviors for road-following and obstacle avoidance.

proportional to a measure of scene quality and the distance to the end of the currently available scene. The other behavior, called *track road edges*, was concerned only with controlling the rate of turn for the vehicle, making it proportional to the scene quality and a heading error measure. These behaviors re-evaluated their speed and turn-rate outputs once every half second so that changes in the vehicle's position relative to the current scene model were constantly taken into account. Updates of the scene models could thus be provided in a completely asynchronous manner with the new data simply replacing the old data whenever it became available. If no new data was provided, the vehicle would naturally slow to a halt since it would be getting close to the end of the available scene model. Many successful road-following tests were performed in simulation using this approach.

The next step was to integrate obstacle avoidance with road-following. General obstacle avoidance was implemented with two simple behaviors. One behavior slowed the vehicle when an obstacle got close, and the other caused the vehicle to turn away from nearby obstacles. Like the road-following behaviors, these behaviors could receive obstacle data and issue control commands in a completely asynchronous manner. In addition, obstacle data could be completely independent of the road data. By designing these behaviors to remain silent when no obstacles were near, and by giving their output commands higher priority than the road-following behaviors, it was fairly straightforward to combine road-following with obstacle avoidance. Although the architecture used allowed for a variety of different modes of command arbitration [18], the priority-based arbitration used in this example may be best illustrated in the equivalent subsumption architecture shown in Fig. 2.

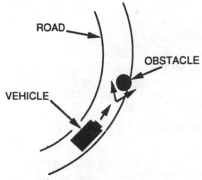

Fig. 3. Obstacle avoidance behaviors do not know which turn choice is most compatible with the needs of road-following.

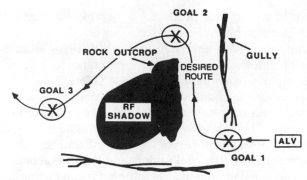

Fig. 4. An ALV route plan expressed as a sequence of intermediate goal points.

The interaction of this collection of behaviors was usually effective in producing desirable actions. As the vehicle, traveling down a road, approached an obstruction, the obstacle avoidance behaviors would begin to slow the vehicle and turn it away from the obstacle. As turning altered the vehicle's path enough to clear the obstacle, the obstacle avoidance behaviors would stop issuing commands, and the vehicle would resume following the road.

Difficulty arose with this approach, however, when the obstacle avoidance behaviors made turn decisions that were not compatible with the needs of road-following. As shown in Fig. 3, for example, an obstacle avoidance behavior may choose to go either left or right around the obstacle, but it is necessary to go left if the vehicle is to stay on the road. The only way to resolve this problem within the architecture shown is to incorporate more road-following knowledge into the obstacle avoidance behaviors. This, however, would violate the goal of maintaining modularity between behaviors.

The difficulty found in combining road-following behaviors with obstacle avoidance behaviors is indicative of a more general problem with priority-based command arbitration. In many cases, the process of selecting an appropriate command requires that several alternatives be weighed. When multiple behaviors are each independently evaluating their alternatives on the basis of different criteria, it may be impossible to arrive at an appropriate compromise. As in the example, the very specific turn command required for road-following was entirely subsumed by a higher priority obstacle avoidance turn command. At the obstacle avoidance level, however, the alternatives for left and right turns were weighted nearly equally, and yet the commands from the road-following behavior could not influence the decision.

This inability to compromise stems from the fact that a great deal of relevant information contained within each individual behavior is not allowed to contribute to the final decision. This results as much from the modularity of behaviors as from the nature of the command arbitration scheme. Modularity typically hides internal state within behaviors. Thus, while some behaviors may weigh several alternatives in order to make an appropriate choice, they cannot express their relative preference for

these alternatives. Similarly, when the commands from several behaviors are combined through arbitration, only a single command will survive. Information is invariably lost when it is hidden within behavior modules or contained in commands which are subsumed or inhibited. The consequences of this information loss are readily apparent.

B. Plan Abstraction Problems

In one of the cross-country experiments performed with the ALV, we witnessed a more dramatic example of how abstraction can interfere with the realization of high-level objectives. In this experiment, a very simple abstraction of a map-based plan was used to provide guidance to sensor-based obstacle avoidance behaviors. As shown in Fig. 4, the basic mission objective was for the vehicle to get from one location to another while maintaining radio contact at all times. The map-based planner generated an appropriate route plan and abstracted a sequence of intermediate sub-goals to represent the critical points along this path. A portion of this sequence is illustrated in Fig. 4 as Goals 1, 2, and 3. Note that the route had to veer specifically around one side of a rock outcrop in order to avoid loss of radio contact. To accomplish the mission, the sensor-based behaviors had primary control of the vehicle so that all obstacles could properly be avoided. The behavior decisions, however, were always biased in favor of selecting a direction toward the current map subgoal whenever possible. As soon as the vehicle got within a specified radius of its current subgoal, that goal would be discarded and the next subgoal would be selected. On paper and in simulation, it seemed that this approach would be effective.

When we attempted to perform this mission with the ALV, the deficiencies of our method became strikingly clear. During the execution of this route, the vehicle achieved Goal 1 but then, because of local obstacles, was unable to turn appropriately to reach Goal 2. Fig. 5 depicts the difference between the desired and actual routes. While this error is clearly apparent from the map data, the control behaviors had only the abstract route description as their guide, and this gave no indication that there was any problem with their action. Fortunately,

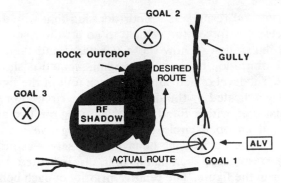

Fig. 5.   Errant vehicle action while executing its route plan.

contrary to our expectations, radio contact was not lost behind the obstacle. This mission could still be completed successfully if the vehicle were to move onward to Goal 3. Despite this new opportunity, however, the vehicle continued to persist toward Goal 2 because the abstract route description failed to give any indication that the original goal sequence was inappropriate.

This example highlights the system's inability to take opportunistic advantage of unexpected situations when such situations are not properly accounted for in the abstract plan. We know from our understanding of the mission constraints that Goal 2 was merely an intermediate waypoint intended to keep the vehicle away from the RF shadow. Looking at the abstract plan in isolation, however, there is no way of knowing why a particular sub-goal has been established. The Goal 2 location could just as easily have been a critical choke point along the only path to Goal 3. It is only through our understanding of the underlying mission constraints that we can both identify the vehicle's failure to turn right and see the opportunity that arose as a result.

The deficiencies of the abstract route plan may at first appear to be due solely to the simplicity of the representation. Certainly a more sophisticated approach could be employed in which further path constraints are added to help prevent the vehicle from straying from the desired route. Should any significant deviation from the plan be detected, the route might then be reevaluated. This strategy, however, focuses on preventing the violation of constraints which may in fact have very little bearing on the successful completion of overall mission objectives. Consider, for example, a case in which the vehicle can get near Goal 2, but cannot get close enough to satisfy the criterion of the abstract plan. The system may expend a great deal of time and energy attempting to reach this arbitrary subgoal when it might otherwise have no difficulty proceeding onward. The problem stems from the fact that the sequence of subgoals is both an overspecification and an underspecification of mission objectives. If the true constraints on vehicle motion relative to a given mission are properly represented, then subgoal locations become immaterial. Therefore, the real deficiency of the abstract route plan lies in the fact that in specifying a predetermined course of action, it fails to supply the information needed for intelligent decisionmaking.

Although dealing with entirely different aspects of system control, both the map planning example and the command arbitration example exhibit the problems that can arise with the inappropriate use of abstraction. In both cases, abstraction causes the loss of information that could otherwise contribute to the decisionmaking process. The abstract route description indicates a desirable path, but it discards a great deal of relevant information that is readily available in the map. Similarly, the priority-based arbitration of behavior commands causes useful information from some behaviors to be ignored when their commands are subsumed by other behaviors. In addition, the output of each individual behavior represents only a small fragment of the available information contained within, thereby hindering the incremental addition of new levels of competence. While abstraction is often useful and necessary, there is a risk that it may be misused, unnecessarily causing obstructions in a system's critical information pathways. In order to determine when the use of abstraction is appropriate and when it creates unwarranted abstraction barriers, we begin by eliminating abstraction wherever possible and studying the implications for a robot control system.

III.   Architectural Solutions

In addressing the aforementioned problems, we have devised a new architecture which attempts to eliminate all abstraction barriers. Previously, we found that internal communication barriers within the lowest level of the system led to difficulties in fusing commands from multiple behaviors. We show how these obstructions can be removed by making behaviors as fine-grained as possible so as to minimize their internal state. We also found that the generation of an abstract route plan for communication between system levels caused the loss of a great deal of meaningful information and did not allow for opportunistic action. This is resolved by creating an architecture in which plans serve as resources for advice within a single system level rather than as constraints imposed upon one level by another. Although these problems are presented within a specific context, we believe they are indicative of difficulties inherent within existing architectures. The architectural solutions presented here attempt to address the underlying causes of those difficulties and thus should provide some insight into how these problems might be solved elsewhere.

A. Fine-grained Behaviors

Our new architecture for robot control removes the communication barriers that existed between behaviors by making them as fine-grained as possible [22]. Instead of modules with internal states and instance variables, behaviors are comprised of atomic functional elements that have no inaccessible internal state. These simple decisionmaking units and their interconnections collectively define a behavior. As shown in Fig. 6, each unit receives

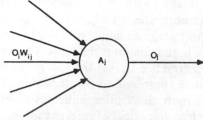

Fig. 6. Unit receives multiple weighted inputs and issues a single output. O_i is the output of unit i. W_{ij} is the weight on the link from unit i to unit j. $A_j = f_A(O_1 W_{1j}, \cdots, O_n W_{nj})$, is the activation level of unit j with n weighted inputs. $O_j = f_O(A_j)$, is the output of unit j.

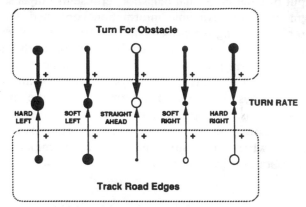

Fig. 7. Connectionist network for fusing turn rate commands.

multiple weighted inputs from other units and from external data sources, computes an activation level, and issues a single output [11], [21]. The activation and output functions for each unit can be any mapping from real numbers to a single real number value. The only constraint is that these functions be defined for inputs between -1 and 1, inclusive, and that the outputs be within this same range. The network is highly structured: in particular there is no attempt to create a network of homogeneous units or to interconnect units so as to form a distributed knowledge representation. Each unit represents a specific concept which the designer establishes through carefully chosen connections, and the functions for each unit are designated based on the desired output characteristics for that unit.

1) Distributed Command Arbitration: Unlike a priority-based arbitration scheme where the first choice of a single behavior is used for controlling the robot, a fine-grained connectionist architecture allows commands to be expressed and arbitrated in a distributed fashion. Command alternatives are represented by the appropriate place-coding of the component units, so that a behavior can explicitly indicate its bias in regards to each of the available choices and each alternative can be evaluated independently. To illustrate, let us consider how a compromise can now be reached in the situation described earlier where the vehicle was following a road and encountered an obstacle, as in Fig. 3. With the subsumption style architecture shown in Fig. 2, the *turn for obstacle* behavior had to override the output of the *track road edges*

behavior; the result was a poor decision that would have unnecessarily forced the vehicle to go off the road.

An illustration of how these behaviors and their command arbitration might now be implemented is shown in Fig. 7. The level of activation for each unit in the figure is roughly indicated by the diameter of the circle representing that unit, with a filled circle signifying positive activation and an open circle denoting a negative activation level. The output of each behavior unit serves as input to the corresponding command unit, as indicated by the arrows in the figure. The relative priority of each behavior is established by the weight on the links to the command units. Positive and negative activations are combined for each command unit, and the one with the highest activation is chosen as the option which best satisfies the needs of both behaviors.

Each behavior is distributed among several units so that the desirability for each choice can be expressed by the output value of the corresponding unit. In addition, since outputs may be negative, a behavior can directly indicate which choices it finds to be undesirable. For the units within the turn for obstacle behavior, a highly negative activation indicates that the corresponding turn would result in collision with an obstacle. The relative activations of the turn for obstacle units express the fact that this behavior most heavily favors turning hard to the left or right, that a soft left or right is less desirable but still acceptable, and that proceeding straight is unacceptable. The collection of units which comprise the track road edges behavior are more activated for turns that would keep the vehicle centered on the road. These units therefore indicate that in order to follow the road a soft left is the best option, a hard left is somewhat less desirable, straight is not favorable, a soft right is bad, and a hard right is very bad.

The desirability of each turn command can now be evaluated by fusing the outputs of the two behaviors. The activation of each turn rate command unit is determined by applying a function such as addition to its inputs. The thicker arrows originating from the turn for obstacle behavior indicate that those links have a larger weight than the links emanating from the units of the track road edges behavior, so that obstacle avoidance has a higher priority than road-following. In this case, the unit representing a hard left turn is the most activated since it fulfills the needs of both behaviors. A winner-take-all network can then select the most activated unit, and the choice that unit represents is then used in sending turn commands to the vehicle. Additional behaviors could also be incrementally added and would influence the decision-making simply by inputting their preferences to the command units. Thus, a connectionist architecture makes it possible to simultaneously satisfy the constraints of multiple independent behaviors.

2) Additional Levels of Competence: Because behaviors are independently capable of producing meaningful action, they can be composed to form *levels of competence* [4], each of which endows the robot control system with a particular capability. As we build up levels of competence, the connections between a new behavior and existing behaviors provide a rich means for expressing alternative actions. New behaviors do not completely subsume the function of existing behaviors, but merely bias decisions in favor of different alternatives. Thus, established levels of competence remain intact and, continue to participate in the decision-making process when higher levels of competence become active.

In keeping with our philosophy of reducing the grain-size of decision-making units, the function computed by each unit in the network may have no internal variables. Although the behaviors in the previous example consisted only of one unit per command option, in general a behavior will consist of several layers of units which together form the behavior's decisionmaking process. Therefore, any intermediate computations or states that a behavior enters are easily accessible to any other behavior and there is no need to define an interface which might need subsequent revision. This greatly simplifies the task of designing new layers for an existing robot control system.

B. Internalized Plans

In providing our fine-grained behaviors with guidance from plans, we must be careful to avoid throwing away information which might otherwise influence local decisions. To minimize the amount of information lost in forming a plan for action, it is best if all relevant knowledge is organized with respect to a given problem and then, without any further abstraction, provided in full for use in real-time decisionmaking. For this to be possible, the plan must no longer be viewed as a program for action, but rather, as a resource to help guide the decisionmaking process [2]. As a resource, plans must serve as sources of information and advice to agents that are already fairly competent at dealing with the immediate concerns of their environment [24]. In this sense, plans are used optionally, and serve only to enhance system performance. This is a significant departure from the conventional view of plans that puts them in the role of specifying a distinct course of action to systems which are often incapable of doing anything without them.

When this viewpoint is adopted, there is no longer a need to represent a plan as a predetermined course of action. Instead, the original state-space in which the plan is formulated can be retained, enabling the plan to provide advice to decisionmaking processes whenever the current state of the system can be identified within that state-space. We refer to plans formulated and used in this manner as *internalized plans*, since they embody the complete search and look-ahead performed in planning, without providing an abstracted account of an explicit course of action [19].

1) Avoiding Unnecessary Abstraction: The difference between the use of internalized plans and conventional plans is best illustrated in the context of the earlier example of a deficient route plan. In contrast to the

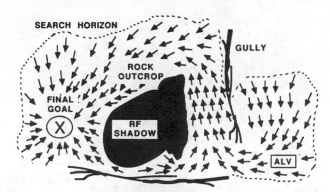

Fig. 8. Gradient field representation provides one form of internalized plan.

abstract route plan, consider a gradient description of a plan to achieve the same objectives. As illustrated in Fig. 8, there is no explicit plan shown, yet one can always find the best way to reach the goal simply by following the arrows. Such a representation would not ordinarily be thought to be a plan because it provides no specific course of action. As a resource for guiding action, however, the gradient field representation is extremely useful. No matter where the vehicle is located, and no matter how it strays from what might have been the ideal path, turn decisions can always be biased in favor of following the arrows.

Closer examination of Fig. 8 reveals not only how the mistake of entering the RF shadow could be avoided, but also how the vehicle could opportunistically continue onward should it happen to enter the shadow. First, upon reaching the bottom of the rock outcrop, the vehicle would find the gradient field strongly biasing it's turn decisions in favor of going to the right. Should it be unable to make this turn, the vehicle will then find itself heading directly against the gradient arrows, indicating that it should attempt to turn around. If circumstances should continue to force the vehicle into the RF shadow area, it would still proceed appropriately. At first, the gradient field data would be unavailable, forcing the vehicle to rely on its robust sensor-based navigation competence to get out of the RF shadow area. Once out of this area, the vehicle would again be directed toward the final goal despite the radical deviation from its expected path. Free of constraints to reach unnecessary pre-established subgoals, the vehicle's actions are opportunistic and directed exclusively toward achievement of mission objectives.

A more dramatic illustration of the difference between a conventional route plan and an internalized plan can be seen in problems requiring the attainment of any of several possible goals. This type of problem is often referred to as the "post office problem" [10] because it can be likened to the task of finding the shortest route to the nearest of several post offices in a neighborhood. In the example shown in Fig. 9, the mission requires that the vehicle reach either of two distinct goal locations. The resultant gradient field is computed by propagating a

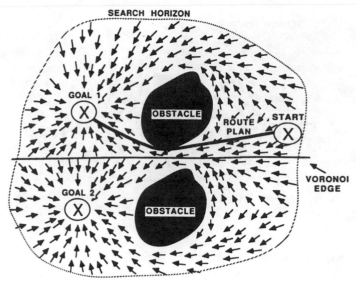

Fig. 9. Gradient field provides a useful internalized plan for reaching either of two goals.

search wavefront simultaneously from each of the two goals. As the wavefronts meet at a Voronoi edge, a ridge is created in the gradient field which will cause the vehicle to be guided toward one goal or the other depending on which side of the ridge it happens to be located.

Clearly, it would be difficult for an abstract route plan to capture the essence of choice contained in the gradient field representation. If we were to produce a route plan, we would invariably have to select a route to the closest goal, as shown in Fig. 9. Once such a choice is made, however, we have discarded all that is known about the alternate goal even though that goal was nearly as close as the one selected. In contrast, by using the gradient field directly as an internalized plan, a goal is reached without ever being explicitly selected. Without making an *a priori* selection of goals, the best choice is pursued at every instant in time, regardless of how the vehicle might stray while avoiding obstacles.

The gradient field is an ideal example of an internalized plan because the map-grid state-space in which the original problem is formulated is the same state-space in which the plan is represented. The gradient field, in fact, is a natural by-product of existing route planning algorithms [16]. These algorithms begin by assigning a cost to each grid cell of a digital terrain map. These costs are usually selected according to mission criteria, with the highest costs associated with untraversable or high threat areas. A search algorithm such as A^* [17], or Dijkstra [9] is then applied to obtain a score for each grid cell, indicating the minimum cost remaining to get from that cell to the goal. The best incremental step to get to the goal from any given cell is obtained by finding the neighboring cell with the lowest score. Ordinarily, to compute a route plan, one begins at the starting point and locally chooses the lowest-score adjacent cell until reaching the goal. The record of steps along the way provides the minimum cost path to the goal. Looking at the scores

from a slightly different perspective, it is easy to see that knowing the best incremental step to the goal from any location is far more valuable than merely knowing a pre-determined path. Thus, without any further abstraction, search in the map-grid can provide a useful resource for action.

2) Using Plans as Resources: The method of use of a gradient field is an important factor in establishing it as an internalized plan representation. Since a digital terrain map generally cannot provide adequate resolution to support detailed maneuvering around small obstacles, there is inevitably a need to incorporate the advice provided by the gradient field into real-time decision-making processes which are attending to immediate sensory data. While, ordinarily, a single abstract route plan is generated, some approaches have taken advantage of a gradient field in order to quickly generate new route plans should the constraints of an initial plan be violated [7, 15]. Problems with establishing and monitoring these constraints, however, are still unavoidable. In contrast, use of the gradient field as an internalized plan requires that the real-time decisionmaking processes continuously attempt to locate the system within the state-space of the plan and bias each decision in favor of the recommended course of action. The absence of an explicit course of action means that no arbitrary plan constraints need be established or monitored. At most, only fairly large discrepancies between the map and the real world will cause a gradient field to be unusable, and such discrepancies should be relatively easy to detect. Consequently, because internalized plans provide only suggestions for preferred action without defining an explicit course of action, they eliminate the need for traditional forms of plan execution monitoring.

Another vector field type of representation, the artificial potential field, appears superficially very similar to the gradient field and it also is used for robot navigation and obstacle avoidance [3], [13], [14]. The basic differences, though, between how these two types of representations are constructed and used sheds further light on what it means for a plan to serve as a resource for action. The computation of potential fields is generally based on a superposition model in which charges are distributed such that repulsive forces are generated near obstacles and attractive forces are generated near goals. Superposition allows the potential field vector at any point to be computed quickly by adding up the contributions from each charge. The resultant field, however, does not represent an optimal path, and may easily contain local minima and traps. In contrast, the gradient field is computed from a more time consuming graph search process. As a result of this search, the gradient field has no local minima and will always yield the set of all optimal paths to the goal.

The most significant distinction between internalized plans and potential field approaches, however, is in how control is achieved. Often, when potential field methods are employed for navigation, the potential field is used for direct control of action. All sensory information is compiled into a single representation which is suitable for modeling an appropriate distribution of charges. The local potential field forces are then continuously computed at the location of the vehicle, and these forces are used directly to compute the desired motion. On the other hand, as internalized plans, gradient fields are never used to provide direct control of the vehicle. Instead, they are merely an additional source of information provided to a set of real-time decisionmaking processes. Since these processes can make use of many disjointed representations of the world in order to control the vehicle, there is never a need for all features of the environment to be abstracted into a single representational framework.

While applications of the gradient field representation are primarily limited to the route planning domain, the broader notion of internalized plans need not be limited to this domain. In his concept of "universal plans," Schoppers has developed an approach to computing symbolic plans which permits opportunistic reaction to a dynamic world [23]. Like a gradient field, these plans are represented in such a way that appropriate actions may be obtained for all possible system states. Consequently, they constitute a significant departure from traditional plan representations. However, universal plans differ from internalized plans in that they are designed to serve as complete programs for action, requiring that all action be determined from knowledge of discrete states and explicitly labeled objects. Universal plans therefore cannot serve as a resource for action, as they do not allow symbolic world knowledge to be used as advice. Nevertheless, universal plans may provide some insights into how to build internalized plans for symbolic domains.

In general, internalized plans should be conceived as representations that allow the raw results of search in an abstract state space to be made available as advice to continuous real-time decision-making processes. In the route planning domain, state is defined in terms of position and orientation, and advice is expressed in terms of an influence to achieve a desired heading. To produce internalized plans for other domains, it must be possible to identify recognizable states, and to determine influences which can guide the system between these states. Search then can be performed, with accessibility between states defined in terms of the likelihood that the system will get from one state to another given the appropriate influences. The result must then serve as advice, identifying the influences to be applied at each recognizable state in order to increase the likelihood of reaching states that are closer to the goal.

To better understand the advisory role of internalized plans, it is helpful to view them as sources of supplementary sensory input data. From this perspective, it is clear that action is not controlled by plans any more than it is by sensory input. Instead, the system must be viewed as an entity which interacts with its environment, responding to both internal and external information sources. The gradient field plan, for example, can be thought of as a phantom compass that always gives a general idea of the right way to go. Just like other sensors, data from this internal sensor influences action but is never used to the exclusion of other sensory data. At any given time, however, a single information source can have significant influence over system behavior if need be. Just as an external sensor can be used to ensure that the vehicle never runs into obstacles, an internalized plan can be used to ensure that mission constraints are not violated. Thus, despite the fact that there is no top-down control, the system can adhere to high level mission requirements.

3) Multiple Internalized Plans: A significant advantage of using internalized plans as resources for action is that it is possible to use multiple internalized plans simultaneously. Each plan can contribute an additional piece of advice which can enhance the overall performance of the system. In this way, different plans may be formulated in incompatible state-spaces without the need to merge these state-spaces through abstraction.

We can consider as an example, the combined use of map-based plans with plans based on symbolic mission constraint data. In the case of the RF shadow problem, a constraint to maintain radio contact may be derived from mission knowledge. If this knowledge is used in conjunction with a signal strength sensor, then whenever the vehicle enters an RF shadow, it can immediately back up in order to regain contact. In the absence of such problems, the gradient field produced from map data can constantly provide advice on which way to go. An unexpected loss of radio contact would then be treated much like an encounter with an obstacle. The vehicle would have to make special maneuvers in order to regain contact and ensure that the same mistake would not be repeated. After this, the map-based plan would regain primary influence.

A great diversity of behavior may also be gained by dynamically combining information from multiple gradient fields. Consider, for example, two independent gradient fields, one that can guide a vehicle along a safe, well hidden route, and another that can lead the vehicle to nearby observation points. We can imagine that the vehicle is guided by the safe gradient field until the time comes for it to make an observation. Then, the gradient field for getting to observation points would become the primary guiding factor. Such a gradient field, formed similar to the field in Fig. 9, would lead the vehicle to the nearest of several possible observation points. Once an observation point had been reached and observation data collected, the safe gradient field would again be used for guidance. Using such a combination of internalized plans allows the performance of tasks that would be difficult to accomplish with a symbolic plan. Without an explicit plan for action, it is the interplay between the vehicle and its environment that determines how the mission will ultimately be carried out.

Although internalized plans can only be used within a system that is already fairly competent at performing most fundamental tasks in its environment, internalized plans do not require a great deal of reasoning power to be

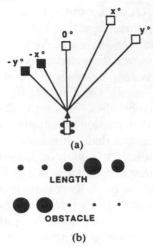

(a)

LENGTH

OBSTACLE

(b)

Fig. 10.   (a) Vehicle Model Trajectory output. (b) Corresponding input units.

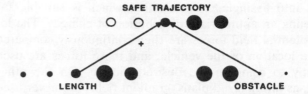

Fig. 11.   *Length* and *obstacle* units represent VMT input data. A combination of these inputs is used to determine the safety of travel in each direction, as represented by the *safe trajectory* units.

interpreted. A properly represented internalized plan gives straightforward advice for a specific set of situations. The breadth of the representation provides an efficiency of interpretation as opposed to an efficiency of space. At the current state of autonomous system technology, this trade-off is necessary to produce efficacious real-time action.

IV.   AN OPERATIONAL SYSTEM

Having addressed two distinct aspects of eliminating information loss within a system, we can now examine how these approaches may be combined to yield a robust system for mobile robot control. To illustrate how a complete system may be constructed through the incremental addition of fine-grained distributed behaviors, we recreate the process of developing an operational planning system that has been used in simulation to navigate a vehicle through cross-country terrain. The system is presented in terms of two distinct levels of competence, the first consisting of the fundamental behaviors required for obstacle avoidance and the second consisting of behaviors which interpret internalized route plans for long-range cross-country navigation.

A.   First Level of Competence—Obstacle Avoidance

The first level of competence allows the robot to avoid obstacles while wandering aimlessly in cross-country terrain. This is accomplished by behaviors which react to processed data from sensors on-board the vehicle.

1) Sensory Input: One type or processed sensory data available to the planner is the Vehicle Model Trajectory (VMT) [8]. A VMT indicates how far the vehicle may safely travel along a linear trajectory in each of several headings, and whether an obstacle is known to exist in that direction. In Fig. 10(a), the distance known to be safe at each azimuth is proportional to the length of the line drawn. While some trajectories terminate at obstacles, designated by the dark squares at $-x°$ and $-y°$, other

trajectories terminate where there is insufficient information about the terrain to determine navigability.

2) Input Units: To make the data available for use by those behaviors that base their decisions on VMTs, the first step is to create input units that represent incoming VMT data. To this end, two groups of units are created, one to represent the trajectory lengths and one to represent trajectory obstacles. These groups of units are labeled *length* and *obstacle*, respectively as shown in Fig. 10(b). Each group contains one unit for each trajectory in a VMT, and the direction that unit represents is indicated by its position within the group. Thus, if the directions of the trajectories were as in Fig. 10, the first unit in each group would represent data concerning the trajectory at $-y°$, the second unit would correspond to the trajectory at $-x°$, the third would be 0° straight ahead, and so on. The activation of a length unit is proportional to the length of the corresponding trajectory, and the activation of an obstacle unit is 1 if the corresponding trajectory terminates at a known obstacle or 0 if it terminates at an unobserved area.

3) Trajectory Selection Behavior: The next step is to use the length and obstacle input units described previously to determine which trajectories pose hazards and which can be followed safely. This is accomplished by the *safe trajectory* group shown in Fig. 11. Each unit in this group combines these inputs to produce an output value that reflects how safe it is to proceed in each direction. Only links for the center unit in a group are shown for clarity; the links for the other units are equivalent. A plus (+) or minus (−) sign next to a link indicates the sign of that link's weight. Thus, the activation of a safe trajectory unit increases with the activation of the corresponding length unit and decreases with the activation of the corresponding obstacle unit.

The safe trajectory group indicates the relative safety of traveling in each direction when considered in isolation, but it is also desirable to avoid traveling adjacent to an unsafe trajectory. Therefore an intermediate set of units is introduced which represents the presence of undesirable neighbors. The *unsafe left neighbor* and *unsafe right neighbor* groups, shown in Fig. 12, provide negative activation whenever a neighbor is more hazardous than the trajectory being evaluated. The *safe path* group can then combine these to indicate how safe each trajectory is when its neighbors are taken into consideration as well. Thus, since the trajectory at 0° is relatively short, the unsafe left neighbor unit for $x°$ is activated. The result is

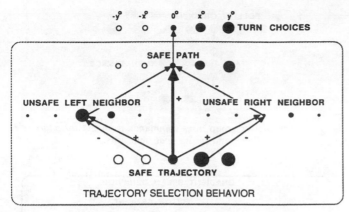

Fig. 12. *Trajectory selection* behavior expresses preference for each turn choice.

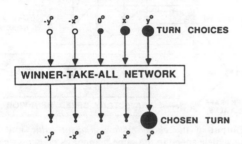

Fig. 13. Winner-take-all network selects the most activated choice.

that the safe path unit for $y°$ is more activated than the one for $x°$, even though the opposite is true for the safe trajectory units. The units in Fig. 12 constitute a behavior which expresses preferences for the turn commands based on VMT input.

The outputs of the trajectory selection behavior are connected to the *turn choices* group of units, which represent the available turn choices. Since only one behavior is inputting preferences to these units, their activations correspond exactly with the activations of the safe path units; however, as other behaviors are added which influence the choice of a heading, the turn choices activations will be influenced by all these behaviors. The activation levels of the turn choices group are then fed into a winner-take-all network, which results in exactly one of the chosen turn units having an activation of 1 and all the rest having an activation of 0, as shown in Fig. 13. The choice this one active unit represents is then used in sending turn commands to the vehicle.

4) Trajectory Speed Behavior: Once a choice of heading change has been made, it is necessary to determine an appropriate speed for travel in that direction. In our previous nonconnectionist implementation of behaviors, the length of the trajectory in the chosen direction was used in computing a speed. In this approach, however, the speed controlling behavior could not take advantage of the relevant information computed by other behaviors. As a result, the vehicle speed was not well suited to the safety of the chosen path.

In our new approach, much more information is now accessible from units within other behaviors; computations that previously were performed within a single mod-

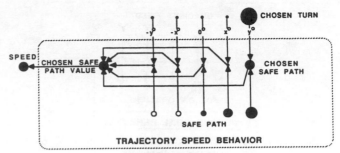

Fig. 14. Output of the selected *safe path* unit is used in determining vehicle speed.

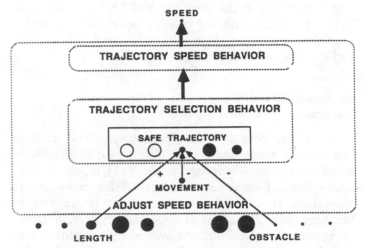

Fig. 15. *Adjust speed* behavior shares units with *safe path* and *trajectory selection* behaviors.

ule are now executed by several groups of units. The activation of a *safe path* unit within the trajectory selection behavior reflects not only the length of the corresponding trajectory, but also the presence or absence of an obstacle and the traversability of the adjacent trajectories. Since all these factors influence the safety of travel along a particular trajectory, the value of the *chosen safe path* unit is a better measure for determining an appropriate speed. This value can be extracted by the subnetwork shown in Fig. 14. The chosen safe path units multiply the inputs from the safe path and chosen turn units, so that a chosen safe path unit is nonzero only if the output of the linked chosen turn unit is 1, in which case the value is equal to the corresponding safe path unit. The outputs of these units are then input to the *chosen safe path value* unit, where a sum yields the desired result. The *speed* unit then sets the vehicle speed as a function of this value.

5) Adjust Speed Behavior: As the vehicle approaches the limit of terrain known to be safe within the current VMT, its speed must decrease until a new VMT is received. This can be accomplished very simply by making use of the existing units and adding in a single unit that represents the distance the vehicle has traveled since receiving the last VMT input. This *movement* unit has a negatively weighted link to the safe trajectory units so that it decreases their activation, as shown in Fig. 15. This leads to a decrease in the activation of the safe path units,

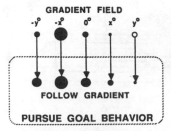

Fig. 16. *Pursue goal* behavior favors heading in direction consistent with the gradient field.

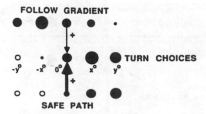

Fig. 17. *Turn choices* group fuses commands from the *trajectory selection* and *pursue goal* behaviors.

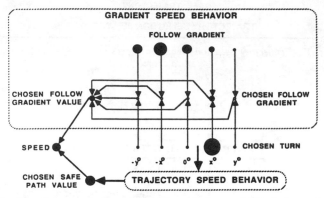

Fig. 18. Output of the *chosen follow gradient value* unit is used in determining vehicle speed in the same manner as *chosen safe path value*.

so that they now represent the safety of travel at all times, rather than just at the time the VMT input is received. Thus, the functionality of existing behaviors can be extended by providing them with additional information, without a need to subsume existing data.

B. Second Level of Competence — Cross-Country Navigation

While the above behaviors constitute a basic level of competence which endows the robot with the ability to avoid any obstacles indicated in the VMT input, an additional level of competence is needed for cross-country navigation. In order for the vehicle to be capable of reaching distant locations in cross-country terrain, something more than local obstacle avoidance is required. Map knowledge must be used so that the vehilce may pursue a path that will ultimately lead to a specified goal location while traversing that terrain which presents the least amount of difficulty to the obstacle avoidance behaviors. As we have seen, internalized route plans such as the one shown in Fig. 8 serve as an ideal representation for this kind of knowledge. Using behaviors which treat the gradient field plan as their sensory input, a new level of competence can be constructed which biases the vehicle's turn decisions in favor of the direction indicated by the gradient field.

The first step in creating this new level of competence is to formulate a representation for the new behavior's input. Since the vehicle has appropriate sensors for estimating its position to an accuracy within the resolution of the map grid, the gradient field plan may be constantly indexed by the current vehicle position in order to determine a desired heading. A set of heading change units can then receive activation according to how closely they correspond with this desired heading.

1) Pursue Goal Behavior: The units that receive the gradient field input are shown in Fig. 16. In this example, the gradient field indicates that $-x°$ is currently the preferred heading change, so the corresponding unit is the most highly activated. The activation for units representing each of the other directions is a function of the angle between that direction and the preferred direction. Note that a heading change of $y°$ significantly diverges from $-x°$, so that input unit has a negative activation. The *follow gradient* units constitute a behavior that uses the *gradient field* units as input. The choice of activation

function for these units is designed so that the sensitivity to the divergence angle is initially small and increases as the divergence increases.

2) Command Fusion: The *pursue goal* behavior endows the robot control system with a higher level of competence by allowing the robot to achieve specified goals. In our earlier architecture, one behavior had to be constructed that combined the gradient field input with VMT input, internally fusing commands in order to reach a decision. In the connectionist architecture, the command fusion can be achieved though the network itself, via the *turn choices* units as shown in Fig. 17. Therefore, the *trajectory selection* behavior can operate as always, utterly unaffected by the addition of the *pursue goal* behavior. The turn choices unit for $x°$ is now the most activated, so that would be chosen instead of $y°$ if both behaviors are used. Note that because the turn choices units have been defined to only update their activation based on new length or obstacle input, a change in the gradient field input will not have an effect on the chosen turn direction while a trajectory is being followed, but will have an effect once a new VMT input arrives.

3) Gradient Speed Behavior: The *follow gradient* unit also influences the vehicle speed in the same manner as the *safe path* units do; i.e., a *chosen follow gradient value* unit is created which has a link to the *speed* unit so that the vehicle speed will vary according to whether the robot is traveling with or against the gradient field. This configuration is shown in Fig. 18.

With the additional level of competence provided by the gradient field behaviors, the vehicle is now capable of traveling to a specific goal rather than wandering aimlessly. In simulation, the vehicle winds its way around a variety of obstacles, always avoiding any path indicated as unsafe by a VMT; but the decisions of the robot control system are also influenced by the gradient field input so that it completely avoids mapped areas which are known to be difficult to traverse, and the vehicle ultimately reaches the goal point.

While the above examples of behaviors for autonomous vehicle control are specific to a given application, many of the sub-networks described can be used, at least conceptually, as the building blocks for other behaviors. The basic techniques for fusing commands as in the safe trajectory layer, comparing neighboring choices as in the unsafe neighbor layers, selecting a chosen action as in the chosen turn layer, and extracting the magnitude of a choice regardless of which choice is made as in the trajectory speed layer, can all be used in a variety of ways to achieve meaningful action. While our fine-grained approach does not eliminate the need to design behaviors which are tailored to suit the autonomous agent's specific task requirements, it does make the job of adding new incremental layers of competence easier by allowing the designer access to the internal state of all existing layers.

V. Conclusion

In our efforts to address the problem of real-time autonomous vehicle control, we have found that many existing architectures may suffer from a loss of critical information between component modules due to the inappropriate use of abstraction. We have seen through experiments on an actual vehicle that serious consequences may result from such a loss. In addressing this problem, we have pursued an architectural approach that attempts to minimize the use of abstraction. While abstraction is sometimes necessary and desirable for the sake of representing available knowledge, our approach has provided a means to clarify the issues surrounding the inappropriate use of abstraction. Two distinct problems were found to capture the essence of these issues.

In the case of the subsumption style behaviors used in our experiments, we observed that the organizational structure of a system can lead to the inaccessibility of important information between its modules. To resolve this problem, we developed a methodology for constructing very fine-grained control behaviors in which the internal states within each module are accessible to all other modules. With the abstraction barriers between behaviors eliminated, we were able to develop a natural and effective way to fuse commands from multiple behaviors in order to obtain coherent action. In the process, we found the ability to represent and combine choices to be an essential aspect of our control approach. Because this architecture truly performs fusion of information, the arbitrary conceptual boundary between sensor fusion and command fusion may be eliminated, recognizing instead that information may be assimilated at a number of stages along multiple pathways from sensing to action.

In the case of plans, we have found that when search results are improperly abstracted, this can often obscure their meaning to the point where serious interpretation failures may occur. In particular, when plans are used only to identify a required course of action, it becomes difficult to take advantage of the valuable information that contributed to the plans' construction. Instead, by viewing plans as resources for action, it is possible to make use of all information derived throughout the planning process. When employing map-based plans for navigation, this change in perspective results in the use of internalized plans as if they were additional sources of sensory input to the real-time behaviors. As an example, we have shown how the use of an unabstracted gradient field provides for decisionmaking that is far more flexible and opportunistic than would be possible with an abstract route plan.

Having gained a better understanding of why some forms of abstraction may be inappropriate for a robot control system, it should be possible to develop new sets of abstractions that are better suited to the construction of intelligent autonomous agents. By recognizing the need for representing alternatives in command fusion and for allowing plans to serve as advice to low-level decision-making processes, we should be able to develop systems that exhibit greater robustness and versatility than existing systems.

Acknowledgment

Many of the concepts presented in this paper came about through fruitful discussions with members of the Hughes AI Center staff.

References

[1] P. Agre, and D. Chapman, "What are plans for?" AI Memo 1050, MIT Artificial Intelligence Laboratory, 1987.

[2] P. Agre, and D. Chapman, "Pengi: An implementation of a theory of activity," in *Proc. Sixth National Conf. Artificial Intelligence*, Seattle. WA, July 1987, pp. 268–272.

[3] R. Arkin, "Motor schema based navigation for a mobile robot: An approach to programming by behavior," *IEEE Conf. Robotics Automat*. pp. 264–271, Mar. 1987.

[4] R. A. Brooks, "A layered intelligent control system for a mobile robot," in *Proc. ISSR; Third Int. Symp. Robotics Res.*, Gouvieux, France. Oct. 1985.

[5] ____, "A robust layered control system for a mobile robot," *IEEE J. Robotics Automat.* vol. RA-2, no. 1, pp. 14–23, Apr. 1986.

[6] ____, "Intelligence without representation," *Preprints of the Workshop on Foundations of Artificial Intelligence*. Dedham, MA: Endicott House, June, 1987.

[7] Y. K. Chan, and M. Foddy, "Real time optimal flight path generation by storage of massive data bases," IEEE Nat. Aerospace Electron. Conf. (NAECON), Dayton, OH, May 1985.

[8] M. Daily, J. Harris, D. Keirsey, K. Olin, D. Payton, K. Reiser, J. Rosenblatt, D. Tseng, and V. Wong, "Autonomous cross-country navigation with the ALV," *Proc. DARPA Knowledge-Based Planning Workshop*, Austin, TX, Dec. 1987, (also appearing in *Proc. IEEE Conf. Robotics Automat.*, Philadelphia, PA., Apr. 1988.)

[9] E. W. Dijkstra, "A note on two problems in connection with graph theory," *Numerische Mathematik*, vol. 1, pp. 269–271, 1959.

[10] H. Edelsbrunner, *Algorithms in Combinatorial Geometry*. Berlin: Springer-Verlag, 1987, pp. 298–299.

[11] J. A. Feldman, and D. H. Ballard, "Connectionist models and their properties," *Cognitive Sci.*, vol. 6, pp. 205–254, 1982.

[12] D. M. Keirsey, D. W. Payton, and J. K. Rosenblatt, "Autonomous navigation in cross country terrain," in *Proc. Image Understanding Workshop*, Boston, MA, Apr. 1988.

[13] O. Khatib, "Real time obstacle avoidance for manipulators and mobile robots," *IEEE Conf. Robotics Automat.*, pp. 500–505, Mar. 1985.

[14] B. H. Krogh, "A generalized potential field approach to obstacle avoidance control," in *Proc. Robotics Int. Robotics Res. Conf.*, Bethlehem, PA, Aug. 1984.

[15] T. A. Linden, J. P. Marsh, and D. L. Dove, "Architecture and early experience with planning for the ALV," in *Proc. IEEE Int. Conf. Robotics Automat.*, pp. 2035–2042, Apr., 1986.

[16] J. S. B. Mitchell, D. W. Payton, and D. M. Keirsey, "Planning and reasoning for autonomous vehicle control," *Int. J. Intell. Syst.*, vol. 2, 1987.

[17] N. J. Nilsson, *Problem Solving Methods in Artificial Intelligence.* New York: McGraw-Hill, 1971.

[18] D. W. Payton, "An architecture for reflexive autonomous vehicle control," in *Proc. IEEE Robotics Automation Conf.*, San Francisco, CA, Apr. 1986, pp. 1838–1845.

[19] D. W. Payton, "Internalized plans: A representation for action resources," in *Proc. Workshop on Representation and Learning in an an Autonomous Agent*, Lagos, Portugal, Nov. 16–18, 1988.

[20] S. J. Rosenschein, and L. P. Kaelbling, "The synthesis of digital machines with provable epistemic properties," in *Theoretical Aspects of Reasoning about Knowledge, Proc. 1986 Conf.*, J. Y. Halpern, Ed. Monterey, CA: Morgan Kauffman Publishers, pp. 83–98.

[21] D. E. Rumelhart, and J. L. McClelland, *Parallel Distributed Processing.* Cambridge: MIT Press, 1986, vol. 1, ch. 2.

[22] J. K. Rosenblatt, and D. W. Payton, "A fine-grained alternative to the subsumption architecture for mobile robot control," in *Proc. IEEE Int. Joint Conf. Neural Networks*, Washington DC, June, 1989, vol. II., pp. 317–323.

[23] M. J. Schoppers, "Universal plans for reactive robots in unpredictable environments," in *Proc. Tenth Int. Joint Conf. Artificial Intell.*, Milan, Italy, Aug., 1987, pp. 1039–1046.

[24] L. Suchman, *Plans and Situated Actions: The Problem of Human Machine Communication.* Cambridge, MA: Cambridge Univ. Press, 1987.

[25] M. A. Turk, D. G. Morgenthaler, K. D. Gremban, and M. Marra, "Video road-following for the autonomous land vehicle," in *Proc. IEEE Int. Conf. Robotics Automat.*, Raleigh, NC, 1987

From 1982 to 1983, he was a software developer at Lisp Machine Inc., Boston, MA. Since December 1983, he has been with the Hughes Research Laboratories, Malibu, CA, working with the Artificial Intelligence Department. Since 1985, he has been head of the Autonomous Systems Section. His research interests have included the development of knowledge representation and control strategies for context-based object recognition. His most recent work has been in the development of software architectures for autonomous vehicle control.

Mr. Payton is a member of Phi Beta Kappa.

J. Kenneth Rosenblatt was born in Santiago, Chile on September 3, 1963. He received the S.B. degree in computer science in 1985 at the Massachusetts Institute of Technology, Cambridge, MA. He is currently a candidate for the Ph.D. degree in robotics at Carnegie Mellon University, Pittsburgh, PA.

From 1985 to 1987, he was a member of the technical staff at the Hughes Image and Signal Processing Laboratory, El Segundo, CA, where he developed an interactive map-based mission planning and monitoring system for semi-autonomous vehicles. Since 1988, he has been with the Hughes Artificial Intelligence Center, Malibu, CA, where he has been actively researching planning systems and control architectures for mobile robots. Other research interests include knowledge representation, machine learning, and cognitive modeling.

Mr. Rosenblatt is a Hughes Fellow.

David M. Keirsey was born in Claremont, CA on January 16, 1950. He received the B.S. degree in information and computer science from the University of California, Irvine, in 1972, the M.S. degree in computer science at University of Wisconsin, Madison, in 1973, and a Ph.D. degree in information and computer science at the University of California, Irvine in 1983.

Since 1980, Dr. Keirsey has been with Hughes Research Laboratories, working with the Artificial Intelligence Department where he is a Senior Computer Scientist. His research interests have included learning new words from natural language text, autonomous vehicles, automated knowledge acquisition, and machine learning. Currently he is involved in developing a knowledge acquisition tool for schema integration of multiple relational databases.

David W. Payton was born in Los Angeles, CA on September 8, 1957. He received the B.S.E.E. degree from the University of California, Los Angeles, in 1979, and the S.M. degree in electrical engineering and computer science from the Massachusetts Institute of Technology, Cambridge, MA, in 1982.

Planning and Active Perception

Thomas Dean*   Kenneth Basye   Moises Lejter
Department of Computer Science
Brown University, Box 1910, Providence, RI 02912

Abstract

We present an approach to building planning and control systems that integrates sensor fusion, prediction, and sequential decision making. The approach is based on Bayesian decision theory, and involves encoding the underlying planning and control problem in terms of a ... mpact probabilistic model for which evaluation is well ...derstood. The computational cost of evaluating such a probabilistic model can be accurately estimated by inspecting the structure of the graph used to represent the model. We illustrate our approach using a robotics problem that requires spatial and temporal reasoning under uncertainty and time pressure. We use the estimated computational cost of evaluation to justify representational tradeoffs required for practical application.

Introduction

In this paper, we view planning in terms of enumerating a set of possible courses of action, evaluating the consequences of those courses of action, and selecting a course of action whose consequences maximize a particular performance (or *value*) function. We adopt Bayesian decision theory [Raiffa and Schlaifer, 1961] as the theoretical framework for our discussion, since it provides a convenient basis for dealing with decision making under uncertainty.[1]

One interesting thing about most planning problems is that the results of actions can increase our knowledge, potentially improving our ability to make decisions. From a decision theoretic perspective, there is no difference between actions that involve sensing or movement to facilitate sensing and any other actions; a decision maker simply tries to choose actions that maximize expected value. In the approach described in this paper, an agent engaged in a particular perceptual task selects a set of sensor views by physically moving about [Bajcsy, 1988, Ballard, 1989].

Having committed to a decision theoretic approach, there are specific problems that we have to deal with. The most difficult concern representing the problem and obtaining the necessary statistics to quantify the underlying decision model. In the robotics problems we are working on, the latter is relatively straightforward, and so we will concern ourselves primarily with the former.

In building a decision model for control purposes, it is not enough to write down all of your preferences and expectations; this information might provide the basis for constructing some decision model, but it will likely be impractical from a computational standpoint. It is frustrating when you know what you want to compute but cannot afford the time to do so. Some researchers respond by saying that eventually computing machinery will be up to the task and ignore the computational difficulties. It is our contention, however, that the combinatorics inherent in sequential decision making will continue to outstrip computing technologies.

In the following, we describe a concrete problem to ground our discussion, present the general sequential decision making model and its application to the concrete problem, show how to estimate the computational costs associated with using the model, and, finally, describe how to reduce those costs to manageable levels by making various representational tradeoffs.

Mobile Target Localization

The application that we have chosen to illustrate our approach involves a mobile robot navigating and tracking moving targets in a cluttered environment. The robot is provided with sonar and rudimentary vision. The moving target could be a person or another mobile robot. The mobile base consists of a holonomic (turn-in-place) synchro-drive robot equipped with a CCD camera mounted on a pan-and-tilt head, and 8 fixed Polaroid sonar sensors arranged in pairs directed for-

---

*This work was supported in part by a National Science Foundation Presidential Young Investigator Award IRI-8957601 with matching funds from IBM, and by the Advanced Research Projects Agency of the Department of Defense and was monitored by the Air Force Office of Scientific Research under Contract No. F49620-88-C-0132.

[1]See Dean and Wellman [1989] for a discussion concerning the use of goals in artificial intelligence and the use of value functions in decision theory.

ward, backward, right, and left.

The robot's task is to detect and track moving objects, reporting their location in the coordinate system of a global map. The environment consists of one floor of an office building. The robot is supplied with a floor plan of the office showing the position of permanent walls and major pieces of furniture such as desks and tables. Smaller pieces of furniture, potted plants and other assorted clutter constitute obstacles that the robot has to detect and avoid.

We assume that there is error in the robot's movement requiring it to continually estimate its position with respect to the floor plan so as not to become lost. Position estimation (*localization*) is performed by having the robot track *beacons* corresponding to walls and corners and then use these beacons to reduce error in its position estimate.

Localization and tracking are frequently at odds with one another. A particular localization strategy may reduce position errors while making tracking difficult, or improve tracking while losing registration with the global map. The trick is to balance the demands of localization against the demands of tracking. The mobile target localization (MTL) problem is particularly appropriate for planning research as it requires considerable complexity in terms of temporal and spatial representation, and involves time pressure and uncertainty in sensing and action.

Model for Time and Action

In this section, we provide a decision model for the MTL problem. To specify the model, we quantize the space in which the robot and its target are embedded. A natural quantization can be derived from the robot's sensory capabilities.

The robot's sonar sensors enable it to recognize particular patterns of free space corresponding to various configurations of walls and other permanent objects in its environment (e.g., corridors, L junctions and T junctions). We tessellate the area of the global map into regions such that the same pattern is detectable anywhere within a given region. This tessellation provides a set of locations $\mathcal{L}$ corresponding to the regions that are used to encode the location of both the robot and its target.

Our decision model includes two variables S_T and S_R, where S_T represents the location of the target and ranges over $\mathcal{L}$, and S_R represents the location and orientation of the robot and ranges over an extension of $\mathcal{L}$ including orientation information specific to each type of location. For any particular instance of the MTL problem, we assume that a geometric description of the environment is provided in the form of a CAD model. Given this geometric description and a model for the robot's sensors, we generate $\mathcal{L}$, S_R, and S_T.

We encode our decision models as a *Bayesian networks* [Pearl, 1988]. A Bayesian network is a directed

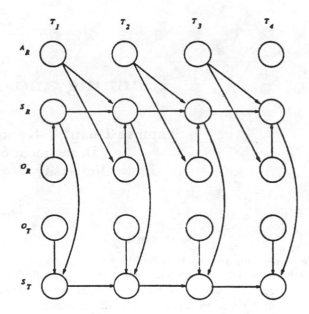

Figure 1: Probabilistic model for the MTL problem

graph $G = (V, E)$. The vertices in V correspond to random variables and are often called *chance nodes*. The edges in E define the causal and informational dependencies between the random variables. In the model described in this paper, chance nodes are discrete valued variables that encode states of knowledge about the world. Let Ω_C denote the set of possible values (*state space*) of the chance node C. There is a probability distribution $\Pr(C = \omega, \omega \in \Omega_C)$ for each node. If the chance node has no predecessors then this is its marginal probability distribution; otherwise, it is a conditional probability distribution dependent on the states of the immediate predecessors of C in G.

The model described here involves a specialization of Bayesian networks called *temporal belief networks* [Dean and Kanazawa, 1989]. Given a set of discrete variables, $\mathcal{X}$, and a finite ordered set of time points, $\mathcal{T}$, we construct a set of chance nodes, $\mathcal{C} = \mathcal{X} \times \mathcal{T}$, where each element of $\mathcal{C}$ corresponds to the value of some particular $x \in \mathcal{X}$ at some $t \in \mathcal{T}$. Let C_t correspond to the subset of $\mathcal{C}$ restricted to t. The temporal belief networks discussed in this paper are distinguished by the following Markov property:

$$\Pr(C_t | C_{t-1}, C_{t-2}, \ldots) = \Pr(C_t | C_{t-1}).$$

Let S_R and S_T be variables ranging over the possible locations of the robot and the target respectively. Let A_R be a variable ranging over the actions available to the robot. At any given point in time, the robot can make observations regarding its position with respect to nearby walls and corners and the target's position with respect to the robot. Let O_R and O_T be variables ranging these observations with respect to the robot's surroundings and the target's relative location.

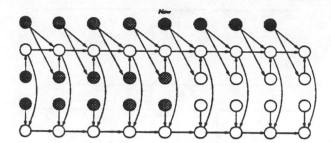

Figure 2: Evidence and action sequences

Figure 1 shows a temporal belief network for $\mathcal{X} = \{S_R, S_T, A_R, O_R, O_T\}$ and $\mathcal{T} = \{T_1, T_2, T_2, T_4\}$. To quantify the model shown in Figure 1, we have to provide distributions for each of the variables in $\mathcal{X} \times \mathcal{T}$. We assume that the model does not depend on time, and, hence, we need only provide one probability distribution for each $x \in \mathcal{X}$. For instance, the conditional probability distribution for S_T,

$$\Pr(\langle S_T, t\rangle | \langle S_T, t{-}1\rangle, \langle O_T, t\rangle, \langle S_R, t\rangle),$$

is the same for any $t \in \mathcal{T}$. The numbers for the probability distributions can be obtained by experimentation without regard to any particular global map.

In a practical model consisting of more than just the four time points shown in Figure 1, some points will refer to the past and some to the future. One particular point is designated the current time or *Now*. Representing the past and present will allow us to incorporate evidence into the model. By convention, the nodes corresponding to observations are meant to indicate observations *completed* at the associated time point, and nodes corresponding to actions are meant to indicate actions *initiated* at the associated time point. The actions of the robot at past time points and the observations of the robot at past and present time points serve as evidence to provide conditioning events for computing a posterior distribution. For instance, having observed σ at T, denoted $\langle O_R{=}\sigma, T\rangle$, and initiated α at $T{-}1$, denoted $\langle A_R{=}\alpha, T{-}1\rangle$, we will want to compute the posterior distribution for S_R at T given the evidence:

$$\Pr(\langle S_R{=}\omega, T\rangle, \omega \in \Omega_{S_R} | \langle O_R{=}\sigma, T\rangle, \langle A_R{=}\alpha, T{-}1\rangle).$$

To update the model as time passes, all of the evidence nodes are shifted into the past, discarding the oldest evidence in the process. Figure 2 shows a network with nine time points. The lighter shaded nodes correspond to evidence. As new actions are initiated and observations are made, the appropriate nodes are instantiated as conditioning nodes, and all of the evidence is shifted to the left by one time point.

The darker shaded nodes shown in Figure 2 indicate nodes that are instantiated in the process of evaluating possible sequences of actions. For evaluation purposes, we employ a simple *time-separable* value function. By

time separable, we mean that the total value is a (perhaps weighted) sum of the value at the different time points. If V_t is the value function at time t, then the total value, V, is defined as

$$V = \sum_{t \in \mathcal{T}} \gamma(t) V_t,$$

where $\gamma : \mathcal{T} \rightarrow \{x | 0 \le x \le 1\}$ is a decreasing function of time used to discount the impact of future consequences. Since our model assumes a finite $\mathcal{T}$, we already discount some future consequences by ignoring them altogether; γ just gives us a little more control over the immediate future. For V_t, we use the following function

$$V_t = -\sum_{\omega_i, \omega_j \in \Omega_{S_T}} \Pr(\langle S_T{=}\omega_i, t\rangle) \Pr(\langle S_T{=}\omega_j, t\rangle) \mathrm{Dist}(\omega_i, \omega_j),$$

where $\mathrm{Dist} : \Omega_{S_T} \times \Omega_{S_T} \rightarrow \Re$ determines the relative Euclidean distance between pairs of locations. The V_t function reflects how much uncertainty there is in the expected location for the target. For instance, if the distribution for $\langle S_T, t\rangle$ is strongly weighted toward one possible location in Ω_{S_T}, then V_t will be close to zero. The more places the target could be and the further their relative distance, the more negative V_t.

The actions in Ω_{A_R} consist of tracking and localization routines (e.g., move along the wall on your left until you reach a corner). Each action has its own termination criteria (e.g., reaching a corner). We assume that the robot has a set of strategies, S, consisting of sequences of such actions, where the length of sequences in S is limited by the number of present and future time points. For the network shown in Figure 2, we have

$$S \subset \Omega_{A_R} \times \Omega_{A_R} \times \Omega_{A_R} \times \Omega_{A_R}.$$

The size of S is rather important, since we propose to evaluate the network $|S|$ times at every decision point. The strategy with the highest expected value is that strategy, $\varphi = \alpha_0, \alpha_1, \alpha_2, \alpha_3$, for which V is a maximum, conditioning on $\langle A_r{=}\alpha_0, Now\rangle$, $\langle A_r{=}\alpha_1, Now{+}1\rangle$, $\langle A_r{=}\alpha_2, Now{+}2\rangle$, and $\langle A_r{=}\alpha_3, Now{+}3\rangle$. The best strategy to pursue is reevaluated every time that an action terminates.

We use Jensen's [1989] variation on Lauritzen and Spiegelhalter's [1988] algorithm to evaluate the decision network. Jensen's algorithm involves constructing a hyper graph (called a *clique tree*) whose vertices correspond to the (maximal) cliques of the chordal graph formed by triangulating the undirected graph obtained by first connecting the parents of each node in the network and then eliminating the directions on all of the edges. The cost of evaluating a Bayesian network using this algorithm is largely determined by the sizes of the state spaces formed by taking the cross product of the state spaces of the nodes in each vertex (clique) of the clique tree.

Following Kanazawa [Forthcoming], we can obtain an accurate estimate of the cost of evaluating a Bayesian network, $G = (V, E)$, using Jensen's algorithm. Let $C = \{C_i\}$ be the set of (maximal) cliques in the chordal graph described in the previous paragraph, where each clique represents a subset of V. We define the function, card $: C \rightarrow \{1, \ldots, |C| - 1\}$, so that card$(C_i)$ is the rank of the highest ranked node in C_i, where rank is determined by the maximal cardinality ordering of V (see [Pearl, 1988]). We define the function, adj $: C \rightarrow 2^C$, by:

$$\text{adj}(C_i) = \{C_j | (C_j \neq C_i) \wedge (C_i \cap C_j \neq \emptyset)\}.$$

The clique tree for G is constructed as follows. Each clique $C_i \in C$ is connected to the clique C_j in adj(C_i) that has lower rank by card$(.)$ and has the highest number of nodes in common with C_i (ties are broken arbitrarily). Whenever we connect two cliques C_i and C_j, we create the *separation set* $S_{ij} = C_i \cap C_j$. The set of separation sets S is all the S_{ij}'s. We define the function, sep $: C \rightarrow 2^S$, by:

$$\text{sep}(C_i) = \{S_{jk} | S_{jk} \in S, (j = i) \vee (k = i)\}.$$

Finally, we define the *weight* of C_i, $w_i = \prod_{n \in C_i} |\Omega_n|$, where Ω_n is the state space of node n. The cost of computation is proportional to $\sum_{C_i \in C} w_i |\text{sep}(C_i)|$. We refer to this cost estimate as the *clique-tree cost*.

The approach described in this section allows us to integrate prediction, observation, and control in a single model. It also allows us to handle uncertainty in sensing, movement, and modeling. Behavioral properties emerge as a consequence of the probabilistic model and the value function provided, not as a consequence of explicitly programming specific behaviors. The main drawback of the approach is that, while the model is quite compact, the computational costs involved in evaluating the model can easily get out of hand. For instance, in our model for the MTL problem, the clique-tree cost is bounded from below by the product of $|\mathcal{T}|$, $|\Omega_{S_T}|^2$, and $|\Omega_{S_R}|^2$. In the next section, we provide several methods that, taken together, allow us to reduce computational costs to practical levels.

Coping with Complexity

To reduce the cost of evaluating the MTL decision model, we use the following three methods: (i) carefully tailor the spatial representation to the robot's sensory capabilities, reducing the size of the state space for the spatial variables in the decision model, (ii) enable the robot to dynamically narrow the range of the spatial variables using heuristics to further reduce the size of the state space for the spatial variables, and (iii) consider only a few candidate action sequences from a fixed library of tracking strategies by taking into account the reduced state space of the spatial variables. In the rest of this section, we consider each of these three methods.

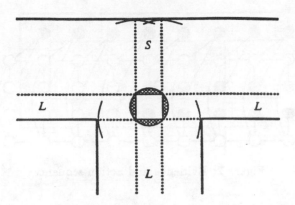

Figure 3: Sonar data entering a T junction

The use of a high-resolution representation of space has disadvantages in the model proposed here: increasing the resolution of the representation of space results in an increase in the sizes of Ω_{S_R} and Ω_{S_T}, and thus raises the cost of evaluating the network. Keeping the sizes of Ω_{S_R} and Ω_{S_T} small makes the task of evaluating the model we propose feasible.

A further consideration arises from the real-world sensory and data processing systems available to our robot. Finer-resolution representations of space place larger demands on the robot's on-board system in terms of both run-time processing time and sensor accuracy. To allow our robot to achieve (near) real-time performance, it seems appropriate to limit the representation to that level of detail that can be obtained economically from the hardware available.

In our current implementation, we have 8 sonar transducers positioned on a square platform, two to a side, spaced about 25 cm. apart. We take distance readings from each transducer, and threshold the values at about 1 meter. Anything above the threshold is "long," anything below is "short." The readings along each side are then combined by voting, with ties going to "long." In this way, the data from the sonar is reduced to 4 bits. Figure 3 shows the result of this scheme on entering a T junction. In addition, we use the shaft encoders on our platform to provide very rough metric information for the decision model. Currently, 2 additional bits are used for this purpose, but only when the robot is positioned in a hallway, which corresponds to only one sonar configuration. So the total number of possible states for O_R is 19, 15 for various kinds of hallway junctions and 4 more for corridors.

This technique results in a tessellation of space like that shown in Figure 4. Our experiments have shown that this tessellation is quite robust in the sense that the readings are consistent anywhere in a given tile. The exception to this occurs when the robot is not well-aligned with the surrounding walls. In these cases, reflections frequently make the data unreliable. One of the tasks of the controllers that underlie the actions described in

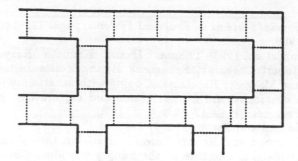

Figure 4: Tessellation of office layout

State space size	Number of time points		
	3	5	8
Constant (6)	40914 (0.58)	78066 (1.11)	133794 (1.90)
Constant (16)	624944 (8.87)	1232176 (17.49)	2143024 (30.42)
Constant (30)	3846330 (54.60)	7669530 (108.86)	13404330 (190.26)
Linear ($2t + 1$)	5844 (0.08)	55088 (0.78)	433759 (6.16)
Quadratic ($t^2 + 1$)	3691 (0.05)	160701 (2.28)	3756559 (53.32)
Exponential (2^t)	2875 (0.05)	107515 (1.53)	4131611 (58.64)

Table 1: Clique-tree costs for sample networks

the previous sections is to maintain good alignment, or achieve it if it is lost.

In addition to reducing the size of the overall spatial representation, we can restrict the range of particular spatial variables on the basis of evidence not explicitly accounted for in the decision model (e.g., odometry and compass information). For instance, if we know that the robot is in one of two locations at time 1 and the robot can move at most a single location during a given time step, then $\langle S_R, 1 \rangle$ ranges over the two locations, and, for $i > 1$, $\langle S_R, i \rangle$ need only range over the locations in or adjacent to those in $\langle S_R, i{-}1 \rangle$. Similar restrictions can be obtained for S_T. For models with limited lookahead (i.e., small $|\mathcal{T}|$), these restrictions can result in significant computational savings.

Consider a temporal Bayesian network of the form shown in Figure 1 with n steps of lookahead. Let $\langle X, i \rangle$ represent an element of $\{S_R, S_T, A_R, O_R, O_T\} \times \{1, \ldots, n\}$. The largest cliques in one possible[2] clique tree for this network consist of sets of variables of the form:

$$\{\langle S_R, i \rangle, \langle S_R, i{+}1 \rangle, \langle S_T, i \rangle, \langle S_T, i{+}1 \rangle\}$$

for $i = 1$ to $n{-}1$, and the size of the corresponding cross product space is the product of $|\Omega_{\langle S_R, i \rangle}|$, $|\Omega_{\langle S_R, i{+}1 \rangle}|$, $|\Omega_{\langle S_T, i \rangle}|$, and $|\Omega_{\langle S_T, i{+}1 \rangle}|$. For fixed state spaces, this product is just $|\Omega_{S_R}|^2 |\Omega_{S_T}|^2$. However, if we restrict the state spaces for the spatial variables on the basis of some initial location estimate and some bounds on how quickly the robot and the target can move about, we can do considerably better.

Table 1 shows the clique-tree costs for three MTL decision model networks of size $n = 3$, 5, and 8 time points. For each size of model, we consider cases in which $\Omega_{\langle S_R, i \rangle}$ and $\Omega_{\langle S_T, i \rangle}$ are constant for all $1 \geq i \geq n$, and cases in which $|\Omega_{\langle S_R, 1 \rangle}| = |\Omega_{\langle S_T, 1 \rangle}| = 1$ and the sizes of the state spaces for subsequent spatial variables, $\Omega_{\langle S_R, i \rangle}$ and $\Omega_{\langle S_T, i \rangle}$, for $1 > i \geq n$ grow by

---
[2] The triangulation algorithm attempts to minimise the size of the largest clique in the resulting chordal graph. There may be more than one way to triangulate a graph so as to minimise the clique size.

linear, quadratic, and exponential factors bounded by $|\Omega_{S_T}| = |\Omega S_R| = 30$. For these evaluations, $|\Omega_{A_R}| = 6$, $|\Omega_{O_T}| = 32$, and $|\Omega_{O_R}| = 19$ in keeping with the sensory and movement routines of our current robot. The number in brackets underneath the clique tree cost is the time in cpu seconds required for evaluation.

Our current idea for restricting the present location of the robot and the target involves using a fixed threshold and the most up-to-date estimates for these locations to eliminate unlikely possibilities. Occasionally, the actual locations will be mistakenly eliminated, and the robot will fail to track the target. There will have to be a recovery strategy and a criterion for invoking it to deal with such failures.

There are certain costs involved with evaluating Bayesian networks that we have ignored so far. These costs involve triangulating the graph, constructing the clique tree, and performing the storage allocation for building the necessary data structures. For our approach of dynamically restricting the range of spatial variables, the state spaces for the random variables change, but the sizes of these state spaces and the topology of the Bayesian network remain constant. As a consequence, these ignored costs are incurred once, and the associated computational tasks can be carried out at design time. Dynamically adjusting the state spaces for the spatial variables is straightforward and computationally inexpensive.

The third method for reducing the cost of decision making involves reducing the size of S, the set of sequences of actions corresponding to tracking and localization strategies. For an n step lookahead, the set of useful strategies of length n or less is a very small subset of $\Omega_{A_R}{}^n$. Still, given that we have to evaluate the network $|S|$ times, even a relatively small S can cause problems. To reduce S to an acceptable size, we only evaluate the network for strategies that are possible given the current restrictions on the spatial variables. For instance, if the robot knows that it is moving down a corridor toward a left-pointing L junction, it can eliminate from consideration any strategy that involves it moving to the end of the corridor and turning right. With appropriate preprocessing, it is computationally simple to dynamically reduce S to just a few possible strategies in most cases.

Related Work

Probabilistic decision models of the sort explored in this paper are just beginning to see use in planning and control. Agogino and Ramamurthi [1988] describe the use of probabilistic models for controlling machine tools. Dean *et al* [1990] show how to use Bayesian networks for building maps and reasoning about the costs and benefits of exploration. Kanazawa and Dean [Kanazawa and Dean, 1989] extend temporal Bayesian networks to handle sequential decision making tasks. Levitt *et al* [1988] describe an approach to implementing object recognition using Bayesian networks that accounts for the cost of sensor movement and inference. Wellman [1987] shows how to integrate qualitative knowledge in probabilistic network models. For some previous approaches to using decision and probability theory in planning, see [Feldman and Sproull, 1977, Langlotz *et al.*, 1987]. For some recent work on temporal reasoning under uncertainty, see [Cooper *et al.*, 1988, Dean and Kanazawa, 1988, Hanks, 1988, Weber, 1989].

References

[Agogino and Ramamurthi, 1988] A. M. Agogino and K. Ramamurthi. Real-time influence diagrams for monitoring and controlling mechanical systems. Technical report, Department of Mechanical Engineering, University of California, Berkeley, 1988.

[Bajcsy, 1988] R. Bajcsy. Active perception. *Proceedings of the IEEE*, 76(8):996–1005, 1988.

[Ballard, 1989] Dana H. Ballard. Reference frames for animate vision. In *Proceedings IJCAI 11*, pages 1635–1641. IJCAI, 1989.

[Cooper *et al.*, 1988] Gregory F. Cooper, Eric J. Horvitz, and David E. Heckerman. A method for temporal probabilistic reasoning. Technical Report KSL-88-30, Stanford Knowledge Systems Laboratory, 1988.

[Dean and Kanazawa, 1988] Thomas Dean and Keiji Kanazawa. Probabilistic temporal reasoning. In *Proceedings AAAI-88*, pages 524–528. AAAI, 1988.

[Dean and Kanazawa, 1989] Thomas Dean and Keiji Kanazawa. A model for reasoning about persistence and causation. *Computational Intelligence*, 5(3):142–150, 1989.

[Dean and Wellman, 1989] Thomas Dean and Michael Wellman. On the value of goals. In Josh Tenenberg, Jay Weber, and James Allen, editors, *Proceedings from the Rochester Planning Workshop: From Formal Systems to Practical Systems*, pages 129–140, 1989.

[Dean *et al.*, 1990] Thomas Dean, Kenneth Basye, Robert Chekaluk, Seungseok Hyun, Moises Lejter, and Margaret Randazza. Coping with uncertainty in a control system for navigation and exploration. In *Proceedings AAAI-90*. AAAI, 1990.

[Feldman and Sproull, 1977] Jerome Feldman and Robert Sproull. Decision theory and artificial intelligence ii: the hungry machine. *Cognitive Science*, 1:158–192, 1977.

[Hanks, 1988] Steve Hanks. Representing and computing temporally scoped beliefs. In *Proceedings AAAI-88*, pages 501–505. AAAI, 1988.

[Jensen, 1989] Finn V. Jensen. Bayesian updating in recursive graphical models by local computations. Technical Report R-89-15, Institute for Electronic Systems, Department of Mathematics and Computer Science, University of Aalborg, 1989.

[Kanazawa and Dean, 1989] Keiji Kanazawa and Thomas Dean. A model for projection and action. In *Proceedings IJCAI 11*, pages 985–990. IJCAI, 1989.

[Kanazawa, Forthcoming] Keiji Kanazawa. *Probability, Time, and Action*. PhD thesis, Brown University, Providence, RI, Forthcoming.

[Langlotz *et al.*, 1987] Curtis P. Langlotz, Lawrence M. Fagan, Samson W. Tu, Branimir I. Sikic, and Edward H. Shortliffe. A therapy planning architecture that combines decision theory and artificial intelligence techniques. *Computers and Biomedical Research*, 20:279–303, 1987.

[Lauritzen and Spiegelhalter, 1988] Stephen L. Lauritzen and David J. Spiegelhalter. Local computations with probabilities on graphical structures and their application to expert systems. *Journal of the Royal Statistical Society*, 50(2):157–194, 1988.

[Levitt *et al.*, 1988] Tod Levitt, Thomas Binford, Gil Ettinger, and Patrice Gelband. Utility-based control for computer vision. In *Proceedings of the 1988 Workshop on Uncertainty in Artificial Intelligence*, 1988.

[Pearl, 1988] Judea Pearl. *Probabilistic Reasoning in Intelligent Systems: Networks of Plausible Inference*. Morgan-Kaufmann, Los Altos, California, 1988.

[Raiffa and Schlaifer, 1961] Howard Raiffa and R. Schlaifer. *Applied Statistical Decision Theory*. Harvard University Press, 1961.

[Weber, 1989] Jay C. Weber. A parallel algorithm for statistical belief refinement and its use in causal reasoning. In *Proceedings IJCAI 11*. IJCAI, 1989.

[Wellman, 1987] Michael P. Wellman. Dominance and subsumption in constraint-posting planning. In *Proceedings IJCAI 10*. IJCAI, 1987.

DYNAMIC CONTROL OF ROBOT PERCEPTION
USING STOCHASTIC SPATIAL MODELS

Alberto Elfes

Department of Computer Sciences
IBM T. J. Watson Research Center
Yorktown Heights, NY 10598
E-Mail (Internet): ELFES@IBM.COM

Abstract

Robot perception has traditionally been addressed as a passive and incidental activity, rather than an active and task-directed activity. Consequently, although sensor systems are essential to provide the information required by the decision-making and actuation components of a robot system, no explicit planning and control of the sensory activities of the robot is performed. This has lead to the development of sensor modules that are either excessively specialized, or inefficient and unfocused in their informational output. In this paper, we develop strategies for the dynamic control of robot perception, using stochastic sensor and spatial models to explicitly plan and control the sensory activities of an autonomous mobile robot, and to dynamically servo the robot and its sensors to acquire the information necessary for successful execution of robot tasks. We discuss the explicit characterization of robot task-specific information requirements, the use of information-theoretic measures to model the extent, accuracy and complexity of the robot's world model, and the representation of inferences about the robot's environment using the Inference Grid, a multi-property tesselated random field model. We describe the use of stochastic sensor models to determine the utility of sensory actions, and to compute the loci of observation of relevant information. These models allow the development of various perception control strategies, including attention control and focussing, perceptual responsiveness to varying spatial complexity, and control of multi-goal perceptual activities. We illustrate these methodologies using an autonomous multi-sensor mobile robot, and show the application of dynamic perception strategies to active exploration and multi-objective motion planning.

1 Introduction

Traditionally, most approaches to robot perception have cast the sensing and perceptual activities of a robot system in what is fundamentally a passive rôle: although the robot planning and action stages depend fundamentally on sensor-derived information, no explicit planning and control of the perceptual activities themselves is performed. Rather, the robot's sensing subsystems acquire data that incidentally happens to be available at the moment of sensory observation during the perception/planning/action cycle. No explicit connection is made between the sensor data acquired and the information required by the robot for the successful execution of a given task (Fig. 1). Robot systems that embody this *passive perception* approach tend to fall into two categories: those where the sensory subsystems are highly specialized and "hardwired" to the feedback control loop that handles the execution of a specific class of robot tasks; and those that embody "general-purpose" sensor understanding and world-modelling subsystems. Typically, the former category cannot reconfigure its sensory subsystems to adapt to different classes of tasks or to accomodate sensor degradation; while the latter category is populated by notoriously large and ineffective systems that are unfocused in their informational output, have slow reaction times, and are inefficient in the use of their sensory and computational resources. Because there is no explicit characterization of the

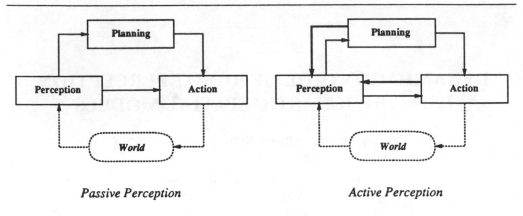

Passive Perception   Active Perception

Figure 1: Passive and Active Control of Robot Perception.

information really needed, over- or undersampling of the data may occur, and data irrelevant to the task at hand may be acquired.

More recently, researchers have identified the limitations intrinsic to the formulation of robot perception as a passive activity, and have argued for viewing it as an active process (Fig. 1). Bajcsy [2] discusses the need for *active sensing*, defined as the application of modelling and control strategies to the various layers of sensory processing required by a robot system. Elfes has developed an approach for *active mapping* using the Occupancy Grid framework [12, 14, 10]; this framework uses estimation-theoretic and Markov Random Field models to compose information from multiple sensors and multiple points of view, thereby addressing in a robust way the underconstrainedness of the sensor data. Aloimonos suggests that the composition of information from multiple views can be used to handle several ill-posed problems in Computer Vision [1]. Related research includes the development of recursive estimation procedures for robot perception [20, 15, 18]; methods for *active sensor control*, where specific parameters of the sensor system (such as camera aperture, focal distance, or sensor placement) can be changed under computer control [19, 25]; development of theoretical foundations for coordination, integration and control of sensor systems [8, 17, 16]; and generation of optimal and adaptive sensing strategies [7, 6, 25].

In this paper, we discuss *Dynamic Perception*, a framework for dynamic and adaptive planning and control of robot perception in response to the information needs of an autonomous robot system as it executes a given mission. The Dynamic Perception framework uses stochastic and information-theoretic sensor and world models to identify the information needs of the robot, plan the acquisition of task-specific data, dynamically servo the robot and its sensors to acquire the information needed for successful execution of the task, update the information acquisition goals as the robot progresses through sequences of tasks, and integrate the sensory tasks with the mission-specific tasks to be performed by the robot. We describe the use of a multi-property tesselated random field model, the Inference Grid, to encode inferences about the robot's environment. We illustrate the explicit characterization of robot task-specific information requirements, and the use of information-theoretic measures to model the extent, accuracy and complexity of the robot's world model. We also discuss the use of stochastic sensor models to evaluate the utility of sensory actions, and to compute the loci of observation of relevant information. We illustrate our approach using an autonomous multi-sensor mobile robot, and show the application of dynamic perception strategies to *active exploration* and *integrated motion planning*, combining perception and navigation goals. A more extensive discussion, with further experimental results, can be found in [13].

2 Dynamic Robot Perception

The work presented here is part of a long-term research effort that addresses the development of *agile* and *robust* autonomous robot systems, able to execute complex, multi-phase missions in unknown and unstructured real-world environments, while displaying real-time performance [10, 13]. For planning,

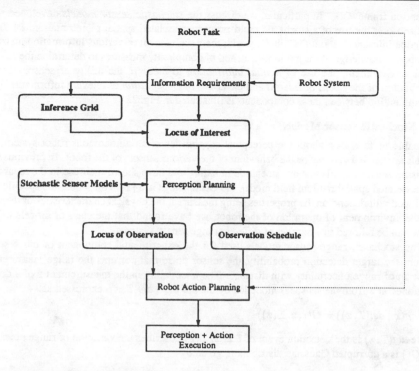

Figure 2: Components of the Dynamic Robot Perception Framework.

scheduling, execution and monitoring purposes, a specific robot *mission* is separated into *phases*, which are in turn decomposed into sets of *tasks*. These tasks can be scheduled in parallel and/or sequentially, depending on the nature of the component activities. As tasks are scheduled for detailed planning and execution by the robot, the information requirements posed by these tasks are used to dynamically update the perceptual goals of the robot. By maintaining explicit models of the task goals and the perceptual goals, the Dynamic Perception framework allows a closer integration between task planning and perception planning, and between task execution and perception execution. Using descriptions of the information required for successful execution of a specific task, appropriate sensors are selected, sensing strategies are formulated, and the robot's locomotion and sensing activities are planned so as to maximize the recovery of relevant information and accomplish the robot's task. Overall, the system's behaviour can be described as *servoing on required information*, as well as on the robotic task itself. We note that in control-theoretic terms this can be phrased as a problem in *dual control* [4, 13].

Some of the issues that have to be addressed in the context of the Dynamic Perception framework include *Sensor Modelling*, *Information Modelling*, *Perception Planning*, and the *Integration of Perception and Action*. This will allow us to enable an autonomous mobile robot with a number of important capabilities, including *attention control*, or the ability to efficiently acquire and process relevant data; *attention focussing* from larger sensing areas to smaller regions of specific interest; and development of *optimal information acquisition* and *exploration strategies*.

3 Components of the Dynamic Perception Framework

We have chosen to address the concerns outlined above through the development of estimation- and information-theoretic models. Contrary to the *ad hoc* AI-based methods still used in much of Computer Vision work, estimation-theoretic models have a long history of success and have been widely applied in signal processing and control tasks [4]. In this section, we describe some of the components of the Dynamic

Perception framework. In particular, we discuss the *stochastic sensor models* developed to describe the robot sensors; the use of a stochastic multi-property tesselated spatial representation, the *Inference Grid*, to encode inferences about the robot's world; the development of various information metrics to measure the robot's knowledge of the environment, and of complexity measures to determine the spatial variability of the environment; the use of *mutual information* to measure the utility of sensory actions; and the computation of the *Locus of Interest* and the *Locus of Observation* of relevant information [13]. The flow of computation between these components is illustrated in Fig. 2.

3.1 Stochastic Sensor Models

To enable us to reason about the perceptual capabilities of an autonomous robot system, mathematical models are needed to describe the behaviour of the various sensors of the robot. In previous work, we have discussed sensor models that are stochastic in nature and have shown their use in the recursive estimation of a tesselated spatial random field model, the Occupancy Grid [14, 10]. While the specific sensor models developed will depend on the properties being measured, the physical characteristics of the sensor systems and the environment of operation of the robot, we have found that the class of models briefly described below can be tailored to a number of different range sensor systems.

The stochastic range sensor models used for the experimental component of our research take into account the target detection probability, the sensor noise that corrupts the range measurements, and the variation of range uncertainty with distance. The uncertainty in the measurement r of a detected target T positioned at z from the sensor is modelled by the pdf $p(r \mid det(T, z))$ expressed as:

$$p(r \mid det(T, z)) = \tilde{G}(r, z, \Sigma(z)) \tag{1}$$

where $det(T, z)$ is the detection event of T at z, $\Sigma(z)$ describes the variation of range noise with distance, and $\tilde{G}()$ is a corrupted Gaussian distribution, given by

$$\tilde{G}() = (1 - \epsilon) G_1() + \epsilon G_2() \tag{2}$$

The distributions $G_1()$ and $G_2()$ are Gaussian, with $G_1()$ modelling the normal behaviour of the sensor, and $G_2()$ occasional gross errors. The parameter ϵ weighs the relative contributions of both terms.

To model the sensor detection behaviour, we use the target detection probability, $P[det(T, z) \mid \exists T$ at $z](z) = P_d(z)$, and the false alarm probability, $P[det(T, z) \mid \not\exists T$ at $z](z) = P_f(z)$ (known as a Type I error [9]). The missed detection probability (Type II error) is of course given by $P[\neg det(T, z) \mid \exists T$ at $z](z) = 1 - P_d(z)$.

The target localization probability, $p(\exists T$ at z $\mid r)$, can be computed using Bayes estimation as:

$$p(\exists T \text{ at } z \mid r) = \frac{p(r \mid \exists T \text{ at } z) P[\exists T \text{ at } z]}{p(r \mid \exists T \text{ at } z) P[\exists T \text{ at } z] + p(r \mid \not\exists T \text{ at } z) (1 - P[\exists T \text{ at } z])} \tag{3}$$

where

$$p(r \mid \exists T \text{ at } z) = p(r \mid det(T, z)) P_d(z) + p(r \mid \neg det(T, z)) (1 - P_d(z)) \tag{4}$$

and

$$p(r \mid \not\exists T \text{ at } z) = p(r \mid det(T, z)) P_f(z) + p(r \mid \neg det(T, z)) (1 - P_f(z)) \tag{5}$$

This class of stochastic sensor models can be applied to a large variety of range sensors, including infrared and laser scanners, sonar sensors, and stereo systems. The required parameters can be obtained through analysis of the physical characteristics of the sensor and through calibration (see, for example, [10, 20, 19, 21]). A typical set of curves showing the dependency of the detection probability and of the range variance on the distance to the target being imaged is given in Fig. 3, while Fig. 4 shows a plot of $p(r \mid \exists T$ at z$)$. These stochastic sensor models are used in perception planning to determine the *Locus of Observation*, from which the robot can acquire spatial information of relevance to specific perceptual goals.

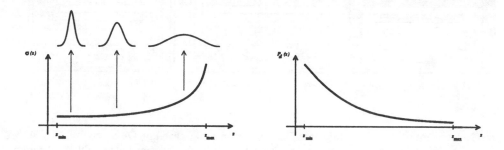

Figure 3: Sensor Range Variance and Detection Probability as a Function of Distance to the Target. These are typical curves for range estimation only.

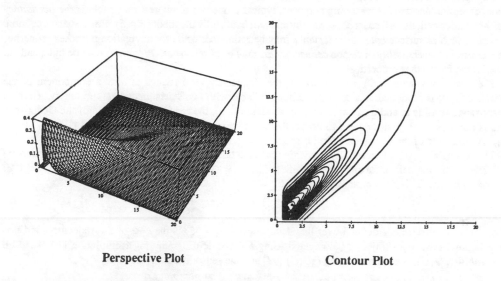

Perspective Plot   **Contour Plot**

Figure 4: Stochastic Range Sensor Model. The function $p(r \mid \exists T \text{ at } z)$ for range estimation only is shown, in both a perspective plot and a contour plot.

3.2 Inference Grids

In previous work, we have developed an approach to spatial robot perception and navigation called the *Occupancy Grid* framework [12, 14, 10]. The Occupancy Grid is a multi-dimensional discrete random field model that maintains probabilistic estimates of the occupancy state of each cell in a spatial lattice. Recursive Bayesian estimation mechanisms employing stochastic sensor models allow incremental updating of the Occupancy Grid using multi-view/multi-sensor data, as well as composition of multiple maps, decision-making, and incorporation of robot and sensor position uncertainty. The Occupancy Grid framework provides a unified approach to a variety of problems in the mobile robot domain, including autonomous mapping and navigation, sensor integration, path planning under uncertainty, motion estimation, handling of robot position uncertainty, and multi-level map generation. It has been successfully tested on several mobile robots, operating in real-time in real-world indoor and outdoor environments [10].

For the Dynamic Perception work we have generalized the Occupancy Grid representation, and have

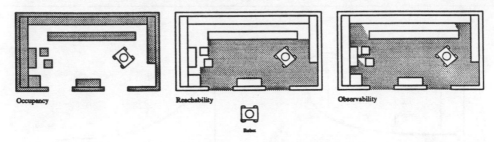

Figure 5: Some Properties of the Inference Grid: Occupancy, Reachability, Observability. The shaded areas of each map indicate regions that are occupied, reachable by the robot, and observable by the robot.

developed a spatial model called the *Inference Grid*. The Inference Grid is a multi-property Markov Random Field defined over a discrete spatial lattice. By associating a random vector with each lattice cell, the Inference Grid allows the representation and estimation of multiple spatially distributed properties. For robot perception and spatial reasoning purposes, typical properties of interest may include the *occupancy* state of a lattice cell, as well as its *observability* and *reachability* by the robot (Fig. 5). For visual perception, properties such as surface *color* or *reflectance* may be estimated, while for navigation purposes properties such as terrain *traversability* or region *connectedness* may be of relevance. Properties may be independent, or derivable from other properties.

The occupancy state, $s(C)$, of a cell C of the Occupancy Grid (which is now a component of the Inference Grid) is modelled as a discrete random variable with two states, *occupied* and *empty*. Given a sensor range reading r, and a stochastic sensor model $p(r \mid z)$, the recursive estimation of the Occupancy Grid is done using Bayes' theorem to obtain the cell state probabilities $P[s(C) = \text{OCC} \mid r]$ (see [14, 10, 11]). An example of the Occupancy Grid is shown in Fig. 6.

The reachability and observability properties can also be treated as binary stochastic variables, and are estimated using the cell occupancy probabilities, $P[s(C) = \text{OCC} \mid M]$, stored in the Occupancy Grid M. Cell *reachability*, $\Xi(C)$, is computed using $P[s(C) = \text{OCC} \mid M]$ as:

$$P[\Xi(C) \mid M] = \prod_{\forall Z \in \eta} (1 - P[s(Z) = \text{OCC} \mid M]) \text{ s.t. } \eta = \arg \min_{\forall \xi(M)} \Gamma(\xi) \tag{6}$$

where $\xi(M)$ is a trajectory defined over the Occupancy Grid, $\Gamma(\xi)$ is the cost of the trajectory, and η is the minimum-cost robot trajectory, computed using a dynamic programming formulation [10, 4]. Cell *observability*, $\Psi(C)$, is estimated using $P[\Xi(C) \mid M]$ and the sensor model of Eq. 3 as:

$$P[\Psi(C) \mid M] = \max_{\forall Z \in M} P[det(C) \mid s(C) = \text{OCC} \wedge \pi(R) = Z] P[\Xi(Z) \mid M] \tag{7}$$

where $\pi(R)$ denotes the position and orientation of the robot.

3.3 Information Measures

To guide the perceptual activities of a robot, we need metrics to evaluate the robot's world knowledge. The specific metric used depends on the robot task and the particular kind of information required for successful execution of the task. For tasks that involve spatial reasoning and navigation, we require measures of the extent and accuracy of the robot's sensory maps, while for target localization or shape recovery, precise surface position information is needed.

Map Uncertainty

For the Inference Grid model discussed above, an intuitive way of expressing the uncertainty in the spatial information encoded in cell occupancy estimates obtained from the sensor data is given by the cell uncertainty function [10]:

$$U(C) = 1 - 4 \left(P[s(C) = \text{OCC} \mid M] - \frac{1}{2} \right)^2 \tag{8}$$

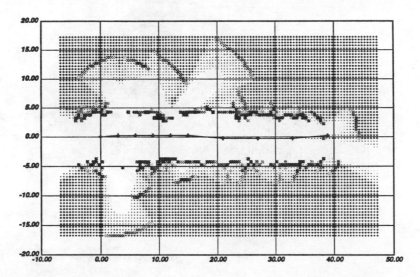

Figure 6: An Example of the Occupancy Grid. The map shows an Occupancy Grid built by a mobile robot using a sonar range sensor array. The robot is moving along a corridor, from left to the right. The gray cells correspond to high occupancy probability areas, while the areas marked with "·" correspond to cells with high emptyness probability. Areas not observed by the robot are identified using "+".

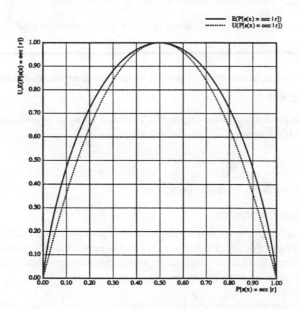

Figure 7: Occupancy Grid Cell Uncertainty Measures. The cell uncertainty function $U(C)$ and the cell entropy function $E(C)$ are shown.

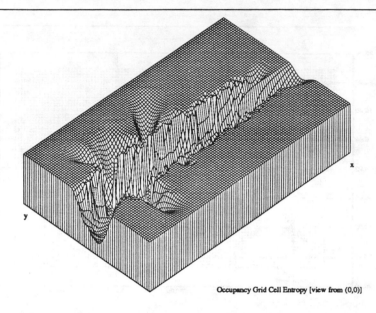

Occupancy Grid Cell Entropy [view from (0,0)]

Figure 8: Entropy of the Occupancy Grid. The cell entropies for the Occupancy Grid of Fig. 6 are shown.

A more generally used metric of uncertainty is the entropy of a random variable [5, 24]. The Inference Grid cell uncertainty can be measured using the entropy of the cell occupancy state estimate as (Figs. 7 and 8):

$$E(C) = -\sum_{s_i} P[s_i(C) \mid M] \log P[s_i(C) \mid M] \tag{9}$$

Using the cell entropy, the uncertainty over a region W of the Inference Grid can be computed as:

$$E(W) = \sum_{\forall C_i \in W} E(C_i) \tag{10}$$

This definition allows us to determine upper and lower bounds on the uncertainty of the region W:

$$0 \le E(W) \le \#(W) \tag{11}$$

where $\#(W)$ is the cardinality of the region W. To obtain an entropy measure that is independent of the region size, the average entropy $\overline{E}(W)$ is defined as:

$$\overline{E}(W) = \frac{E(W)}{E_0(W)} \tag{12}$$

where $E_0(W) = \#(W)$ is the maximum entropy of W.

Target Localization Uncertainty

For precise localization or shape recovery tasks, the error probabilities for target detection and the variance in the position estimate of detected features, such as surfaces or vertices, can be directly used as quantitative uncertainty measures. Consider a range sensor whose measurements are corrupted by Gaussian noise of zero mean and variance σ^2, modelled by the pdf:

$$p(r \mid z) = \frac{1}{\sqrt{2\pi}\sigma} \exp\left(\frac{-(r-z)^2}{2\sigma^2}\right) \tag{13}$$

The uncertainty in surface localization is given directly by σ, and the Type II error probability $1 - P_d(z)$ provides a direct measure of detection uncertainty.

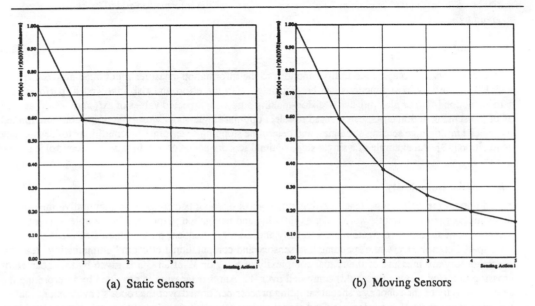

| (a) Static Sensors | (b) Moving Sensors |

Figure 9: Occupancy Grid Entropy Change With Sequential Sensing Actions. Graphs (a) and (b) both show the change in OG uncertainty, measured as the average entropy $\overline{E}(M)$, as range sensor data is acquired sequentially and used to update the Occupancy Grid M. For graph (a), the robot and its sensors are static, and data is being acquired from a single view. In graph (b), the robot and its sensors are moving, and data is being acquired from multiple views. For this case, it can be seen that the uncertainty decay rate is faster, and that the asymptotic uncertainty limit is lower.

3.4 Information Provided by a Sensing Action

The information $\Delta I(\alpha_k)$ added to the Inference Grid by a sensing action α_k can be defined directly in terms of the decrease in uncertainty caused by that sensing action:

$$\Delta I(\alpha_k) = -(E_k(W) - E_{k-1}(W)) \tag{14}$$

where $E_{k-1}(W)$ is the entropy of a region W of the Occupancy Grid before the sensing action α_k, and E_k is the entropy of W after the sensing action α_k. This measure is known as *Mutual Information* [24].

Eq. 14 serves as an indicator of the efficiency of a specific sensor or sensing action, and can be used to implement "stopping rules" such as those used in statistical analysis [3]. Fig. 9(a) shows how the average entropy $\overline{E}(M)$ of an OG M decreases for the case of a static robot, where multiple scans are performed from a fixed location. In contrast, Fig. 9(b) shows how the average entropy decreases for a mobile robot that is taking sensor scans from multiple locations, as it moves around in the environment. The graphs make explicit and express in quantitative terms what would be intuitively expected, namely, that it is more useful in terms of world knowledge acquisition to integrate data obtained from multiple sensing locations than from a single view. Note that, for non-biased robot sensors, we have $\lim_{k \to \infty} I(\alpha_k) = 0$. In the case of a static robot, the average entropy $\overline{E}_k(W)$ will tend towards $\lim_{k \to \infty} \overline{E}_k(W) = 1 - \#(\omega)/\#(W)$, where $\omega \subset W$ is the region observable by the sensors from the robot's location, and W is the total extent of the region of interest. For the case of a moving robot, the average entropy $\overline{E}_k(W)$ will tend towards $\lim_{k \to \infty} \overline{E}_k(W) = 1 - \#(\Omega)/\#(W)$, where $\Omega \subset W$ is the region observable by the robot as its explores its environment, and W is again the total extent of the region of interest. This behaviour can be observed in Figs. 9(a) and (b).

As we already mentioned, in localization problems the quality of an estimate is usually measured by its variance. If n measurements are taken using a sensor described by the model of Eq. 13, the sample variance of the estimated parameter $\hat{r}$ will be $\sigma_r^2 = \sigma^2/n$, and the information added by sensing action α_k is given by:

$$-\Delta\sigma_k^2 = \frac{\sigma^2}{k\,(k-1)} \tag{15}$$

It is straightforward to associate utility functions to the information measures mentioned above (Eqs. 14 and 15). Similarly, cost functions can be associated with the effort and risk involved in performing a sensing action [4]. For planning and decision-making purposes, expected values of ΔI_k and $-\Delta\sigma_k^2$, as well as of the sensing action costs, can be used to select optimal single-stage sensory actions, compute limited lookahead multi-stage sensing strategies, and determine stopping or termination conditions for sensor data acquisition [13]. An example of a single-stage optimal sensory action choice for attention control is shown in Fig. 13.

3.5 Spatial Complexity

An additional metric of importance for dynamic control of robot perception is a quantitative measure of the complexity of the robot's world. For exploration and navigation purposes, for example, it is intuitive that the robot should investigate more carefully and do more frequent measurements of areas that have high spatial complexity, while spending less sensory and computational effort in "uninteresting" regions. A straightforward measure of the spatial complexity of a region W of an OG is given by the mean zero-crossing rate (for a given threshold) computed over W. Another measure is provided by interpreting the sensing activity of the robot as a spatial sampling process performed over the robot's environment. Given the tesselated nature of the Inference Grid model and its representational similarity to images [10, 11], a position-dependent Fast Fourier Transform (FFT) $\mathcal{F}_W[M(x,y)]$ can be computed over a finite-size window W of the Occupancy Grid M to obtain the spatial frequency spectrum of a specific region. Appropriate window functions include the rectangular window and the Hamming window [22]. The spatial frequency spectrum gives us a metric of the spatial complexity of the regions being explored by the robot, and allows us to derive an optimal sensing strategy, by performing the sensing actions at spatial intervals determined using the Nyquist (or optimal sampling) rate [22]:

$$\Delta x = \frac{\pi}{\omega_x} \qquad \text{and} \qquad \Delta y = \frac{\pi}{\omega_y} \tag{16}$$

where ω_x and ω_y are the band limits of the spatial frequency spectrum. An example of the use of spatial complexity measures to plan the locomotion of a mobile robot is shown in Fig. 6. Using spatial variability estimates of its immediate surroundings as constraints on the distances between data acquisition stops, the robot is able to respond to the complexity of its environment: it stops more frequently in regions of high spatial variability, such as the two open doors on either side of the left portion of the corridor, and speeds up when the corridor becomes "dull".

4 Strategies for Dynamic Control of Robot Perception

We now turn to the application of the stochastic and information-theoretic models, discussed in the previous sections, in the development of strategies for dynamic control of robot perception. We will discuss the *Locus of Interest* and the *Locus of Observation*, outline methods for attention control and attention focussing, and illustrate the application of these strategies in autonomous robot exploration and in the integrated planning of robot navigation and robot perception.

4.1 Task-Directed Perception

Application scenarios for autonomous mobile robots require the execution of a variety of tasks related to spatial perception, reasoning and navigation, such as motion planning, detection and inspection of spatially distributed features, object recognition and pose determination, grasp planning, etc. In the work discussed here, the connection between robot task and robot perception is done by explicitly mapping the information needs of the task on the Inference Grid. Typical perceptual tasks may include observing a spatial feature with some minimum detection probability, localizing a spatial feature with some bound on the positional uncertainty, or selecting specific regions of interest, so that non-pertinent sensor data can be ignored.

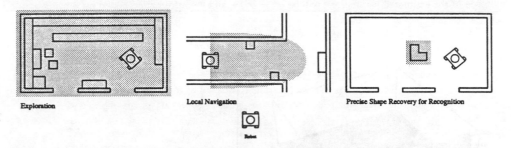

Figure 10: Locus of Interest for Several Robot Tasks. The regions of interest to the robot for *exploration*, *local navigation* and *precise shape recovery* tasks are shown as shaded areas.

Perception Constraints and the Locus of Interest

We define the *Locus of Interest* as a region specified on the Inference Grid that is fundamentally relevant for a specific robot task. It is determined by having the task define a utility function $R(W)$ over a region W of the Inference Grid, which measures the relevance of knowledge about W for successful task execution. Consequently, the Locus of Interest LI defined by task τ_i is computed as the region of the Inference Grid that exceeds a utility threshold u_i:

$$LI(M, \tau_i) = \{\forall C \in M \text{ s. t. } R(C) \geq u_i\} \tag{17}$$

Examples of the Locus of Interest for some specific robot tasks are shown in Fig. 10. In addition to region selection, tasks can impose additional constraints on the information to be acquired. For example, *Detection Constraints* and *Localization Constraints* of the form $det(T, z) \geq D_t$ and $\sigma(z) \leq \sigma_t$ can be defined, where D_t and σ_t are detection and range uncertainty thresholds (see example in Fig. 12). Note that if multiple goals are being pursued by the robot, the corresponding task-defined Loci of Interest and perceptual constraints can be merged and simultaneously represented on the same Inference Grid. It should also be mentioned that the robot system itself may impose LIs derived from robot-specific operational tasks.

The Locus of Observation

After determining the Locus of Interest, we can compute the *Locus of Observation*. The Locus of Observation is the configuration space region where the robot has to position a selected sensor or set of sensors to acquire the information needed by the robot's tasks and specified in the LI. The LO is computed as:

$$LO(M, \Theta[LI(M)]) = \{\forall C \in M \text{ s. t. } \pi(R) \in C \wedge \Theta[LI(M)] \geq \lambda\} \tag{18}$$

where $\Theta[LI(M)]$ is a predicate or utility function defined over the perceptual constraints, and which has to be above a threshold λ to be satisfactory. Fig. 11 shows the limits imposed on the region of operation of the stochastic sensor model of Eq. 4 by detection and range uncertainty constraints. An illustration of the Locus of Observation is given in Fig. 12. Industrially used mobile platforms, such as AGVs or sentry robots, frequently rely on the detection of specific landmarks, such as active beacons radiating on a known signature frequency, to determine the location of the vehicle or select the next segment of the path to be traversed. The shaded areas in Fig. 12 correspond to the Loci of Observation, and indicate the regions from which the beacons can be observed with $det(T, z) \geq D_t$.

4.2 Attention Control and Attention Focussing

In the Dynamic Perception framework, *Attention Control* is performed through the selection of regions in the Inference Grid that have both *high current relevance* as measured by the Locus of Interest and are *observable* as measured in the Inference Grid. An example of Attention Control is given in Fig. 13. The

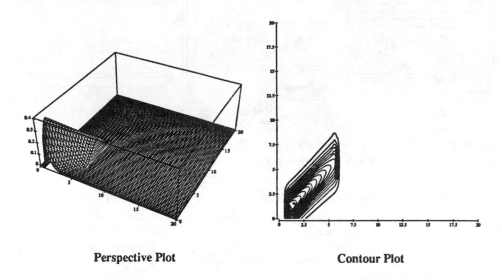

Perspective Plot   **Contour Plot**

Figure 11: Imposing Perceptual Constraints on the Stochastic Sensor Model. The stochastic sensor model of Eq. 4 is shown with the operational limits imposed by the perceptual constraints $det(T, z) \geq D_t$ and $\sigma(z) \leq \sigma_t$.

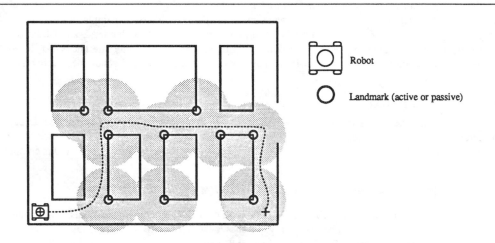

Figure 12: Locus of Observation for Landmark-Based Navigation. An environment instrumented with beacons (active landmarks) is shown. The shaded regions correspond to the Loci of Observation.

task is exploration, and the robot has already partially mapped the region of interest. After a perception planning cycle during which the observability of the cells of the Inference Grid was estimated, four regions were selected for further exploration. These regions correspond to windows that have both high average Occupancy Grid entropy and high average probability of being observable.

Given the uncertainty intrinsic to sensory observations and execution of robot actions, it is generally not possible to compute exact *a priori* measures of observability, sensor information, etc., and many

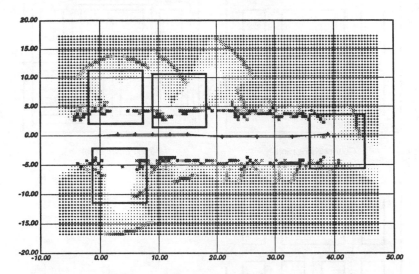

Figure 13: Controlling the Attention of the Robot. The same Occupancy Grid of Fig. 6 is shown. The robot's task is exploration, and the four areas to be investigated next are shown superimposed on the map.

of the measures used for perception planning are estimates or expected values of the actual parameters. Therefore, most scenarios require an iterative refinement approach, since at the beginning of the sensing and world modelling activity the robot will have only an incomplete map and partial information to work with. This situation leads naturally to the need for *Attention Focussing*, where the Locus of Interest covers a large area during the preliminary reconnaissance stage, but is narrowed down to more specific regions as more information becomes available. This is again illustrated in the exploration scenario presented in Fig. 13, where after a general reconnaissance phase the attention of the mapping system is now being narrowed to specific regions to be explored further.

5 Applications

Robot Exploration

We now outline two concrete scenarios. The first is the *exploration scenario*, whose components were already discussed in previous sections. As illustrated in Figs. 6 and 13, the Dynamic Perception framework allows the robot to react to the complexity of its environment, as well as to determine optimal exploration strategies, by reasoning explicitly about what the robot needs to know, what it does know, and what it doesn't know about its environment. A second exploration example is given in Fig. 14. It should be mentioned that the exploration problem we address can be seen as a stochastic generalization of the deterministic *Art Gallery Problem* [23].

Integration of Navigation and Perception

The second scenario involves *integrated perception and navigation planning*. As discussed in section 1, robot perception and robot task planning have generally been treated as separate stages of the robot's cycle of operation. This is clearly seen in the area of robot motion planning. Path planning methods have generally been limited to planning safe trajectories from a given starting point to a given goal, avoiding obstacles and taking into account the kinematic and dynamic characteristics of the robot. Other concerns, such as robot registration constraints, perceptual requirements, or environment complexity, are ignored.

The Dynamic Perception framework allows integrated navigation planning, where both perceptual and locomotion requirements can be taken into account. Simple obstacle-avoidance path planning is performed

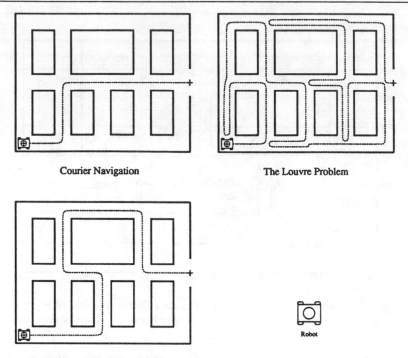

Courier Navigation   The Louvre Problem

Robot

The Tourist in the Louvre Problem

Figure 14: Integration of Perception and Navigation Tasks. The maps show the behaviour of a mobile robot in three cases: 1. Courier navigation tasks, when the goal is finding the fastest route. 2. Exploration tasks, when the primary goal is careful mapping of the robot's area of operation. 3. Integration of courier navigation and exploration, when both activities are given comparable importance, and the robot automatically adjusts its behaviour accordingly.

---

on the Inference Grid as the minimization of a dual-objective cost function [10]:

$$\min_{\mathbf{P}} f(\mathbf{P}) = w_d \operatorname{length}(\mathbf{P}) + w_c \sum_{\forall C \in \mathbf{P}} \Gamma(C) \tag{19}$$

where $\mathbf{P}$ is the robot path, w_d is the path length weight, w_c is the cell cost weight, and $\Gamma(C) = f_c(P[s(C) = \mathrm{OCC}])$ is the cell traversal cost, defined directly in terms of the Occupancy Grid cell state estimates.

Integrated navigation planning for perceptual and locomotion tasks can be performed on the Inference Grid as the minimization of a multi-objective cost function:

$$\min_{\mathbf{P}} f(\mathbf{P}) = \sum_i w_i c_i(M) \tag{20}$$

where the $c_i(M)$ are cost functions representing various perception and locomotion requirements, computed on the Inference Grid M, and w_i is the weight vector. An example of this approach is shown in Fig. 14, where different behaviours of a mobile robot are obtained by varying the relative importance of two tasks: *exploration*, where extent and accuracy of the resulting map are important, and *courier navigation*, where finding the shortest distance to the goal is essential.

6 Conclusions

The *Dynamic Robot Perception* framework outlined in this paper stresses the active and adaptive control of the perceptual activities of an autonomous robot. This is done by explicitly determining the evolving information requirements of the different tasks being addressed by the robot, and by planning appropriate sensing strategies to recover the information required for successful completion of these tasks. We discussed the development of strategies for dynamic control of robot perception that emphasize the use of stochastic and information-theoretic sensor interpretation and world modelling mechanisms, and explored the connection between specification of a task and its information requirements.

We have performed an initial experimental validation of the components of the Dynamic Perception framework discussed in this paper. Currently, our research group is finishing the software structure and sensor interfaces for a more powerful mobile robot. This vehicle is based on an omni-directional platform, and is equipped with a number of sensors, including infrared proximity sensors, a sonar sensor array, and an optical rotating range scanner. It will be used to conduct more extensive experimental work.

Acknowledgments

The author wishes to thank José Moura for useful insights into the information control literature, and Ingemar Cox for making me aware of the Art Gallery literature. The research discussed in this paper was performed at the Intelligent Robotics Laboratory, Computer Sciences Department, IBM T. J. Watson Research Center. It incorporates some results from research performed by the author during his association with the Mobile Robot Lab, Robotics Institute, Carnegie-Mellon University.

The views and conclusions contained in this document are those of the author and should not be interpreted as representing the official policies, either expressed or implied, of the sponsoring organizations.

References

[1] J. Aloimonos, I. Weiss, and A. Bandyophadyay. Active Vision. *International Journal of Computer Vision*, 1(4), January 1988.

[2] R. Bajcsy. Active Perception. *Proceedings of the IEEE*, 76(8), August 1988.

[3] J. O. Berger. *Statistical Decision Theory and Bayesian Analysis*. Springer-Verlag, Berlin, 1985. Second Edition.

[4] D. P. Bertsekas. *Dynamic Programming: Deterministic and Stochastic Models*. Prentice-Hall, Englewood Cliffs, NJ, 1987.

[5] R. E. Blahut. *Principles and Practice of Information Theory*. Addison-Wesley, Reading, MA, 1988.

[6] T. Dean, K. Basye, and M. Lejter. Planning and Active Perception. In *Proceedings of the DARPA Workshop on Innovative Approaches to Planning, Scheduling, and Control*, DARPA, 1990.

[7] T. Dean et al. Coping with Uncertainty in a Control System for Navigation and Exploration. In *Proceedings of the Eight National Conference on Artificial Intelligence*, AAAI, Boston, MA, July 1990.

[8] H. F. Durrant-Whyte. *Integration, Coordination, and Control of Multi-Sensor Robot Systems. Kluwer International Series in Engineering and Computer Science*, Kluwer Academic Publishers, Boston, MA, 1988.

[9] J. L. Eaves and E. K. Reedy. *Principles of Modern Radar*. Van Nostrand Reinhold, New York, 1987.

[10] A. Elfes. *Occupancy Grids: A Probabilistic Framework for Robot Perception and Navigation*. PhD thesis, Electrical and Computer Engineering Department/Robotics Institute, Carnegie-Mellon University, May 1989.

[11] A. Elfes. Occupancy Grids: A Stochastic Spatial Representation for Active Robot Perception. In *Proceedings of the Sixth Conference on Uncertainty and AI*, AAAI, Cambridge, MA, July 1990.

[12] A. Elfes. Sonar-Based Real-World Mapping and Navigation. *IEEE Journal of Robotics and Automation*, RA-3(3), June 1987.

[13] A. Elfes. *Strategies for Dynamic Robot Perception Using a Stochastic Spatial Model*. Research Report, IBM T. J. Watson Research Center, 1991. In preparation.

[14] A. Elfes. A Tesselated Probabilistic Representation for Spatial Robot Perception and Navigation. In *Proceedings of the 1989 NASA Conference on Space Telerobotics*, NASA/Jet Propulsion Laboratory, JPL, Pasadena, CA, January 1989.

[15] E. Grosso, G. Sandini, and M. Tistarelli. 3-D Object Reconstruction Using Stereo and Motion. *IEEE Transactions on Systems, Man, and Cybernetics*, 19(6), November/December 1989.

[16] G. Hager. *Information Maps for Active Sensor Control*. Technical Report MS-CIS-87-07, GRASP Lab, Department of Computer and Information Science, University of Pennsylvania, Philadelphia, PA, February 1987.

[17] G. Hager and M. Mintz. Estimation Procedures for Robust Sensor Control. In *Proceedings of the 1987 AAAI Workshop on Uncertainty in Artificial Intelligence*, AAAI, Seattle, WA, July 1987.

[18] M. Hebert, T. Kanade, and In So Kweon. 3-D Vision Techniques for Autonomous Vehicles. In *Analysis and Interpretation of Range Images*, Springer-Verlag, Berlin, 1990.

[19] E. P. Krotkov. *Active Computer Vision by Cooperative Focus and Stereo*. Springer-Verlag, Berlin, 1989.

[20] L. H. Matthies. *Dynamic Stereo Vision*. PhD thesis, Computer Science Department, Carnegie-Mellon University, 1989.

[21] G. L. Miller and E. R. Wagner. An Optical Rangefinder for Autonomous Robot Cart Navigation. In *Proceedings of the 1987 SPIE/IECON Conference*, SPIE, Boston, MA, November 1987.

[22] A. V. Oppenheim and R. W. Shafer. *Discrete-Time Signal Processing. Prentice Hall Signal Processing Series*, Prentice Hall, Englewood Cliffs, NJ, 1989.

[23] J. O'Rourke. *Art Gallery Theorems and Algorithms*. Volume 3 of *International Series of Monographs on Computer Science*, Oxford University Press, Oxford, 1987.

[24] A. Papoulis. *Probability, Random Variables, and Stochastic Processes*. McGraw-Hill, New York, 1984.

[25] H.-L. Wu and A. Cameron. A Bayesian Decision Theoretic Approach for Adaptive Goal-Directed Sensing. In *Proceedings of the Third International Conference on Computer Vision*, IEEE, Osaka, Japan, December 1990.

Consideration on the Cooperation of Multiple Autonomous Mobile Robots

Suparerk PREMVUTI and Shin'ichi YUTA

Institute of Information Science and Electronics,
University of Tsukuba, Tsukuba 305, JAPAN

Abstract

The paper proposed basic concepts of the cooperation of multiple autonomous mobile robots. The authors suggested to divide cooperation into two types, active and non-active, according to setting goal of the system. And in non-active cooperation there must be "modest cooperation" to prevent any collision of resource access by robots. The way to design multiple robots system also discussed.

1.Introduction

Human-being can do his work independently. But at the same time, he can cooperate with others. "Intelligent" of an autonomous mobile robot means including this kind of action or not?. Up until now, researches on cooperation of robots have been focusing on arm robots or manipulators. But this kind of cooperation generally uses centralized control method that means all movements of robots are controlled directly by one controller.

On the other hand, self organizing system or distributed cooperation system contains many simple functional cells which is independent from each other. These cells cannnot be said that they are independent from each other or have intelligence. The cooperation of intelligent robots may be something different from cooperation we known. The authors surveyed factors concerning cooperation of autonous mobile robots and proposed concepts of modest cooperation which is necessary for this type of robot.

2.Consideration of Autonomous and Cooperation of Robots

2.1 Scanning other researches

Grossman[1] considered movement of multiple AGV's system using indidual planning by each robot and proposed the method using high ways for high throughput. But his research does not discuss any factors that are essential to mobile robots such as the use of sensors. Fukuda[2] proposes a new concept of combined robot which called Cellular Robot. The concept is excellent but with regard to decision making method and planning of movement of the whole system are not discussed sufficiently.

2.2 Classification of multiple robots system

There is no doubt that if the capability of a robot is improved, the robot can work efficiently and flexibly. But in some case a robot cannot do all works solely. So, many researches focuses on how to make several robots work together. But detail consideration of essence of cooperation has not been scrutinized yet. The authors proposed consideration concerning the process of designing multiple autonomous mobile robots system. Techniques used in the system are also discussed.

First, we divide cooperation of robots into two type as follows,
(1) Active cooperation, and
(2) Non-active cooperation

The word, "Active", means all robots in the system take up a positive attitude to join together for doing something that create a progressive result, such as increasing performance, save time. On the contrary, an attitude which not consists such nature is called "Non-active".

2.3 Multiple robots system that has one goal

When several robots work in the same space, the first thing we must think about is goals of each robot and their ties. In most multiple robots system, usually industrial arm robots, system's main goal is to manufacture products. All robots in the system do their works cooperatively to reach the goal of the system. When we know exactly the goal of the system, it is better to make the system that use centralized decision structure. Because, all parameters of the systems can be optimized easily to obtained the best performance.

Here, an example for showing how to consider the problem of cooperation with one common goal is given.

Comparison between distributed control method and centralized control method to solve a problem of multiple robots system

Problem:

There are 4 mobile robots in a room(fig.1). Robots are at arbitrary positions. Wanting all robots move to the coners of the room. A corner can be accessed by only one robot.

EH0342-6/91/0000/0219$01.00 © 1990 IEEE

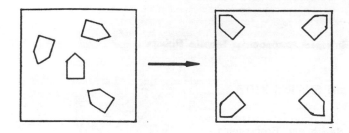

FIG.1   MULTIPLE ROBOTS IN A ROOM

If men are confronted with the same situation what will they do?. When four men are told to go the corner to have one person at one corner. Men will notice actions of each other and decide which corner to go to with out any consultation needed. This kind of movement seem to be easy for human-being which has excellent sensor, his eyes and ears. But not easy for a robot because the sensor of robot is not so powerful. Existing sensor cannot detect movements of other robots immediatly, then it is impossible for robots to act like men. Robots will act as follows.
(1) Use sensor to sense movements of other robots and decide its movement. Sometime there may be wrong dicisions, a corner is selected by more than one robot. Robots must do the method of trial and error.
(2) Using communications path for communicate each other for exchange information about position or plan of movement.

The way robots decide their movement can be done in two different ways, distributed control method or centralized control method. Each robot's hardware and software can be listed as follows.

In case of distributed control method
Each robot contains
(1) Decision making algorithm
(2) Map
(3) Environment sensor
(4) Inter-robot communications capability
(5) Actuator

In case of centralized control method
Each robot contains
(1) Environment sensor
(2) Communications to the center capability
(3) Actuator

and the center (may be one of robots) contains
(1) Decision making algorithm
(2) Communications to robots capability
(3) Map

Let's compare processes done by the two methods.

Distributed control method
In this method decision making part of the system is divided in to several sub parts which are implemented in each robot. Each robot is responsible for its motion. But all robots's movements must lead to the main goal of the system, then communications between all robots is indispensable. Messages

exchanged between robots would be plans of movement sended to others for adjusting if there will be any collision.
Diagram of decision making process is shown in fig.2.

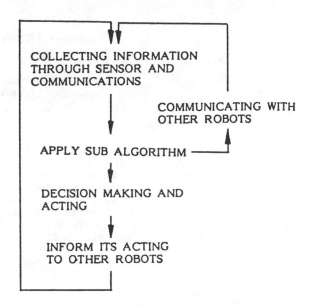

FIG.2 DECISION MAKING PROCESS USING DISTRIBUTED CONTROL METHOD

Decision making algorithm of the whole system divided into sub algorithms implemented in each robot. Sensory information and other information through communication network are used to determine movement of the robot according to sub algorithm. Since the algorithm of the whole system is divided into several sub parts, all sub algorithm's connection and mutual reaction must be considered carefully. Thus, the algorithm tends to be complicated.

Centralized control method
Decision making part of the system is placed at the center, which may be a ground base or one of robots. Since all movements of robots in the system are controlled from the single place, conflicts among robots can be solved inside one processor. Communication inside a CPU requires less time than communication between processors installed inside mobile robots. And the performance of system can be worked out relatively easily than distributed system, then the centralized control method is suitable for system aim at high performance.
Diagram of decision making process of centralized control method is shown in fig. 3.

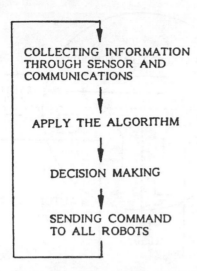

FIG.3 DECISION MAKING PROCESS USING
CENTRALIZED CONTROL METHOD

Decision making is done by the center's CPU. Real-time sensor data and status information data of all robots in the system are sended to the center through center-robot communication network. The center uses these informations to determine movements of robots and send appropriate commands to robots.

2.4 Multiple robots system that each robot has its own goal

In this case, the centralized control method can be used also. But when we think of autonomy of each robot, the way each robot makes decision by itself should be a better method. But when robots determine their movements by themself, conflicts such as collision between robots may occur. Then, when the distributed control method is applied to multiple robots system, robots must try not to interfere with each others. This behavior, that the authors called "modest cooperation", can be seen as an essential action of multiple autonomous robots system.

Considering human society, a man is autonomous. He can do anything he want by himself. We can say that men can live together peacefully because they respect to modest cooperation rule, not interfere with others.

2.5 Modest cooperation in details

The main goal of modest cooperation is to make the multiple autonomous robots sytem can work smoothly through peacefully sharing of resources among robots.

Resource sharing

In system there will be several resources for robots to use. Accessing to the same resource by several robots at the same time causes a conflict or collision. Then resource sharing can be said in another way "collision avoidance of access to resource". For autonomous mobile robots, there are many resources used.

(1) Space resoure for movement of robots

This resource is basic resource of a mobile robot. Sharing space for movement among robots in multiple robots system is being done by many researchers under the theme "collision avoidance among mobile robots"(see[3] for example).

(2) Tools

Tools means all physical objects shared by robots in the system. The tools may be placed at some position and can be accessed by any robot.

(3) Band of active sensor

Environment sensor is very important for movement of a mobile robot. In general, sensing method can be divided into two categories; active sensor and nonactive sensor. Some active sensor may use ultrasonic sound, laser, or radio wave to project to the object and sense reflection. The sensor cannot detect reflection correctly if there is another same type of sensor operating near by. Then, rules for using sensors or modulation-demodulation of radiated wave must be considered.

What is needed for modest cooperation?

When a mobile robot moves solely in a space, it just senses the environment, mostly static, and adopts proper movement. But, in multiple robots system, a robot must can sense other robots which also moving. This is difficult than sensing only environment. At first, we need a powerful sensor that can perceive dynamic movement of other robots. But, at present, such kind of sensor is not powerful enough. For this reason, we also need communication network among robots to provide supplementary information to data from sensor. Exchange of messages between robots needs time. And we must remind this while design algorithm for cooperation of robots.

How to design modest cooperation of autonomous mobile robot?

For achieve high throughput and high efficiently using of resources. The authors proposed dividing motions of robots into two modes, autonomous mode and modest cooperation mode(fig.4).

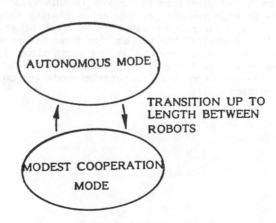

FIG.4 MODE TRANSITION CENCEPT

When robots come close together at a specific distance, the robots may form a group, called "Modest cooperation group", and the robots in this group will act under "Modest cooperation mode". The robots in the group are provided with different priority which may be used in some cooperation rules. When distance between robots in cooperation group is far enough, robots will exit the group and return to act autonomous mode.

Merits of dividing actions of robots into the two modes.
(1) Independent movement of each autonomous mobile robot which is main action can be maximized.
(2) Due to clearly defining boundary of cooperative motion, designing of cooperative algorithms and rules can be done easily.

Modest cooperation mode
Basic policy of this mode is to avoide interferance of resources accessing. A robot will refrain itself from action which may cause damage to other robots' movement. Rules for access resources may be established for properly sharing of resources among robots in the system without deadlock state or collision occurred.

Autonomous mode
In multiple autonomous mobile robots system although there are many robots, but it does not mean that all robots always interfere each other. In modest cooperation mode, some measures must be taken for sharing resources among robots. Under the measures, action or movement of all robot shoud be decreased. A robot cannot use resources freely. That means the performance of an individual robot also declined. To achieve high performance, robots must work under autonomous mode as long as possible.

Example of how to design modest cooperation of autonomous mobile robots
In multiple robots system that robots move along roads or corridors, the changing of modest cooperation mode and autonomous mode depends on distance and direction of robots. In this example, roads is wide enough for one robot running along. However, two robots can pass by each other if running carefully. We assume that three robots are coming to a cross section at the same time. The three robots that are running in autonomous mode will change to modest cooperation mode and cross the junction safely.

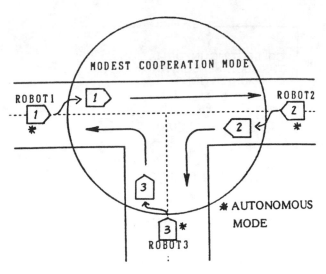

FIG.5 MULTIPLE ROBOTS AT A CROSS SECTION

Features of robot's movements under each mode

Autonomous mode
(1) Moving with high velocity
(2) Freely access resource
   -Running in the middle of the roads
   -Always using active sensors
(3) Exchange position information with other robots through communications network

Modest cooperation mode
(1) Moving with low velocity
(2) Access resources only when necessary
   -Running close to left side of the roads
   -Using sensors according to the rule
(3) Exchange high volume of information with other robots
   -Position information
   -Planned movement
(4) Act according to modestcooperation rules

About modest cooperation rules
Rules used in modest cooperation mode must be designed while keeping in mind the following items.
(1) Considering rules applied to local movement
(2) There is loosely link communication network among robots. Exchange of messages between robots needs time.
(3) Robots must use refrain itself strategy in accident or collision avoiding.

2.6 More complicated multiple robots system
When the goal of multi robots system is unique, the authors propose using centralized decision making method in cooperate with modest cooperation. A robot may be decided to be a leader which provide moving plans to all other robots in the system. The leader does not control movements of robots always. Rather, the leader sends moving plans to other

robots and let those robots perform movement by themself. The method of centralized decision making that provide good performance is used in dividing task and local collision avoidance or resources sharing is done under modest cooperation rules.

3. Example of modest cooperation and experiment
3.1 Collision Avoidance Problem
Typical modest cooperation is collision avoidance between robots at a cross section. The authors will take an example of implementation of modest cooperation to multiple robots system under environment of road network.
3.2 Experiment
How to detect collision?
The authors use autonomous mobile robot 'YAMABICO.M (Level-12)'(Photo 1) in the experiment. At present, the robot is implemented with four directions ultrasonic range sensors which can measure distance between the robot and objects in front, back, left or right. These active sensors cannot be used to detect position of others robots. But these sensors can be used to detected environment and compare to map implemented inside each robot.

Yamabico also have deadreckoning system to obtain coordinate of the present position. Yamabico's can communicate each other through inter-robot network,CAR-Net[5], which can be used to transfer any kind of message including position information.
Then, collision detection method of Yamabico robots can be done using,
(1) Map and deadreckoning system for detect the present position.
(2) Ultrasonic sensors for comparing and adjust position coordinate detected by deadreckoning and for sensing obstacles.
(3) Communication network for sending the position coordinate to others and receiving the position coordinate of other robots. And compare positions of robots if there will be any risk of collision or not.

PHOTO 1 YAMABICO.M(LEVEL 12)

Resource sharing
Yamabico autonomous mobile robot uses ultronic range sensors for sensing environment. The radiation of ultrasonic can be seen as using of sound media resources. Two robots cannot use ultrasonic sensor at the same time. Then the ultrasonics sensors of Yamabico is improved so that can be activated or turned to be inactive by robot's user program.

Implementation of Modest Cooperation mode and Autonomous mode
Modest cooperation mode and autonomous mode can be expressed well by the robot language ROBOL/0[4] which introduce using of "action mode". In ROBOL/0 language all movements of robots are translated into action modes. Modest cooperation mode or autonomous mode are a group of ROBOL/0's action modes.

4. Conclusion
Cooperation is said to be an important behavior of robots. But understanding this behavior has not been analyzed much. The paper proposed steps to approch cooperation of robots systematically. Bring several robots to work in the same space without preparing infrastructures for cooperation will cause interference between robots in many ways. And robots cannot operate properly. The authors proposed divides cooperation into two types, active cooperation and non-active cooperation, according to purpose of the system. Moreover, in non-active cooperation robot system, robots must perform "modest cooperation" for preventing collision of accessing resources. And using of centralized decision making method with modest cooperation in complicated multiple robots system is proposed. For autonomous mobile, modest cooperation is an essential action.

5. References
[1] D.D.Grossman,"Traffic Control of Multiple Robot Vehicles",IEEE J. Robotics and Automation, vol.4 No.5, pp.491-497, Oct.1988
[2] Fukuda, T., Nakagawa, S., Kawauchi, Y. and Martin, B.:"Self Organizing Robots Based on Cell Structures - CEBOT", Proc. 1988 IEEE Internation Workshop on Intelligent Robots and Systems (IROS'88),pp.145-150
[3] M. Saito, T. Tsumura,"Collision Avoidance Between Mobile Robots", Proc. 1989 IEEE IROS'89,pp.473-478
[4] S.Suzuki,S.Yuta and J.Iijima,"A Consideration on the Programming Method of Sensor-Driven Based on the Action Mode Representation", Proc. 20th ISIR, Tokyo, Dec.1989
[5] S.Premvuti,S.Yuta,"Radio Communication Network on Autonomous Mobile Robots For Cooperative Motions", Proc. IECON'88, Singapore,1988

Learning to Coordinate Behaviors

Pattie Maes & Rodney A. Brooks

AI-Laboratory
Massachusetts Institute of Technology
545 Technology Square
Cambridge, MA 02139
pattie@ai.mit.edu
brooks@ai.mit.edu

Abstract

We describe an algorithm which allows a behavior-based robot to learn on the basis of positive and negative feedback when to activate its behaviors. In accordance with the philosophy of behavior-based robots, the algorithm is completely distributed: each of the behaviors independently tries to find out (i) whether it is *relevant* (i.e. whether it is at all correlated to positive feedback) and (ii) what the conditions are under which it becomes *reliable* (i.e. the conditions under which it maximizes the probability of receiving positive feedback and minimizes the probability of receiving negative feedback). The algorithm has been tested successfully on an autonomous 6-legged robot which had to learn how to coordinate its legs so as to walk forward.

Situation of the Problem

Since 1985, the MIT Mobile Robot group has advocated a radically different architecture for autonomous intelligent agents (Brooks, 1986). Instead of decomposing the architecture into functional modules, such as perception, modeling, and planning (figure 1), the architecture is decomposed into task-achieving modules, also called *behaviors* (figure 2). This novel approach has already demonstrated to be very successful and similar approaches have become more widely adopted (cfr. for example (Brooks, 1990) (Rosenschein & Kaelbling, 1986) (Arkin, 1987) (Payton, 1986) (Anderson & Donath, 1988) (Yamaushi, 1990) (Zhang, 1989)).

One of the main difficulties of this new approach lies in the control of behaviors. Somehow it has to be decided which of the behaviors should be active and get control over the actuators at a particular point in time. Until now, this problem was solved by precompiling the control flow and priorities among behaviors either by hand (Brooks, 1986), or automatically, using a description of the desired behavior selection (Rosenschein & Kaelbling, 1986). In both cases the result is some "switching circuitry" among the behaviors which is completely fixed at compile time by the designer.

However, for more complicated robots prewiring such a solution becomes either too difficult or impractical. As the number of behaviors goes up, the problem of control and coordination becomes increasingly complex. Additionally, it is often too difficult for the programmer to

Figure 1: Classical decomposition of an autonomous robot.

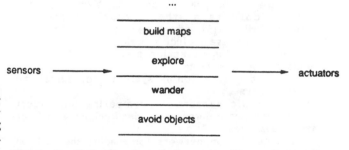

Figure 2: Behavior-based decomposition of an autonomous robot.

fully grasp the peculiarities of the task and environment, so as to be able to specify what will make the robot successfully achieve the task (Maes, 1990).

We therefore started developing an algorithm for learning the control of behaviors through experience. In accordance with the philosophy of behavior-based robots, the learning algorithm is completely distributed. There is no central learning component, but instead each behavior tries to learn when it should become active. It does so by (i) trying to find out what the conditions are under which it maximizes positive feedback and minimizes negative feedback, and (ii) measuring how relevant it is to the global task (whether it is correlated to positive feedback).

We hope that ultimately, this learning algorithm will allow us to program the behavior of a robot by selecting a number of behaviors from a library of behaviors, connecting these to the actual sensors and actuators on the robot, defining positive and negative feedback functions, and making each of the behaviors learn from experience when it is appropriate for it to become active.

224

The Learning Task

The learning task we are concerned with is defined as follows. Given a robot which has:

- a vector of binary perceptual conditions which are either being perceived (or "on") or not being perceived (or "off") at every instant of time,

- a set of behaviors; where a behavior is a set of processes involving sensing and action; a behavior has a precondition list, which is a conjunction of predicates testing a specific value (on or off) for a certain perceptual condition; a behavior may become active when all of its preconditions are fulfilled; an active behavior executes its processes,

- a positive feedback generator, which is binary and global, i.e. at every time t, the robot (and therefore all of the behaviors) either receive positive feedback or not,

- a negative feedback generator, which is again binary and global.

The learning task is to incrementally change the precondition list of behaviors so that gradually only those behaviors become active that fulfill the following two constraints:

1. they are *relevant*,
   where a relevant behavior is a behavior that is positively correlated to positive feedback (i.e. positive feedback is more often received when the behavior is active then when it is is not active) and not positively correlated to negative feedback (i.e. either not correlated at all or inversely correlated),

2. they are *reliable*,
   where a reliable behavior is defined as a behavior that receives consistent feedback (i.e. the probability of receiving positive (respectively negative) feedback when the behavior is active is close enough to either 1 or 0).

An additional requirement, imposed by our philosophy, is that we want the algorithm to be *distributed*. It should allow individual behaviors to change their precondition list in response to certain feedback patterns so that the global behavior of the set of behaviors converges towards a situation where maximal positive feedback and minimal negative feedback is received. Finally, three additional constraints are related to the fact that this algorithm has to be useful for real robots in unconstrained environments: (i) the algorithm should be able to deal with *noise*, (ii) the algorithm should be *computationally inexpensive*, so that it can be used in real-time, and (iii) the algorithm should support *readaptation*, i.e. if the robot changes (e.g. some component breaks down), or its environment changes (e.g. the feedback generators change) the algorithm will make the robot adapt to this new situation (which possibly involves forgetting or revising learned knowledge).

We adopted some simplifying, but realistic, assumptions. One is that, for every behavior, there exists at least one conjunction of preconditions for which the probability of positive feedback as well as the probability of negative feedback are within some boundary (which is a parameter of the algorithm) from either 0 or 1. Another important assumption is that feedback is immediate (there is no delayed feedback) and does not involve action sequences. And finally only conjunctions of conditions (including negations) can be learned. The last section of the paper sketches how the algorithm could be extended to deal with the more general problem[1]. Nevertheless, even after adopting these simplifying assumptions, the learning task is still far from trivial. More specifically, the global search space for a robot with n behaviors, and m binary perceptual conditions is $n * 3^m$, since every behavior possibly has to learn an "on", "off" or "don't-care" value for every perceptual condition.

The Algorithm

The learning algorithm employed by each behavior is the following. Each behavior starts from a "minimal" precondition list. More specifically, only conditions that are necessary in order to be able to execute the processes of the behavior are present. Behaviors maintain data about their performance. A first set of data is related to whether the behavior is relevant or not. A behavior measures i, j, k and l:

	active	not active
positive feedback	j	k
no positive feedback	l	m

Where, j is the number of times positive feedback happened when the behavior was active, k is the number of times positive feedback happened when the behavior was not active, l is the number of times positive feedback did not happen when the behavior was active, and m is the number of times negative feedback did not happen when the behavior was not active. The same statistics are maintained for negative feedback. The statistics are initialized at some value N (for example $N = 10$) and "decayed" by multiplying them with $\frac{N}{N+1}$ every time they are updated. This ensures that impact of past experiences on the statistics is less than that of more recent experiences. The *correlation* (the Pearson product-moment correlation coefficient) between positive feedback and the status of the behavior is defined as

$$corr(P, A) = \frac{j * m - l * k}{\sqrt{(m + l) * (m + k) * (j + k) * (j + l)}}$$

[1] Some additional assumptions being made are that the robot is first of all, able to do experiments, and second, that these do not involve too big risks (the environment can not be too hostile nor the robot too fragile).

225

This gives a statistical measure of the degree to which the status of the behavior (active or not active) is correlated with positive feedback happening or not. $corr(P, A)$ ranges from -1 to 1, where a value close to -1 represents a negative correlation (feedback is less likely when the behavior is active), 0 represents no correlation and 1 represents a positive correlation (feedback is more likely when the behavior is active). In an similar way, $corr(N, A)$, i.e. the correlation between the status of the behavior and negative feedback is defined. The *relevance* of a particular behavior is defined as

$$corr(P, A) - corr(N, A)$$

It is used by the algorithm to determine the probability that the behavior will become active (which is related to the effort which will be put into doing experiments in order to improve the behavior, i.e. making it more reliable). The relevance of a behavior ranges from -2 to +2. The more relevant a behavior is, the more chance it has of becoming active. A behavior that is not very relevant has little chance of becoming active. So, these statistics makes it possible to determine which behaviors are the interesting ones (relevant ones). The relevant behaviors are not necessarily very reliable yet: they might only receive positive feedback in a minority of the times they are active. They also might still cause a lot of negative feedback to be received. All that is "known" is that positive feedback will be more likely received when these behaviors are active, than when they are not active (respectively negative feedback being less likely). The *reliability* of a behavior is defined as (where index P stands for positive feedback and index N stands for negative feedback)

$$min(max(\frac{j_P}{j_P + l_P}, \frac{l_P}{j_P + l_P}), max(\frac{j_N}{j_N + l_N}, \frac{l_N}{j_N + l_N}))$$

The reliability of a behavior ranges from 0 to 1. When the value is close to 1, the behavior is considered very reliable (i.e. the feedback is very consistent: the probability of receiving feedback is either close to 0 or to 1). The reliability of a behavior is used by the algorithm to decide whether the behavior should try to improve itself (i.e. learn more conditions or modify the existing preconditions). If the behavior is not reliable enough, i.e. if either the negative feedback is inconsistent or the positive feedback is inconsistent or both, then one or more additional preconditions are relevant. In this case, the behavior will pick a new perceptual condition to monitor in order to determine whether this condition might be related to the inconsistent feedback [2]. An additional set of statistics is related to the specific condition being monitored (if there is one):

	cond. on	cond. off
positive feedback	n	o
no positive feedback	p	q

Where n is the number of times positive feedback happened when the behavior was active and the condition was on, o is the number of times positive feedback happened when the behavior was active and the condition was off (not on), p the number of times positive feedback did not happen when the behavior was active and the condition was on, and q is the number of times negative feedback did not happen when the behavior was active and the condition was off. Again the same statistics are maintained for negative feedback. If the behavior notices a strong correlation between the condition being monitored and positive and/or negative feedback, it will adopt this condition as a new precondition. More specifically, if the correlation

$$corr(P, on) = \frac{n * q - p * o}{\sqrt{(q + p) * (q + o) * (n + o) * (n + p)}}$$

becomes close to 1 (respectively -1), then the condition will be adopted in the precondition list, with a desired value of "on" (respectively "off"). And similarly, if the correlation for negative feedback becomes close to 1 (respectively -1), then the condition will be adopted in the precondition list, with a desired value of "off" (respectively "on"). If the values for positive and negative feedback are incompatible, the one suggested by the negative feedback dominates. From the moment a new condition has been learned, a behavior only becomes active when this condition has the desired value. If after monitoring a condition for a while, the behavior doesn't notice any correlation between the value of the condition and positive or negative feedback, and the behavior is still not reliable enough, it will start monitoring another condition.

After learning a new condition, the behavior will not necessarily be completely reliable. There might still be other conditions related to the feedback. Until the behavior is reliable enough, it will try to find extra preconditions [3]. Notice that the list of conditions being monitored/evaluated is circular. When all of the conditions have been evaluated and feedback is still inconsistent, the behavior will start monitoring conditions from the start of the list again, reevaluating also those conditions which have already been taken up in the precondition list. A behavior might "forget" something it learned and reevaluate the relevance of that condition. This guarantees that if the environment (e.g. the feedback) or the robot changes, the behaviors are able to adapt to the new situation.

The control strategy of the algorithm is as follows. Behaviors are grouped into groups which control the same actuators. At every timestep the selectable behaviors in every group are determined (those behaviors which are

---

[2] Currently there is a complete connectivity between behaviors and perceptual conditions. The connectivity will be decreased in a subsequent implementation through the use of a switchboard. In applications involving a large vector of perceptual conditions, one could restrict the subset of conditions that a particular behavior considers for learning.

[3] We make the assumption here that for every condition that has to be learned the correlation to feedback is independently detectable. This is likely to be true, because we are not dealing with the problem of learning disjunctions.

not yet active and whose preconditions are fulfilled). For each of these groups, one or zero behaviors are selected probabilistically according to (in order of importance)
- the relative relevance of behaviors,
- the reliability of behaviors, and
- the "interestingness" of the current situation for behaviors, where a situation is more interesting for a behavior if the condition being monitored by the behavior appears in the situation with a value (on or off) that has been experienced a lesser number of times.

The selected behaviors are then activated. The probabilistic nature of the selection process ensures that there is a balance between behaviors being selected (i) because they are successful (are relevant and reliable) and (ii) because of experimentation purposes (to learn). Notice that the learning algorithm is biased: if behaviors are not very relevant (in comparison with other behaviors) they have very little chance of becoming active, which means that little effort is put into making them more reliable (learn new preconditions).

Finally, there are a number of global parameters which can be varied to change the algorithm
- how strong a condition has to be correlated to adopt it as a new precondition,
- how long a condition is monitored before it is dropped,
- how reliable a behavior should try to become,
- how adaptive the behavior is (the relative importance of new data versus data of past experiences).

These parameters have to be tuned to the particular circumstances of task and robot at hand.

A Robot that Learns to Walk

The Task

The described algorithm is being tested on a six-legged robot, called Genghis (see figures 4 and 5). The goal of the experiment is to make Genghis learn to walk forward. This task was chosen because of its complexity. The current version consists of 12 behaviors learning about 6 perceptual conditions, which corresponds to a search space of $12 * 3^6 = 8748$ nodes. Another reason for choosing this task was the availability of a 6-legged robot with a lot of degrees of freedom (12 to be precise) (Angle, 1989). The final reason is that a lot is known both about insect walking (Wilson, 1966) (Beer, 1989) and about 6-legged robot walking (Donner, 1987) (Brooks, 1989). The results reported in this literature demonstrated that the task was feasible (that the completely distributed walking which would result from our learning was indeed robust and successful). This literature also made it possible to compare and interpret our results.

The Robot

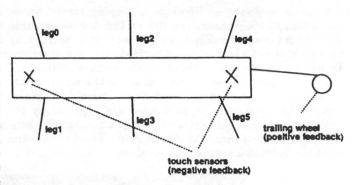

Figure 3: Schematic representation of Genghis, its positive and negative feedback sensors and its distributed collection of learning behaviors.

Genghis is an autonomous six-legged robot with twelve servo motors controlling the two degree of freedom legs (Angle, 1989). It has 4 on board 8-bit microprocessors linked by a 62.5Kbaud token ring. The total memory usage of the robot is about 32Kbytes. Genghis has been programmed before to walk over rough terrain and follow a person passively sensed in the infrared spectrum (Brooks, 1989). Our experiments were programmed using the Behavior Language and Subsumption Compiler (Brooks, 1989b). The entire learning program runs on board.

The sensors used in this experiment are two touch sensors on the bottom of the robot (one in the front and one in the back) and a trailing wheel which measures forward movement. Negative feedback is received by all of the behaviors every time at least one of the touch sensors fires. Positive feedback is received every time the wheel measures forward movement. We equipped Genghis with a dozen behaviors: 6 swing-leg-forward behaviors (which move a leg that is backward, up, forward and then down again), 6 swing-leg-backward behaviors (which move a leg that is forward, up, backward and then down again) (figure 3). Further there is one horizontal balance behavior, which sums the horizontal angles of the legs and sends a correction to all of the legs so as to reduce that sum to 0 (i.e. if one leg is moved forward, all of the legs are moved backwards a little). The 12 swing behaviors try to learn what the conditions are under which they should become active. The vector of binary perceptual conditions has 6 elements, each of which records whether a specific leg is up (not touching the ground).

Results

In a first experiment only six swing forward behaviors plus the balance behavior were involved. Genghis was able to learn to activate behaviors safely (avoiding negative feedback or "falling on its belly") and successfully (producing positive feedback or moving forward). More specifically, it learned to adopt a tripod gait, keeping three legs on the ground at any moment: the middle leg on one side and the front and back leg on the other side. Notice that this task is not trivial: negative feedback is not related to particular behaviors (none of the behaviors by itself causes negative feedback), but rather to the way they are coordinated (or uncoordinated). It is therefore a necessity in this application that actions are explored by the robot in parallel. This extra difficulty is successfully handled by the algorithm: the distributed learning behaviors are able to learn a task which requires their coordination.

This experiment has been successfully demonstrated on the robot. Using a non-intelligent search through the condition vector (i.e. every behavior monitors the conditions in the same order, starting with the status of the first leg, then the status of the second leg, etc) this takes in the order of 10 minutes. Using an intelligent search (i.e. every behavior starts by monitoring the status of legs that are nearby, then the ones that are further away, and so on), this experiment takes approximately 1 minute and 45 seconds.

Figure 4: Genghis learning to walk. Initially, the behaviors do not know yet how they are supposed to coordinate (under what conditions they should become active). Since the two front leg "swing forward" behaviors are being activated at the same time, Genghis falls down and receives negative feedback from the touch sensor mounted on the front of its belly.

Figure 5: Gradually, a more coherent "walk" emerges. In this case the global behavior converges towards a tripod gait: two groups of three legs are being swung forward alternately.

The reason why this is so much faster is that in this case, behaviors learn and converge more or less simultaneously (in the non-intelligent search case all of the behaviors have to "wait" for leg 4 and 5 to go through the whole condition vector before finding correlated conditions). The preconditions that are learned are that a swing-forward behavior is only allowed to become active when the neighboring legs (e.g. leg3 and leg0 in the case of leg1) are down. Actually, we noticed that not all of the behaviors learn that both of the neighboring legs have to be down. If for example leg0 and leg4 learned not to be active at the same time as leg2, then leg2 doesn't have to learn how to avoid negative feedback, because its two neighbors are taking care of the coordination problem.

In a second experiment, which has been demonstrated in simulation (it is therefore difficult to compare the resulting convergence times), six swing-backward behaviors were added. The robot now also had to learn that only certain behaviors are relevant for receiving positive feedback (in the first experiment, positive feedback didn't play much of a role, because every behavior was correlated to positive feedback under all conditions). More specifically, the robot had to learn that even though the swing-leg-backward behaviors do not cause negative feedback to be received (when coordinated), they should never become active because they are not correlated to positive feedback. In our simulation, the "non-relevant" swing-backward behaviors slowly die out, because they are not correlated to positive feedback. Their probability of becoming active gradually goes down, so that they have less opportunities to find out what the conditions are under which they can minimize negative feedback. Most of them "die out" before they are able to find the optimal list of preconditions so as to avoid negative feedback.

The gait that emerges in the experiments is the *tripod gait*, in which alternatively 2 sets of 3 legs are simultaneously swung forward. As reported by Wilson (Wilson, 1966) and confirmed in simulations by Beer (Beer, 1989), the gait that emerges in a distributed 6-legged walking creature is an emergent property of the time it takes to push a leg backwards during the "stance phase". One of the experiments we are working on right now is to try to obtain different gaits as a result of varying the speed of the stance phase and by disconnecting one of the legs at run time. The circular monitoring scheme should take care that the behaviors can adapt to this new situation and modify their precondition lists accordingly. Another experiment we are currently working on is to make Genghis learn to walk backward, by adding a switch which inverts the movement feedback.

Related Learning Work

This work is related to Drescher's PhD thesis on "translating Piaget into LISP" (Drescher, 1989). The main differences are: that our behaviors (corresponding to his "schemas") do not maintain statistics for all of the perceptual conditions in the environment, but instead only

for one or zero conditions at the time. As a consequence our algorithm is less computationally complex. Drescher's algorithm would not be usable in real time. A second important difference is that the algorithms are concerned with different learning tasks. In Drescher's case the task is to discover the regularities of the world when taking actions (a condition-action-effect kind of representation of the world is built up), while in the work presented here the only things learned are the conditions which optimize positive feedback and minimize negative feedback. Our system evolves towards a task-dependent (goal-oriented) solution, while in Drescher's case, generally useful knowledge is built up.

There is further also some relation to Classifier Systems and Genetic Algorithms (Holland et al., 1986) (for an application to control problems cfr. (Greffenstette, 1989)). The main difference is that our learning technique is basically constructivist, while theirs is selectionist. In our algorithm the right representations are built up (in an incremental way) instead of being selected. An advantage is that our algorithm is faster because it does not perform a "blind", unstructured search. It further also uses memory more efficiently because there is no duplication of information (all the information about one action is grouped in one behavior) and because we only monitor/explore a certain condition when there is a need for it (the behavior is not reliable yet).

Finally, the problem studied here is related to a class of algorithms called Reinforcement Learning Algorithms (Sutton, 1984)(Sutton, 1988)(Sutton, 1990)(Kaelbling, 1990)(also related are (Narendra & Thathachar, 1989) and (Berry & Fristedt, 1985)). The main differences are that (i) the algorithm discussed here is distributed and parallel: several actions an be taken at once, and asynchronously (this is even crucial to learn the tripod gait, for example), (ii) this algorithm is action- oriented, whereas reinforcement learning algorithms are state-oriented (utility functions are associated with states, while here relevance and reliability are associated with actions) and (iii) here feedback is binary and dual (positive and negative), whereas in reinforcement learning the utility function is real valued (both have advantages and disadvantages: the former's advantage is that positive and negative goals/feedback are treated separately, while the latter's advantages is that there is a more gradual evaluation).

Conclusion and Future Work

We have developed a learning algorithm which allows a behavior-based robot to learn when its behaviors should become active using positive and negative feedback. We tested the algorithm by successfully teaching a 6-legged robot to walk forward. In future work we plan to test the generality of the algorithm by doing more experiments with the same and different robots for different tasks and by studying the properties and limitations of the algorithm with a mathematical model.

We further plan some extensions to the algorithm, the first one being the addition of a mechanism to deal with delayed feedback (or learning action sequences). Three possible solutions to this problem will be investigated: (i) the usage of some Bucket Brigade Algorithm or Temporal Difference method (Sutton, 1988) (Holland et al., 1986), (ii) the extension of the perceptual condition vector with conditions representing the past actions taken and (iii) composing actions into macro-actions (so that feedback is still immediate for such an action sequence).

Acknowledgements

Grinell Moore did most of the mechanical design and fabrication of Genghis. Colin Angle did much of the processor design and most of the electrical fabrication. Olaf Bleck built the sensors providing positive and negative feedback and replaced the broken servos every time our buggy programs "tortured" the robot. Leslie Kaelbling, Maja Mataric and Paul Viola provided useful comments on an earlier draft of this paper. Richard Lathrop and David Clemens helped with the statistics.

Supported by Siemens with additional support from the University Research Initiative under Office of Naval Research contract N00014–86–K–0685, and the Defense Advanced Research Projects Agency under Office of Naval Research contract N00014–85–K–0124. The first author is a research associate of the Belgian National Science Foundation. She currently holds a position as visiting professor at the M.I.T. Artificial Intelligence Laboratory.

Bibliography

Anderson T.L. and Donath M. (1988) A computational structure for enforcing reactive behavior in a mobile robot. In: Mobile Robots III, Proc. of the SPIE conference, Vol. 1007, Cambridge, MA.

Angle C.M. (1989) Genghis, a six-legged autonomous walking robot. Bachelors thesis, Department of EECS, MIT.

Arkin R. (1987) Motor schema based navigation for a mobile robot: An approach to programming by behavior. IEEE Conference on Robotics and Automation '87.

Beer R.D. (1989) Intelligence as Adaptive Behavior: An Experiment in Computational Neuroethology. Technical Report 89-118, Center for Automation and Intelligent Systems Research, Case Western Reserve University.

Berry D.A. and Fristedt B. (1985) Bandit problems: Sequential allocation of experiments. Chapman and Hall, London.

Brooks R.A. (1986) A robust layered control system for a mobile robot. IEEE Journal of Robotics and Automation. Volume 2, Number 1.

Brooks R.A. (1989) A robot that walks: Emergent behavior from a carefully evolved network. Neural Computation, 1(2).

Brooks R.A. (1989b) The behavior Language; User's - Guide. Implementation Note. AI-laboratory, MIT.

Brooks R.A. (1990) Elephants don't play chess. In: P. Maes (ed.) Designing Autonomous Agents, Bradford-MIT Press, in press. Also: special issue of Journal of Robotics and Autonomous Systems, Spring '90, North-Holland.

Donner M.D. (1987) Real-time control of walking. Progress in Computer Science series, Vol. 7, Birkhauser, Boston.

Drescher G.L. (1989) Made-Up Minds: A Constructivist Approach to Artificial Intelligence, PhD Thesis, Department of EECS, MIT.

Greffenstette J.J. (1989) Incremental learning of control strategies with genetic algorithms. Proceedings of the - Sixth International Workshop on Machine Learning, Morgan Kaufmann.

Holland J.H., Holyoak K.J. Nisbett R.E. and Thagard P.R. (1986) Induction: Processes of inference, learning and discovery. MIT-Press, Cambridge, MA.

Kaelbling L. (1990) Learning in embedded systems, PhD thesis, Stanford Computer Science Department, forthcoming.

Maes P. (1990) Situated agents can have goals. In: P. Maes (ed.) Designing Autonomous Agents, Bradford-MIT Press, in press. Also: special issue of Journal of Robotics and Autonomous Systems, Spring '90, North-Holland.

Narendra K, and Thathachar M.A.L. (1989) Learning Automata, an Introduction. Prentice Hall, New Jersey.

Payton D.W. (1986) An architecture for reflexive autonomous vehicle control. IEEE Robotics and Automation Conference '86, San Francisco.

Rosenschein S.J. and Kaelbling L. (1986) The synthesis of digital machines with provable epistemic properties, in Joseph Halpern, ed, Proceedings of the Conference on Theoretical Aspects of Reasoning About Knowledge, Monterey, CA.

Schlimmer J.C. (1986) Tracking Concept Drift, Proceedings of the Sixth National Conference on Artificial Intelligence '86.

Sutton R. (1984) Temporal credit assignment in reinforcement learning. Doctoral dissertation, Department of Computer and Information Science, University of Massachusetts, Amherst.

Sutton R. (1988) Learning to predict by the methods of Temporal Differences. Machine Learning Journal, Vol. 3, 9-44.

Sutton R. (1990) Integrated architectures for learning, planning and reacting based on approximating dynamic programming. Proceedings of the Seventh International Conference on Machine Learning.

Wilson D.M. (1966) Insect walking. Annual Review of Entomology, 11: 103-121.

Yamaushi B. (1990) Independent Agents: A Behavior-Based Architecture for Autonomous Robots. Proceedings of the Fifth Annual SUNY Buffalo Graduate Conference on Computer Science '90, Buffalo, NY.

Zhang, Y. (1989) Transputer-based Behavioral Module for Multi-Sensory Robot Control. 1st International Conference in Artificial Intelligence and Communication Process Architecture '89, London, UK.

Chapter 4: Systems and Applications

This chapter discusses successful systems in various domains, including systems currently under development. Examples include the Mars Rover planetary exploration projects, autonomous underwater vehicles, autonomous vehicles for hazardous environments (mining and nuclear facilities, for example), road following, and aids for the handicapped. Nilsson's paper is of historical significance, since it discusses Shakey — the first mobile robot — developed at SRI. Thompson describes the JPL robot's control structure and the operation of its navigational subsystem, paying special attention to path planning.

Giralt et al. focus on aspects of the Hilare project on autonomous-mobile-robot development, discussing issues of system integration, multisensor navigation, motion control, and learning related to automatic environment modeling.

Weisbin et al. describe the design of an autonomous robot at the Oak Ridge National Laboratory, focusing on the development and experimental validation of intelligent control techniques that can plan and perform various assigned tasks in unstructured environments.

Cox describes the position estimation system for an autonomous robot called Blanche, designed for use in unstructured office or factory environments, paying particular attention to a method for constructing an environmental model from sensor information.

In two papers, Dickmanns and Graefe discuss the integration of real-time monocular vision in a control-theoretic approach to the visual navigation of roads, which constitutes a seminal contribution to autonomous mobile robotics. Kriegman et al. discuss MOBI, a mobile robot system that integrates stereo and sonar information to navigate inside buildings.

Fujie et al. discuss the major technologies needed to develop a lightweight high-speed flexible moving mechanism, a high-speed intelligence guidance technique for self-controlled movement, and a dextrous manipulator that can deal with movement.

Madarasz et al. provide first steps in the design and fabrication of a completely autonomous vehicle for patients with physical and sensory disabilities, providing an overview of project objectives and a detailed description of the components currently at work.

Tachi and Komoriya delineate the design concepts of Meldog, a guide-dog robot, describing a navigational method that uses an organized map and landmarks, an obstacle detection/avoidance system based on ultrasonic environment measurements, and man/machine communication via electrocutaneous stimulation. Ishikawa et al. discuss an AGV system that uses visual line recognition for navigation.

Rembold describes the architecture and functions of an autonomous mobile robot, explaining individual functions — including knowledge-based planning, execution, and supervision of such a vehicle.

Turk et al. describe VITS, the vision system for Alvin (an autonomous land vehicle). Alvin's road following is of particular interest: The authors describe various road segmentation methods for video-based road following, boundary extraction, and boundary transformation from the image plane into a vehicle-centered three-dimensional-scene model.

Thorpe et al. present a mobile robot that includes perception-and-navigation tools for outdoor environmental applications, and discuss two types of vision algorithms — color vision for road following, and three-dimensional vision for obstacle detection and avoidance.

Sharma and Davis introduce a road-following system based on image analysis for an autonomous vehicle, and discuss the three-dimensional boundary recognition process in two parts — low-level data-driven analysis, followed by high-level model-directed search. Harmon discusses the architecture of a vehicle designed for operation in rough terrain.

Wilcox et al., discuss ongoing research and field testing involving the vision system for a Mars Rover and describe the stereo-correction technique, including the use of more than two cameras to reduce correspondence errors and possibly limit the processing overhead.

Bares et al. overview Ambler, the NASA planetary-rover project at Carnegie Mellon University that focuses on locomotion, perception, planning, and control. The authors summarize goals and the requirements to satisfy those goals.

Blidberg and Chappell discuss the development of autonomous underwater vehicles.

A MOBILE AUTOMATON: AN APPLICATION
OF ARTIFICIAL INTELLIGENCE TECHNIQUES

by

Nils J. Nilsson
Stanford Research Institute
Menlo Park, California

Summary

A research project applying artificial intelligence techniques to the development of integrated robot systems is described. The experimental facility consists of an SDS-940 computer and associated programs controlling a wheeled vehicle that carries a TV camera and other sensors. The primary emphasis is on the development of a system of programs for processing sensory data from the vehicle, for storing relevant information about the environment, and for planning the sequence of motor actions necessary to accomplish tasks in the environment. A typical task performed by our present system requires the robot vehicle to rearrange (by pushing) simple objects in its environment.

A novel feature of our approach is the use of a formal theorem-proving system to plan the execution of high-level functions as a sequence of other, perhaps lower-level, functions. The execution of these, in turn, requires additional planning at lower levels. The main theme of the research is the integration of the necessary planning systems, models of the world, and sensory processing systems into an efficient whole capable of performing a wide range of tasks in a real environment.

Key Words

Robot
Robot System
Visual Processing
Problem-Solving
Question-Answering
Theorem-Proving
Models of the World
Planning
Scene Analysis
Mobile Automaton

Acknowledgment

At least two dozen persons at the Stanford Research Institute have made substantial contributions to the project that the author has the good fortune to describe in this paper. All of us express our appreciation to the Rome Air Development Center and the Advanced Research Projects Agency, who supported this research under Contract No. F 30602-69-C-0056.

I   Introduction

At the Stanford Research Institute we are implementing a facility for the experimental study of robot systems. The facility consists of a time-shared SDS-940 computer, several core-loads of programs, a robot vehicle, and special interface equipment.

Several earlier reports[1] and papers[2-4] describing the project have been written; in this paper we shall describe its status as of early 1969 and discuss some of our future plans.

The robot vehicle itself is shown in Fig. 1. It is propelled by two stepping motors independently driving a wheel on either side of the vehicle. It carries a vidicon television camera and optical range-finder in a movable "head." Control logic on board the vehicle routes commands from the computer to the appropriate action sites on the vehicle. In addition to the drive motors, there are motors to control the camera focus and iris settings and the tilt angle of the head. (A motor to pan the head is not yet used by present programs.) Other computer commands arm or disarm interrupt logic, control power switches, and request readings of the status of various registers on the vehicle. Besides the television camera and range-finder sensors, several "cat-whisker" touch-sensors are attached to the vehicle's perimeter. These touch sensors enable the vehicle to know when it bumps into something. Commands from the SDS-940 computer to the vehicle and information from the vehicle to the computer are sent over two special radio links, one for narrow-band telemetering and one for transmission of the TV video from the vehicle to the computer.

The purpose of our robot research at SRI is to study processes for the real-time control of a robot system that interacts with a complex environment. We want the vehicle to be able to perform various tasks that require it to move about in its environment or to rearrange objects. In order to accomplish a wide variety of tasks rather than a few specific ones, a robot system must have very general methods. What is required is the integration in one system of many of the abilities that are usually found separately in individual Artificial Intelligence programs.

We can group most of the needed abilities into three broad classes: (1) problem-solving, (2) modelling, and (3) perception:

(1)   Problem-Solving

A robot system accomplishes the tasks given it by performing a sequence of primitive actions, such as wheel motions and camera readings. For efficiency, a task should first be analyzed into a sequence of primitive actions calculated to have the desired effect. This process of task analysis is often called planning, because it is accomplished before the robot begins to act. Obviously, in order to plan, a robot system must "know" about the effects of its actions.

(2) Modelling

A body of knowledge about the effects of actions is a type of model of the world. A robot problem-solving system uses the information stored in the model to calculate what sequence of actions will cause the world to be in a desired state. As the world changes, either by the robot's own actions or for other reasons, the model must be updated to record these changes. Also, new information learned about the world should be added to the model.

(3) Perception

Sensors are necessary to give a robot system new information about the world. By far the most important sensory system is vision, since it allows direct perception of a good sized piece of the world beyond the range of touch. Since we assume that a robot system will not always have stored in its model every detail of the exact configuration of its world and thus cannot know precisely the effects of its every action, it also needs sensors with which to check predicted consequences against reality as it executes its plans.

The integration of such abilities into a smoothly-running, efficient system presents both important conceptual problems and serious practical challenges. For example, it would be infeasible for a single problem-solving system (using a single model) to attempt to calculate the long chains of primitive actions needed to perform lengthy tasks. A way around this difficulty is to program a number of coordinating "action-units," each with its own problem-solving system and model, and each responsible for planning and executing a specialized function. In planning how to perform its particular function, each action-unit knows the effects of executing functions handled by various of the other action-units. With this knowledge it composes its plan as a sequence of other functions (with the appropriate arguments) and leaves the planning required for each of these functions up to the action-units responsible for executing them at the time they are to be executed.

Such a system of interdependent action-units implies certain additional problems involving communication of information and transfer of control between units. When such a system is implemented on a serial computer with limited core memory, obvious practical difficulties arise connected with swapping program segments in and out of core and handling interrupts in real time. The coordinated action-unit scheme serves as a useful guide in explaining the operation of our system, even though practical necessities have dictated occasional deviations from this scheme in our implementation. In the next section we shall discuss the problem-solving processes and models associated with some specific functions of the present SRI robot system.

II SOME SPECIFIC FUNCTIONS OF THE ROBOT SYSTEM AND THEIR ASSOCIATED PROBLEM-SOLVING PROCESSES AND MODELS

A. Low Level Functions

The robot system is capable of executing a number of functions that vary in complexity from the simple ability to turn the drive wheels a certain number of steps to the ability to collect a number of boxes by pushing them to a common area of the room. The organization of these functional action-units is not strictly hierarchical, although for descriptive convenience we will divide them into two classes: low level and high level functions.

Of the functions that we shall mention here, the simplest are certain primitive assembly language routines for moving the wheels, tilting the head, reading a TV picture, and so on. Two examples of these are MOVE and TURN; MOVE causes the vehicle to roll in a straight line by turning both drive wheels in unison, and TURN causes the vehicle to rotate about its center by turning the drive wheels in opposite directions. The arguments of MOVE and TURN are the number of steps that the drive wheels are to turn (each step resulting in a vehicle motion of 1/32 inch) and "status" arguments that allow queries to be made about whether or not the function has been completed.*

Once begun, the execution of any function proceeds either until it is completed in its normal manner or until it is halted by one of a number of "abnormal" circumstances, such as the vehicle bumping into unexpected objects, overload conditions, resource exhaustion, and so on. Under ordinary operation, if execution of MOVE results in a bump, motion is stopped automatically by a special mechanism on the vehicle. This mechanism can be overridden by a special instruction from the computer, however, to enable the robot to push objects.

The problem-solving systems for MOVE and TURN are trivial; they need only to calculate what signals shall be sent to registers associated with the motors in order to complete the desired number of steps.

At a level just above MOVE and TURN is a function whose execution causes the vehicle to travel directly to a point specified by a pair of (x,y) coordinates. This function is implemented in the FORTRAN routine LEG. The model used by LEG contains information about the robot's present (x,y) location and heading relative to a given coordinate system and information about how far the vehicle travels for each step applied to the stepping motors. This information is stored along with some other special constants in a structure called the PARAMETER MODEL. Thus for a given (x,y)

---

* Our implementation allows a program calling routines like MOVE or TURN to run in parallel with the motor functions they initiate.

destination as an argument of LEG, LEG's problem-solving system calculates appropriate arguments for a TURN and MOVE sequence and then executes this sequence. Predicted changes in the robot's location and heading caused by execution of MOVE and TURN are used to update the PARAMETER MODEL.

Ascending one more level in our system, we encounter a group of FORTRAN "two-letter" routines whose execution can be initiated from the teletype. Our action-unit system ceases to be strictly hierarchical at this point, since some of the two-letter commands can cause others to be executed.

One of these two-letter commands, EX, takes as an argument a sequence of (x,y) coordinate positions. Execution of EX causes the robot to travel from its present position directly to the first point in the sequence, thence directly to the second, and so on until the robot reaches the last point in the sequence. The problem-solving system for EX simply needs to know the effect caused by execution of a LEG program and composes a chain of LEG routines, each with arguments provided by the successive points specified in the sequence of points. Under ordinary operation, if one of these LEG routines is halted due to a bump, EX backs the vehicle up slightly and then halts. A special feature of our implementation is the ability to arm and service interrupts (such as caused by bumps) at the FORTRAN programming level.

Another two-letter command, PI, causes a picture to be read after the TV camera has been aimed at a specified position on the floor. The problem-solving system for PI thus calculates the appropriate arguments for a TURN routine and a head-tilting routine; PI then causes these to be executed, reads in a picture from the TV camera, and performs processing necessary to extract information about empty areas on the floor. (Details of the picture processing programs of the robot system are described in Sec. III below.)

The ability to travel by the shortest route to a specified goal position along a path calculated to avoid bumping into obstacles is provided by the two-letter command TE. Execution of TE involves the calculation of an appropriate sequence of points for EX and the execution of EX. This appropriate sequence is calculated by a special problem-solving system embodied in the two-letter command PL.

The source of information about the world used by PL is a planar map of the room called the GRID MODEL. The GRID MODEL is a hierarchically organized system of four-by-four grid cells. Initially the "whole world" is represented by a four-by-four array of cells. A given cell can be either empty (of obstacles), full, partially full, or unknown. Each partially full cell is further subdivided into a four-by-four array of cells, and so on, until all partially full cells represent areas of some suitably small size. (Our present system splits cells down to a depth of three levels, representing a smallest area of about 12 inches.)

Special "model maintenance" programs insure that the GRID MODEL is automatically updated by information about empty and full floor areas gained by either successful execution or interruption of MOVE commands.

The PL program first uses the GRID MODEL to compute a network or graph of "nodes." The nodes correspond to points in the room opposite corners of obstacles; the shortest path to a goal point will then pass through a sequence of a subset of these nodes. In Fig. 2 we show a complete GRID MODEL of a room containing three objects. The robot's position, marked "R," and the goal position, marked "G," together with the nodes A,B,C,D, E,F,H,I,J and K are shown overlaid on the GRID MODEL. The program PL then determines that the shortest path is the sequence of points, R,F,I, and G by employing an optimal graph-searching algorithm developed by Hart, et al.[5]

If the GRID MODEL map of the world contains unknown space, PL must decide whether or not to treat this unknown space as full or empty. Currently, PL multiplies the length of any segment of the route through unknown space by a parameter k. Thus if k=1, unknown space is treated as empty; values of k greater than unity cause routes through known empty space to be preferred to possibly shorter routes through unknown space.

Execution of TE is accomplished by first reading and processing a picture (using PI with the camera aimed at the goal position) and taking a range-finder reading. The information about full and empty floor areas thus gained is added to the GRID MODEL. A route based on the updated GRID MODEL is then planned using PL, and then EX is executed using the arguments calculated by PL. If the EX called by TE is halted by a bump, a procedure attempts to manuever around the interfering obstacle, and then TE is called to start over again. Thus, vision is used only at the beginning of a journey and when unexpected bumps occur along the journey.

Although our present robot system does not have manipulators with which to pick up objects, it can move objects by pushing them. The fundamental ability to push objects from one place to another is programmed into another two-letter FORTRAN routine, called PU. Execution of PU causes the robot to push an object from one named position along a straight line path to another named position. The program PU takes five arguments: the (x,y) coordinates of the object to be pushed, the "size" or maximum extent of the object about its center of gravity, and the (x,y) coordinates of the spot to which the object is to be pushed. The problem-solving system for PU assembles an EX, a TURN, and two MOVE commands into a sequence whose execution will accomplish the desired push. First a location from which the robot must begin pushing the object is computed. Then PL is used to plan a route to this goal location. The sequence of points along the route serves as the argument for EX that is then executed. (Should EX be stopped by a bump, PU is started over again.) Next, PU's problem-solving system (using the PARAMETER model) calculates an argument for TURN that will point the robot in the direction that the object is to be pushed. A large argument is provided for the first MOVE command so that when it is executed, it will bump into the object to be pushed and automatically halt. After the bump and halt, the automatic stopping mechanism on the vehicle is overridden and the next MOVE command is executed with an argument calculated to push the object the desired distance.

B. Higher Level Functions

As we ascend to higher level functions, the required problem-solving processes must be more powerful and general. We want our robot system to have the ability to perform tasks possibly requiring quite complex logical deductions. What is needed for this type of problem-solving is a general language in which to state problems and a powerful search strategy with which to find solutions. We have chosen the language of first-order predicate calculus in which to state high level problems for the robot. These problems are then solved by an adaptation of a "Question Answering System" QA-3, based on "resolution" theorem-proving methods.[6-9]

As an example of a high level problem for the robot, consider the task of moving (by pushing) three objects to a common place. This task is an example of one that has been executed by our present system. If the objects to be pushed are, say, OB1, OB2, and OB3, then the problem of moving them to a common place can be stated as a "conjecture" for QA-3:

$$(\exists p,s)\{\text{POSITION } (OB1,p,s)$$
$$\land \text{ POSITION } (OB2,p,s)$$
$$\land \text{ POSITION } (OB3,p,s)\} \quad .$$

(That is, "There exists a situation s and a place p, such that OB1, OB2, and OB3 are all at place p in situation s.") The task for QA-3 is to "prove" that this conjecture follows from "axioms" that describe the present position of objects and the effects of certain actions.

Our formulation of these problems for the theorem-prover involves specifying the effects of actions in terms of functions that map situations into new situations. For example, the function PUSH (x,p,s) maps the situation s into the situation resulting by pushing object x into place p. Thus two axioms needed by QA-3 to solve the pushing problem are:

$$(\forall x,p,s)\{\text{POSITION } (x,p, \text{ PUSH } (x,p,s))\}$$

and

$$(\forall x,y,p,q,s)\{\text{POSITION } (x,p,s) \land \sim \text{SAME } (x,y)$$
$$\Rightarrow \text{POSITION } (x,p,\text{PUSH } (y,q,s))\} \quad .$$

The first of these axioms states that if in an arbitrary situation s, an arbitrary object x is pushed to an arbitrary place p, then a new

situation, PUSH (x,p,s), will result in which the object x will be at position p. The second axiom states that any object will stay in its old place in the new situation resulting by pushing a different object.

In addition to the two axioms just mentioned, we would have others describing the present positions of objects. For example, if OB1 is at coordinate position (3,5) in the present situation, we would have:

$$\text{POSITION } (OB1, (3,5), \text{PRESENT}) \quad .$$

(This information is provided automatically by routines that scan the GRID MODEL, giving names to clusters of full cells and noting the locations of these clusters.)

In proving the truth of the conjecture, the theorem-prover used by QA-3 also produces the place p and situation s that exist. That is, QA-3 determines that the desired situation s is:

$$s = \text{PUSH } (OB3, (3,5), \text{PUSH } (OB2, (3,5), \text{PRESENT})).$$

All of the information about the world used by QA-3 in solving this problem is stored in the form of axioms in a structure called the AXIOM MODEL. In general, the AXIOM MODEL will contain a large number of facts, more than are necessary for any given deduction.

Another LISP program examines the composition of functions calculated by QA-3 and determines those lower level FORTRAN two-letter commands needed to accomplish each of them. In our present example, a sequence of PU commands would be assembled. In order to calculate the appropriate arguments for each PU, QA-3 is called again, this time to prove conjectures of the form:

$$(\exists p,w)\{\text{POSITION } (OB2,p,\text{PRESENT}) \land \text{SIZE } (OB2,w)\} \quad .$$

Again the proof produces the p and w that exist, thus providing the necessary position and size arguments for PU. (Size information is also automatically entered into the AXIOM MODEL by routines that scan the GRID MODEL.)

In transferring control between LISP and FORTRAN (and also between separate large FORTRAN segments), use is made of a special miniature monitor system called the VALET. The VALET handles the process of dismissing program segments and starting up new ones using auxiliary drum storage for transferring information between programs.

The QA-3 theorem-proving system allows us to pose quite general problems to the robot system, but further research is needed on adapting theorem-proving techniques to robot problem-solving in order to increase efficiency.* The generality of

---

* We can easily propose less fortunate axiomatizations for the "collecting objects task" that would prevent QA-3 from being able to solve it.

theorem-proving techniques tempts us to use a single theorem-prover (and axiom set) as a problem-solver (and model) for all high level robot abilities. We might conclude, however, that efficient operation requires a number of coordinating action-unit structures, each having its own specialized theorem-prover and axiom set and each responsible for relatively narrow classes of functions.

Another LISP program enables commands stated in simple English to be executed.[10,11] It also accepts simple English statements about the environment and translates them into predicate calculus statements to be stored as axioms. English commands are ordinarily translated into predicate calculus conjectures for QA-3 to solve by producing an appropriate sequence of subordinate functions. For some simple commands, the theorem-prover is bypassed and lower level routines such as PU, TE, etc., are called directly.

The English program also accepts simple English questions that require no robot actions. For these, it uses QA-3 to discover the answer, and then it delivers this answer in English via the teletypewriter. (Task execution can also be reported by an appropriate English output.)

III VISUAL PERCEPTION

Vision is potentially the most effective means for the robot system to obtain information about its world. The robot lives in a rather antiseptic but nevertheless real world of simple objects--boxes, wedges, walls, doorways, etc. Its visual system extracts information about that world from a conventional TV picture. A complete scene analysis would produce a description of the visual scene, including the location and identification of all visible objects. Currently, we have two separate operating vision programs. One of these produces line drawings, and has been used for some time to identify empty floor space, regions on the floor into which the robot is free to move. The other, which is more recent, locates and identifies the major, non-overlapping objects. In this section we shall give brief descriptions of how these programs operate.

A. Line Drawing Program

The line drawing program produces a line drawing representation of a scene by a series of essentially local operations on a TV picture.* Fig. 3a shows a typical digitized picture, which is stored in the computer as a 120 x 120 array of 4-bit (16-level) intensity values. The scene shown is fairly typical, and includes some of the problems of mild shadows and reflections, some faint edges, and objects not completely in the field of view.

---

* Most of these operations were adaptations of earlier work by Roberts.[12] Details of our procedures, together with a description of special hardware for doing them efficiently, are given in Refs. 1 and 4.

The first of the local operations is a digital differentiation used to find points where there are significant changes in light intensity. These changes usually occur at or near the boundaries of objects, as can be seen from Fig. 3b. The next step is to determine the local direction of these boundaries. This is done by systematically placing small masks over the differentiated picture, and looking for places where the masks line up well with the gradient. Fig. 3c shows the locations and orientations of masks that responded strongly.

The next step is to fit these short line segments with longer straight lines. This is done by first grouping the short line segments, and then fitting a single straight line to all of the segments in a group. Grouping is a systematic procedure in which short segments are linked if they are sufficiently close in location and orientation. Fig. 3d shows the results of fitting longer lines to the segments in each group. The final step is to join the endpoints of these long lines to produce a connected line drawing. This is done by considering the endpoints one at a time and creating candidate connections--straight connections to neighboring endpoints, extrapolations to points of intersection, extrapolations to T-junctions, etc. The candidate that best fits the corresponding part of the derivative picture is the one selected. The final line drawing produced by this procedure is shown in Fig. 3e.

While the line drawing preserves much of the information in the quantized picture in a compact form, it often contains flaws due to missing or extra line segments that complicate its analysis. Currently, the only information we extract from the line drawing is a map of the open floor space. A program called floor boundary first finds a path along those line segments that bound the floor space in the picture. These lines are typically the places where the walls or objects meet the floor, or where sides of objects obscure our view of the floor. Fig. 3f shows the floor boundary extracted from the line drawing.

Now corresponding to any point in the picture is a ray going from the lens center of the camera through the picture point and out into space. This ray is the locus of all points that can produce the given picture point. If we follow the rays going through points on the floor boundary to the points at which they pierce the floor, we obtain an irregular polygon on the floor that bounds space known to be empty. In this way the line drawing is used to identify empty floor space, and the vision system enters information about open area into the GRID MODEL.

B. Object Identification Program

Were the line drawing program able to produce a perfect line drawing, the analysis needed to locate and identify objects in the scene would be relatively straightforward. However, the line drawing often contains flaws that seriously complicate its analysis. Some of these flaws could be corrected by more elaborate local processing.

However, there is a limit to how well local processing can perform, and when significant edges cannot be told from insignificant edges on the basis of local criteria, the goal of producing a perfect line drawing in this way must be abandoned.

The object identification program locates and identifies non-overlapping objects by gathering and interpreting evidence supplied by local operators. The program consists of two parts: a repertoire of local operators and an executive. The local operators process the gradient picture to perform tests such as deciding whether or not there is a line between two points, or finding all lines leaving a given point. Each operator returns not a single answer, but a set of possible answers with associated confidences (ideally, probabilities) that each answer is in fact correct. The executive explores the scene by calling local operators and evaluating the results in the light of both prior test results and built-in knowledge of the world.

The executive program is organized as a decision tree. Each node in the tree specifies that a particular test is to be performed. The branches leaving a node correspond to the possible test outcomes, and since each outcome has an associated confidence, these confidences are attached to the branches.

A given node in the tree can be thought of as representing a hypothesis about the contents of the scene. This hypothesis is simply that the scene is partially described by the test results specified by the path from the start node to the given node. The hypothesis is given a confidence by combining the confidences of these test results. The test called for at the given node is designed to provide an answer that will tend to confirm or infirm the hypothesis.

An analysis of a scene proceeds as a search of the decision tree described. At any stage in the search we have a partially expanded tree corresponding to the tests already performed. The nodes at the tips of this partial tree, which we shall call open nodes, present us with a choice of possible next tests (or, alternatively, hypotheses to be further investigated). The open node with highest associated confidence is selected for expansion, i.e., the test called for by that particular node is performed. The search proceeds until the open node with highest confidence is a terminal node of the tree. A terminal node typically represents a complete description of at least a portion of the scene, and hence constitutes at least a partial "answer." (Certain terminal nodes correspond to impossible physical situations; in this event, the search is resumed at the next most confident node.) After returning a partial answer the portion of the scene containing the object found is deleted and the analysis begins again. Thus the tree search is iterated until the scene contains no further objects.

The decision tree itself embodies the strategy for searching the scene. The basic ideas behind this strategy are simple, and will be illustrated by following the operation of the program on the

scene of Fig. 3a. The first operation called for is a search for vertical lines, since these are usually both reliably detectable and significant. Fig. 4a shows that the appropriate operator found three vertical lines, which happened to rank in confidence as numbered.

Starting with the highest confidence line and checking to see that its lower endpoint was within the picture, the program next looked for other lines leaving that endpoint. A failure here would have led to the conclusion that something was strange, and therefore to a transfer to the next most confident node in the tree. However, a "spur" was detected, as shown in Fig. 4b. Hypotheses that the lower endpoint was connected to the lower endpoints of other verticals were rejected because the direction of the spur was not correct. Thus, at this point attention was shifted to the top of the vertical. A spur was found there, as expected, and that spur was followed to its endpoint as shown in Fig. 4c. The fact that its endpoint was on the picture, coupled with the fact that the program had failed in its attempt to connect the vertical to other verticals, provided strong evidence that a wedge had been found. A further check of the angle at the top of the vertical confirmed the wedge hypothesis, and the same calculations used in the floor boundary program were used to locate the lower vertices.

At this point one object had been found and identified, and a search for other objects began. On the second iteration, the remaining verticals were successfully joined at their lower endpoints, as shown in Fig. 4d, and various spurs were found, as shown in Fig. 4e. A similar attempt to spot a wedge failed to produce strong evidence, as shown in Fig. 4f, and the final output indicated the existence and location of an object partly out of view without specifying what it was. At this point there were no more vertical lines, and the analysis was completed.

The object identification program is capable of locating and often identifying non-overlapping objects on the basis of partial information. There are a number of obvious ways in which it can be improved and extended, and further research will be devoted to these tasks. However, even as it stands it can provide the robot with much valuable information about the robot's world.

IV CONCLUSIONS

There are several key questions that our work has helped to put into focus. Given that a robot system will involve the successful integration of problem-solving, modelling, and perceptual abilities, there are many research questions concerning each of these. Let us discuss each in turn.

A. Problem-Solving

Our somewhat hierarchical organization of problem-solvers and models seems a natural, even if ad hoc, solution to organizing complex behavior. Are there alternatives? Will the use of theorem-proving techniques provide enough generality to permit a single general-purpose problem-solver, or will several "specialist" theorem-provers be needed to gain the required efficiency?

Other questions concern the use of theorem-proving methods for problem-solving. How do they compare with the "production methods" as used by the General Problem Solver (GPS) or with the procedural language approach as developed by Fikes?[13] Perhaps some combination of all of these will prove superior to any of them; perhaps more experience will show that they are only superficially different.

Another question is: To what level of detail should behavioral plans be made before part of the plan is executed and the results checked against perceptual information? Although this question will not have a single answer, we need to know upon what factors the answer depends.

Our problem-solving research will also be directed at methods for organizing even more complex robot behavior. We hope eventually to be able to design robot systems capable of performing complex assembly tasks requiring the intelligent use of tools and other materials.

B. Modelling

Several questions about models can be posed: Even if we continue to use a number of problem-solvers, must each have its own model? To what extent can the same model serve several problem-solvers? When a perceptual system discovers new information about the world, should it be entered directly into all models concerned? In what form should information be stored in the various models? Should provisions be made for forgetting old information? Can a robot system be given a simple model of its own problem-solving abilities? Ensuing research and experience with our present system should help us with these questions.

C. Visual Perception

The immediate vision problems involve including more tests in the object identification program to complete unfinished analysis, and removing the restriction to non-overlapping objects. Beyond these improvements there are still longer range problems to be solved. The scene analysis programs implicitly store information about the world in their structure. Changes in the robot's world can require extensive changes to the whole program. What program organization would minimize these problems? How can the scene analysis program interrogate and use facts stored in the model to advantage? Since "facts" obtained from either the model or the subroutines are subject to error, it is natural to accompany them by a confidence measure. How should these confidences be computed and how should they be combined, since, loosely speaking, we operate under conditions of strong statistical dependence? How can we augment the current repertoire of subroutines with others to make use of such properties as color, texture and range? Future vision research will be devoted to answering questions such as these.

The main theme of the project has been, and will continue to be, the problem of system integration. In studying robot systems that interact with the real world, it seems extremely important to build and program a real system and to provide it with a real environment. Whereas much can be learned by simulating certain of the necessary functions (we use this strategy regularly), many important issues are likely not to be anticipated at all in simulations. Thus questions regarding, say, the feasibility of a system of interacting action-units for controlling a real robot can only be confronted by actually attempting to control a real robot with such a system. Questions regarding the suitability of candidate visual processing schemes can most realistically be answered by experiments with a system that needs to "see" the real world. Theorem-proving techniques seem adequate for solving many "toy" problems; will the full generality of this approach really be exploitable for directing the automatic control of mechanical equipment in real-time?

The questions that we have posed in this section are among those that must be answered in order to develop useful and versatile robot systems. Experimenting with a facility such as we have described appears to be the best way to elicit the proper questions and to work toward their answers.

REFERENCES

1. N. Nilsson, et al, "Application of Intelligent Automata to Reconnaissance," Contract AF30(602)-4147, SRI Project 5953, Stanford Research Institute, Menlo Park, California (four Interim Reports and one Final Report dated December 1968).

2. C. A. Rosen and N. J. Nilsson, "An Intelligent Automaton," IEEE International Convention Record, Part 9 (1967).

3. B. Raphael, "Programming a Robot," Proc. IFIP Congress 68, Edinburgh, Scotland (August 1968).

4. G. E. Forsen, "Processing Visual Data with an Automaton Eye," in Pictorial Pattern Recognition (Thompson Book Company, Washington, D.C., 1968).

5. P. Hart, N. Nilsson, and B. Raphael, "A Formal Basis for the Heuristic Determination of Minimum Cost Paths," IEEE Trans. on Systems Science and Cybernetics, Vol. SSC-4, No. 2, pp. 100-107, (July 1968).

6. C. Green and B. Raphael, "The Use of Theorem-Proving Techniques in Question-Answering Systems," Proc. 1968 ACM Conference, Las Vegas, Nevada (August 1968).

7. B. Raphael, "Research on Intelligent Question-Answering Systems," Final Report, Contract AF 19(628)-5919, SRI Project 6001, Stanford Research Institute, Menlo Park, California (May 1968).

8.   C. Green, "Theorem-Proving by Resolution as a Basis for Question-Answering Systems," _Machine Intelligence 4_, B. Meltzer and D. Michie, Eds. (Edinburgh University Press, Edinburgh, Scotland; to appear 1969).

9.   C. Green, "Applications of Theorem-Proving to Problem-Solving," _Proc. of the International Joint Conference on Artificial Intelligence_, Washington, D.C., (May 1969).

10. L. S. Coles, "An On-Line Question-Answering System with Natural Language and Pictorial Input," _Proc. 1968 ACM Conference_, Las Vegas, Nevada (August 1968).

11. L. S. Coles, "Talking with a Robot in English," _Proc. of the International Joint Conference on Artificial Intelligence_, Washington, D.C., (May 1969).

12. L. G. Roberts, "Machine Perception of Three-Dimensional Solids," _Optical and Electro-Optical Information Processing_ (MIT Press, Cambridge, Massachusetts, 1965).

13. R. Fikes, "A Study in Heuristic Problem-Solving: Problems Stated as Procedure," _Proc. of Fourth Systems Symposium_, held at Case Western Reserve University, Cleveland, Ohio, November 1968 (to be published).

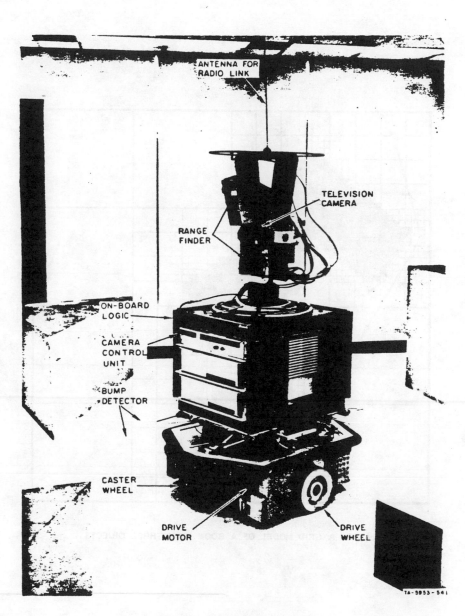

ANTENNA FOR
RADIO LINK

TELEVISION
CAMERA

RANGE
FINDER

ON-BOARD
LOGIC

CAMERA
CONTROL
UNIT

BUMP
DETECTOR

CASTER
WHEEL

DRIVE
MOTOR

DRIVE
WHEEL

TA-5953-541

FIG. 1   THE ROBOT VEHICLE

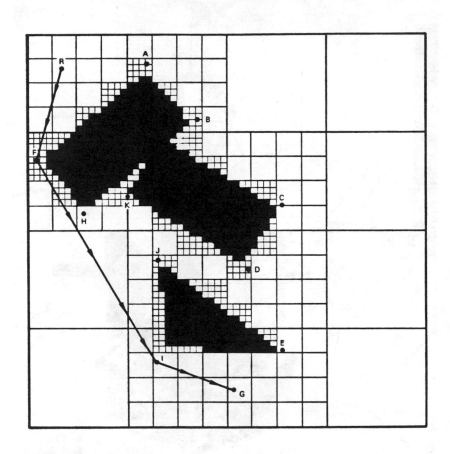

FIG. 2   A GRID MODEL OF A ROOM WITH THREE OBJECTS

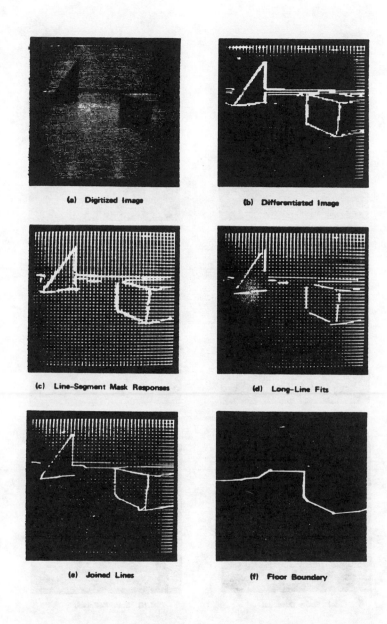

(a) Digitized Image   (b) Differentiated Image

(c) Line-Segment Mask Responses   (d) Long-Line Fits

(e) Joined Lines   (f) Floor Boundary

FIG. 3   PICTURE PROCESSING BY THE LINE-DRAWING PROGRAM

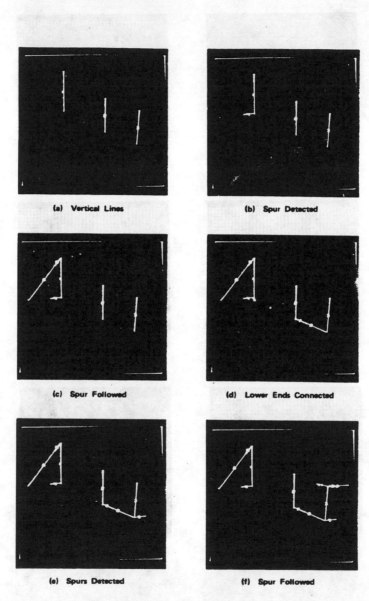

(a) Vertical Lines   (b) Spur Detected

(c) Spur Followed   (d) Lower Ends Connected

(e) Spurs Detected   (f) Spur Followed

FIG. 4   PICTURE PROCESSING BY THE OBJECT IDENTIFICATION PROGRAM

THE NAVIGATION SYSTEM OF THE JPL ROBOT*

Alan M. Thompson
Jet Propulsion Laboratory
Pasadena, California

ABSTRACT

The control structure of the JPL research
robot and the operations of the navigation subsystem are discussed. The robot functions as a
network of interacting concurrent processes distributed among several computers and coordinated
by a central executive. The results of scene
analysis are used to create a segmented terrain
model in which surface regions are classified by
traversibility. The model is used by a path-
planning algorithm, PATH*, which uses tree search
methods to find the optimal path to a goal. In
PATH*, the search space is defined dynamically as
a consequence of node testing. Maze-solving and
the use of an associative data base for context-
dependent node generation are also discussed.
Execution of a planned path is accomplished by a
feedback guidance process with automatic error
recovery.

1. Introduction

The Robotics Research Program at the Jet
Propulsion Laboratory is aimed at developing the
capabilities in machine intelligence systems
required for a semi-autonomous vehicle to be used
in remote planetary exploration. To achieve this
end, a "breadboard" robot has been constructed
(Fig. 1) to serve as test bed and demonstration
tool for the programs and hardware. Research is
being conducted in the areas of robot vision,
problem-solving and learning, hardware and system
architecture, motor-function control in manipulation and locomotion, as well as the terrain
modeling and navigation tasks described herein.

In the JPL experiments, the robot is deposited
in an unknown laboratory environment consisting of
many arbitrary obstacles (rocks, walls, and
other objects) and is given tasks such as finding
and collecting selected rock samples. As a robot
subsystem, the navigation system has the responsibility of finding an unobstructed path to a
designated goal and then controlling the vehicle's
movement along the path. To do this, it must
maintain an internal representation of its environment from sensory input for use in the planning
phase and then use additional sensory input to
monitor execution of movement. The environment is
initially completely unknown, and the path planner
requests updates to the terrain model as needed
for planning.

Figure 1. JPL Research Robot

A terrain model was chosen for the robot that
simplifies the task of path planning while simultaneously providing a means of representing large
areas of terrain in a compact, segmented, hierarchical structure that is easily updated or
extended. Having a numeric description of the
location and shape of obstacles allows the path
planner to accurately model the characteristics of
the vehicle while conducting the optimal path
search, so that the resulting path is in a form
that is readily executable by the guidance programs
to within a known error tolerance.

2. Robot System Structure

The JPL robot operates as a hierarchy of
separate concurrent processes which are distributed
among three computers. The main control structure
(Fig. 2) consists of a Robot Executive (REX) which
communicates with the operator via the "ground
system." Other processes, whose functioning is
coordinated by the executive (though not necessarily determined by it), perform the tasks of
vision, manipulation, and navigation. The control
hierarchy is not strictly enforced, however, as
processes may interact freely in such functions as
hand-eye coordination, etc. Recent additions to
the system include processes for error recovery
and problem-solving, which will be the nucleus of

---

*This paper presents the results of one phase of research carried out at the Jet Propulsion Laboratory,
California Institute of Technology, under Contract No. NAS 7-100, sponsored by the National Aeronautics
and Space Administration.

Keywords: optimal path planning, terrain modeling, parent backup, subgoal stacking, and maze-solving.

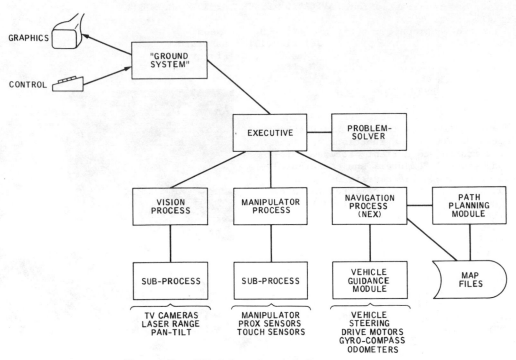

Figure 2. JPL Robot Control Structure

a system for automatic planning, error correction and learning. The sensory-motor processes have sub-processes on the minicomputer containing the actual vehicle interfaces. Processes suspend themselves when not needed.

Communication between the processes is handled by a shared program segment "mailbox" method, similar to that of the Stanford Hand-Eye System (FS1). Messages passed within one computer are merely stored in the appropriate slot in the shared segment, whereas inter-computer messages are transmitted by a separate communications process to the appropriate machine and then deposited. This structure is extensible to multiple processors. A process need not know on which CPU another process runs. At present, three CPU's are connected: a DEC PDP-10, a GA SPC-16 mini-computer, and an IMLAC PDS-1D graphics system. The "high-level" processes run on the PDP-10, and are implemented in the SAIL language and, recently, in LISP. In the minicomputer, FORTRAN and assembly code are used. All interfaces to the vehicle hardware are currently contained in the mini, but since it has limited capacity for parallel processes, the use of a microcomputer network is being investigated.

The navigation system runs as three concurrent processes: the navigation executive (NEX), the path planner module (PPM), and the vehicle guidance module (VGM). Both NEX and PPM access the terrain model files. NEX is the controlling process for all navigation functions. It contains the command interface to the robot executive which translates acceptable commands into the appropriate action. The NEX process invokes the path planner upon request and processes map update requests generated by PPM. Map requests are forwarded to the vision system, and replies from the vision system are

processed by NEX procedures into the terrain model format and added to the data base. The required transformations will be discussed below. NEX also invokes the VGM either to move the vehicle along planned paths or to execute movement primitives.

3. The Terrain Model

In order to perform path planning, the navigation system must maintain a model of the robot's environment in which features that would affect the vehicle's movement are represented. Since the area explored by the robot may be large and many such obstructions encountered, it is desirable to have a terrain model that is partitioned into segments of a convenient size, within which the features have a compact numerical representation. The map segments should normally reside in bulk storage and should have an access structure that allows rapid loading when needed. The model used in the navigation system was designed to meet these requirements. In addition, since the testing of proposed path links is the major computational expense of path-planning, the representation is optimized for this purpose. The segments, when loaded, form a hierarchy for accessing the barrier descriptions and, as discussed below, the structure of the descriptions was chosen to facilitate the process of path search.

The territory represented in the terrain model is partitioned into map sectors by a fixed lattice of grid lines. The grid lines are drawn parallel to the axes of the robot's absolute (lab-based) coordinate system and are equally spaced, so that the sectors are square and may be numbered relative to the origin. The sector number may thus be used to compute the absolute coordinates of the sector. The map sector is defined by providing to the model a file containing the terrain description for the area covered; the resultant directory of sectors is thus analogous to a catalog of charts. At present, a three-meter grid spacing is used for within the lab.

The primary source of input to the model is the vision system. When a description of a sector is requested by NEX for use by the planner, the request is sent to the vision system, which performs the required terrain analysis. Here the perceptual problem is that of producing from stereo TV and laser rangefinder inputs a segmentation of the area covered by the requested sector into regions described as traversible, obstructed (non-traversible) or unknown (YC1, YC2). The information available from a new viewpoint must be combined with that already in the model to produce a new sector description. This map maintenance process is subject to errors which will be discussed later.

It is of interest to note that what is described in the map is the traversibility of the terrain surface. This is adequate for the purposes of path-planning and allows a two-dimensional representation for all types of obstructions, as shown in Fig. 3. As a greater variety of terrain descriptors are added to the model, such as slope and altitude, information pertaining to the third dimension may influence the cost of a path on the goals of the robot. Should it become necessary to describe multi-level structures, new description types may easily be added to the model.

Within a map sector, terrain regions are described by polygonal boundaries which are represented as lists of the vertices (corners) of the polygon. At present, regions are classified either as obstacle (non-traversible) or unknown. All else is presumed clear and traversible. Some generality is lost by dividing barriers that overlap a sector boundary between the adjacent sectors, but the path planner detects this case and regards the parts as a single obstacle. There is also a capability for a region at sector level to represent a cluster of objects, in which case the description contains a list of other boundary or cluster descriptors. When loaded into memory, the descriptors for regions within a sector become the "datums" of SAIL "items," so that the associative search features of SAIL may be used in the path-planning process (VL1).

In new map sector descriptions provided by the vision system, the borders of the terrain regions are represented by lists of predefined unit vectors, in which consecutive vectors describe (15 cm) unit steps along the boundary. The navigation system must translate this representation into the polygon vertex list used in the map. The translation procedure locates the minimum number of corners necessary to describe the boundary to within a known tolerance. The operations include smoothing, elimination of inaccessible regions in the interior of closed boundaries and finally, an iterative polygon approximation to the boundary. The resulting

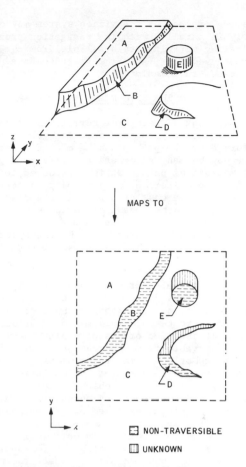

MAPS TO

☐ NON-TRAVERSIBLE

☐ UNKNOWN

Figure 3. A Terrain Map Sector

description is a list of corners in order of their connection, plus a centroid and clearance radius used to simplify the path testing process (T1, F1).

In the normal mode of operation, when the PPM is assigned a planning task, its first operation is to determine which map sectors lie along the straight line path from start to goal. Queries for the indicated sectors are sent to the vision system, which then determines if a map update is possible for any requested sector. Each sector update is dated, so that if the vehicle has moved since the last update, additional information may be available from the new point of view. In the latter case, the vision system would provide a new sector update; otherwise it would indicate that the planner should use the existing model. As the queries are answered the corresponding sector maps are loaded into memory for use by the path-planning process. The boundary between the loaded and not-loaded sectors is represented in the map as a special obstacle. If, during the course of planning, the map border is encountered, the map must be expanded by adding additional sectors. The appropriate queries are then generated and the process is suspended until the replies are received. Thus, the sector loading mechanism forms a sort of "virtual memory" for the segmented map. The system is structured so that the map updating process of the vision system may operate freely in parallel with the navigation process, collecting terrain data available from the current viewpoint, even though the updates may not be needed immediately.

4.   The Path-Planning Process

The task of using the terrain model to find an unobstructed path to a selected goal is performed by the path planning module. Naturally, there may be many alternative routes to a goal, so a measure of path cost is introduced to define a selection criterion for optimal path search. At present, the cost of a path is the distance along it, but other measures, such as time or energy, may be used in the future. The cost metric could also be redefined by the system from one path search to the next to solve specific problems.

True optimal search is possible only when all obstructions in the area encountered by the search are known. The planner, however, has access only to terrain information represented in the map at the time of planning plus those features observable from the robot's current location. From the robot's initial point of view, much of the terrain may be obscured (by occlusion or distance), but since the obscured terrain is also represented in the model by unknown regions, the planner can detect the case where a proposed traverse intersects the unknown and may terminate the search at that point. An optimistic executive could move the robot to that point (or near enough to classify the unknown area), update the map, and plan a new path to the original goal. A pessimistic executive could regard unknown terrain the same as an obstruction and look for a (possibly) longer path. The planning algorithm is capable of functioning in the latter mode, and may be told to switch to this mode after detecting (and remembering) the first (possibly optimal) partial path.

With the selection of a cost metric and a function for defining nodes in the search space, traditional methods of optimal path search may be used (HNR1). The decision to use the energy required for the traverse (which in a zero slope lab environment translates to distance) as a metric was motivated by the need to demonstrate a system suitable for application in an actual robot planetary exploration vehicle, as compared to, for example, SRI's robot, SHAKEY, which simply used search depth as the cost, resulting in a path with the fewest number of links (R1). Although, as one would expect, the strategy of avoiding an obstruction by generating candidate paths to either side of it is a feature common to previous path planners, the minimum distance requirement, as shown below, demands a more complex node generation scheme than that required for simpler cost measures. Also, proper choice of successor nodes, combined with pruning, keeps the problem one of tree search rather than graph search.

Path Planning Defined

For the purposes of this discussion, path planning will be defined for a point vehicle and then elaborated for the finite case. The map will be defined in terms of a set of vertices, $\underline{V}$, and a set of non-traversible walls, $\underline{W}$, where

$$\underline{W} = \left\{ A_1B_1,\ldots,A_mB_m; \ A_i, B_i \ \varepsilon \ \underline{V}, \text{ and the} \right.$$
$$\text{segment } A_iB_i \text{ is part of}$$
$$\left. \text{some closed polygon} \right\}.$$

The set of vertices, with the addition of two points defining the start and the goal, define $\underline{P}$, the set of all points in the map. Then for each point P_i in $\underline{P}$ we define a set $\underline{L}_i$, where

$$\underline{L}_i = \left\{ P_iP_j; \ j \neq i, \ P_iP_j \ \varepsilon \ \underline{W}, \text{ or} \right.$$
$$P_j \ \varepsilon \ \underline{V} \text{ and } P_iP_j \text{ cuts no polygon}$$
$$\left. \text{in the map} \right\}.$$

$\underline{L}_i$ is called the link set of P_i and is composed of the walls adjacent to P_i (if P_i is a vertex) plus all line-of-sight links from P_i to other members of $\underline{P}$. A path from the start to some point $Q \ \varepsilon \ \underline{P}$ is a list of links

$$(P_{k_1}P_{k_2}, P_{k_2}P_{k_3}, \ldots P_{k_{n-1}}P_{k_n}),$$

where P_{k_1} = start, P_{k_n} = Q, and $P_{k_i}P_{k_{i+1}} \ \varepsilon \ \underline{L}_{k_i}$, etc.

Path search may be defined in these terms. We define a successful node in the search as a point (in $\underline{P}$) to which there is a known path. The link of a node is the path segment from the parent node to the given node. Similarly, successor links go from a node to successor nodes. A goal link is the link from a node to the goal. For each node P_{k_i} in the search, we select successor links from L_{k_i} until the (optimal) path is found.

In our map definition, however, the link sets must be derived, since only the walls are represented. By deriving the link set as needed, we avoid the combinational explosion that would result from representing the link set of each point in the map. The link set from a node is found by proposing candidate links to be tested for membership in the set. Normally, the first candidate is the goal link from the node. A candidate link $P_kP_{k'}$ is tested by examining the set $\underline{W}$ for intersections. If $P_kP_{k'} \ \varepsilon \ \underline{W}$, the membership assertion is true. If $P_kP_{k'}$ cuts no wall, the assertion is true, and $P_{k'}$ becomes a node in the search with P_k as its parent and (typically) the goal link from P_k, as its successor candidate. However, if there is an intersection with some $A_i B_i$ in $\underline{W}$ the assertion is false, and lines $P_k A_i$ and $P_k B_i$ then become candidate links for future testing. Note also that lines $P_j A_i$ and $P_j B_i$, $j=1,2,\ldots k-1$, may also be candidates for their own link sets, but, in an optimal path search, all but one of the successor candidates to A_i or to B_i may be pruned as discussed below.

With the goal of finding the minimum cost path from start to finish, the A* algorithm (HNR1) may be applied, with modifications, to perform optimal path search in the space defined above. Given a node generating function, Γ, and an admissible node cost criterion, the A* method is guaranteed to find the lowest cost path through the space defined by Γ, if it exists. In PATH*, the algorithm used in the JPL robot, a node in the search space is actually what is described here as a path link, since it is the link record that is tested for success or failure. Also, the notions of parent and successor are different from those of A*. After a path link is tested, PATH* may select one or more points in the map as destinations for candidate links, but, by a procedure discussed below, the search tree is traced back to find the optimal parent for each chosen destination. The destination then becomes a candidate for the link set of that parent. New candidate links generated as a consequence of the failure of a link are said to be "engendered" by the failed link and are associated with it for possible use in subgoal generation. The algorithm uses these and other relations between links to form an associative data base describing the search context for use in node generation and pruning. These techniques will be illustrated in the examples below.

The cost function for a link is the actual distance along the (unique) path from the start to its endpoint plus the straight line distance remaining to the goal. Since the straight-line distance from the endpoint to the goal is the lower bound on the actual path cost of reaching the goal, this heuristic estimate satisfies the admissibility requirement of A*. In certain maze-like configurations, this heuristic estimate may be increased to improve search efficiency. As in A*, untested candidate links are kept in a list ordered by this cost estimate, so that the main loop may always select the least-cost link for the next test.

The question remains, of course, whether the successor generation procedure described above is capable of covering the link set from a node or, in the case of optimal search, of generating that member of the link set that lies on the optimal path from the given node. This turns out to be the heart of the planning problem, because there are maze-like configurations of concave barriers where the search must move away from the goal in

order to reach it. Barriers with concave boundaries must be avoided on the edge-by-edge (wall following) basis discussed above, rather than by generating links to the extreme tangent points of the barrier (which is adequate for convex barriers). Also, as will be shown, it is often necessary to choose as the successor candidate link from a new node some link other than the goal link. This alternative to the goal link is called a subgoal link, whose end node (the subgoal) is the end node of the link whose failure (obstruction) led to the generation of the given (successful) link.

To illustrate these principles, consider the search space shown in Fig. 4a. The straight path from start to goal is obstructed by wall CD. Candidates SC and SD are generated. In the search tree notation shown in Fig. 4b, the line cutting the tree link from parent to node implies an obstruction of the physical link, as well as indicating that the successor nodes have the same parent as the failed link. The parent node is mentioned in the box for the successor for convenience, and also illustrates the interchangeability of the notions of node and physical link.

The successors would be tested in order by cost, but for the purpose of discussion we will consider a more depth-first approach. Link SD is unobstructed, so the goal link is generated. This link is in turn obstructed by wall FH. Note that the successor H is to the right of the parent link SD, and that SD is avoiding wall CD on the right. This implies that H may be reached from the parent of D (in this case the start point S) and is guaranteed to avoid CD on the right, so the link SH is generated instead of DH. In practice the successor generator function will trace back perhaps several generations to find the oldest ancestor that does not satisfy the "parent-backup" condition, thus selecting the parent with the shortest path to the successor that will avoid on the same side those same walls avoided by the intervening links. Of course, the new link from the backed-up parent is not guaranteed to be unobstructed, just that it will not hit those walls avoided by the intervening links. If the shortest path to the successor lies on the opposite side of one of the intervening walls, it would be found by the normal search process proceeding from the nodes on the opposite ends of the walls.

Continuing with the example, the new link SH is obstructed by wall IJ. The links SJ and SI are generated as usual. However, note that link SI hits the wall near D again. This repetition is detected by an associative mechanism (discussed below), and since SD was previously found to be successful, the link DI may be generated at once.

Returning to consider link SC, other features of the algorithm may be shown. SC fails, suggesting SB as a candidate avoiding B on the right. SB succeeds, but note that now the goal is on the right of the line containing SB. This state would normally indicate parent backup, but since the

249

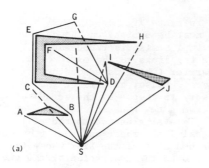

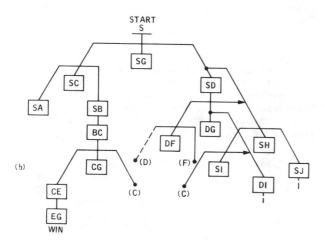

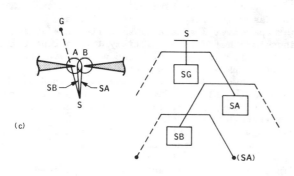

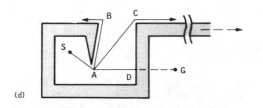

Figure 4. Path Search Examples

goal link from any parent would have already been considered, the destination of the failed link that engendered SB is proposed, in this case C, so BC is generated to avoid CD on the left. This remembering of subgoals is accomplished by associating with the successor links the link whose failure led to their generation. In this case, the failed link SC is associated with both links SA and SB. Note that SB could be obstructed as well, and new links from S would be engendered with B as the subgoal. Such an occurrence would represent the "pushing" of a new level of subgoal onto an implied "stack". In general, once a node is successfully reached and if the successor link is generated to a subgoal (instead of the goal), the subgoal of the link associated with the successful link is then passed along (associated) to the successor, i.e., if the successor is a subgoal, it inherits the subgoal of that subgoal. This represents a "pop" of the implied subgoal stack.

Pruning The Search Space

One of the advantages of performing optimal cost-directed search is that the first path found to a node is the optimum (HNR1). This allows a node marked as having been successfully visited to be used for pruning the search. No different path to that node need be considered later in the search. This eliminates the need for a graph search process in which a lower cost to a node may be discovered later in the search, requiring updating of all successor node costs. Thus, for example, when the goal link from A hits wall CD in the figure, neither AC nor AD should be generated. Pruning is indicated in the tree of Fig. 4b by a dot in place of the successor. The requirement for barriers to be closed polygons is dictated by this pruning consideration, since if the same vertex could be reached from both sides of a barrier it would be necessary, when testing a candidate for pruning, to determine if the candidate is on the same side of the wall as the successful link. That would not always be a simple test, so considerable time is saved by the requirement that barriers have "thickness".

The other category of pruning deals with the detection of duplicate links with the same parent, which can occur as a consequence of repetitive failure configurations, or due to parent backup (as shown above), or in cases where the search originates within a concave barrier. Whenever a new link is to be proposed, the destination of the link is compared with that of every other link proposed from the parent. The parent-successor associations are used to derive this set. If no match is found, the proposed link may be generated. However, if the link had been previously generated (i.e., a match is found), the link could be either unobstructed, obstructed, or not yet tested. For each of these cases, action is taken that results in the generation of the appropriate link required to guarantee continuation of the optimal search. Required subgoals may be associated with untested nodes, or, as in the example, the tree below the parent may be examined for the proper node from which to generate a link to the subgoal. Also, the repetition detection will recognize those barrier configurations in which a gap is too narrow for a finite-sized vehicle. In Fig. 4c the circles around the vertices A and B indicate the radius by which the (finite-sized) vehicle must avoid the corner. When link SA attempts to avoid A on the right it encounters the wall at B. Then when SB attempts to avoid B on the left, A is encountered again, but this is detected, and since SA engendered SB, the repetition is suppressed, effectively treating the gap as closed.

Maze-Solving

Another case that requires special treatment is that of maze-solving, where wall-following is needed if the shortest (or even the only) path is to be found. In Fig. 4d the starting point is contained within a concave polygon. From the vertex B, the goal link is generated, but is obstructed by wall CD. Since there is no path to the right of D, the search would proceed to the left of C, perhaps indefinitely, if there were no other rule. However, as mentioned in the definition of the search space, the walls connected to vertex B are contained in the link set of B, and in this case the shortest path is along the wall AB. It should be noted that unless an adjacent wall is encountered by the normal search that always proceeds toward the goal, wall-following is needed only if it becomes necessary to circle back around the starting point of the search. This allows the normal heuristic estimate of the remaining distance to the goal (used in computing the node cost that determines the order of testing) to be increased by the straight-line distance from the start to the node in question, since that is a lower bound on the path length back around the enclosing obstruction. This increase in total node cost results in fewer unnecessary tests. Thus, when a new successful node is found, its total cost is increased by the defined amount and then reinserted in the list of untested nodes as a candidate for wall-following which would not be tested unless the observed search cost reached its new cost estimate.

Using these rules, the search will continue until a terminating state is reached. If a successful goal link is found, or if an obstructing wall is the border of an unknown region, normal termination occurs. If the goal is enclosed within a barrier, or if the list of untested nodes is exhausted due to repetition pruning, the goal is declared inaccessible. Also, the search could run out of memory, in which case, the path to the successful node nearest the goal is returned.

Real-World Considerations

It is useful for the path planner to conduct the path search in accordance with the actual vehicle size and maneuvering constraints. The size and shape of the vehicle must of course be considered in detecting collisions, and modelling the vehicle turning capability during path search eliminates the need for adjusting and re-testing the planned path. Modelling the vehicle's turning geometry is easily done in the path link generator by storing in each link record the center of a turning circle at (or near) the subgoal and the straight line path that is the tangent between that circle and the turning circle at the parent link endpoint (Fig. 5a). A link is then defined as a turn from the parent's endpoint to the link heading followed by a straight traverse ending at the tangent point of the subgoal turning circle. The sign of a turn, indicating left or right, is dictated by which side of an obstructing edge the link is avoiding, i.e., if the parent link ends on the left of an edge, its successors will begin with a right turn about the vertex, etc. The turning center at a vertex need not be located on the vertex. As shown in Fig. 5b, the tangent is found between the turning circle at the parent node and the avoidance circle (of radius r_a) centered on the vertex. The solution is obtained by solving the geometrically equivalent problem for the right triangle shown, where S_p is the sign of the turn at the parent node and S_v is the desired side of the destination vertex. ($S_v=0$ indicates $r_a=0$.) The endpoint and direction of the link then determine the location of the new turning center near the vertex. Turns in reverse may be represented for those situations where a shorter path may be obtained by backing up (initial or terminal heading constraints) or where normal forward movement is restricted (Fig. 5c). The vehicle's length and width must also be considered by the link-testing procedure which searches the map for the first obstructing wall (if any) encountered by either the turn toward the link heading or the straight part of the link. The testing procedure is discussed in the appendix.

PATH* also has several special purpose move generators for such cases as confined spaces requiring complicated maneuvers or for special goal categories. A goal may be specified as a requirement that the robot's manipulator be positioned near enough to a selected object to reach it, etc. Also, goals near an edge may impose heading constraints on the vehicle at the goal. The use of tree search methods and state-space representation by node records is an improvement over recursive reduction in this domain. In fact, examples may be constructed in which the success of the search depends upon the ability of PATH* to abandon attempts to reach an inaccessible subgoal in favor of proceeding toward the goal directly from some intermediate node. A recursive algorithm that required reaching the subgoal would fail. In PATH*, the search tree and the other associations become a data base for a variety of operators with all levels accessible through the parent-successor and other relations.

5. Planned Path Execution and Error Recovery

Upon command to execute a planned path, NEX invokes the VGM and sends it the path links in

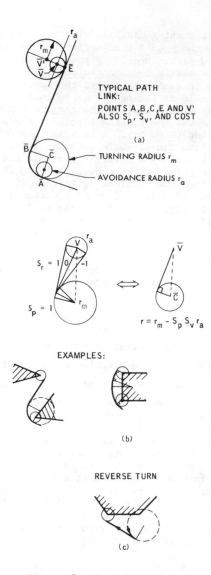

TYPICAL PATH
LINK:

POINTS A,B,C,E AND V'
ALSO S_p, S_v, AND COST

(a)

TURNING RADIUS r_m

AVOIDANCE RADIUS r_a

$S_r = 1 \quad 0 \quad -1$

$S_p = 1$

$r = r_m - S_p S_v r_a$

EXAMPLES:

(b)

REVERSE TURN

(c)

Figure 5.   Path Link Examples

succession. The VGM has a system of feedback
control loops for translating the movement
commands sent by NEX into vehicle steering and
drive signals. Vehicle odometer and gyro-compass
heading feedback is used to maintain an estimate
of the vehicle's current location, which is then
used in the guidance loops to keep the vehicle on
the planned path. Front and rear wheels steer in
opposite directions, placing the turning radius
through the vehicle center. It is desirable to
make heading changes without stopping the vehicle,
and since the rolling turn is not a circle due to
the finite steering rate, this creates a systematic tracking error. The path planner requires
a clearance along the planned path that is
actually larger than the vehicle width, so that
this error is tolerable.

The vehicle will be equipped with proximity
and tilt sensors and already possesses a scanning
laser rangefinder to aid in the detection of
unexpected obstacles. Limited evasive maneuvers

by the VGM are allowed, but if avoidance of the
obstructing region requires substantial deviation
from the planned route, the path is aborted, and
error recovery procedures are invoked.

There are numerous error sources having direct
impact on the robot's performance. The uncertainty
in vehicle position as determined by dead
reckoning grows with distance from a known
location and may be reduced only by external
references such as landmarks. The sensory limitations of the vision system result in uncertainty
in the relative position of terrain features
which at present are added to the map by using the
vehicle's location as an absolute reference point,
thus increasing error. Terrain classifications
are themselves probabilistic in nature, and in an
unstructured environment, mistakes will be made.
The end result is that the robot will eventually
encounter a rock it never saw, and update the map
by remembering the rock in the wrong location
relative to a lost robot! After an intervening
sojourn it may even repeat the process on the same
rock. Of course, laser and proximity sensors
should prevent actual collision, but both the
position and map errors remain.

Landmark navigation, when perfected, will
allow the system to reduce the robot's positional
uncertainty below some upper bound. The map
updating process provides another opportunity for
error reduction. Knowing the sensory uncertainties, the position of perceived terrain features,
and given the locations of previously detected
features, the new perspective may be matched
against the old by varying the estimated vehicle
position (within the error bounds) until the best
fit is found. Data structures have been proposed
(M1) that record a robot's perceptual history and
associated uncertainties, so as to facilitate such
a process.

Until such features are implemented, error
recovery will consist of a simple map update from
the current estimated position, followed by
execution of a replanned path.

6. Future Work

It is expected that the navigation capabilities of the robot will be expanded in the following areas, more or less in order:

1. In confined quarters it is necessary for
    the path planner to generate moves that
    simultaneously avoid several obstacles
    and that make heading changes by a com-
    bination of forward and reverse turns or
    movements. In some situations reverse
    search is useful. Such features will be
    provided either as an improvement to the
    current algorithm or else be integrated
    with the general problem solver.

2. A landmark location function, to be
    provided as part of the vision system,
    will be used to reduce position uncer-
    tainty either when error estimates in

the dead-reckoning position exceed a maximum or else continuously by landmark tracking feedback during vehicle motion. Similarly, real-time visual feedback would be used to assist obstacle avoidance.

3. The terrain model will be expanded to categorize areas by slope, texture, etc. Objects may be given functional properties such as "pushable," or "fuel-source," etc. to be used in conjunction with "high-level" planning. Previously executed paths may be remembered, forming a sort of "road map."

Appendix. Collision Testing in the Terrain Map

The vehicle is approximated as a rectangle, with variables VL2 and RS representing half the length and width respectively. The value RS defines a "safety radius" on either side of the path. The actual coordinates of path endpoints, etc. contained in a path link record refer to the position of the vehicle center point, so that actual collision testing is performed relative to that point. The test consists of detecting whether any edge of an obstacle intersects the area swept out by the vehicle along the path (Fig. 6). Turns are tested by testing successive 0.5 radian chord lines until the turn is completed.

When a path line is to be tested, the endpoint is extended VL2 units along the path and used to locate the vehicle front edge line as shown. Collision with an obstacle is detected if either of the following tests is true:

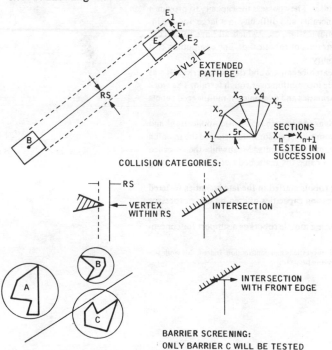

Figure 6. Collision Testing

a) Any vertex is $\leq$ RS units from the extended path-line.

b) Any edge intersects either the path line or the vehicle front edge.

The nearest collision to the start of the path is found, and the left and right (of path) vertices of the obstructing edge are offered as possible subgoals.

It is useful to limit the search to only those obstacles that lie near the path line to avoid performing the detailed search of every obstacle in the map. Included in the obstacle record is a centroid and clearance radius (RC, the radius of the superscribed circle about the centroid). Barriers within the map sectors that contain the path are tested to see if the distance from the centroid to the path-line is $\leq$ RC + VL2. If so, then the detailed test is performed.

References

F1   Freeman, H., Computer Processing of Line-Drawing Images, Computing Surveys, Vol. 6, No. 1, March 1974.

FS1   Feldman, J.A., and Sproull, R.F., System Support for the Stanford Hand-Eye System. 2nd International Joint Conference on Artificial Intelligence, 1971.

HNR1 Hart, P.E., Nilsson, N.J., and Raphael, R., "A Formal Basis for the Heuristic Determination of Minimum Cost Paths," IEEE Transactions on Systems Science and Cybernetics, July 1968.

M1   Merriam, E.W., Robot Computer Problem Solving System. Bolt Beranek and Newman Inc., Cambridge, Mass., Progress Report (in publication).

R1   Raphael, et al., Research and Applications - Artificial Intelligence, Report on SRI Project 8973, Dec. 1971.

T1   Thompson, A.M., The Navigation System of the JPL Robot, JPL Pub. No. 77-20, (May 1977).

VL1   Van Lehn, K.A. (Ed.), SAIL User Manual. Stanford Artificial Intelligence Laboratory Memo AIM-204, Stanford University, Stanford, Ca. (July 1973).

YC1   Yakimovsky, Y. and Cunningham, R., Data Base for Image Analysis with Non-Deterministic Inference Capability. Jet Propulsion Laboratory Technical Memorandum TM 33-733 (FEB. 1976). (Also printed in Pattern Recognition and Artificial Intelligence, C.H. Chen, ed., Academic Press, N.Y., 1976.

YC2   Yakimovsky, Y. and Cunningham, R., A System for Extracting 3-D Measurements from a Stereo Pair of T.V. Cameras. JPL TM 33-769 (March 1976).

An Integrated Navigation and Motion Control System for Autonomous Multisensory Mobile Robots

Georges Giralt, Raja Chatila, and Marc Vaisset

An essential task for a mobile robot system is navigation and motion control. The characteristics of perception required by environment modeling or motion control are very different. This may be basically obtained using several sensors. The described NMC system integrates the elementary data acquisition, modeling, planning, and motion control subsystems. A set of rules determines the dynamic structure and the behavior of the system and provides a man/machine and system to system interface.

1 Introduction

Research on mobile robots began in the late sixties with the Stanford Research Institute's pioneering work. Two versions of SHAKEY, an autonomous mobile robot, were built in 1968 and 1971. The main purpose of this project was "to study processes for the real-time control of a robot system that interacts with a complex environment" ⟨NIL 69⟩. Indeed, mobile robots were and still are a very convenient and powerful support for research on artificial intelligence oriented robotics. They possess the capacity to provide a variety of problems at different levels of generality and difficulty in a large domain including perception, decision making, communication, etc., which all have to be considered within the scope of the specific constraints of robotics: on-line computing, cost considerations, operating ability, and reliability.

A second and quite different trend of research began around the same period. It was aimed at solving the problem of robot vehicle locomotion over rough terrain. The work focus was the design and the study of the kinematics and dynamics of multilegged robots ⟨McG 79⟩.

During the seventies various reasons, such as too remote real-world applications and lack of efficient on-board instrumentation (computers, sensors, etc.), slowed the research thrust in the field and even lead to important funding cuts. Meanwhile the so-called industrial robots, i.e., manipulator robots, became the main body of a fast expanding field of robotics.

The present renewal of interest in mobile robots started in the late seventies fostered by powerful on-board signal and data processing capacities offered by microprocessor technology.

Today, in 1983, the scientific reasons for using mobile robots as a support for concep-

The authors would like to thank Malik Ghallab for his suggestions and contribution. Indeed, this work was only made possible thanks to the group efforts of the entire project.

Reprinted from *International Symposium on Robotics Research*, 1984, pages 191-214, "An Integrated Navigation and Motion Control System for Autonomous Multisensory Mobile Robots" by G. Giralt, R. Chatila, and M. Vaisset, by permission of The MIT Press, Cambridge, Massachusetts.

tual and experimental work in advanced robotics hold more than ever. Furthermore a number of real-world applications can now be realistically envisionned, some for the near future. These applications range from intervention robots operating in hostile or extremely dangerous environments to day-to-day machines in highly automated factories using flexible manufactoring systems (FMS) technology.

In this paper we focus on aspects of the HILARE project's current research that we believe are the key to autonomous mobile robots development: system integration, multisensory driven navigation, and motion control.

2 Overview of HILARE, A Mobile Robot

The HILARE project started by the end of 1977 at LAAS ⟨GIR 79⟩. The project's goal is to perform general research in robotics and robot perception and planning. A mobile robot was constructed to serve as an experimental means.

The environment domain considered is a world of a flat or near flat smooth floor with walls which include rooms, hallways, corridors, various portable objects, and mobile or fixed obstacles.

2.1 The Physical Infrastructure
The vehicle has three wheels as shown in figure 1. The two rear wheels are powered by stepping motors and the front wheel is free. This structure is simple but allows the robot to perform such trajectories as straight lines, circles and clothoids.

size ≃ 1.10 x 1.10 x 0.70 m
weight ≃ 400 kg
moving velocity ≃ 1 m / sec max

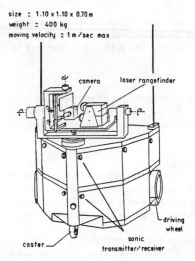

Figure 1

The perception system is composed of two separate subsystems serving different purposes: an ultrasonic system and a vision-based system. To improve odometry path-control, two optical encoders are also used. A manipulator is to be put on the robot in the future. The computer system supporting the various robot functions has a distributed multilevel architecture (figure 2): several (currently six) robot-borne microprocessors are radio-linked to a 32-bit computer accessing one or more other larger or similar processors.

2.2 The Perception System
A multisensory system provides the robot with the information it needs about its environment. The various sensors are used independently or in concert.

Ultrasonic Perception
A set of 14 ultrasonic emitter-receivers distributed on the vehicle provides the range data up to 2 m. The system has two functions:

1. An alarm function that warns the robot of the near vicinity of some object. The reaction of the robot is usually to come to a full stop if moving, but in some circumstances it will try to avoid the detected object.

2. A closed-loop local obstacle avoidance function. In this mode the robot uses the range data to move along an object, maneuvering to stay at a fixed distance from its surface.

Vision
A camera and a laser range-finder are the main perception system. They are mounted on a pan and tilt platform. The laser can be used in scanning mode, or it can measure ranges within the camera's field using a retractable mirror. This provides the robot with 3-D data about its environment ⟨FER 82⟩.

Position Referencing
HILARE's position can be obtained either relatively to objects and specific environment patterns or in a constructed frame of reference. To do this, HILARE is equipped with an infrared triangulation system which operates in areas where fixed beacons are installed.

2.3 Decision Making and Execution Monitoring
One important question that arises in the research area of decision-making system organization is the extent of decision distribution in the system, the degree of decomposition of the system in independent modules, and their level of abstraction. The issues of synchronization and communication between modules have to be addressed to answer this question.

The robot decision-making system is composed of several specialized decision modules (SDMs) some of which are implemented on different processors. Several architectures are being examined wherein either distribution or centralization is enhanced. One ap-

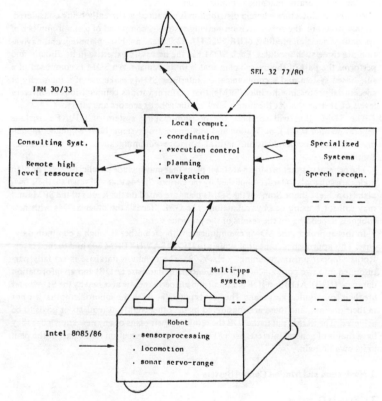

Figure 2

proach is to consider a large number of SDMs organized in a fixed hierarchical structure corresponding to a predetermined ordering. The hierarchy expresses the fact that a lower level module is a primitive of the next one in the structure. On top is a general planner and a plan execution monitor. The execution monitor is a central system through which modules interact. It also controls the robot's interaction with the outer world, i.e., sensing and operator-machine communication.

Another architecture wherein distribution is enhanced is currently being considered. In this structure, the robot decision-making system is composed of a small number of specialized decision modules ⟨GIR 79, CHA 82⟩. Scene analysis, planning, and navigation are three such modules. Each SDM has the necessary expertise in its domain and performs the part of the overall plan that is in its domain. An SDM is composed of a rule-based system and a communication interface. SDMs make use of a hierarchy of specialized processing modules (SPMs) that perform various computations at different levels of abstraction. At the lowest level are the robot's sensors and effectors.

The SDMs that will compose the decision-making system of HILARE include general planner (GPL), navigation and motion control system (NMCS), object modeling and scene comprehension, natural language communication, and manipulator control.

The general planner produces abstract plans that may include parallel nodes. The plan nodes are subgoals to be accomplished by the various SDMs that will generate their own plans to achieve them. Some of the SDMs plans are based on the results of the SPMs and may call for sensing or physical action. The GPL "feeds" the other SDMs with new subgoals depending on the results of the previous steps.

In this approach, the SDMs communicate with each other through a common database. This approach is somewhat similar to HEARSAY II ⟨ERM 80⟩ and to the Hayes-Roths' "opportunistic planning" ⟨HAY 79⟩ The common database is actually partitionned into two components: an announcement database (ADB) and an information database (IDB). ADB and IDB are small and are permanently accessed by the SDMs and are their communication means. The SDMs put in the ADB the subproblems that are not in their domain, and these are considered by the other SDMs as requests or goals to be achieved. The SDMs put in the IDB the results of their plans or any new knowledge they have that is of general interest. Each SDM controls the execution of the part of the plan in its own domain.

3 Navigation and Motion Control System

3.1 System Overview
Environment dependent sensory driven navigation is at present the key issue in mobile robots research. And this is still more the case when we consider real-world applications.

The navigation and motion control system (NMCS) is one of HILARE's specialized decision modules. Its domain is all that concerns the robot's mobility activity. The basic procedures it makes use of are routing, navigation, low-level vision, locomotion, position finding, and local obstacle avoidance. The sensors and effector are camera, laser and

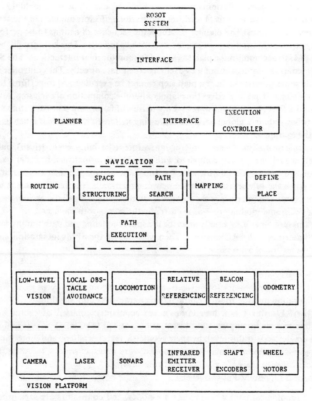

Figure 3
Navigation and motion control system.

vision platform positionning, ultrasonic transducers, infrared transmitters, shaft encoders, and wheel motors (figure 3).

NMCS will use all these resources to produce a general plan of action in the navigation domain and to control its validity and its execution using the sensory information.

NMCS is composed of a planner, an execution controller (expressed in two separate rule systems), and an interface. The rules are the operational specialized knowledge that the system possesses in its domain, and they are fired according to the factual knowledge in its database. The various procedures and sensors/effectors are used in the action part of some of the rules. The planner produces the sequence of actions to be performed by the system's SPMs and sends it step by step to the controller. This last actually has to call the SPMs to accomplish the plan steps, and to monitor their interactions. The SPMs are activated by the controller but they exchange their data directly. The controller is able to handle some situations where a plan step cannot be executed and to perform the necessary mending. It acknowledges the planner after the completion of each step, and in case of failure, the robot current state and the reason of failure are sent back to the planner. The robot state is a global variable expressing its location and some other parameters (see the example).

The rule formalism has some interesting features (flexibility, event driven), but it could be rather inefficient when compared with procedural formalisms. In order to keep the power of a rule system and increase its efficiency, we compile the rules into a decision tree. When the rule set is extended, a new compilation is required, but this is not a frequent event. The interface is the part of NMCS that handles all the exchanges with the other decisional modules or—as is currently the case—with the user.

In this section we first briefly describe the space modeling and path finding techniques investigated in HILARE's project. Then we focus on the SPMs integration and on the execution control aspects.

3.2 Space Models

The knowledge necessary for navigation and motion control concerns mainly the space configuration and the robot's characteristic parameters and location.

HILARE's world is a human-made environment composed of rooms and work stations and containing several objects to be avoided (obstacles) or manipulated. Objects are defined, at this level, by their geometric features (shape and dimensions).

Space is composed of places, e.g., rooms, work areas, corridors, hallways, and connectors between places (e.g., doors and gateways).

Space is represented at two levels:

1. the topological level where places are nodes, and connectors are arcs in a connectivity graph (figure 4) and

2. the geometric level, which assigns dimensions to the elements of the connectivity graph: width to the connectors and boundary dimensions to the places.

Every place has an explicit operator provided or an implicit frame of reference. The latter is set by the robot while learning about its world.

Places can also contain specific locations, such as landmarks and work stations, which

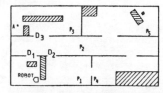

Figure 4
Topological representation of space: a connectivity graph whose nodes are places (rooms, ...) and arcs are traversible boundaries (doors, ...) between places.

can be used in similar fashion to the frame of reference to define at a relational level the robot position.

The space structure is further refined considering the internal topology of places. These may contain a variety of static or movable objects (e.g., obstacles and elements of the internal space structure). An internal topological model is created by the definition of polygonal cells:

1. polygonal cells O_i which are floor polygonal projections of the objects and the walls, and

2. convex polygonal cells C_i, which represent empty space.

C_i cells are bounded by segments of O_i and segments of other C_j, $j \neq i$. Segments of space within a place which are either partially undefined or nonconvex are noted C_i^* cells.

Hence, a connectivity graph between cells is created whose nodes are cells and arcs are the connectors i.e., the traversible frontiers between cells (figure 5).

3.3 Path Determination

Let us consider the case, which can be shown to be general, where the robot possesses information on the general setting of its environment (e.g., connectivity graph between places) and no information on the internal structure of places (e.g., obstacles). Now a goal as "Move to place P_3 location $A(X, Y)$" (figure 4) is first interpreted using ROUTING a specialized processing module (SPM).

Routing

Based on the connectivity graph of figure 4, this expert module transforms "Move to P_3 location $A(X, Y)$" into a sequence of elementary point-to-point moves.

For instance,

"Move to door D_1 (X, Y),"
"Move to D_3 (X, Y),"
"Move to A (X, Y)."

The routing expert database contains only topological knowledge and the graph search would consequently provide the best path in terms of search depth only (number

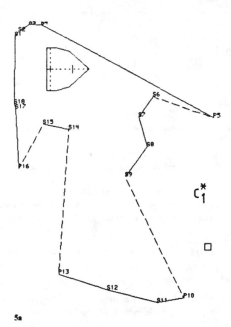

5a

Figure 5 (pp. 199–204)
Example of robot navigation: (a) first vision; (b) first cell structuring and graph; (c) the robot moves to the best frontier of the unknown space; (d) second vision; (e) second cell structuring and modified graph; (f) goal is reached.

of crossed places between start and goal nodes). In order to avoid this limitation, a cost function has to be defined taking into account other information of the robot's world.

The concept of place allows for a nested decomposition of space since a place j can be itself described as a set of places $\frac{i}{j}$. Hence in HILARE's decision-making structure a ROUTING expert can be included both at the NMCS and at the GPL levels.

Everyone of the elementary moves is now to be executed within a place whose internal structure is unknown. This task, at the very core of the navigation system, is performed by two cooperating subsystems: space structuring and path search.

Space Structuring
Let us consider now the general problem facing HILARE: the task of going from an initial location R to the goal G. Nothing else but the reference frame is given to the robot.

The first action to be carried out is a perception of the environment (e.g., using the laser system), which provides the robot with an outline of the perceived space.

In figure 5 an example of environment perception is shown and the S_i and P_i types of vertices are defined.

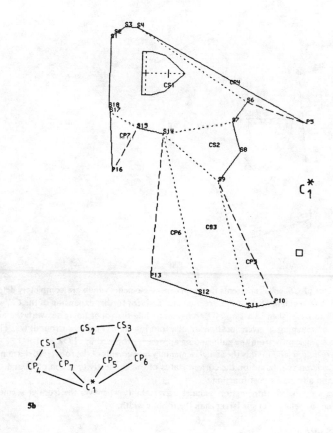

5b

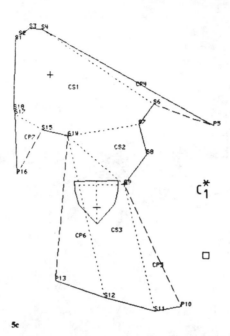

5c

The set of CS_i cells represents the empty space elements which are completely determined. The segments $S_j P_i$ or $P_j S_i$ are used as a basis for further expansion of the CS_i set. This is accomplished in a step-by-step manner while the robot navigates with the aim either of reaching a given location or of exploring its unknown surroundings. This differs significantly from the path-finding approach proposed in ⟨LOZ 81, BRO 82⟩.

Every action, and mainly the actual elementary movements of the robot, is determined by a decision rule based on the current status of the connectivity graph of C_i and C_i^* cells, and a heuristic cost function.

Arcs are labeled. Information includes traversability whenever the frontier segment between the related cells is larger than the robot's width.

Path Search

Together with the connectivity graph previously defined a second graph is constructed whose nodes and arcs are, respectively, the frontier segments and cells.

The cost function used has the following general form:

$$F = \sum_i (F_i^1 + F_i^2);$$

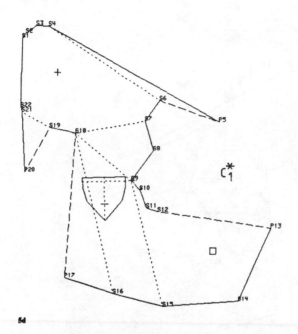

F_i^1 measures every cost associated with determined trajectory travel (CS_i and CP_i cells);
F_i^2 measures uncertainty and is evaluated over C_i^* cells.

After giving either the robot-centered or absolute coordinates of the goal to the robot, the path search is performed on the segment graph. We use an A^* algorithm ⟨HAR 68, NIL 80⟩, guided by F. For the evaluation of a minimum distance path, a segment is currently represented by its center. The robot and goal, represented by points, are added to the graph after determination of the cells in which they are.

The robot is obviously always within a CS-cell. If the goal is inside a CS or CP cell all the path is in the known part of the robot's world and the heuristic function reduces to F^1. Otherwise, we introduce in F^2 an evaluation of the information that may be gathered on the unknown world at the selected frontier segment and an evaluation of obstacle density in the region of the unknown world to be traversed, in addition to a lower bound of the distance to be covered ⟨CHA 81, 82⟩. The search procedure is performed using the whole graph at each stage so that the best frontier segment with the C^* containing the goal is selected. Furthermore, if we assume that obstacles do not change place, the cell structuring is definitive and enables the robot to select another frontier segment when it reaches a dead end, for example, without any need for new perceptions and structurings.

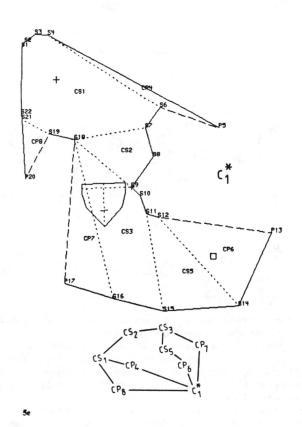

5e

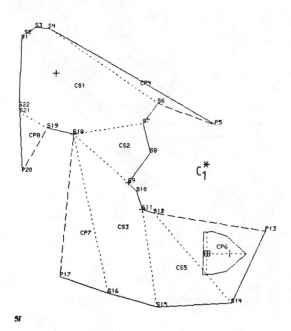

5f

The example given in figure 5 shows how robot movements and space structuring actually interact.

Path execution is accomplished by a lower module which has to take into account the robot maneuvering capabilities and a smoothed near-optimal time trajectory is finally obtained composed of straight lines, clothoids, and circles.

3.4 Automatic Environment Modeling

To date, two aspects which concern automatic environment modeling have been investigated: (1) environment mapping and (2) environment structure acquisition.

Environment Mapping

In the previous section, when HILARE had completed its task, it had obtained a partial map of its environment. This map could become complete after execution of several random displacements. To systematize this mapping function, a specific operator, MAP SPACE (X_0, Y_0), has been defined which associates a decision rule based on segments of the type $S_i P_j$ or $S_i S_j$ to a cost function Fm similar to F. Such an operator produces a map of a place in an optimized way (figure 6). This map will either be referred to the previously given frame of reference (X_0, Y_0) or to a coordinate system automatically defined by the robot and transmitted to the operator.

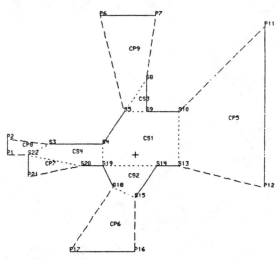

Figure 6a
Space mapping: first view and cell structure. The robot will explore until no more *SP* segments (or *P* vertices) remain.

Environment Structure Acquisition

Notice that automatic mapping can be generalized when the environment explored by the robot is not a simple place but rather a more complex arrangement of places whose structure is not known to the robot. The mapping obtained either in a systematical way or through random moves will produce eventually, together with a map of objects O_i, a set of cells representing empty space. Over this topological model of the robot environment, an automatic partition into places can be performed by use of graph decomposition techniques (figure 7).

Current work at LAAS solves the problem whenever the global connectivity graph can be decomposed in a tree of biconnected graphs ⟨LAU 83⟩.

On this basis a DEFINE PLACE operator can be defined which allows the robot automatically to acquire the overall structure and through subgraph labeling to produce an internal coding for the places.

3.5 The Specialized Processing Modules

In sections 3.3 and 3.4 several specialized processing modules have been introduced, i.e., routing, space structuring, path search, map space, and define place.

All of them concern HILARE's high level understanding of space and mainly control, at a high level, the robot navigation. More specialized processing modules are needed actually to control robot motion.

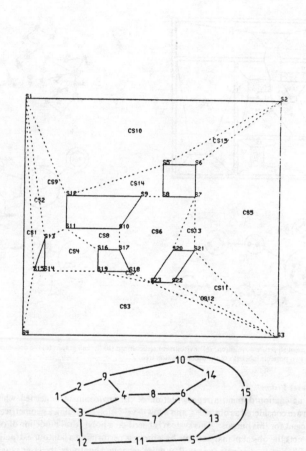

Figure 6b
Space mapping: final cell structure and graph.

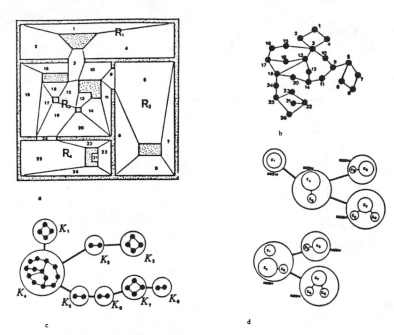

Figure 7
Environment structure acquisition: (a) environment structuring; (b) the cell graph; (c) the decomposition tree; (d) two possible labelings of the decomposition tree.

Low Level Vision
In the navigation domain a representation of the environment is needed wherein all objects are considered as obstacles, and a low-level vision providing a geometrical model is sufficient for this purpose. This system is based on a polyhedral modeling of obstacles and detects the object planar surfaces by a process performing a contour extraction, and after a filtering to eliminate too small or noise generated contours, the laser range-finder is used to determine the planar surfaces slope ⟨FER 82⟩. Objects ground projection or trace is computed from this information and used to build or update the geometrical world model.

Local Obstacle Avoidance and Maneuvering
This module uses the ultrasonic range data and provides for local maneuvering and obstacle avoidance. It is used, for example, when an obstacle is detected on the robot's path whereas none was in that location in the current space model.

Three categories of obstacles are defined: fixed (e.g., walls and internal structure elements), movable, and mobile. Movable objects can, in some conditions defined by

GPL, be considered as occasional obstacles and be disregarded for space structuring purposes. This means they can be avoided by the robot in a way analogous to when it encounters a mobile obstacle.

Thus if the obstacle is identified as occasional it will be merely avoided, taking into account the local space configuration. The rule system contains the conditions under which this may be accomplished without questioning and modifying the world model by a new costly 3-D perception. Another situation where this module may be used is when the robot moves in a corridor: no obstacles are to be avoided and it has only to move along a wall. Proximity navigation and sonar servoranging are also useful in the approach phase of docking maneuvers.

This processing module is also activated by two operators—Move Along, $(Z; v; C_1, C_2)$; Dock, $(Frame; Z; C_1, C_2)$—where Z is the list of servoing parameters, v is the robot speed, and C_1, C_2 are two predicates which define stop conditions for success and failure.

Position Finding

Position finding decomposes into three modules:

1. Absolute position measurement using external beacons that determine a frame reference. The beacons are infrared diodes and the camera is used to detect them.

2. Position computing using the shaft encoders to integrate the robot's movements and deduce its current position knowing it initial state (odometry).

3. Relative position measurement using some environment or characteristic object features.

Two important implemented cases are docking, where specific proximity sensors are used (e.g., sonar), and fixed obstacle edge referencing, using a laser.

Locomotion

This low-level module operates the wheel motors to produce the robot's moves. It computes and controls the accelerations and rotation speeds of each wheel to execute trajectories (straight lines, circles, clothoids) consistent with the elementary move orders it receives.

3.6 The Execution Control Rule System

Plan execution is mainly monitored through sensory information. The multisensory approach is very rich in the execution control context since some sensors, such as the laser or the ultrasonic range-finders, have a good real-time response and can thus be used to check rapidly the robot's world model, whereas the camera is used to build a new one. On the other hand, one sensor, such as the laser, may be used differently in the acquisition phase, where what is important is to provide a complete perception, and in execution monitoring, where quick measurements are often sufficient.

In the execution control mode, the system has to take into account the various inaccuracies induced by its sensors and effectors and the uncertainties in its world model.

This important problem in robotics has been addressed by several authors since the birth of the field (e.g., ⟨MUN 71⟩ and ⟨BRO 82⟩), and its solution is not straightforward. Our approach to this problem is based on the use of the multisensory data to check the information of one sensor by another and of the system's ability to react rapidly to changes in its factual knowledge. Thus some rules contain the necessary actions to be performed in specified failure conditions in the current plan step.

Sensor Integration

To date HILARE's perception system consists of five independent sensing devices: 1 TV camera, 1 laser range-finder, 14 ultrasonic emitter-receivers, 1 infrared emitter-receiver, and 2 optical shaft encoders.

Through the SPM structure and the execution control rule system the sensing devices operate in three modes:

1. *Cooperative* That is, for instance, the case of the TV camera and the laser range-finder jointly used to obtain 3-D scene analysis.

2. *Monitoring* In this very important mode one of the sensors monitors the operation of a module involving other independent sensing devices. This is, for instance, the case of the Move Along operator, which involves the sonars and can be monitored by odometry or laser position referencing.

3. *Cross Checking* In several configurations data can be obtained from two or more independent measures. For instance, after a "Move to" step both odometry and triangulation using laser telemetry information will allow one to compute the current robot location.

Given accuracy information on the different sensors, the cross checking mode provides

i. a dynamic ranking between measures and
ii. a consistency test (both measures match within accuracy ranges).

Whenever the latter fails, validation and, if necessary, higher level intervention will be requested by the system.

The Rule System

It is necessary to coordinate and organize the interactions of the various system's modules, and this can be achieved in several ways ⟨ELF 83⟩. We have chosen to investigate the rule approach, which appears to be very suitable to describing the actions that a system has to undertake in a set of situations that might occur in a specific domain ⟨VAI 83⟩.

The controller's rules are expressed in a LISP-based system. Each rule has a condition field and an action field. The condition field has 3 parts: ⟨pattern in database⟩, ⟨pattern not in database⟩, and ⟨evaluable expression⟩.

The action field directly modifies the database or is an explicit call of one or more of the system's procedures whose results will in turn change the robot's state description.

Rule examples

R_1: ⟨MONITORP-Si MOVEP⟩NIL NIL → STOP

Actions are monitored through sensors. This function is expressed by the predicate MONITORP-Si which concerns in this rule the sonar sensors. MOVEP is a predicate that expresses the motion state of the robot. The action STOP actually stops the robot. Hence it modifies the state parameters and deletes MOVEP.

R2: ⟨MONITORP-Si AVOIDP⟩⟨MOVEP⟩NIL
     → MOVE ALONG $(Z; v; C_1, C_2)$ MONITOR $S_i, C_1)$

The move orders from the planner contain the predicate AVOIDP that enables the local avoidance of occasional obstacles.

 The MOVE ALONG order is executed using the sonar system with a termination success condition.

 In general, AVOIDP is added by the planner in places known to be usually empty and where the robot's trajectories are straight lines. Occasional obstacles are stated to be rather small (boxes). Hence the termination success condition is that the robot aims again toward the goal. The execution of the MOVE ALONG order is thus monitored through odometry and if necessary absolute position measurement.

R3: ⟨MONITORP-Si NIL ⟨POSITION-UNCERTAINTY.⟩.Z⟩
     → MEASURE-POSITION

R4: ⟨MONITORP-Si NIL ⟨POSITION-UNCERTAINTY.≤.Z⟩
     → MOVE TO (X, Y) MONITOR (S_i, C_i)

MONITORP-Si is added to the database after the avoidance successful completion. POSITION-UNCERTAINTY is compared with a preset value Z and a measure of the absolute position is made whenever it is necessary (R3). Otherwise the next MOVE order can be executed monitored by the ultrasonic system (R4).

 The controller is an asynchronous sequential process that reacts to interruptions coming through its interface and according to its database contents, which includes the current state of the system. An interruption corresponds to a stimulus that has to be processed. A stimulus can be an explicit order to be executed or a particular event occurrence reported by a sensor which has been requested to do so earlier. This sequential interruption-driven process allows for parallelism in the execution of some specified orders if they do not share the same primitive processing modules (considered as ressources) and the sensors and effectors they operate.

 The system has to react in real time to rapid changes in its environment. Hence, efficiency aspects become proeminent, and inefficiency is one of the most important disadvantages of rule system ⟨DAV 77⟩. To increase efficiency, we compile the rules into a decision tree. This is possible only if some constraints are met in the rule: at each cycle, there must be at most one rule that matches the context. The tree nodes are the conditions in the left-hand side of the rules. A path from the root to a leaf in the tree corresponds to one rule to be executed. For the rule compilation, we use algorithms developed by Ghallab ⟨GHA 82⟩.

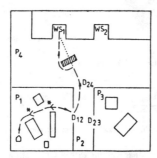

Figure 8
Example of robot task: "MOVE TO WS1." The general layout is known. Place 1 is initially partially known. Place 2 is a corridor. WS1 and WS2 are known work stations. • denotes that 3-D vision is performed.

3.7 The Execution Controller Interface
The interface deals with the input to the controller from the planner and the various other modules. Three different types of orders can reach the interface:

1. Imperative orders that must be processed immediately by the controller. This type of command can be produced by the planner or by the user and is transmitted directly in this case to the controller's interface. It supersedes any other order or sequence of orders in the interface and is considered in the next controller cycle, interrupting any current action.

2. Sequenced commands that have to be executed in a specified order. This sequence is usually a plan or a part of a plan. A given step of the sequence cannot be executed before the successful completion of the preceding one.

3. Commands that can be executed whenever the system's state enables their execution, without any specified ordering.

4 Example

Let us consider the situation (figure 8) in which the robot is initially in a room (place 1) and receives the order to go to work station 1:

"MOVE TO WS1."

In this initial state, the robot's world model contains a partial knowledge of the obstacle configuration in place 1 and the complete layout of the other places. Place 2 is known to be a corridor, and place 4 is an empty room with work station of known location and docking features.

This order is to be processed by the robot decision system and, at the NMCS level, a sequence of actions to be executed will eventually be produced. In this sequence, the robot will first navigate in place 1, avoiding the obstacles it gradually discovers, to reach the door D12 to place 2, where the next order MOVE ALONG is processed. This is an efficient step since place 2 is a corridor where there is no need for costly 3-D vision. The MOVE ALONG order is achieved on a success termination condition when the robot's position measured by a monitoring sensor (in this case odometry), reaches the set value of the entrance D24 of place 4. At this stage, two sequenced orders remain to be processed:

(MOVE TO (X, Y) (DOCK (frame (WS1); Z; C_1, C_2))

where X, Y are coordinates in the access area of WS1 where the robot has to dock. Place 4 is empty of obstacles, but occasional movable objects may be present in it. Thus the MOVE TO order will authorize to avoid such objects without any need to modify the current world model and plan.

Let us examine the system's behavior when the event of the existence of an unpredicted object on the robot straight line path is reported by the sonar system.

The database contents is

DB1:   (MONITORP-Si MOVEP AVOIDP)

Rule R1 is valid and executed producing

DB2:   (MONITORP-Si AVOIDP)

which tiggers rule R2. The MOVE ALONG order parameters are computed from the robot's state description.

The execution of MOVE ALONG deletes MONITORP-Si and adds MOVEP and AVOIDANCE:

DB3:   (MOVEP AVOIDP AVOIDANCE)

After completion of this order we have

DB4:   (MONITORP-Si AVOIDP)

Then either rule R3 or R4 is valid depending on the evaluation of position uncertainty. In any case, R4 will be eventually fired. The execution of R4 deletes MONITORP-Si and adds MOVEP,

DB5:   (MOVEP AVODIP)

The completion of this order deletes MOVEP and AVOIDP since the goal is reached:

DB6:   NIL

The next order, DOCK, is processed according to the robot's state:

DB7:   (DOCK MOVEP)

The execution proceeds till the final completion of the sequence.

5 Conclusion

Research on mobile robots is today an important and fast growing field ⟨GIR 83⟩. Autonomous mobility is, we believe, the central problem at hand.

Current research on HILARE emphasizes this aspect. The navigation and motion control system which has been presented in this paper is now partially integrated and *in situ* experimentation is being carried out. Complete integration is in progress ⟨VAI 83⟩ and extensive experimentation to test fully the execution control rule system will then be achieved.

Among the other subjects which are investigated in the HILARE project we devote also an important effort to high level decision making, i.e., the general planner and the distributed decision structure and learning.

Learning is indeed among the most important research topics in robotics and very much a wide-open field. In the HILARE project, learning plays a central role, although we intend to keep ambitions and claims on the subject at a precisely defined and tractable level. In this paper learning was considered in two aspects related to automatic environment modeling.

References

⟨BRO 82⟩ R. A. Brooks, "Symbolic error analysis and robot planning," Intern. Jal of Robotics Research, vol. 1, Winter 1982.

⟨BRO 83⟩ R. A. Brooks and T. Lozano-Perez, "A subdivision algorithm in configuration space for findpath with rotation," Proc. of the 8th IJCAI Karlsruhe, W. Germany, 1983.

⟨CHA 81⟩ R. Chatila, "Système de navigation pour un robot mobile autonome: modélisation et processus décisionnels," Thesis Dissertation, U. P. S. Toulouse, France, July 1981.

⟨CHA 82⟩ R. Chatila, "Path planning and environment learning in a mobile robot system," Proc. of the European Conference on Artificial Intelligence, Orsay, France, July 1982.

⟨DAV 77⟩ R. Davis and J. King, "An overview of production systems," in Machine Intelligence, vol. 8, Chichester: Ellis Horwood, 1977.

⟨ELF 83⟩ A. Elfes and S. Talukdar, "A distributed control system for the CMU Rover," Proc. of the 8th IJCAI, Karlsruhe, W. Germany, 1983.

⟨ERM 80⟩ L. D. Erman et al., "The Hearsay-II speech-understanding system: integrating knowledge to resolve uncertainty," ACM Computing Surveys, vol. 12, June 1980.

⟨FER 82⟩ M. Ferrer, "Système multisenseur de perception 3D pour le robot mobile HILARE," Thesis Dissertation, U.P.S., Toulouse, France, December 1982.

⟨GHA 82⟩ M. Ghallab, "Optimisation de processus décisionnels pour la robotique," Thèse d'Etat, U.P.S., Toulouse, France, October 1982.

⟨GIR 79⟩ G. Giralt, R. P. Sobek, and R. Chatila, "A multi-level planning and navigation system for a mobile robot; a first approach to HILARE," Proc. 6th IJCAI, Tokyo, Japan, August 1979.

⟨GIR 83⟩ G. Giralt, "Mobile Robots," NATO Advanced Study Institute on Robotics and Artificial Intelligence, Castelvecchio Pascoli (BARGA), Italy, 26 June–8 July 1983.

⟨HAR 68⟩ P. E. Hart, N. J. Nilsson, and B. Raphael, "A formal basis for the heuristic determination of minimum cost paths," IEEE Trans. on System Science and Cybernetics, vol. 4, 1968.

⟨HAY 79⟩ B. Hayes-Roth and F. Hayes-Roth, "A cognitive model of planning," Cognitive Science, vol. 3, 1979.

⟨LAU 83⟩ J. P. Laumond, "Model structuring and concept recognition: two aspects of learning for a mobile robot," Proc. 8th IJCAI, Karlsruhe, W. Germany, 1983.

⟨LOZ 81⟩ T. Lozano-Perez, "Automatic planning of manipulator transfer movements," IEEE Trans. on Syst. Man and Cyb., vol. SMC-11, 1981.

⟨McG 79⟩ R. B. McGhee et al., "Adaptive Locomotion of a multilegged robot over rough terrain," IEEE Trans. Syst. Man and Cyber., vol. SMC 9, 1979.

⟨MUN 71⟩ J. H. Munson, "Execution, and monitoring in an uncertain environment," 2nd IJCAI, London, September 1971.

⟨NIL 69⟩ N. J. Nilsson, "A mobile automaton: an application of artificial intelligence techniques," Proc. of the 1st IJCAI, Washington, D.C., May 1969.

⟨NIL 80⟩ N. J. Nilsson, "Principles of artificial intelligence," Tioga Pub., Palo Alto, California, 1980.

⟨VAI 83⟩ M. Vaisset, "Intégration des structures décisionnelle et informatique pour le robot mobile HILARE," Thesis Dissertation, December 1983.

Autonomous Mobile Robot Navigation and Learning

C.R. Weisbin, G. de Saussure, J.R. Einstein, and F.G. Pin

Oak Ridge National Laboratory

E. Heer

Heer Associates

The rapid development of solid-state computer technology has brought intelligent systems within the realm of possibility. Striking changes have occurred in the way all types of modern systems are monitored, controlled, and operated. Machines now accomplish many of the tasks formerly performed by humans. The complexity and speed of decisions required by many modern control systems suggest augmenting traditional automation with machine intelligence to handle such functions as problem solving, perception, and learning.

Autonomous mobile robots, designed to govern themselves and make decisions, are experimental testbeds for research in intelligent machines. In accomplishing given objectives, they manage their resources and maintain their integrity. In addition, robots must be able to communicate with humans at an appropriately high level.

Autonomous robots find a wide variety of potential applications. These include undersea operations, space exploration, mining operations, and hazardous-waste disposal, where environments are characteristically more unstructured and unpredictable. In most of these areas, the appeal of autonomous mobile

> **Having prototyped an autonomous robot that can perceive, plan, navigate, handle contingencies, learn, and manipulate, researchers are looking ahead to machines that will approach human-scale performance.**

robots is their proposed capabilities for maintenance, surveillance, and repair in dangerous and/or otherwise inaccessible places, allowing humans to remain in a safe environment while acting in a supervisory capacity. In addition to safety considerations, there are related factors of economy and a desire to minimize unnecessary communication with human

operators. The major goal of research in autonomous robot systems is to move humans out of the control loop as much as possible, so that communication between robots and humans can occur at a high level -- possibly at the human-speech level. This requires the development of machine intelligence technologies and their integration at the systems level.

The Center for Engineering Systems Advanced Research (CESAR) at the Oak Ridge National Laboratory focuses on the development and experimental validation of intelligent control techniques for autonomous mobile robots able to plan and perform a variety of assigned tasks in unstructured environments. The assignments originate with human supervisors in a remote control station. The robots then perform detailed implementation planning and execute the tasks.

Since the operational environment is generally dynamic, a robot must be in sensory contact with its surroundings to capture and recognize changes relevant to task objectives and, if necessary, re-plan its behavior. This implies cognitive capabilities that enable the robot to form and modify a model of the world around it and relate this world model causally to the task objectives. This world model

Reprinted from *IEEE Computer* Magazine, June 1989, pages 29-35. Copyright © 1989 by The Institute of Electrical and Electronics Engineers, Inc. All rights reserved.

Figure 1. HERMIES-IIB and the process control panel used for concept demonstrations at CESAR. Wooden boxes like the one in the background are used as obstacles during navigation experiments.

includes the robot itself as an element of the environment. We also want to enable the robot to learn from its past experience. This can require acquisition of explicit knowledge, development of skills, theory formation, and inductive inference so that the robot can systematically improve its performance.

Currently we conduct research at CESAR in a controlled indoor environment. However, actual system implementation and operation must ultimately cope with many additional realities that are either avoided or not present in the laboratory. Theoretical idealizations in the laboratory usually get distorted when building and operating real systems. While it is possible to take an algorithmically oriented approach in the laboratory, in practical implementations all necessary data may not be available, some may be suspect, and some of the knowledge for interpreting the data may be unreliable. Input from human operators during operation may contain errors, and sensory input may be inaccurate, spotty, and fragmentary. The bewildering problem of capturing reasoning processes with, and drawing inferences from, uncertain or incomplete data has led to a variety of approaches based on heuristic techniques and multivalued logic.

Robot navigation and exploration

An autonomous mobile robot. HERMIES-IIB (hostile-environment robotic machine intelligence experiment series) is a self-powered, wheel-driven platform containing an on-board 16-node Ncube hypercube parallel processor interfaced to effectors and sensors through a VME-based system containing a Motorola 68020 processor, a phased sonar array, dual manipulator arms, and multiple cameras (see Figure 1). It has been described in more detail in an earlier article.[1]

The first HERMIES robots provided valuable experience in planning, world modeling, mobility, sensor perception, and communication; but intensive computations and high-level decision making were performed off-board in computers linked to the robots by radio. With HERMIES-IIB the emphasis is on computational autonomy -- hence, the need for powerful on-board computing capabilities. HERMIES-IIB's position can be controlled in an open-loop fashion to a precision of about 0.1 percent for each linear movement and about 0.1 degree per rotation using real-time monitoring of the wheel encoders.

Earlier research[2] determined that using expert systems combined with modular procedures provided a convenient and powerful method of controlling the robot's behavior. The expert systems make high-level decisions and diagnose unexpected occurrences. When a standard procedure is required -- such as avoiding or removing an obstacle, manipulating an object, or mapping an area -- the navigation or learning expert systems can call the appropriate routine, which executes until completed or until an unexpected event generates an interrupt that returns control to the expert system. For HERMIES-IIB, Lisp-type format rule bases were prepared in a text editor. These rule bases control the high-level decisions and can call on C-compiled navigation and manipulation procedures. The rule bases are loaded in an expert system shell, Clips, and linked to the navigation and manipulation procedures. Clips and the navigation and learning routines run on one of the Ncube nodes.

Research in robot navigation. Considerable work has been reported on the problem of robot navigation in known static terrains (see Whitesides[3] and the references therein). Algorithms have been proposed and implemented to search for an optimum path to the goal, taking into account the finite size and shape of the robot. Whitesides analyzed many of these algorithms. Not as much has been reported on robot navigation in unknown, unstructured, or dynamic environments.[4]

A robot navigating an unknown environment must explore with its sensors, construct an abstract representation of the environment to plan a path to the goal (world modeling and planning), and monitor its performance by comparing expected and actual sensor responses. The source of a discrepancy must be diagnosed, and real-time replanning may be necessary.

Treatment of sensor data. HERMIES-IIB is equipped with sonar sensors and cameras. Each of these sensors has limitations. The sonars provide depth information but have poor angular resolution and suffer from specular reflection. The vision system does not provide direct depth information and requires extensive processing. A desirable goal is to improve the knowledge of the environment by combining data from the different sensors into a consistent representation. This sensor fusion requires a good knowledge of each detector's response function to provide accurate resolution of systematic errors that result from inappropriate interpretation of the sensor data.

Beckerman and Oblow[5] have developed techniques for treating systematic errors in the processing of sonar sensor data. Pixels of the world model (a two-dimensional map of the robot's environment) are assigned one of several labels during initial processing. One of the labels flags conflict among interpretations from two or more sensor measurements. This happens whenever there are erroneous interpretations of the data (for example, obstacles detected within a conical region are assumed to be located along the sonar central axis, while interpretation of a video image indicates no obstacle at this location). To remove errors, the data are then reinterpreted, using pattern analyses and consistent labeling operations based on the response function of each detector.

Mann et al.[6] proposed a methodology for the fusion of sonar and vision data. They described an iterative strategy that incorporates range information from the sonar sensors into the vision feature extraction algorithms (to find edges, for example) and vice versa.

World modeling and path planning. If the robot operates in a dynamic environment and precise positioning of the obstacles is unlikely because of the intrinsic uncertainties and resolution limits of the sensors, it may not be feasible to obtain a complete map of the environment. In that case a digitized representation of the environment, such as a quadtree (or octree) representation, may best suit rapid path-planning algorithms. R.C. Fryxell[7] successfully demonstrated navigation planning and execution using quadtrees.

A laser range finder provides far more precise range data than commonly used sonar transducers. Representing large amounts of precise range data using a Cartesian map or quadtree may be impractical because of the amount of memory required. Goldstein et al.[8] investigated three-dimensional world modeling using combinatorial geometry

and developed algorithms for autonomous robot navigation with this representation. This combinatorial geometry technique for describing complex 3D objects is efficient and widely used in Monte Carlo simulation of particle transport and in the computer graphics literature.

Other world model techniques have been used in connection with robot exploration and terrain acquisition. Using acquired sensor data, Rao et al.[9] constructed a model of the terrain comprising polygonal boundaries approximating perceived obstacles. As more sensor data is collected, the polygonal boundaries shrink to better approximate the obstacles' actual surfaces, free space for transit is correspondingly enlarged, and additional nodes and edges are recorded on the basis of path intersections and stop points. Elfes[10] successfully used a probabilistic approach to map sonar-based data.

Mission navigation for an autonomous security robot. A particularly attractive application of autonomous navigation in environments where unexpected obstacles may exist is that of autonomous safeguards and security robots (for intrusion prevention, for example). A major characteristic of this application field is the requirement for minimal communication between the robot and its home base (to prevent external interference or tampering with the robot's mission). This requirement means that we need an on-board autonomous decision-making capability. Because of its demonstrated autonomous sensing and on-board reasoning capabilities, the HERMIES-IIB robot was selected as a testbed for development and verification of navigation algorithms for an autonomous robot mission in secured environments.

A network of nodes and node connections is prescribed a priori for HERMIES as an initial navigation path graph structure. Any divergence of the actual environment from the robot's expectations is treated as an unanticipated obstacle requiring on-line replanning during the mission. The locations of several robot workstations are specified, corresponding to areas where the robot is to perform predetermined tasks (testing security devices, maintaining process control devices, etc.). Each workstation is associated with the approximate time needed to complete the predetermined task. A mission is defined as a series of stations

to be visited and a series of time constraints for these visits (for example, latest allowed arrival times, latest absolute time for completion of a task, etc.).

When requested to perform a mission, the robot first examines the time constraints associated with each workstation to be visited. The station with the shortest time constraint is selected as the first subgoal. Knowing its own motion parameters (acceleration, cruise speed, time for turning, etc.), the robot estimates the minimum time needed to reach the station, including a fixed overhead to account for initial planning activities and some contingencies (parameter uncertainties). If the subgoal is not reachable within the time constraint, a record is generated on the mission log and the station with the next shortest time constraint is examined as a subgoal; otherwise, the robot starts moving on its path. If during navigation an unexpected obstacle is detected blocking the planned path, the robot returns to the intersection (node) previously crossed and restarts its path-planning from this new origin. If the subgoal is still reachable under the time constraints with a new plan, the robot continues toward it. If not, the current plan to reach the subgoal is abandoned and the station with the next shortest time constraint is examined as the next subgoal.

Automatic learning by an autonomous robot

The literature includes a considerable amount of research on automatic learning.[11] Although most of this research is not directed to robotics, the development of learning systems would profoundly enhance the effectiveness of an autonomous robot operating in an unstructured environment. The learning system would compensate for the lack of a priori information (which would have to have been provided explicitly by the human programmer) by assessing the current state of the environment and, on the basis of past experience, proposing the next action. Furthermore, a learning system could provide a more flexible and robust means to insert knowledge into a robot; teaching with a set of training examples can be easier than reprogramming a more conventional system.

Learning by interacting with the environment. To investigate autonomous

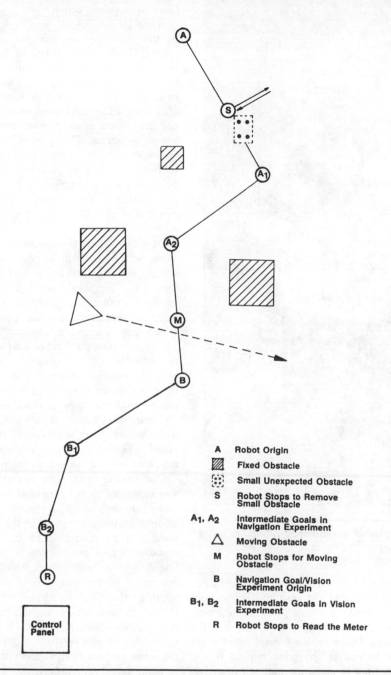

A   Robot Origin

▨   Fixed Obstacle

⊞   Small Unexpected Obstacle

S   Robot Stops to Remove
        Small Obstacle

A₁, A₂ Intermediate Goals in
        Navigation Experiment

△   Moving Obstacle

M   Robot Stops for Moving
        Obstacle

B   Navigation Goal/Vision
        Experiment Origin

B₁, B₂ Intermediate Goals in Vision
        Experiment

R   Robot Stops to Read the Meter

Figure 2. Layout of the experimental area for HERMIES-IIB. The robot navigates from point A to point B, stopping at intermediate goals A_1 and A_2 and responding to dynamic obstacles at points S and M. The robot moves from point B to the control panel, stopping at intermediate goals B_1 and B_2. At point R the robot is close enough to read the meter.

learning, we constructed a mock-up control panel with large push buttons and big levers that could be manipulated with the primitive Hero arms of HERMIES-IIB. The panel was equipped with four buttons (each with a pilot light), two levers, and a "danger" light. The status (on/off) of the pilot lights and danger light and the position of the meter needles, all functions of the push-button status and the lever positions, allowed HERMIES-IIB to observe the consequences of its actions.

The particular function (relating button/lever status to meter/danger light) was programmed into a PC attached to the control panel. The function was unknown to the robot and had to be learned. The learning system consisted of production rules programmed on the expert system shell Clips. The production rules had access to C-coded external routines, which controlled the sensor and effector primitives.

In a preliminary training activity, the robot was to discover by trial and error the sequence of push-button and lever moves that would turn off the danger light for a number of different panel configurations. For this task the robot was provided with domain knowledge about which primitive panel moves were possible and what feedback might be expected from correct and incorrect moves. For instance, a correct push-button move would turn on the associated pilot light.

Inference and concept formation. Following the training activity, the robot was confronted with a new panel configuration. From the initial training set, sampled configurations of the panel were associated into categories according to the description of the type of initial states leading to a common response sequence. The new panel state was then categorized to best match past experience, and the response sequence associated with the selected category was attempted as a possible solution. The learning system was provided with a set of heuristics about principles of classification (analogous to the generalization language in Mitchell's "version-space learning" program[12]).

Different strategies of category formation were investigated and their efficiency in optimizing the learning was examined with respect to generating the most general hypothesis consistent with the training set. The hypothesis so generated was continually modified by specialization from experimental feedback of erroneous classification.

A demonstration of navigation and learning

To focus our research activities, prove the correctness of our general approach, test and verify the sensor-based reasoning methodologies and algorithms, and better identify areas for further investi-

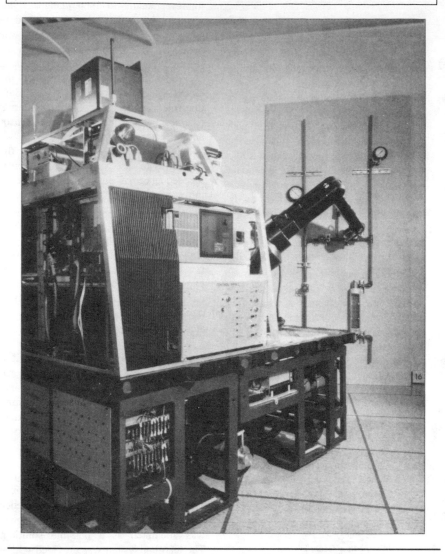

Figure 3. HERMIES-III, the latest in a series of autonomous robots, is being used to approach human-scale performance.

gation, we developed several experimental scenarios and carried out associated demonstrations using HERMIES-IIB. Figure 2 illustrates such a demonstration. It was conducted to test HERMIES-IIB's capabilities in world modeling, autonomous navigation in dynamic environments, handling of contingencies, sensor-guided exploration and goal recognition, reading and understanding of complex control devices, vision-guided manipulation, and innovative problem-solving on the basis of prior experience.

The robot starts from some arbitrary initial location, denoted by A on the figure. The coordinates of this initial location A and of a subgoal location B are sent to the robot via the radio frequency link. HERMIES-IIB's task is to navigate from A to B, avoiding or removing several types of static and moving

obstacles. From location B, HERMIES-IIB is to find and position itself in front of a mock-up control panel. There are no obstacles between location B and the panel. However, the control panel could be located at any point beyond B and need not be along the wall. All this navigation is done autonomously. After observing a lighted danger signal and the status of the control panel levers and meters, HERMIES-IIB is to determine, autonomously, the appropriate sequence of button and lever motions on the basis of prior experience, thereby removing the danger status warning.

HERMIES-IIB starts by making a wide-angle sonar scan of the environment and planning a collision-free path to the reachable point closest to B. As the robot approaches its destination, a sonar scans the area ahead of it. If the sonar

detects an unexpected obstacle within two feet of its path, the robot stops, diagnoses the nature of the obstacle, and takes appropriate action.

When HERMIES-IIB reaches location B, the vision system searches for the panel.[1] This is accomplished as follows: Gray-value images of the scene are obtained via the camera and a wide-angle lens. They are converted to binary images using a threshold based on the average gray-value as determined from the image histogram. This proved to be a reasonably robust method, since the control panel and its immediate vicinity do not produce complicated gray-value distributions. The binary image is searched to identify groups of connected pixels (contiguous and like-colored). Each binary-image component is labeled along with component features such as size, aspect ratio, and moment of inertia. This list is compared with a description of the control panel generated from a priori information about the geometric relationship among the objects (meters and switches) on the front of the panel. If the binary-image components approximate the expected features of the panel, the robot moves toward this image component, and the steps just outlined are repeated. The robot continues forward until the size of the binary-image components identified as meters indicates that the robot is close enough to the panel.

One camera with a wide-angle lens was calibrated using the simple pinhole-camera model, which is based on the assumption that each point in the image plane can be connected to the corresponding world point using a straight line through a "pinhole" displaced from the center of the image by a distance equal to the focal length of the lens. Hence, in this demonstration, distance information is inferred from a priori knowledge of relative sizes of objects in the scene and not from stereo vision, as in some other experiments.

With the robot centered in front of the panel, the vision system is used to determine the location of all devices on the panel face and evaluate their state (attribute value). In the case of the analog meters, the region of the binary image identified as a meter is searched to find groups of pixels that form lines. Since the meters have two needles (one is a preset, or limit, needle), the images are searched to find the most prominent pair of lines. A Hough transform converts the needle position data from Cartesian to polar coordinates. The meter reading is determined by comparing the needle angle to a table of angle-versus-ampere values, and a corresponding attribute value is assigned. The knowledge gained during a previous training session with the panel emulator is used to infer the correct sequence of actions on the control devices to turn off the danger light, and vision-guided manipulation is used to perform these tasks.

We have carried out this complete experiment successfully for a number of difficult sets of obstacle arrangements, vectors from the intermediate goal to the panel, and panel configurations. Thus we have validated our research approach in world modeling, autonomous navigation in dynamic environments, handling of contingencies, sensor-guided exploration and goal recognition, reading and understanding of complex control devices, vision-guided manipulation, and innovative problem solving based on prior experience.

Although HERMIES-IIB is a powerful and versatile research tool, it has limited manipulative capabilities. To approach human-scale performance, CESAR has designed and is assembling a much larger testbed, HERMIES-III, to be used in future experiments.[13]

HERMIES-III is an autonomous robot comprising a seven-degree-of-freedom manipulator designed for human-scale tasks, a laser range finder, a sonar array, an omnidirectional wheel-driven chassis, multiple cameras, and a dual computer system containing a 16-node hypercube expandable to 128 nodes (see Figure 3). The experimental program involves performance of human-scale tasks (for example, valve manipulation and use of tools), integration of a dexterous manipulator and platform motion in geometrically complex environments, and effective use of multiple cooperating robots (HERMIES-IIB and HERMIES-III). The environment the robots will operate in has been designed to include valves, pipes, meters, obstacles on the floor, valves occluded from view, and paths of differing navigational complexity. The ongoing research program supports the development of autonomous capability for HERMIES-IIB and III to perform complex navigation and manipulation under time constraints while dealing with imprecise sensory information. □

Acknowledgments

This article describes the work of many members of the CESAR team. We wish to acknowledge in particular the contributions of W.R. Hamel, W.W. Manges, S.M. Killough, R.R. Feezell, and D.H. Thompson to the design and construction of HERMIES-IIB and HERMIES-III. D.L. Barnett, B.L. Burks, R. Fryxell, M. Goldstein, N.S.V. Rao, A. Sabharwal, and N. Sreenath performed the navigation research and contributed to the design of the robots' architecture. P.S. Spelt, E. Lyness, and G. Oliver developed the learning program. M. Beckerman, J.P. Jones, and R.C. Mann conducted the research in sensor interpretation and concurrent computation. R.D. Lawson and L.C. Whitman typed the article, and E.S. Howe did the coordination.

We also acknowledge the support of O. Manley of the Engineering Research Program of the Office of Basic Energy Sciences, US Department of Energy, which sponsored the described research under contract DE-AC05-840R21400.

References

1. B.L. Burks et al., "Autonomous Navigation, Exploration and Recognition," *IEEE Expert*, Vol. 2, No. 4, Winter 1987, pp. 18-27; also Tech. Report CESAR-87/25, Oak Ridge National Laboratory.

2. C.R. Weisbin, G. de Saussure, and D.W. Kammer, "Self-Controlled: A Real-Time Expert System for an Autonomous Mobile Robot," *Computers in Mechanical Engineering*, Vol. 5, No. 2, Sept. 1986, pp. 12-19; also Tech. Report CESAR-86/25, Oak Ridge National Laboratory.

3. S.H. Whitesides, "Computational Geometry and Motion Planning in Computational Geometry," G.T. Toussaint, ed., Elsevier Science Publishing Co., New York, 1985 (see references therein).

4. "Intelligent Systems and their Applications," theme feature articles, *IEEE Expert*, Vol. 2, No. 4, Winter 1987.

5. M. Beckerman and E.M. Oblow, "Treatment of Systematic Errors in the Processing of Wide Angle Sonar Sensor Data for Robotic Navigation," to be published in *IEEE J. Robotics and Automation*.

6. R.C. Mann, "Multi-Sensor Integration Using Concurrent Computing," *Proc. SPIE Symp. Infrared Sensors and Sensor Fusion*, Vol. 782, Int'l Soc. Optical Eng., Bellingham, Wash., 1987, pp. 83-90; also Tech. Report CESAR-87/18, Oak Ridge National Laboratory.

7. R.C. Fryxell, "Navigation Planning Using Quadtrees," Tech. Memorandum 10481, Oak Ridge National Laboratory, Nov. 1987.

8. M. Goldstein et al., "3-D World Modeling Based on Combinatorial Geometry for Autonomous Robot Navigation," *Proc. 1987 IEEE Int'l Conf. Robotics and Automation*, Vol. 2, CS Press, Los Alamitos, Calif., pp. 727-733; also Tech. Report CESAR-86/51, Oak Ridge National Laboratory.

9. S.V.N. Rao et al., "Robot Navigation in an Unexplored Terrain," *J. Robotic Systems*, Vol. 3, No. 4, 1986, pp. 389-407; also Tech. Report CESAR-86/28, Oak Ridge National Laboratory.

10. A. Elfes, "Sonar-Based Real-World Mapping and Navigation," *J. Robotics and Automation*, Vol. RA-3, No. 3, June 1987, pp. 249-265.

11. R.S. Michalski, J.G. Carbonell, and T.M. Mitchell, *Machine Learning, An Artificial Intelligence Approach*, Vol. 2, Morgan Kaufmann, Los Altos, Calif., 1986.

12. T. Mitchell, "Generalization as Search," *Artificial Intelligence*, Vol. 18, No. 2, 1982.

13. C.R. Weisbin et al., "HERMIES-III: A Step Toward Autonomous Mobility, Manipulation and Perception," NASA Conf. Space Telerobotics, Jan. 31 - Feb. 2, 1989, to be published by Jet Propulsion Laboratory, Pasadena, Calif., and by *Robotica J.*

Charles R. Weisbin is director of the Robotics and Intelligent Systems Program at Oak Ridge National Laboratory. He is also section head of the laboratory's Mathematical Modeling and Intelligent Control Section and an associate professor at the University of Tennessee. Weisbin received his degree in nuclear engineering from Columbia University. He has extensive experience in data and computer model validation, and he contributed to and edited *Sensitivity and Uncertainty Analysis of Reactor Performance Parameters* (Plenum Press, 1982). An IEEE Computer Society member, he was program chairman for the IEEE Second International Conference on Artificial Intelligence Applications and is an *IEEE Expert* editorial board member.

Gerard de Saussure, a staff scientist for Oak Ridge National Laboratory's Engineering Physics and Mathematics Division, has worked for over 30 years in experimental nuclear physics. Currently he divides his time between nuclear physics and research in strategy planning and machine intelligence at the Center for Engineering Systems Advanced Research. He is also an honorary professor at the University of Tennessee. He received his PhD in experimental physics from the Massachusetts Institute of Technology.

Francois G. Pin heads the Machine Reasoning and Automated Methods Group at Oak Ridge National Laboratory. He also leads the machine intelligence and advanced computing systems activities of the Robotics and Intelligent Systems Program and is a principal investigator at the Center for Engineering Systems Advanced Research. His technical interests include high-level planning, reasoning, problem solving and learning for autonomous mobile systems, and man-machine symbionts. He received his MS and PhD degrees in mechanical engineering from the University of Rochester, New York.

J. Ralph Einstein is a member of the Advanced Computer and Integrated Sensor Systems Group of the Center for Engineering Systems Advanced Research at Oak Ridge National Laboratory. His principal current interest is concurrent computing for autonomous robotics. He received a BS in physics from Yale University and a PhD in biophysical chemistry from Harvard University. Einstein is a member of the IEEE Computer Society.

Ewald Heer is president of Heer Associates, an engineering consulting group. He organized two international conferences on robotic systems, in 1972 at Caltech and in 1975 at USC. In addition, he organized the NASA Study Group on Machine Intelligence and Robotics in 1977. He is a fellow of the American Society of Mechanical Engineers and has published numerous articles and books on machine intelligence and robotics. He received an MS from Columbia University and a doctorate in engineering science from the University of Hannover, Germany. Heer is a member of the IEEE Computer Society.

The authors can be contacted at the Robotics and Intelligent Systems Program, Oak Ridge National Laboratory, PO Box 2008, Oak Ridge, TN 37831-6364.

Blanche: Position Estimation for an Autonomous Robot Vehicle

Ingemar J. Cox

AT&T Bell Laboratories
Murray Hill, New Jersey 07974

ABSTRACT

This paper describes the position estimation system for an autonomous robot vehicle called Blanche, which is designed for use in structured office or factory environments. Blanche is intended to be low cost, depending on only two sensors, an optical rangefinder and odometry. Briefly, the position estimation system consists of odometry supplemented with a fast, robust matching algorithm which determines the congruence between the range data and a 2D map of its environment. This is used to correct any errors existing in the odometry estimate. The integration of odometry with fast, robust matching allows for accurate estimates of the robot's position and accurate estimates of the robot's position allow for fast, robust matching. That is, the system is self sustaining.

The vehicle and associated algorithms have all been implemented and tested within a structured office environment. There is no recourse to passive or active beacons placed in the environment. The entire autonomous vehicle is self contained, all processing being performed on board. We believe this vehicle is significant not just because of the sensing and algorithms to be described, but also because its implementation represents a high level of performance at very low cost.

1. INTRODUCTION

A key to autonomy is navigation, i.e. an accurate knowledge of position. By position, we mean the vehicle's (x,y,θ) configuration with respect to either a global or local coordinate frame, *not* topological position, e.g. to the left of the wall. Dead reckoning using inertial guidance and odometry sensors drift with time so that the estimated position of the vehicle becomes increasingly poor. This complicates the process of position estimation. In order to correct for these cumulative errors, the vehicle must sense its environment and at least recognize key landmarks. Using sensory information to locate the robot in its environment is the most fundamental problem to providing a mobile robot with autonomous capabilities.

This paper describes the position estimation system employed by the vehicle. Section (2) begins with a brief overview of the vehicle and its guidance system. Section (3) then discusses in detail the navigation (position estimation) subsystem employed by the vehicle. Included in this is a description of the robot's map representation, Section (3.1), the algorithm used to match the range data to this map, Section (3.3), as well as some implementation details. Section (3.4) discusses how the odometry position estimate is combined with the correction estimated by the matcher. The system has been implemented on the vehicle and experimental results are included in Section (4).

2. OVERVIEW OF BLANCHE

Blanche [3] is an experimental vehicle intended to operate autonomously within a structured office or factory environment. It is designed to be low cost, depending on only two sensors, an optical rangefinder and odometry. Blanche, shown in Figure (1), has a tricycle configuration consisting of a single steerable drive wheel at the front and two passive rear wheels. The vehicle is powered by two sealed 12V 55Ah batteries which, in the current configuration, provide a useful lifetime of approximately seven hours. Control of the cart is based on a Multibus system consisting of a MC68020 microprocessor with MC68881 math coprocessor, 2 Mbyte of memory, an ethernet controller, a custom two-axis motor controller and an analogue-to-digital convertor. The Motorola 68020 runs a real-time UNIX® derived executive called NRTX [10].

The cart is equipped with two primary sensors: odometry on each of

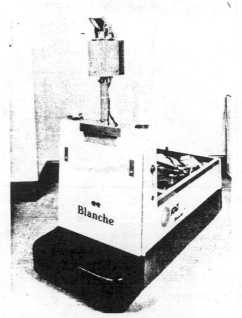

Figure 1: Blanche, an autonomous robot vehicle.

the two rear wheels and an optical rangefinder. Both sensors are extremely low cost (under $1000 each for components), and together provide all the navigational sensing information available to the cart. The advantage of odometry is, of course, that it is both simple and inexpensive. However it is prone to several sources of errors. First, surface roughness and undulations may cause the distance to be over estimated. Second, wheel slippage can cause distance to be under estimated. Finally, variations in load can distort the odometer wheels and introduce additional errors. If the load can be measured, then the distortion can be modelled and corrected for [17]. Where appropriate, a simple and more accurate alternative is to provide a pair of knife edge, non-load bearing wheels solely for odometry. Blanche uses this approach. In addition, even very small errors in the vehicle's initial position can lead to gross errors over a long enough path. Consequently, it is imperative that the vehicle's environment be sensed.

A simple low cost time-of-flight optical rangefinder has been developed specifically for cart navigation [12]. The rangefinder uses an approximately 1″ diameter beam and a rotating mirror to provide 360° polar coordinate coverage of both distance and reflectance out to about 15 feet. Both radial and range resolution correspond to about 1 inch at a ranging distance of 5 feet, with an overall bandwidth of approximately 1 kHz. Figure (2) shows a typical range map of a room obtained from a single scan of the rangefinder. A scan typically takes about one second. Each point is represented by its corresponding region of uncertainty, denoted by a circle of radius twice the standard deviation in the measurement. It should be pointed out that the error due to assuming that the range output is linear with distance, may sometimes exceed the error due to noise in the rangefinder. This systematic error can be removed by using a table look up technique to accurately map range output into distance.

The control of Blanche can be classified into three main components; path planning, guidance (trajectory generation and low level control) and navigation (position estimation). These components are illustrated in Figure (3).

Reprinted from the *IEEE/RSJ International Workshop on Intelligent Robots and Systems*, 1989, pages 432-439. Copyright © 1989 by The Institute of Electrical and Electronics Engineers, Inc.

EH0342-6/91/0000/0285$01.00 © 1989 IEEE

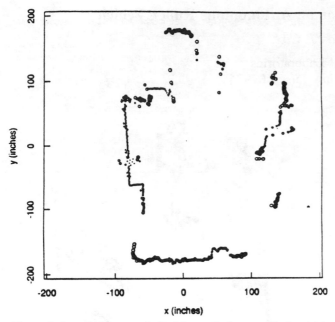

Figure 2: A range data scan obtained in a typical room. (Each point is denoted by twice its standard deviation).

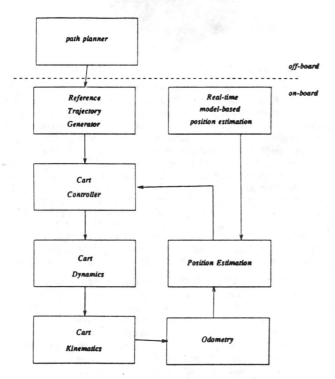

Figure 3: Block diagram of the overall control system.

The path planner [19] is an off-line program which generates a series of collision free maneuvers, consisting of line and arc segments, to move the vehicle from a current to a desired position. Since the cart has a minimum turning radius (approximately 2 feet), it is not possible to simply turn on the spot and vector to the desired position as is the case for differentially driven vehicles.

This path is downloaded to the vehicle, which then navigates along the commanded route. The line and arc segments specifications are sent to control software consisting of low level trajectory generation and closed-loop motion control [13]. Briefly, the reference state generator takes each segment specification from which it generates a

reference vector at each control update cycle (every 0.1 secs). The cart controller controls the front steering angle and drive velocity using conventional feedback compensation to maintain small errors between the reference and measured states.

3. NAVIGATION (POSITION ESTIMATION)

Navigation can be broadly separated into two distinct phases, reference and dead reckoning, as discussed in [6]. Reference guidance refers to navigation with respect to a coordinate frame based on visible external landmarks. Dead reckoning refers to navigation with respect to a coordinate frame that is an integral part of the guidance equipment. Dead reckoning has the advantage that it is totally self contained. Consequently, it is always capable of providing the vehicle with an estimate of its position. Its disadvantage is that the position error grows without bound unless an independent reference is used to periodically reduce the error. Reference guidance has the advantage that the position errors are bounded, but detection of external references or landmarks and real-time position fixing may not always be possible. Clearly inertial and external reference navigation are complementary and combinations of the two approaches can provide very accurate positioning systems.

Position estimation based on the *simultaneous* measurement of the range or bearing to three or more known landmarks is well understood [15]. However, recognizing naturally occurring reference points within a robot's environment is not always easy due to noise and/or difficulties in interpreting the sensory information. Placing easy to recognize beacons in the robots workspace is one way to alleviate this problem. Many different types of beacons have been investigated including (i) corner cubes and laser scanning system, (ii) bar-code, spot mark or infra-red diodes [16] and associated vision recognition systems [9] and (iii) sonic or laser beacon systems. We chose not to rely on beacons, believing that the ability to operate in an unmodified environment was preferable from a user standpoint.

There have been many efforts to use high level vision to navigate by, particularly stereo vision [1], [8]. However, conventional vision systems were ruled out because of the large computational and associated hardware costs: We want the vehicle to be economic.

Figure (4) is an overview of Blanche's position estimation system. It consists of:

1. An *a priori* map of its environment.

2. A combination of odometry and optical range sensing to sense its environment.

3. An algorithm for matching the sensory data to the map [4].

4. An algorithm to estimate the precision of the corresponding match/correction [5] which allows the correction to be optimally (in a maximum likelihood sense) combined with the current odometric position to provide an improved estimate of the vehicle's position.

Provided the error models are accurate, the combined position estimate is less noisy than any one of the sets of individual measurements. The sensor integration process can, of course, be routinely mechanized by use of the Kalman filter [11]. The Kalman filter is not explicitly used in this system, but equivalent results are obtained.

3.1 Map Representation

Many spatial representations have been proposed. However, it is worth reflecting on the purpose of a map representation. Our purpose is to compare sensed range data to a map in order to refine our position estimate. The map is *not* intended to be used for path planning, it is not even necessarily intended to be updated by sensory data. It's sole purpose is for position estimation in an absolute coordinate frame. While many spatial representations have been proposed few appear to have been tested on a real vehicle. One major exception is occupancy grids [7]. Occupancy grids represent space as a 2- or 3D array of cells, each cell hold an estimate of the confidence that it is occupied. A major reason given for not using a more geometric representation is that sensor data is very noisy making geometric interpretation difficult.

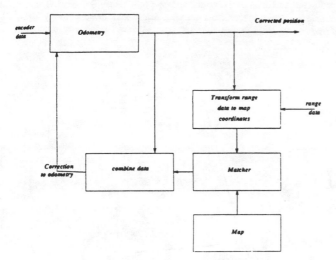

Figure 4: Block diagram of the navigation system.

This is especially true of sonar data (which Elfes and others used) and is one reason why sonar was not used on Blanche.

The infrared range data is much less noisy. This, combined with the fact that factory or office buildings are easily described by collections of line segments also influenced our choice of representation. We represent the environment as a collection of discrete line segments in the plane. A 2D representation was chosen because (i) much of the robot's environment is uniform in the vertical direction; there is not much to be gained from a 3D representation, (ii) the range sensor currently provides only 2D information (r, θ), (iii) matching sensor data to 2D maps is significantly simpler than matching to 3D maps. But above all else, a line segment description was chosen because a matching algorithm had been developed that used line segments. Moreover, the matching algorithm use (see Section 3.3) does not require the explicit extraction of any features from the image.

3.2 Sensor data

Sensing of the environment is quite straightforward, relying on odometry and an infrared rangefinder, as previously mentioned. The choice of a rangefinder was based primarily on the belief that ultrasonic rangefinders were poor, with severe problems due to specular reflection. A characteristic of the position estimation system described here is that the vehicle does not move under odometric control for a period, then stop to locate beacons, update its position and continue. Rather, its position is constantly being updated as the vehicle moves. Since range data is collected while the vehicle is moving, there is a need to remove any motion distortion that may arise. This is done by reading the odometric position at each each range sample. The current position and range value are then used to convert the data point into world coordinates for matching to the map.

If (r, α) is the range, angle pair from the rangefinder's local coordinate frame, it's position in world coordinates is given by

$$\begin{bmatrix} x \\ y \\ 1 \end{bmatrix} = \mathbf{C} \, \mathbf{R} \begin{bmatrix} r\cos(\alpha) \\ r\sin(\alpha) \\ 1 \end{bmatrix}$$

where $\mathbf{R}$ is the homogeneous transformation describing the relative position of the rangefinder with respect to the cart and $\mathbf{C}$ describes the vehicle's position with respect to the base coordinate system.

3.3 Matcher

We now describe how matching of range data to the map is achieved. First, an example. The solid line in Figure (5a) is the line segment description of a laboratory room in which Blanche is moved. The model, consisting of 24 line segments, was constructed very simply from measurements based on the one foot square floor tile grid. Item (a) is a tall desk, item (b) is a ventilation duct, item (c) is some cardboard posters, item (d) is a small refrigerator and item (e) a large

cupboard. It should be pointed out that the model is very simple; some items in the room are not modeled, and others are only roughly approximated, such as item (c), which in fact is made up of several cardboard sections all at different distances from the wall. The dotted points show the image based on range data acquired from a single scan of the sensor. In this example there is a small degree of rotation and a large translation, of the order of nine feet in x and eight feet in y. Figure (5b) shows the results of applying the algorithm. It is evident that the correct congruence has been found.

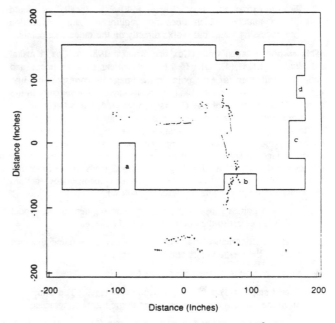

Figure 5a: Range image and associated map of room.

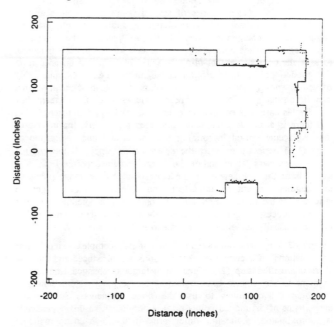

Figure 5b: Registered range image and associated map of room.

The general matching problem has been extensively studied by the computer vision community. Most work has centered upon the general problem of matching an image of arbitrary position and orientation relative to a model. Matching is achieved by first extracting features followed by determination of the correct correspondence between image and model features, usually by some form of constrained search.

Once the correspondence is known, determination of the congruence is straightforward.

The following observations motivated the derivation of the matching algorithm described below.

1. The displacement of the image relative to the model is small, i.e. we roughly know where we are at all times. This assumption is almost always true for practical autonomous vehicle applications, particularly if position updating occurs frequently.

2. Feature extraction can be difficult and noisy. Ideally, we would like a matcher which does not require a feature extraction preprocessing stage, but works directly on the range data points.

This section is taken from (Cox and Kruskal 1988 & Cox, Kruskal and Wallach 1988) [4], [5]. If the displacement between image and model is small, then for each point in the image its corresponding line segment in the model is very likely to be its closest line segment in the model. Determination of the (partial) correspondence between image points and model lines reduces to a simple search to find the closest line to each point. The central feature of the original conception was devising a method to use the approximately correct correspondence between image *points* and model *lines* to find a congruence that greatly reduces the displacement. Iterating this method leads to the following algorithm (for actual use, the algorithm incorporates certain computational simplifications as described below):

1. For each point in the image, find the line segment in the model which is nearest to the point. Call this the *target*.

2. Find the congruence that minimizes the total squared distance between the image points and their target lines.

3. Move the points by the congruence found in (2).

4. Repeat steps (1-3) until the procedure converges. The composite of all the step (3) congruences is the desired total congruence.

The rationale for our method of performing step (2) is explained with the aid of Figure (6). The model shown there consists of two line segments. The image consists of several points from both segments which have been displaced by translation up and to the right and by a slight rotation counterclockwise. For every point in the image, its target segment is the correct segment, as one would hope. It is natural to seek the congruence that minimizes the total squared distance from the model points to their target segments, i.e., the minimizing procedure tries to move the image points from each segment so that they lie on that segment, but it doesn't care where on the segment they go. In this case it is possible to reduce the total squared distance to 0, and there is a unique congruence that does so, namely, the inverse of the displacement used in forming the image, and one application of steps (1-3) perfectly recovers the desired congruence. If the original displacement was larger and/or the image contained points from near the ends of the line segments, the correspondence of image points to line segments might be imperfect, and several iterations might be required. If the image points contain some error, a potentially infinite iteration process might be required, though in practice only a few iterations usually achieve convergence to sufficient accuracy.

Step (2) is computationally the most complex step. For computational efficiency, two modifications are introduced and the new version is called step (2'). First, each target is changed from a line segment to the infinite line containing the segment. Note, however, that step (1) continues to use the finite segments. Second, the dependence of the moved points on θ is non-linear, so this dependence is approximated by the first order terms in θ. Such an approximate congruence is called a pseudo-congruence. Note, however, that step (3) continues to use a real congruence, namely, the congruence with the same $(t_x, t_y, 0)$ as the pseudo-congruence found in step (2'). Since the algorithm is iterative and the final iterations involve vanishingly small displacements, the approximations involved in these two modifications do not cause error in the final result.

Any congruence can be described as a rotation by some angle θ followed by a translation by t, and can thus be denoted by (t, θ).

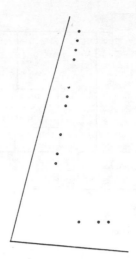

Figure 6: Simple model and displaced image.

However, instead of taking rotations around an arbitrary origin, we will take them around the center of gravity c of the image, for reasons that will be noted shortly. Thus the congruence (t, θ) maps any point

$$x \rightarrow R(\theta)(x-c)+(c+t),$$

where $R(\theta)$ is a clockwise rotation of angle θ, and can be denoted by the matrix

$$R(\theta) = \begin{bmatrix} \cos(\theta) & -\sin(\theta) \\ \sin(\theta) & \cos(\theta) \end{bmatrix}.$$

In practice, c will be the mean point (center of gravity) of the image. Using a center that moves with the image (such as its mean point or its k-th point) has the advantage that (t_1, θ_1) followed by (t_2, θ_2) is the same motion of the image as $(t_1+t_2, \theta_1+\theta_2)$. This does *not* hold for rotations around a point, such as the origin, that does not move with the image. Furthermore, an approximation that will be introduced shortly has an error that is proportional to the distance from the center to the point being rotated. Use of the mean point of the image is desirable to minimize this error.

We wish to find the value of (t, θ) that minimizes

$$S = \sum_i ([R(\theta)(v_i-c)+(c+t)]'\, u_i - r_i)^2.$$

where v_i are the image points, u_i their corresponding target lines and ' denotes transpose. However, because θ enters the image of v_i in a non-linear way, it is not evident how to do this in a rapid simple manner. Hence we introduce a linear approximation, namely,

$$R(\theta) = \begin{bmatrix} 1 & -\theta \\ \theta & 1 \end{bmatrix}.$$

Geometrically, this means that a point x is not moved in a circle around c, but is instead moved along the tangent to the circle. If x is at distance d from c, then instead of moving along an arc of angle θ (in radians), it moves along the tangent a distance of $d\theta$.

Now we proceed as we did above. The vector derivative of S with respect to t and the derivative of S with respect to θ are set equal to 0. After simplification, the following equations are obtained,

M_2	m_1		t		d_1
m_1'	m_0		θ	$=$	d_0

where

$$M_2 = \sum u_i\, u_i' \text{ (matrix)},$$

$$m_1 = \sum u_i\, (u_i'\, v_i) \text{ (vector)},$$

288

$$m_0 = \sum (\mathbf{u}_i{}' \mathbf{v}_i)^2 \text{ (scalar)},$$

$$\mathbf{d}_1 = \sum \mathbf{u}_i (r_i - \mathbf{u}_i{}' \mathbf{v}_i) \text{ (vector)},$$

$$d_0 = \sum (r_i - \mathbf{u}_i{}' \mathbf{v}_i)(\mathbf{u}_i{}' \mathbf{v}_i) \text{ (scalar)}.$$

3.3.1 Implementation details

The algorithm as described so far is intrinsically robust against incompleteness of the image, i.e. missing data points. To make it robust against spurious data, e.g. people walking by, and incompleteness of the model, it is modified further by deleting from consideration in step (2') points whose distance to their target segments exceed some limit. Note that for the general matching problem, in which the assumption of small displacement is not necessarily valid, such an approach would not be possible. Our experience is that robustness is very desirable, yet very hard to achieve. The robustness of this algorithm is one of its major features.

Step (1) of the algorithm requires that for each point, its corresponding (closest) line segment be found. Since the map is currently unordered, this entails exhaustively calculating the distance of *every* line from each point in order to find the closest line segment. If there are n point and m line segments, the complexity of this operation is $O(nm)$. However, if the map is first preprocessed to compute the corresponding Voronoi diagram (a step which takes $m\log(m)$ time) determination of the closest line segment can then be achieved in $O(n\log(m))$ [14] Presently, we have not implemented the Voronoi solution.

Since determination of the partial correspondence can quickly dominate the processing time, it pays to keep N and M small. We therefore restrict the number of image points to approximately 180 out of a possible 1000. Originally the map was small enough that we did not have to worry about the size of M, the number of line segments. However, this is no longer the case. Fortunately, our knowledge of the vehicle's position allows us to extract only those lines within say a 20' radius. It is this subset of the entire map, which often only numbers four or five, that is then used by the matching algorithm.

Currently, position updates occur approximately every 8 seconds for an image size of 180 points and a map of 24 line segments. This is sufficient for our requirements. However, it should be noted that the matching is the *lowest* level priority process running on the uniprocessor system. A considerable improvement in speed would be gained by simply dedicating a processor board to this task.

Finally, we note that the matcher returns an (x, y, θ) correction based on an origin at the centroid of the image/range scan. A coordinate transformation is therefore required to convert the matcher's centroid centered correction to a correction based on the vehicle's current position.

3.4 Integrating Odometric and Matched Positions

It is assumed that the x, y and θ position and orientation are independent of one another. We therefore describe how the x position is updated, y and θ being identical. We first need to estimate the standard deviation in the measurement of x for both the matcher and odometry.

3.4.1 Estimating matcher accuracy

This section is taken from (Cox, Kruskal and Wallach 1988) [5]. Let $\mathbf{b}$ be the vector of parameters $\mathbf{b} = (t_x, t_y, \theta)'$ that describes a congruence or a pseudo-congruence. An approach to estimating the variances of the three parameter estimators arises from the method described in Section 3 for forming the estimators. This method can also estimate the covariances as well.

The covariance matrix of the estimator $\hat{\mathbf{b}}$ of $\mathbf{b}$ is a 3×3 symmetric matrix whose diagonal elements are the variances of the three parameter estimators and whose off-diagonal elements are their covariances.

Now consider the method described in Section 3. Each time step (2') is performed, an estimate $\hat{\mathbf{b}}$ must be found. This estimate is calculated by solving a matrix equation

$$X'\mathbf{y} = (X'X)\mathbf{b} ,$$

where X and $\mathbf{y}$ are calculated from the data in an elementary way.

It is explained in Section (3.3) that after convergence is complete, the desired congruence is the composite of the $\hat{\mathbf{b}}$ vectors found in all repetitions of step (2'). The covariance matrix of this composite $\hat{\mathbf{b}}$ vector is the same as the covariance matrix of the final $\hat{\mathbf{b}}$ vector. This covariance matrix is estimated by $s^2(X'X)^{-1}$, where s^2 is the usual estimate of σ^2,

$$s^2 = (\mathbf{y} - X\hat{\mathbf{b}}) \cdot (\mathbf{y} - X\hat{\mathbf{b}})/(n-4) .$$

The covariance matrix is calculated as part of the overall calculation for x [5].

3.4.2 Estimating odometry accuracy

Estimation of the standard deviation for odometry is less rigorous. Presently, it is assumed to be directly proportional to the distance moved. Empirically, we have set the constant of proportionality to 0.01, i.e. after 100' the standard deviation is 1'. For a more rigorous treatment of odometric error see (C. Ming Wang 1988) [18].

3.4.3 Integration

Given (x_o, σ_o) and (x_m, σ_m) the position and standard deviation from odometry and matching, respectively, we can optimally combine these using the formulas [2], [11]

$$x_c = x_o + \frac{\sigma_o}{\sigma_o + \sigma_m}(x_m - x_o)$$

$$\frac{1}{\sigma_c^2} = \frac{1}{\sigma_o^2} + \frac{1}{\sigma_m^2}$$

where (x_c, σ_c) are the updated values for the x position and its corresponding standard deviation. It is clear that if the standard deviation in the odometry is small compared with the matcher, the error term $(x_m - x_o)$ has little effect. Alternatively, if the standard deviation in the odometry is large compared with the matcher, almost all the error term is added to the corrected value. This is intuitively correct.

This updated value is fed back to the odometry where it is used as the new value from which the current position is estimated. This is referred to as an *indirect feedback* configuration [11] in comparison to an *indirect feedforward* configuration, since the error correction is fed back into the odometry subsystem. In this way, the odometry errors are not allowed to grow without bound. Further, since the odometry estimate is corrected after each match or registration, failure of the rangefinder or matching subsystem does not lead to immediate failure of the robot system. Instead, the robot is free to continuing navigating for some period using only odometry until such time as the odometric estimates of the standard deviation in position exceed some unacceptable limits.

4. EXPERIMENTAL RESULTS

Figure (7) is a sequence of range images taken by the robot vehicle as it moved along a predetermined path. At initialization the vehicle was given its initial position. In practice, the vehicle had been displaced approximately 3" in its (x, y) position. Its orientation is approximately correct.

Table (1) tabulates the partial set of corrections made by the vehicle as it traverses its path.

TABLE 1. A sequence of corrections estimated by the matcher

correction	x (inches)	y (inches)	θ (degrees)
1	0.0	2.8	0.7
.	.	.	.
.	.	.	.
.	.	.	.
8	0.0	-0.1	0.3
9	-1.2	0.1	-0.6
10	6.7	-0.1	-0.5
11	-3.0	0.1	0.3

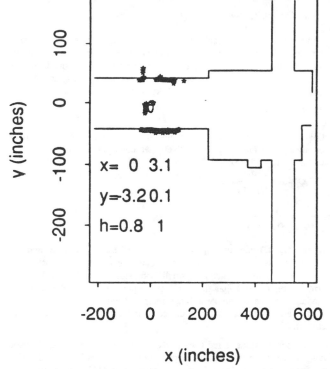

Figure 7a: Sequence (1) of range images taken as vehicle navigated along a corridor.

The first correction correctly determines the error in the y direction. A correction of 0.7° is also applied to the heading, but, examination of the heading corrections suggests that this is noise. However, there is no correction in the x direction. This is because the vehicle cannot tell where along the corridor it is[1]. The matcher has an infinite standard deviation along the x direction. Only when the vehicle comes out into the elevator bay — see scan (9) of Figure (7) — is it able to begin to refine its x position.

Large collections of range points not associated with the model usually indicate people walking by. In addition, maintenance and cleaning personnel may occasionally leave unmodeled objects in the environment. However, almost all these data points are rejected as spurious because their vicinity to a modelled line segment is not within preset thresholds.

The matcher's estimates of the standard deviations in (x, y, θ) assume a gaussian noise form with zero mean. However, the prototype rangefinder appears to have systematic errors: range values have been seen to depend on the strength of the received signal. For example,

very bright objects appear further away. We believe that the large corrections in the x direction at scans 9 through 11 are due in part to this problem. An improved rangefinder is currently under test which we hope will solve this problem. Other potential sources of systematic errors include errors in the map and errors in the alignment of the rangefinder relative to the vehicle. A misalignment of only 1° would cause an almost 2″ error in the y direction as the vehicle travelled 9′ along the x direction. This may account for why the sum of the corrections in the y direction totals to 7.8″ rather than to the approximately 3″ error expected.

5. CONCLUSIONS

This paper described the position estimation system of an autonomous robot vehicle for use in structured office or factory environments. The navigation system consists of (i) an *a priori* map of its environment consisting of a collection of discrete line segments in the plane, (ii) a combination of odometry and optical range sensing to sense its environment, (iii) an algorithm for matching the sensory data to the map. The precision of the corresponding match/correction is determined which allows the correction to be optimally combined with the current odometric position to improve the estimate of the vehicle's position. There is no recourse to passive or active beacons placed in the environment.

A 2D representation based on collections of line segments in the plane was chosen because much of the robot's environment is uniform in the vertical direction, the range sensor currently provides only 2-D information and matching sensor data to 2-D maps is significantly simpler than matching to 3-D maps.

The matching algorithm assumes that the displacement between image and model is small. This is a very reasonable assumption since odometry is providing the vehicle with a good estimate of position and matching occurs very frequently. The algorithm is intrinsically robust against incompleteness of the image, i.e. missing data points. Spurious data, e.g. people walking by, and incompleteness of the model, is dealt with by deleting from consideration any points whose distance to their target segments exceed some limit. Note that for the general matching problem, in which the assumption of small displacement is not necessarily valid, such an approach would not be possible.

In contrast to many papers presenting results of simulations, the vehicle and associated algorithms have all been implemented and tested within a structured office environment. The entire autonomous vehicle is self contained, all processing being performed on board. We believe this vehicle is significant not just because of the sensing and algorithms described, but also because its implementation represents a high level of performance at low cost. There also appears to be a self sustaining property to this configuration: Accurate knowledge of position allows for fast robust matching which leads to accurate knowledge of position.

There are several areas of in which the vehicle might be improved. First, any map has so far been assumed to be perfectly accurate, all errors being considered due to sensor noise or errors in the matcher. In practice, the map also has an associated accuracy which should also be modeled. Second, the need to provide a map of the environment can become tedious and it would therefore be desirable to automate this step, i.e. allow the vehicle to construct its own map. Third, the covariance matrix estimated by the matcher assumes there is no correlation between scans of the rangefinder. This is not necessarily true in practice because of (i) errors in the map and (ii) sensor noise may be correlated with particular wall surfaces. Future work is directed to addressing these problems.

Acknowledgements: The development of "Blanche" would not have been possible without the help of many individuals. It is a pleasure to thank J. B. Kruskal whose collaboration with the matching algorithm was critical to the success of the project. Also special thanks to G. L. Miller, and E. R. Wagner for the optical ranger. Finally, R. A. Boie, W. J. Kropfl, D. A. Kapilow, J. E. Shopiro, W. L. Nelson, F. W. Sinden, and G. T. Wilfong all provided help and assistance of one kind or another. Thank you.

---

1. Remember that doorways are not modeled. Locally, the environment appears as two parallel lines.

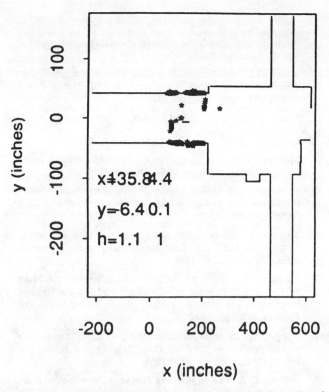

x=35.8 1.4
y=-6.4 0.1
h=1.1 1

Figure 7b: Sequence (8) of range images taken as vehicle navigated along a corridor.

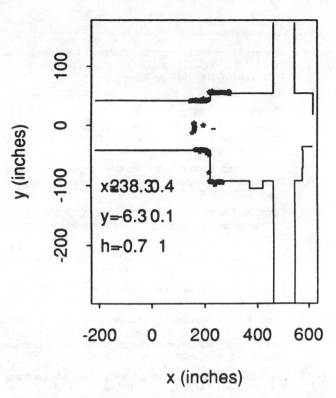

x=238.3 0.4
y=-6.3 0.1
h=0.7 1

Figure 7d: Sequence (10) of range images taken as vehicle navigated along a corridor.

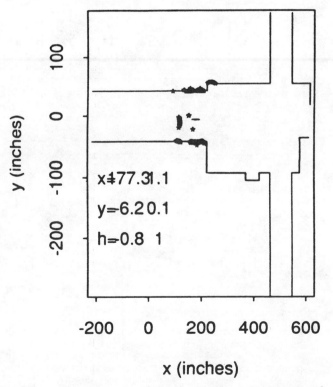

x=177.3 1.1
y=-6.2 0.1
h=0.8 1

Figure 7c: Sequence (9) of range images taken as vehicle navigated along a corridor.

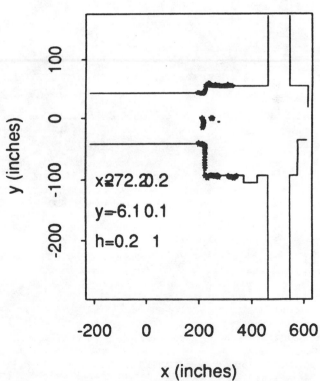

x=272.2 0.2
y=-6.1 0.1
h=0.2 1

Figure 7e: Sequence (11) of range images taken as vehicle navigated along a corridor.

REFERENCES

1. Ayache, N. and Faugeras, O. Building, Registrating, and Fusing Noisy Visual Maps. In *Int. Conf. Computer Vision*, IEEE, London, UK, 1987, pp. 73-79.

2. Bar-Shalom, Y. and Fortmann, T.E. *Tracking and Data Association*. Academic Press, 1988.

3. Cox, I.J. Blanche: An Autonomous Robot Vehicle for Structured Environments. In *IEEE Int. Conf. on Robotics and Automation*, IEEE, 1988, pp. 978-982.

4. Cox, I.J. and Kruskal, J.B. On the Congruence of Noisy Images to Line Segment Models. In *Int Conf. Computer Vision*, IEEE, 1988.

5. Cox, I.J., Kruskal, J.B., and Wallach, D.A. Predicting and Estimating the Performance of a Subpixel Registration Algorithm. 1988.

6. Cox, I.J. and Wilfong, G.T., Eds. *Autonomous Robot Vehicles*. Springer-Verlag, New York, to be published.

7. Elfes, A. Sonar-Based Real-World Mapping and Navigation. *IEEE J. of Robotics and Automation RA-3*, 3 (1987), 249-265.

8. Elfes, A. and Matthies, L. Sensor Integration for Robot Navigation: Combining Sonar and Stereo Range Data in a Grid-Based Representation. In *IEEE Conference on Decision and Control*, IEEE, 1987.

9. Giralt, G., Chatila, R., and Vaisset, M. An Integrated Navigation and Motion Control System for Autonomous Multisensory Mobile Robots. In *1st Int. Symp. on Robotics Research*, Bretton Woods, NH, USA, 1983, pp. 191-214.

10. Kapilow, D.A. Real-Time Programming in a UNIX Environment. *1985 Symposium on Factory Automation and Robotics* (1985), 28-29.

11. Maybeck, P.S. *Stochastic Models, Estimation, and Control*. Vol. 1. Academic Press, 1979.

12. Miller, G.L. and Wagner, E.R. An Optical Rangefinder for Autonomous Robot Cart Navigation. In *SPIE Mobile Robots II*, Vol. 852, SPIE, 1987, pp. 132-144.

13. Nelson, W.L. and Cox, I.J. Local path Control of an Autonomous Vehicle. In *IEEE Int. Conf. Robotics and Automation*, IEEE, 1988, pp. 1504-1510.

14. Preparata, F.P. and Shamos, M.I. *Computational Geometry: An Introduction*. Springer-Verlag, New York, 1985.

15. Torrieri, D.J. Statistical Theory of Passive Location Systems. *IEEE Trans. on Aerospace and Electronic Systems AES-20*, 2 (1984), 183-198.

16. Tsumura, T. Survey of Automated Guided Vehicle in Japanese Factory. In *IEEE Int. Conf. on Robotics and Automation*, IEEE, 1986, pp. 1329-1334.

17. Tsumura, T., Fujiwara, N., and Hashimoto, M. An Experimental System for Self-Contained Position and Heading Measurement of Ground Vehicle. In *Int. Conf. Advanced Robotics*, 1983, pp. 269-276.

18. Wang, C.M. Location Estimation and Uncertainty Analysis for Mobile Robots. In *IEEE Int. Conf. Robotics and Automation*, Vol. 2, IEEE, 1988, pp. 1230-1235.

19. Wilfong, G.T. Motion Planning For An Autonomous Vehicle. In *IEEE Conf. Robotics and Automation*, IEEE, 1988, pp. 529-533.

Dynamic Monocular Machine Vision

Ernst Dieter Dickmanns and Volker Graefe

Fakultät für Luft- und Raumfahrttechnik (LRT), Universität der Bundeswehr München, W-Heisenberg-Weg 39, 8014 Neubiberg, West Germany

Abstract: A new approach to real-time machine vision in dynamic scenes is presented based on special hardware and methods for feature extraction and information processing. Using integral spatio-temporal models, it bypasses the nonunique inversion of the perspective projection by applying recursive least squares filtering. By prediction error feedback methods similar to those used in modern control theory, all spatial state variables including the velocity components are estimated. Only the last image of the sequence needs to be evaluated, thereby alleviating the real-time image sequence processing task.

Keywords: 4-D machine vision, real-time image sequence processing, automatic visual motion control, vehicle guidance

1. Introduction

Dynamic vision is more than fast processing of static image sequences. The dynamics aspect rests primarily in the scene observed or in the motion of the sensor and is independent of the image frequency; as in any sampled measurement process, high sampling rates are necessary for recovering highly dynamical changes. In vision, however, in addition to this, high sampling rates reduce the so-called correspondence problem, that is, keeping track of special image features or objects in space from one frame to the next.

Address reprint requests to: Prof. Ernst D. Dickmanns, Steuer-und Regelungstechnik, Universität der Bundeswehr München, Fakultät für Luft- und Raumfahrttechnik, Institut Für Systemdynamik und Flugmechanik, Werner-Heisenberg-Weg 39, D-8014, Neubiberg, West Germany.

This research has been partially supported by the German Federal Ministry of Research and Technology (BMFT), the Deutsche Forschungsgemeinschaft (DFG), the Daimler-Benz AG, and by Messerschmitt-Bölkow-Blohm GmbH (MBB).

Note that humans, when talking about dynamic scenes, do not converse in image terms but do prefer spatial interpretations, both in position and velocity, whenever possible. They try to see motion of objects in space. Motion properties of objects are an integral part of a person's knowledge base like possible shapes and colors. Similarly in the approach described below, a direct spatial interpretation of image sequences is achieved by using spatial and temporal models in conjunction. This unified approach in space and time is the core of the 4-D method developed and tested for machine vision. Applications are discussed in a companion paper (Dickmanns and Graefe 1988, this issue; p. 241).

The immediate inclusion of temporal aspects is very essential since it allows a proper definition of state variables and the introduction of temporal continuity conditions for image sequence interpretation by exploiting differential equations. Geometric shape descriptions and generic models for motion *together* constitute the basis for an integrated spatio-temporal approach, which may be termed "4-D vision" or "dynamic vision."

This means that not just objects are being seen but motion processes of objects in space and time. Note that unlike "static" image sequence processing, dynamic vision has no separation between spatial object recognition from one frame to the next as a first step and motion reconstruction afterwards as a second one. Instead, object and motion are treated as a unit and the least squares fit for determining the best estimate for the object motion state, based on noise corrupted image sequences, is done in space and time simultaneously.

As a very beneficial side effect, the need for storing past images (e.g., for computation of displacement vector fields or optical flow) is reduced. The state of the scene observed is represented on a very high symbolic level by the shape descriptors and the spatio-temporal state variables including spatial velocity components as an integral part (state vector components).

This approach provides an efficient framework for data fusion and active control of the viewing direction. Angular rates are state variables directly and translational velocity components of the egomotion are time integrals of the corresponding accelerations; both may easily be sensed by inertial sensors. Vision and inertial sensors have complementary properties when used for state recognition under egomotion: High angular rates, causing motion blur in the imaging process, are easily measured inertially; slow drift rates, hard to detect inertially, are easily discovered optically. For this reason, many organic species have developed this sensor combination and corresponding control facilities, for example, in vertebrates, the vestibular/ocular measurement and control system (Dichgans et al. 1973; Bizzi 1974). Active gaze control, in addition, allows the anchoring of the viewing direction on relatively fast moving prominent features of an object and thereby reduces motion blur for this object. If this object happens to be of special interest, the deterioration induced for the viewing conditions of most other objects may be acceptable. This fixation mode of vision also is very common in biological systems. In machine vision, as of course in biological vision too, a precisely servoed gaze anchoring allows the reading of object angular position from the measurement of mechanical angles while object shape may be determined from a quasistationary image.

For these reasons, active fast control of the viewing direction by the interpretation process is considered essential for dynamic vision. Therefore, it has been included in the systems design from the beginning.

The observation that in biological systems the sense of vision seems to be intimately linked to active motion control has lead us to consider motion *control* as the proper entry point for developing machine vision.

In hindsight, this turned out to be the right decision since the dynamical models of modern control theory proved to become the cornerstone of the new method for dynamic vision.

Motion control in space requires spatial or stereo vision. How many cameras are most appropriate for stereo vision? This question is still unresolved. Putting emphasis on motion and temporal integration, we decided to use just one camera. Spatial ambiguities may be resolved through motion stereo over time. In the case of active motion control, movements may be planned and executed in a way allowing to disambiguate a situation. For vehicle con-

trol as discussed in the companion paper (this volume) it seems more favorable to devote a second camera to high resolution imaging for better farsight than to direct stereo.

The considerations described above have lead to an approach to machine vision different from the mainstream of vision research originating from digital image processing and artificial intelligence. Gennery (1981, 1982) has taken a similar approach. In recent years Broida and Chellappa (1986) and Rives et al. (1986) seem to be heading in the same direction. For a literature survey on image sequence processing see Nagel (1983).

The next section summarizes some basic considerations on computer architectures for dynamic vision. In section 3 the nonuniform low level image processing schemes are described, upon which the approach is based. The general method for the higher levels of dynamic vision is developed and explained in section 4. Section 5 very briefly presents application results. All four application examples treated so far have been performed with real image sequence processing hardware in the real-time loop. More details on the implementation of the 4-D method, together with a discussion of the hardware developed, are given in the companion paper.

Finally, in section 6, development perspectives for the future are discussed.

2. Computer Architecture for Dynamic Vision

There are basically two approaches to the design of a real-time vision system. One is what might be called the brute force approach, using extremely fast hardware elements and possibly a massively parallel structure, yielding a supercomputer with an impressive power in terms of the notorious MIPS (million instructions per second). The other one is to look for the inherent structure and, possibly, simplicity of the problem of dynamic vision, and to find a computer architecture which is well matched to the task of visual motion control.

The second approach is, indeed, feasible and has led to the construction of a family of multiprocessor systems specialized for dynamic vision. The architecture of these real-time systems is very different from that of a typical image processing system. In spite of their relative simplicity, they have proven to be a very powerful hardware basis for various real-world experiments where mechanical systems or vehicles were controlled by dynamic vision.

An important concept upon which to base the design of a dynamic vision system is temporal continuity. Usually, natural scenes change only gradu-

294

ally, and if two pictures of such a scene are taken within a few milliseconds they will normally be very similar to each other.

In order to understand how the temporal continuity of natural scenes can facilitate dynamic vision, assume that a first TV image of such a scene has just been interpreted. It is then rather easy to interpret the immediately following image, as the differences between the two are very small. This observation has important consequences for the design of a real-time vision system. It means that the task of dynamic scene interpretation becomes easier if the time spent on each image is reduced, and that the task becomes more difficult if the system is slower. Therefore, the cycle time of the low level vision subsystem should ideally be less than one frame period of the TV signal used, making it possible to evaluate every single image as it is delivered by the camera. (The higher levels of the vision system which operate on symbolic descriptions of the scene may use longer cycle times, depending on the dynamics of the machine to be controlled and of the objects in the scene.)

Another important aspect on which to base the architecture of hardware for dynamic vision for motion control is the desired output of the system: it is the behavior of a visually controlled machine, and not, as often in traditional static image processing, either another image or a fairly complete, perhaps even verbal, description of the image.

The appropriate behavior of a vision controlled machine typically depends on the presence and location, or absence of certain objects in its environment. The vision task is then clearly goal directed, the first subtask being to locate features in the image which are indicative of the presence and location of important objects. It seems obvious that such features in many typical situations occupy only a small fraction of the total area of each image (Figure 1). It suffices then to process only those areas of each image which actually contain relevant features.

In dynamic scene interpretation the location of all important features is usually known in advance and with fairly good precision from the interpretation of previous images. This means that, when interpreting the next image in the sequence, the search space in which the feature of interest should be looked for is small, and the feature can be rediscovered rather quickly if the search is indeed focused on this small search space. This leads to the probably most important point in the design of hardware for real-time vision: since nearly all the relevant information in the image is contained in a limited number of small regions the combined size of

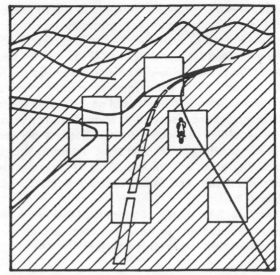

Figure 1. Small regions of an image contain almost all information relevant for motion control.

which is only a small fraction (often less than 10%) of the whole image, much will be gained if all the available computing power can be concentrated on those regions. Moreover, since each region may contain a different type of feature, it is important to be able to use different algorithms in each region.

This shows that a conventional image processing system which is designed to treat all pixels in an image in the same way does not have the proper structure for dynamic vision. The same is true for some massively parallel computers of the single instruction, multiple data (SIMD) type. Because these machines, too, must treat all pixels of an image in the same manner, they may waste 90% or more of their computing power on processing parts of the image which are known in advance to contain no relevant information. In the worst case, additional computing power is needed to delete all the irrelevant data which are produced in the process.

The concept of processing only a limited number of well defined regions within an image is also the key to a natural division of the problem into subtasks which can be executed in parallel on a coarsely grained multiprocessor system. Each parallel processor in such a system can be assigned one relevant region, and it can locate—independently of all other processors—the associated features in that region. Such a system not only has a very clear structure (one region—one group of features—one subtask—one processor), but it can also be very efficient, since the parallel processors do not have to spend time synchronizing or coordinating each other.

An important key to this concept is that the size, shape, and location of each region may be varied during the interpretation process in a data dependent way. Each region will normally be continuously adjusted in such a way as to completely contain a relevant feature or object. If the regions were fixed, the system would be much less efficient for two reasons. First, some regions would contain no relevant information but would nevertheless absorb computing power; secondly, some features or objects would be dissected by the borders between regions, creating the difficulty of detecting and interpreting arbitrarily dissected parts, representing them internally, and finally recombining them into objects.

Architectural details of a family of vision systems designed according to these concepts are given in the companion paper (this volume).

3. Feature Detection and Tracking Algorithms

In a dynamic vision system as discussed here, the time available for feature extraction is limited to about 50 ms per image. In order to evaluate every available image, the time should in fact be limited to less than one frame period of the TV camera (17 ms). An image contains (roughly) 10^5 pixels, and standard image processing methods require many operations per pixel [Reddy (1978) has estimated that 1000 operations per pixel are required for segmentation]. It is therefore obvious that speed is a most critical characteristic of any feature extraction algorithm for dynamic vision.

Two powerful approaches are available to maximize the speed of feature extraction in dynamic vision: application of advance knowledge, and a strict concentration on obtaining that, and *only that*, information which is necessary to accomplish the given task. In other words, only a relatively small number of carefully selected relevant features should be extracted depending on the situation and on the requirements of the task; knowledge should be applied to maximize the efficiency in processing the selected features.

Both approaches emphasize the difference between static image processing and dynamic vision. In static image processing very little is often known in advance of the image presented to the system. The task then is to extract as much information from the image as possible. This is very different from dynamic vision, where each new image is known to be a natural continuation of a sequence of images which the system has interpreted already; differences relative to the previous image are to be expected in small details only. Most of the time the feature extraction has to answer only a small number of precise questions relating to one or another of the small differences, such as how much and in which direction did a certain feature move in a small fraction of a second.

These two approaches will be discussed in more detail in the sequel.

3.1 Task Specific Feature Extraction

Limiting the number of features to be processed is not meant to exclude useful redundancy, which is absolutely necessary for any robust system, but rather to avoid wasting time or computing resources on processing irrelevant parts of a scene. All the available resources in a dynamic vision system should be concentrated on obtaining that information which is necessary, or at least helpful, to accomplish a certain task, usually the control of a moving system. The point is to extract only *task relevant* features and not every conceivably extractable feature. In the applications described in the companion paper, never more than about 10 features were needed, and 1000 features will probably be sufficient to handle rather complex scenes.

One of the problems which must be solved in the design of a dynamic vision system for a specific task is defining and selecting the relevant features. This is best done in a top-down approach, starting from the task the vision system is supposed to execute.

For balancing an inverted pendulum on an electric cart, which is a simple but typical example, it is sufficient to know the coordinates of two points on the rod as a function of time. If the coordinates of more points are available, this provides valuable redundancy which can be used to make the system more robust. On the other hand, nothing can be gained for the performance of the system by analyzing the background or the floor. All the available computing power should, therefore, be concentrated on locating various points of the rod.

Similarly, in the docking experiment discussed in the companion paper, the docking partner can be recognized and its relative position can be estimated by localizing corners of its contour in the image. Analyzing the entire contour might provide useful redundancy, but certainly all background features are irrelevant and should be ignored in the interest of efficiency.

Defining all features relevant for an autonomous vehicle in a natural environment is more difficult because of the great variety of situations it may encounter. It is, however, easy to see that large parts of a typical scene will never contain any relevant information, such as the mountains and the sky in Figure 1. Certain features will always be relevant, for instance, grey level edges which are characteristic of the borders of the road or lane, while it is not clear yet what kinds of other features may be relevant for the detection and classification of obstacles in certain situations. A pragmatic approach is to start with relatively simple scenes, such as an empty freeway, where the number of relevant features is small and their nature is obvious (borders of the road or lane), and then, as experience is accumulated, admit more complexity, like obstacles, other vehicles or intersections.

In any case, the key point is that the low level part of a dynamic vision system should process only those features which yield information actually required by the higher levels. The requirements of the higher levels are derived from the desired performance of the machine to be controlled.

3.2 Knowledge Based Feature Extraction

Typically, although there are a few exceptions, the appearance of a dynamic scene changes only gradually; this is due to the inertia and limited energy of all massive objects. In a sequence of TV images of such a scene it is, therefore, possible to predict the appearance of each new image from the previous images. Predicting the entire image would be expensive, and usually it suffices to predict the location, and possibly the appearance, of a limited number of features, like edges, corners, etc. The prediction will not be perfectly correct, but that is not necessary. All that is required is that the remaining search space, which corresponds to the uncertainty area of the prediction, be sufficiently small to complete the actual search for the feature within one video cycle. As most features in typical scenes move by at most one or two pixels between two successive images, a "zeroth order prediction," where it is assumed that the feature will reappear at the same location as in the last image, will often be sufficient. In exceptional cases with very fast moving objects (e.g., the inverted pendulum in the start-up phase or after an extreme external disturbance) a "first order prediction," which also takes the estimated velocity of the feature into account, may sometimes be more appropriate. It should be noted, however, that the motion blur caused by all normal TV cameras places a natural limit on the admissible velocity of features in an image.

Such a prediction-and-correction method has been the key to the success in balancing the inverted pendulum. It was used there in combination with an ad hoc method for locating the pendulum in the image, based on an anisotropic, nonlinear filter. The filter took into account the effects of motion blur, caused by the sometimes rapid motions of the system (Haas and Graefe 1983). Such a fairly sophisticated method was necessary to cope with such problems as camera shading, irregular lighting, camera and electronic noise, and the high speed of the mechanical system. Applying the filter in the entire image in real time would have exceeded the capabilities of even very powerful computers. By predicting the location of the pendulum in the image using either a zeroth or a first order prediction, depending on momentary velocity, the search space was reduced to less than 1% of the image. Two eight-bit microprocessors were then sufficient for the task.

As a more general realization of the prediction-correction concept, the method of "controlled correlation," sometimes also referred to as "intelligent correlation," has been developed (Kuhnert 1986a, 1986c, 1988). First real-time results obtained with this method (road tracking and obstacle detection from within a simulated autonomous vehicle) were reported in Kuhnert and Zapp (1985).

Correlation is the basis of many visual trackers. It is, however, not often used in computer vision, probably because it is considered computationally expensive. In its discrete non-normalized form, the 2-D correlation function C is defined as

$$C_{i,j} = \sum_k \sum_l I_{i+k,j+l} \cdot M_{k,l}$$

$$k = -K \ldots K; l = -L \ldots L$$

The essence of correlation is that a 2-D reference pattern M (the "mask"), which is usually much smaller than the image, is laid over the image I (2-D array of gray-level values), where each element of the mask is multiplied with the corresponding pixel of the image, and the products are summed, yielding a correlation value corresponding to the position (i,j) of the mask. The process is then repeated for all positions of the mask relative to the image, yielding a correlation function. If a region in the image resembles the reference pattern, the correlation function has a peak at that location. Thus, correlation can be used to find such regions.

If there are n pixels in the image and m elements in the mask ($m \ll n$) the computation of the correlation function requires almost $m \cdot n$ multiplication, which usually is a very large number. After the correlation function has been computed, it has to be searched for the relevant peaks. Because of the large number of correlation values this search is expensive, too.

On the other hand, correlation has some very desirable properties. It is very flexible in the sense that any pattern can be used as a mask and thus be looked for in the image. Most importantly, however, correlation is robust against noise. From communication theory it is known that correlation is the best linear method to detect a signal in the presence of additive ergodic white noise.

Noise is a severe problem in feature extraction. It may exist in many forms, for example, camera noise (time varying and fixed pattern), electronic noise in the analog part of the vision system, rounding errors in the digital part, irregular lighting, shadows, dirt covering parts of a scene, or branches of trees moving in the wind. It causes many feature extraction methods, which work well in noise-free synthetic images, to break down in natural scenes. Not all the kinds of noise which impede the processing of natural scenes can be considered additive, ergodic, and white. But, nevertheless, correlation is certainly a good candidate for a noise resistant feature extraction method. This is supported by results of an investigation (Kuhnert 1984) where several edge detectors were compared with the human visual system with respect to their ability to detect edges in synthetic pseudo-noise images. The correlation-based operators reproduced the abilities of the test persons more closely than any of the other ones. This observation adds to the attractiveness of correlation, since the human visual system is one of the best vision systems known.

Several types have been taken to reduce the computational cost of feature extraction by correlation, leading to "controlled correlation" and making it a very efficient method for dynamic vision. The first step is to correlate the mask only with a small region of interest which includes the predicted location of the feature, rather than with the entire image. This alone can reduce the required effort by more than two orders of magnitude.

If only elementary features such as edge elements are searched, the correlation masks can be small and, moreover, a small set of masks is sufficient to cover all possible orientations of short edge elements. If, on the other hand, complex features or even images of entire physical objects are looked for, the masks, in general, will be larger and many different ones will be needed to cover various sizes, orientations, aspect angles, and illumination conditions of a single class of feature or object. In regard to efficiency, short edge elements are therefore excellent features to base an image analysis on. They may be used in subsequent steps to construct longer edges, while still higher levels in the system may use knowledge to combine several edges, reconstructing 2-D images of objects and finally the 3-D objects and their motions. This result perhaps bears

a relationship with the findings of Hubel and Wiesel (1959) indicating that vertebrates also have receptive fields in their visual system which are tuned to short edge elements with specific orientations.

Depending on the situation, choosing an appropriate path along which to look for a correlation peak often helps to gain additional advantages. It should be remembered that one is not really interested in the entire correlation function, but only in the location of that peak which corresponds to the correct feature, or, considering the effects of noise and of possibly existing false features in the vicinity of the desired one, of a small number of candidates for the correct peak. If a good search path is chosen, all good candidates for the correct peak can be found quickly and the search can then be discontinued.

Correlation masks often contain elements whose absolute magnitude is much smaller than that of others. The correlation function, and in particular the locations of its peaks, do not change much if these small values are replaced by zero. It is possible to implement the correlation method in such a way that mask elements of value zero are skipped in the execution of the program and do not cause any operation of the computer at all. Setting many mask coefficients to zero will then reduce the number of operations to be executed, making the resulting algorithm faster. Masks of this type are called "sparsely populated." It is important to notice that, when sparsely populated masks are used, the computational cost of the algorithm does not depend on the extent of the mask, but only on the number of its nonzero elements.

Most correlation algorithms cause the computer to spend much time generating addresses and checking loop termination conditions. If all addresses are computed during compilation and if loops are avoided altogether, a much faster program results, consisting of only one (very long) linear piece of code. An additional step of optimization applies only to computers like the BVV 2, which require much more time for a multiplication than for an addition (the coprocessor of the BVV 3 is different, it multiplies and adds simultaneously). The fact is utilized that even very coarsely quantized correlation masks, if they meet certain symmetry conditions, are almost as effective as finely quantized masks (Kuhnert 1988). This is true even for ternary masks which contain only elements with values of -1, 0, and $+1$. Such masks can be realized by algorithms which perform subtractions and additions only.

Figure 2 shows an example of controlled correlation as it is used to track the left shoulder of a road

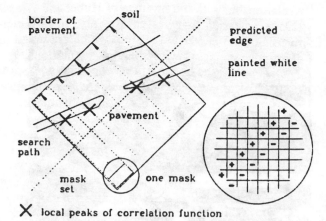

border of pavement

soil

predicted edge

painted white line

pavement

search path

mask set

one mask

✕ local peaks of correlation function

Figure 2. Using controlled correlation in order to find the left shoulder of a road. [The figure shows a very small section of the total image where the dark road surface borders the brighter soil. In the mask, "+" represents a value of +1 and "−" a value of −1. The five search paths together constitute the region of interest.]

from within a vehicle; the figure shows a small section of the image. A white line is painted on the road near the border of the pavement. The feature looked for is the right edge of the painted white line. Its *predicted* location and direction are indicated as a dashed line. The region of interest is centered around the predicted edge; its size is 2016 pixels (3% of the entire image). The mask is also indicated in Figure 2; it contains 14 nonzero elements. The correlation function is initially computed along a search path beginning near the lower corner of the region of interest and ending near its left corner. The search path is a straight line perpendicular to the predicted edge. Of the peaks found on the search path either the first peak of significant size or the "best" peak is accepted as a border point.

In natural scenes it is sometimes difficult to discriminate between a valid peak and noise effects or irregularities in the scene. Therefore, several algorithms are available to select a "best" peak of the correlation function; one of them may be chosen during runtime according to the situation.

All masks constituting the initial search path form a mask set. The search is repeated four more times on parallel search paths, using the same mask set, shifted to the upper right by six pixels each time, yielding altogether five border points.

In Figure 2 it is assumed that the white line is not clearly visible everywhere. It is, therefore, missed by the third search path, and the border of the pavement is found instead. It happens frequently in dynamic vision that a feature is missed temporarily and it is questionable whether an error-free feature extraction is at all possible. Fortunately, such er-

rors usually do not persist and in one of the next few images the lost feature or an equivalent one will be available again. In reality any robust system will have to use redundancy and world knowledge to handle such errors as a matter of routine. In any case it is up to the higher levels of the system to handle such a situation, for example, to recognize the outlying point and to eliminate it, or to tolerate the error. A more sophisticated version of the feature extractor could be designed which would not accept the edge of the pavement instead of the white line (Kuhnert 1986c); it would, however, run more slowly. It should be realized that, because of the transient nature of such errors, in dynamic vision the total system performance may well be better with a fast algorithm for feature extraction which occasionally makes errors, than with a less error prone, but slower algorithm.

The controlled correlation as described requires less than 33 ms on the 8086 microprocessor of the BVV 2. A similar algorithm employing only three paths instead of five runs in less than 16 ms. Various versions of this method have been used in different applications, among them the automatic aircraft landing and the autonomous road vehicle described in the companion paper. They have also been tested extensively in laboratory simulations where videotapes and 8 mm films taken from within a landing airplane and from cars driving on freeways or in cities, were played back and analyzed in real time (Kuhnert 1986a; Eberl 1987). These experiments were greatly simplified by the robustness of the algorithms. It was not even necessary to synchronize the film projector with the TV camera of the vision system which picked up the projected images from a screen; the features could be tracked reliably in spite of the severe fluctuations in brightness of the digitized TV images caused by the lack of synchronism.

Controlled correlation can also be used to track features other than edge elements, for instance, corners. Figure 3 shows an example (Kuhnert 1988) based on a very simple mask. Applying the mask once takes about 50 μs on an 8086 microprocessor. The mask works quite well, even if a corner is rotated slightly or if the enclosed angle is not exactly 90 deg. The problem is that the response it gives for a perfect match is only twice as high as the response for, say, a straight horizontal edge. This lack of selectivity is both an advantage and a disadvantage. If a mask is not very selective it will yield many false responses in a natural scene. If the mask is very selective (such a mask can easily be constructed by increasing its size), very large mask sets are needed to cover all possible angles and orienta-

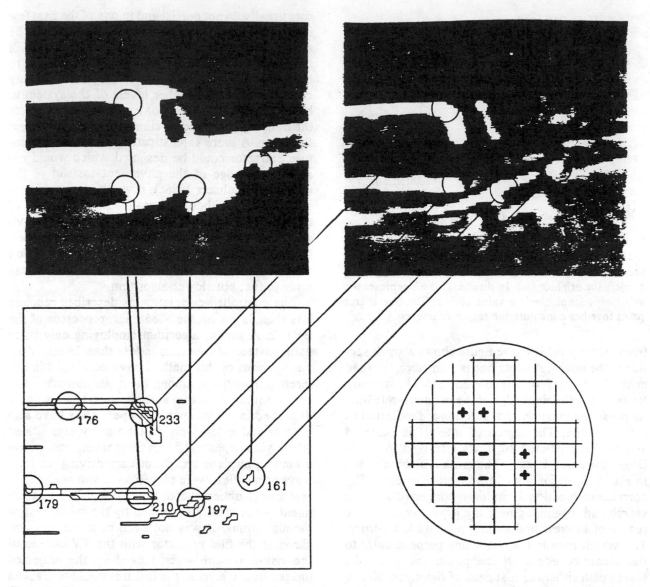

Figure 3. Detection of corners in a noisy image of a natural scene by correlation with a small ternary mask. Upper left: Original image (photo from the monitor of an image processing system); the figure shows only a small section of the entire image. Lower right: The correlation mask; "+" represents a value of +1 and "−" a value of −1; the shading shows the theoretical prototype corner of the mask. Upper right: Visualization of the correlation function; bright areas correspond to positive values and dark areas to negative values. Lower left: selected contours and the greatest maxima of the (non-normalized) correlation function; the numbers indicate the magnitudes. [The image stems from an 8-mm movie, taken from within a moving car; it has been projected onto a screen, picked up with a TV camera (262 TV lines) and digitized with 8 bits of resolution.]

tions of corners. This problem has to be investigated in greater detail, but probably it is better to look for two intersecting edges separately and then compute the point and angle of intersection, rather than looking for each specific type of corner directly.

Corners have been used as features in the vehicle docking experiment (Wünsche 1987; Kuhnert 1988). The method was knowledge based; however, it was not based on correlation. The contours of visible objects were extracted and maxima in their

"curvature" were taken as corner points. This method was implemented on the older BVV 1 and cannot be generalized easily. It is, however, also based on the concept of prediction and correction, predicting the location of the corners in the image and analyzing only those parts of the contours which are close to the predicted locations.

3.3 Feature Detection
So far it has been assumed that the position of a feature in the image can be predicted using the

knowledge gained from processing the immediately preceding images. This is usually true, but there are at least three exceptions: the initialization phase, the recovery after a feature has been lost, and the appearance of a new object in the scene. What makes the initialization phase manageable, is the fact that it is not time critical. It is difficult to make general statements about the initialization phase, but enough scene- or task-specific knowledge can usually be built into the system to avoid searching the entire image for each single feature that is needed. In the case of the inverted pendulum, a horizontal search path through the center of the image will find the pendulum, and in the case of a road vehicle it can, perhaps, be assumed that the vehicle is initially standing on a road, oriented nearly parallel to it. This limits the regions in the image, where the borders of the lane can be expected, sufficiently for efficient search.

It is normal for a feature being tracked to get lost once in a while, for example, the center edge element in Figure 2 or the border of the road when passing under a bridge while the auto-iris has not had time to adjust to the darkness. Usually, a lost feature will soon reappear near the location where it has been found last or near a location which can be easily predicted using adjacent features. If this does not happen soon enough, or if the feature tracker locks on to a wrong feature, higher levels in the system, which have a more comprehensive knowledge of the situation, must guide the feature extraction level to reacquire the lost feature. Wünsche (1983, 1987) has shown that this can be done very effectively, yielding a remarkably robust system.

The discovery of objects which suddenly enter the scene is of a different nature and often difficult, in particular if little is known regarding the visual appearance of the new objects and the region of the image where they will first be visible. An example is the discovery of obstacles which might obstruct the path of a vehicle. What makes this class of problems particularly difficult is that, unlike in the initialization phase, only a small amount of time is available.

4. Feature Based 4-D Dynamic Scene Analysis Using Integral Spatio-Temporal World Models

In the previous section it has been shown how simple elementary features can be extracted efficiently in a robust manner at video rate with relatively modest computing power. But what are these simple features good for? They receive their significance from a method which is able to provide a link from 3-D features on objects moving in space to 2-D features in the image. Straight contour elements are especially suited for this purpose since on a proper scale many objects have straight or nearly straight contour elements and zero curvature is an invariance property under perspective projection.

As caricatures show in a most impressive way, lines and curves do carry most of the information characterizing a scene. If the linear features (contour or edge elements) detected are considered to be tangents to curves, they constitute very general elements for shape description in differential geometry terms (Dickmanns 1985a). If sets of features in the image move in conjunction, it can be hypothesized that they belong to the same object, though there are exceptional cases where this may not be true.

Objects in the real world may be described as 3-D shapes realized by a massive substance having a center of gravity. The dynamical models for motion of and around the center of gravity (CG) of a physical object are combined with representations of its 3-D shape, emphasizing the position of (contour element) features relative to the CG. Feature groupings (aggregations) in the image, interpreted as a perspective map, are used to recognize objects in a 3-D scene. However, only in the initialization phase, if at all, is this done in the usual nonunique inverse way. If the type of scene is known a priori, forward perspective mapping of generic models and adjustment of model and relative position parameters is done until the measured image is matched or until the model pool is exhausted. Once the real-time phase has been initiated, only the model based approach is applied. Up to now, features used have been limited to linear contour elements (edge elements) and corners. A theory for efficient curvilinear shape representation has been developed and is presently under investigation (Dickmanns 1985a, 1985b).

Figure 4 shows a juxtaposition of the conventional method in image sequence processing (top) with the integrated 4-D approach (bottom). In the former, a considerable amount of computation is done in image coordinates and in inter-image comparisons, whereas in the latter a 4-D representation of a model world is being maintained in the interpretation process and servo-controlled by measurement data from the last image of the sequence only.

This model based prediction-error-feedback of feature positions has several advantages: no differencing between noise corrupted images of a sequence is required for obtaining velocity components (as in optical flow computation). Motion interpretation does not have to be done in image coordinates first with a reinterpretation in space af-

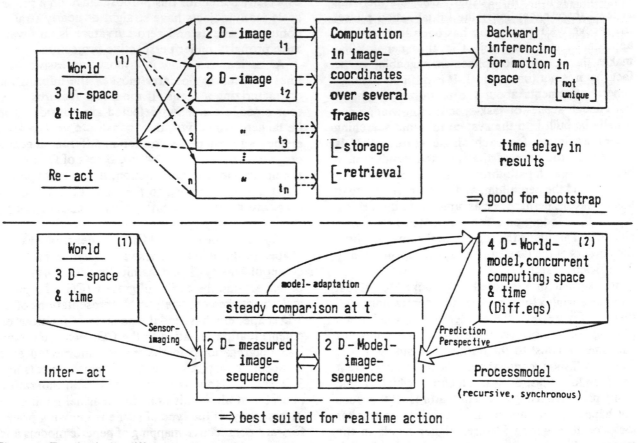

Figure 4. Two basically different methods of image sequence processing.

terwards; instead, all interpretation is immediately performed in the 4-D space–time continuum.

Similar to using shape models for object recognition, temporal models appear to be of great advantage for motion parameter recognition: As the term *recognition* tells, the interpretation process does have background knowledge of what it is going to "see," at least as a generic class from which special objects are being instantiated through data based hypothesis generation. This usual approach for shape recognition has been augmented by associating the object with its environment and the viewing conditions for image sequence taking: If the object is at rest and the camera moves, a generic dynamical model with state variables x for the camera motion is introduced; if the camera is at rest and the object moves, a model for this motion is selected. In both cases physical motion constraints and optional control or disturbance inputs are included. (The case where both camera and object are moving is much more difficult and presently under investigation.)

The general standard form of a generic dynamical model is a set of n differential equations for n state variables, usually nonlinear, sometimes with time-varying coefficients. As in modern control theory for sampled systems, locally linearized approximations with transition matrices for the sampling period T and influence coefficients for the control are being used. All coefficients are assumed to be constant over T. This basic cycle of period T for model based measurement interpretation and control action has been selected around 0.1 s (10 Hz); the more complex situation analysis on a higher level may be slower.

The goal of basing visual process recognition on an integral spatio-temporal world model is threefold:

1. Eliminate the need to access past images
2. Determine spatial velocity components by smoothing numerical integration
3. Bypass the nonunique inversion of the perspective projection by doing recursive least squares state estimation exploiting the Jacobian matrix of the measured image features (their partial derivatives with respect to the state variables of the dynamical model).

This matrix contains rich information for the direct interpretation of feature prediction errors in spatial coordinates; this information becomes accessible through the measurement model including perspective projection and sensor properties.

One might say that the problem of dynamic scene analysis is solved by doing a servo-controlled synthesis, based on feature–prediction–error feedback, using generic world models for shape, motion, and the measurement process. This is in contrast to today's common approaches that try to achieve geometric reconstructions from several images (shown schematically in the top part of Figure 4).

The recursive estimation based on the smoothing numerical process of integration proceeds as follows: An estimate $\hat{x}$ of the complete state vector x describing the process to be interpreted is assumed to be given; this hypothesis generation in the initialization phase is a hard task in partially or fully unknown environments. Knowing the n vector $\hat{x}$ and a dynamical model of the process in the discrete form for sampled measurement and control, a state prediction x^* for the next measurement at time $(k + 1)T$ can be made

$$x^*[(k + 1)T] = A[k]\hat{x}[kT] + Bu[kT] \qquad (1)$$

A is a $n \cdot n$ state transition matric over one sampling period T and B is the $n \cdot q$ control effectiveness matrix for the q components of the control vector u assumed to be constant over one period T.

If the shape of the objects observed and the relative geometry is known, maybe even described in terms of state components x_i^*, the predicted position y^* of features in the next image can be derived by forward application of the laws of perspective projection exploiting a model of the actual camera used for measurement. In general, this will be a nonlinear relationship containing measurement parameters p

$$y^* = f(x^*,p) \qquad (2)$$

Both the process modeled in Eq. (1) and the measurements will be corrupted by noise, designated $v(kT)$ for Eq. (1) and $w(kT)$ for the measurements. The problem is to determine best estimates for the state

$$x = x^* + \delta x \qquad (3)$$

given the measured quantities

$$y = f(x,p) + w \qquad (4)$$

Assuming that the influence of the noise is small

and its average is zero, a linearized relationship between y, y^* and x, $\hat{x}$ may be a good approximation to reality

$$
\begin{aligned}
\delta y &= y - y^* \\
&= f(x^* + \delta\hat{x},p) + w - f(x^*,p) \\
&\approx \frac{\partial f(x^*,p)}{\partial x^*}\delta\hat{x} + w \\
&= C\delta\hat{x} + w \qquad (5)
\end{aligned}
$$

Note that the Jacobian matrix C contains the $m \cdot n$ partial derivatives of the m measurement quantities y^* as predicted, relative to the state variables x in 3-D space including the spatial velocity components.

Figure 5 shows the physical meaning of the C matrix (some components) for a simple top-down view by the reader onto a camera at point P imaging a rectangular box O_o. The present state is given by the distance r and the polar angle coordinate v to the center and the angular orientation ψ of the object around its center. In the image plane (normal to r and to the plane containing the figure as shown) the edges of the box, designated M_{10}, M_{20}, and M_{30}, may be tracked laterally by three line element features. In the two subfigures at the bottom of Figure 5 (exploded view with 90 deg plane change), the feature position in the image is shown for the nominal state vector (index 0) and for state vectors with perturbed components, one each individually:

1. Radial translation (index r)
2. Azimuthal translation (index v)
3. Rotational displacement (index ψ)

the left subfigure shows the feature displacements δy_{2i} for feature 2 for each perturbation in one state variable keeping the other ones at the nominal value; the right subfigure shows the same for feature 1.

Combining the state variable change, for example, δr, with the feature position change (δy_{1r}), one obtains the partial derivative: for example

$$C_{y1r} \approx \delta y_1/\delta r$$

Performing this analytically, based on the 4-D model for all features and all feature coordinates (besides the image coordinates y_B and z_B, a rotational feature angle Φ_B may also be measurable) with respect to all state components yields the C matrix. It is rich information provided by the integral spatio-temporal world model which allows the bypassing of the nonunique inverse perspective

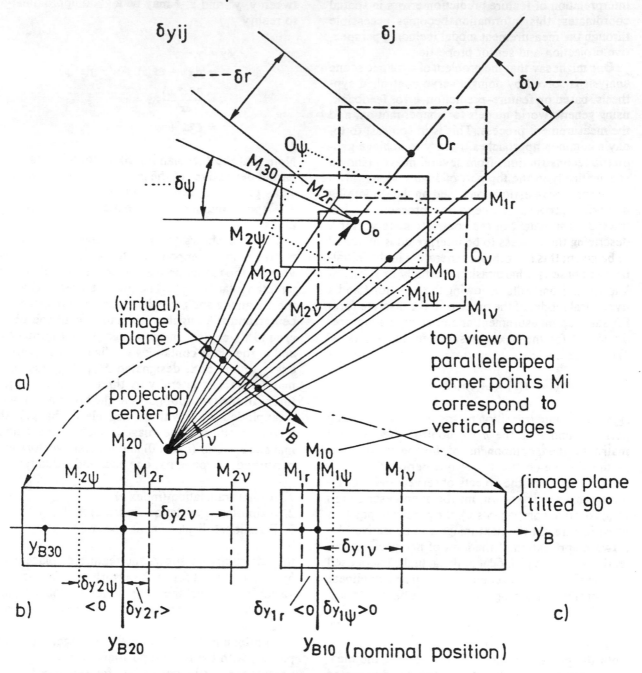

feature displacements due to state changes

δy_{ij}

δj

δr

δv

$O\psi$

O_r

M_{30}

M_{2r}

$\delta\psi$

M_{1r}

$M_{2\psi}$

O_0

M_{20}

O_v

r

M_{10}

M_{2v}

$M_{1\psi}$

(virtual) image plane

M_{1v}

top view on parallelepiped corner points Mi correspond to vertical edges

a)

projection center P

M_{20}

ν

y_B

M_{10}

P

$M_{2\psi}$ M_{2r} M_{2v}

M_{1r} $M_{1\psi}$ M_{1v}

image plane tilted 90°

$\leftarrow \delta y_{2v} \rightarrow$

$\leftarrow \delta y_{1v} \rightarrow$

y_B

y_{B30}

$\rightarrow \delta y_{2\psi} \rightarrow$

<0 $\delta y_{2r}>$

$\delta y_{1r} <0$ $\delta y_{1\psi}>0$

b)

c)

y_{B20}

y_{B10} (nominal position)

Determination of C-Matrix elements

Figure 5. Physical meaning of C matrix elements for recursive least squares state estimation by 4-D vision. (a) Object and imaging geometry, nominal and perturbed. (b) y position shifts for feature 2. (c) y position shifts for feature 1 due to singular state changes.

transformation. In fact, it allows going to a 4-D representation of the scene observed from the image data in just one step. This is performed recursively, based on data from the last image only. Besides exploiting this information for state estimation in a

numerical (procedural) way, it may also be used for spatial reasoning and general inferencing in a supervisory process on hierarchically higher levels. The recursive least squares state estimation is done using well-known techniques: If the covariance matrices of the noise processes v and w are known, Kalman filter techniques or derivatives thereof may be used (Bierman 1975; 1977; Maybeck 1979). The new best estimate $\hat{x}$ then becomes

$$\hat{x} = x^* + K(y - y^*) \tag{6}$$

where the gain matrix K for innovation is determined depending on the method used; sequential formulations well suited for time varying measurement vector lengths, such as those due to occlusion, are available (Wünsche 1987). One set of equations for the K update showing the role of the Jacobian matrix C is given below:

$$K(k) = P^*(k)C^T(k)\,[C(k)P^*(k)C^T(k) + R(k)]^{-1}$$

$$P^*(k) = A(k-1)P(k-1) + Q(k-1) \tag{7}$$

$$P(k-1) = P^*(k-1) - K(k-1)C(k-1)P^*(k-1),$$

where Q = covariance matrix of process noise
  R = covariance matrix of measurement noise
  $P^*(0)$ is selected according to the confidence in the initial values $x^*(0)$.

Figure 6 shows the recursive 4-D image sequence interpretation method in the form of a block diagram. At the top left the real world is sensed by a television camera (TV) the video signal of which is digitized by the image sequence processing system BVV and given onto a video bus. In the initialization mode, the processors of the BVV are coordinated to do a feature search over the entire search space; based on the distribution and orientation of the features found, object hypotheses are generated consisting of shape, relative position, and angular orientation components (upper right). The 3-D shape instantiation proposed is transferred into the "geometric reasoning" block of the 4-D real-time world representation (center of Figure 6, "world 2" in the interpretation process); the relative position and orientation estimates are installed as the initial conditions for state prediction together with the proper dynamical model (lower center left). The predicted state of the CG motion is combined with the shape description of the object (shown as a circled "and" (Λ) sign in the geometric reasoning box) to yield the internal representation of this object at the point in time when the next measurement is going to be taken. According to this state the visibility of various features is checked and those which are best suited for relative state estimation are selected for tracking (upper line of geometric reasoning block). This automatic dynamic allocation of processing power to meaningful areas of interest, decided upon at a high interpretation level

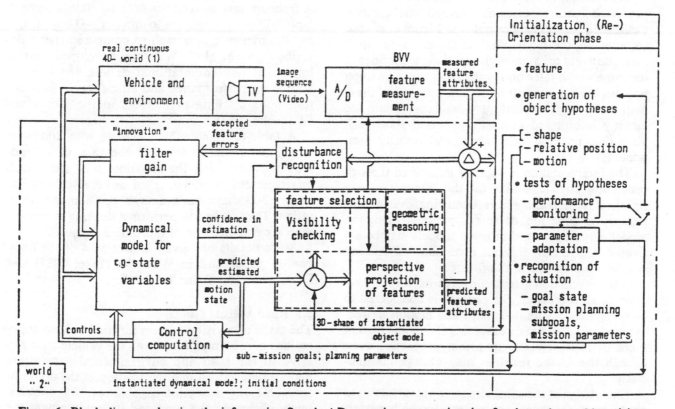

Figure 6. Block diagram showing the information flow in 4-D recursive state estimation for dynamic machine vision.

with a rich world representation as a basis for inferencing, is considered to be most beneficial for efficient dynamic scene analysis. It has been developed in (Wünsche 1987): The determinant of the pseudo-inverse of the properly scaled and weighted C matrix as occurring in a Gauss/Markov estimator formulation is maximized by selecting different feature combinations (global version); in a local version suited for real-time application, single features are substituted by other ones, one at a time, and the local maximum always is adopted.

The features selected are communicated to two locations:

1. To the BVV system for attention focussing and nonuniform image processing
2. To the perspective mapping module within the geometric reasoning block for computing the "model image" as the reference for prediction–error–feedback.

Note that at this point the Jacobian matrix C [Eq. (5)] is computed which is instrumental for bypassing the nonunique inversion of the perspective 3-D → 2-D mapping by recursive least squares filtering; the matrix C contains all the essential first order dependencies needed for intelligent interpretation of the measured feature data, given the scene model.

The difference between the measured and the predicted feature positions, formed at the circle designated with $\triangle$ (upper right) in Figure 6, is used for adapting the model state to the measurements. First, outlyers are removed exploiting a confidence measure which later on results from the estimation process itself. Note that through proper use of the C matrix the accepted feature measurement data are directly interpreted in spatio-temporal state variables (3-D position, orientation, and velocity components).

The interpretation process is monitored through prediction error checking (at the $\triangle$ circle, upper right). If systematic errors occurring over longer periods in time are detected, parameters of the model may be changed or other models may be activated. Up to now this has been done off line but will be done on line in the future.

If the prediction errors converge initially and remain small thereafter, the process is considered to be recognized. Note that shape and motion are recognized simultaneously assuming the model parameters to represent the invariant properties even though the image features may change continuously.

Knowledge of the complete state vector allows to apply state feedback controllers for high performance (lower left). There may be a direct feedback for fast reflex-like behavior. By checking the state against a sequence of goal states for coordinated mission performance, and by superimposing a control model switching, very flexible and highly efficient behavioral competences may be achieved. The controls actually output are fed both into the real-world machine being controlled and the (world 2) model in order to generate "expected" motion state components [see Eq. (1), last term; signal flow in Figure 6 left].

5. Applications

The general method as described above evolved during application to four problem areas.

5.1 Balancing of an Inverted Pendulum

The first real-time hardware application was in the early 1980s in the balancing of rods of various lengths in one degree of freedom on an electrocart. Rods from 0.4 to 2 m length have been investigated (Haas 1982; Meissner 1982; Meissner and Dickmanns 1983). Closed-loop eigenvalues of up to 1 Hz have been achieved. More details are given in section 3 of the companion paper.

5.2 Vehicle Docking

A frequent task in robotics is to position a controllable 3-D vehicle relative to another 3-D object. Using the dynamic approach to computer vision described above, H. J. Wünsche developed several important implementational details and demonstrated its performance and efficiency in fully autonomous docking maneuvers in the laboratory (Wünsche 1987).

A table-top air cushion vehicle with computer controlled reaction jets has the task of autonomously recognizing the situation in its (technical) environment, consisting of several objects of known 3-D shape but unknown position and orientation. Then, it has to perform a docking maneuver with a particular one of these objects.

More details are given in section 4 of the companion paper and in Wünsche (1986; 1987) and Dickmanns and Wünsche (1986b).

5.3 Road Vehicle Guidance

The tasks of a vision system for road vehicle applications may be manifold. Both the vehicle state relative to the road and environmental parameters may be determined in order to support the driver or for achieving autopilot capabilities. By continuously observing the road and its environment the following items relevant to safe road vehicle guidance can be estimated or recognized in principle: road curvature, both horizontal and vertical, lane width and number of lanes, surface conditions such as smoothness, surface states such as dry, wet, or dirt or snow covered, presence of obstacles or other vehicles and traffic signs. A discussion of possibilities and problems related to the application of machine vision to road vehicle guidance is given in Dickmanns (1986).

Only the guidance task proper will be discussed in the companion paper, section 5, demonstrating the specific application of the integrated 4-D approach described in general terms in section 4. The following results have been achieved: Fully autonomous runs starting from rest have been performed under various road and weather conditions, including bright sunshine and light rain as well as road surfaces with and without lane markings. Increasing the maximum speed limit step by step, in August 1987 the maximum speed of the vehicle ($V_{max} \approx 96$ km/h on a level surface) has been reached. In order to obtain some results with respect to reliability, the total run length of more than 20 km has been driven several times (in both directions) without the need for intervention by the safety driver in the driver's seat.

Lane changes from right to left and back have been performed on free lanes as well as highway entry maneuvers from the acceleration strip.

The following conclusions for road vehicle guidance can be drawn: The capability of real time image sequence processing for guiding high speed road vehicles along well structured roads is becoming a reality. The method derived has been shown to allow autonomous vehicle guidance even at high speed with a relatively small set of today's microcomputers.

More computing power will be needed to improve the checking for obstacles and other objects in a less restricted environment. In principle, the vision system considered has the growth potential to allow the development of autonomous vehicles that fit neatly into the traffic system developed up to now for the human driver. Both gradual deployment and mixed human and automatic traffic seem to be possible.

5.4 Aircraft Landing Approach

This is the most complex real-time motion control task solved by computer vision by our group up to now. Aircraft motion occurs simultaneously in six degrees of freedom: three translatory and three rotatory ones. In each degree of freedom, according to Newton's law, one differential equation of second order is required in order to model the dynamical behavior. So 12 state variables are necessary to describe the rigid body motion.

An aircraft is controlled by selecting four control variable time histories: elevator for pitch and altitude, aileron for roll, rudder for yaw and sideslip, and throttle for thrust level control; in addition, diverse flaps may be set for certain flight regimes (takeoff and landing). To direct such a vehicle in a well controlled maneuver requires skill and concentration even for a trained human pilot; he has to acquire this capability in an extended learning process.

Exactly this knowledge, coded in differential equations as side constraints to the development of trajectories, should be available to an automatic system for recognizing and controlling landing approaches by machine vision.

Simultaneously exploiting spatial and temporal models as shown in section 4 and Figure 6, Eberl (1987) has shown that the problem of controlling landing approaches by computer vision may be tackled successfully relying on present-day microprocessors. In a six degree of freedom fixed base simulation with real-time image sequence processing hardware in the loop, complete landings starting from 2 km distance have been performed fully autonomously with airspeed V being the only quantity not determined from vision. It seems unlikely that such a complex task can be handled by computer vision without using integrated spatio-temporal process models.

Space does not allow us to go into more detail here. A somewhat more extended discussion of machine vision for flight vehicles is given in Dickmanns (1988) containing a summary of the dissertation (Eberl 1987) in English.

6. Development Perspectives

The following principles have been found to be essential for real-time dynamic machine vision.

1. Tangency detection by correlation of linear contour elements
2. Active control of the viewing direction, both top-down (feature search, mode switching) and bottom-up (feature tracking, fixation), coupled with nonuniform image processing
3. Analysis by synthesis, that is, building up internal spatio-temporal representations via 4-D models, the parameters of which are adjusted exploiting the sensor data input.

They will be discussed in the sequel with respect to future applications in machine vision.

6.1 Tangency Detection for Shape Recognition

In Kuhnert (1988) it has been shown that by performing interpolations over the correlation values of shifted and rotated bar masks of about 10 pixel length, angular resolutions of about 1–2 deg and edge position localizations of subpixel resolution can be achieved. Taking this type of data as the basic input for shape description in differential geometry terms, a very efficient signal-to-symbol transition has been proposed in Dickmanns (1985a, 1985b). Using local coordinates only, by simple weighted summing and differencing of the slopes, the two parameters of a linear curvature element can be determined. For direction changes smaller than about $\pi/6$ (30 deg), this element closely corresponds to a spline curve element of third order. Whole contours may be pieced together and corners can be incorporated as curvature impulses. These can be isolated by locating slope discontinuities using the same tangency operations.

This representation is by itself position and rotation invariant. It can be made scale invariant by normalizing the contour length; this may be achieved by some natural scale length, such as normalizing the contour length to the range 0, 1 (or 0, 2π) or dividing it by the square root of the area enclosed. The resulting very compact representation of smooth contours has been termed normalized curvature function (NCF). The interesting point is that NCFs yield a nice basis for shape idealizations, symmetry detection, and the definition of shape terms (e.g., circle, n angle, concavity). Complex features of a 2-D object may be decomposed easily on a local basis into an aggregation of simple edge element features that will allow object tracking and motion interpretation based on edge element tracking using the 4-D model based approach. Thus, it is in combination with point 3 mentioned above, that edge element correlation appears as a powerful tool in 3-D dynamic scene analysis.

The hardware under development presently for low level vision will increase performance by two orders of magnitdue for correlation based feature extraction, yielding the capability of measuring many dozens of edge element features per video cycle by a single processor. This will allow us to tackle visually much more complex scences.

6.2 Active Gaze Control

When egomotion and object motion occur simultaneously, signals from inertial sensors may help considerably in discriminating the motion components. High angular relative speeds will lead to motion blur. It may result from two components: egomotion and/or object motion. Angular egomotion may be compensated by stabilizing the camera platform inertially. Angular velocities due to external object motion may be cancelled for the imaging process by active feature tracking and corresponding viewing direction control by the camera platform. This allows (close to static) shape analysis in the image and motion measurement via the platform orientation.

The tuning of the feedback control loops both with inertial and visual signals is presently under investigation in a simulation facility. The quality of stabilization achievable will influence the range of the teleoptics useful with the second camera (for high resolution). During periods when the viewing direction is quickly changed, the image data are blurred and evaluation has to be suppressed; in these short intervals, motion control is done purely based on the internal models. Experience will have to show what are the best strategies for compromising foveal fixation and the consequent motion in the wide angle image.

6.3 Representing the World by a Servo-Maintained Internal "World 2"

The basic feedback approach chosen (see bottom, Figure 4), using integral spatio-temporal world models including perspective (forward) projection, may be expanded by incorporating active gaze control and long term memory for models. A somewhat unconventional block diagram of such a system is shown in Figure 7. The basic arrangement is the same as in Figure 4; the viewing direction control has been introduced as the central horizontal bar. It actively controls the camera pointing and takes care of the corresponding geometrical transformations. For orientation relative to the dominating gravity vector, inertial and other conventional sensors (upper left) are being used, yielding a stabilized internal 4-D world model (center right). The actual models instantiated there are installed by a hypothesis generator (center top) basing its decisions on bottom up feature data and on rules implemented for this purpose. Models for both shapes and motion processes may be selected from a long term memory called model store (upper right). Instantiations are first done separately per object (second block from top, upper right) and then integrated into the world model as a representation of the situation as recognized by the interpretation process (world 2).

If the interpretation process has a set of goals against which it checks the situation recognized, it may start or continue action planning in order to achieve the goals. Actions may be both attention focussing through active vision (lower center and right) taking other sensory modalities into account

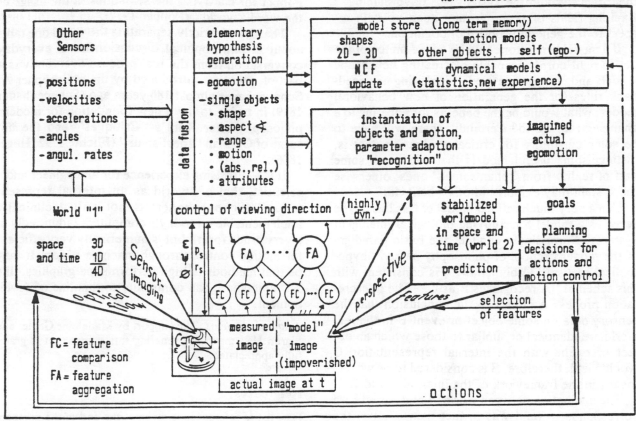

Figure 7. Block diagram of 4-D feature based vision concept including active gaze control, long term model storage, and goal driven activity planning.

(upper left), and application of motion control through effectors (bottom). Since dynamical models are available, direct state variable feedback may be used in order to achieve reflex-like behavioral competences. Thus, for example, in road vehicle guidance, lane-keeping and proper speed control may be achieved without continuously running cumbersome planning activities. Monitoring subprocesses just have to provide "road recognized" and "road free of obstacle" signals. As long as these are true and the goal is not yet achieved, the system continues in this mode. All logical variables required for mode continuation form the set of continuation control tags.

Their value, in turn, may be changed either by sensory data including situation variables derived therefrom or by decisions taken in the continuously active mission planning and monitoring subprocesses. Depending on the particular continuation control tag becoming false, specific other behavioral modes with proper sensing activities and feedback control laws, if necessary adaptable by situation dependent parameters, may be invoked, taking care of a gradual transition from the old mode to the new one.

A sufficiently rich set of behavioral modes including smooth transitions has to be developed and stored in long term memory. In addition, knowledge has to be implemented in the interpretation process as to which behavioral competences should be activated with which set of parameters, depending on the situation and the goals to be achieved.

In the long run, the system should be able to learn from statistics it accumulates during each mission. This is, however, far off in the future.

The systems we have developed up to now only have very simple reflex-like behavioral competences. Some interesting questions arise when we try to imagine what kind of behavior much more complex systems might display (in a not very near future), if they continue to be based on the general principles explained in the previous sections.

The actual world 2 instantiated in the interpretation process is forced to remain close to the real world by critical feature comparison and corresponding model adaptation based on the measured image data and the data from other real-world sensors (left column in Figure 7). What could happen if all these sensory inputs would be cut off and the central and right blocks would continue working on their own? Would the system be "daydreaming"? Exciting fields of research may open up with respect to the general problem of cognition.

If a model generator could be added on top of the upper right corner, capable of creating new shape models and new motion models including new feedback rules for the generation of new behavioral modes, what would be the benefits for adpating to a changing real world? Certainly, there will have to be some capability for critical evaluation, that is, distinguishing useful models that can catch some part of reality from "phantasmal" ones; otherwise the system may exhibit "idiotic" behavior.

Can a very much refined model of this basic type serve as a model for studying biological intelligent systems? Surprisingly enough, the basic paradigm of the modern school of philosophy named "hypothetical realism" (Vollmer 1975) is consistent with this scheme: "A recognizes B as C." The interpretation process A comes to the conclusion that the sensory data on some object or event B in the real world are identical or similar to those which an object or event with the internal representation C would yield; therefore, B is considered to be what C means in the framework of the internal world 2.

The distinction between worlds 1 and 2 used here corresponds to (and was in fact named after) the terminology introduced by the philosopher Karl Popper for clarifying the semantics in the usage of the word "world" (Popper 1977).

The idea of strictly separating the world one talks about in philosophical discussion or in everyday conversation from the real process everybody is a part of, was first introduced by the philosopher A. Schopenhauer almost 180 years ago (Schopenhauer 1819) in trying to lay a new foundation for modern philosophy after Kant's "Critiques" and the upsurge of German "Idealismus" (Flichte 1792, Hegel 1806).

It is a surprising experience for an engineer and a scientist, that a "world as an internal representation" (Schopenhauer) is not only technically implementable on today's computers (admittedly in a very crude form), but is numerically very efficient for motion control through machine vision. It may be that methods in high performance graphics, like 3-D animation, can contribute to this line of development.

Acknowledgments. The support by Madeleine Gabler and Sigrun Hausmann in preparing this manuscript is gratefully appreciated.

References

[The list of references is given at the end of the companion paper—see pp. 260–261.]

Applications of Dynamic Monocular Machine Vision

Ernst Dieter Dickmanns and Volker Graefe

Fakultät für Luft- und Raumfahrttechnik (LRT), Universität der Bundeswehr München, W-Heisenberg-Weg 39, 8014 Neubiberg, West Germany

Abstract: The 4-D approach to real-time machine vision presented in the companion paper (Dickmanns and Graefe 1988, this volume) is applied here to two problem areas of widespread interest in robotics. Following a discussion of the vision hardware sed, first, the precise position control for planar docking between 3-D vehicles is discussed; second, the application to high speed road vehicle guidance is demonstrated. With the 5 ton test vehicle VaMoRs, speeds up to 96 km/h (limited by the speed capability of the basic vehicle) have been reached. The test run available, of more than 20 km length, has been driven autonomously several times under various weather conditions.

Key Words: 4-D machine vision, real-time vision hardware, automatic visual motion control, vehicle docking, road vehicle guidance

1. Introduction

In the companion paper (this volume) a set of principles and methods emphasizing the dynamic aspects of machine vision have been introduced. In this paper, four applications of these concepts in the form of demonstration experiments are described, together with some technical details of the real-time vision hardware used. These experiments not only served for developing the concepts but also for convincing both ourselves and others that the concepts are indeed practical.

Address reprint requests to: Prof. Ernst D. Dickmanns, Steuer-und Regelungstechnik, Universität der Bundeswehr München, Fakultät für Luft- und Raumfahrttechnik, Institut Für Systemdynamik und Flugmechanik, Werner-Heisenberg-Weg 39, D-8014, Neubiberg, West Germany.

This research has been partially supported by the German Federal Ministry of Research and Technology (BMFT), the Deutsche Forschungsgemeinschaft (DFG), the Daimler-Benz AG, and by Messerschmitt-Bölkow-Blohm GmbH (MBB).

Simple motion control tasks in the dynamic range of a human operator, albeit confined to a visually well structured environment, were considered to be both sufficiently demanding and rewarding for the initial demonstration of practical applicability. In order to emphasize the real-time aspect, we deliberately imposed a cycle time limit of about 0.1 s (100 ms) on the interpretation and control process. This was intended to and succeeded in spurring the group to think in different terms than when time constraints do not play any role, hoping for processors to become fast enough to run any algorithm in real time. With the "real time" concept we understand two things in the context of computer vision: that the low level part of the system should sample the scene at the highest rate practical and thus process every image delivered by the TV camera, which limits the processing time for each image to 17 or 40 ms, depending on the TV standard used; and that the response time of the entire vision system (the time between some visible event in the scene and the output of a control signal which has been caused or influenced by the event) should not be much longer than for a human. Both software and hardware concepts have developed in different directions than they would have done without this side constraint; but especially the methods for information processing have shaped up differently. The real-time constraint called for recursive methods as have been developed in modern control theory. Gradually, it became clear that temporal coherence is as important as spatial coherence for visual dynamic scene analysis.

This has lead to the 4-D method presented which appears to be a natural extension of modern control engineering tools. The availability of temporally dense time histories of the spatial states including velocity components of all relevant objects in the scene provides a rich basis for inferencing and applying AI methods in the future in order to perform a more complex analysis of the situation.

Having all relevant state variables of the process to be controlled available from the model based vision process, state feedback control for achieving well defined goal functions is readily determined and implemented. This provides the system (e.g., a mobile robot) with basic behavioral competences on a reflex-like level for fast response. It is this lower stratum of visual based behavior which has been developed for the different application areas.

More intelligent behavioral programs can be developed on top of this by combining these basic behavioral competences and adapting them to the different situations encountered.

The inclusion of time and of the cause and effect sequences into the world model right from the beginning, through the proper use of the dynamical models, yields an appropriate frame for dealing with the temporal development of actions. This seems to be another big advantage as compared to conventional AI approaches.

In the next section the computer architecture and hardware as developed for the feature based 4-D approach will be described. In section 3, a short survey on our entry point to dynamic vision will be given: the inverted pendulum balanced on an electrocart. The vehicle docking problem, from which we learned many implementational details for the general method, is discussed next to some extent. Section 5 then is devoted to a problem of much current interest: the autonomous guidance of road vehicles. Here, a promising field of practical applications may open up rather soon.

Because of space limitations, the simulated aircraft landing approach is not discussed here. References are given in section 5.4 of the companion paper.

2. Computer Architecture and Hardware

In this chapter a family of real-time vision systems will be introduced, the members of which have been designed especially for dynamic vision and motion control. A key factor in their design has been to optimize the structure of the hardware according to the requirements of the methods and algorithms used for feature extraction in dynamic scenes. Letting methods and algorithms influence the design of the hardware resulted in systems which are rather different from, and in a sense simpler than, many computer vision systems which have been described in the literature. A good synopsis of the field, including numerous references, is given in Duff (1986).

Having powerful hardware available was of utmost importance, since it allowed the development

and verification of all concepts, methods, and algorithms for dynamic vision in *real-world experiments* where the vision system was part of closed loops controlling the motion of *real* mechanical systems. This experimental approach has proven very expensive, time consuming, and laborious, but also an extremely effective means for learning to understand the nature of dynamic vision and for finding at least partial solutions to some of its problems.

2.1 The Original Hardware Realization

The first computer vision system which has been designed according to the architectural concepts set forth in the companion paper was the BVV 1 (Haas 1982; Graefe 1983a). It could contain up to 15 microcomputers: one system processor (SP), and up to 14 parallel processors (PP) (Figure 1). All processors use the Intel 8-bit microprocessor 8085 A as a CPU and have 16 kB of memory (the design was begun in 1977). The video signal is digitized and distributed in digital form via the videobus to all PPs. Each PP copies that part of the image which constitutes its momentary region of interest into an internal private memory. Having a private image memory in each PP avoids delays which would exist if a central image memory would be shared by all PPs. Each PP is a self-contained computer executing its own task. Messages can be sent from any PP to any other PP or to an external host computer. The SP transports all messages, relieving the PPs completely from this task; they merely have to de-

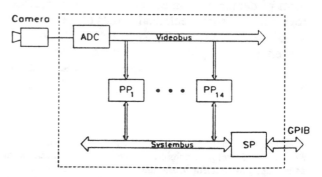

Figure 1. The BVV 1, a multiprocessor system for dynamic vision.
ADC: Analog to digital converter.
GPIB: General purpose interface bus (IEEE 488) connecting the BVV 1 to other equipment, e.g. a host computer.
PP: Parallel processor; a specially designed single-board computer, based on the microprocessor Intel 8085 A (8 bit), equipped with a videobus interface for inputting and storing pixel data and with a buffered systembus interface.
SP: System processor; it coordinates all internal and external communication.

posit outgoing messages into their private "out basket" and periodically check their private "in basket."

Typically, each PP will produce one result message per video cycle and it will occasionally receive a message from higher levels, instructing it to enter a different mode of operation. The total message traffic is, therefore, so weak that the systembus is never a bottleneck which might slow down the operation of the system. The videobus creates no bottleneck either, since all PPs can receive video data from it simultaneously without disturbing each other.

The structure of the BVV 1 is clear and simple. It is, however, more flexible than might be apparent at a first glance. Although all PPs are equal and are connected to the systembus in parallel, it is, nevertheless, possible to give the system a hierarchical structure by merely letting the application programs in the PPs communicate with each other in the appropriate way. For instance, PP_1 through PP_4 could send their results to PP_5, and PP_6 through PP_8 could send results to PP_9. PP_5 and PP_9 could then be considered to be on a higher hierarchical level. If they, in turn, produced results and sent them to, say, PP_{12}, this would constitute still another level in the hierarchy. The interesting point is that the grouping of the parallel processors into a parallel, hierarchical, or mixed structure is perfectly flexible. The structure can be freely rearranged at any time, even during runtime.

The first major demonstration experiment performed with the BVV 1 was the balancing of an inverted pendulum (see chapter 3). It may appear surprising that the computing power of three 8-bit microprocessors (two for feature extraction and one for state estimation and control) is sufficient for visually controlling such a highly dynamic, fast moving mechanical system. It underlines, however, the points made in the companion paper, that an adequate system architecture and proper data processing schemes are the key to real-time vision, rather than sheer computing power.

The low level vision for the planar docking maneuver of an air cushion vehicle described in chapter 4 was performed on the BVV 1, too.

2.2 The Second Generation

The successor of the BVV 1, known as the BVV 2, has a very similar architecture and the same degree of flexibility (Figure 2). When the design of the BVV 2 was started in 1982, single-board computers based on the 16-bit processor Intel 8086 had become available. They were used for both PPs and SP. Compared with the 8-bit processors of the BVV 1, they are about 10 times faster and have much more memory.

The main difference in a conceptual sense is, however, the videobus system which lets the BVV 2 transmit in a first version two, and now four, independent image sequences simultaneously (Graefe 1983b, 1984). There are several reasons for having more than one image available simultaneously. In autonomous vehicles, for instance, a wide-angle camera is needed to cover a large sector of the environment immediately in front of the vehicle, and a telecamera (on a controllable platform) is required to scrutinize more distant objects which might be obstacles. Other conceivable uses of the videobus system include color cameras and several cameras pointing into different directions, such as to the front and to the rear. Also, when the BVV 2 is used as a test bed for pixel processors which generate a preprocessed image out of the original one and make it available to the PPs, more than one image must be transmitted simultaneously.

The BVV 2 is different from the BVV 1 in a number of technical details, mainly as a consequence of more modern semiconductor components. No special hardware is necessary for interprocessor communication because each PP has a dual port memory which can be accessed by the SP via a standard multibus which is the systembus. In order to gain access to the videobus system, a PP needs a special hardware which we designed, the videobus interface. (If a particular PP is used only in such a way that it does not require direct access to the digitized images, for example, as the control processor for the camera platform or for processing the results from other PPs, it does not have to be equipped with a videobus interface. In this role one or two of the newer and more powerful single-board computers with the 80286 or 80386 CPU, which are not fully compatible with the BVV 2, may also be used.)

If more than one camera is connected to the system, they must run synchronously. A clock generator in the BVV 2 provides the control signals necessary to synchronize the cameras and all other parts of the system. In the BVV 1, where only one camera could be used, the camera provided the synchronization signals.

The BVV 2 has proven to be an efficient hardware basis for dynamic computer vision. For instance, it allows tracking three to five contour elements in every 60 Hz frame (17 ms) per PP, using a special correlation method described in the companion paper.

Again, the main reason for the relatively good system performance as evidenced in the applica-

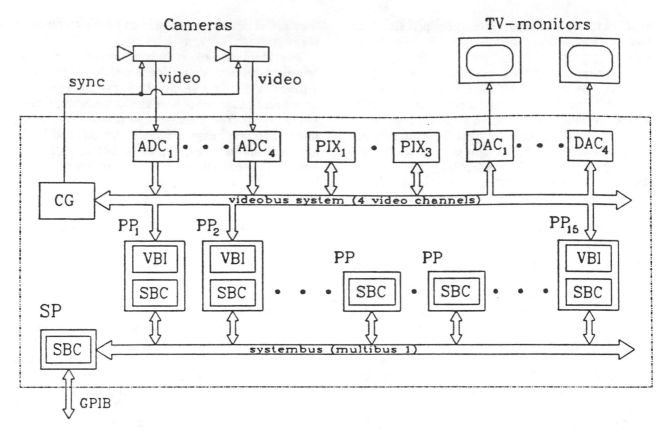

Figure 2. The second generation vision system BVV 2.
ADC: Analog to digital converter
CG: Central clock generator
DAC: Digital to analog converter
GPIB: General purpose interface bus connecting the BVV 2 to other equipment, e.g. a host computer
PIX: Pixel processor (for preprocessing digital images on the fly)
PP: Parallel processor
SBC: Single board computer, based on the Intel 8086 processor
SP: System processor (it coordinates all internal and external communication)
VBI: Videobus interface; it reads and stores pixel data from the videobus and makes them available to the PP (not all
     PPs need a VBI; see text)
The total number of analog to digital converters and pixel processors is limited to 4, the number of parallel processors is
limited to 15.

tions is not the sheer computing power of the processors of the BVV 2 but rather its flexibility, which allows the entire power to be concentrated on the relevant regions of the image. Optimizing the hardware by itself is, however, only part of the solution. What is really important is to optimize several things as a whole and, if possible, simultaneously: the basic methods for feature extraction, the way these methods are cast into algorithms, the hardware on which the algorithms are to be executed, and the feature based 4-D interpretation of image sequences. In reality it is difficult to optimize so many different interdependent things at the same time; a sequence of iterations is a more realistic approach.

2.3 The Future

By performing experiments in dynamic vision using the BVV 1 and BVV 2 (see chapters 3 through 5), much has been learned regarding methods and algorithms for low level vision. In 1986, as a result, a new vision system, the BVV 3, has been designed (Kuhnert 1986b; Graefe and Kuhnert 1987). The goal is to build a system which is optimized for those classes of methods and algorithms which have been found particularly useful for low level vision.

The BVV 3 will have the same structure as the BVV 2; the parallel processors will, however, be different (Figure 3). The single-board computer will now be based on the microprocessor Intel 80286; it is about two to three times more powerful and it has

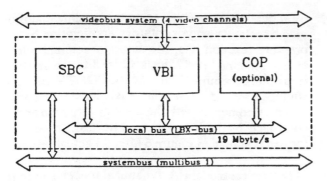

Figure 3. The parallel processor of the real-time vision system BVV 3.

COP: Coprocessor (optional)
SBC: Single-board computer, based on the Intel 80286 processor, equipped with 1 MB of RAM
VBI: Videobus interface with 2 frame memories in a double buffer arrangement

more memory (1 MB) and a larger address space (16 MB). As discussed in chapter 3 of the companion paper it is possible to gain speed by having a very large memory available; the 128 kB which are available on the PP of the BVV 2 are sometimes insufficient. The videobus interface of the new PP includes two memories, each of them large enough to hold an entire image, in a double buffer arrangement. While the region of interest in which each PP of the BVV 2 can work is limited to 4 KB, such a limitation will no longer exist in the BVV 3.

The most significant innovation is the coprocessor. It is a special RISC (reduced instruction set computer) which can perform 10^7 complex operations per second. It has a pipeline structure with seven stages; therefore, each complex operation can include seven elementary operations like address generation, addition, multiplication, maxi-

mum detection, etc. Originally, the coprocessor was intended specifically for data controlled correlation, a method we have found to be particularly effective in low level vision. The coprocessor is, however, equally well suited for certain other methods which may become important, too. This includes linear and nonlinear filtering and resolution pyramids. It can compute a resolution pyramid over the entire image in 12 ms, but, more importantly, it can also compute a pyramid over ¼ image in 3 ms or over ¹⁄₁₆ image in 0.8 ms.

A breadboard model of the coprocessor has been built and tested with correlation based programs equivalent to the ones used in the BVV 2. The results indicate that such programs run about 100 times faster on the coprocessor than on the PP of the BVV 2. This lets us hope that the BVV 3 will be an adequate tool for developing and evaluating vision algorithms which are required to handle much more complex situations than have been studied by our group so far.

3. Balancing of an Inverted Pendulum

The first real-time hardware application (in the early 1980s) was the balancing of rods of various lengths in one degree of freedom on an electrocart (Figure 4). The system has a mass of 7 kg and is capable of linear accelerations of about 6 m/s².

The scene is observed with a TV camera and the resulting image sequence is fed into the BVV 1. Two (or more) parallel processors track one segment of the rod each, and transmit the coordinates of the segments to a control computer. The control computer may either be an external general purpose computer (Meissner 1982) or it may be one of the parallel processors within the BVV 1 which is not

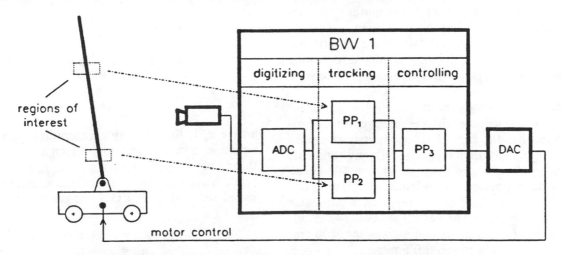

Figure 4. Balancing an inverted pendulum on an electrocart. (Reduced version with BVV 1 only.)

used for low level vision (Haas 1982; Graefe 1983a). In the control computer the information on the rod segments is combined into an integrated system state, and the control signal necessary to balance the rod is computed and transmitted to the cart.

Balancing the inverted pendulum is a very demanding task because of the instability of the mechanical system. It requires both accuracy of measurement and speed of computation. The position of the rod has been measured to within 0.5 pixel accuracy, and a cycle time of 35 ms for the measurement process was achieved. The rod sometimes moves very rapidly, for instance when compensating for an external disturbance or during the initialization phase. Its image is then severely blurred and loses most of its contrast. Detecting and precisely locating the rod under such conditions and in the presence of noise requires a careful modeling of the appearance of the *moving* rod (Haas and Graefe 1983; Graefe 1983c).

Rods from 0.4 to 2 m length have been investigated (Meissner and Dickmanns 1983). Closed-loop eigenvalues of up to 1 Hz have been achieved. Cycle times have been 80 ms in general and 40 ms for the very short rods having large eigenvalues (Wünsche 1983). By dealing with this relatively simple fourth order system, many of the basic features of our method have been developed. The dynamical model, initially introduced for control computation, has become the cornerstone of the integrated approach. It also has been used in this application for robust process recognition and control under severe perturbations (Dickmanns and Wünsche 1986a).

4. Vehicle Docking

Using the dynamic approach to computer vision described above, H.J. Wünsche developed several important implementational details and demonstrated its performance and efficiency in fully autonomous docking maneuvers in the laboratory (Wünsche 1987).

Figure 5 shows the general arrangement. A tabletop air-cushion vehicle with computer controlled reaction jets has the task of autonomously recognizing the situation in its (technical) environment, consisting of several objects of known 3-D shape but unknown position and orientation. Then, it has to perform a docking maneuver with a particular one of these objects. The vehicle is about 64 cm tall, has a mass of about 20 kg and a diameter of 36 cm; its maximum speed is about 0.5 m/s.

The main computer, a VAX 750, can control the position and the angular orientation of the satellite by activating control valves on the satellite that provide translatory and/or rotatory acceleration by air pressure reaction jets. Maximum acceleration levels are 3.8 cm/s^2 or 12 deg/s^2. Fixed to the satellite is a charge coupled device (CCD) TV camera equipped with an 8 mm wide-angle lens. Its viewing direction coincides with the docking direction and is controlled by body rotation. The task to be performed is shown in Figure 6 after an actual test run.

Dynamical model A dynamical model is used to describe all rigid body motions which lead to motions in the image sequence. In the experimental setup as shown in Figure 5, the only body moving is the air cushion vehicle carrying the camera, whereas all other objects in the scene are at rest. Therefore, the dynamical model consists of the equations of motion (EOM) of the air cushion vehicle only, describing the motion of this vehicle relative to the rendezvous partner. Here, a formulation in polar coordinates centered at the rendezvous partner's assumed center of gravity (CG) is chosen, with r being the range from the satellite CG to the partner CG, v being the aspect angle of the partner, and Ψ being the satellite yaw angle (see Figure 7). This reference system was chosen not only because the orientation of the partner (i.e., the aspect angle v) might not be visually detectable at greater distances while r and Ψ already can be detected and controlled, but also because the measurement equations turn out to be better conditioned in this coordinate system rather than in a cartesian coordinate system.

Because of the rotatory disturbances, a noise moment s_ψ is introduced into the rotatory equations of motion which are given by

$$\dot{\psi} = \omega - \dot{v} \tag{1}$$

$$\dot{\omega} = a_\psi u_\psi + s_\psi, \quad u_\psi = -1, 0, +1 \tag{2}$$

$$\dot{s}_\psi = z_\psi \tag{3}$$

where a_ψ is the rotatory acceleration, u_ψ is the reaction jet control computed through an s_ψ-adaptive three point bang–bang controller, z_ψ is a Gaussian random noise, and ω is the inertial angular rate. The radial and tangential EOM are given with

$$a_{rx} = -ax\cos\psi,$$
$$a_{ry} = ay\sin\psi,$$
$$a_{tx} = -ax\sin\psi,$$
$$a_{ry} = -ay\cos\psi$$

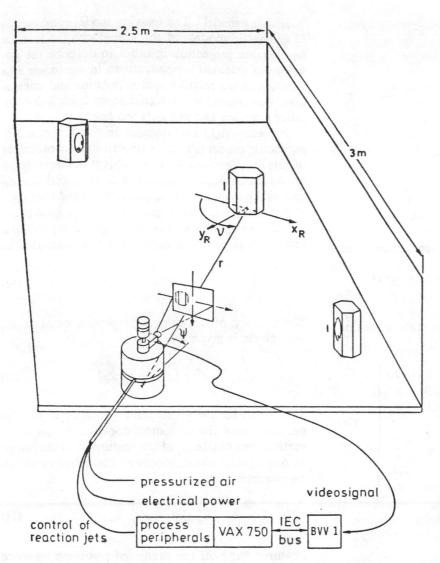

by

$$\dot{r} = v_r \qquad (4)$$

$$\dot{\nu} = v_t/r \qquad (5)$$

$$\dot{v}_r = v_t^2/r + a_{rx} + a_{ry} \qquad (6)$$

$$\dot{v}_t = -v_t v_r/r + a_{tx} + a_{ty}. \qquad (7)$$

The camera line of sight is aligned with the positive body x-axis of the satellite; however, it may be manually pitched up and down before a test run to center the rendezvous partner in the image. The camera pitch angle as an unknown parameter has been included in the EOM through $\dot{\Theta} = z_\Theta$, where z_Θ is a Gaussian random noise virtual acceleration. These nonlinear equations are written in vector notation as

$$\dot{x}(t) = f[x(t), u(t), z(t)]$$

Figure 5. Automatic docking of an air cushion vehicle propelled by reaction jets and controlled by visual object and relative state recognition (planar satellite model plant).

where

$$x = [\psi, \omega, s_\psi, \Theta, r, v_r, \nu, v_t]^T \qquad (8)$$

is the system state vector to be estimated by computer vision. Given an estimate for the state variables at some time instant t_{k-1}, called $\hat{x}_{k-1}$, a prediction of the state to the current time t_k, called x_k^*, has to be determined.

With some simplifying assumptions for small sampling periods $T = t_k - t_{k-1} = 0.13$ s, an approximate discrete algebraic equation for state prediction results [for more details see Wünsche (1987)]

$$x_k^* = A_{k-1} \cdot \hat{x}_{k-1} + B_{k-1} \cdot u_{k-1}, \qquad (9)$$

where $u = [u_\psi, u_x, u_y]^T$ is the vector of control commands, held constant from time t_{k-1} to t_k, A is

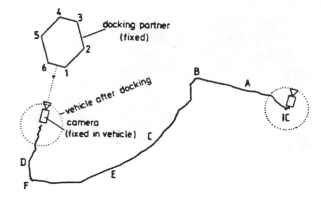

Figure 6. Mission sequence for docking phases:
A: approach to good visibility.
B: interpretation of relative orientation, mission planning.
C: circular arc path to final docking direction D.
IC: starting position of vehicle.

the time variable (state dependent) state transition matrix

$$A = \begin{bmatrix} 1 & T & T^2/2 & 0 & 0 & 0 & 0 & -T/r \\ 0 & 1 & T & 0 & 0 & 0 & 0 & 0 \\ 0 & 0 & 1 & 0 & 0 & 0 & 0 & 0 \\ 0 & 0 & 0 & 1 & 0 & 0 & 0 & 0 \\ 0 & 0 & 0 & 0 & 1 & T & 0 & 0 \\ 0 & 0 & 0 & 0 & \omega_r^2 & 1 & 0 & 0 \\ 0 & 0 & 0 & 0 & 0 & 0 & 1 & Te/r \\ 0 & 0 & 0 & 0 & 0 & 0 & 0 & e \end{bmatrix} \quad (10)$$

and B is the input gain matrix

$$B = \begin{bmatrix} T^2 a/2 & 0 & 0 \\ aT & 0 & 0 \\ 0 & 0 & 0 \\ 0 & 0 & 0 \\ 0 & T^2 a_{rx}/2 & T^2 a_{ry}/2 \\ 0 & T a_{rx} & T a_{ry} \\ 0 & T^2 e\, a_{tx}/(2r) & T^2 e\, a_{ty}/(2r) \\ 0 & Te\, a_{tx} & Te\, a_{ty} \end{bmatrix} \quad (11)$$

with

$$r = \hat{r}_{k-1}$$
$$\omega_r = \hat{v}_t/r|_{k-1}$$
$$e = \exp(-v_r T/r)$$

Because of the unknown modelling errors and the disturbances acting on the vehicle, the predicted state x_k^* will be off the true state $x(t_k)$. Therefore x_k^* needs to be updated using some measurements.

Imaging model The imaging model combines 3-D geometric models of the scene with the laws of perspective projection in order to describe the positions of relevant scene features in the image as a function of the relative spatial position and orientation; this procedure is well known from 3-D computer graphics and not detailed here.

For each rigid body object in the scene, a 3-D geometric model is used to describe the coordinates of relevant features Φ_i in an object-centered coordinate frame X_R, Y_R, Z_R, for example, with the Z_R axis as the symmetry axis (if existent) and the origin at an assumed center of gravity or in the base plane (see Figure 8). This yields for the position p_i of each feature Φ_i three space components, summarized as

$$P_i^T = [X_{Pi}, Y_{Pi}, Z_{Pi}] \quad (12)$$

The position of the camera projection center F on the vehicle is given by

$$d^T = [X_K, Z_K, X_F, Z_F] \quad (13)$$

This yields for the given state vector x_k^* of the dynamical model the horizontal coordinate y_{ci} and the vertical coordinate z_{ci} of the feature Φ_i in the image as one special two-component contribution to the measurement vector:

$$[\dot{y}_{ci}, \dot{z}_{ci}]^T = g(x_k^*, p_i, d) \quad (14)$$

Features lying off the predicted positions by more than a threshold ϵ_{yi} or ϵ_{zi} (which may be adjusted as a byproduct of the Kalman filter), or which cannot

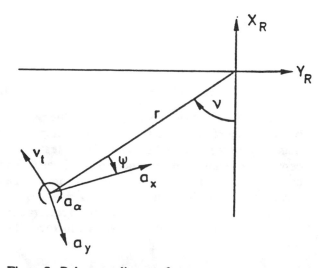

Figure 7. Polar coordinate reference system.

be aligned with some other feature in the vicinity (disturbances may cause the vision system to track a similar looking feature) are ignored. The remaining features define the measurement vector

$$y_k = [y_{c1}, z_{c1}, y_{c2}, z_{c2}, \ldots, y_{cj}, z_{cj}]^T$$
$$= g_k(x_k, p, d) + w_k \qquad (15)$$

with the time variable dimension $m_k = 2j$. The measurement disturbance vector w_k is assumed to have a Gaussian distribution and a diagonal measurement covariance matrix R_k.

Recursive state estimation The main computational effort in a time-variable Kalman filter used for updating the state variables is the computation of the gain matrix K, requiring the inversion of an $m \cdot m$ matrix and the update and time propagation of the covariance matrix P of the state estimation errors (see Bierman 1977; Maybeck 1979). However, the state update can be performed efficiently through sequential scalar processing of the m_k measurements. Space, here, does not allow us to go into details (see Wünsche 1986, 1987).

Several Kalman filter approaches have been implemented and tested numerically. The best and most time-efficient results were obtained using the U–D factorized Kalman filter due to Bierman (1977), where the covariance matrix P is factorized into UDU^T, with the upper unit triangular matrix U and the diagonal matrix D to be updated and time-propagated instead of P. With the extensions for vector-stored U and A matrices, as described in Thornton and Bierman (1980), and with modifications for more efficient indexing, a FORTRAN program used for real-time motion estimation takes only 22 ms on a VAX 750 for the complete state- and U–D-update and prediction when tracking three features in the scene ($m_k = 6$).

The new state estimate $\hat{x}_k$ is then used for determining the vehicle control. However, because of the delay time caused by the feature extraction algorithms (corner positions taking about 70 to 80 ms on the parallel processors of the BVV 1 using Intel 8085 CPUs), the feature positions transmitted at time t_k originally belong to t_{k-1}. Thus, $\hat{x}_k$ is the estimated vehicle motion state of time t_{k-1}, too. Therefore the motion state is extrapolated via Eq. (9), in order to use x^*_{k+1} for computing the reaction jet control commands for the three-point bang–bang controllers.

Feature selection Usually there will be more features perceptible in the scene than there are processors available in the BVV for tracking and more than are necessary for the motion state vector to be completely observable (which is a well defined term in modern systems theory).

Therefore an algorithm is needed for selecting those m features out of the s available ones, the measurement equations of which allow the best state update possible; for instance those which are not only well perceptible but which also give the best conditioned measurement equations. This condition may be computed by evaluating the determinant of the pseudo-inverse of the properly scaled and weighted Jacobian C, as appearing in the Gauss-Markov estimator, which is given by

$$\delta \hat{x}_R = S \cdot \delta \hat{x}_{NR}$$
$$= S\{(C_R S)^T R^{-1}(C_R S)\}^{-1}(C_R S)^T R^{-1}(y - y^*) \quad (16)$$

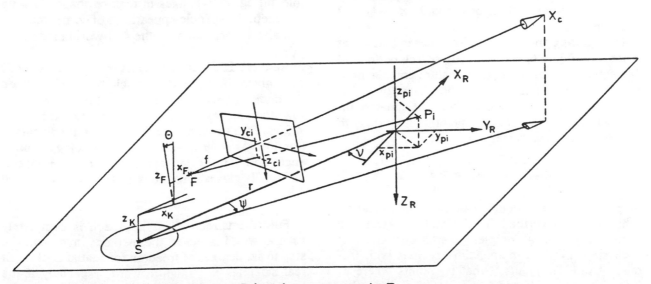

Figure 8. Perspective projection of feature P_i into the camera at point F.

Here the index R corresponds to the reduced state vector $[\psi, \Theta, r, \nu]^T$ with $n_R = 4$, C_R being the Jacobian of Eq. (15) with the zero columns removed, R being the diagonal measurement covariance matrix to be determined off line beforehand, and S being a diagonal scaling matrix for balancing the state variables such that the Gauss-Markov estimator minimizes the errors $(\delta \bar{x}_N)^T (\delta \bar{x}_N)$ instead of $(\delta \bar{x})^T (\delta \bar{x})$, with $\delta \bar{x} = (\delta x_R - \delta \hat{x}_R)$. With the definition

$$C_N = \sqrt{R^{-1}} \, C_R S \qquad (17)$$

the performance index to be maximized is written as

$$J = |C_N^T \, C_N| \qquad (18)$$

which involves computation of a $4 \cdot 4$ determinant. Rather than having to evaluate Eq. (18) for each of the $\binom{s}{m}$ possible feature combinations, an expression is sought giving the change in J if one feature at a time is replaced by another, allowing a gradient search technique. This is found by expanding Eq. (16) into a recursive Gauss-Markov estimator, which transforms Eq. (18) into

$$J = |\sum_i C_i^T \, C_i| \qquad (19)$$

where C_i is the $2 \cdot 4$ matrix with those two rows of C_N that correspond to the y_{ci} and z_{ci} measurements of feature Φ_i. Using further assumptions and matrix identities, Eq. (19) is reduced to a $2 \cdot 2$ determinant as detailed in Wünsche (1986). Due to efficient implementation, evaluation of Eq. (19) requires only 16 multiplications, 2 divisions, and $9\,(m-1) + 7$ additions.

Only during the initialization phase, that is, at point B of Figure 6, is a complete search performed evaluating J for each possible feature combination. In the following real-time motion phase, only that feature is looked for which gives the greatest increase in J when replacing some other feature, with the additional requirement of

$$J_{new} = \alpha \cdot J_{old}, \quad \alpha > 1 \qquad (20)$$

to prevent too frequent switching. In simulation runs it has been verified that this method selects the same features as a complete search performed at each step, even over long time periods with frequent "singular points," that is, points where at least two of the selected features start disappearing

or become occluded. Figure 9 shows the time history of the performance index and corresponding feature distributions selected automatically for a real-time test run. In the initial approximation phase ($t \leq 48$ s) the performance index increases steadily; then, the circumnavigation phase begins. At about 55 s a new surface area becomes visible at the left hand side causing two corner features to lie close to each other in the image; this deteriorates perceptibility. Therefore, first the upper feature tracker is shifted to the corner right next to it. Since the reassignment process takes two interpretation cycles, the performance index drops significantly for the next cycle (downward needle peaks). Shortly afterwards, the lower left window is shifted to the right for the same reason. At around 65 s the circumnavigation has progressed so far that the lower left corner of the newly visible surface has become clearly distinguishable; therefore, the lower left window is shifted to this feature (see third box in top row of Figure 9). At around 100 s, when the image of the docking partner has become almost symmetric, an additional feature tracker is invoked: it is identical to the one used for pole balancing (see section 3) and is assigned the task of searching and tracking a rod-like pointer standing out in front of the docking demarcation. This added feature introduces a quantum jump in the value of the performance index (transition from region 6 to 7 in Figure 9).

If a sudden occlusion of a feature, such as by an unexpected object occurs, after a short period of repeated trials at the old feature position a new combination of features is selected, yielding the next best performance index. Due to the strong perspective distortion of imaged features at small distances, the corner extraction algorithms have to be tolerant against changes in feature shape. In a future step, the specific appearance of corner features will also be predicted by the 4-D world model.

Results In Figure 10 results of a test run over 90 s are given, in which the vehicle first turns toward the docking partner (ψ-reduction for $t < 9$ s, first row) then moves towards it (R-reduction for $13 < t < 20$ s, fifth row), circumnavigates it at constant R (for 20 s $< t < 60$ s, last two rows, ν being the polar angle and VT the tangential velocity component) and finally closes in for docking ($t > 60$ s, see R and VR).

Future extension This approach is completely general with respect to the methods involved. The step to six degrees of freedom for spatial docking of real satellites is straightforward, especially if the total motion can be decomposed into decoupled, only slightly interfering submotion groups; this holds true for usual docking maneuvers dictated by safety aspects.

However, the method may also be readily adapted to docking of ships or land vehicles, such as in berthing or car parking.

5. Road Vehicle Guidance

In this section the guidance task for high speed road vehicles will be discussed demonstrating the specific application of the integrated 4-D approach described in general terms in the companion paper.

5.1 Concept for Autonomous Road Vehicle Guidance

In a coordinate frame fixed to the vehicle when moving along the road, the geographically fixed road curvature appears in the car as a temporally variable state of the environment. Since the generic law, according to which highways are built (clothoids), and the ranges for the parameters yielding reasonable results are known, effective filtering

methods may be based on this road model for smoothing noisy image processing data. By these methods, the relevant road parameters are estimated recursively, exploiting the dynamical model for the road while driving on it at speed V. These parameters are (a) the actual curvature and (b) the rate of curvature change with arc length (differential geometry road skeleton model). The image of the road in a look-ahead distance, however, also depends on the position and the orientation of the camera relative to the road. If the camera position and its orientation in the vehicle are fixed, then the state variables of the vehicle relative to the road (lateral position y_v and heading angle ψ) determine its perspective image (see Figure 11). This holds true spatially. Temporal continuity conditions result from the vehicle having only limited mobility: Its wheels revolve in a plane normal to their axis; the sliding angle β due to soft tires and slipping is small, but not negligible. From this, side constraints in the form of differential equations for vehicle motion result: If the vehicle does have an orientation in azimuth relative to the road not equal to zero, a lateral offset y_v will result in the future in general;

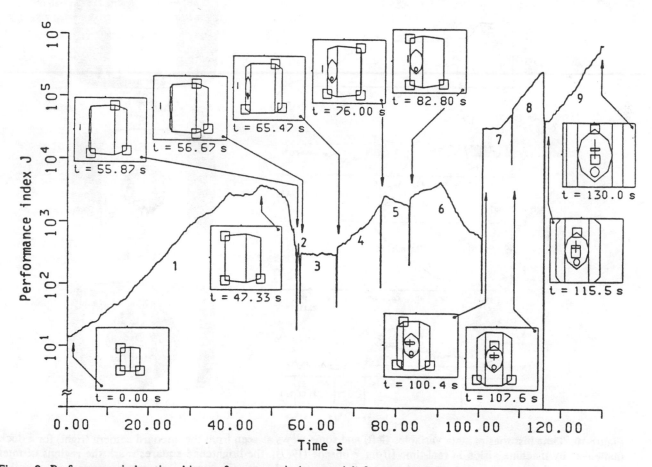

Figure 9. Performance index time history for automatic (sequential) feature selection.

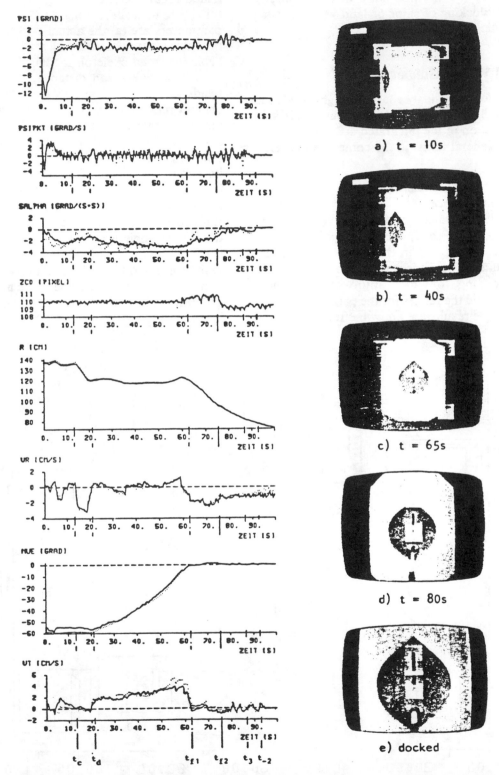

Figure 10. Time histories of state variables (left) and some views as seen from the onboard camera (right) for a docking maneuver by machine vision in real-time [from Wünsche (1987)]; the brightened squares mark the regions of interest evaluated by the system at that moment.

this, in addition, depends on the centrifugal force ($\sim V^2$) and the steering control. Introducing the knowledge of these interactions into the process of image sequence interpretation yields again a very efficient recursive approach for estimating the entire state vector of the vehicle. Though only some feature positions are being measured, all position and speed components of the vehicle relative to the road are determined, exploiting always the last image of the sequence only (Dickmanns and Zapp 1985, 1986, 1987; Zapp 1988). Figure 12 shows the cooperation of the two dynamical models for curvature determination (upper part) and for vehicle state estimation relative to the road (lower part) in the feedback loop.

Road curvature c determined in the look-ahead range is not only used for driving the anticipatory part of the lateral control u_{alat}, but also for automatically adapting longitudinal speed V. This is adjusted in such a way that the lateral acceleration $a_y = cV^2$ stays below a preset limit value (e.g., 0.1 of Earth gravity g, i.e., ~1 m/s^2). Both in simulations with real sensors in the loop and in real experiments with our test vehicle for autonomous mobility and computer vision, VaMoRs, this method has proven to be very efficient.

Other vehicles as partners in road traffic may be observed and tracked using similar methods as in section 4. This is presently under development. For achieving both wide-angle coverage in the near range and good resolution farther away for early obstacle recognition, a combination of two cameras with different focal lengths has been selected. In order to further increase visual flexibility, the two cameras are mounted on a two-axis, fast pan and tilt platform ZP which is controlled by a microprocessor ZPP integrated into the image sequence processing system BVV 2 (see Figure 13). This allows optical gaze anchoring on features or feature combinations (fixation) via image sequence processing. Active gaze control is considered to be an essential aspect of efficient real-time vision systems since it allows flexible control and allocation of high resolution sensing capability combined with reduction of motion blur during fixation periods.

In the upper right of Figure 13 a typical road scene is depicted; only the wide angle image will be treated in the sequel. Those areas of utmost importance to the task are covered by rectangular regions of interest ("windows") marked 1 through 6. The large number of windows has been selected for redundancy reasons to obtain robust road recognition under perturbations. Each rectangular window may contain up to 4K pixel data and have any shape, position, size and sparseness; these parameters may be adjusted by software control during runtime. In the BVV 2, PPs 1 to 6, each independently programmable, are ordered to acquire image data in windows 1 to 6. Each PP searches for linear contour elements under a preferred slope in its window using correlation methods as described in section 3.2 of the companion paper. The slope is prescribed by

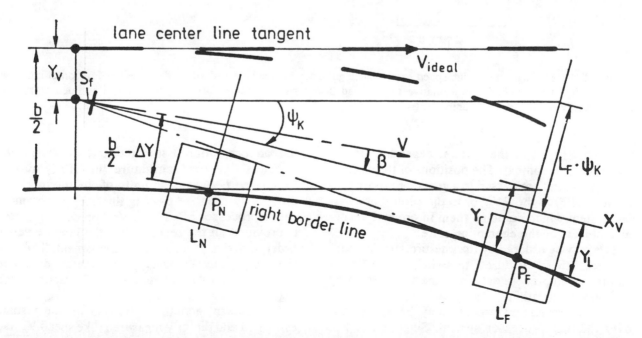

Figure 11. Top view of general imaging situation with lateral offset y_v, nontangential viewing direction ψ_K and road curvature y_c; two preview ranges L_N and L_F.

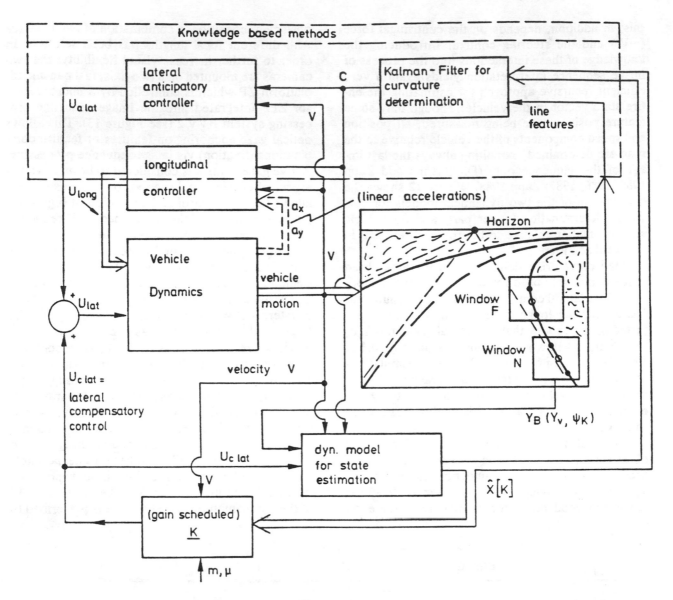

Figure 12. Block diagram for high speed road vehicle guidance by machine vision: model based estimation of curvature C (upper right) and vehicle state relative to the road (lower right); anticipatory (upper left) and compensatory state feedback (lower left) control components.

the higher levels of the system, depending on the interpretation context. The positions of the contour elements found are passed to a general purpose processor (GPP) in the BVV or to the central processing system which combine them in order to recognize objects or the egomotion state.

PPs may be ordered by the interpretation process to track "their" features, to switch to a different feature, or to process data from the same region by a different algorithm.

The interpretation process may be spread out over the multiprocessor system: Single GPPs will usually be allocated to the recognition and parameter or state estimation of one object, such as GPP1

to road recognition. Estimated state variables or parameters like road curvature (in GPP1) may be exchanged between the units of the interpretation process by message passing through the communication processor SP. The central processing system is provided all the results from the lower levels in order to arrive at a situation assessment. This may then be used as the basis for parameter adjustment in the behavioral mode running, or for mode switching.

The control actuations depend on the visually estimated and directly measured (like speed V) state variables. They are implemented by an input/output controller.

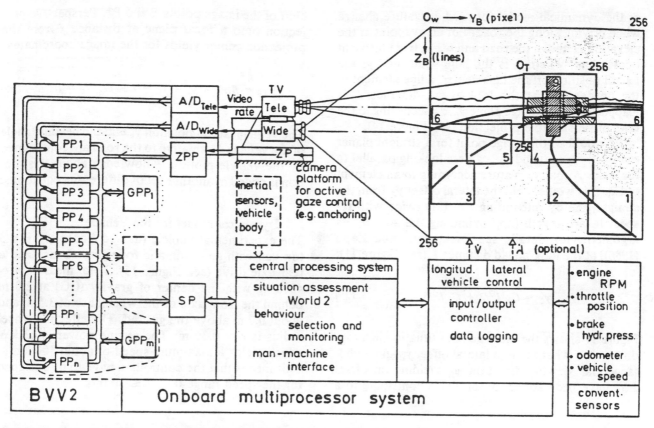

Figure 13. System architecture for a vision controlled autonomous road vehicle.

In the interpretation process, exploiting the predicted state and feature positions y^*, the incoming measurement data are screened, outliers are removed, and the image sequence analysis performed by the PPs is supervised and adjusted to the present interpretation status.

5.2 Road Model for Recursive Estimation

Here, only those parts of the visual recognition process necessary for guiding the vehicle along a free lane are given, omitting other objects and traffic information. High speed roads are modelled as a sequence of N arcs with linear curvature models defining the skeleton R of a band with constant width b. With λ as absolute and l as relative run length in each segment this may be written

$$R = \sum_{i=1}^{N} (C_{oi} + C_{li}l_i)$$

$$0 \le l_i \le \lambda_i - \lambda_{i-1} \qquad (21)$$

$$C_{ji} = 0 \text{ outside}$$

C_o is the constant curvature part of the model and $C_1 = dC/dl$ is the linearly varying part over arc length. The parameters of this structural model (C_{oi}, C_{li} and λ_i) have to be determined from the visual input for a certain range L in front of the vehicle.

By definition of the road model Eq. (21), using the chain rule, there follows for the curvature C_V at the location of the vehicle traveling at speed V tangentially to the road

$$\dot{C}_V = \frac{d}{dt}(C_V) = \frac{d}{dl}(C_V)\frac{dl}{dt} = C_1 V \qquad (22)$$

$$\dot{C}_1 = \frac{d}{dt}(C_1) = \begin{cases} 0 \text{ on one segment} \\ V\delta(1 - \lambda_i), \text{ a Dirac-impulse} \\ \text{at a transition point } \lambda_i \end{cases} \qquad (23)$$

for practical purposes $\dot{C}_1$ is considered to be random noise $n_R(t)$.

These equations may be written in state space form

$$\begin{bmatrix} \dot{C}_V \\ \dot{C}_1 \end{bmatrix} = \begin{bmatrix} 0 & V \\ 0 & 0 \end{bmatrix} \begin{bmatrix} C_V \\ C_1 \end{bmatrix} + \begin{bmatrix} 0 \\ n_R(t) \end{bmatrix} = F_R x_R + v_R$$

$$(24)$$

as the dynamical model for road curvature change while driving along the road. For an eyepoint at the elevation H above the road and $b/2-y_v$ from the right border line, where b is the lane width and y_v the lateral position off the lane center, a line element of the border line at the look-ahead distance L is mapped onto the image plane y_B, z_B according to the laws of perspective projection (see Figure 14).

P^* would be the image point for a straight planar road with $y_v = 0$ and the camera looking parallel to the road. A road curvature according to an element of Eq. (21) would yield the lateral offset y_C from the straight line by integrating Eq. (21) twice with respect to arc length [approximating the sine by its argument; for details see Dickmanns and Zapp (1986)] at the look-ahead distance L (see Figure 11):

$$y_C \approx C_V L^2/2 + C_1 L^3/6 \qquad (25)$$

Figure 11 shows the more general situation in a top-down view where both a lateral offset y_v and a non-tangential viewing direction ψ_K yielding an offset $L\psi_K$ have been added, all of which contribute to a shift of the image points P and P^*. Perspective projection onto a focal plane at distance f from the projection center yields for the image coordinates

$$y_B = f(b/2 - y_v + y_C - L\psi_K)/L; \; z_B = fH/L \qquad (26)$$

The vehicle lateral position y_v and the heading angle of the vehicle ψ_K relative to the road are constrained by the vehicle motion. Knowledge about this motion is coded in another set of differential equations.

5.3 Dynamical Model for the Vehicle

The experimental vehicle (named VaMoRs), is a 5 ton van with an automatic four gear change and a rear axle drive (see Figure 15). Total mass is $m \approx 4000$ kg with the center of gravity (CG) at 2.0 m behind the front axle of the wheelbase of 3.5 m and about 0.7 m above the ground. The effective wheel radius is $r \approx 0.36$ m. The engine power is 63 kW which yields a maximum speed of about 100 km/h. This means that the controllers have to be designed for the speed range $0 < V \leq 30$ m/s.

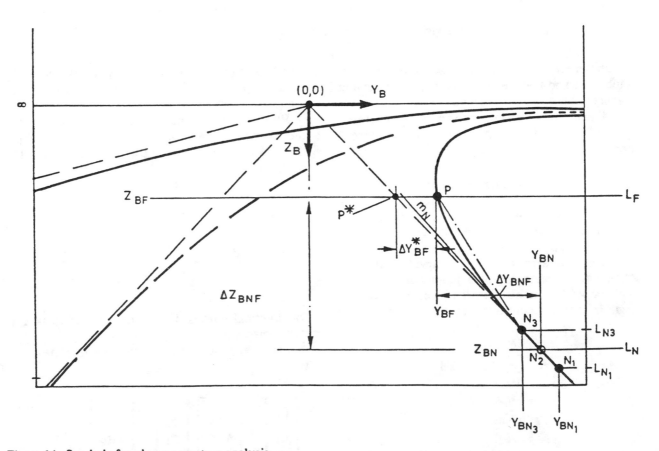

Figure 14. Symbols for planar curvature analysis.

Lateral dynamics. A planar "bicycle" substitution model according to Strackerjan (1975) has been applied. Two degrees of freedom are the sideslip angle β and the inertial yaw rate $\dot{\psi}_v$, described by the following equations of motion:

$$mV\dot{\beta} + (U_f + U_r + \mu c_f + \mu c_r)\beta$$
$$- \left(mV + \mu c_f \frac{l_f}{V} - \mu c_r \frac{l_r}{V}\right)\dot{\psi}_v$$
$$= -(U_f + \mu c_f)\delta \qquad (27)$$

$$I_z \ddot{\psi}_v + \frac{\mu}{V}(c_f l_f^2 + c_r l_r^2)\dot{\psi}_v - \mu(c_f l_f - c_r l_r)\beta$$
$$= (U_f l_f + \mu c_f l_f)\delta \qquad (28)$$

where I_z is the moment of inertia around a vertical axis, $c_{f,r}$ are normal side force coefficients applicable to front and rear axles. $l_{f,r}$ are the distances of the CG to the front and rear axle respectively, $U_{f,r}$ are the circumferential forces on the wheels, and μ is the friction coefficient between wheels and road.

δ is the steer angle of the front wheels. In the vehicle a computer controlled stepping motor serves as actuator for steering. It has been modelled as an integrator

$$\dot{\delta} = k_\delta \cdot u \qquad (29)$$

where $\dot{\delta}$ is limited to 15 deg/s.

The lateral position y_v on the road is constrained by the differential equation

$$\dot{y}_v = V(\Delta\chi + \delta \cdot l_r/d) = V(\psi_K - \beta + \delta \cdot l_r/d) \quad (30)$$

where $\Delta\chi$ is the path azimuth angle relative to the

Figure 15. VaMoRs, the experimental vehicle for autonomous mobility and machine vision.

road, $d = l_r + l_f$ is the axle distance, and the road-oriented vehicle yaw angle ψ_K is linked to the inertial yaw angle ψ_v by

$$\dot{\psi}_K = \dot{\psi}_v - C_V V \qquad (31)$$

The last term represents the temporal road heading change due to curvature C_V and vehicle speed V.

5.4 Integrated State Space Model for the 4-D Approach

The imaging Eq. (26) contains contributions both from the road (b and y_C) and from the vehicle state (y_v and ψ_K).

Combining Eqs. (26) to (31) one obtains a state space model for lateral dynamics relative to a road with visual measurements taken at the look-ahead distance L. It is of fifth order with the state variables

$$x^T = (\dot{\psi}_v, \beta, \psi_K, y_v, \delta)$$
$$\dot{x} = F_v x + gu + hC_V \qquad (32)$$

where

$$F_v = \begin{bmatrix} f_{11} & f_{12} & 0 & 0 & f_{15} \\ f_{21} & f_{22} & 0 & 0 & f_{25} \\ 1 & 0 & 0 & 0 & 0 \\ 0 & -V & V & 0 & f_{45} \\ 0 & 0 & 0 & 0 & 0 \end{bmatrix} \quad g = \begin{bmatrix} 0 \\ 0 \\ 0 \\ 0 \\ k_\delta \end{bmatrix} \quad h = \begin{bmatrix} 0 \\ 0 \\ -V \\ 0 \\ 0 \end{bmatrix}$$

The component $\dot{\psi}_K$ contains a contribution due to road curvature C_V. The elements f_{ij} depend upon the parameters V, m, μ, the side force coefficients, the vehicle CG location, and the circumferential forces on the wheels.

The dominant effects caused by parameter changes are due to speed variations, which also are the most frequent ones. Adaptations to changes in mass m are—if at all—only necessary at the beginning of a mission. Changes in friction coefficient μ may be due to road surface parameters including weather conditions. Since speed is the only easily measurable variable and has the largest influence on vehicle behavior under normal conditions, its effect on the model is always taken into account, that is, the nonlinear physical model is approximated by a time-varying mathematical model with speed V governing the coefficients (Zapp 1985). In the experimental vehicle VaMoRs, speed is derived from the digital odometer system.

The two systems of Eqs. (24) and (32) may be

combined to yield a 7 · 7 matrix which nicely shows the interactions from the road onto the vehicle dynamics

$$
\begin{array}{c}
\text{Column no.} \quad 6 \quad 7 \\
\begin{bmatrix}
& & \vdots & 0 & 0 \\
& & \vdots & 0 & 0 \\
F_v & & \vdots & -V & 0 \\
(5 \cdot 5) & & \vdots & 0 & 0 \\
& & \vdots & 0 & 0 \\
\hline
\underline{0} & & \vdots & 0 & V \\
(2 \cdot 5) & & \vdots & 0 & 0
\end{bmatrix}
\begin{bmatrix}
\dot{\psi}_v \\
\beta \\
\psi_K \\
y_v \\
\delta \\
\hline
C_V \\
C_1
\end{bmatrix}
= F_t x_t
\end{array}
\qquad (33)
$$

The main diagonal blocks are treated by separate Kalman filters. The effect of column 6 in the upper right block is handled separately as an anticipatory part for the lateral control (steer angle to achieve a curved trajectory according to C_V). The filter for road curvature estimation is a complete Kalman filter while for vehicle state estimation a quasi-stationary filter with velocity dependent coefficients is being used. In addition, dependencies on other parameters like m and μ may be taken into account. The measured output variables are the borderline positions y_B in the image, corresponding to the near and far windows (index N, F in figures 11 and 12). The look-ahead distances L_N and L_F for a flat road model are directly related to the vertical position z_B in the image.

For sampled measurement and control the state transition matrix A_t corresponding to F_t and the control coefficients b corresponding to g [Eq. (32)] for a given sampling period T ($=100$ ms for VaMoRs) are obtained by standard state space methods.

Measurement equations. Fixing the look-ahead range L by choosing the z_B coordinate for feature tracking correspondingly, Eq. (26) becomes, with (25):

$$
y_B/f = \frac{b}{2L} - \psi_K - y_v/L + C_V L/2 + C_1 L^2/6. \qquad (34)
$$

Since the mapping factor f and the elevation H of the camera above the road are known, L is given by Eq. (26) (right side) for the viewing direction parallel to the road. With the state vector x_t according to Eq. (33) the measurement equation for the y component of one feature is linear in the state variables and is obtained as

$$
y_i = c_i x_t = f [0 \; 0 \; -1 \; -1/L_i \; 0 \mid L_i/2 \; L_i^2/6] x_t \qquad (35)
$$

where

$$
x_t = [\dot{\psi}_v \; \beta \; \psi_K \; y_v \; \delta \mid C_V \; C_1]^T
$$

This clearly shows that the lateral offset y_v (fourth component) is best determined from small values L_i (near range), that is, large values of z_B at the bottom of the image. On the contrary, curvature effects are best detected in the far range (large L_i or small values z_B); however, since measurement accuracy decreases with distance and visibility becomes poorer, a compromise has to be found.

5.5 Results with Test Vehicle VaMoRs

In the last quarter of 1986 VaMoRs became available for autonomous driving tests. The first task performed is shown in Figure 16. Initially the vehicle is parked at the start position (upper left) in the vicinity of a lane boundary marking according to the German standard. After searching and detecting this line the two real-time tracking windows in Figure 12 are positioned in the near and far look-ahead range. Then the vehicle starts moving while the line marking is tracked laterally. The system steers the vehicle in such a way that it moves at a preset distance beside the line with its velocity vector approximately tangential to it, exploiting curvature prediction. As long as the line is straight and the maximum speed, set in advance, is not yet achieved, the vehicle accelerates. When the curve (a tightening spiral) is approached, curvature becomes nonzero, the anticipatory lateral control u_a is computed, and speed is adjusted to the estimated curvature.

At point E the line marking ends at a border line between the bright concrete pavement (outer rink) and the dark basalt pavement of the inner rink. The vehicle continues its path following this border line at a speed of ≈ 25 km/h which corresponds to a preset limit for the lateral acceleration of 1 m/s^2 (0.1 g).

Further tests were performed in 1987 on the construction site of a new stretch of Autobahn close to Munich. Fully autonomous runs starting from rest have been performed under various road and weather conditions, including bright sunshine and light rain as well as road surfaces with and without lane markings. Increasing the maximum speed limit step by step, in August 1987 the maximum speed of the vehicle ($V_{max} \approx 96$ km/h on a level surface) was reached. In order to obtain some results with respect to reliability, the total run length of more than 20 km was driven several times (in both directions) without a need for intervention by the safety driver in the driver's seat.

Lane changes from right to left and back have been performed on free lanes as well as highway entry maneuvers from the acceleration strips, even under the condition of light rainfall.

Tracking of lower standard roads (without autonomous vehicle control) with a group of up to seven cooperating processors in the BVV 2 has been demonstrated in Mysliwetz and Dickmanns (1986, 1987).

Driving on these types of roads with small radii of curvature was investigated in 1988. For this purpose a window configuration similar to the one shown in the upper right of figure 13 was used. Each of the windows 1 to 6 tracks its feature combination independently; in addition, the platform ZP is controlled in the azimuth direction in such a way that the centroid of the set of windows remains in the center of the image. Thus, the camera "looks into the curve" automatically.

Test runs on normal roads without any lane markings, and even beneath trees casting shadows on the road, have been performed. Speeds up to about 50 km/h have been driven autonomously under these conditions. On the other extreme, the capability of following this type of road has been demonstrated under the condition of light rainfall with the wipers operating.

This shows that road vehicle guidance by machine vision is maturing to a state where practical applications in simple tasks may come into reaching distance. More computing power will be needed to handle visually more complex scenes with obstacles and other vehicles.

Combining the results described in the previous and the presently discussed areas, tests for road vehicle guidance in the presence of other objects and vehicles are in preparation.

6. Conclusions from Application Examples

The 4-D approach offers a sufficiently rich embedment for the interpretation of many different processes in space and time. The approach is computationally efficient. The models may be time varying (e.g., vehicle dynamics as function of speed); even then, the method works reasonably well provided the model rate of change is slow as compared to the state variable dynamics and the image frequency. The results achieved up to now in four application areas are encouraging; they have been obtained with sampling rates from 8 to 25 Hz. Development steps are being investigated presently toward the capability of handling more complex situations

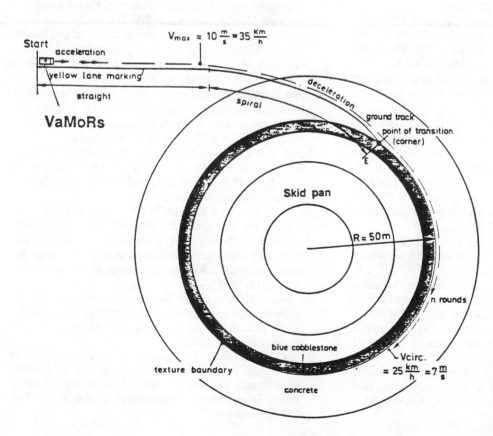

Figure 16. Initial test run for VaMoRs autonomous visual guidance.

where (several) objects of unknown shape may occur having unknown dynamical characteristics.

In order to be able to handle more complex tasks efficiently, the introduction of artificial intelligence methods on top of the 4-D state representation seems to be favorable. Note that in this approach actions are an integral part of the internal representation.

Observing and analyzing state variable time histories may provide a direct access to temporally deep reasoning including frequency domain methods.

REFERENCES

Bierman GJ (1975) Measurement updating using the $U–D$ factorization. Proc. IEEE Control and Decision Conf., Houston, Texas, pp 337–346

Bierman GJ (1977) Factorization methods for discrete sequential estimation. Academic Press, New York

Bizzi E (1974) The coordination of eye-head movement. Scientific American 231, pp 100–106

Broida TJ, Chellappa R (1986) Estimation of object motion parameters from noisy images. IEEE Transactions on Pattern Analysis and Machine Intelligence, Vol. PAMI-8, No. 1, pp 90–99

Dichgans J, Bizzi E, Morasso P, Tagliasco V (1973) Mechanisms underlying recovery of eye head coordination following bilateral labyrinthectomy in monkeys. Exp. Brain Res., 18, pp 548–569

Dickmanns ED (1985a) 2D-object recognition and representation using normalized curvature functions. In: Hamza MH (ed) Proc. IASTED Int. Symp. on Robotics and Automation '85. Acta Press, pp 9–13

Dickmanns ED (1985b) Vermessung und Erkennung von Figuren mit linearen Krümmungsmodellen. UniBwM/ LRT/WE 13/FB/85-2

Dickmanns ED (1986) Computer vision in road vehicles— chances and problems. Preprint, ICTS Symposium on Human Factors Technology for Next-Generation Transportation Vehicles, Amalfi, Italy

Dickmanns ED (1988a) Object recognition and real-time relative state estimation under egomotion. In: Jain AK, (ed) Real-Time Object Measurement and Classification. Springer-Verlag, Berlin, pp 41–56

Dickmanns ED (1988b) Computer vision for flight vehicles. Zeitschrift für Flugwissenschaft und Weltraumforschung (ZFW) 12, pp 71–79

Dickmanns ED, Eberl G (1987) Automatischer Landeanflug durch maschinelles Sehen. (DGLR-Jahrbuch 1987) Jahrestagung der DGLR, Berlin, pp 294–300

Dickmanns ED, Wünsche HJ (1986a) Regelung mittels Rechnersehen. Automatisierungstechnik (at) 34, 1/ 1986, pp 16–22

Dickmanns ED, Wünsche HJ (1986b) Satellite rendezvous maneuvers by means of computer vision. Jahrestagung DGLR München. Jahrbuch 1986 Bd 1, DGLR, Bonn, pp 251–259

Dickmanns ED, Zapp A, Otto KD (1984) Ein Simulationskreis zur Entwicklung einer automatischen Fahrzeugführung mit bildhaften und inertialen Signalen. Proc. 2. Symp. Simulationstechnik, Wien, Informatik-Fachberichte 39. Springer-Verlag, Berlin

Dickmanns ED, Zapp A (1985) Guiding land vehicles along roadways by computer vision. Proc. Congres Automatique 1985, AFCET, Toulouse, pp 233–244

Dickmanns ED, Zapp A (1986) A Curvature-based scheme for improving road vehicle guidance by computer vision. In: Mobile Robots, SPIE-Proc. Vol. 727, Cambridge, MA, pp 161–168

Dickmanns ED, Zapp A (1987) Autonomous high speed road vehicle guidance by computer vision. Preprint, 10th IFAC-Congress, München, Vol. 4, pp 232–237

Duff MJB (1986) Intermediate-level image processing. Academic Press, London

Eberl G (1987) Automatischer Landeanflug durch Rechnersehen. Dissertation, Fakultät für Luft- und Raumfahrttechnik der Universität der Bundeswehr München

Fichte JG (1792) Versuch einer Kritik aller Offenbarung. Hartung, Königsberg (appeared anonymous at first)

Gennery DB (1981) A feature-based scene matcher. Proceedings 7th International Joint Conference on Artificial Intelligence, pp 667–673

Gennery DB (1982) Tracking known three-dimensional objects. Proceedings American Association for Artificial Intelligence, Pittsburgh, pp 13–17

Graefe V (1983a) A Pre-processor for the real-time interpretation of dynamic scenes. In: Huang TS (ed) Image sequence processing and dynamic scene analysis. Springer-Verlag, Berlin, pp 519–531

Graefe V (1983b) Ein Bildvorverarbeitungsrechner für die Bewegungssteuerung durch Rechnersehen. In: Kazmierczak H (ed) Mustererkennung 1983, NTG Fachberichte. VDE-Verlag, pp 203–208

Graefe V (1983c) On the representation of moving objects in real-time computer vision systems. In: Tescher AG (ed) Applications of digital image processing VI. Proceedings of the SPIE, Vol. 432, pp 129–132

Graefe V (1984) Two multi-processor systems for low-level real-time vision. In: Brady JM, Gerhardt LA and Davidson HR (eds) Robotics and artificial intelligence. Springer-Verlag, Berlin, pp 301–308

Graefe V, Kuhnert KD (1987) Low-level vision for advanced mobile robots. In: Martin T (ed) International advanced robotics programme—proceedings of the first workshop on manipulators, sensors and steps towards mobility. Kernforschungszentrum Karlsruhe, KfK 4316, pp 239–246

Haas G (1982) Meßwertgewinnung durch Echtzeitauswertung von Bildfolgen. Dissertation, Fakultät für Luft- und Raumfahrttechnik der Universität der Bundeswehr München

Haas G, Graefe V (1983) Locating fast-moving objects in TV-images in the presence of motion blur. In: Oosterlinck A and Tescher AG (eds) Applications of digital image processing V. Proceedings of the SPIE, Vol. 397, pp 440–446

Hegel GWF (1806) Phänomenologie des Geistes. In: Glockner H (ed) Hegel, Sämtliche Werke. Stuttgart, 1927

Kalman RE (1960) A new approach to linear filtering and prediction problems. Trans. ASME, Series D, Journal of Basic Engineering, pp 35–45

Kuhnert KD (1984) Towards the objective evaluation of low level vision operators. In: O'Shea T (ed) ECAI 84, Proceedings of the Sixth European Conference on Artificial Intelligence, Pisa, p 657

Kuhnert KD (1986a) A modeldriven image analysis system for vehicle guidance in real time. In: Wolfe WJ (ed) Proceedings of the Second International Electronic Image Week, CESTA, Nizza, pp 216–221

Kuhnert KD (1986b) A vision system for real-time road and object recognition for vehicle guidance. In: Marquino N (ed) Advances in intelligent robotics systems. Proceedings of the SPIE, Vol. 727, Cambridge, MA, pp 267–272

Kuhnert KD (1986c) Comparison of intelligent real-time algorithms for guiding an autonomous vehicle. In: Hertzberger LO (ed) Proceedings of the Conference on Intelligent Autonomous Systems, Amsterdam

Kuhnert KD (1988) Zur Echtzeit-Bildfolgenanalyse mit Vorwissen. Dissertation, Fakultät für Luft- und Raumfahrttechnik der Universität der Bundeswehr München

Kuhnert KD, Zapp A (1985) Wissensgesteuerte Bilfolgenauswertung zur automatischen Führung von Straßenfahrzeugen in Echtzeit. In: Niemann H (ed) Mustererkennung 1985. Springer-Verlag, Berlin, pp 102–106

Maybeck PS (1979) Stochastic models, estimation and control, Vol. 1, Academic Press

Meissner HG (1982) Steuerung dynamischer Systeme aufgrund bildhafter Informationen. Dissertation, Fakultät für Luft- und Raumfahrttechnik der Universität der Bundeswehr München

Meissner HG, Dickmanns ED (1983) Control of an unstable plant by computer vision. In: Huang TS (ed) Image sequence processing and dynamic scene analysis. Springer-Verlag, Berlin, pp 532–548

Mysliwetz B, Dickmanns ED (1986) A vision system with active gaze control for real-time interpretation of well structured dynamic scenes. In: Hertzberger LO (ed) Proceedings of the Conference on Intelligent Autonomous Systems, Amsterdam

Mysliwetz B, Dickmanns ED (1987) Distributed scene analysis for autonomous road vehicle guidance. Proc. SPIE Conf. on Mobile Robots, Vol. 852, Cambridge, MA, pp 72–79

Mysliwetz B, Dickmanns ED (1988) Ein verteiltes System zur Echtzeitinterpretation von Straßenszenen für die autonome Fahrzeugführung. In: Lauber R (ed) Prozeßrechnersysteme '88. Informatik-Fachberichte 167, Springer-Verlag, Berlin, pp 664–673

Nagel HH (1983) Overview on image sequence analysis. In: Huang TS (ed) Image sequence processing and dynamic scene analysis. Springer-Verlag, Berlin, pp 2–39

Popper KR, Eccles JC (1977) The self and its brain—an argument for interactionism. Springer International, Berlin

Reddy R (1978) Pragmatic aspects of machine vision. In: Hanson A and Riseman E (eds) Computer vision systems. Academic Press, New York, pp 89–98

Rives, P, Breuil E, Espian B (1986) Recursive estimation of 3D features using optical flow and camera motion. In: Hertzberger LO (ed) Proceedings of the Conference on Intelligent Autonomous Systems, Amsterdam, pp 522–532

Schopenhauer A (1819) Die Welt als Wille und Vorstellung. In: Löhneysen W (ed) Arthur Schopenhauer, Sämtliche Werke. Suhrkamp, Frankfurt a.M., (Nachdruck 1986)

Strackerjan B (1975) Theoretische Untersuchungen des dynamischen Lenkverhaltens von Personenkraftwagen. Automobil-Industrie 3/75, pp 49–56

Thornton CL, Bierman GJ (1980) UDU^T Covariance factorization for Kalman filtering. In: Leondes CT (ed) Control and dynamic systems, advances in theory and application, Vol. 16, Academic Press, New York, pp 177–248

Vollmer G (1986) Was können wir wissen? Bd 1: Die Natur der Erkenntnis, Bd 2: Die Erkenntnis der Natur. S. Hirzel Verlag, Stuttgart

Wünsche HJ (1983) Verbesserte Regelung eines dynamischen Systems durch Auswertung redundanter Sichtinformation unter Berücksichtigung der Einflüsse verschiedener Zustandsschätzer und Abtastzeiten. Report HSBwM/LRT/WE13a/IB/83-2.

Wünsche HJ (1986) Detection and control of mobile robot motion by real-time computer vision. In: Marquino N (ed) Advances in intelligent robotics systems. Proceedings of the SPIE, Vol. 727, Cambridge, MA, pp 100–109

Wünsche HJ (1987) Erfassung und Steuerung von Bewegungen durch Rechnersehen. Dissertation, Fakultät für Luft- und Raumfahrttechnik der Universität der Bundeswehr München

Zapp A (1985) Automatische Fahrzeugführung mit Sichtrückkopplung; Erste Fahrversuche mit einer Fernsehkamera als Echtbauteil im Simulationskreis. Universität der Bundeswehr München, LRT/WE 13/FB/85-1

Zapp A (1988) Automatische Straßenfahrzeugführung durch Rechnersehen. Dissertation, Fakultät für Luft- und Raumfahrttechnik der Universität der Bundeswehr München

Stereo Vision and Navigation in Buildings for Mobile Robots

DAVID J. KRIEGMAN, ERNST TRIENDL, AND THOMAS O. BINFORD

Reprinted from *IEEE Transactions on Robotics and Automation*, Vol. 5, No. 6, December 1989. Copyright © 1989 by The Institute of Electrical and Electronics Engineers, Inc. All rights reserved.

Abstract— A mobile robot that autonomously functions in a complex and previously unknown indoor environment has been developed. The omnidirectional mobile robot uses stereo vision, odometry, and contact bumpers to instantiate a symbolic world model. Finding stereo correspondences across a single epipolar line is adequate for instantiating the model. Uncertainty in sensor data is represented by a multivariate normal distribution, and uncertainty models for motion and stereo are presented; uncertainty is reduced by extended Kalman filtering. To execute a high-level command such as "Enter the second door on the left," a model is instantiated from sensing and motions are planned and executed. Experimental results from the fast, running system are presented.

I. INTRODUCTION

FOR MOBILE ROBOTS to be successful, they must quickly and robustly perform useful tasks in a complex, dynamic, and previously unknown environment. If there is going to be a mass proliferation of mobile robots, they must be easy to use and should not require programming. Instead, a set of tasks should be presented to the robot which would determine how to achieve them based on some understanding of the environment. The robot, with the aid of some general knowledge, may build as detailed a model of its environment as necessary. This paper presents results in that direction, particularly building a model from sensor information.

The goal of this work is to enable a mobile robot to autonomously operate in unknown surroundings, in particular buildings, without explicit cues or markers (i.e., the real world). The robot should function without *a priori* knowledge or map of the building, but should know what to expect in any building and use sensing to fill in the details. With the acquired knowledge and a task to achieve, the robot should determine what actions are necessary. Additionally, the system should be robust and reasonably fast so more than a few experiments can be performed; we are not happy to wait 15 min/m [23].

These goals are achieved by a combination of modeling, sensing, uncertainty reduction, planning, and action. Generic modeling describes what to expect in the environment (e.g., a

Manuscript received May 16, 1988; revised November 21, 1988. This research was supported in part under a Subcontract to Advanced Decision Systems S1093 S-1 (Phase II) "Knowledge Based Vision Task B," from a Contract to the Defense Advanced Research Projects Agency. Partial support was also provided by the Air Force Office of Scientific Research under Contract F33615-85-C-5106 "Basic Research in Robotics." D. J. Kriegman was supported by a fellowship from the Fannie and John Hertz Foundation.

The authors are with the Robotics Laboratory, Department of Computer Science, Stanford University, Stanford, CA 94305.

IEEE Log Number 8930461.

building contains walls, floors, doors, rooms, etc.). Given the constraints of this model, sensing (primarily vision) is used to instantiate the model of a particular building. Stereo correspondences across a single epipolar line adequately instantiate the model; details of the fast vision system are presented. Because neither sensors nor the interpretation of their data are perfect, uncertainty must be considered and hopefully reduced when building the model, making decisions, and planning actions. The planning system then determines what actions and further sensing are needed to attain a goal. Finally, a robot implements the plan through action.

The system has been implemented and runs in the hallways of our building using two motion planning strategies. It is fast, taking between 8 and 12 s per 1-m step on conventional computer hardware.

In the past couple of years, a large number of mobile robot projects have tackled many of these research issues including the use of multiple sensors, environmental modeling, and uncertainty reduction. At one extreme, world knowledge is assumed to be complete, and all possible actions are known *a priori* as with automated guided vehicles (AGV) which follow stripes, wires, or other markings [13]. In Moravec's pioneering work, the Stanford Cart built a world model of obstacle points found by nine-eyed stereo vision [23]. More recently, uncertainty grids fuse information from multiple sensors: stereo vision and ultrasonics. Though motion uncertainty can be taken into account when building the grid, no method for reducing that uncertainty is presented nor are symbolic models created [10]. Hilare [11] uses a laser range finder to find the 2-D boundary of free space which is partitioned into a graph of convex polygons. Based on the graph structure, nodes are grouped into rooms. A beacon determines robot location, easing problems with motion uncertainty; this issue has been considered more recently in [8]. With better sensing such as INRIA's trinocular stereo [2] along with methods for extracting geometric features from uncertain data [3], more detailed models can be created. An alternative approach to fast mobile robot vision is the work on Flakey at SRI [14]; line segments are sequentially tracked in monocular images yielding their 3-D location. On a larger scale, the problems of outdoor navigation have been attacked by the Autonomous Land Vehicle [29], [28] and the CMU Navlab [25], [12] which use multiple sensors to build a model of the environment. These approaches as well as ours build a model of the surroundings from sensor data; in some of these, sensor and/or motion uncertainty has been considered. At the other extreme, the MIT approach avoids any map or model building and opts for a

Fig. 1.   The robot.

more decentralized, behavioral approach to the production of complex behaviors [7].

II. The Experimental Testbed

The Stanford Mobile Robot (Fig. 1), known as the Mobile Autonomous Robot Stanford (MARS) or more affectionately, Mobi, is an omnidirectional vehicle (3DOF) with a novel three-wheel configuration built by Unimation West. The robot is essentially cylindrical with a diameter of 65 cm and height of 170 cm. The wheels, forming the edges of an equilateral triangle, have six contoured passive rollers along their chords, allowing the wheels to move sideways while driven. With a combination of driving the wheels about their major axis, and passive turning of the rollers, the vehicle can translate while simultaneously rotating. With this form of locomotion, the vehicle is well suited for indoor travel and maneuvering about tight obstacles. However, because of the small roller diameter, the vehicle is confined to traveling over relatively smooth terrain.

Each wheel motor contains a shaft encoder that is used for odometry and trajectory following. Trajectories are specified by two alternative methods: B-splines connect knot points to form a path with smoothly changing velocity. Alternatively, points along with a motion direction (postures) are connected by straight-line segments and clothoid curves to create trajectories having continuous, linearly changing curvature and smooth acceleration along the path [16].

The robot has four sensing modalities: vision, acoustics, tactile, and odometry. A stereo pair of cameras is mounted on a pan/tilt head, giving it two degrees of freedom. The acoustic system is composed of twelve Polaroid ultrasonic sensors, equally spaced about the circumference of the robot. These sensors return a distance that is proportional to the time of flight of an echoed acoustic chirp. To a first approximation, the acoustics find the nearest object within a 30° cone. Though the acoustics have very low angular resolution, they measure depth quite accurately, and are useful for measuring proximity during guarded moves. As a last line of defense, if the vision and acoustic systems miss an object, a ''tactile'' sensing system, composed of twelve bumpers with internal tape switches, detects collisions. The bumpers also aid navigation through tight areas such as doorways where the door frame is

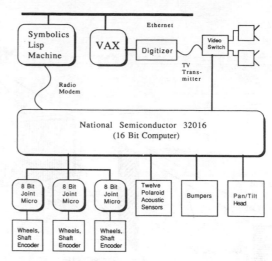

Fig. 2.   Computational architecture and subsystems.

within the minimum ultrasonic range and not within the region of stereo vision. Finally, odometry determines the robot's position.

The robot's distributed computational system resides both on and off the vehicle; Fig. 2 presents an overview of the architecture and subsystems. On-board, a 16-bit computer (National Semiconductor 32016) is responsible for trajectory planning, sensor data acquisition, and communication. Because of the burden of on-board program development and access to mass storage of images, most of the ''intelligence'' is off-board. The on-board computer executes simple motion and sensing commands. Additionally, it handles real-time emergencies such as stopping when a bumper is unexpectedly pressed (a crash!).

The robot communicates, via a digital radio link, with a (necessarily offboard) Symbolics Lisp Machine, which is responsible for operations such as sensor interpretation (except early vision), sensor fusion, uncertainty reduction, model building, and path planning. The Lisp Machine graphically displays the robot's status, including position and sensor values. For safety, planning algorithms are developed and tested on a simulator. A TV transmitter sends the video signal from the cameras to the digitizer. The image is transferred via DMA to a VAX which performs low-level image processing and determines stereo correspondence points that are sent to the Lisp Machine over an ethernet.

III. Modeling the Environment

To autonomously perform useful tasks, the robot must possess knowledge about its environment. In particular, a framework is needed to represent sensor information and extract meaningful features. As a further requirement, the robot should build the map or model of its environment without human intervention; the map should not come from a CAD database. Ultimately, a symbolic description is desired so high-level goals can be specified. The model should be hierarchical, built bottom-up with sensor data at the lowest level and more abstract symbolic and topological information at higher levels. In this way, reasoning and planning can initially occur at higher levels with details from lower levels filled in later.

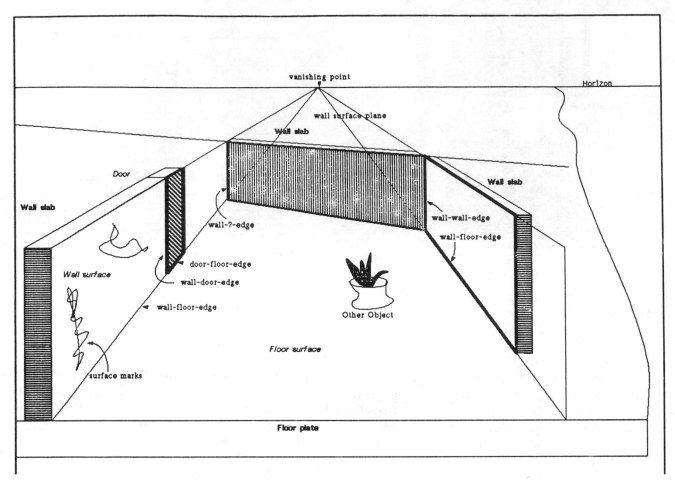

Fig. 3. A possible instantiation of the generic hallway model showing all of the observables.

We refer to two types of models: the generic building model and an instance. The generic model should describe the structure of all (or at least most) buildings. From the generic model and sensor models, observables can be predicted [5]. Now given sensed data, a model can be instantiated to create an instance describing a particular building. The instance is used for further sensing and action planning. In the implementation described here, the generic building model is implicit within the code instantiating the model though we are currently developing an explicit representation of generic models [18]. The sensing and methods for instantiating the model that are presented should be a subset of those derivable from the explicit generic model.

Fig. 3 presents a drawing of a possible instance of the generic model of a hallway. It shows some of the observable edges that can be determined from the generic hallway model such as the wall–door edge, wall–floor edge, and the door–floor edge. In addition to the model observables, there are other edges that might be sensed but do not fit the model; these might correspond to markings on the wall, obstacles, parts of unmodeled objects, or part of a modeled object that has not been instantiated due to incomplete information. Though these observations cannot be described in terms of known models, they must be included in the instance for obstacle avoidance and sensing planning. At some later time in the instantiation process, more information may be known so these observations can be explained in terms of the generic models.

For motion planning, the robot is modeled as a cylinder with a circular projection onto the ground plane. For sensing planning, the cameras are modeled by their separation and field of view; the projection onto the ground plane of the region of stereo overlap can be calculated.

Additionally, uncertainty in sensor data and model abstraction must be considered. Because of sensor noise and wheel slippage, uncertainty in spatial location may lead to an inconsistent model. A method for reducing uncertainty of sensor data as well as coping with inconsistencies in the model is needed. Because of uncertainty, object location is represented relative to other locations in a graph structure rather than locating them in an absolute coordinate system or grid [6]. Measurement uncertainty is described by a multivariate normal distribution (Gaussian).

Currently, the world is represented as two-dimensional (2-D); all objects are projected into the ground plane. Since a mobile robot is constrained to travel along a surface, a 2-D representation is reasonable for motion planning, but when more sophisticated actions (such as manipulation) and sensing planning are desired, a full 3-D model is needed [18].

IV. THE VISION SYSTEM

The structure of the environment as represented by the generic model has implications for the vision system; all observable edges of the model are either horizontal or vertical in the world. Notice the edges in Fig. 4. Vertical edges in the world project to vertical edges in a vertical image plane

Fig. 4. A stereo pair seen by Mobi while roaming through our building.

and provide sufficient information to instantiate the model. On the other hand, nonvertical lines that fit the model (e.g., floor–wall and floor–door edges) are often outside the field of view of the camera, too low in contrast, ill-defined, or obstructed by furniture. All important information, except edge length, is provided by any edge of a vertical. In fact, looking for edges at the horizon suffices, since all vertical edges in the hallway model cross a horizontal plane at camera height. So finally we arrive at our present solution: fast epipolar stereo vision across the horizon.

A. Edge Detection

A vertical edge detector with an aperture of 5 columns by 10 rows proceeds in two stages:

First, a 1 by 10 vertical averaging filter is applied to both images at the horizon, which is known from calibration. This vertical smearing has the following effects:

1) Vertical edges retain their acuity.
2) Slanted edges are blurred; horizontal edges vanish.
3) The effect of image noise on verticle edges is reduced.
4) The effect of small obscurations and markings is reduced.
5) Tilt and roll angle misalignment have less effect in the matching process because a larger effective image area will be compared.

Second, a 5 by 1 version of the edge appearance model is applied to the filtered image line. The edge appearance model [26] compares a local patch of image to the image that would have been created if the camera were looking at an ideal step edge. It uses the spatial filter created by the lens, camera, digitizer, and averaging filter pipeline for this purpose. The operator returns edge quality, position, direction, left and right gray levels, and an estimate of the localization error (1/8 pixel for good edges).

B. Stereo Algorithm

From the edges found in the two images, stereo correspondences are determined. The stereo mechanism uses edges, gray levels, correlation of intensities, constellations of edges and constraint propagation. It tries to preserve epipolar ordering of edge matches but allows ordering violations (e.g., caused by a pillar in the middle of a room). Stereo correspondences are determined in three stages: First, matches are proposed based on local edge information. These matches are either confirmed or denied by considering the intervals between neighboring matches. More globally, consistent series of neighborhoods are linked together.

With all possible pairs of edges within stereo constraints, the first stage proposes those matches that are similar on the left or right side of the edge or have a similar intensity profile.

These matches are ordered according to an initial estimate of match quality. Note that this includes matches of occluding edges where the object side of the edge is similar but the backgrounds differ. Fig. 5 shows the intensity profile of the right (top) and left (bottom) epipolar lines with matching edges superimposed. The diagonal lines between the edges in each image represent candidate matches at this stage.

Next, the gray-level intensity curves between neighboring matches in each image are compared. The intensity curves are interpolated and resampled to a constant number of samples; the normalized cross correlation is then determined. The gray-level comparison function returns this result weighted by the standard deviation of gray levels, difference in mean gray level, and difference in interval length. A large value of the gray-level comparison function confirms proposed matches and establishes a link between these matches. As a side effect, high correlation in an interval between edges often results from a solid object in the model, a wall or a closed door.

Ultimately, a set of directed graphs is produced where the nodes are proposed matches, and the arcs are the neighborhood links determined above. All of the paths through each graph form a set of consistent matches. When there is no incoming or outgoing arc for a node, another graph is formed; this may be caused by an occlusion, the boundary of the image, or an oversight in an earlier stage. The quality of a path within a graph is determined by summing up the match qualities along the path. Finally, paths in multiple graphs are linked together if they are consistent; each edge only matches once. Again for each linked path, a quality measure is determined by summation. A checkerboard that is entirely visible will be matched correctly. If there is more than one consistent, high-quality path, the stereo algorithm returns a set of potential matches. Motion correspondences, if available, might resolve any remaining ambiguity.

Fig. 6 shows the result of this algorithm applied to the stereo pair in Fig. 4. The edge detector, first-stage stereo matcher, and gray-level comparison function are implemented in C; the rest is implemented in Lisp. Running on a VAX 11/750, processing time is about 1 s per stereo pair.

C. About Calibration

Clearly, when mounting a stereo pair of cameras, they cannot be exactly configured with coplanar vertical focal planes and a precise horizontal baseline. Instead, they are mounted as close as possible to the ideal configuration and then calibrated with the aid of a calibration chart which is hung on a wall at a precise height as shown in Fig. 7. There are three reasons that the cameras must be calibrated. First, the epipolar lines corresponding to the horizon for each camera must be known for matching. Second, the relative position of the cameras and quadratic distortion are needed to calculate the three-dimensional location of correspondences. Third, to compare intensities between the two cameras, the relative intensity transfer function between the two cameras must be determined.

For each camera, the horizon line is determined from the row of four squares at the top of the chart which is hung so their centers lie on the robot's horizon plane. Roll error

335

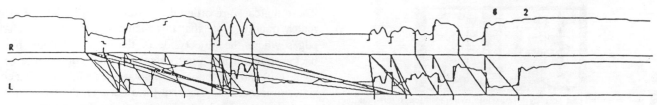

Fig. 5. The intensity profile of the left and right epipolar lines of a hallway: Possible matching edges are superimposed and the diagonal lines represent candidate matches.

Fig. 6. An overhead view of the correspondence points found in the scene of Fig. 4.

Fig. 7. The chart used for stereo calibration.

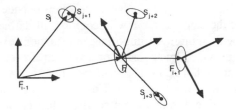

Fig. 8. Relationship of frames (F_i), transformations $(T_{i,i+1})$, sensor readings (S_j), and their uncertainty ellipses.

sensing, a frame F_i or a local coordinate system with the robot at the origin is created. The frame is related to the previous frame F_{i-1} by a transformation $\vec{T}_{i-1,i} = (x_i y_i \phi_i)^T$. An example of these relationships is shown in Fig. 8. Locations are drawn as ellipses because they are the equi-probability density contours of a multivariate (2-D) normal distribution. In the following, models of motion uncertainty and stereo uncertainty are developed; this is followed by a tool for reducing uncertainty, the extended Kalman filter.

Other researchers have used similar uncertainty models including Smith and Cheeseman [24], Durrant-Whyte [9], Ayache and Faugeras [2], and Mathies and Shafer [21]. Alternative approaches include using uncertainty manifolds [6], uncertainty grids [10], fading [8], and statistical decision theory [22].

A. Motion Uncertainty

As the robot moves, motor velocity $\dot{\vec{\varphi}}$ is mapped through the kinematics K to determine the robot's velocity relative to itself. The relative velocity is rotated yielding world velocity and then integrated to calculate the vehicle's location. So the 2-D transformation from the origin is

$$\vec{p}(t) = \begin{pmatrix} x(t) \\ y(t) \\ \phi(t) \end{pmatrix} = \int_0^t \begin{pmatrix} \cos \phi(u) & -\sin \phi(u) & 0 \\ \sin \phi(u) & \cos \phi(u) & 0 \\ 0 & 0 & 1 \end{pmatrix} K(\dot{\vec{\varphi}}) \, du. \tag{2}$$

Odometry is generally accurate in distance traveled, but the robot's orientation may be off by a few degrees. For linear motion, there is more uncertainty in the direction orthogonal to the travel direction. For a commanded straight line motion of

$$\vec{p} = \begin{pmatrix} x \\ y \\ \phi \end{pmatrix} = \begin{pmatrix} d \cos \theta \\ d \sin \theta \\ \phi \end{pmatrix} \tag{3}$$

the mean $\vec{\mu}_p$ is computed by (2), and in the direction of motion, the covariance matrix Σ_r is modeled as a linear operator

for each camera can also be determined, however, the vertical filter reduces its significance. The important parameters of the relative transform are the baseline separation and vergence angle because errors in these could lead to large errors in the location of correspondences. The other parameters do not significantly contribute to location errors and are assumed to be mounted correctly. By knowing the size and horizontal spacing of the white squares, the baseline separation and vergence angle along with second-order lens distortion are calculated. Finally, because the narrow baseline leads to nearly identical lighting/reflectance geometry when viewing the calibration chart, the relative intensity transfer function between the cameras is determined by averaging the gray levels in the same black and white regions of the chart. The two parameters of a linear relation are retained. After calibration, measured pixel coordinates map into those of an ideal epipolar stereo configuration.

V. Uncertainty

As a mobile robot moves, it gathers sensor data relative to its current position. When comparing the spatial location of objects sensed from different positions, the uncertainty in the position of the robot as well as sensor data must be considered. Uncertainty is represented by a multivariate normal distribution with mean vector $\vec{\mu}$ and covariance matrix Σ, and so the probability that a sensed point s is at $\vec{x}$ is

$$f_s(\vec{x}) = \frac{\exp\left[-\frac{1}{2}(\vec{x} - \vec{\mu}_s)^T \Sigma_s^{-1} (\vec{x} - \vec{\mu}_s)\right]}{2\pi \sqrt{|\Sigma_s|}}. \tag{1}$$

All sensor data are gathered at discrete locations, and for each of distance d

$$\Sigma_r = \begin{pmatrix} K_r & 0 & 0 \\ 0 & K_\perp & 0 \\ 0 & 0 & K_\phi \end{pmatrix} d \qquad (4)$$

where the linear coefficients are: K_r in the direction of motion, $K_\perp$ perpendicular to the direction of motion, and K_ϕ in orientation.

Σ_r can be rotated by θ yielding Σ_p giving the covariance relative to the previous frame.

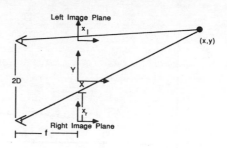

Fig. 9.   The stereo camera configuration.

$$\Sigma_p(d, \theta) = R_\theta \Sigma_r R_\theta^T = \begin{pmatrix} (K_r \cos^2 \theta + K_\perp \sin^2 \theta)d & \cos \theta \sin \theta (K_r - K_\perp)d & 0 \\ \cos \theta \sin \theta (K_r - K_\perp)d & (K_r \sin^2 \theta + K_\perp \cos^2 \theta)d & 0 \\ 0 & 0 & K_\phi d \end{pmatrix} \qquad (5)$$

where R_θ is the rotation matrix of angle θ. Thus the transformation between two successive frames is given by $\vec{\mu}_p$ and Σ_p.

B. Stereo Uncertainty

The stereo vision system locates vertical edges in space. These project into the ground plane as points, so only their 2-D location is considered. As with other measurements, a normal distribution describes the position uncertainty of a vertical.

Given a stereo vision system of two cameras with parallel image planes and horizontal epipolar lines, the uncertainty in the location of a stereo correspondence point can be found. Consider the overhead view of the stereo arrangement in Fig. 9 during this derivation. The focal length of the cameras f, the baseline distance $2d$ between cameras are known. The location of the edges in the image plane x_l and x_r are measured and described by a normal distribution with a mean μ_{x_i} and variance σ_{x_i}. From [4] the location of the point in space is

$$\vec{p} = \begin{pmatrix} x \\ y \end{pmatrix} = \begin{pmatrix} F(x_l, x_r) \\ G(x_l, x_r) \end{pmatrix} = \begin{pmatrix} \dfrac{d(x_r + x_l)}{x_r - x_l} \\ f - \dfrac{2df}{x_r - x_l} \end{pmatrix}. \qquad (6)$$

This is a nonlinear function of x_l and x_r so $\vec{p}$ will not be normally distributed. However, when σ_{x_i} is small, F and G can be linearized about the mean of x_l and x_r with a Taylor series expansion and $\vec{p}$ will be Gaussian.

$$x = F(\mu_{x_l}, \mu_{x_r}) + \left. \frac{\partial F(x_l, x_r)}{\partial x_l} \right|_{\substack{x_l = \mu_{x_l} \\ x_r = \mu_{x_r}}} (x_l - \mu_{x_l})$$

$$+ \left. \frac{\partial F(x_l, x_r)}{\partial x_r} \right|_{\substack{x_l = \mu_{x_l} \\ x_r = \mu_{x_r}}} (x_r - \mu_{x_r}) + \cdots$$

$$\approx d \frac{\mu_{x_r} + \mu_{x_l}}{\mu_{x_r} - \mu_{x_l}} + \frac{2d\mu_{x_r}}{(\mu_{x_r} - \mu_{x_l})^2} (x_l - \mu_{x_l})$$

$$- \frac{2d\mu_{x_l}}{(\mu_{x_r} - \mu_{x_l})^2} (x_r - \mu_{x_r}) \qquad (7)$$

and similarly

$$y \approx f - \frac{2df}{\mu_{x_r} - \mu_{x_l}} - \frac{2df}{(\mu_{x_r} - \mu_{x_l})^2} (x_l - \mu_{x_l})$$

$$+ \frac{2df}{(\mu_{x_r} - \mu_{x_l})^2} (x_r - \mu_{x_r}). \qquad (8)$$

Now, the Jacobian J of the transform is

$$J = \begin{pmatrix} \dfrac{\partial x}{\partial x_l} & \dfrac{\partial x}{\partial x_r} \\ \dfrac{\partial y}{\partial x_l} & \dfrac{\partial y}{\partial x_r} \end{pmatrix}$$

$$= \begin{pmatrix} \dfrac{2d\mu_{x_r}}{(\mu_{x_r} - \mu_{x_l})^2} & \dfrac{-2d\mu_{x_l}}{(\mu_{x_r} - \mu_{x_l})^2} \\ \dfrac{-2df}{(\mu_{x_r} - \mu_{x_l})^2} & \dfrac{2df}{(\mu_{x_r} - \mu_{x_l})^2} \end{pmatrix} \qquad (9)$$

and the covariance is determined by

$$\Sigma_p = \begin{pmatrix} \sigma_{xx} & \sigma_{xy} \\ \sigma_{xy} & \sigma_{yy} \end{pmatrix} = J \begin{pmatrix} \sigma_{x_l}^2 & 0 \\ 0 & \sigma_{x_r}^2 \end{pmatrix} J^T. \qquad (10)$$

The off-diagonal terms are zero because the locations of the edges in the two images are independent. Assuming that $\sigma_{x_l}^2 = \sigma_{x_r}^2 = \sigma^2$, the covariance matrix of a stereo correspondence point is

$$\Sigma_p = \frac{4d^2}{(\mu_{x_r} - \mu_{x_l})^4} \sigma^2 \begin{pmatrix} \mu_{x_l}^2 + \mu_{x_r}^2 & -f(\mu_{x_r} + \mu_{x_l}) \\ -f(\mu_{x_l} + \mu_{x_r}) & 2f^2 \end{pmatrix}. \qquad (11)$$

So the mean of the location of the point in space $\vec{\mu}_p$ is determined from (6)

$$\vec{\mu}_p = \begin{pmatrix} \mu_x \\ \mu_y \end{pmatrix} = \begin{pmatrix} F(\mu_{x_l}, \mu_{x_r}) \\ G(\mu_{x_l}, \mu_{x_r}) \end{pmatrix} \qquad (12)$$

and the covariance Σ_p is determined from (11).

Fig. 10 shows the resulting error ellipses of a stereo image looking down a hallway. Similar results have also been

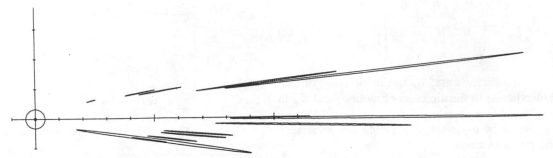

Fig. 10.   Uncertainty in the location of the stereo correspondence points.

reported in [2], [21]. The ellipses grow quickly in length with respect to the mean distance (proportional to distance squared). Also, notice the one large nearby ellipse caused by poor localization of a low resolution edge. Interestingly, even at moderate distances, the uncertainty in distance measurement from stereo becomes larger than angular uncertainty; this complements the measurements from the acoustic sensors which have broad angular resolution and good depth accuracy. Using both sensors could, for example, accurately pinpoint the location of a pillar in the middle of a room.

One should note that the derivation of the covariance matrix required linearizing the stereo equation. For distant correspondence points the linearization is no longer valid because $\mu_{x_r} - \mu_{x_l} \to 0$, and the higher order terms of the expansion dominate. Thus the location is no longer normally distributed.

C. Uncertainty Reduction

Instantiated models are built and updated from sensor data gathered at a series of locations. From experience, the most influential source of uncertainty when fusing information from different scenes is the motion transform between the robot's two positions because uncertainty is compounded across multiple transforms. Thus uncertainty in position caused by uncertainty in odometry must be reduced.

The tool used for uncertainty reduction is the extended Kalman filter. Consider an example of two verticals seen from two positions in Fig. 11. In Fig. 11(a), two verticals (C_a and C_b) were viewed from two different locations (P_1 and P_2), and their uncertainty ellipses were determined from (11). This figure shows that there is an error in the motion transform $T_{1,2}$ and sensor data because correspondence points C_{a_1} and C_{a_2} should be coincident as well as C_{b_1} and C_{b_2}. The uncertainty in robot location P_2 is determined from (2) and (5) and represented by the ellipse about its mean in Fig. 11(b). The uncertainty in the location of the stereo points C_{a_2} and C_{b_2} with respect to location P_1 is found by compounding [9], [24]. Motion correspondence between vertical edges from two successive views can be determined from the edge characteristics and relative location. In this case, $C_{a_1} \equiv C_{a_2}$ and $C_{b_1} \equiv C_{b_2}$, so the extended Kalman filter is applied to reduce the uncertainty in the stereo correspondences and motion transform. The odometry transform and the correspondence points as seen from P_1 become the initial state vector along with their covariances. The stereo measurements as well as the transform are assumed to be independent. The correspondence points as seen from P_2 becomes the measurement

vector, and the Kalman filter [15], [24] is applied to obtain a better estimate of the state as shown in Fig. 11(c). Note that motion uncertainty has been reduced as has the location of the correspondence points as viewed from both positions.

Because of the linear approximation to the nonlinear equations relating the state to sensor values in the extended Kalman filter, we experimented with the iterative extended Kalman filter. This only led to a very small (about 3-percent) change in the difference between sensor values and estimated values between the extended Kalman filter and the iterative extended Kalman filter. So we decided that the computational expense was not worthwhile.

This technique can be used to integrate motion sequences and build models from multiple views. Fig. 12(a) shows a motion sequence composed of correspondence points found from eight positions. The scale on the axes is 1 m per tick. The shape of the correspondence and the mark at the center of the robot indicates the robot's position when sensing that vertical. In Fig. 12(a), the robot's location was determined purely by odometry. Between successive locations, motion correspondences were determined along with the error models of odometry and stereo. The Kalman filter was applied and led to the verticals in Fig. 12(b). Notice that the verticals now form the two walls of a hallway, demonstrating that uncertainty was greatly reduced. The circles were seen only once while the squares depict motion correspondences. The observant reader will notice that one of the correspondence points is touching Mobi's perimeter; Mobi crashed in this run.

VI. INSTANTIATING THE MODEL

From the stereo correspondences, the hallway model is instantiated and updated with further sensing. See Fig. 15 for an example of the model building process. First the direction of the walls are determined, and then doors and pieces of the walls are added.

Along the epipolar line in the image, the intervals between pairs of neighboring correspondence edges are tentatively considered the image of a wall or door when their appearance is similar in the left and right image and when the intensity varies smoothly. The direction of the prospective walls are clustered with weights proportional to their lengths. The most prominent direction is considered to be the direction of a wall in the scene.

From the clustered intervals, the probable locations of the walls (wall spines) are determined. A wall spine is represented as a straight line (i.e., a vertical plane in three dimensions)

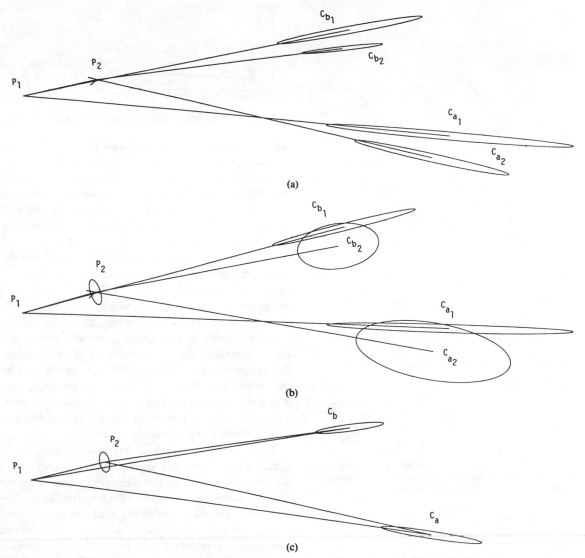

Fig. 11. Uncertainty reduction, merging two views. (a) Two points seen from different locations. (b) Uncertainty of motion and stereo with respect to P_1. (c) Reduced uncertainty.

determined by a least mean square fit of a cluster of correspondences. Given the location of the spine relative to the robot, it can be labeled *leftwall* or *rightwall*. For planning, a wall spine is not actually a wall but the expected locations of a wall. So before crossing a spine, the planner should confirm, through further sensing, that there is actually a gap in the spine (e.g., an open door).

Potential doors and pieces of the wall are added to the model. A door is seen as an interval between two not necessarily neighboring verticals that 1) are between 60 to 140 cm apart (wider than a person, but not too wide), 2) have edges with high contrast, 3) have complementary gray-level changes across the two edges (e.g., dark on the inside, bright on the outside) indicating a uniform door on a uniform wall, 4) has at least one side near the spine. Walls are seen as intervals near a spine that are not doors and have fairly uniform intensity between the bounding edges. If there are spines (walls) on opposite sides of the robot that are further apart than twice that of a typical person, the robot is in a hallway.

Finally, the verticals that are not recognized as the bound of a door or wall are added to the model. The ones that are near a wall are added as wall markings, and those that are away from the walls are considered obstacles. These points may be seen from other positions and must be avoided during motion.

Fig. 13 shows an instance of the hallway model that was built from 20 images during a run down a hallway. All of the doors in the hallway were correctly identified except some very large accordion doors at the far left end; these do not fit the doorway model. The gaps in the spine that were not recognized as solid walls were either the accordion doors, two bulletin boards, fuse boxes, or a region where both edges of the interval were not seen in the same image. Eliminating the first three of these gaps would require modeling more objects. To recognize the last missing section of the wall requires interpolating between edges across multiple images or considering observables in other areas of the image such as the floor–wall edge.

VII. MOTION PLANNING WITH AND WITHOUT THE MODEL

From the sensor information, quite a bit is known about the free space around the robot even without modeling. Two

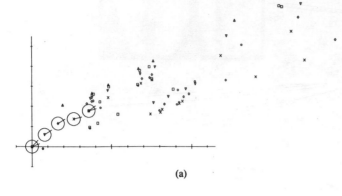

(a)

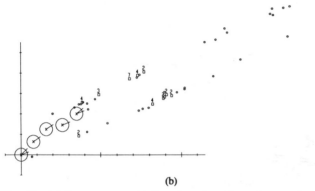

(b)

Fig. 12. Motion sequence. (a) Location of correspondence points with odometry determining robot position. (b) Location of the same points after uncertainty reduction and merging of matched points.

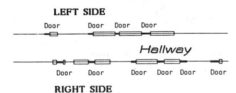

Fig. 13. An instance of the hallway model built from 20 images in our building.

motion planning strategies have been implemented, one using the model and one not.

A. Exploring Free Space

From the projection of stereo correspondences and a small set of assumptions, a polygon bounding free space is determined; a simple motion planning strategy enables the robot to safely explore. At each step, the polygon is updated, and a new motion is planned and executed. The bounding polygon contains two types of segments, ones that are candidate obstacle boundaries and those that are inferred (e.g., the edge of the field of view). For each correspondence, there must be free space between each camera and the vertical, the visibility constraint [6]. As a simplification, because the cameras are close together relative to the distance to the points, the visibility constraint is satisfied with respect to the center of the two cameras instead of each camera. For a single view, the points are ordered by their polar angle about the center and

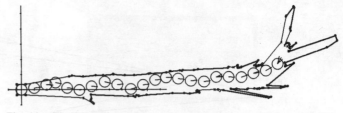

Fig. 14. The free space and robot motion during a run without modeling.

adjacent ones are connected with line segments to form the bounding polygon. This is the only possible set of boundaries or walls that could connect all of the points without violating the visibility constraint. The first and last points in the polar ordering are connected to a polygon bounding the robot, which is obviously in free space; these segments are inferred boundaries.

Now, the next motion of the robot is determined. For each step, a 1.5-m radius circle is divided into arcs according to its intersection with the bounding polygon. Candidate goals are the center of the arcs that are within free space. Excluding the direction to the previous position, the linear path where the robot is entirely in free space is chosen. If free space bifurcates, a random choice is made. If there is no safe move, the robot rotates in place. For our omnidirectional robot, this leaves one degree of freedom unspecified, orientation. The same motion planning algorithm is applied from the goal point, and the robot looks in the direction of the second goal; this is where it will probably go next.

From the new position, the correspondence points are matched to the previous ones to reduce uncertainty. Any new verticals are added to the free-space map so as not to violate the visibility constraint. Known free space is generally expanded, and the robot safely explores its surroundings.

The computational cost of adding a new vertical to the bounding polygon is $O(n^2)$ where n is the number of vertices; every edge between the vertical and all of the vertices is considered, which is $O(n)$, along with checking for violations of the visibility constraint for each candidate edge, $O(n)$ per candidate. Our approach is similar to that mentioned in a remark in [1]. Overall, the algorithm for building the polygon-bounding free space is $O(n^3)$; as the robot moves and integrates new verticals into the map, it slows substantially. Recently, a related algorithm has been reported that may allow the bounding polygon to be built in $O(n \log n)$ time [1].

When constructing the polygon, it is possible to have inconsistencies where it is not possible to add a point and satisfy the visibility constraint. The inconsistencies may be due to mismatched correspondences or sensor noise. To remove the inconsistency, nearby points with high relative uncertainty are removed from the polygon until the visibility constraint can be satisfied for the remaining points and the new point. The points with higher uncertainty were often seen from further away or were found after traversing a large distance and returning near the inconsistency. Unfortunately, this heuristic method only considers sensor noise and may remove valid points. Another option for dealing with inconsistencies might include retaining multiple hypotheses. However, this has a high computational/memory cost.

Fig. 14 shows the free space and motion of the robot from

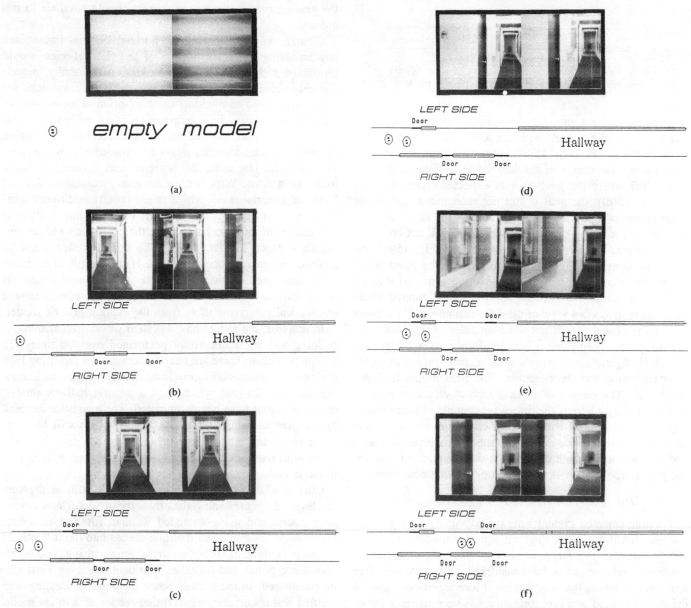

Fig. 15. The stereo images and models for the command: **Enter the second door on the left**.

a successful run reported in [27] where the dark edges are obstacle boundaries, and the light edges are inferred boundaries. The robot successfully ran through 45 stereo pairs.

This simple motion planning strategy guarantees that the robot will safely explore. In a hallway the robot will move to one end of the hall, turn around and navigate through the previously explored free space to the other end (and so on). If an open door or another hallway is encountered, the robot will randomly choose whether to turn. This algorithm does not generate a path to a goal. A two-dimensional path planner (e.g., [19]) could be invoked to find such a path from the boundary representation. However, because the goal and all of the obstacles may not have been seen, local obstacle avoidance and navigation methods may be more appropriate [17], [20].

B. Model-Based Motion Planning

More interesting goals can be attained if they are specified symbolically with respect to the model (e.g., enter the last door on the right side of the hallway). Given a goal, our system builds a model and then plans motions and sensing to achieve that goal. A simple planner based on a finite-state machine is used; state transitions occur whenever the model is updated from new sensor information. From the current state, if certain conditions hold (e.g., both walls of the hallway are seen), then the robot can perform a particular action (e.g., move down the center of the hall) and enter a new state. A small number of states and transitions were actually implemented; the results of a run are displayed in Fig. 15 and described below. The figure displays the stereo pairs, the state of the model, the current position of the robot, and the planned next position.

In this run, the robot wakes up in a state with an empty model and is commanded to **enter the second door on the left**. Since it is viewing a blank wall, nothing can be added to the model, and the only safe move is performed, rotation (Fig. 15(a)). This continues for two more rotations until image pair 4 (Fig. 15(b)) where the robot sees enough of the hallway

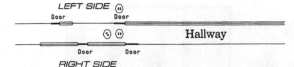

Fig. 16. The final model and motion through the door.

to recognize at least one wall. The planner enters a new state, and the robot aligns itself with the wall. Now, the first door on the left is visible (Fig. 15(c)) but not the goal, so the robot moves down the center of the hall while looking at the side of the hall where the goal door is expected. After the next step (Fig. 15(d)), the goal is still not recognized, and so the same action is repeated; the state remains the same. At each step, the planner checks that the swept path will not cross any viewed edges. Finally, after the seventh step (Fig. 15(e)), the modeling system recognizes the goal door; the robot heads toward the door, stopping a couple more times to visually update its position. Now at Fig. 15(f), a state is entered where a motion to give a last view of the door is planned. If the robot moves any closer, it will not see both sides of the door.

Finally, the robot plans a path to align itself with the door (which is slightly ajar), pushes it open, and enters using only dead reckoning and the bumpers. The final model is shown in Fig. 16. The robot has about 1 inch of clearance on each side and actually bumps the doorway a couple of times before entering. Each contact with the door constrains the location of the robot relative to the door. The robot's relative position is updated, and a new path through the door is planned. Finally, the goal is accomplished, and the robot is in a new room.

VIII. CONCLUSIONS AND FUTURE DIRECTIONS

The implemented system runs quickly using only general-purpose hardware. Planning without the model, it is also slower because the large number of line segments in the bounding polygon must be considered when growing free space and planning the next motion. More significantly, when the model is used, the representation has become more structured; experimentally, we have found that cycle times remain nearly constant, only depending on the complexity of the current scene. The cycle time ranges between 8 and 12 s for the model-based planner including stereo, uncertainty reduction, model building, planning, data communication, and robot motion.

In the described system, the generic hallway model is implicitly represented within the code; we are currently developing explicit models of object class, such as a generic building model, and hope to model outdoor environments as well as other objects [18]. Observable features are determined from the model, and sensing strategies can be planned to quickly instantiate the model. For example, it can automatically be determined from the generic hallway model along with the robot model that vertical lines in space project to verticals in the image as in this paper; furthermore, these edges can be tracked to ascertain if they contact the floor. Additional edges found in the image provide confirming evidence of walls, doors, windows, and other features that are described in the explicit generic model. By planning sensing strategies from

the generic model, the vision system should continue to run quickly.

Clearly, only considering the horizon line has limitations; any low-lying obstacles outside of the field of view would be missed such as waste paper baskets. Additionally, modeling only provides necessary but not sufficient conditions for sensed data to be admitted as an observation of an instance of a model. Since very little information is currently used to instantiate the model, the necessary conditions may be satisfied by other objects; a model might be erroneously instantiated. For example, just using the horizon line, a fuse box might look like a door. With further necessary conditions derived from the generic model which might require additional sensing (e.g., looking at a larger portion of the image), there is less chance of an object satisfying the conditions and not being an instance of that model. The fuse box would not be admitted as an instance of a door if the length of verticals were considered. Rather than further embed these conditions in the current system, we have directed our efforts toward automatically deriving them from the explicit generic model.

In practice, the model-based system works well within our building and has successfully performed over 50 times. To recognize a door, there are heuristic conditions requiring high contrast, complementary gray-level changes across the bounding verticals. In [18] which used a generic hallway model, these conditions have been eliminated. The heuristics can lead to a degree of brittleness in the system which will be overcome by having a good representation of the generic model along with methods for interpreting sensor data with respect to the model.

Currently, uncertainty is primarily dealt with at the sensor level. By representing data by a mean vector and covariance matrix and using extended Kalman filtering to reduce uncertainty, integration of multiple scenes into the instantiated model is facilitated. However, gross errors such as bad correspondence points and mistakes in model matching must also be considered. In the future, better methods for dealing with conflict and uncertainty when fitting sensor data to the model are needed. Also, issues related to robust decision making should be explored further [22].

Finally, action planning just using the generic model should be possible. Further details of the plan can be determined as the model is instantiated. This contrasts the current system where planning is embedded in the finite-state planner. With such a system, the robot will be able to perform quickly, robustly, and flexibly in a complex and previously unknown environment.

ACKNOWLEDGMENT

The authors would like to thank S. Y. Kong, J. Ponce, R. Fearing, G. Gorali, S. Fishman, L. Dreshler-Fischer, R. Rise, and R. Rubenstein for all their help throughout this work.

REFERENCES

[1] P. Alevizos, J. D. Boissonnat, and M. Yvinec, "An optimal $O(n \log (n))$ algorithm for contour reconstruction from rays," in *ACM Symp. on Computational Geometry*, June 1987.
[2] N. Ayache and O. D. Faugeras, "Building, registering, and fusing

noisy visual maps," in *Int. Conf. on Computer Vision*, pp. 73–82, 1987.

[3] ——, "Maintaining representations of the environment of a mobile robot," Tech. Rep. 789, INRIA, 1988.

[4] D. H. Ballard and C. M. Brown, *Computer Vision.* Englewood Cliffs, NJ: Prentice-Hall, 1982.

[5] T. O. Binford, "Generic surface interpretation: Observability model," in *Int. Symp. on Robotics Research*, 1987.

[6] R. A. Brooks, "Visual map making for a mobile robot," in *IEEE Int. Conf. on Robotics and Automation*, 1985.

[7] R. A. Brooks, "A robust layered control system for a mobile robot," *IEEE J. Robotics Automat.*, vol. RA-2, no. 1, pp. 14–23, Mar. 1986.

[8] R. Chatila and J. P. Laumond, "Position referencing and consistent world modeling for mobile robots," in *IEEE Int. Conf. on Robotics and Automation*, 1985.

[9] H. F. Durrant-Whyte, "Uncertain geometry in robotics," *IEEE J. Robotics Automat.*, vol. 4, pp. 23–31, Feb. 1988.

[10] A. Elfes and L. Matthies, "Sensor integration for robot navigation: Combining sonar and stereo range data in a grid-based representation," in *IEEE Conf. on Decision and Control*, 1987.

[11] G. Giralt, R. Chatila, and M. Vaisset, "An integrated navigation and motion control system for autonomous multisensory mobile robots," in Brady and Paul, Eds., *Robotics Research.* Cambridge, MA: MIT Press, 1983, pp. 191–214.

[12] M. Hebert, "3-D vision for outdoor navigation by an autonomous vehicle," in *Proc. Image Understanding Workshop* (Cambridge, UK, 1988).

[13] R. H. Hollier, *Automated Guided Vehicle Systems.* New York, NY: Springer, 1987.

[14] William M. Wells, III, "Visual estimation of 3-D line segments from motion—A mobile robot vision system," in *Proc. Amer. Assoc. Artificial Intell.*, 1987.

[15] T. Kailath, *Lectures on Wiener and Kalman Filtering.* New York, NY: Springer-Verlag, 1981.

[16] Y. Kanayama and N. Miyake, "Trajectory generation for mobile robots," in *Int. Symp. on Robotics Research*, 1985.

[17] O. Khatib, "Real-time obstacle avoidance for manipulators and mobile robots," *Int. J. Robotics Res.*, vol. 5, no. 1, pp. 90–98, 1986.

[18] D. J. Kriegman and T. O. Binford, "Generic models for robot navigation," in *IEEE Int. Conf. on Robotics and Automation*, 1988.

[19] T. Lozano-Perez, "Automatic planning of manipulator transfer movements," in M. Brady *et al.*, Eds., *Robot Motion: Planning and Control.* Cambridge, MA: MIT Press, 1982.

[20] V. Lumelsky, "Algorithmic issues of sensor-based robot motion planning," in *IEEE Conf. on Decision and Control*, 1987.

[21] L. Mathies and S. A. Shafer, "Error modeling in stereo navigation," *IEEE J. Robotics Automat.*, vol. RA-3, no. 3, pp. 239–248, 1987.

[22] R. McKendall and M. Mintz, "Robust fusion of location information," in *IEEE Int. Conf. on Robotics and Automation*, 1988.

[23] H. P. Moravec, "The Stanford cart and the CMU rover," *Proc. IEEE*, vol. 71, no. 7, pp. 872–884, July 1983.

[24] R. Smith and P. Cheeseman, "Estimating uncertain spatial relationships in robotics," in *AAAI Workshop on Uncertainty in Artificial Intelligence*, Aug. 1986.

[25] C. E. Thorpe, M. Hebert, T. Kanade, and S. Shafer, "Vision and navigation for the Carnegie-Mellon Navlab," *IEEE Trans. Pattern Anal. Machine Intell.*, vol. 10, no. 3, pp. 362–373, May 1988.

[26] E. Triendl, "How to get the edge into the map," in *Int. Conf. on Pattern Recognition*, 1978.

[27] E. Triendl and D. J. Kriegman, "Stereo vision and navigation within buildings," in *IEEE Int. Conf. on Robotics and Automation*, 1987.

[28] M. A. Turk, D. G. Morgenthaler, K. D. Gremban, and M. Marra, "VITS—A vision system for autonomous land vehicle navigation," *IEEE Trans. Pattern Anal. Machine Intell.*, vol. 10, no. 3, pp. 342–361, May 1988.

[29] A. M. Waxman, J. J. LeMoigne, L. S. Davis, B. Srivivasan, T. R. Kushner, E. Liang, and T. Siddalingaiah, "A visual navigation system for autonomous land vehicles," *IEEE J. Robotics Automat.*, vol. RA-3, no. 2, pp. 124–141, 1987.

David J. Kriegman received the B.S.E. degree (summa cum laude) in electrical engineering and computer science from Princeton University, Princeton, NJ, in 1983; the M.S. degree in electrical engineering from Stanford University, Stanford, CA, in 1984; and is currently completing the requirements for the Ph.D. degree there.

He was awarded the Charles Ira Young Award for undergraduate electrical engineering research at Princeton and, while at Stanford, he was studying under the Hertz Foundation Fellowship. His research interests include mobile robots, computer vision, shape representation, sensor fusion, and computer architectures for high-speed vision.

Mr. Kriegman is a member of Tau Beta Pi and Phi Beta Kappa.

Ernst Triendl received the Ph.D. degree from the University of Innsbruck, Innsbruck, Austria, in 1966.

From 1966 to 1983 he worked at DFVLR (Deutsche Forschungs- und Versuchsanstalt für Luft- und Raumfahrt), where he was Head of the Image Processing Division. From 1983 to 1987 he was a Research Associate at Stanford University, Stanford, CA. He is now an independent consultant. His research interests include stereo vision, mobile robots, artificial intelligence, image processing, and remote sensing.

Thomas O. Binford is a Professor of Computer Science (Research) at Stanford University, Stanford, CA. From 1967 to 1970 he was a Research Scientist at the MIT Artificial Intelligence Laboratory, Cambridge, MA. He has been at Stanford since 1970. His research interests include computer vision (model-based systems, shape representation, mobile robots, stereo vision, edge segmentation), manufacturing, robotics, and applications to surgery.

343

Mobile Robot System with Transformable Crawler, Intelligent Guidance, and Flexible Manipulator

Masakatsu Fujie, Yuji Hosoda, Taro Iwamoto, Koji Kamejima, and Yoshiyuki Nakano

Mechanical Engineering Research
 Laboratory
Hitachi, Ltd.
502, Kandatsu-machi, Tsuchiura-shi,
 Ibaraki-ken, 300 Japan

Reprinted from *International Symposium on Robotics Research*, Vol. 3, pages 341-347, "Mobile Robot System with Transformable Crawler, Intelligent Guidance, and Flexible Manipulator" by M. Fujie, Y. Hosoda, T. Iwamoto, K. Kamejima, and Y. Nakano, by permission of The MIT Press, Cambridge, Massachusetts.

The intelligent mobile function must be small, light, and fast moving. We have developed new robot technologies aimed at achieving this function. The new technologies developed consist of 1) a crawler-type moving mechanism, in which robot posture is controlled to fit road conditions, 2) a guidance system, in which viewing and proximate information related to surrounding structures such as doors, walls, etc., is compared with stored map information to determine the speed of the crawler, and 3) a manipulator mechanism which has six degrees of freedom and force feedback control.

In addition, to evaluate robot performance, we set up a model terrain consisting of a corridor with vertical angle corners, a door, stairs, a step, and an obstacle which the robot must avoid. We can gauge the robot's properties in this model terrain, and then estimate its functions for general purpose.

By incorporating these technologies, an autonomous robot can detour or overstep stairs or obstacles to reach its destination. Through comprehensive testing, the robot proved to be practical for near-future use.

1 INTRODUCTION

Since 1980, the robot has established itself in manufacturing; now new demands for robots are spreading to other industrial fields. Robot market trends are shown in Fig.1; also shown are those breakthrough technologies already perfected, and those under investigation, which will be necessary in meeting these demands. In the past few years, new robots have become prominent in the nuclear energy field. Within the next seven or eight years, demands in the off-shore bottom, construction, transportation, medical, and other fields will increase to a total production of ¥100 billion(=$400 million)[1].

However, these fields have requirements that are difficult to meet at present: a) diversification, b) non-repetition, c)instability, and d) wide fields[2]. Furthermore, areas of applications may be inaccessable to humans. Therefore, a self-controlled moving function should be developed, in addition to conventional manipulation functions. Table 1 summarizes technical improvements that are part of the steps toward realization of a new type of robot. A wide range of new technologies must be developed in the areas of mechanism and control. Mobile and intelligence functions are essential, unlike the instance of current conventional robots.

Our aim was to develop a robot system capable of intelligence and self-controlled movement at almost human speed[2]. This paper describes some of the major technologies needed in developing a light-weight, high-speed flexible moving mechanism, a high speed intelligent guidance technique for self-controlled movement, and a dexterous manipulator which can deal with movement. In addition, experimental results of the robot

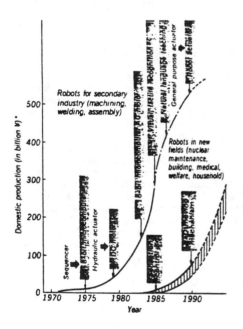

Fig.1. Trends of Robot Demand and
                 Breakthrough   Technologies.

system in which the technologies described above are synthesized and the model environment by which the robot system is estimated are also described here.

2 MOVING ENVIRONMENTS AND TARGET SPECIFICATIONS OF THE ROBOT

A new type of robot is needed for such use as movement within a building for trouble shooting and inspection work in nuclear plants or other plants. Here the robot must move in areas together with men. Along its way, there are stairs, curbs, slopes, and other routing facilities, which are convenient for humans, but obstructive for the robot. To make the robot more practical, a moving mechanism was developed which allows it to maintain steady motion despite obstacles. Generally, passageways in buildings are narrow and the permissible floor load, small. Therefore, robots need to be light and compact, these two factors also important for high speed movement[1].

Table 1. Technical Demands for
Next Generation Robots.

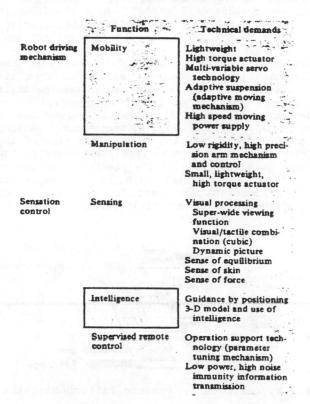

Function		Technical demands
Robot driving mechanism	Mobility	Lightweight High torque actuator Multi-variable servo technology Adaptive suspension (adaptive moving mechanism) High speed moving power supply
	Manipulation	Low rigidity, high precision arm mechanism and control Small, lightweight, high torque actuator
Sensation control	Sensing	Visual processing Super-wide viewing function Visual/tactile combination (cubic) Dynamic picture Sense of equilibrium Sense of skin Sense of force
	Intelligence	Guidance by positioning 3-D model and use of intelligence
	Supervised remote control	Operation support technology (parameter tuning mechanism) Low power, high noise immunity information transmission

The robot must immediately understand any changes in environment as it moves along. Available technology, however, has not resolved the problems of complicated environmental information ; the robot was obliged to stop continually to allow sufficient time to deal with comprehension of moving and stopping cycles. We have shortened this comprehension time for the robot destined for use within a building. First, before operation, the environment of robot movement is stored as a limited number of patterns corresponding to doors, corridors, stairs, etc.. Next, 3-D shape of the building is prepared as a map on which feature patterns are stored. The robot can thus maintain its pace at the speed of a slow human walk. In addition, the light, dexterous manipulator, which can be mounted on the mobile mechanism, must be developed for achieving the task at the destination. This requires lightening of the multi-joint mechanism and force control technology.

3 MOVING MECHANISM OF ROBOT AND ITS CONTROL

3.1 Principles of moving mechanism

Among the three types of moving mechanisms; wheel-type, crawler-type, and leg-type, the crawler-type is best suited to our needs in this case, as the wheel-type cannot overstep an obstacle larger than its wheel diameter and the leg-type cannot be easily applied to high-speed robots. The crawler-type can travel even on rough surfaces with high controllability, at high speed.

However, the conventional crawler-type has no means of controlling posture; it sometimes falls down when negotiating high steps on stairs or curbs. In the past, one proposed type had four crawler units, and by swinging each unit as was necessary to maintain posture, it adapted to the road surface[3]. However, this system was not so enough. We developed the new variable shape crawler car with a compact, light-weight structure that can move around easily within a building. As shown in Fig.2, main and sub wheels are mounted on both sides of the car body as part of the crawler unit supported with arms and composed of planet wheels. By moving these planet wheels, the wheel tracks vary to adapt to road conditions, thereby enhancing travelling capability.

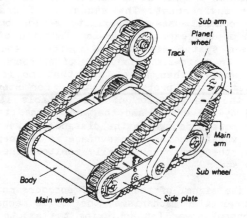

Fig.2. Variable Shape Crawler Car.

3.2 Crawler shape control

Crawler shape is altered by changing planet wheel position. At that time, the tracks should not become loose. Constant tension control was one solution to the problem of preventing slack in the tracks by extending and contracting the arm. However, this requires a heavier, complicated control mechanism. Another method was developed by studying the locus of the motion for the planet wheels when the tracks are spanned.

As the three wheels have the same diameter, the locus becomes an oval with focal points at the centers of both the main and sub wheels[4], as shown in Fig.3(a). A sub arm is mounted at the top of the main arm and rotated at the same angle of main arm rotation in the reverse di-

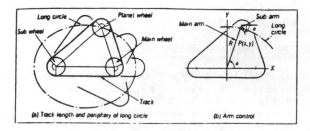

Fig.3. Control of Crawler Shape.

rection as shown in Fig.3(b). The sub arm moves in the oval when the main arm rotates around the center point between the main and sub wheels. Both arms can be connected by a simple gear mechanism while maintaining constant track tension which makes for a small, light crawler.

In addition, the robot features high-speed and high-torque, so that most of its weight is apportioned to the driving mechanisms. One pair of reduction gears and a DC servo motor are used for speed control, creating higher torque upon obstacles and higher speed on flat surfaces.

3.3 Crawler motion control

To overstep obstacles and stairs, the relative positions of the car and the obstacles should to be accurately detected. The sensors which are arranged on the mobile mechanism for observation of the environment. The sequence of crawler motion control uses a proximity sensor on the bottom of the car. The viewing function of the guidance system detects an obstacle and the controller decides on the angle of approach. The front wheels then rise, and that motion is detected by the proximity sensor. For stepping down, the end position of the obstacle is detected by the sensor and the arm, turned to the lower front side, while the distance between the planet wheel and the road surface is reduced. On stairs, the sensor continually monitors the edge of the stair.

Using the signal of the pair of encoders mounted on both crawler mechanisms, the mobile mechanism can control speed for achieving the smooth curb movement, and can generate forecast patterns for navigation.

4 PERCEPTIVE GUIDANCE OF ROBOT

4.1 Principles of guidance

Robot guidance means understanding environment and recognizing position. An obstacle of proximity, e.g. stairs, curb, etc. can be over-riden through recognition its position by a proximity sensor. In perceptive guidance of robot, the position relative to distant objects is perceived using a viewing function.

As moving environments are limited in buildings, a 3-D guide can be prepared as a "map". This map is used as preliminary information for observation planning. In addition, data to be observed can be forecast.

According to the guidance system, the forecast image is calculated on the bases of the map and the moving distance signal from the travelling car. This is compared with the actual image sighted by the viewing function: the steering angle and speed signal are then calculated. A diagram of this dynamic guidance procedure for pattern matching is shown in Fig.4.

The map is created in the wire frame model to represent characteristic patterns of buildings and, computed as a forecast pattern stored in the memory. In addition, the edge pattern is extracted from the detected image and arranged as a measured pattern which is also stored in the memory. A displacement vector pattern is obtained by connecting patterns of all corresponding points. The displacement pattern is calculated by pattern processors I and II, while the forecast pattern is replaced with a potential gradient vector from the source of the forecast pattern.[5]

The displacement vector is created when the viewpoint of the robot is dislocated from its estimated position. In that case, steering angle reference signals and speed are controlled to decrease this displacement while moving toward the object point. In Fig. 4, the measured pattern (solid line) is shifted to the left side of the forecast pattern (broken line). Accordingly, the robot is guided to move the viewing point toward the right. The vertical and holizontal components of displacement are used for calculation of directional and distant dislocations.

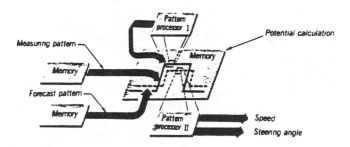

Fig.4. Guidance by Dynamic Pattern Matching.

4.2 Computer simulation

The sample room shown in Fig.5, with an area of 5mx5m and one door, was devised using a

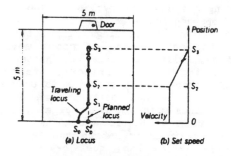

Fig.5. Simulation of Viewing Guidance.

computer. Robot guidance was simulated and the following conclusions were obtained:[6]

(1) Instruction signals for robot steering and speed can be calculated in 1/60 seconds, allowing the robot to move at a practical speed (2.5 km/h).

(2) Forecast patterns may be updated intermittently (marked ⊙ in the figure), sufficient for practical purposes.

(3) Vertical or holizontal edges can be acknowledged in one second. The robot can be safely guided as long as the edge is within the field of view during this time.

(4) For stable system operation, moving speed of the robot shall be decelerated as it approaches its destination.

4.3 Equipment configuration

A fisheye lens with a picture angle of 162° was used for the TV camera for image pick-up to allow for a wide viewing field and large focal distance. Hence, the measuring pattern will not be removed from the field of view. In addition, the detecting unit of the robot's guidance system is made smaller and lighter. However, the pick-up image is usually deformed and compressed at the periphery. A new corrective method was devised to correct this deformation in which the pattern of lens distortion was stored in a memory; reference to it, deleted the distortion from the image.

A block diagram of the developed system is shown in Fig.6, while Fig.7 presents the configuration

Fig.6. Mobile Robot Guidance System.

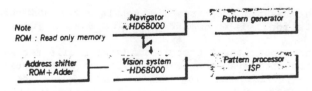

Fig.7. Configuration of
Dynamic Image Processing System.

of a dynamic image processing system in which guidance signals are computed during movement. The system consists of a pattern processor provided with a frame memory of 64 kbytes and an image processor LSI (ISP[7]). Processing speed is 16 ms per image screen. Pattern generation is processed by the perspective projection of the wire frame component vector. To ensure high speed processing, only edge point calculations are performed by the software. Picture description is carried out by a special LSI.

5 MANIPULATOR WITH MULTI JOINTS

5.1 Manipulator mechanism

The manipulator mounted on the mobile mechanism must satisfy the following requirements.

(1) High degree of freedom
(2) Compactness
(3) Light in weight
(4) Real-time treatment by the on-board computer
(5) Energy saving
(6) Toughness for the vibration with the mobile

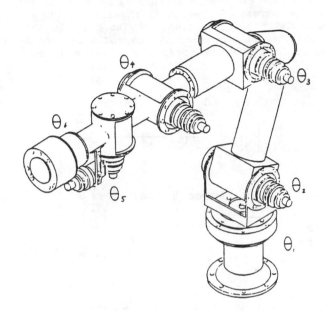

Fig.8. Exterior of Manipulator.

The exterior of our manipulator designed in consideration of these requirements is shown in Fig.8. This manipulator has six joints in each of which a driving DC servo motor, a reduction gear, a encorder, and a tachogenerator are stored.[7] Brakes are stored only in three bottom joints. To control in good performance, the axes of the joints of the wrist must not be separated.

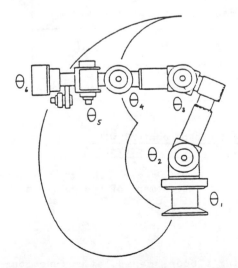

Fig.9. Working Area of Manipulator.

Therefore sensors of the rotating end joint of the wrist are attached outside of the joint and operated using pullies and a timing belt.

The working area of this manipulator is shown in Fig.9. This area expands only vertically; the mobile mechanism can move horizontally; this manipulator has a large working area.

Each joint and each structual part, a thin shell made of CFRP, is a flexible structure. The manipulator can handle workpieces of 3 kg. Layout of the degrees of freedom is shown in Fig.10 and manipulator specification, in Table 2. In this table the maximum speed of the end point of the manipulator is the speed for opening a door.

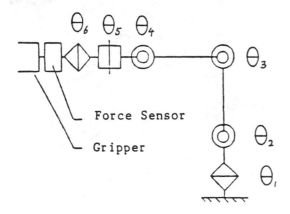

Fig.10. Layout of Six Joints.

Table 2. Manipulator Specifications.

Freedom	6
End max. speed	817mm/s
Positionning accuracy	±0.23mm
Handling weight	3 kg
Arm length	840mm
Weight	26.6kg

5.2 Force feedback control with force-displacement conversion

A manipulator which can achieve the task in cooperation with movement of the mobile mechanism must have following control functions.
(1) Teaching and play back
(2) Shifting the co-ordinate axis
(3) Force control

In opening a door, manipulator functions are accomplished as follows.

The position data of normal motion, when the mobile mechanism is standing at normal position by the door, is taught to the manipulator. When the mobile mechanism arrives in front of the door, the difference from the normal position is detected and the co-ordinate axis of the position data of the motion, shifted to that degree. While the manipulator is opening the door, including knob gripping, the control function of the force at the end of the manipulator is added to the teaching play back function.

The block diagram of the force feedback system with force-displacement conversion which we developed is shown in Fig.11. The manipulator is modeled on the system of spring, mass, and damper; force is converted to displacement, and this displacement is added to reffered position data.

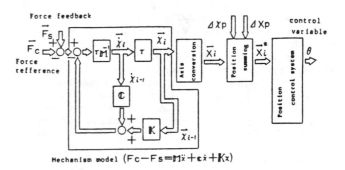

Fig.11. Block Diagram of Feedback System.

6 ROBOT SYSTEM

6.1 Mobile robot system

By a synthesis of developed technologies described in chapters 3, 4, and 5, the mobile robot system, shown in Fig.12, is developed for the purpose described in chapter 2.

This system comprises the mobile robot and the station with command and data base systems.

The robot itself comprises (1)the mobile mechanism with crawler, (2)the navigator with visual and proximate sensing, (3)the multi joints force feedback manipulator with cooperation to the movement, (4)the communicator, which exchanges data between station and robot, and (5)the power source.

The purpose of our development is to achieve a robot light enough to maintain its pace at the speed of a slow human walk, predescribed. We achieved a robot system whose specifications are shown in Table 3.

6.2 The control system of the mobile robot

The control system of the predescribed mobile robot system, in which the mobile control system plays the main role and the visual recognition system and manipulator control system connected to it, achieve the following.

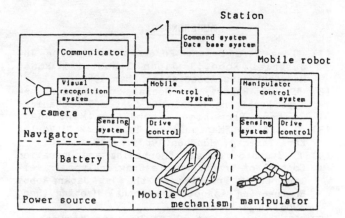

Fig.12. Mobile Robot System.

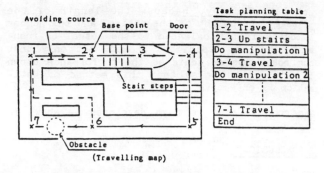

Fig.13. High Level Control Information.

Table 3. Robot System Specifications.

Element	Sprcification	
	L ∎∎ W ∎∎ H ∎∎	
Body size	1200×900×1500	
Weight	~280 kg	
Speed	2.0(2.2) km/h	Mean(Max)
	0.8 km/h	Stairs up
Navigator		
Power	150 w	
Speed	1/60 sec/Frame	High speed
	1 sec/Frame	Low speed
Mobile mechanism		
Weight	~160 kg	
Power	1500 w (max)	High speed
	100 w (max)	Low speed
Manipulator		
weight	~30 kg	Handling weight:
Length	800 mm	3 kg
Power source		
Weight	~40 kg	2.2 kw 20 min.

The map containing straight passages, a curbed corridor, stair steps, obstacles in the passage, and doors, is prepared in advance. Information which is stored in high level control system is shown in Fig.13.[7] Elements of the mobile environment are described on the map. The robot always has the environment information. On the task planning table, each task of each base point is described. The task of the mobile control system is shown in Fig.14. The manipulator task is classified as manipulator control system alone and in co-operation with the manipulator control and mobile control systems, for example, the task of opening the door. In the latter case, the manipulator control system is the main source of operation and the operating status of the mobile mechanism is regarded as the manipulator's environment.

6.3 Model mobile environment

Further development in the robot system will require development of estimation methods. In our

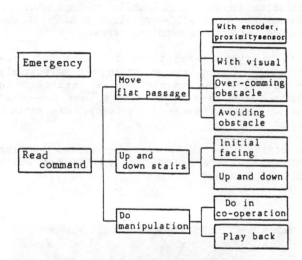

Fig.14. Task of Mobile.

robot development we try to estimate our system using model mobile environment which we set up. This environment, shown in Fig.15,[6] is designed with characteristic construction of usual buildings, a passage like a corridor with curb, stair steps, doors, steps, and obstacles be avoided. If experimental results in such model environment are estimated, the performance in general environment can easily be supposed.

Our model environment, shown in Fig.15, is

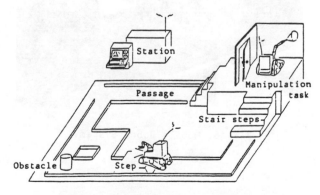

Fig.15. Model Mobile Environment.

not satisfactory, because the length of the passage, the height of the stairs, and the number of curbs is not balanced to the actual environment. But our experiment showed, not only that our robot system achieved the performance shown in Table 3 but also that this method of estimation is effective.

The robot system we have developed is shown in Fig.16.

7 CONCLUSIONS

We developed a moving mechanism, perceptive guiding technique, and dexterous manipulation technology to detour or overstep stairs or other obstacles in buildings, while moving toward a destination in a self-controlled manner; it can achieve the various handling tasks.

Also we synthesized these features and constructed a robot system which can effectively be used for the inspection of equipment in plants, such as nuclear plants, and for other various growing needs in primary and tertiary industries.

In addition, with these experimental results we can evaluate robot performance and robot estimating method.

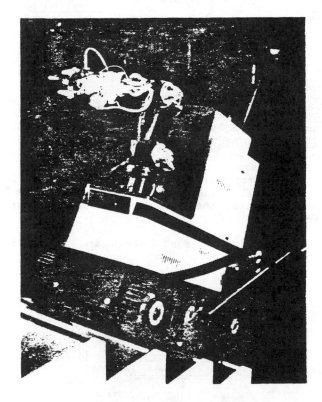

Fig.16. Mobile Robot System.

REFERENCES

(1) "Long Term Demand Forecast Report for Industrial Robot," Japan Industrial Robot Manufacturers Association (Mar.1981).
(2) "Survey Report for Social Stock Retention-Type Repairing Machinery System," Mechanical System Promotion Foundation(Mar.1982).
(3) G.W.Köhler."Manipulator Type Book," Verlag Karl Thiemig, München(1981),p.570.
(4) T. Iwamoto, et al. "Mechanism and Operation Control of Transformable Crawler-Type Travelling Car," Preprint of the 1st Japan Robot Engineering Association Annual Meeting,(Dec. 1983),pp.1-4
(5) K. Kamejima, et al."A Pattern Kinematics Concept and Its Application to Mobile Robot Navigation," 6th SICE Symp.on DST,(1983),pp. 87-90.
(6) Y. Ogawa, et al."A Method of Perceptive Guiding of Moving Robot," Preprint of Communication and Electronic Engineers Institute Annual General Meeting,(1984),pp. 6-131.
(7) T. Fukushima, et al. "Development of Image Processing LSI-ISP," Preprint of Information Processing Engineers Institute Annual General Meeting 5B-2,(1983).
(8) M. Satoh, et al "Robot Arm Mechanism with Joint Driving," Preprint of JAACE Flexible Automation Symp. Annual meeting, (May 1984), pp29-32.
(9) Y. Hosoda, et al. "Travelling Control of Mobile Robot," Preprint of the 2nd Japan Robot Engineering Association Annual Meeting, (Nov.1984),pp.227-228.
(10)M. Fujie, et al. "Organization of Autonomous Mobile Robot" Preprint of the 2nd Japan Robot Engineering Association Annual Meeting,(Nov. 1984),pp221-224.
(11)Y. Nakano, et al. "A Concept for An Autonomous Mobile Robot and A Prototype Model," Preprint of 2nd. ICAR Annual meeting,(Sep. 1985).
(12)M. Fujie, et.al. "Mobile Robot with Transformable Crawler and Intelligent Guidance," Preprint of International Conference on the Manufacturing Science and Technology of the Future,(Oct.1984),pp. 81-83.

The Design of an Autonomous Vehicle for the Disabled

RICHARD L. MADARASZ, MEMBER, IEEE, LOREN C. HEINY, ROBERT F. CROMP, AND NEAL M. MAZUR

Reprinted from *IEEE Journal of Robotics and Automation*, Vol. RA-2, No. 3, September 1986, pages 117-126. Copyright © 1986 by The Institute of Electrical and Electronics Engineers, Inc. All rights reserved.

Abstract—The first steps in the design and construction of an autonomous vehicle for the physically and sensory disabled are described. The vehicle is basically a self-navigating wheelchair, which is designed to transport a person to a desired room within an office building given only the destination room number. The wheelchair has been equipped with an on-board microcomputer, a digital camera, and a scanning ultrasonic rangefinder. The goal of the project is to develop a vehicle that can operate without human intervention within a crowded environment with little or no impact on the building or its inhabitants. An overview of the project objectives and a detailed description of the components that are currently working are presented.

I. INTRODUCTION

THE ABILITY to move about freely in the world is highly valued by all persons. However, this ability is sometimes hindered by a physical disability. A person with a mobility impairment must rely on an external transport mechanism. A blind person must be able to navigate in the world without the obvious ability to visually sense the state of his or her immediate environment. For some, the physical coordination necessary for the control of a moving device is lacking. Combinations of these disabilities greatly increase the difficulty of achieving the desired mobility. These problems can often be overcome only at the expense of considerable personal energy and the risk of possible harm.

The technology being developed for autonomous machines that are capable of moving about without human intervention may offer certain multi-handicapped persons the assistance needed to achieve this freedom of mobility. These are not preprogrammed devices but are machines capable of performing the complex tasks of navigation, locomotion, and problem solving. Current research in mobile autonomous vehicles has been limited to very constrained environments, such as a deserted nuclear power plant, or contain details that can be ignored, as is the case of a weapon system [1]-[4]. In many cases the operation of the system has been simulated. A device that will be used to transport people must be capable of functioning in unconstrained dynamic environments with complete regard for the safety of its passenger, its surroundings, and itself.

II. GOALS AND CONSTRAINTS

The goal of this project is to develop an autonomous vehicle which can be used to transport a person from one location to

Manuscript received July 31, 1985; revised February 24, 1986.
The authors are with the Computer Science Department, Arizona State University, Tempe, AZ 85287, USA.
IEEE Log Number 8608950.

another without outside assistance. While this is similar in concept to other autonomous vehicle projects, there are certain objectives which make it unique. First, the machine is designed to function inside an office building. This makes it easier to plan strategies and to model the working environment. It also places constraints on the types of problems that may be encountered. The objective of the system is to plan its own path from its current location to a particular room in the building, and then travel to that location. The problem is made more difficult by the fact that the destination may be on another floor, requiring the successful operation of an elevator. Another difficulty is the fact that the machine may not know its current location due to reinitialization or an accident. In this case it must determine its starting point and then replan its strategy to the goal.

The second unique feature of the system is that it must function with minimum impact on the building in which it will be used. This means that the building cannot be equipped with a guidance mechanism, such as embedded wires in the floor or painted stripes that can be followed. It also means that the machine must operate along with the current inhabitants of the building, without danger to itself, its passenger, the other people in the hallways, or the building itself. This is in contrast to an autonomous vehicle, such as a tank or mining machine, which can ignore those elements of the world that will not impede its operation.

In addition to the above constraints, the vehicle must be self-contained. All of the sensing and decision making should be performed by the on-board equipment. Previous designs for mobile robots have relied upon two-way communication with another computer to perform some of the tasks [5]. The cost and reliability of a communication link as well as the cost and availability of an additional computer make this an unacceptable solution for an autonomous wheelchair. While it can be argued that the techniques need to be perfected before a final design is developed, it is hard to separate the practical considerations of the target system from the conceptual design process.

III. SCOPE AND JUSTIFICATION

The major emphasis of this project is to experiment with the basic concepts of autonomous machines. A general approach to this problem was initially outlined, but the actual work is evolutionary. The intent is to develop a working system that can deal with a very restricted environment, namely the interior of an office building. "Real-time" in our context means a period of time that is compatible with the speed of the vehicle. There are no claims made about the generality of the techniques we are using, nor for the vehicle's operation outside of the target domain. However, this restriction does not trivialize the problem in any way, and in some cases actually adds unnecessary difficulty to the problem.

Our other objective for the project is to make progress towards assisting disabled persons to live as normal a life as possible. The mere suggestion of relinquishing one's control to a machine generates criticism of the approach; and yet we do it all the time, with autopilots for airliners and highway traffic controllers. It is true that a system such as this will directly benefit only a small number of people (the blind and physically disabled) in its complete form. But the individual concepts being developed here will hopefully spawn ideas with a broader scope. Since the projects inception, we have been in contact with and supported by local professionals who work with the disabled. We have also received help and encouragement from disabled students on campus.

IV. System Overview

The system is an on-going project, which was started in February 1985, and is intended to continue for several years. It is being designed in a modular fashion, and at present it is not an integrated system. Everything that is described in Section IV-A through IV-D has been implemented, except for the mechanism to operate the elevator buttons. There are still many problems that must be solved before the system meets its design objective: completely autonomous operation.

The current vehicle is a standard motorized wheelchair equipped with an on-board computer, a digital camera, and an ultrasonic rangefinder (Fig. 1). All processing for planning, navigation, collision detection and avoidance, object location, and motion control is to be performed on the vehicle at the time it is required. The components of the system were chosen on the basis of cost and availability, rather than being the best choice for the task. No structural or functional changes have been made to the wheelchair itself.

The on-board processor is an IBM Portable PC with 320 Kbytes of memory, two floppy disk drives, parallel I/O, analog-to-digital converters, and a digital camera interface. The entire unit is mounted in a vertical position above the rear axle of the wheelchair. At present, the system must run from standard line current, but the plans are to provide for operation from the two 12 volt batteries which power the wheelchair.

The vision system consists of a single digital camera with a resolution of 128 × 128 picture elements. The camera is mounted in a fixed position on the upper right corner of the wheelchair frame. A wide angle lens provides a viewing angle of approximately 35°. This combination of lens and mounting was chosen to give the best compromise for all of the tasks the vision system must perform. The camera is connected to the processor via a custom DMA interface, which allows images to be recorded at a maximum rate of 11 frames per second.

Range information for the vehicle is provided via a Polaroid ultrasonic rangefinder which is mounted on a platform above the centerline of the wheelchair [6]. A scanning device can position the sensor at 3° intervals through a 360° sweep.

Fig. 1. Autonomous wheelchair.

Resolution of the sensor is 0.1 feet over a maximum range of 35 feet.

The computer controls the motion of the vehicle through the standard wheelchair joystick. This requires the system to perform the same motion-control functions that a human passenger would have to provide. Low-level feedback of the vehicle's motions is provided via optical encoders mounted on the driving wheels. Wheel motions are encoded in 1-inch increments of travel; however, they are not relied upon to keep track of the position of the vehicle. The wheelchair itself is an imprecisely constructed device (more imprecise in this case due to a bent frame), which means that it often does not move exactly as it is commanded to. The optical encoders are used to keep the wheels moving in approximately the desired direction.

In the current configuration, all of the various operations that are performed by the system are done sequentially as needed by single-purpose modules. To manage the flow of control between these modules, a special operating system has been developed that allows the individual software modules to be written in any language supported by the PC. In this way the software modules can be developed by several different people in parallel, without regard for the details of program interaction. The operating system provides for control and data to be passed between any of the modules, which themselves are stand-alone programs. Operation is similar to a batch processing system, in which the commands to be executed are stored in a file, and then executed in sequential order. The difference here is that the command file is initially empty. As the system runs, the command file is continually rewritten by the executing modules. In this way, one system module can call another without having to worry about the internal process-control mechanisms. This process is also necessitated by the limited size of the processor's memory. Individual modules can either be stored on one of the disk drives, or within a small portion of the memory, which is reserved for use as a "RAM disk."

One of the goals of the system is to be able to operate a standard elevator. A quick survey of elevator control panels indicates that the buttons are usually arranged in a grid pattern. At this time there are no plans to incorporate a manipulator arm into the system to push the buttons. Instead, a linear array of solenoids will be mounted on a vertical support at the front corner of the wheelchair. A button is selected by positioning the wheelchair in front of the elevator panel and activating the closest solenoid. This part of the system has not yet been implemented.

A. Planning and Problem Solving

An autonomous vehicle must be able to plan its own actions. At first this may seem unnecessary, but it must be remembered that the passenger of such a vehicle may not be able to ascertain the current status of the environment or have had the opportunity to previously explore the building. Therefore, the machine must be able to generate a series of primitive operations to get from its current location to the goal. The goals and constraints of this system allow the use of a simple yet effective planning mechanism.

The inclusion of a world model is essential to a system that must plan a strategy for navigation. It is highly desirable to be able to load the model of a particular building into the system as the vehicle enters the building. This allows the user to be transported through a building about which he has no previous knowledge. The world model, in this case, is a symbolic structure representing the significant features of the building, such as relative locations of offices, hall intersections, and elevators. The model contains no explicit recording of separating distances because such detail was found inappropriate with respect to the robot's limited sensory ability and the imprecision of its movements. Our design requires instead that the system have the capability to detect and identify both static landmarks and dynamic objects.

The major data structure employed is a three-dimensional array that can be indexed to retrieve a room number. The first dimension corresponds to the floor number, the second dimension represents the wall number, and the final refers to the absolute position of the room as measured from one end of the building. This structure captures the relative locations between all offices in the building (See Fig. 2).

Hallway intersections by definition are the first and last landmarks along a corridor. The existing program handles elevators as special cases, though methods have since been devised that include them in the three-dimensional building array.

These features are static. However, since the wheelchair is navigating in a public building the model must contend with a dynamically changing environment. Knowledge about the location of obstacles (people, furniture, ladders, etc.) is essential for intelligent planning. When the wheelchair encounters an obstacle that it cannot maneuver around, its location (in terms of the room to which it is closest) is entered into the world model. In addition, an age is associated with the obstacle, which indicates the duration of time it has inhabited the environment. Disabled elevators are handled in an analogous manner.

Given the input of the destination room number and the knowledge of the starting location, the planning system uses an algorithm to determine the shortest path to the destination. This algorithm has several built-in heuristics to handle obstacles and disabled elevators.

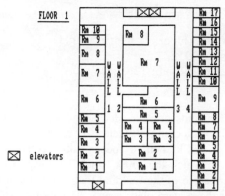

Fig. 2. Typical office building floorplan.

The basic shortest path algorithm separates the plan up into movements on a single floor, either traveling from the start location to a destination, or in the case where movement between floors is necessary, from the start location to an elevator and the elevator to the destination. On a single floor, the wheelchair can move either in a clockwise or counterclockwise fashion from the source to the destination. The proper path is chosen by observing the world model and choosing the direction, which contains the fewer intervening rooms from source to destination. This may not always be the shortest path in distance due to the nonuniformity in room sizes, but if it is not, the difference is negligible.

When obstacles and disabled elevators are included, the shortest route is not always the most desirable. An obstacle-free path is always chosen if it is available. If all paths contain obstacles, the path containing obstacles of the larger average age is chosen (the age of an obstacle is incremented each planning session). This follows the reasonable assumption that an obstacle tends to be moved after a period of time (people leave the hall, equipment is brought into rooms, etc.). An elevator that is marked as disabled is not used since it is assumed it will not be fixed for a good length of time. However, after the age of the disabled elevator reaches a preset threshold value, it is removed from the disabled list in the hope that it has been fixed.

Before executing the plan, the obstacles that exist along the chosen path are removed from the obstacle list. If the plan is properly executed, it indicates that the obstacles are no longer there. If, however, one is encountered, it is placed back on the obstacle list with a minimum age assigned to it. The planning algorithm will then cause the wheelchair to select another route during subsequent planning operations due to the youthful age of the newly added obstacle.

The output of the planning system is a series of primitive operations, such as MOVE UNTIL (an intersection is reached, a room is located, etc.), ROTATE (execute a turn of a specified number of degrees), and ELEVATOR (enter the elevator and travel from the current floor to a specified floor). A plan to get from room 389 to room 129 might look as follows:

MOVE UNTIL (Floor: 3, Intersection: 4)
ROTATE (− 90°)
MOVE UNTIL (Floor: 3, Elevator: 1)
ROTATE (90°)

```
ELEVATOR (Floor: 3, Floor: 1)
ROTATE (90°)
MOVE UNTIL (Floor: 1, Intersection: 1)
ROTATE (−90°)
MOVE UNTIL (Room: 129)
END OF PLAN
```

These commands are executed sequentially until the goal is reached or a blocking obstacle is encountered. If the current location is not known, a special robot behavior module is invoked, in which the robot "feels" its way around until it locates a feature of the building that it is familiar with. From this point the system plans a new strategy to reach the goal.

As has been alluded to, there are cases where the primitive operations cannot be executed successfully. The strategy of error recovery has been distributed throughout the system. That is, a primitive operation attempts to recover from its own errors rather than defer back to the planning system. For example, if obstacles are found to be blocking a hallway, the primitive movement routine will invoke a collision avoidance routine (not yet implemented) that will try to find a path with enough clearance between the obstacles for the wheelchair to pass. It will then oversee execution of the motions to get the wheelchair past the obstacles, and, upon completion, relinquish control to the movement routine. If this fails, an error status is passed back up through the hierarchy of primitive operations until it finally reaches the planning routine. The obstruction is added to the world model and a new plan is generated to reach the destination using the current location as the starting point.

Like all of the modules of the system, the planning module is stored on a disk until it is needed. The loading of the module is the most time-consuming part of its operation. Once the module is loaded, a plan of operation is generated within 1 to 2 s. Using the RAM disk speeds up the process considerably. Since the complete system is not fully integrated, the planner has only been tested in isolation.

B. Sensing

Autonomous operation of the wheelchair is made possible by the ability to sense the differences between what is expected in the environment and what is actually there. The two primary sensing mechanisms in this design are vision and ultrasonic ranging. The two systems provide very different types of information, and are used in conjunction with each other in very specific instances, such as checking elevator doors.

Vision can provide a very detailed description of the spatial layout of the environment for both humans and machines. There are, however, some major difficulties to the use of vision as a primary sense for an autonomous machine. First, depth cannot be measured directly from a single image. The use of multiple cameras can be used to implement a stereo vision approach, but this is computationally expensive [5]. The second disadvantage is the processing time required to analyze the images, even though only a single camera is used. The third problem is the fact that images are intrinsically ambiguous. That is, there are an infinite number of combinations of factors that can produce the same image.

In this design, vision will be used for some very specific purposes. It is used to locate and verify known objects (such as room numbers and elevators), determine the status of elevator indicator lights, detect changes in the scene (which indicates that people or other moving objects are present), and to keep the wheelchair centered in the hallways. By limiting the use of the vision system to these tasks, the processing can be optimized. Rather than trying to understand the layout of the environment, a specific set of independent tasks with a high probability of success are utilized.

Illumination within a building may vary greatly at different times of the day. Although there are usually lights in the hallways, a window in an inopportune place may allow sunlight that is much brighter than the artificial light to enter. Since the camera does not have an automatic exposure control mechanism, an on-board light source can be used to balance the difference between the areas of high and low illumination. A structured light system (line projector) is also being tested to facilitate object location for the purpose of collision avoidance.

The ultrasonic rangefinder is used to determine the distance between the wheelchair and the objects surrounding it. The sensor can be scanned to generate a depth map of the immediate area [7]. This may sound like an ideal sensor, but there are several factors in our working environment that make this data highly questionable. The sensor has a very broad angle of acceptance (25°). Therefore, any object within the field of view of the sensor can be located anywhere along a spherical surface at the indicated distance from the sensor. This also means that if the sensor is scanned at an angular increment less than the angle of acceptance, the same object may be detected in successive readings, making it appear larger than it actually is.

A more serious problem with the use of the ultrasonic rangefinder concerns false readings. Due to the behavior of sound waves, any hard surface will act as a reflector. If the sensor is oriented relative to a smooth, untextured surface at an angle greater than half the angle of acceptance, the sound has a tendency to bounce off of the surface at an oblique angle, rather than returning to the transducer. Eventually, it may be reflected back by another surface along the same path. The result of this is an indicated distance to a particular surface that is much greater than the true distance. Since the wheelchair is designed to operate within hallways, the problems associated with reflections make the raw sensor data very unreliable. This is analogous to a person trying to walk through a maze of mirrors, as can be seen in Fig. 5 and will be discussed in detail later.

The ultrasonic rangefinder is used primarily to orient the wheelchair with respect to the walls of the hallways. This is necessary whenever the vehicle becomes lost or trapped. The second use of the rangefinder is for the detection of conditions that result in a change in depth. This may be the detection of a moving obstacle in the halls, or the opening of an elevator door. This sensor is used also in conjunction with the vision system to provide gross information about distances.

C. Motion Control

Precise control of moving machines has traditionally relied upon accurate construction of the machine elements [8]. Control systems are designed for a particular machine configuration with known and predictable dynamic behavior. Yet it can be argued that these are not necessary preconditions for precise control of mechanisms. Most humans can accomplish the same tasks, such as locomotion and manipulation, with an adequate degree of precision even though their physical constructions vary greatly. The ability to sense the difference between the desired and affected motions seems to be essential.

A wheelchair is a very imprecise mechanism, yet, with a little practice, it can be precisely controlled by a novice user. The human user relies heavily on vision to provide the feedback to make the necessary corrections to the machine's motions. Vision is used in this vehicle to control motion relative to the walls. The task is to keep the vehicle moving in a straight line aligned with the center of the hallway, and accommodate the imprecision that is inherent in the construction of the wheelchair.

The typical paradigm for visual control of machines is to determine the spatial layout of the environment, plan a strategy, and then move a desired increment in that direction [9]. However, existing systems, which integrate sensing with machine control, often suffer from the inability to process the sensory information fast enough to suit the needs of the process. Another disadvantage is the inability to deal with moving objects in the working environment.

An alternate approach is to reduce the amount of information that is sensed and implement a heuristic approach for its interpretation. (A similar approach for use on sidewalks was described by Deering and Collins [10].) This eliminates the need for time-consuming scene analysis. The computational savings result in the ability to deal with a dynamic environment while the vehicle is moving. The objective is to use significant features of a known environment for visual guidance. Restricting the operation of the vehicle to the interior of a building allows the use of the junction between the floor and the wall as one such significant feature. In this case, the baseboards on either side of the hallway are utilized to center the moving vehicle.

The basic strategy is to search each image for the occurrence of dark converging edges. By determining the slope and position of the edges, the centerline of the hall can be estimated. A correction value can then be determined to bring the vehicle's trajectory back to the center of the hall (Fig. 3). This is all accomplished without stopping the vehicle.

The scanning is done in a windowed region in the lower portion of each image. Starting at the bottom center of the window, the image is searched toward both sides until a transition to a dark region is found. The process is repeated for each line of the window. The locations of these points are saved and later interpreted to reconstruct the baseboards.

The continuity of the baseboards is interrupted at various points along the hall, such as at doorways. The presence of people and other obstacles also interferes with the detection of the edges. To handle these situations, the system is designed

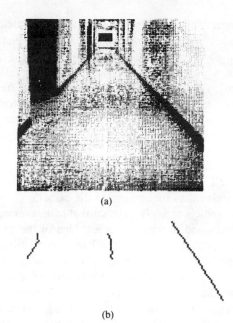

(a)

(b)

Fig. 3. (a) Image of hallway as seen by wheelchair. (b) Baseboard edges and desired path (center).

with a preference to track one wall at a time. Usually the right wall is chosen. If the non-preferred wall is momentarily lost, the vehicle is still able to continue as long as the slope and location of the preferred wall stays within an acceptable range. Should the preferred wall be lost, the vehicle will transfer preference to the other side. The loss of both walls invokes another process which tries to realign the vehicle in the hall or determine if a hallway intersection has been reached.

Obstacles most often appear as sudden deviations in the detected slope and position of the wall. In these cases, the vehicle is usually able to continue. However, the vehicle may have to alter its path to avoid a collision with the obstruction. This is achieved by directing the vehicle towards the midpoint between the wall and the obstacle. Because of the restricted field of view, this is usually the best plan of action. If the width of the path is too small, the vehicle is stopped, and after a short delay, movement is again attempted. This wait state has been added to handle the common case of a person crossing the path of the wheelchair. If the problem still exists after the delay, assistance is sought from other routines to handle the problem, as explained in Section IV.

A very important factor in the performance of the visual servoing is the ratio of the frame rate to the speed of the vehicle. Our algorithms and hardware restrict the image processing rate to slightly less than two frames per second while in motion. To compensate for this image rate, the speed of the wheelchair is usually kept under ten inches per second. The actual time it takes the vehicle to traverse a hallway is dependent upon the occurrences of obstacles, doors, and poor lighting conditions.

The visual motion control mechanism has been tested for approximately 60 h of operation under various conditions. Most of these test runs have been made over a distance of less than 70 feet (the distance from our lab to the nearest intersection). In "best-case" situations, where there are no obstacles, the wheelchair averages 98 s to traverse this distance. This time increases if the wheelchair must avoid obstacles or call other modules (ie: when the lighting is poor or the baseboards have been completely lost).

The wheelchair has also been run in several other halls of the target building. Some of these have different lighting, widths, and patterns of doors and alcoves. Variance in lighting conditions causes the greatest degree of difficulty. During the day time, windows at the end of the hallways create the most problems. Usually, we must manually adjust the intensity of our on-board light to compensate for the lighting conditions of each hall. A high-priority item for future work is to add an aperture-control mechanism to the camera.

The vehicle is fairly successful at manuevering around both stationary and moving obstacles. One of the problems in the current implementation is the handling of blind spots. Since the wheelchair moves based on what is in its field of view, it may collide with an obstacle it is passing once it is out of view. The problem is more pronounced with obstacles on the left side of the wheelchair since the camera is mounted on its right.

In general, people move much faster than the wheelchair can react. Many times people will move beyond the wheelchair before it has had a chance to respond to their presence. Likewise, the wheelchair may commit itself to a move just before a person comes into the field of view. This can only be resolved by using a faster image processing rate.

D. Orientation

The problem of machine mobility would be greatly simplified if the world were perfect. Unfortunately, deviations in the construction of a building from the plans, imprecision in the movement of the vehicle, noise in sensor signals, hardware failures, and tampering by others may all cause the system to lose its current position. It is essential that an autonomous machine be able to recover if it becomes disoriented. In this case, the strategy is for the vehicle to realign itself relative to a wall and then proceed until it finds a recognizable landmark. Whenever it is determined that the wheelchair is lost, a procedure is invoked that attempts to reorient the vehicle parallel to a wall, facing the direction of the longest open area. The procedure is one of iterative sensing and motion and continues until an acceptable solution has been found.

The process begins by taking multiple distance measurements with the scanning ultrasonic rangefinder. The distances from the center of the wheelchair are measured at small increments through 360° of rotation. The increment which seems to provide the best compromise between coverage and scanning speed is 6°. The actual scan is made in two passes to minimize the effect of people moving through the area of measurement. Since each scan takes approximately 30 s, it is unlikely that a moving person will appear in two consecutive measurements.

The next step is to try to determine the location and orientation of a hallway wall. A situation in which the vehicle has become disoriented in a clear hallway is shown in Fig. 4. The measured distances clearly illustrate the problem which is caused by the reflections off of the smooth walls. The nearest wall is detected at an orientation of 180° from the front of the vehicle at a distance of approximately two feet. However, as

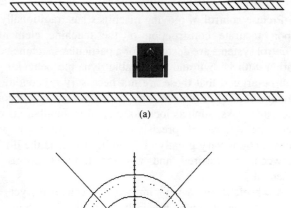

(a)

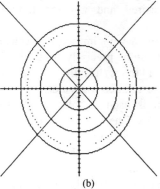

(b)

Fig. 4. (a) Disorientation in clear hallway. (b) Ultrasonic rangefinder scan (each circle represents five feet).

soon as the sensor is positioned beyond the angle of acceptance relative to the wall, the reading jumps to its maximum value. (Through experience it was determined that the maximum reading in a confined space was limited by the height of the ceiling. In this particular building, the maximum rangefinder measurement is 13.5 feet.) There are also considerable noise points in the scan.

The poor quality of the rangefinder signals in this working environment makes it difficult to reconstruct the spatial layout of the area [7]. Instead, it was decided to take a heuristic approach [11]. The chosen heuristic is to look for the farthest distance that is approximately perpendicular to the closest distance. The farthest and closest distances are determined by first examining the distribution of range measurements and picking the highest and lowest values with a sufficient number of occurrences, determined by a preset noise threshold. The scan is then examined to find the longest contiguous run of the chosen highest and lowest values. The center of each of these runs is used as the direction of the closest and farthest distances. If the longest run in each case is shorter than an acceptability threshold, the process is repeated for the next highest or lowest (respective) value. When the orientation of the wall is determined, the wheelchair is rotated until it is parallel to it and facing the farthest distance. To accomodate inaccuracies in sensing and motion, the process is repeated until the current orientation is within an acceptable range of the path parallel to the wall. At this point, the process terminates and control is returned to the procedure, which visually guides the vehicle down the hall.

The situation may arise where the vehicle is not lost in a clear hallway, but is actually trapped in a corner. This occurs in the test building when the wheelchair becomes stuck in one of the many alcoves in the hallways (Fig. 5). In this case, the

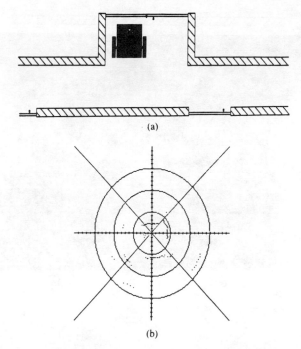

(a)

(b)

Fig. 5. (a) Disorientation in a corner. (b) Ultrasonic rangefinder scan.

ultrasonic scan does not find a suitable wall–hallway combination. When this happens, the vehicle attempts to maneuver to a better location and tries again. The wheelchair is rotated to the direction of the farthest distance and moved to the minimum of one half of the measured distance or three feet. Prior to rotation, the minimum distance is checked to see if it is greater than the turning radius of the wheelchair. If it is not, then a direction and distance is chosen which will get it clear of the obstacle. Once it has moved, the process is repeated until the desired wall is found. In the case of Fig. 6, the wheelchair would move backwards one foot, and scan again. A subsequent motion would be a rotation 135° to the right and a move of approximately three feet into the clear space. A final move would then position it parallel to the wall and facing down the right hallway.

The two situations just described, and their associated sonar scans, are typical of what is encountered in the test building. The orientation module has been tested and demonstrated in the hallways over 100 times. The test runs were performed at various times of the day and night, with normal levels of building activity. In only a few instances were the hallways free of people and obstacles. The obstacles usually consisted of inanimate objects, stationary observers, moving pedestrians that were no longer interested in this activity, and those that actively tried to fool the system (ie: waving hands in front of the sonar). In clear hallway situations, such as shown in Fig. 4, the vehicle is usually able to orient itself in two iterations (two scans and one movement), or a total of approximately 2.5 min. The alcove situation (Fig. 5) is usually accomplished in three to five iterations (3–6 min), depending on the type of corner and its actual initial placement. Each situation may require more time and trials if there is heavy traffic in the hallways, particularly if there are a lot of spectators close to the vehicle. At present there is a limit of 15 on the number of iterations that are allowed, and in some cases, this limit is

reached. When this happens, the machine stops moving and starts beeping for someone to come and help it.

E. Typical Test Run

A typical test run of the vehicle is shown in the series of photographs in Fig. 6. The test was run at 11:00 AM on a Friday and is representative of the hallway conditions between class changes. The test started by pushing the wheelchair into the alcove and plugging it in. This is a "worst case" situation; a cold start in an unknown orientation. The sequence shows the vehicle orienting itself in the hallway and then traveling under visual guidance. The orientation of the front wheels gives a clue to the movement that was just executed. Note that people in the vicinity will confuse the sonar, but that it will recover on subsequent moves. The vehicle traveled approximately 50 feet from its starting position before it was stopped. The elapsed time for this particular run was 6 min.

V. CURRENT CAPABILITIES

As was previously stated, the system is currently capable of planning a path through the building, orienting itself in a hallway, and performing visual guidance down a hallway. These functions are working slowly but satisfactorily. The detection of intersections still poses a problem. If it is unambiguous, such as a right angle, the vehicle has no trouble. The opposing wall is simply treated as a large stationary obstacle, and the turn is negotiated successfully. The detection of the other types of intersections is accomplished by a failure to track the sides of the hallways. The identification of the type of intersection encountered is still being investigated.

In addition to these functions, we have working modules for a variety of special purposes, such as locating and determining the condition of elevator doors (by looking for vertical lines in a particular configuration). Door numbers are determined visually by locating and reading a "circle-coded" marker. This is simply a black disk with a series of white dots that encode the room numbers. So far, this is our only modification to the building and was done to avoid performing character recognition.

The software and hardware for the various functions is being developed in parallel and has not been fully integrated. The elevator operating mechanism has been designed but has not been implemented, nor is there a user interface in place. The operating system that was developed for the machine is functioning, but has not been used extensively. Obviously, this means that the system does *not* run autonomously at present. When the individual modules are performing reasonably well, they will be integrated into one system. A suitable on-board power source is still a major problem.

Since the system was designed to operate in a dynamic environment, it has always been tested and demonstrated during normal working hours in the candidate building. The initial reaction of the inhabitants of the building to this vehicle was fascination. This soon turned to irritation at having to walk around the machine. Finally, the machine has become accepted as a normal part of the building. If a device such as this is ever going to be useful, it must be accepted as normal by everyone.

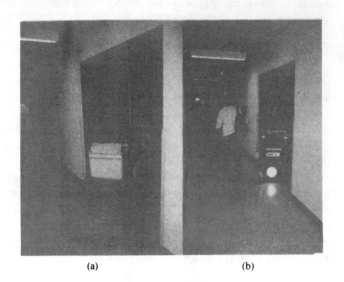

(a) (b) (c) (d)

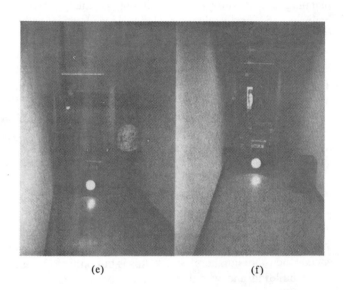

(e) (f) (g) (h)

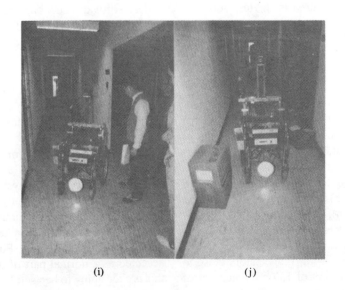

(i) (j)

Fig. 6.   Typical test run.

VI. Future Work

The next step in the design of an autonomous vehicle for the disabled is the integration of the various modules into a single working system. The modules that have been previously described are running on the wheelchair's PC. Other modules are running on different computers and must be made to operate successfully given the time and memory constraints of the on-board computer. In addition, a power source must be found that will provide more freedom than 130 feet of extension cord.

A considerable amount of work must still be done to make the sensing process more useful and robust. A better approach to the use of the vision system would be to identify general features in the scene rather than looking for specific objects, such as baseboards. The precise reconstruction of the physical environment is not necessary for this task, but a symbolic representation of the layout of the world may be determinable in a period of time which is compatible with the speed of the vehicle. The usefulness of the raw data from the ultrasonic rangefinder is questionable at best. Much work must still be done to identify techniques to reconstruct the surrounding environment from the raw data.

At present, all of the processing within the system is done on a single processor. Current plans are to develop a multiprocessor architecture that would separate the tasks to allow concurrent operation. The low-level motion control functions will be moved to a separate processor. It would also be desirable to move the rangefinder control to a separate processor so that it could be used while the vehicle is moving. Since the vision processing is the most time-consuming of the processes that must be done in real time, it will stay on the PC.

Once the system is working sufficiently within the particular building in which it is being developed, it will be extended to other buildings. These solutions to the problems of navigation and guidance within hallways may be extended to an outdoor environment which is also constrained to well-defined paths. While the vehicle may never be capable of taking a walk in the woods, it may be extended to operation on sidewalks. The outdoor environment, however, provides challenges which have not been considered here.

VII. Summary

This work represents the first step in the design of a completely autonomous vehicle for the transport of people. In addition to being an aid to the disabled, it is a test bed for basic research in robotics. The uniqueness of this project comes from the goal of being able to work in a dynamic and crowded environment. It also represents an attempt to actually implement a working system which is capable of functioning in an imprecise environment. The difficulties of autonomous machine operation are presently not very well understood. This work has given some insight into what the problems really are.

The reader should keep in mind that this paper describes only the work performed in the first three months of a multi-year project. It is not yet an integrated system, and does not perform all of the tasks that it is being designed to do. However, we feel that significant progress has been made given the time and monetary constraints.

It should be realized that completely autonomous operation of such a vehicle may not ultimately be practical, nor desirable. However, several concepts may be applicable in a more limited form which would also be an aid to the disabled. A wheelchair that is capable of visual tracking and collision avoidance would be useful to a sighted operator who has difficulty in directly controlling the motion. High-level visual and ranging systems may also provide blind persons with more information about the environment than can be currently obtained from manual tactile sensing.

Acknowledgment

The autonomous vehicle described in this report was designed in part as a class project for CSC591 Robotics at Arizona State University in the Spring of 1985. Much of the work was done as a group effort by the members of the class. The wheelchair was provided by the Disabled Student Services Center at ASU. We would also like to thank the ASU Manufacturing Technology Department for the use of their machining and welding facilities.

References

[1] F. Mavaddat, "WATSON/I: Waterloo's Sonically Guided Robot," J. Microcomputer Applications, vol. 6, p. 37–45, 1983.

[2] B. Bullock, D. Keirsey, J. Mitchell, T. Nussmeier, and D. Tseng, "Autonomous vehicle control: An overview of the Hughes project," in Proc. 1983 Trends and Applications: Automating Intelligent Behavior, 1983.

[3] G. Giralt, R. Sobek, and R. Chatila, "A multilevel planning and navigation system for a mobile robot," in Proc. 6th Int. Joint Conf. Artificial Intelligence, 1979, pp. 335–338.

[4] Y. Ichikawa, N. Ozaki, and K. Sadakane, "A hybrid locomotion vehicle for a nuclear power plant," IEEE Trans. Syst., Man, Cybern., vol. SMC-13, no. 6, pp. 1089–1093, Nov. 83.

[5] H. P. Moravec, "Rover visual obstacle avoidance," in Proc. 7th Int. Joint Conf. Artificial Intelligence, 1981, pp. 785–790.

[6] C. Biber, S. Ellin, E. Shenk, and J. Stempeck, "The polaroid ultrasonic ranging system," presented at the Audio Engineering Society Convention, New York, 1980.

[7] J. Crowley, "Navigation for an intelligent mobile robot," IEEE J. Robotics Automat., vol. RA-1, no. 1, pp. 31–41, Mar. 1985.

[8] L. Shih, "Automatic guidance of mobile robots in two-way traffic," Automatica, vol. 2, no. 2, pp. 193–198, 1985.

[9] G. J. Agin, "Servoing with visual feedback," in Proc. 7th Int. Symp. Industrial Robots, pp. 551–560, 1977.

[10] M. Deering and C. Collins, "Real-time natural scene analysis for a blind prosthesis," Proc. 7th Int. Joint Conf. Artificial Intelligence, pp. 704–709, 1981.

[11] R. Chattergy, "Some heuristics for the navigation of a robot," Int. J. Robotics Res., vol. 4, no. 1, pp. 59–66, Spring 1985.

Richard L. Madarasz (M'84) is an Assistant Professor of Computer Science at Arizona State University. He received the B.S. degree in manufacturing engineering technology from Arizona State University, Tempe, in 1975. He also received the M.S. and Ph.D. degrees in computer science from the University of Minnesota in 1981 and 1983, respectively.

Prior to graduate school, he worked as a manufacturing engineer with General Dynamics, Ft. Worth Division, Ft. Worth, TX, and with Ball Corp., Electronic Display Division, St. Paul, MN. His research interests include autonomous systems, machine perception, manufacturing automation, advanced aids for the disabled, and sensory motion control.

Dr. Madarasz is currently the Chairman of the Phoenix Chapter of the IEEE Computer Society. He is also a member of the Society of Manufacturing Engineers and the American Association for Artificial Intelligence.

GUIDE DOG ROBOT

Susumu Tachi and Kiyoshi Komoriya

Mechanical Engineering Laboratory, MITI
Tsukuba Science City
Ibaraki, 305 Japan

The Guide Dog Robot Project started in the 1977 fiscal year at MEL. The project's goal is to enhance mobility aids for the blind by providing them with the functions of guide dogs, i.e., obedience in navigating or guiding a blind master, intelligent disobedience in detecting and avoiding obstacles in his/her path, and well-organized man-machine communication which does not interfere with his/her remaining senses. In this paper the design concept of the Guide Dog Robot MELDOG is described first. Next, the navigation method using an organized map and landmarks, obstacle detection/avoidance system based on the ultrasonic environment measurement and man-machine communication via electrocutaneous stimulation system are presented. The results of the feasibility studies using MELDOG MARK I, II, III and IV test hardwares are discussed. Future problems are also elucidated.

INTRODUCTION

Independent travel is one of the strongest desires of about three hundred and forty thousand blind or severely visually impaired individuals in Japan. Since the concept of technological assistance for the blind is of recent origin (after World War II), they have been largely on their own, depending upon more sensitive and subtle utilization of their remaining senses, and extending them through the use of the cane, or relying upon human or dog guides.

Ideal mobility aids for the blind should support the three necessay functions for mobility; i.e., (1) the blind person's next step, (2) his/her directional orientation, and (3) his/her navigation along reasonably long travel path on both familiar and unfamiliar terrain [Mann, 1974]. However, existing mobilty devices; e.g., the Pathsounder [Russell, 1971], the Sonic Glasses [Kay, 1973], the Laser Cane [Farmer et al., 1975], the Mowat Sensor [Morrissette et al., 1981] and the Nottingham Obstacle Detector [Dodds et al., 1981], have only functions (1) and (2). The information processing system employed by the existing devices is very simple and crude so that the blind user must concentrate on the devices, resulting in the fatigue of the user or loss of other information which otherwise might be obtained through the remaining senses.

It is quite desirable to design more intelligent mobility aids for the blind which combine the above three functions with the enhancement of functions (1) and (2) by increasing the information processed by the device or the machine. These devices should warn only if the blind persons are in danger, thereby not distracting the attention of the blind traveler from other potential cues through their remaining senses. This design concept is very similar to traveling with a guide dog (Seeing-eye).

Reprinted from *International Symposium on Robotics Research*, Vol. 2, pages 333-340, "Guide Dog Robot" by S. Tachi and K. Komoriya, by permission of The MIT Press, Cambridge, Massachusetts.

The purpose of the Guide Dog Robot Project (dubbed MELDOG) which started in 1977 is to enhance mobility aids for the blind by providing them with the functions of guide dogs; i.e., obedience in navigating a blind master, intelligent disobedience in detecting and avoiding obstacles in his/her path, and well-organized man-machine communication which does not interfere with his/her remaining senses.

In this paper the design concept of MELDOG is first described. Next, the navigation using an organized map and landmarks, obstacle detection/avoidance system based on the ultrasonic environment measurement and man-machine communication via an electrocutaneous stimulation system are presented. While theoretical consideration has been done for the realization of these functions by machines, feasibility studies of the proposed methods have been conducted both by computer simulation and field tests using the test hardwares.

The results of the feasibility experiments using MELDOG MARK I, II, III and IV test hardwares are discussed and the future problems are elucidated.

GUIDE DOG ROBOT

In order to realize a robot that can assist a blind master's mobility, the following three fundamental control and communication problems of man-machine systems must be solved.

(a) How a robot guides itself by using an organized map of the environment and registered landmarks in the environment.

(b) How the robot finds obstacles which are not registered on the map and avoids them.

(c) How the robot informs its blind master about the route and the obstacles detected.

Two main functions of real guide dogs are obedience and intelligent disobedience, which corresponds to the navigation and obstacle detection, respectively. Adding to these is the necessary communication between the blind master and the dog is necessary. In order to realize these main functions by solving the above three problems we have set the following specifications for the guide dog robot:

(1) In principle, the master takes the initiative. The master commands the robot by control switches connected by a wired link. The robot precedes the master and stops at each intersection, waits for the master's next order (right, left, straight, or stop) and obeys it. If the master does not know the area and wants full automatic guidance, all he has to do is assign the starting code and the destination code. The robot determines whether there is a route to the destination. If more than one route exists, it chooses the optimal route and guides the master accordingly. The robot stops at each intersection as a safety precaution (See Landmark Sensor Subsystem of Fig. 1).

(2) When the robot detects a dangerous situation on the road, it no longer obeys the master's command but gives him a warning. If the obstacle is moving toward the master, it stops and alerts the moving object and the master. If the obstacle is moving in the same direction but slower than the master, it asks the master to reduce his/her speed to follow the preceding object, probably a human traveler. If something is crossing in front of the robot, the robot waits till it passes. If it detects an obstacle which does not move, it tries to determine if it is possible to find space that will permit the safe transport of the master around the obstacle. If space exists, it safely guides the master around the obstacle. If not, it tries to find a new route to the destination without using an undesirable path (See Obstacle Detection Subsystem of Fig. 1).

(3) In general, the speed of the robot is controlled so that it coincides with that of the blind master's gait. Thus, if the master walks slowly or rapidly, the robot moves accordingly, keeping the distance between them almost constant. As long as the master is considered to be safe by the robot he is not warned, so that (s)he may concentrate on his/her remaining senses and his/her own decisions. Only when (s)he fails to detect an obstacle or is out of the safety zones, is (s)he warned by the robot (See Man-Machine Communication Subsysytem of Fig. 1).

NAVIGATION

The fundamental data base of the robot is its navigation map stored in the auxiliary memory; e.g., cassette tapes, and transferred into the main memory of the robot when in use. The navigation map consists of information about intersections, i.e., names and types of intersections, distance between two adjacent intersections, and orientation to the adjacent intersections. Information on the landmarks to identify the intersections and other essential points of navigation are also included in the navigation map. This map is represented as an automaton as shown in Fig. 2.

The next step the robot should take is to identify the real intersection as specified on the map and correct its position and orientation so that it can travel farther. In order to do so, specific landmarks are chosen for each intersection or other essential points of navigation. In the initial phase (from 1977 to 1982) white painted lines on the streets with a length of about 2m and a width of 0.15m were adopted as the landmarks. These marks had to be set at every crossing at this stage of development. The automaton represention map for the robot could be automatically produced by an off-line computer from an ordinary map using picture processing techniques. Landmark laying

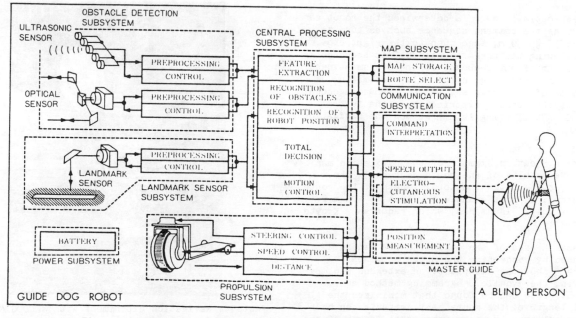

Fig. 1 Schematic diagram of the guide dog robot system (MELDOG).

instructions which would be used to place the landmarks on the streets could be provided at the same time [Kaneko et al., 1983].

At the second stage (from 1983 to the present) registered natural landmarks such as poles and walls are being used as markers for the correction of the robot's position and orientation. However, the navigation method is fundamentally the same.

Navigation Map

Figure 2 shows an example of landmarks set on the streets and the automaton representation of the map of landmarks in the memory. Landmark codes which contain information on intersection identification number, intersection type, i.e., crossings, forked roads, straight roads, etc., and stop information, i.e., it should stop at the landmark or not, correspond to the states of the automaton. Commands from the blind traveler (or Central Processing Subsystem (CPS) in automatic guidance mode) such as turn to the left, right, or go straight correspond to the input of the automaton, while information to the CPS and/or the blind master such as the steering angle to be used to reach the next landmark, the distance between two landmarks, and intersection attributes correspond to the automaton outputs.

The same map can be interpreted as the tree-structure shown in Fig. 3. In this representation landmark codes correspond to the nodes of the tree and commands from the user correspond to the branches. Each branch has an attribute represented as the output in the automaton representation. If the user assigns a starting landmark code and a destination code, the robot can find whether there is a route to reach the destination or not, by using the tree-structure representation of the map and searching techniques commonly used in artificial intelligence study, and can find an optimal route if plural routes exist.

Once an optimal route is determined the robot can determine the command sequence such as turn to the left, go 30 m, then turn to the right etc., by following the tree-structure. This sequence is used as the input sequence to the aforementioned automaton, resulting in the fully automatic guidance of the traveler. Photo 1 shows a general view of the outdoor experiments of the test MELDOG MARK II using landmarks and the navigation map.

Figure 4 shows an example of the navigation map and some results of the route search. In the figure, s indicates the total length of the route in meters. In this example the area is 500 m x 500 m with 276 landmarks, which requires 2 K byte memories.

The search area of the optimal route can be extended by connecting the above sub-maps. Figure 5 shows an example of an extended map. By applying the dynamic programming method an optimal route can be found that minimizes the total length of the route. Any criteria can be chosen arbitrarily, e.g., the total length of the

route, minimum number of intersections encountered, etc., or a combination of these [Tachi et al., 1980].

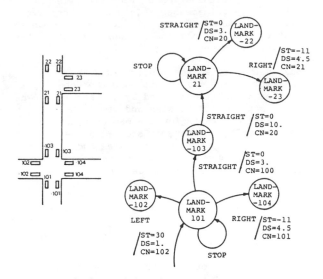

Fig. 2 An example of the landmarks and the automaton representation of the navigation map.

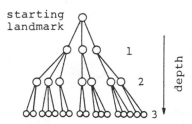

Fig. 3 Tree-structure representation of the navigation map.

Photo. 1 Navigation experiment with MELDOG MARK II using landmarks and navigation map.

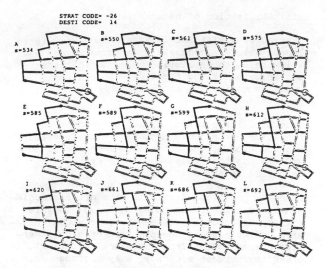

Fig. 4 Results of the route search.

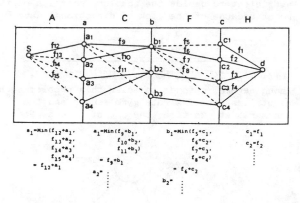

Fig. 5 Optimal route searching by connecting sub-maps.

Navigation between Landmarks

The robot travels from one landmark to another using landmark information in the navigation map to generate a desirable path.

Figure 6 shows an example of path generation when the starting vector and the destination vector are assigned. The designed path, which connects the current position with an arbitary intermediate destination, consists of two arcs and their common tangent. After determining a path the robot travels along it using the encoders of the steering shaft and the rear wheels. Each arrow of Fig.6 indicates the final experimental position of the robot after following the path. In the figure, b) and c) show better navigation results through controlled steering compensation [Komoriya et al., 1984].

In navigation using internal sensors, accumulation of error from a course is inevitable. In order to guide the robot along the path accurately it is necessary to compensate for this. Three methods are studied to solve this problem.

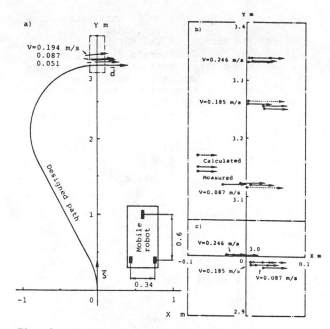

Fig. 6 Path generation and the result of the path following experiments.

i) landmark tracking method

In the first method landmarks are used to compensate for course error. When the robot reaches a landmark, it adjusts its orientation and position by moving along the landmark (See Fig.7). The robot has two landmark sensors, one at its front and one at its rear end, which optically detect landmark edges. After lateral course error Δy and orientational error $\Delta\phi$ are measured, equation (1) gives the steering angle which enables the robot to follow the landmark [Komoriya et al., 1983, Tachi et al., 1980].

$$\theta = K_1 \cdot \Delta y + K_2 \cdot \Delta\phi \qquad \ldots\ldots(1)$$

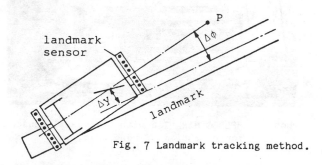

Fig. 7 Landmark tracking method.

ii) Utilization of road edges

A landmark tracking method is effective if the deviation from the course can be within the detection area of landmark sensors when the robot reaches a landmark. This condition restricts the distance between landmarks. Using the road edges as an auxiliary method, supports landmark tracking, and enables the distance between landmarks to be longer.

Fig.8 shows a general view of this method. The robot detects the road edge, shown as the x-coordinate axis, from the points where the road edge crosses the CCD camera's field of view by processing the visual data using the road edge attributes of the navigation map. After calculating ϕ_R robot's orientation to the road edge, and Y_R distance from its course by equations (2), the steering angle is given by equation (3).

$$\phi_R = \tan^{-1} \frac{X_2 - X_1}{Y_1 - Y_2}$$

$$Y_R = X_1 \cos \phi_R + Y_1 \sin \phi_R \qquad \ldots\ldots(2)$$

$$\theta(t+\tau) = K_1 \phi_R(t) - K_2 (y_R(t) - y_s) \quad \ldots\ldots(3)$$

where τ is sampling time.

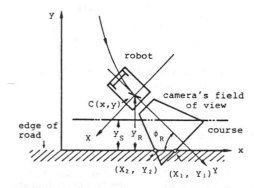

Fig. 8 Utilization of road edges.

Photo. 2 MELDOG MARK IIb with CCD visual sensor for detection of road edge.

Photo.2 shows the CCD camera assembled for this purpose and installed on MELDOG MARK IIb. Its field of view can be changed to the front, right and left side of the robot by turning the table which supports the camera and far and near by tilting the mirror which alters the vision line. One micro-computer is mounted on board the robot to control these operation and to process visual data exclusively [Tachi et al., 1982 a].

iii) Utilization of natural landmarks

Instead of artificial landmarks such as painted lines, natural landmarks such as poles and walls which have rather simple shapes so that the sensor on board the robot can measure their position easily are more desirable for navigation.

From the view point of signal processing in real-time, ultrasonic sensors are preferable. The construction of the ultrasonic sensor used here and the position measurement algorithm is described in the next section.

Landmark position using this kind of ultrasonic sensor is measured as shown in Fig.9. In this figure the robot is assumed to move along the x-axis and to measure the position of a cylinder-like shaped object from the plural points Pi. In order to increase accuracy, only distance data si is used. Using the radius of the object, relative position of the object (xm, ym) can be calculated by equations (4). However this information is not sufficient to decide the absolute position of the robot because of the lack of directional data.

$$x_m = \frac{1}{2n(n+1)d} \left\{ \sum_{i=1}^{n} s^2_{-i} - \sum_{i=1}^{n} s^2_i - 2r \sum_{i=1}^{n}(s_i - s_{-i}) \right\}$$

$$\ldots\ldots(4)$$

$$y_m = \sqrt{\frac{1}{2n} \left\{ \sum_{i=1}^{n}(s_i+r)^2 + \sum_{i=1}^{n}(s_{-i}+r)^2 \right\} - \frac{1}{6}(n+1)(2n+1)d^2 - x^2_m}$$

Among several methods to solve this problem such as utilization of a rate gyro sensor and two landmarks at one time, the use of flat surfaces

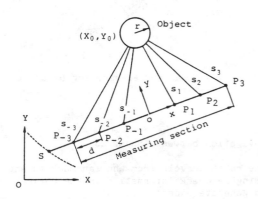

Fig. 9 A method to measure an object as a landmark.

Photo .3 MELDOG MARK IV used to demonstrate the navigation using walls as landmarks.

such as walls is practical if the navigation is inside buildings. Photo.3 shows the test hardware MELDOG MARK IV which has this ability to use walls as landmarks and with ultrasonic sensors at both sides of its body [Komoriya et al., 1984].

Obstacle Detection

It is important for the robot to find various obstacles while it guides the master, which appear in front of it, such as obstacles which block its path, objects and humans that come toward the robot and the master, steps or uneven streets, overhanging objects like awnings, etc.

In order to detect these obstacles, an ultrasonic sensor, which can determine not only the distance from the obstacle but also its direction by the traveling time measurement of ultrasound was developed.

Fig.10. shows its construction with one transmitter and plural receivers arranged in a array d distance apart from each other. A tone burst of frequency 40KHz and duration time 25msec is sent by the transmitter. Each of the receivers detects the reflected signal by obstacles t_i seconds later from the transmission, which corresponds to the distance S_i from T to R_i through the obstacle surface. Detected signals are amplified, processed by band-pass-filters and compared with the appropriate threshold so as to make stop-pulses for the counters which measure t_i in order to get the aforementioned S_i.

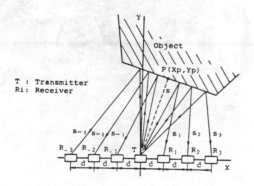

Fig. 10 Construction of ultrasonic sensor.

Location of the object can be calculated from S_i by equations (5), which assumes that the ultrasound is reflected at a flat surface as shown in the figure. According to the numerical calculation this algorithm gives almost correct position and direction with less than two percent error if the obstacle has a cylindrical surface or is a circle in a two dimensional figure [Komoriya et al., 1984].

$$s = \sqrt{\frac{1}{2n} \left(\sum_{i=-n}^{n} s_i^2 - \sum_{i=-n}^{n} i^2 d^2 \right)} \qquad \ldots (5)$$

$$x_p = \frac{1}{2n(n+1)} \left(\sum_{i=-1}^{-n} s_i^2 - \sum_{i=1}^{n} s_i^2 \right) \qquad y_p = \sqrt{s^2 - x_p^2}$$

When the robot detects an object in its sensing area, it can determine the relative speed V of the object by measuring distance at more than two instances. If V is positive, it means that the object is moving away from the robot and it will not bother the robot. Therefore the robot needs not take any action.

If V is negative, the robot behaves as follows. If the absolute value of V is larger than the speed of the robot Vr, i.e. the object is coming towards the robot, the robot quickly stops and warns the master and the object in order to avoid a collision.

If the absolute value of V is smaller than Vr, i.e. the object is moving in the same direction of the robot and the master at a slower speed, the robot asks the master to slow down and tries to follow the object keeping a safe distance between them (See Photo.4).

If the absolute value of V is equal to Vr, i.e. the object is standing still, the robot modifies its path to avoid the obstacle. Fig.11 shows the path to avoid an obstacle in front of it. At the

Photo. 4 MELDOG MARK III in the experiment to demonstrate an obstacle detection.

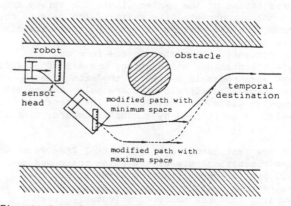

Fig. 11 Example showing the avoidance of an obstacle by modifying the robot's path.

time the robot detects a stationary obstacle, it generates a path with maximum space between the obstacle and the robot as shown by the dotted line.

While it moves along this modified path, the ultrasonic sensor continues to detect obstacles along the road by turning the sensor head. If the obstacle leaves the detection area of the sensor and the robot has open space in front of it, the robot generates its path again to return to the initial path with minimum space as shown by the solid line after avoiding the obstacle.

When the robot doesnot have open space along the path with maxmum space, i.e. the road is blocked, the robot turns back generating a new path to its final destination.

COMMUNICATION

In order to guide a blind individual in accordance with the information acquired, an information communication channel between the master and the robot must be established.

When a robot which directs or guides a blind individual has somehow acquired information about the direction of, and width of, an unobstructed path along which it should lead the blind individual, the problem is the choice of sensory path display and its safe margins appropriate for presentation to the remaining exterior receptive senses of the blind individual. Quantitative comparison method of display scheme has been proposed and an optimal auditory display scheme has been sought [Tachi, et al., 1983].

In the MELDOG system the location of the master is measured by the robot in real-time by the triangulation of the ultrasonic oscillator put on the belt of the master and the two receivers on board the robot (See Fig.12). The result of the measurement is used to control the robot's speed to coincide with that of the blind master.

A safety zone is set behind the robot in which the master is supposed to walk (See Fig.13). The triangulation is also used to transmit warning signals from the robot to the master. When he is outside the zone he is warned by the robot, while he receives no feedback when he is safe. When the orientation of the master within the safety zone is not appropriate, the Master Guide detects the condition and informs the master. These signals are transmitted through a wired link and presented to the master in the form of electrocutaneous stimulation. One set of Ag-AgCl wet electrodes is placed on the skin of each brachium. The signals used are pulse trains with a pulse width of about 100 μs, the energy of which is controlled by a constant energy circuit [Tachi, et al., 1982 b].

In the test hardware MELDOG MARK I (See Photo.5) the repetition rate of the pulse train was set at 100 pps for normal warning stimuli that the master was outside the safety zone and 10 pps for warning that the master's orientation was inappropriate. For example, the signal presented to the right arm with 100 pps means the master

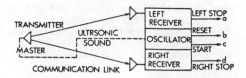

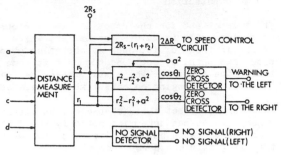

Fig. 12 Block diagram of measurement system of the master location using ultrasounds.

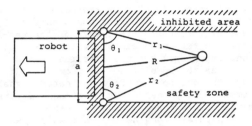

Fig. 13 Safety zone set behind the robot.

Photo. 5 Communication experiment with MELDOG MARK I.

should step to the right to come back to the safety zone and with 10 pps means (s)he should turn his/her body counterclockwise to correct his/her orientation [Tachi, et al., 1978 and 1981a].

CONCLUSIONS

The idea of guiding a blind person using an autonomous robot and a method for the realization of the idea were proposed (Photo.6). The robot processes both the information stored in the memory of the robot and environmental information acquired by the sensors on board the robot and passes the processed information to the blind master.

Photo. 6 General view of the guidance by MELDOG (MARK IV).

A navigation map is given to the robot prior to guidance. The robot travels according to the information on the navigation map. The error between the internal representaion of the environment (navigation map) and the real world is compensated by detecting landmarks in the real environment and correcting the robot's position and orientation according to the landmark location measurement. The landmark location data is already stored on the navigation map.

The feasibility of the navigation method was demonstrated both by computer simulation and outdoor experiments using the test hardware called MELDOG MARK II and MARK IV.

Some of the obstacle detection and avoidance functions were considered theoretically and the feasibility of the method was demonstrated by the test hardware called MELDOG MARK III and MARK IV.

Experiments concerning the transmission of course information and obstacle information via electrocutaneous stimulation were conducted using the test hardware called MELDOG MARK I and MARK IV.

The remaining problems include:

(1) The selection of general criteria for environmental objects as navigation landmarks, the detection method of the landmarks, and organization and utilization method of the navigation map with the information of the selected landmarks.

(2) Finding a more general obstacle detection and avoidance method.

(3) Finding an optimal choice of sensory display method of the navigation information acquired by the robot appropriate for presentation to the remaining exterior receptive senses of the blind individual.

ACKNOWLEDGEMENTS

The authors would like to express their thanks to the members of the Mechanical Engineering Laboratory who are concerned with this project for their kind support and valuable advice.

REFERENCES

Dodds, A.G., Armstrong J.D. and Shingledecker, C.A. 1982. The Nottingham Obstacle Detector: Development and evaluation, J. Visual Impairment & Blindness, 75, pp. 203-209.

Farmer, L.W., Benjamin, J.M. Jr., Cooper, D.C., Ekstrom, W.R., and Whitehead, J.J. 1975. A teaching guide for the C-5 laser cane: An electronic mobility aid for the blind, Kalamazoo: College of Education, Western Michigan University.

Kaneko, M., Tachi, S., and Komoriya, K. 1983. A constructing method of data base for mobile robot navigation, J. of Mechanical Engineering Laboratory, 37, pp. 160-170.

Kay, L. 1973. Sonic glasses for the blind - Presentation of evaluation data, Research Bulletin of the American Foundation for the Blind, 26.

Komoriya, K., Tachi, S., Tanie, K., Ohno, T., and Abe, M. 1983. A method for guiding a mobile robot using discretely placed landmarks, J. Mechanical Engineering Laboratory, 37, pp. 1-10.

Komoriya, K., Tachi, S., and Tanie, K. 1984. A method for autonomous locomotion of mobile robots, J. Robotics Society of Japan, 2, pp. 223-232.

Mann, R.W. 1974. Technology and human rehabilitation: Prostheses for sensory rehabilitation and/or sensory substitution, Advances in Biomedical Engineering, 4, pp. 209-353.

Morrissette, D.L., Goodrich, G.L. and Hennessey, J.J. 1981. A follow-up study of the Mowat Sensor's applications, frequency of use, and maintenance reliability, J. Visual Impairment & Blindness, 75, pp. 244-247.

Russell, L. 1971. Evaluation of mobility aids for the blind, Pathsounder travel aid evaluation, National Acad. Eng., Washington, D.C.

Tachi, S., Komoriya, K, Tanie, K., Ohno, T., Abe, M., Hosoda, Y., Fujimura, S., Nakajima, H., and Kato, I. 1978. A control method of a mobile robot that keeps a constant distance from a walking individual, Biomechanisms, 4, pp. 208-219.

Tachi, S., Komoriya, K., Tanie, K., Ohno, T., Abe, M., Shimizu, T., Matsuda, K. 1980. Guidance of a mobile robot using a map and landmarks, Biomechanisms, 5, pp. 208-219.

Tachi, S., Tanie, K., Komoriya, K., Hosoda, Y., and Abe, M. 1981 a. Guide Dog Robot - Its basic plan and some experiments with MELDOG MARK I, Mechanisms and Machine Theory, 16, pp. 21-29.

Tachi, S., Komoriya, K., Tanie, K., Ohno, T., and Abe, M. 1981 b. Guide Dog Robot - Feasibility experiments wit MELDOG MARK III, Proceeding of the 11th International Symposium on Industrial Robots, Tokyo, Japan, pp. 95-102.

Tachi, S., Komoriya, K., Tanie, K., Ohno, T., Abe, M., Hosoda, Y. 1982 a. Course control of an autonomous travel robot with a direction-controlled visual sensor, Biomechanisms, 6, pp. 242-251.

Tachi, S., Tanie, K., Komoriya, K., and Abe, M. 1982 b. Electrocutaneous communication in Seeing-eye Robot (MELDOG), Proceedings of IEEE/EMBS Frontiers of Engineering in Health Care, pp. 356-361.

Tachi, S., Mann, R.W., and Rowell, D. 1983. Quantitative comparison of alternative sensory displays for mobility aids for the blind, IEEE Trans. on Biomedical Engineering, BME-30, pp. 571-577.

Visual Navigation of an Autonomous Vehicle Using White Line Recognition

SHIGEKI ISHIKAWA, HIDEKI KUWAMOTO,
AND SHINJI OZAWA

Abstract—We present a method for an autonomous, visually navigated vehicle using white guide line recognition. In our approach, the vehicle follows a white guide line on flat ground or floor, contrasted with its background, sensing through a forward looking TV camera. The white line recognition algorithm presented here, which uses a state transition algorithm, performs well. We also present a field pattern monitoring method for vehicle guidance using a state transition scheme. When the vehicle comes to a branch, it selects the appropriate direction according to a path planner output. It can also avoid collisions with obstacles in front of it by monitoring the field patterns and their changes. An experimental moving vehicle system was constructed, and tests confirmed the effectiveness of this approach.

Index Terms—Autonomous vehicle, edge detection, field pattern, machine vision, scan line pattern, state transition, visual navigation, white line recognition.

I. INTRODUCTION

Research in robot sensing technology has gained attention and made rapid progress during the last few years. Using these sensing technologies, various sensor based robot systems have been developed. Some of these systems are experimental autonomous vehicles. Because the environment of an AGVS can change dramatically, it is necessary for the vehicle to monitor the environment measuring its position by real-time processing using sensors.

There are two main tasks for navigation, which are to guide a vehicle from a certain starting point to a destination correctly, and to avoid collision with any obstacle in the vehicle's path. The existing schemes for autonomous vehicle systems have various sensors, such as sonar, buried wire, infrared, optical line sensors, radar, TV-camera, etc., to achieve the tasks. The typical example for practical use is the buried wire AGVS [1]. This system is very robust, but difficult to install, alter, or repair, because the wire is usually buried in a concrete floor. Therefore white line or metal line guided systems have been developed [2]. Many systems using ultrasonic sensors to obtain local environment information have been reported recently (for example [3], [4]). The path planning problem is also frequently discussed [5], [6].

Recently, progress in image processing has made it possible to use some visual functions in the field of robotics. However, visual systems have many limitations compared to the human vision. Generally speaking, it takes so much time to process images that, although many earlier efforts have tried to solve this problem [7]–[9], most of the system which have been constructed so far have some difficulties. A visual navigation system, however, has the advantage that it can be easily used for both navigation and collision avoidance.

As the earliest research of visual navigation, the "Stanford Cart" and the "Robot Rover" are famous for their approach, described

Manuscript received December 15, 1986; revised May 15, 1987.
S. Ishikawa is with Tokyo Research Laboratory, IBM Japan, 5-19, San-ban-cho, Chiyoda-ku, Tokyo 102, Japan.
H. Kuwamoto and S. Ozawa are with the Department of Electrical Engineering, Faculty of Science and Technology, Keio University, 3-14-1, Hiyoshi, Kouhoku-ku, Yokohama-shi, Kanagawa 223, Japan.
IEEE Log Number 8820098.

Reprinted from *IEEE Transactions on Pattern Analysis and Machine Intelligence*, Vol. 10, No. 5, September 1988, pages 743-749. Copyright © 1988 by The Institute of Electrical and Electronics Engineers, Inc. All rights reserved.

in [10], in which multiple stereo view is used to identify obstacles in the path of the robot. However, the robots moved in "lurches" of up to 1 m once every 10 min. Perhaps ultimate goal in the visual navigation of autonomous vehicle, is that they can move without any guide, as if they were human beings. Recent research in this direction has provided a number of sophisticated methods [13]–[16]. Herbert and Kanade [13] presented an active vision system using range data from a laser range scanner for outdoor scene analysis. Wallace *et al.* [14] discussed the CMU Terregator using color image segmentation. Waxman *et al.* [15] presented a navigation system using rule-based reasoning. Turk [16] discussed a method of detecting obstacles using video data on the Martin Marietta ALV. Although many different techniques have been investigated for visually based, autonomous navigation with collision avoidance, none has been shown to be ready for practical use. We have discussed a method of visual navigation using white guide lines [17] intended for factory transportation. Our approach does not include complex image processing, but performs well in a lower vision system.

The use of a white guide line [11], [12] is considered to be more flexible than the buried cable guided system. In [11] and [12], a geometrical model of the guide line is used to navigate the vehicle more accurately.

In this correspondence, we present a method for an autonomous, visually navigated vehicle using white guide line recognition. The vehicle follows a white guide line, which resides on the flat ground or floor and contrasts with its background, sensing through a forward looking TV camera. The white line recognition algorithm presented uses a state transition algorithm, which performs well. We also present a field pattern monitoring method for vehicle guidance using the state transition scheme. When the vehicle comes to branches, it selects the appropriate direction according to a path planner output. It can also avoid collision with obstacles in front of it by monitoring the field patterns and their changes. An experimental moving vehicle system was constructed and tests confirmed the effectiveness of this approach.

II. VISUAL NAVIGATION USING WHITE LINE RECOGNITION

A. Conception

We assume that the major intended use of this vehicle is in the factory and the path layout of the vehicle system includes straight lines, branches, crosses, and junctions as shown in Fig. 1. In Fig. 1, a node means a branch, a junction, or a station where the vehicle will stop. The station is represented by a landmark. A path is defined as a course between nodes.

The TV camera, mounted on the front of the vehicle, looks forward and down, at an area between 0.5 and 2.5 m ahead. Images of white guide lines, which are determined from the field of the camera view, vary according to the vehicle movement. The system recognizes them as field patterns from the input images (described in Section III). Because our approach can quickly extract positional information from the scene, it is possible to continuously navigate the vehicle by monitoring recognition results.

In our system, if the vehicle has a path layout map, the system will plan a path route from a certain station node to destination according to any request. The vehicle can realize its position on the path route by comparing field pattern changes along with the path route order predetermined in the path planner. A detailed algorithm for monitoring the field pattern is described in Section IV.

In this correspondence, we assume that obstacles have unexpected edges. The vehicle can also avoid collision with obstacles in front of it, by monitoring field patterns and their changes. First, the system must check 110 field patterns to be recognized. The

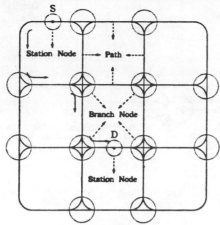

Fig. 1. Path layout of white guide lines.

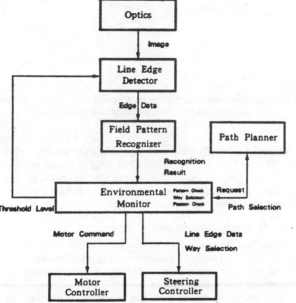

Fig. 2. Visually navigated vehicle system using white guide line recognition.

system checks the field patterns against 110 possible solution. Other field patterns encountered may imply:

1) Obstacles may exist in front of the vehicle.

2) Optical parameters such as thresholding level must be updated.

Therefore, when unknown field patterns are recognized, the vehicle stops, avoiding collisions with obstacles. Next, the system judges whether it should update optical parameters and restart the vehicle or remain stopped until the obstacles are moved away.

The reliability of avoiding collisions is increased when field pattern changes are monitored because correct field patterns are limited by the vehicle's position and the vehicle realizes its position of the path route by monitoring field pattern transitions.

Thus, the visual navigation system can control the vehicle by monitoring the recognition results and comparing the current results to the anticipated values.

B. System Structure

The visually navigated vehicle system discussed in this paper is composed of several processes as shown in Fig. 2. First of all, the white line edge detector detects line edges from an input image. The recognizer recognizes a field pattern from the detected line

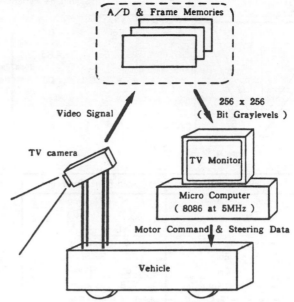

Fig. 3. Experimental vehicle system.

edges in real time (Section III), and also extracts the line edge positions in the image plane to control the steering.

The process of environment monitoring (Section IV) handles recognition results of field patterns. It checks whether field patterns and their changes are correct or not, monitors the vehicle position, selects a direction on a branch, and judges if the vehicle reaches its destination. If the field pattern is unknown, or the field pattern transition is illegal, it directs the motor controller to stop the vehicle and updates its thresholding level.

The steering process controls the steering angle based on edge positions. In our experimental system, we assume that the vehicle runs on a flat ground or floor, which greatly simplifies the steering process. The steering angle is determined to be proportional to the distance between an edge position on a raster scanning line and the center point of the scanning line. Since several edges are detected when the vehicle comes to branches or junctions, the steering process will select a suitable edge position according to the orders of the environmental process monitor. To choose suitable edge data, scan line patterns extracted from the recognition process (Section 3-A) are used.

The path planner, which has not been implemented yet on the experimental system, searches and directs a path route before the vehicle begins.

C. Experimental Vehicle System

Fig. 3 shows an overview of the experimental vehicle system. The vehicle is driven by an electric motor, controls various mechanical configurations, and runs at 20(cm/s). The navigation system is composed of a TV-camera, an A/D converter, frame memories, and a microcomputer, in which the CPU is an 8086 running at 5 MHz. The frame memory consists of 256 * 256 pixels, monochrome with 8 bits (256) gray level, which are mapped to the main memory of the microcomputer. The navigational system is implemented on the microcomputer. It takes about 300 ms per image for recognition and navigation.

III. FIELD PATTERN RECOGNITION METHOD

Our approach for field pattern recognition uses a structural analysis method combined with a state transition algorithm. This method is composed of two steps. First, an input image is scanned horizontally to detect edges and to extract scan line patterns. The second step scans the scan line patterns extracted in the first step and assigns a single field pattern value to the input image.

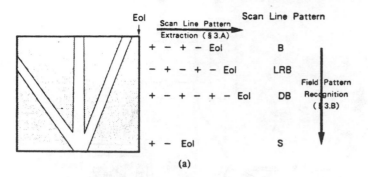

Eol → Scan Line Pattern Extraction (§3.A) → Scan Line Pattern

+ − + − Eol   B

− + − + − Eol   LRB

+ − + − + − Eol   DB

Field Pattern Recognition (§3.B)

+ − Eol   S

(a)

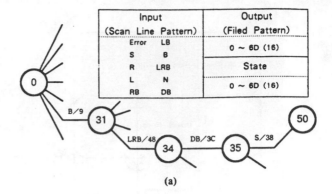

Input (Scan Line Pattern)		Output (Filed Pattern)
Error	LB	0 ~ 6D (16)
S	B	
R	LRB	State
L	N	
RB	DB	0 ~ 6D (16)

(a)

Detected Edges	Scan Line Pattern
Eol	N (Nothing)
+ Eol	R (Right)
− Eol	L (Left)
+ − Eol	S (Straight)
+ − + Eol	RB (R−Branch)
− + − Eol	LB (L−Branch)
+ − + − Eol	B (Branch)
− + − + Eol	LRB (L·R−Branch 1)
+ − + − + Eol	LRB (L·R−Branch 2)
− + − + − Eol	LRB (L·R−Branch 3)
+ − + − + − Eol	DB (Dual Branch)
Others	Error

(b)

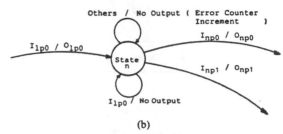

(b)

Fig. 5. State transition diagram for field pattern recognition. (a) State transition example. (b) State transition mechanism.

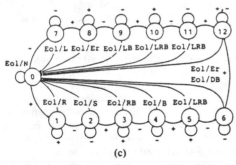

(c)

Fig. 4. Field pattern example and scan line pattern extraction. (a) Example of field pattern and its edges. (b) Scan pattern definitions. (c) State transition diagram for scan line pattern extraction.

A. Line Edges Detection and Scan Line Pattern Extraction

We have paid particular attention to edges of white lines, in order to recognize field patterns efficiently. Since white lines in an image plane exist along the vertical direction, line edges are detected by searching gray level changes along the raster (horizontal) direction. In this paper, we classify line edges according to the gray level change, such as plus edges and minus edges. Plus edges are defined as gray level changes from the background to white, and minus edges are defined as the position where gray levels change from the white line to the background. In addition, we include an Eol (end of scan line) for efficient recognition. Plus edges and minus edges are detected by thresholding spatial gray level difference values which are calculated by comparing gray levels of two distant pixels on a scan line. In the experimental system, the pixel separation is a balance between image resolution and sensitivity.

Fig. 4(a) shows an example of a three-way branch and its scan line patterns. On the top scan line, two white lines are observed, on which edges of +, −, +, −, and Eol are included. The next scan line contains three white lines which is composed of −, +, −, +, −, and Eol. We define 12 scan line patterns as shown in Fig. 4(b).

We apply the state transition algorithm shown Fig. 4(c) to extract the scan line pattern. The initial state is 0 and inputs are +, −, and Eol. Regarding detected edges on the scanning line as the input data sequence, the scan line patterns are extracted as outputs of transitions when the inputs are Eols.

B. Field Pattern Recognition

The second step of the white line recognition scans the line patterns vertically. The field patterns are regarded as occurrence patterns of the scan line patterns. We again apply a state transition algorithm to this information.

Fig. 5 shows an example of the recognition process. The scan line patterns shown in Fig. 4(a) are B, LRB, DB, and S. Fig. 5(a) shows the state transition path in this case. The initial state at the beginning of a scan line pattern is always 0. When the first input B (6) comes, the transition outputs 9 which means field pattern No. 9. Each output corresponds to a unique field pattern. The latest output of all the outputs is taken as the recognition result. In this case, other inputs will follow. Finally, the recognition result becomes 38 which represents the field pattern shown in (a). The input image is assigned to the field pattern value.

The state transition algorithm applied here differs somewhat from the one described in the previous section. The scan line patterns as inputs at state-n are classified into three types. The first type is permitted to transit state-n with an output which means a field pattern [Inp0/Onp0, Inp1/Onp1 in Fig. 5(b)]. However, the transition is acknowledged only when the same input continues above a certain threshold count for a given number of lines. If the same inputs do not continue for more than M lines, the error counter is incremented. The second type is the same as the input which comes into State-n (Ilp0). This type has no output. Finally, the last one is not permitted to transit State-n. This type also has no output but increase the error counter. If the error counter is beyond a certain threshold count, the recognition result should end in failure.

Field patterns on an input image depend on the TV-camera field view, the height of the TV-camera, the camera tilt angle, the type of the steering control, and so on. In this correspondence, we get

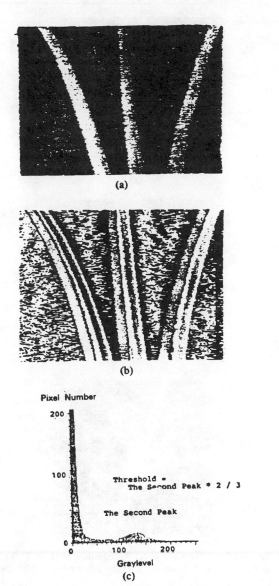

(a)

(b)

Pixel Number

200

100

Threshold =
The Second Peak * 2 / 3

The Second Peak

0

0   100   200

Graylevel

(c)

Fig. 6. Threshold calculation for edge detection. (a) An input image. (b) The spatial difference image. (c) Histogram of the spatial difference image.

different 110 patterns, as experimental results, which the system can recognize quickly and efficiently.

C. Threshold Calculation

The threshold level is calculated before the vehicle runs, and it is held constant while the vehicle is moving. The threshold is updated when the vehicle is stopped by the environment monitor shown in Fig. 6. Fig. 6(a) shows an example of the input white line image, and Fig. 6(b) shows a spatial gray level difference image, where absolute value $(D(i, j) = |F(i + n, j) - F(i, j)|)$.

Fig. 6(c) shows the histogram of the (b) image. As shown in this figure, the histogram has two peaks. One is a background peak, and the other is an edge peak. The threshold level is determined to be 2/3 of the second peak value.

IV. FIELD PATTERN MONITORING

In this section, we propose a model for running a vehicle running a model that represents actual field pattern changes. Field patterns are always changing while the vehicle is moving. However, the field pattern changes in a fixed way according to the course the vehicle is expected to follow. Fig. 7 shows an example of the view area change and a sequence of the field pattern changes according to the movement when the vehicle selects a right fork at a three-

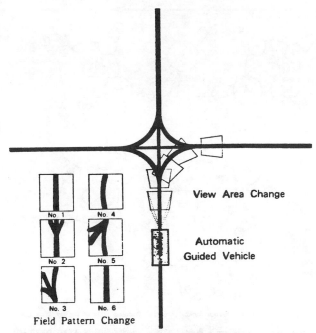

View Area Change

Automatic Guided Vehicle

No. 1   No. 4

No. 2   No. 5

No. 3   No. 6

Field Pattern Change

Fig. 7. Change of view area according to vehicle movement.

way branch. Considering the above, this system can monitor environmental conditions while it is running. We apply this to the environmental monitor process, which generates the orders of vehicle start/stop and selections of branch directions to follow.

Fig. 8 shows a state transition diagram we propose. There are six basic states, which describe the kind of nodes the vehicle is moving towards, or is currently at. For example, S-state means that the vehicle is moving toward a Station node, or leaving a Station node, and BBR means that the vehicle is moving toward a two-way branch and will select the right fork. These states have substates of their own, which are illustrated by field pattern changes in the nodes.

In this system, we assume that the path planner searches for a path before the vehicle starts. Fig. 1 shows an example of the vehicle moving from a station S to a station D, in which case the path planner selects left direction at the first branch, the right direction at next one, and finally runs along a left guide line. In the case that the vehicle selects a right direction at the three-way branch, the field pattern monitoring is processed as shown in Fig. 8(b) which corresponds to a TBR-state [Fig. 8(a)]. TBR0 is an entrance state of TBR, and TBR1 is an exit state. A new recognition result of a field pattern triggers a transition between states. Whenever the environmental monitor gets a recognition result, it checks whether the result is suitable for the current state.

E-state is an error state, which means that an unknown field pattern or an illegal field pattern change is detected, and the environmental monitor stops the vehicle and updates the threshold level for edge detection. If any obstacle exists in front of it, the vehicle does not start again, because the next recognition results or field pattern transitions will return an error state.

When a focus of attention state comes to an end state such as TBR1, a path order pointer is incremented and points to the next path order. According to this, the environment monitor generates another selection of a direction in which to progress. The vehicle position at a node can be extracted from this state transition model.

V. EXPERIMENTS

To prove the effectiveness of our visual navigation approach, we constructed and tested an experimental vehicle system.

A. Field Pattern Recognition Experiments

Fig. 9 depicts some experimental results of field pattern recognition. These figures show the scan line pattern and the recognition

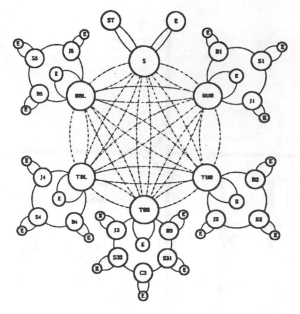

BBR : Running State of 2-way Branch Right   BBL : Running State of 2-way Branch Left

TBR : Running State of 3-way Branch Right   TBS : Running State of 3-way Branch Left

TBL : Runnig State of 3-way Branch Straight   S : Running State towards Station Node

E : Error State

(a)

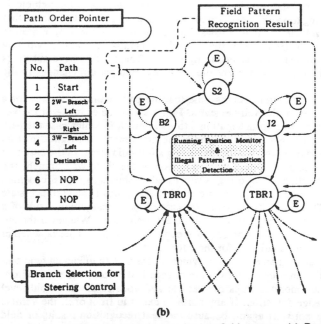

(b)

Fig. 8. State transition diagram for monitoring field pattern. (a) Basic states. (b) Substates.

results. Fig. 9(a) shows a straight field pattern image and its recognition. A straight line was recognized correctly. Fig. 9(b) shows a case when an obstacle exists in front of the vehicle. The recognition result shows that it is illegal, that is, the obstacle was detected. In these figures, our approach works well for vehicle navigation; however, in the case of Fig. 9(c), the recognition shows an illegal pattern, although the correct pattern is straight, as in Fig. 9(a). The reason is a shadow in the field of view. Since the system detects wrong edges by shadow, the recognition fails. We will discuss this more in the following section.

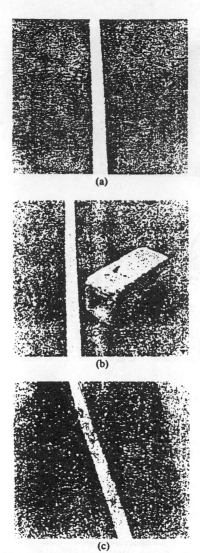

Fig. 9. Recognition experiments. (a) Recognition result = 1. (b) Recognition result = FFFF (fail). (c) Recognition result = FFFF (fail).

B. Guiding Experiments

Fig. 10(a) shows an experiment in avoiding collision with an obstacle, which was put on the course. The vehicle detected the obstacle and stopped in front of it, to avoid collision. After removing the obstacle, the vehicle started again.

Fig. 10(b) shows a navigation experiment. A white guide line path was laid out on the asphalt pavement, which included two station nodes and two branch nodes. In this experiment, the planned path was input manually as follows:

"Start" ➡ "3-way Branch Left"
"3-way Branch Right" ➡ "Destination"

The vehicle selected the correct way when it came to branches, and stopped correctly when it reached its destination.

VI. DISCUSSION

In this section, we discuss some considerations of our approach. Our approach has not only advantageous points but also some disadvantages.

A. Recognition Processing

Since the reliability of guidance depends on the recognition results which are extracted from edge data, it is important to detect edges correctly and necessary to determine a suitable threshold level. The optical environment can change rapidly with various

(a)

(b)

Fig. 10. Navigation experiments. (a) Collision avoidance. (b) Vehicle guidance.

conditions, and the TV camera should adjust diaphragm automatically and hold a constant threshold level.

The number of field patterns needs to be as small as possible. The system stops the vehicle, when a recognition result or a field pattern change fails. This means that the fewer field patterns the system has, the more reliably the system is able to avoid collisions and false stops.

B. Running Environments

The environment changes constantly while the vehicle is moving. Some environments are not desirable for running the vehicle. These conditions are summarized as follows:

1) An environment which causes a failed recognition even though no obstacle exists, e.g., shadows, pictures on the ground.

2) An environment which causes a successful recognition result from its image structure even though an obstacle exists.

The latter case is the dangerous one but it is very rare, because the number of field patterns which the system holds are as small as possible and the patterns are quite simple. Besides, the field pattern change monitoring limits the number of correct field patterns.

The system can eliminate small noisy edges because the state transition algorithm used in Section III absorbs them. However, it is necessary to consider other suitable algorithms under severe conditions such as a shadow. We have developed some ideas [18] considering height information to solve this problem. Another simple and useful method to solve the shadow problem is lighting ahead, because the field of view is narrow enough to remove shadows.

C. Field Pattern Transition Monitoring

The method of field pattern change monitoring gives a visual navigation system useful and effective information. We believe that the recognition process can be improved under severe conditions, if the field pattern prediction is fed back to the recognition process.

Another topic for future research is the multiple autonomous vehicle system. In such a system, the system needs to know other vehicle movements, or the center control system is needed. However, even in these cases, the field pattern monitoring method would be successful without any change.

VII. CONCLUSIONS

It is necessary for a moving vehicle system to monitor the surrounding environment and use the local information to guide the vehicle. In this correspondence, we presented a method of visual navigation for an autonomous vehicle using white line image recognition. The vehicle is guided by white line images and the system recognize 110 white line image patterns.

The recognition method consists of two state transition algorithm steps. The first step is concerned with horizontal scan line patterns. The second step manages these scan line patterns in order to recognize a whole image pattern. This method can process an image quickly and efficiently, and making it possible to guide a moving vehicle by image processing.

The ability to avoid collision with obstacles can be added to the system by monitoring the white line image recognition results. The field pattern transition monitoring gives useful information to help navigate the vehicle. This was confirmed experimentally.

ACKNOWLEDGMENT

Our special thanks go to E. Takinami in Toyoda Automatic Loom Works, Ltd. and the members of Ozawa Lab. in Keio University for their great help.

REFERENCES

[1] Y. Oshima *et al.*, "Control system for automatic automobile driving," in *Proc. IFAC Tokyo Symp.*, 1965, p. 347.

[2] S. Ando, "Unattended travelling vehicle guided by optical means," in *Proc. 4th ISPR*, 1974, p. 385.

[3] R. A. Cooke, "Microcomputer control of free ranging robots," in *Proc. 13th Int. Symp. Indust. Robots Robot 7*, vol. 2, Apr. 1983.

[4] J. L. Crowley, "Dynamic world modeling for an intelligent mobile robot," in *Proc. 7th Int. Conf. Pattern Recognition*, Aug. 1984, p. 207.

[5] R. Ruff and N. Ahuja, "Path planning in a three dimensional environment," in *Proc. 7th Int. Conf. Pattern Recognition*, Aug. 1984, p. 188.

[6] K. Kant and S. W. Zucker, "Trajectory planning problems, I: Determining velocity along a fixed path," in *Proc. 7th Int. Conf. Pattern Recognition*, Aug. 1984, p. 196.

[7] T. Yatabe *et al.*, "Driving control method for automated vehicle with artificial eye," in *Proc. IAM*, 1978, p. 29.

[8] Y. Okawa, "Vehicle guidance by digital picture processing and successive state estimation," in *Proc. IFAC*, 1981, p. 2419.

[9] J. L. Moigne and A. M. Waxman, "Projected light grids for short range navigation of autonomous robots," in *Proc. 7th Int. Conf. Pattern Recognition*, Aug. 1984.

[10] H. P. Moravec, "The Stanford Cart and the CMU Rover," *Proc. IEEE*, vol. 71, no. 7, pp. 872–884, July 1983.

[11] K. C. Drake *et al.*, "Sensing error for a mobile robot using line navigation," *IEEE Trans. Pattern Anal. Machine Intell.*, vol. PAMI-7, no. 4, pp. 485–490, July 1985.

[12] E. S. McVey *et al.*, "Range measurements by a mobile robot using a navigation line," *IEEE Trans. Pattern Anal. Machine Intell.*, vol. PAMI-8, no. 1, pp. 105–109, Jan. 1986.

[13] M. Herbert and T. Kanade, "Outdoor scene analysis using range data," in *Proc. IEEE Int. Conf. Robotics and Automation*, San Francisco, CA, Apr. 1986, pp. 1426–1432.

[14] R. Wallace, "Robot road following by adaptive color classification and shape tracking," in Proc. AAAI-86.
[15] A. Waxman *et al.*, "A visual navigation system," in *Proc. IEEE Int. Conf. Robotics and Automation*, San Francisco, CA, Apr. 1986, pp. 1600-1606.
[16] M. Turk, "Color road segmentation and video obstacle detection," in *Proc. SPIE Mobile Robot Conf.*, Cambridge, MA, Oct. 1986.
[17] S. Ishikawa *et al.*, "A method of image guided vehicle using white line recognition," in *Proc. IEEE Conf. Computer Vision and Pattern Recognition*, June 1986, pp. 47-53.
[18] E. Takinami, S. Ishikawa, *et al.*, "A method of detecting obstacles for the automatic image guided vehicle," *Trans. Inst. Electron. Commun. Eng. Japan*, vol. J68-D, no. 10, pp. 1789-1791, Oct. 1985.

The Karlsruhe Autonomous Mobile Assembly Robot

by

Prof. Dr.-Ing. Ulrich Rembold

Institute for Realtime Computer Systems and Robotics

University of Karlsruhe 7500 Karlsruhe F. R. G.

Summary

In this paper the architecture and functions of an autonomous mobile system are described. For the operation of such a system knowledge-based planning, execution and supervision modules are necessary which are supported by a multi-sensor system. The individual functions of such a vehicle are explained with the help of an autonomous mobile assembly robot which is being developed at the University of Karlsruhe. The vehicle contains a navigator, a docking module and an assembly planner. Navigation is done with the help of a road map under the direction of the navigator. The docking maneuver is controlled by sensors and the docking supervisor. The assembly of the two robot arms is prepared by the planner and controlled by a hierarchy of sensors. The robot actions are planned and controlled by several expert systems.

1 Introduction

For several years, various autonomous mobile robots are being developed in Europe, Japan and the United States. Typical areas of application are mining, material movement, work in atomic reactors, inspection of under-water pipelines, work in outer space, leading blind people, transportation of patients in a hospital, etc. The first results of these research endeavors indicate that many basic problems still have to be solved until a real autonomous mobile vehicle can be created; e.g. the development of an integrated sensor system for the robot is a very complex effort. To recognize stationary and moving objects from a driving vehicle is several orders of magnitude more complex than the identification of workpieces by a stationary camera system. In most cases the autonomous system needs various sensors. For processing of multi-sensor signals, science has not found a solution to date. An additional problem imposes the presentation and processing of the knowledge needed for planning and following a route or trajectory which is necessary to execute an assignment. Unexpected obstacles have to be recognized, and if necessary an alternate coarse of action has to be planned.

At the University of Karlsruhe an autonomous mobile robot for the performance of assembly tasks is being developed. The assignment of the system is to retrieve parts from a storage, to bring them to a work table and to assemble them to a product. All assignments have to be done autonomously, according to a defined manufacturing plan which is given to the system.

With autonomous mobile robots it is possible to develop manufacturing plants of great flexibility. Any combination of machine tools may be selected according to a virtual manufacturing concept. E.g., an autonomous assembly system equipped with robot arms is capable of working at various assembly stations. For welding or riveting tasks, the robot can move along a large object, such as the hull of a ship and perform the desired operations. An increase in flexibility can only be obtained by the use of knowledge based planning, execution and supervision modules which are sensor supported. In addition, omnidirectional drive systems have to be conceived, capable of giving the vehicle a three-dimensional flexibility, including turning on a spot.

2 Components of an Autonomous Mobile System

An autonomous system must be capable of planning and executing a task according to a given assignment. When a plan is available, its execution can start. A complex sensor system must be activated which leads and supervises the travel of the vehicle. Furthermore, it is necessary to recognize and solve conflicts with the help of a knowledge processing module. The basic components of an autonomous intelligent robot are shown in Fig. 1. To conceive and build these components,

```
1. Mechanics and drive system

2. Sensor system
     internal sensors
     external sensors

3. Planner and navigator
     planner
     navigator
     expert system
     knowledge base
     meta knowledge

4. World model
     static component
     dynamic component

5. Knowledge acquisition and world modeling

6. The computer system
```

Fig 1:   Components of an autonomous robot

expertise of many disciplines such as physics, electronics, computer science, mechanical engineering, etc. is required. It is very important to design good interfaces between the functional components of the system.

The most difficult task is building the software components. This is a universal problem with automation efforts involving computers. Designing software for autonomous vehicles is, however, complicated by the fact that very little is known about their basic concepts.

Reprinted from the *Proceedings of the IEEE International Conference on Robotics and Automation*, 1988, pages 598-603.

3 Mechanics and Drive System

Many present research efforts on autonomous vehicles use experience gained from guided vehicle experiments. Most of these vehicles are not suitable for autonomous operations. Their mobility is limited because they employ a drive system with 2 degrees of freedom and a rather simple control system.

For autonomous mobile robots drive systems with 3 degrees of mobility are of interest. However, there are no matured control systems available. This was a major consideration when the Karlsruhe autonomous robot was designed. Fig. 2 shows the robot and its drive systems. It has

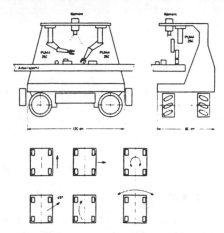

Fig. 2: The Karlsruhe autonomous mobile assembly robot

four active Mecanum wheels. Along the circumference of each wheel, passive rollers are fastened in a fixed angular orientation with regard to the main axis. The rollers of the front wheels are arranged in a positive, and those of the rear wheels in a negative herringbone pattern. In case the vehicle drives forward, all active wheels are rotated in the same direction. For a sideward motion the front wheels are rotated in the opposite direction of the rear wheels. Both sets of wheels rotate at the same speed. Thus the vector of the drive force acts perpendicular to the centerline of the vehicle and pushes the vehicle sidewards via the passive rollers. This vehicle can also be moved in any direction or it is able to rotate on a spot under the guidance of a proper control circuit.

4 The Sensor System

A sensor system of the Karlsruhe autonomous mobile robot consists of various sensors which are interconnected by a hierarchical control concept. The sensors furnish the planning and supervision modules with information about the status of the robot world. For each of the three major tasks of the vehicle, the navigation, docking and assembly an own sensor system is provided.

The sensoric has the following assignments:
- locating workpieces in storage
- supervising the vehicle navigation
- controlling the docking maneuver
- identifying the workpieces and their location and orientation on the assembly table
- supervising the assembly
- inspecting the completed assembly

For the navigation of the autonomous vehicle, a multisensor system is necessary. A distinction is made between vehicle based internal and external sensor and world based sensors.

Internal sensors are increment decoders in the drive wheels, the compass, inclinometer, etc. External sensors are TV cameras, range finders, approach and contact sensors, etc. World based sensor systems use sonic, infrared, laser or radiotelemetry principles. The Karlsruhe robot will be guided by several cameras, sonic sensors and a ranging device. The vehicle must constantly monitor its path with its sensor system and must be able to react to unforeseeable events. Recognition of objects is done by extracting from the sensor data of the scenario specific features and comparing these with a sensor hypotheses obtained from a world model. For scenes with many and complex objects the support of an expert system is needed for the sensor evaluation.

The docking maneuver will be supported by optical and mechanical proximity sensors. Thereby, for coarse positioning a vision system is used and for fine positioning mechanical feelers.

Recognition of parts for assembly will be done with a 2D vision system which also determines the position and orientation of the object. For the supervision of the assembly process a 3D vision system is being developed. Operations such as parts mating, fastening and aligning are monitored by force/torque, approach and touch sensors. Thereby, the vision system is the most important sensor; it is connected with the other sensors to a multisensor module to supervise the complex assembly operation.

5 Planning and Supervision

5.1 Global Planning and Supervision

The task of the global planning and supervision system is the conception of an action plan for the autonomous mobile robot and the supervision of the execution of the plan. With the autonomous mobile robot of the University of Karlsruhe, there are three functions to be performed, Fig. 3. For each function, an individual planning module is required.

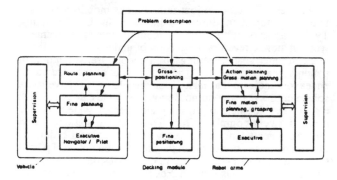

Fig. 3: Global planning system of the Karlsruhe autonomous mobile robot

A typical task for this system may be the instruction to perform an assembly of a product. All necessary information about the part, obtained from the assembly drawing and the bill of materials, is sent to the action module, Fig. 4. The action module retrieves knowledge from the world model about the inventory, product delivery date, the configuration of the material storage, the assembly workstation and the layout of the plant. The vehicle must be brought into a suitable fixed position for the retrieval of the parts and for assembly. For both of these operations knowledge about the docking maneuver must be available. An additional task is the supervision and synchronization of the navigation, docking

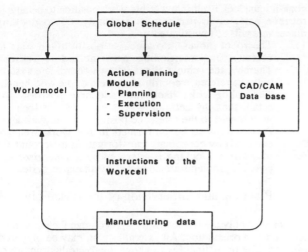

Fig. 4: Global action planning module for an autonomous mobile system

and assembly. For these operations the corresponding parameters, transfer procedures, interfaces and synchronization protocols must be provided.

For the assembly of the product the necessary operational steps must be determined and ordered in proper sequence, Fig. 5. The number of products to be manufactured determines

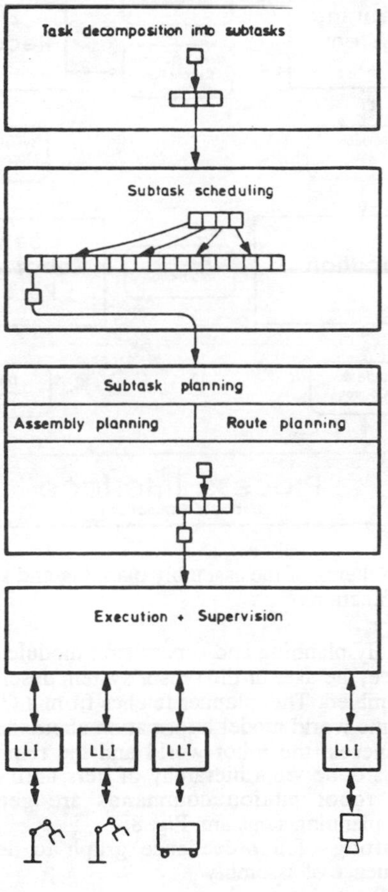

Fig. 5: Schema of planning, execution and supervision functions

determines the required number of parts. With this knowledge the retrieval of the material, the material flow and the navigation route are determined. In case of conflicts, it may be necessary to perform the planning repeatedly or to divide the material flow into several subactivities.

The majority of the tasks has to be executed sequentially. However, many tasks can be solved in quasi-parallel operations and must be prepared for parallel execution by the task scheduler. The task scheduler is also the executive which initiates the processing of the tasks and which controls asynchronous events. The scheduler employs algorithms for path and time optimization under the boundary conditions that the precedence requirements of the task are obeyed, duedates and priority assignments are met and the load capacity of the robot is not exceeded.

The scheduler performs the route planning under global considerations with the help of a very abstract model. The geometric model is presented by a graph, whereby the nodes represent locations and the edges segments of the path. The edges are directed and represent distances. At stochastic intervals a data acqustion system reports the actual state of the world, for example about the traffic density, jams, road blocks, station times, queues, etc. With the help of online information, the scheduler selects the proper workstation in case there are several alternatives. The execution planner reports to the scheduler at periodic intervals the completion of a task or possible problems which had arisen. In case of a problem, it may be necessary to do the task at another station or to abort the whole endeavor and to remove the task from the queue.

The scheduler gives the task to the task planner for execution. There are three task planners, one each for the navigation, docking and assembly.

The output of the planning modules is entered into the execution and supervision modules which are preparing the instruction for the low level interpreters. Characteristic of these modules is the ability to perform their own planning operations. The plans are further refined, brought to execution and the actions are supervised. In case of deviations from the plan, a situation analysis will be made and a new plan will be generated.

If a task is not solvable, the system has to give information about the cause of the problem; e.g. if a sensor has failed and a corrective action has to be initiated. For this purpose, it is required to store knowledge for the recognition and solution of problems. Thus, the system must have an expert system which is able fo responding to any unusual situation.

5.2   Path Planning and Navigation System

With the help of the path planning and navigation module, the vehicle will be capable of determining its path along the plant floor. It will try to plan an optimal route between the start and goal position, and it will attempt to stay on the path, avoiding obstacles and possible collisions. The navigation system will be tailored to the application. Since unexpected obstacles can be met on the path, the navigation becomes very complex. In this case the world has to be monitored constantly and the vehicle must be capable of recognizing dangerous situations and performing corrective actions. For the solution of conflicts, knowledge about new strategies is needed. The planner must have access to a map to know the action space of the vehicle and to interpret the map. For the evaluation of online sensor data, it is necessary to store in a world model sensor hypotheses for every situation to be expected.

Path planning is done according to a hierarchical scheme, whereby different tasks are solved at different levels, Fig. 6.
1.   At the planning level the planner draws up a plausible and collision free path by using expert knowledge.

However, at first the vehicle has to determine its start position. This can be done by starting at a known position or by sensors. The planner may try to search for an optimal plan in case there are alternative solutions. Optimization criteria may be the shortest path or the route with the least number of obstacles to be encountered.

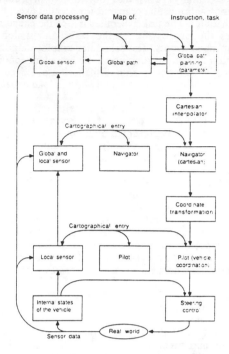

Fig. 6: The hierarchical navigation system

2. On the navigation level there is a local planning module which has knowledge about the path, its objects and possible obstacles. Fine planning of the navigation is done on the basis of the plan which was drafted on the previous level, thereby all local situations must be considered. It is important to recognize unknown obstacles by the use of sensors and to report any abnormal situation to the planning system. All problems must be considered for further planning.

3. On the pilot level the elementary control functions of the vehicle are performed. For trajectory planning a map of the world is used showing the path to be travelled in detail. For simple operations only a two-dimensional map of the routes, branches and crossings is needed. With the help of the map trajectory segments, like straight lines or curves are calculated. The passage of doorways or underpaths must be investigated with a three-dimensional map. The navigator corrects the map with the aid of realtime sensor data. A wrong way or obstacles entering or learing the path must be recognized by the sensors. The information is processed quantitatively and qualitatively in order to plan a necessary corrective action. Possible alternative moves are checked with the planner at the higher level. If the action is approved detailed move instructions are sent to the pilot level for execution.

5.3 Planning and Supervision of the Docking

As soon as the autonomous mobile robot has reached its

goal, it must lock itself into a stable work position to be able to retrieve parts or to perform an assembly. Two docking maneuvers will be investigated:

1. The robot moves into a coarse position and tries to perform fine positioning with the aid of sensors. Thereby, the vehicle aligns its own coordinate system with the reference coordinate system of the workbench.

2. The robot locks itself into a coarse position and establishes, by activating sensors, its exact location with regard to the reference position of the work area. Any actions to be performed in the work area are corrected by coordinate transformation to account for the offset. Docking is overseen by a supervisor and possible problems are corrected by an expert system.

5.4 Planning and Supervision of the Assembly

For assembly the robot will try to recognize the parts, their location and orientation. An assembly plan may be prepared, either online or offline. Fig. 7 shows the overall structure of

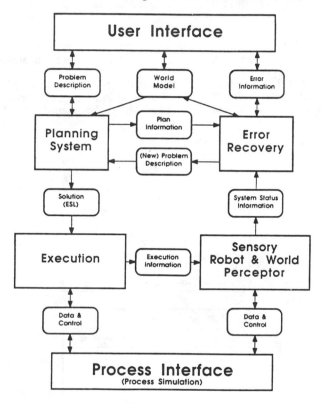

Fig. 7: Schema of the assembly planning and supervision functions

the assembly planning and supervision module. By means of an interface, the user or the sensor system describes the parts to be assembled. The planner fetches from a CAD databank and from the world model information about the geometry of the workpieces, the robot world and the required fixtures. Planning is done via a hierarchy of tiers until at the lowest level the robot motion commands are generated. The individual planning steps are, Fig. 8:

1. Drafting of a precedence graph to determine the sequence of assembly

2. At the strategic planning tier the assembly motions, fixtures and auxiliary peripherals are determined according to a virtual assembly cell concept. From a set

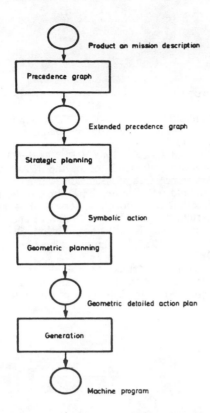

Fig. 8: Different tiers of assembly planning module

of alternative solutions a strategy is selected which will render the best solution for the described problem.

3. At this level the equipment to be used is selected. The virtual assembly plan is mapped into the real world by defining the robots, grippers and fixtures for the operations. At this level a distinction between coarse and fine planning is made. For coarse planning the start and goal positions and the trajectories are determined. The fine positioning is concerned with parts mating and the required fine motions.

4. A collision investigation is made at this level. No collision may occur between cooperating robots, equipment or workpieces. In case of a possible collision, a new planning cycle is necessary.

5. The instruction code for robot joint motors and the peripherals is generated at this level. For example, the sensor system must obtain the control information to monitor the world and report back the world status. This information is sent to the planning module via the error recovery module, Fig. 7.

The planning module sends the code to the execution module for processing, Fig. 7. Here, every line of code is checked by the monitor to verify that it can be executed. The monitor contains an up-to-date report on the state of the system from the sensors. In case the state of the assembly area has changed, e.g. by the displacement of an object, the code is not consistent with the present situation and will not be transmitted to the robot. A message is sent to the error recovery module to find the cause of the problem. Thereafter, the planning module is activated anew in order to suggest an alternative assembly method. Simultaneously, a message is sent to the user interface to report the problem. Planning, supervision and error recovery are supported by various expert systems.

6 The World Model

A world model can be considered in a broad sense as the extension of the CAD databank. The CAD databank contains information about the workpiece, its geometry, physical data and information about the manufacturing process. However, an autonomous robot must know more about its world. The Karlsruhe robot needs information from the world model for the navigation, docking and assembly. Thus, three different types of world models are needed in which the following information is stored.

- geometry of the objects and their geometric relations
- functions, operation parameters and capability of the objects
  (e.g. of the vehicle, robots and auxiliary equipment)
- interpolation routines for the vehicle motions and the robot trajectories
- motion parameters for all moving parts (speed, acceleration and deceleration)
- sensor hypotheses for planning and supervision of the action sequences
- information about the effect of tolerances for the different operations
- preferred routes and standard motions
- behavior rules for the avoidance and removal of errors and collision
  The world model for the assembly robots needs the following additional information:
- assembly parameters such as fits, mating forces and torques, as well as approach and work speeds

Since the world may change during the execution of a task, part of the world model has to be conceived as a dynamic world model. For this reason, the world is constantly monitored by sensors. For example, if a part falls over, the event must be reported for updating of the world model. Thus, information for the following tasks has to be supplied to the dynamic world model:

- evaluation result of the online sensor system
- reports about changes in the world
- updates about the world model

Part of the world model for the navigator is shown in Fig. 9.

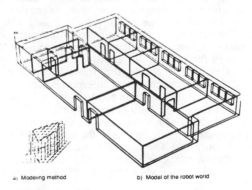

a) Modeling method   b) Model of the robot world

Fig. 9: Modeling of the robot work area using the oct-tree cube space method

The modeling method employed divides the work space recursively into cubes until there are only free and occupied cubes left. Thus, the obstacles are presented by an occupied cube structure. This information is stored in the computer. With the aid of the cube structure, the computer is able of calculating the optimal path the vehicle should take in a three dimensional space.

7 Knowledge Acquisition and World Modeling

To plan, execute and supervise its actions, an autonomous mobile vehicle must be able to make decisions on several levels. Thus, is must be equipped with a hierarchy of knowledge based modules and submodules. A special problem presents the acquisition and storage of expert knowledge. For some functions, knowledge may be stored in independent databanks, in other cases knowledge has to be shared via a blackboard architecture.

The Karlsruhe autonomous vehicle will have the following knowledge-based modules.

1. The global action planner and supervisor on the level of the manufacturing cell
2. The route planner and supervisor for the navigation
3. The action planner and supervisor for the assembly
4. The sensor data processor for the navigation
5. The sensor data processor for the assembly
6. The error recognition and recovery module for the navigation
7. The error recognition and recovery module for the assembly
8. The expert systems are presently being implemented with the aid of OPS5 and installed in various computers of the robot. Other expert shells are being investigated.

An interactive modeler for the robot and the workspace is used for the system design and the investigation about the robot motions. It is possible to observe the actions of all components of the assembly system and to watch for possible conflicts.

8 The Computer System

With the presently available computers two different computer systems are necessary for the planning and control of the autonomous vehicle. Planning is done offline on a powerful scientific computer and the execution and supervision is done online by a vehicle based realtime computer. The main computer contains the global planner and various expert systems for the planning and supervision of the subfunctions of the vehicle. For example, the main computer gives the action program to the vehicle computer for execution. The vehicle computer interpretes stepwise the instructions and executes them. In addition, expert knowledge is given to the vehicle computer to process sensor information and to solve conflicts which may arise during the navigation, docking or assembly. Since the size of the vehicle computer is restricted, it only can solve simple problems. In serious situations the main computer will be notified and it in turn tries to find a solution. It will also prepare and issue a situation report for the operator.

Acknowledgement

The autonomous mobile robot of the University of Karlsruhe is being developed by the Institute of Realtime Computer Systems and Robotics (Prof. Dr.-Ing. U. Rembold and Prof. Dr.-Ing. habil. R. Dillmann). The work is funded by the Deutsche Forschungsgemeinschaft as part of a cooperative research project on artificial intelligence (Sonderforschungsbereich Künstliche Intelligenz).

Literature

1. Rembold, U. and Levi, P.: "Development Trends for Expert Systems for Robots", Der Betriebsleiter, March April 1987
2. Rembold, U. and Dillmann, R.: "Artificial Intelligence in Robotics", Proceedings of the IFAC SYROCO 85 Conference, Oct. 85, Barcelona, Spain
3. Rembold, U., Blume, C., Dillmann, R. and Levi. P.: "Intelligent Robots", a series of 4 articles, VDI Zeitschrift, Vol. 127, No. 19, 20, 21 and 22, 1985
4. Frommherz, B.: "Robot Action Planning", CIM EUROPE 1987 Conference, Cheshire, England, May 1987
5. Soetadji, T.: "Methods for Solving the Rout Planning for a Navigation System for an Autonomous Mobile Robot", Dissertation at the University of Karlsruhe, 1987

VITS—A Vision System for Autonomous Land Vehicle Navigation

MATTHEW A. TURK, MEMBER, IEEE, DAVID G. MORGENTHALER, KEITH D. GREMBAN,
MARTIN MARRA

Reprinted from *IEEE Transactions on Pattern Analysis and Machine Intelligence*, Vol. 10, No. 3, May 1988, pages 342-361.

Abstract—In order to adequately navigate through its environment, a mobile robot must sense and perceive the structure of that environment, modeling world features relevant to navigation. The primary vision (or perception) task is to provide a description of the world rich enough to facilitate such behaviors as road-following, obstacle avoidance, landmark recognition, and cross-country navigation. We describe VITS, the vision system for Alvin, the Autonomous Land Vehicle, addressing in particular the task of road-following. The ALV has performed public road-following demonstrations, traveling distances up to 4.5 km at speeds up to 10 km/hr along a paved road, equipped with an RGB video camera with pan/tilt control and a laser range scanner. The ALV vision system builds symbolic descriptions of road and obstacle boundaries using both video and range sensors. We describe various road segmentation methods for video-based road-following, along with approaches to boundary extraction and transformation of boundaries in the image plane into a vehicle-centered three dimensional scene model.

Index Terms—Autonomous navigation, computer vision, mobile robot vision, road-following.

I. INTRODUCTION

TO achieve goal-directed autonomous behavior, the vision system for a mobile robot must locate and model the relevant aspects of the world so that an intelligent navigation system can plan appropriate action. For an outdoor autonomous vehicle, typical goal-directed behaviors include road-following, obstacle avoidance, cross-country navigation, landmark detection, map building and updating, and position estimation. The basic vision task is to provide a description of the world rich enough to facilitate such behaviors. The vision system must then interpret raw sensor data, perhaps from a multiplicity of sensor and sensor types, and produce consistent symbolic descriptions of the pertinent world features.

In May of 1985, "Alvin," the Autonomous Land Vehicle at Martin Marietta Denver Aerospace, performed its

Manuscript received December 15, 1986; revised May 15, 1987. This work was performed under the Autonomous Land Vehicle Program supported by the Defense Advanced Research Projects Agency under Contract DACA76-84-C-0005.

M. A. Turk was with Martin Marietta Denver Aerospace, P.O. Box 179, M. S. H0427, Denver, CO 80201. He is now with the Media Laboratory, Massachusetts Institute of Technology, Cambridge, MA 02139.

D. G. Morgenthaler and M. Marra are with Martin Marietta Denver Aerospace, P.O. Box 179, M.S. H0427, Denver, CO 80201.

K. D. Gremban is with the Department of Computer Science, Carnegie-Mellon University, Pittsburgh, PA 15213, on leave from Martin Marietta Denver Aerospace, P.O. Box 179, M.S. H0427, Denver, CO 80201.

IEEE Log Number 8820101.

first public road-following demonstration. In the few months leading up to that performance, a basic vision system was developed to locate roads in video imagery and send three-dimensional road centerpoints to Alvin's navigation system. Since that first demonstration, VITS (for Vision Task Sequencer) has matured into a more general framework for a mobile robot vision system, incorporating both video and range sensors and extending its road-following capabilities. A second public demonstration in June 1986 showed the improved road-following ability of the system, allowing the ALV to travel a distance of 4.2 km at speeds up to 10 km/hr, handle variations in road surface, and navigate a sharp, almost hairpin, curve. In October 1986 the initial obstacle avoidance capabilities were demonstrated, as Alvin steered around obstacles while remaining on the road, and speeds up to 20 km/hr were achieved on a straight, obstacle-free road. This paper describes Alvin's vision system and addresses the particular task of video road-following. Other tasks such as obstacle detection and avoidance and range-based road-following are discussed elsewhere [10], [11], [36].

A. A Brief Review of Mobile Robot Vision

SRI's Shakey was the first mobile robot with a functional, albeit very limited, vision system. Shakey was primarily an experiment in problem solving methods, and its blocks world vision system ran very slowly. The JPL robot [32] used visual input to form polygonal terrain models for optimal path construction. Unfortunately, the project halted before the complete system was finished.

The Stanford Cart [25], [26] used a single camera to take nine pictures, spaced along a 50 cm track, and used the Moravec interest operator to pick out distinctive features in the images. These features were correlated between images and their three dimensional positions were found using a stereo algorithm. Running with a remote, time-shared computer as its "brain," the Stanford Cart took about five hours to navigate a 20 meter course, with 20 percent accuracy at best, lurching about one meter every ten to fifteen minutes before stopping again to take pictures, think, and plan a new path. The Cart's "sliding stereo" system chose features generally good enough for navigation in a cluttered environment, but it did not provide a meaningful model of the environment.

Tsugawa *et al.* [34] describe an autonomous car driven

up to 30 km/hr using a vertical stereo pair of cameras to detect expected obstacles, but its perception of the world was very minimal, limiting its application to a highly constrained environment. The "intelligent car" identified obstacles in an expected range very quickly by comparing edges in vertically displaced images. A continuous "obstacle avoidance" mode was in effect, and a model of the world was not needed.

A vision system for a mobile robot intended for the factory floor was presented by Inigo et al. [18]. This system used edge detection, perspective inversion, and line fitting (via a Hough transform) to find the path, an a priori road model of straight lines, and another stereo technique using vertical cameras, called motion driven scene correlation, to detect obstacles. The Fido vision system [33] uses stereo vision to locate obstacles by a hierarchical correlation of points chosen by an interest operator. Its model of the world consists of only the 3-D points it tracks, and it has successfully navigated through a cluttered environment and along a sidewalk. Current work in multisensory perception for the mobile robot Hilare is presented by de Saint Vincent [7], describing a scene acquisition module, using stereo cameras and a laser range finder, and a "dynamic vision" module for robot position correction and tracking world features. Another stereo vision system based on matching vertical edges and inferring surfaces is described by Tsuji et al. [35].

The goal of a mobile robot project in West Germany is to perform autonomous vehicle guidance on a German Autobahn at high speeds [8], [22], [29]. The current emphasis is on control aspects of the problem, incorporating a high-speed vision algorithm to track road border lines. The system has performed both road-following and vehicle-following in real-time.

Other mobile robots have been or are being developed that use sensors particularly suited to an indoor environment (e.g., [4], [19]). The project headed by Brooks [2] implements a novel approach to a mobile robot architecture, emphasizing levels of behavior rather that functional modules; much of the current vision work may be incorporated into such a framework.

B. ALV Background

The Autonomous Land Vehicle project, part of DARPA's Strategic Computing Program, is intended to advance and demonstrate the state of the art in image understanding, artificial intelligence, advanced architectures, and autonomous navigation. A description of the project and the initial system configuration is found in [24]. Related vision research is proceeding concurrently by a number of industrial and academic groups, as is work in route and path planning, as well as object modeling and knowledge representation. The ALV project is driven by a series of successively more ambitious demonstrations. The ultimate success of the project depends on coordination among the different groups involved to enable rapid technology transfer from the research domain to the application domain. As the ALV is intended to be a national testbed for autonomous vehicle research, various vision systems and algorithms will eventually be implemented. Some of the current work is briefly described in the remainder of this section.

Vision research areas currently being pursued in relation to the ALV program include object modeling, stereo, texture, motion detection and analysis, and object recognition. An architecture for terrain recognition which uses model-driven schema instantiation for terrain recognition is presented by Lawton et al. [23]. Such representations for terrain models will be important for future cross-country navigation. Waxman et al. [40], [41] present a visual navigation system that incorporates rule-based reasoning with image processing and geometry modules. The system, developed at the University of Maryland, finds dominant linear features in the image and reasons about these features to describe the road, using bootstrap and feed-forward image processing phases. In the feed-forward phase, previous results are used to predict the location of the road in successive images. A subset of this system has been used to autonomously drive the ALV for short distances. DeMenthon [6] describes an alternative geometry module for the above visual navigation system.

Significant ALV-related work is proceeding at Carnegie-Mellon University (CMU). A review of recent results from the CMU program is presented by Goto and Stentz [13]. Outdoor scene analysis using range data from a laser range scanner is presented by Hebert and Kanade [16], describing methods for preprocessing range data, extracting three dimensional features, scene interpretation, map building, and object recognition. Fusion of video and range data is also discussed. Range data processing has been used on the CMU Navlab to demonstrate obstacle avoidance capabilities. Vision algorithms used for successful outdoor navigation of the CMU Terregator are described by Wallace et al. [37]–[39]. The Terregator has achieved continuous motion navigation using both edge-based and color-based sidewalk finding algorithms.

Hughes Artificial Intelligence Center is developing knowledge-based vision techniques for obstacle detection and avoidance using the concept of a virtual sensor which blends raw sensor data with specialized processing in response to a request from the planning system [5], [30]. Work at SRI International is focused on object modeling and recognition, and on modeling uncertainty in multiple representations [1].

FMC Corporation and General Dynamics have demonstrated successful transfer of ALV technology to mission-oriented scenarios of mixed teleoperation and autonomous navigation, performed at the Martin Marietta test site in 1986. Kuan et al. [20], [21] describe FMC's research in vision-guided road-following. Other university and industrial laboratories which are engaged in vision research related to ALV include Advanced Decision Systems, Columbia University, General Electric, Honeywell Research Center, MIT, University of Massachusetts at Amherst, University of Rochester, and USC. The Proceedings of the February 1987 Image Understanding

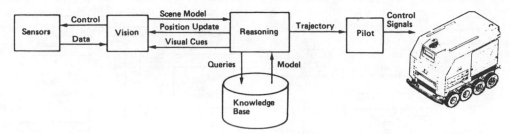

Fig. 1. The ALV system configuration.

Workshop, sponsored by DARPA, contains descriptions and status reports of many of these projects.

The vision system described in this paper (VITS) is the system meeting the perception requirements for testing and formal demonstrations of the ALV through 1986. Section II gives a system overview, briefly describing the various subsystems; it is important to understand the vision system in its context. Video-based road following is discussed in Section III, describing sensor control, road segmentation, road boundary extraction, and geometric transformation to three dimensional world coordinates.

II. ALV System Overview

It is important to view Alvin's vision subsystem as an integral part of a larger system, which can affect and be affected by the performance of the system as a whole. Fig. 1 illustrates the basic system configuration of the ALV, including the interfaces to the major modules. In the paragraphs below, each of Alvin's major components will be discussed in the context of the interaction as a complete system.

A. Hardware Components

The primary consideration behind selection of the hardware components was that Alvin is intended to be a testbed for research in autonomous mobility systems. Consequently, it was necessary to provide Alvin with an undercarriage and body capable of maneuvering both on-road and off-road, while carrying on board all the power, sensors, and computers needed for autonomous operation. In addition, the requirements of autonomous operation directed the selection of sensors and processing hardware.

1) Vehicle: Fig. 2 is a photograph of Alvin. The overall vehicle dimensions are 2.7 m wide by 4.2 m long; the suspension system allows the height of the vehicle to be varied, but it is nominally 3.1 m.

Alvin weighs approximately 16 000 pounds fully loaded with equipment, yet is capable of traveling both on-road and off-road. The undercarriage is an all-terrain built by Standard Manufacturing, Inc. The basic vehicle is eight-wheel drive, diesel-powered, and hydrostatically driven. Alvin is steered like a tracked vehicle by providing differential power to the two sets of wheels.

Alvin's fiberglass shell protects the interior from dust and inclement weather, and insulates the equipment inside. The shell provides space for six full-size equipment racks, as well as room for service access. The electronics within the ALV are powered by an auxiliary power unit.

Fig. 2. Alvin.

An environmental control unit cools the interior of the shell.

2) Sensors: In order to function in a natural environment, an autonomous vehicle must be able to sense the terrain around it, as well as keep track of heading and distance traveled. The ALV hosts a number of sensors to accomplish these tasks.

Alvin's sense of direction and distance traveled is provided by odometers on the wheels coupled to a Bendix Land Navigation System (LNS). These sensors enable Alvin to follow a trajectory derived from visual data or read from a prestored map. The LNS provides direction as an angle from true North, while distance traveled is provided in terms of horizontal distance (Northings and Eastings), and altitude.

Two imaging sensors are currently available on the ALV for use by VITS. The primary vision sensor is an RCA color video CCD camera, which provides 480 × 512 red, green, and blue images, with eight bits of intensity per image. The field of view (38° vertical and 50° horizontal) and focus of the camera are kept fixed. The camera is mounted on a pan/tilt unit that is under direct control of the vision subsystem.

The other vision sensor is a laser range scanner, developed by the Environmental Research Institute of Michigan (ERIM). This sensor determines range by measuring the phase shift of a reflected modulated laser beam. The laser is continuously scanned over a field of view that is 30° vertical and 80° horizontal. The output of the scanner is a digital image consisting of a 64 × 256 array of pixels with 8 bits of range resolution.

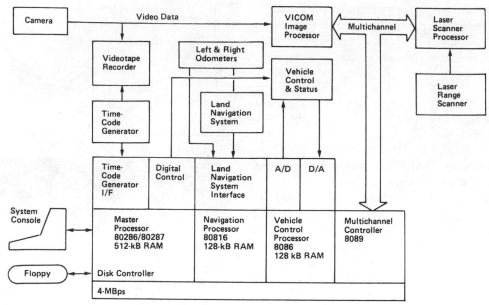

Fig. 3. The first-generation ALV processor configuration.

3) Computer Hardware: Alvin currently uses a variety of computers, resulting from the range of processing requirements of the different software subsystems. The diverse processing requirements were met by designing a modular multiprocessor architecture. VITS is hosted on a Vicom image processor, while the other software subsystems are hosted on an Intel multiprocessor system. VITS communicates with the other subsystems across a dedicated communication channel, while the other subsystems communicate across a common bus. Fig. 3 depicts the processor configuration.

The special capabilities of the Vicom hardware were important to the development of the ALV vision subsystem (VITS). The Vicom contains video digitizers, and can perform many standard image processing operations at near video frame rate (1/30 second). For example, 3 × 3 convolution, point mapping operations (such as thresholding, or addition and subtraction of constants), and image algebra (such as addition or subtraction of two images) are all frame rate operations. The Vicom also contains a general purpose microcomputer for additional, user-defined operations.

As stated above, Alvin is intended to be a testbed for autonomous systems. In fulfilling this charter, plans have been made to integrate a number of advanced experimental computer architectures in future generations of the ALV system. This will begin with a new architecture in early 1987.

B. Vision

The vision subsystem is composed of three basic modules: VITS, the vision executive, which handles initialization, sets up communication channels, and "oversees" the processing; VIVD, the video data processing unit; and VIRD, the range data processing unit. Range data processing has been implemented on the ALV, and results of

range-based road-following and obstacle avoidance are presented in [10], [11].

The vision system software resides entirely on the Vicom image processor, which also houses a board dedicated to camera pan/tilt control and a board to enable communication with the Intel system. Nearly all of the application code is written in Pascal and uses the Vicom-supplied libraries for accessing high-speed image operations. Some low level control routines have been implemented in Motorola 68000 assembly language.

The responsibility of the vision subsystem in road-following is to process data in the form of video or range images to produce a description of the road in front of the vehicle. This description is passed to the reasoning subsystem, which uses additional data such as current position, speed, and heading to generate a trajectory for Alvin to follow. Communication between the vision subsystem and Reasoning takes place in three different forms: the scene model, the position update, and visual cues. A special communication control processor, part of the utilities subsystem, mediates communication between VITS and the other subsystems. The control processor shares memory with VITS, and handles communication by examining the content of key memory locations every 100 ms and modifying them as appropriate.

1) Scene Model: The *scene model*, a description of the observed road, is the output of the vision subsystem after each frame of images is processed. The scene model contains a record of Alvin's position and heading at the time of image acquisition, a description of the road found in the imagery, consisting of lists of vehicle-centered 3-D points denoting left and right road edges, and an optional list of points surrounding an obstacle. The reasoning subsystem must then transform the road description into a fixed, world coordinate system for navigation. VITS may optionally specify the scene model in world coordinates;

this is more efficient when data acquired from multiple sensors or at different times is used to create the scene model.

Since the time needed to compute a scene model is non-deterministic, VITS sets a "scene model ready" flag indicating that a new scene model is ready to be processed. The communication controller examines this flag, and, when set, transfers the scene model to the reasoning subsystem and clears the flag.

Fig. 4 illustrates the format of a scene model. Fig. 5 is an example of a hypothetical road scene and the corresponding scene model.

2) Position Update: VITS must know the position and heading of the vehicle at the time of image acquisition to integrate sensor information acquired at different times, and to transform vehicle-centered data into world coordinates. In addition, VITS must be able to predict the location of the road in an image, given its location in the preceding image (see Section III-B-1-d).

Communication of vehicle motion and position information is effected by means of a *position update* message passed from Reasoning to the vision subsystem. The position update specifies the current vehicle speed, position, and heading. Synchronization of position update and image acquisition is mediated by a position update request. At the time VITS digitized an image, the "position update request" flag is set. When the communication controller finds the flag set, it sends a message to Reasoning which immediately (within 100 ms) generates the required information, builds a position update message, and sends it to VITS.

3) Visual Cues: The reasoning subsystem interfaces to a knowledge base which contains information about the test area. Some of this information can be used by VITS to specify behavior (find road, locate obstacles, pause, resume) or to optimize processing, much as the information on a road map can guide a driver. When Reasoning determines that a visually identifiable feature should be within the field of view, a *visual cue* is sent to VITS enabling vision processing to be modified. In the future, when Alvin's domain becomes more complex, these cues will be used to guide the transition from one road surface to another, from on-road to off-road and vice versa, or to guide the search for a landmark. In the current version of the system, stored knowledge about the shape of the road shoulder has been used to guide a transition between range-based and video-based road-following. Apart from this, the cue facility has been used to date only to notify VITS that the vehicle is approaching a curve (which causes the camera panning mechanism to be enabled) and to send pause and resume commands to VITS.

C. Reasoning

The Reasoning subsystem is the executive controller of the ALV; Vision is a resource of Reasoning. At the highest level, Reasoning is responsible for receiving a plan script from a human test conductor and coordinating the

```
type scene_model = record
    time: array[1..4] of word; { time stamp }
    count: word; { # of road edge records }
    x,y,psi: real; { vehicle position }
    SM_rec: array[1..10] of record
            tag: string[2]; { left or right }
            numpts: word; { # of points }
            pts: array [1..10]
                of array[1..3] of real;
        end;
    version: string[10]; {current SW version }
    num: word; { scene model # }
end;
```

Fig. 4. The scene model format.

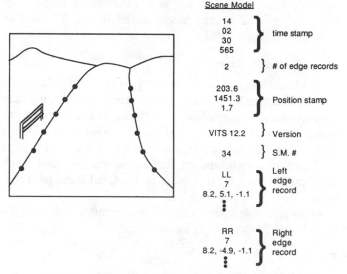

Fig. 5. Road scene and corresponding scene model.

other subsystems on Alvin in order to accomplish the goals specified in the script.

Because the processing involved in creating a visual description of the environment is beyond the real-time capability of present computers, the scene model is not used directly in the vehicle's control servo loop. Instead, the Navigator (part of the reasoning subsystem) pieces together scene models from the vision system and builds a reference trajectory that is sent to the Pilot for control. The reasoning subsystem accepts a position update request from VITS, generates the appropriate data, and sends back a position update. Upon receipt of a scene model, Reasoning evaluates it and plots a smooth trajectory if the data is acceptable. The new trajectory is computed to smoothly fit the previous trajectory.

Evaluation of scene models is a powerful capability of the reasoning subsystem. Small environmental changes, such as dirt on the road, or the sudden appearance of a cloud, can significantly affect the output of the vision subsystem. Reasoning uses assumptions about the smoothness and continuity of roads to verify data from VITS. Every scene model is evaluated based on the smoothness of the road edges, and on how well they agree with previous edges. A scene model evaluated as "bad" is discarded.

Reasoning creates a new trajectory by minimizing a cost function based on current heading, curvature of the scene model, attraction to a goal, and road edge repulsion. The final trajectory is a sequence of points that lie near the center of the road. Each point is tagged with a reference speed. The reference speeds are computed so that, if no new scene models are received, the vehicle will stop at the end of the trajectory. The trajectory is then sent to the Pilot.

The reasoning subsystem also interacts with the knowledge base to locate features significant for vision processing. As each new trajectory is generated, the knowledge base is searched to determine if any features are within the field of view of the vehicle. Features that are both within a maximum distance and a maximum angle from the current heading are incorporated into a Visual Cue which is passed to VITS.

D. Knowledge Base

The knowledge base consists of *a priori* map data, and a set of routines for accessing the data. Currently, the map data contains information describing the road network being used as the ALV test track. The map data contains coordinates which specify the location of the roadway, as well as various significant features along the road, such as intersections, sharp curves, and several local road features.

At present, the vision subsystem communicates with the knowledge base through Reasoning.

E. Pilot

The Pilot performs the actual driving of the vehicle. Given a trajectory from Reasoning, the Pilot computes the error values of lateral position, heading and speed by comparing LNS data with the target values specified in the trajectory. The Pilot uses a table of experimentally obtained control gains to determine commands needed to drive the errors toward zero; these commands are output to the vehicle controllers.

The vision subsystem has no direct communication with the Pilot.

III. VIDEO-BASED ROAD-FOLLOWING

The task of the vision system in a road following scenario is to provide a description of the road for navigation. Roads may be described in a variety of ways, e.g., by sets of road edges, a centerline with associated road width, or planar patches. We have chosen to represent a road by its edges, or more precisely, points in three space that, when connected, form a polygonal approximation of the road edge. Road edges are intuitively the most natural representation, since they are usually obvious (to humans, at least) in road images. Often, however, the dominant linear features in road images are the shoulder/vegetation boundaries rather than the road/shoulder boundaries. The difficulties in extracting the real road boundary from the image led us to adopt a segmentation algorithm to first extract the road in the image, track the road/nonroad boundary, and then calculate three dimensional road edge points.

The current video data processing unit (VIVD) uses a clustering algorithm to segment the image into road and nonroad regions. A detailed description of image segmentation by clustering can be found in [3]. After producing a binary road image, the road boundaries are traced and select image points are transformed into three dimensional road boundary points. The complete cycle time, from digitization to producing a symbolic description of the road, is currently just over 2 seconds. The algorithm is summarized in the following steps, which are discussed in detail in the following sections: 1) digitize the video images; 2) segment road/nonroad regions; 3) extract road boundaries by tracing the binary road edges; and 4) transform 2-D road edge points to 3-D coordinates and build the scene model. Fig. 6 depicts the flow of control in a complete scene model cycle.

A. Sensor Control and Image Acquisition

1) Camera Panning: The position of the road with respect to the vehicle may change due to a curving road, vehicle oscillation, or a sudden path correction. Consequently, the position of the road within the field of view of a fixed camera may change. Because the video segmentation algorithm requires sampling a population of road pixels, two methods were developed to maintain knowledge of the road position from frame to frame: camera panning and *power windowing*. Power windowing, a "software panning" technique, is described in Section III-B-1-d.

Control of the pan/tilt mechanism is a function of vehicle orientation and desired viewing direction. During road-following, we would like the camera to point "down the road," regardless of the vehicle orientation, keeping the road approximately centered in the image. This requires the vision system to know global position information and relate the vehicle-centered road description to present vehicle location and orientation, and then to calculate and command the desired pan angle. If only one road boundary is detected, then VITS will attempt to pan the camera to the right or left to bring both road edges into view in the next image. The activation of planning is also controlled by cues from the reasoning subsystem that indicate when panning would be useful (e.g., going around a sharp corner), and when it would not be helpful (e.g., passing a parking lot).

In the initial implementation of camera panning, the camera was allowed to assume only three positions, *left*, *mid*, and *right*, with simple rules for switching from one to another based on road location in the image. With this technique we were able to successfully navigate a sharp corner; however, this was not very repeatable. After studying the failure symptoms we rejected this simplistic approach to panning in favor of a technique allowing "continuous" pan positions based on the difference between vehicle orientation and the perceived road orientation, and constraints on minimum and maximum incremental panning and on the maximum absolute pan angles.

We have not yet needed the camera tilting capability. The tilt angle is fixed at approximately 17° below the horizon.

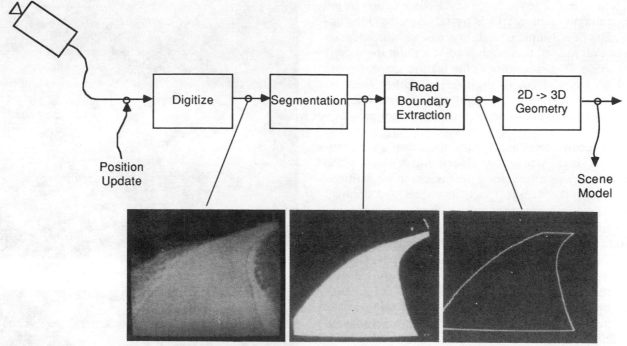

Fig. 6. VITS flow of control.

2) Image Acquisition: The image processing computer digitizes red, green, and blue images, directly into memory from the video camera. Typically, the images are blurred to reduce noise; since convolution is distributive, blurring is performed after the images are combined. Our camera has an optional automatic iris which compensates for global intensity changes. Calibration is performed on the camera before a test run to calculate the exact tilt angle and focal length, and the proper color response.

Along with the raw images, the *position update* is requested and received from the communication control processor. The position update includes a time stamp and the two dimensional position and orientation of the vehicle—this information allows conversion from a vehicle-centered road description to world coordinates.

B. Road Segmentation

There are many techniques used in computer vision systems to segment images into regions of similarity. Segmentation of natural outdoor scenes is a particularly complex problem [15], but it is simplified when a predominant feature in the scene (i.e., the road) is the main focus of the segmentation. *A priori* information is available about roads, as is information from previous frames and vehicle movement.

Road segmentation is rather simple to accomplish on a stored road image, given enough time to experiment and modify parameters. In a real-time, outdoor environment with a mobile robot, however, road segmentation is complicated by the great variability of vehicle and environmental conditions. Changing seasons, weather conditions, time of day, and manmade changes complicate the video segmentation, as do the variable color response of the cameras, the vehicle suspension system, performance of the navigation and control subsystems, and other changes in the vehicle system. Because of these combined effects robust segmentation is very demanding. Particular conditions that have proven difficult to handle are the presence of dirt on the road, spectral reflection when the sun is at a low angle, shadows on the road, and tarmac patches (used to repair road segments). These will be discussed further in Section IV.

The segmentation methods used by VITS are motivated by the hardware supporting it, speed requirements, and assumptions about road and nonroad image characteristics. We have proposed and tested various segmentation techniques, all based on knowledge of road characteristics in color images. Section III-B-1 describes the specific techniques developed for the segmentation algorithms. The algorithms discussed in the following sections are somewhat cryptically called *red minus blue*, *color normalization*, and *shadow boxing*.

1) A Classification Problem: To understand the road segmentation algorithms discussed below, it is best to view road segmentation in the broader context of a general classification problem with only two classes: road and nonroad. A typical classification problem involves five basic steps: feature extraction, feature decorrelation, feature reduction, clustering, and segmentation [3]. Feature extraction is the process of computing the features used to distinguish between classes. Feature decorrelation is formally a multidimensional transformation that results in a new, orthogonal set of features. Feature reduction discards those features that are unnecessary or yield little information. Clustering partitions the feature space into K mutually exclusive regions, and segmentation assigns each pixel in the image to one of the regions.

a) VITS Clustering: The clustering algorithm used

in VITS is a real-time (i.e., abbreviated) solution to the classification problem. Step 1 is trivial. A color video image consists of a 3-tuple of red, green, and blue intensity values at every pixel; thus each pixel is a point (or vector) in RGB space. By using color as the feature space, feature extraction is equivalent to digitization. Other features, such as texture, saturation, and reflectance data from the laser scanner, have been considered but are not presently used.

Feature decorrelation and feature reduction are accomplished in a single step in our algorithm. A single plane in three-space does a very good job under most conditions of partitioning the feature space into road and nonroad regions. The original features are, in effect, combined into a single linear feature whose axis is perpendicular to the separating plane. The projection of image points onto this perpendicular line is equivalent to a dot product of the pixel vector and the normal of the plane. Feature decorrelation and reduction then reduce to the problems of finding the orientation of this plane and projecting image points onto a line normal to the plane; these are discussed in Section III-B-1-b.

Clustering (step 4) now involves determining the threshold along the "feature line" that separates road from nonroad clusters, discussed in Sections III-B-1-c and III-B-1-d. Segmentation (step 5) is just the creation of a binary road/nonroad image (since $K = 2$ for simple road-following) from the clustering information, labeling each pixel as *road* or *nonroad* according to whether it projects above or below the feature threshold.

b) Color Parameter Selection: The projection of image points in RGB space onto a line is equivalent to taking the dot product of every pixel vector (R, G, B) with a vector in the direction of the line (r, g, b). This is accomplished by a *tricolor* operation, a linear weighted combination of the red, green, and blue images:

$$I(i, j) = rR(i, j) + gG(i, j) + bB(i, j)$$

$$= (r, g, b) \cdot (R, G, B).$$

The vector (r, g, b) represents the red, green, and blue coefficients of the tricolor operation. When this vector is normal to the plane that separates road and nonroad clusters in the RGB space, the outcome I is the single band "feature enhanced" image. The orientation of the separating plane is relatively consistent under given weather and camera conditions. It can often be chosen by hand at the beginning of a run and not modified thereafter. However, we have found that some segmentation failures have occurred because of a change in the orientation of the separating plane, due to changing weather conditions, seasonal changes, road surface variations, and the camera color response. We have therefore developed a method to dynamically compute the optimal plane orientation based on data from the image currently being processed.

Experience has shown that a good segmentation is achievable without using the green band, so we can reduce the problem conceptually to finding the slope of a

(a)

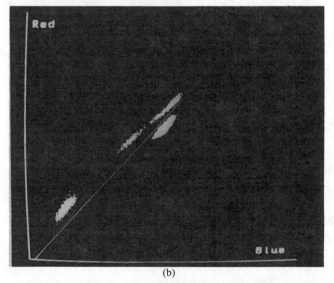

(b)

Fig. 7. Road image. (a) Original. (b) Red/blue scatter diagram of image. Line in (b) depicts road/nonroad boundry.

line in two dimensional Red/Blue space, rather than finding the normal of a plane in RGB space. (The techniques presented are easily generalized to include the green band, however.) Fig. 7(b) shows a *scatter diagram* of the red and blue components of Fig. 7(a); this can be thought of as a projection of the RGB space onto the Red/Blue plane, or as a two-dimensional histogram. Road pixels cluster nicely, distinct from nonroad pixels, and it should be clear that the line drawn in the figure will successfully separate road and nonroad clusters and therefore segment the image. This line is the *linear discriminant function* in Red/Blue space.

The slope of the linear discriminant function determines the red and blue components of the tricolor operation, with the green component equal to zero. In the scatter diagram, the road cluster is consistently an elliptical shape whose principal axis is parallel to the linear discriminant function. Hence the discriminant function can be found by calculating the principal axis of the road cluster. The angle the principal axis makes with respect to the red axis (θ) is defined by the equation [17]

$$\theta = 0.5 \tan^{-1}\left(\frac{b}{a-c}\right) \quad (1)$$

where

$$a = \sum_i (r - \bar{r})^2$$

$$b = 2\sum_i (r - \bar{r})(b - \bar{b})$$

$$c = \sum_i (b - \bar{b})^2$$

and ($\bar{r}, \bar{b}$) is the mean of the cluster. From θ the red, green, and blue color parameters are calculated as

$$(r, g, b) = (\cos\theta, 0, \sin\theta). \quad (2)$$

The method to dynamically choose color parameters, then, proceeds as follows: sample road points in the image, calculating the red and blue means ($\bar{r}, \bar{b}$) and a, b, and c which lead to the calculation of θ and then r, g, and b. This provides the normal of the plane in RGB space, or equivalently the line in Red/Blue space, that separates the road and nonroad clusters. This automatic color parameter selection technique is yet to be fully integrated into VITS—the current version still uses preset constant color parameters.

c) Threshold Selection: Once the color parameters are known (either calculated or preset), the tricolor operation is performed, creating an image for which each pixel represents the distance (which may be positive or negative) from the original RGB pixel to the plane $rR + gG + bB = 0$. At this point, the image is blurred to reduce noise. Choosing a value with which to threshold the new image, then, is equivalent to translating the plane in RGB space. The segmentation is described by the following equation which makes explicit the use of both color parameters (r, g, b) and the threshold (λ):

$$I'(i,j) = \begin{cases} 1 & \text{if } rR(i,j) + gG(i,j) \\ & \quad + bB(i,j) + \lambda < 0 \quad (3) \\ 0 & \text{otherwise.} \end{cases}$$

The resulting binary image I' is a function of the threshold λ, which is selected by sampling a population of road pixels from the current image. In the original version of VITS, we did a histogram equalization of the feature image (the tricolor result) and used a constant threshold to segment. This assumed that the road occupied a constant percentage of the image pixels from image to image. This worked well over most of the initial test track, but is not generally true. As an alternative, we opted for a more robust and faster method of calculating the threshold by sampling road pixels in each image. The road sampling technique is described in the next section.

Our original threshold calculation involved finding the mean and standard deviation of the road samples; the threshold was calculated as the mean of the road cluster plus a constant number of standard deviations. This proved to be very sensitive to the presence of shadows, dirt on the road, potholes, and patches of new tarmac used in road repair; these often caused the calculated road mean to be unreliable. (See Section III-B-4 for a proposed solution to this problem.) To overcome some of these problems, we chose the *median* of the top M sampled values (typically $M = 15$) rather than the mean of the whole sample population as the nominal threshold value. This takes advantage of the knowledge that true road pixels are brighter in the feature image than most of these problem areas (with the nagging exception of dirt). This normally prevents small shadows, cracks and potholes, and patches on the road from causing erroneous road sampling. The most positive sampled road points define the approximate boundary of the road cluster along the feature line. The median of the top values is used in case spurious nonroad values are sampled, and a constant offset is added to the nominal value to move the threshold just above the cluster boundary. The threshold calculation is critical to a good segmentation; Fig. 8 shows the results of thresholding at different values.

d) Road Sampling and Power Windows: Sampling road pixels in a dynamic environment is not straightforward. Our original implementation sampled at 125 fixed image locations, as shown in Fig. 9(a), assuming that the road covered these points. Because the road can drastically change position within the field of view from frame to frame, the sampling "windows" sometimes fell partially off the road, sampling dirt or grass. Since dirt and grass tend to fall above the road cluster boundary, this upset the calculation of the threshold.

To prevent sampling portions of the scene outside of the road boundary, then, *power windowing* was developed for road sampling. Instead of sampling at *fixed* image locations, the sampling window is projected onto the *predicted* road position in the image. Because of vehicle movement, this involves projecting a trapezoid representing the boundary of the road found in the previous scene model into world coordinates and then back into the new image plane location. This trapezoid, the prediction of the location of the road in the new image, is now the bounding window for image sampling, given the new position and orientation of the vehicle. This relies on the position update information (described in Section II) to do the geometric calculations between the previous and the present vehicle positions. Fig. 9(b) shows the sampling window computed as Alvin travels around a curve.

Power windowing is used along with or independent of pan/tilt control. Without the pan/tilt mechanism, it gives the ALV a software panning ability; with it, power windowing provides fine adjustment for road sampling. In relatively straight portions of road terrain, power windowing alone is preferred, because of the time involved in panning the camera. Even small angular panning is significant because of the acceleration and deceleration times of the pan/tilt mechanism. On a straight road the vehicle will often oscillate slightly as it traverses its path, so rather than allowing many small panning motions the pan mechanism is disabled and power windowing takes care of

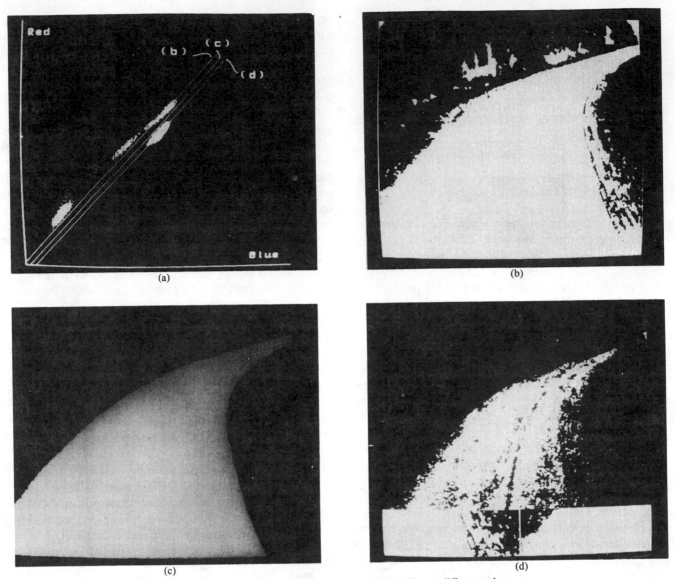

Fig. 8. (a) Scattergram. (b)–(d) Result of thresholding at different values.

"panning" the sample window. In extreme conditions such as a bad segmentation suggesting a window that is unreasonably small or large, a default window is used, similar to the old fixed image sampling points.

Because Alvin may travel as much as 10 m between successive scene acquisitions at top speeds, the projection of the old road model into the new road image may be small and fill only the lower portion of the image. To compensate for this, we use the speed from the previous position update to extend the top of the window forward so that it reaches a fixed distance in front of the vehicle. A larger sample area reduces the danger of sampling solely on a patch of dirt, shadow or stained road. The extension of the window also allows us to sample on shadowed pavement, which is helpful in the *shadow boxing* algorithm (Section III-B-4).

2) Red Minus Blue Segmentation: The segmentation algorithms are implemented on the Vicom image processing computer and take advantage of its frame rate convolution and lookup table operations. The feature reduction is accomplished by a tricolor operation, a weighted sum of the red, green, and blue images. These weights are chosen by hand or more recently calculated based on road image statistics (Section III-B-1-b). Typical values for (r, g, b) are $(0.5, 0.0, -0.5)$; hence the name "Red minus Blue." In practice, for the Martin Marietta test site these values are nearly constant. The dynamic parameter selection described above makes it possible to handle changes in road surface type, for example, during a run.

This segmentation method was originally motivated by noticing that the road appears darker than the dirt on the road shoulder in the red image and brighter than the dirt in the blue image. Since the spectral content of the pavement is "mostly blue" and the dirt bounding the road is "mostly red," subtracting the images became an obvious way to enhance the road/nonroad boundary. (Actually, "Blue minus Red" enhances the pavement, while "Red minus Blue" enhances the dirt!) An edge-based road-following algorithm would best work on the enhanced image rather than a normal intensity or single-band image.

A threshold value is chosen from the road statistics to differentiate road and nonroad clusters, as discussed in

(a)

(b)

Fig. 9. (a) Original fixed sampling points. (b) Sampling window around sharp curve, calculated by power windowing module.

Section III-B-1-c. The resulting image is then thresholded to produce the binary road/nonroad image; this is simply one pass of the image through a lookup table. The steps of the algorithm are depicted in Fig. 10 for two different road scenes.

3) Color Normalization: Another clustering algorithm involves segmenting a color normalized image, rather than the "Red minus Blue" feature image. Color intensities vary quite a bit within the road surface in a road image with shadows; intensities from the shaded regions are much smaller than those from the sunny region of the road. Intuitively, normalizing the color components will allow both shaded and sunny regions to cluster together. This assumes that the ambient illumination is identical or similar in spectral content to the incident illumination of the scene. Gershon *et al.* [12] discuss this assumption and propose a tool that can be used to classify whether discontinuities in an image are due to material changes or shadows.

We have found that using a normalized blue feature im-

age enhances the pavement/dirt boundary and therefore gives a good road/nonroad segmentation. The calculation of this feature and the resulting segmentation is described by

$$I'(i, j) = \begin{cases} 1 & \text{if } \dfrac{B}{R + G + B} - \lambda > 0 \\ 0 & \text{otherwise.} \end{cases} \quad (4)$$

The threshold equation of (4) can be rewritten as

$$B - \lambda(R + G + B) > 0$$

or

$$\lambda R + \lambda G + (\lambda - 1)B < 0.$$

This can also be implemented as a tricolor operation followed by a threshold, as described by the equation:

$$I'(i, j) = \begin{cases} 1 & \text{if } \lambda R(i, j) + \lambda G(i, j) \\ & \quad + (1 - \lambda)B(i, j) < 0 \quad (5) \\ 0 & \text{otherwise.} \end{cases}$$

Equation (5) is equivalent to a plane segmentation of the RGB color space, where the dynamically chosen threshold actually varies the *orientation* of the plane, rather than its translation as in the "Red minus Blue" algorithm. Calculation of the threshold λ proceeds as described in Section III-B-1-c.

4) Shadow Boxing: Segmenting the road using a single threshold on a combined RGB image supposes that the road cluster is the only significant cluster in an RGB half-space. If there are significant nonroad regions inside of the half-space defined by the plane normal (r, g, b) and the threshold λ, they will also be labeled as "road" and perhaps cause faulty scene models to be output. Fig. 11 shows a scatter diagram of such a case: a large region labeled "shaded nonroad," caused primarily by ditches and shadows of bushes off the road, falls in the *road* half-space. Shadows that fall on the road are close to this region in the scatter diagram; the cluster labeled "shaded road" must be distinguished from the "shaded nonroad."

This could perhaps be solved by segmenting twice, corresponding to the two threshold lines in Fig. 11(b), and performing a logical AND of the resulting binary images. The dynamic threshold calculation for the boundary between the sunny and shaded road regions is very sensitive to noise, however, because there is very little information in the shaded regions, even when digitized from a camera with a reasonably good dynamic range. Another technique that we have considered is to use a nonlinear discriminant function for segmentation. The problem is to find an adequate nonlinear function that reliably corresponds to the bends observed in the boundaries of road features in Red/Blue color space. On inspection of a wide variety of road scenes, no general pattern could be observed.

Rather than segmenting complete half-spaces, then, the road regions may be bounded by rectangles in the scatter

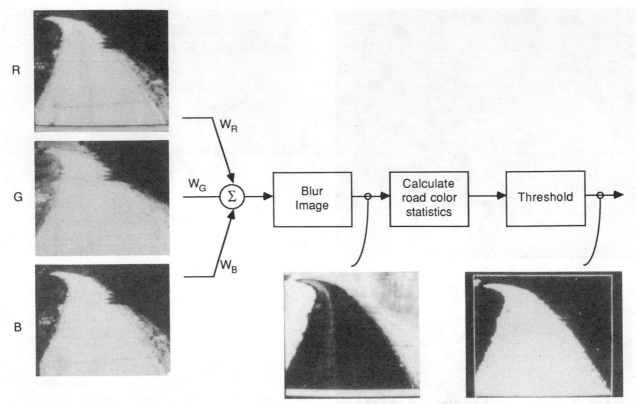

Fig. 10. Incremental results from video segmentation algorithm.

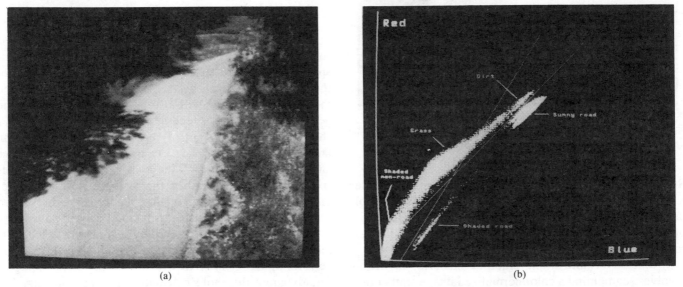

(a)   (b)

Fig. 11. (a) Original image. (b) Scatter diagram of road scene with shadows.

diagram, and only the *boxed* regions are segmented and labeled as road. This is particularly helpful in conditions with significant shadows; hence the name "Shadow Boxing."

The bounding boxes are again motivated by the current hardware, as the segmentation can be implemented quickly by two global lookup table operations per boxed region. The boxes must be oriented with the axes of the two dimensional feature space, however, so the first step is to perform a rotation of the Red/Blue axes so that the red axis is aligned with the road cluster in the scatter diagram. A rotation of θ about the origin is desired, where θ is defined in (1) as the angle between the red axis and the principal axis of the road cluster in the scatter diagram. As shown in the Appendix, this rotation is performed by replacing the red image with the result of a tricolor operation using $(r, 0, b)$ and replacing the blue image with a tricolor result using $(-b, 0, r)$, where r and b are defined in (2). Fig. 12(a) shows the new scatter diagram with the bounding boxes, from the original in Fig. 11.

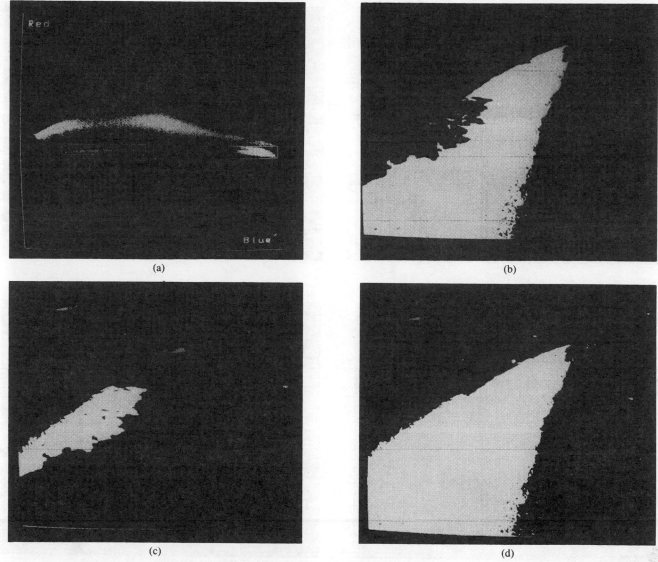

Fig. 12. Scatter diagram from Fig. 11, (a) rotated to align with the road cluster. (b) Sunny road result. (c) Shaded road result. (d) Final road image.

The extents of the bounding boxes in the rotated space are calculated as the road image points are sampled by keeping track of the maximum and minimum sampled values within expected ranges for both shaded road and sunny road regions. The threshold calculation uses only values in the expected range for sampling the road, making that stage less sensitive to noise. Lookup tables are built to perform the segmentation; two binary images are produced from each bounding box, one per axis. For each box, a logical AND of its resulting binary images is performed, giving the road region corresponding to the box. Figs. 12(b) and (c) show the image regions corresponding to the sunny road and shaded road boxes, respectively. The results from each box are then combined with a logical OR operation, resulting in a binary road/nonroad image, as in Fig. 12(d).

Shadow boxing is similar to a dynamic Bayesian classifier [9] with three decision regions and rectangular decision boundaries. In summary, the steps involved in the algorithm are: 1) calculate the color parameters (r, g, b); 2) rotate the Red/Blue space by performing two tricolor operations; 3) sample the road points, keeping track of the extents of the bounding boxes; 4) build the lookup tables; and 5) perform the segmentation for each box by passing the rotated images through the corresponding lookup tables and combining the images with the proper sequence of logical operations.

C. Boundary Extraction

The segmentation algorithms produce a binary road/nonroad image. From this image, the road edges are extracted and transformed into three dimensional coordinates to fill in the scene model. The boundary extraction is an edge-tracking process in which the road boundary is found and then traced while keeping track of the pixel locations.

The initial task is to find the road/nonroad boundary in the image. To facilitate easier boundary tracing and to

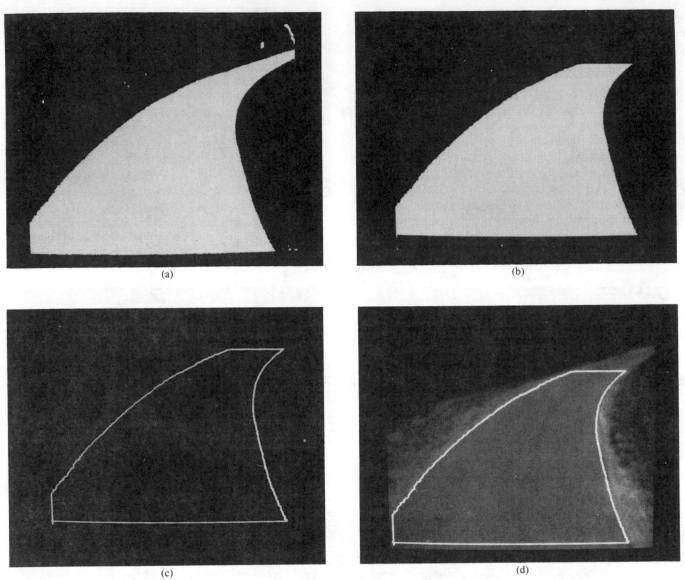

Fig. 13. (a) Binary road image. (b) False boundary added. (c) Road edges.
(d) Road edges overlaid on original image.

avoid looking for the road on or above the horizon, a false road boundary is added around the image, creating an artificial horizon and borders, as in Fig. 13(b)—this prevents following the road edges up into the sky or over into the next segment of image memory. In order to find an initial boundary, we start in the bottom quarter of the image and step upwards until a boundary is detected. The border is then traced in both directions, using an 8-neighbor nonroad, 4-neighbor road connectivity rule, and image coordinates of boundary points are saved. The boundary detection and tracing uses preset ''skip factors'' in both row and column directions to speed processing. This effectively reduces the image size by the row and column skip factors. The boundary tracking method allows for reasoning on the fly—''bubbles'' are properly ignored, and globally nonlinear segments, such as corners of intersections, can be detected and noted. When the right or left false boundary is detected, the corresponding road edge is known to be out of the camera's field of view.

Fig. 13(c) and (d) show the boundary traced for the segmentation of Fig. 7(a).

A completely general approach to boundary extraction is prohibitively expensive with the current hardware. Such an approach, however, would allow more extensive reasoning about road shape, obstacles in the road, and choosing road sampling regions. Kuan and Sharma [21] have demonstrated model-based reasoning about road boundaries. Future hardware improvements should allow more sophisticated reasoning capability in road boundary extraction.

Once the image coordinates of both right and left road edge points are found, we choose a small number of points (up to ten) on each edge to form a polygonal representation of the road edge. Image coordinates of a small neighborhood of edge points are averaged to avoid sending ''stray'' points. The row locations of these points are spaced by a quadratic function so that the three dimensional locations of the road edge points will be approximately equal distances apart. These points are then sent to the geometry module for conversion to three dimensional road edge point.

D. Three Dimensional Geometry Transformations

Once road edge points are selected in the image, a three-dimensional description, called the scene model, must be sent to Reasoning for trajectory calculation. This process of recovering the three dimensional information projected onto the two dimensional image plane is the "inverse optics" problem of vision. As Poggio [31] and others have pointed out, this is an under-constrained (formally "ill-posed") problem that requires the introduction of generic constraints to arrive at a unique solution. In our case, such constraints are assumptions about the structure of the road environment. This is the *forward-geometry* problem.

VITS also uses the solution to the *inverse-geometry* problem, determining the location within the image plane of a point whose three-dimensional location is known. Unlike the forward-geometry problem, the inverse-geometry problem has an exact solution; no assumptions need to be made in order to constrain the problem and make it well-posed. The combination of the forward-geometry and inverse-geometry processes allow for frame-to-frame registration of features as well as predictions about, for example, the continuation of the road.

The original forward-geometry module used in VITS was essentially a model driven "shape from contour" method developed at the University of Maryland [40], based on calculating the vanishing point of parallel lines projected onto the image plane. Experiments soon showed that assuming a *flat-earth* road model allows for a much faster forward-geometry module and performs better for the roads Alvin encounters and the speeds attained during the demonstrations through 1986. While flat-earth geometry is clearly an assumption that is very useful in certain circumstances, it is not accurate enough for all road-following applications. Work is proceeding to incorporate a *hill-and-dale* geometry module [28] that uses a fast "shape from contour" method to solve the forward-geometry problem.

These various techniques for recovering the three dimensional descriptions are discussed in the remainder of this Section. The flat-earth model is presented first because it is used by the other techniques. Next, the hill-and-dale technique, a slight modification of the flat-earth model, is presented. The *vanishing point* method and a *zero-bank* method, both developed at the University of Maryland, are mentioned for comparison.

1) Flat-Earth Geometry Model: In the *flat-earth* geometry model we assume that the road is planar, and that the plane containing the visible portion of the road is the same plane which is giving support to the vehicle. Thus, the three dimensional location of an edge point found in the image at (*col*, *row*) can be determined by finding the point of intersection of the vector from the focal point of the camera through this point with the ground plane.

The flat-earth geometry model has several advantages over the other forward-geometry models. The first of these is its speed: a straightforward calculation gives the three dimensional location for a given image point. Second, this model can be applied to any single image point, even those which are not edges of the road; there is no need for multiple image points as in the vanishing point geometry model. This property also implies that this algorithm is the only one of the four discussed here which is presently capable of handling intersections. Third, the error in the output three dimensional locations is only a function of the extent to which the flat-earth assumption is violated, and not additionally a function of the goodness of the segmentation.

In practice there are a number of problems which limit the applicability of this technique. First is its sensitivity to inaccuracies in the assumed tilt angle formed by the camera to the road plane. The camera is in a fixed position relative to the body of the ALV, but the body of the vehicle is able to rock forward and backward on the undercarriage. When traveling uphill the vehicle body rocks backwards, decreasing the effective tilt angle of the camera. When traveling downhill the vehicle body rocks forwards, increasing the effective tilt angle of the camera. These changes in the effective tilt angle cause parallel road edges to be output as converging or diverging three dimensional edge segments. If the convergence or divergence is too severe it is difficult to connect road edges from one scene model with those of the next scene model. Because of the problems caused by this rocking motion of the vehicle, we are adding sensors which will measure the angle formed between the vehicle body and the vehicle undercarriage. Knowledge of this angle will enable a more accurate forward-geometry transformation.

A second problem with the flat-earth model occurs at inflection points, such as at crests of hills and at bottoms of valleys. These situations cause the description of the road to both converge and diverge within the same scene model. This problem is addressed by each of the geometry models described below.

Examples of both of these problems can be seen in Fig. 14; plots of successive scene models reveal herringbone patterns caused by a combination of a nonplanar road and inaccuracy of the camera tilt angle.

2) Hill-and-Dale Geometry Model: The "hill-and-dale" geometry model was developed to address the problems of converging and diverging scene model edges, as illustrated in Fig. 14. The essence of this technique is to use the flat-earth geometry model for the two roadway points nearest the vehicle in the image, and then to force the road model to move up or down from the flat-earth plane so as to retain a constant road width.

Let $p(i, j)$ be the world coordinates of the points which appear in the image as indicated in Fig. 15. The first step of the algorithm is to use flat-earth geometry to solve for $p(0, 1)$ and $p(0, 2)$. From this it is possible to compute the width of the road W, where $W = \| p(0, 1) - p(0, 2) \|$.

One way to maintain a constant road width in the scene

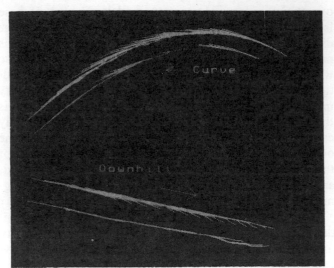

Fig. 14. Scene model plots from a test run/(em road edges viewed from above). Herringbone patterns are seen in the overlap of scene models.

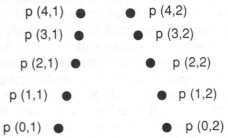

p (4,1) ●　　　● p (4,2)

p (3,1) ●　　　● p (3,2)

p (2,1) ●　　　　● p (2,2)

p (1,1) ●　　　　● p (1,2)

p (0,1) ●　　　　● p (0,2)

Fig. 15. Road edge points to be converted into the vehicle coordinate system.

model is to intersect the successive pair of edge points $(i, 1)$ and $(i, 2)$ with a different plane than the ground plane. To see this, note that the rays from the camera origin through the image locations for these points are diverging; thus using a plane above the ground plane will produce a narrower road than using a plane below the ground plane. For each successive pair of edge points $(i, 1)$ and $(i, 2)$ we can compute the elevation of a plane containing these points, perhaps above or below the assumed ground plane, such that the road maintains the same width. The elevation is then used in a flat-earth geometry calculation to produce scene models for which the road is of constant width when measured at paired scene model points.

Testing has shown that this algorithm produces more accurate scene models than the flat-earth algorithm on straight or slightly curved roads which go up and down hill, and when the segmentation is good. The algorithm is, however, very dependent upon good segmentation, as a slightly wider road segmentation will cause the road to appear to travel uphill, and a slightly narrower road segmentation will cause the road to travel downhill. While this is not particularly important to vehicle behavior when the road is straight, it can cause the distance to a curve to be in error by a significant amount.

A potentially larger drawback to this algorithm is its performance near and in curves and intersections. Many curves exhibit banking which this algorithm is unable to reproduce; the apparent location of the lower edge of the banked road will always be too high, and that of the outer edge will always be too low. This problem is potentially addressed by the vanishing-point algorithm described below. Also, the selection of opposing pairs of edge points is critical, since the width of the road is measured between these pairs of points. Thus, if the road is curving it is necessary to select pairs of edge points such that the resulting tiles of the road are pie shaped. Finally, it should be noted that the constant width assumption of this algorithm is violated at intersections, and may be violated at other roadway areas as well.

We are currently investigating heuristics for selecting matched edge points. This is the "tiling problem" of road geometry.

3) Other Geometry Models: The vanishing point geometry model [40] uses flat-earth geometry to determine the location of the nearest visible right and left edge points within the image, and allows for the road to slope uphill or downhill relative to the vehicle ground plane. This algorithm uses constraints based on assumptions of parallel road edge segments, continuity, and local flatness of the road. These assumptions allow the computation of a set of tiles that approximate the road in a viewer-centered coordinate frame. The technique involves solving for the image coordinates of the vanishing point of each pair of matched left and right road edge segments within each tile. This geometry model was used in the initial May 1985 ALV demonstration.

The zero-bank algorithm [6] models the road as a centerline spine and horizontal line segments cutting the spine at their midpoint at a normal to the spine. Modeling the road in this way constrains tiles of the road to be "warped isosceles trapezoids." As in the other algorithms, the three-dimensional locations of the closest point on each edge are found using the flat earth algorithm. In each successive step the three-dimensional location of the next point on one side of the road is parameterized by its distance from the camera. The image location of the paired point on the other side of the road is constrained by the warped isosceles trapezoid as well as the edge curve within the image plane. This leads to a cubic equation in the parameter; when the roots of this equation (if any) are found, the root which yields minimum reasonable slope difference with respect to the previous road direction is kept.

While the zero-bank algorithm is computationally the most costly, it addresses both the tiling problem and the problem of hills and dales. None of the algorithms can reproduce banking in curves, and all but the flat-earth algorithm rely heavily on good segmentation and cannot handle intersections.

IV. DISCUSSION

The ALV public demonstrations in 1985 and 1986 have displayed Alvin's road-following capabilities. Fig. 16 shows a schematic map of the current Martin Marietta test track. In May of 1985, Alvin traversed from points 2 to

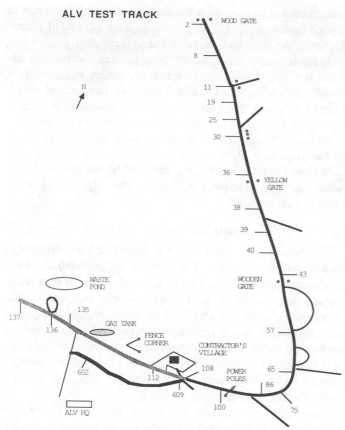

ALV TEST TRACK

Fig. 16. Martin Marietta ALV Test Track (Denver, CO).

Fig. 17. Representative image of the test track.

65 (a distance of approximately 1 km) at a speed of 3 km/hr. In June of 1986, Alvin traversed the entire test track (4.2 km) at speeds up to 10 km/hr. Additional capabilities demonstrated in 1986 include:

1) switching back and forth between video-based and range-based road-following
2) obstacle avoidance based on a fusion of video and range data
3) high-speed runs of up to 20 km/hr
4) varying vehicle speed as a function of scene model length
5) slowing to a stop and turning completely around, then traveling back in the opposite direction
6) negotiating a hairpin curve
7) traveling over two different types (and colors) of pavement.

Fig. 17 illustrates some of the various conditions encountered on various portions of the test track. A typical test run over the entire track will involve the acquisition and processing of hundreds of images. These images vary from day to day, since weather, sun angle, shadows, and tire tracks all alter the appearance of the road and its surroundings. To compensate for occasional failures in VITS, the reasoning subsystem builds trajectories so that the vehicle will halt at the end of the most recent scene model if no additional data is received. This convention makes Alvin very forgiving of occasional vision failures, usually causing a slight slowing of the vehicle when the process-

ing of a single image frame fails. During one test run, a cloud passed overhead and changed the road appearance enough so that no road was found in the images. The vehicle ramped down to a stop and waited until the cloud passed by and VITS was able to segment the road; Alvin then resumed safe travel.

Current vision efforts are concentrated on speed and robustness of road-following, obstacle detection and location, and vision for off-road navigation. Long term research areas include object modeling, landmark recognition, terrain typing, stereo, and motion analysis. Researchers at many groups are currently working on these problems for future ALV application. Their efforts are critical to meeting future demonstration requirements, and to the success of the program in general. Interaction with these groups has influenced our present system to a large degree.

To travel at higher vehicle speeds, the vision system must not only produce scene models more rapidly but also provide longer scene models to allow for the distance needed to slow down for a detected obstacle or to stop in case of an emergency. We are presently investigating methods to more accurately model the road at far distances (25 meters and beyond) and to extend calculated road edges based on road history and assumptions about road curvature. Because of limited field of view and accuracy in the range of the current range scanner, we are working on fast methods to detect obstacles at a distance using video data [36].

To meet the demands of future demonstrations, implementation of a new architecture begin in early 1987. Each subsystem of the ALV will have significantly more processing power available. For the vision subsystem, the second generation hardware consists of two Vicom image processors, a Sun 3/180, and a Warp machine [14]. We are also investigating the use of a FLIR sensor and special sensors for obstacle detection, and other advanced computer architectures (such as the Butterfly and the Connection Machine).

Of the video road-following algorithms described, the "Red minus Blue" algorithm has proved to be the most dependable so far, and it has been used (at different stages of development) in the formal demonstrations to date. The color normalization algorithm performs well in very sunny conditions when shadows present a problem to "Red minus Blue." "Shadow boxing" is designed to deal with shadows and obstacles. It has been tested but not yet used to drive the ALV in a formal demonstration.

The evolution of the vision system has been largely motivated by failures, caused by deviations from "ideal" road-following conditions. Dirt and tire tracks on the road, unexpected tarmac patches, changing weather conditions (from a momentary cloud overhead to the presence of snow for weeks) and seasonal variations (such as increased shadows and spectral reflection caused by a lower sun angle in the winter months) have all caused failures of the vision system. In analyzing these failures, we have learned much about the vision/navigation interplay, sensor control needs, and the dynamic nature of video parameters.

The Autonomous Land Vehicle is intended not only as a development project to meet specified demonstration requirements, but also as a national testbed for research in image understanding, AI planning, and advanced architectures. As different parts of the system change, we anticipate learning more about how system interaction affects the individual components. A vision system for a mobile robot does not stand alone; we have spent much effort analyzing the vision/navigation interaction to discover the causes of certain vehicle behaviors or failures. Likewise, the algorithm level is not totally independent of the implementation or hardware level. Much of the current vision system has been motivated by the chosen hardware—as the hardware changes, the algorithms may change substantially.

V. Summary and Conclusion

The time is ripe for fruitful research and experimentation in mobile robotics. Early work suffered from a lack of processing power, but with the availability of relatively cheap, fast machines, particularly special-purpose image processing hardware and lots of memory, an important step in mobile robot research is now realizable: experimentation with fully autonomous, continuous motion, self-contained systems. The dynamics of system interaction provides important insight into the development of sophisticated autonomous systems.

Experiments in mobile robot road-following prove the importance of an evolving, robust vision system to model the environment for navigation. Such a system must exhibit intelligent behavior under varying vehicle and environmental conditions: seasonal variation in scene characteristics, diverse and changing weather conditions, unexpected visual information (e.g., obstacles, shadows, potholes), changes in navigation and control systems, and changing sensor characteristics. Our experiments with Alvin have driven the development of such a vision framework. As mobility may be a key to the development of intelligence [27], real-time interaction with the environment is a significant step in the development of intelligent machines. For this reason, research in mobile robot vision is largely an incremental process of hypothesis and test. Analyzing the failure of a particular test is often much more informative than a success.

We have presented the vision system for Alvin, the Autonomous Land Vehicle, discussing in particular the task of video road-following. The system provides an effective level of behavior in both speed and performance. As the processing power of the ALV increases and the performance requirements become more ambitious, the system must become more robust, faster, and more "intelligent." Work is progressing at a number of institutions to meet these goals.

Appendix
Red/Blue Axes Rotation

We want to show that the rotated coordinate system used for shadow boxing is accomplished by replacing the Red image with the tricolor image with parameters $(r, 0, b)$ and replacing the Blue image with the tricolor image with

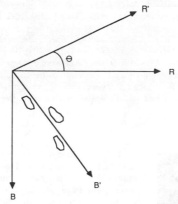

Fig. 18. (R, B) to (R', B') rotation for shadow boxing.

parameters $(-b, 0, r)$. The coordinate systems are shown in Fig. 18, where (R', B') is rotated to align with the major axis of the blob. The transformation is described by:

$$R' = R \sin \theta - B \cos \theta$$
$$= (R, G, B) \cdot (\sin \theta, 0, -\cos \theta)$$
$$B' = R \cos \theta + B \sin \theta$$
$$= (R, G, B) \cdot (\cos \theta, 0, \sin \theta)$$

From (2) in Section III-B-1-b, this is equivalent to

$$R' = (R, G, B) \cdot (r, 0, b)$$
$$B' = (R, G, B) \cdot (-b, 0, r)$$

which describes a pair of tricolor operations. In practice, a *translation* of the coordinate system may also be necessary to avoid negative values.

ACKNOWLEDGMENT

T. Dunlay, S. Hennessy, B. Bloom, J. Bradstreet, Z. Shinno, and G. Arensdorf all contributed to the development of the vision system. The support of the whole ALV team is greatly appreciated.

REFERENCES

[1] S. Barnard, R. Bolles, D. Marimont, and A. Pentland, "Multiple representations for mobile robot vision," in *Proc. SPIE Mobile Robot Conf.*, Cambridge, MA, Oct. 1986, pp. 143–151.
[2] R. A. Brooks, "A robust layered control system for a mobile robot," *IEEE J. Robotics and Automation*, vol. RA-2, no. 1, pp. 14–23, Mar. 1986.
[3] G. B. Coleman and H. C. Andrews, "Image segmentation by clustering," *Proc. IEEE*, vol. 67, no. 5, pp. 872–884, May 1979.
[4] J. L. Crowley, "Navigation for an intelligent mobile robot," *IEEE J. Robotics and Automation*, vol. RA-1, no. 1, pp. 31–41, Mar. 1985.
[5] M. J. Daily, J. G. Harris, and K. Reiser, "Detecting obstacles in range imagery," in *Proc. DARPA Image Understanding Workshop*, Feb. 1987, pp. 87–97.
[6] D. DeMenthon, "A Zero-bank algorithm for inverse perspective of a road from a single image," in *Proc. IEEE Int. Conf. Robotics and Automation*, Raleigh, NC, Apr. 1987, pp. 1444–1449.
[7] A. R. de Saint Vincent, "A 3D perception system for the mobile robot Hilare," in *Proc. IEEE Int. Conf. Robotics and Automation*, San Francisco, CA, Apr. 1986, pp. 1105–1111.
[8] E. D. Dickmanns and A. Zapp, "A curvature-based scheme for improving road vehicle guidance by computer vision," in *Proc. SPIE Mobile Robot Conf.*, Cambridge, MA, Oct. 1986, pp. 161–168.
[9] R. O. Duda and P. E. Hart, *Pattern Classification and Scene Analysis*. New York: Wiley-Interscience, 1973.
[10] R. T. Dunlay and D. G. Morgenthaler, "Robot road-following using laser-based range imagery," in *Trans. SME Second World Conf. Robotics Research*, Scottsdale, AZ, Aug. 1986.
[11] R. T. Dunlay and D. G. Morgenthaler, "Obstacle detection on roadways from range data," in *Proc. SPIE Mobile Robot Conf.*, Cambridge, MA, Oct. 1986, pp. 110–116.
[12] R. Gershon, A. D. Jepson, and J. K. Tsotsos, "The effects of ambient illumination on the structure of shadows in chromatic images," Dep. Comput. Sci., Univ. Toronto, Tech. Rep. RBCV-TR-86-9, 1986.
[13] Y. Goto and A. Stentz, "The CMU system for mobile robot navigation," in *Proc. IEEE Int. Conf. Robotics and Automation*, Raleigh, NC, Apr. 1987, pp. 99–105.
[14] T. Gross, H. T. Kung, M. Lam, and J. Webb, "Warp as a machine for low-level vision," in *Proc. IEEE Int. Conf. Robotics and Automation*, St. Louis, MO, Mar. 1985, pp. 790–800.
[15] A. R. Hanson and E. M. Riseman, "Segmentation of natural scenes," in *Computer Vision Systems*, A. R. Hansen, and E. M. Riseman, Eds. New York: Academic, 1978.
[16] M. Hebert and T. Kanade, "Outdoor scene analysis using range data," in *Proc. IEEE Int. Conf. Robotics and Automation*, San Francisco, CA, Apr. 1986, pp. 1426–1432.
[17] B. K. P. Horn, *Robot Vision*. Cambridge, MA: MIT Press, 1986.
[18] R. M. Inigo, E. S. McVey, B. J. Berger, and M. J. Mirtz, "Machine vision applied to vehicle guidance," *IEEE Trans. Pattern Anal. Machine Intell.*, vol. PAMI-6, no. 6, pp. 820–826, Nov. 1984.
[19] M. Julliere, L. Marce, and H. Place, "A guidance system for a mobile robot," in *Proc. 13th Int. Symp. Indust. Robots and Robots7*, Chicago, IL, Apr. 1983.
[20] D. Kuan, G. Phipps, and A. Hsueh, "A real-time road following and road junction detection vision system for autonomous vehicles," in *Proc. AAAI-86*, Philadelphia, PA, Aug. 1986.
[21] D. Kuan and U. K. Sharma, "Model based geometric reasoning for autonomous road following," in *Proc. IEEE Int. Conf. Robotics and Automation*, Raleigh, NC, Apr. 1987, pp. 416–423.
[22] K. Kuhnert, "A vision system for real time road and object recognition for vehicle guidance," in *Proc. SPIE Mobile Robot Conf.*, Cambridge, MA, Oct. 1986, pp. 267–272.
[23] D. T. Lawton, T. S. Levitt, C. McConnell, and J. Glicksman, "Terrain models for an autonomous land vehicle," in *Proc. IEEE Int. Conf. Robotics and Automation*, San Francisco, CA, Apr. 1986, pp. 2043–2051.
[24] J. M. Lowrie, M. Thomas, K. Gremban, and M. Turk, "The autonomous land vehicle (ALV) preliminary road-following demonstration," in *Intelligent Robots and Computer Vision (Proc. SPIE 579)*, D. P. Casasent, Ed., Sept. 1985, pp. 336–350.
[25] H. P. Moravec, "Obstacle avoidance and navigation in the real world by a seeing robot rover," Ph.D. dissertation, Stanford Univ., Stanford, CA, Sept. 1980. (Reprinted as *Robot Rover Visual Navigation*. Ann Arbor, MI: UMI Research Press, 1981.)
[26] ——, "The Stanford Cart and the CMU Rover," *Proc. IEEE*, vol. 71, no. 7, pp. 872–884, July 1983.
[27] ——, *Mind Children*. Cambridge, MA: Harvard University Press, 1986.
[28] D. Morgenthaler, "Hill and dale geometry," Martin Marietta Internal Memo, May 1986.
[29] B. Mysliwetz and E. D. Dickmanns, "A vision system with active gaze control for real-time interpretation of well structures dynamic scenes," in *Proc. Intelligent Autonomous Systems*, Amsterdam, The Netherlands, Dec. 1986.
[30] K. E. Olin, F. M. Vilnrotter, M. J. Daily, and K. Reiser, "Developments in knowledge-based vision for obstacle detection and avoidance," in *Proc. DARPA Image Understanding Workshop*, Feb. 1987, pp. 78–86.
[31] T. Poggio, "Early vision: From computational structure to algorithms and parallel hardware," *Comput. Vision, Graphics, and Image Processing*, vol. 31, pp. 139–155, 1985.
[32] A. M. Thompson, "The navigation system of the JPL robot," in *Proc. Fifth IJCAI*, 1977, pp. 749–757.
[33] C. Thorpe, "FIDO: Vision and navigation for a mobile robot," Ph.D. dissertation, Dep. Comput. Sci., Carnegie-Mellon Univ., Dec. 1984.
[34] S. Tsugawa, T. Yatabe, T. Hirose, and S. Matsumoto, "An automobile with artificial intelligence," in *Proc. Sixth IJCAI*, 1979, pp. 893–895.
[35] S. Tsuji, J. Y. Zheng, and M. Asada, "Stereo vision of a mobile robot: World constraints for image matching and interpretation," in

Proc. IEEE Int. Conf. Robotics and Automation, San Francisco, CA, Apr. 1986, pp. 1594–1599.

[36] M. A. Turk and M. Marra, "Color road segmentation and video obstacle detection," in *Proc. SPIE Mobile Robot Conf.*, Cambridge, MA, Oct. 1986, pp. 136–142.

[37] R. S. Wallace, "Robot road following by adaptive color classification and shape tracking," in *Proc. AAAI-86.*

[38] R. Wallace, A. Stentz, C. Thorpe, H. Moravec, W. Whittaker, and T. Kanade, "First results in robot road-following," in *Proc. IJCAI-85.*

[39] R. Wallace, K. Matsuzaki, J. Crisman, Y. Goto, J. Webb, and T. Kanade, "Progress in robot road-following," in *Proc. IEEE Int. Conf. Robotics and Automation*, San Francisco, CA, Apr. 1986, pp. 1426–1432.

[40] A. M. Waxman, J. LeMoigne, and B. Scrinivasan, "Visual navigation of roadways," in *Proc. IEEE Int. Conf. Robotics and Automation*, St. Louis, MO, Mar. 1985.

[41] A. M. Waxman, J. J. LeMoigne, L. S. Davis, B. Srinivasan, T. R. Kushner, E. Liang, and T. Siddalingaiah, "A visual navigation system for autonomous land vehicles," *IEEE J. Robotics and Automation*, vol. RA-3, no. 2, pp. 124–141, Apr. 1987.

David G. Morgenthaler received the B.S. degree in electrical engineering from Princeton University, Princeton, NJ, in 1975, and the M.S. and Ph.D. degrees in computer science from the University of Maryland in 1978 and 1981, respectively. His dissertation was on three dimensional image processing.

He joined Martin Marietta Denver Aerospace, Denver, CO, in 1982 as the Computer Vision Unit Head. He has led the computer vision research and development activities on several robotic related programs. He is currently the Software Lead and the Perception Lead on the Autonomous Land Vehicle program.

Keith D. Gremban received the B.S. degree in mathematics and the M.S. degree in applied mathematics from Michigan State University, East Lansing, in 1978 and 1980, respectively. He is currently working toward the Ph.D. degree in computer science and robotics at Carnegie-Mellon University, Pittsburgh, PA.

Since 1980, he has been with the Advanced Automation Technology Section of Martin Marietta, and is currently a visiting scientist at the Robotics Institute, Carnegie-Mellon University. His primary research interests are in the perception and interpretation of natural scenes, and mobile robot navigation. He has worked on several mobile robot projects, including the NAVLAB at CMU, and the ALV at Martin Marietta.

Matthew A. Turk (S'81–M'84) received the B.S. degree in electrical engineering from Virginia Polytechnic Institute and State University, Blacksburg, in 1982 and the M.S. degree in electrical and computer engineering from Carnegie-Mellon University, Pittsburgh, PA, in 1984.

In 1984 he joined the Advanced Automatic Technology Section at Martin Marietta Denver Aerospace, Denver, CO, where he was involved in computer vision research and development for autonomous vehicles and aerial photointerpretation. He is currently pursuing the Ph.D. degree in computational vision, affiliated with the Media Laboratory at the Massachusetts Institute of Technology, Cambridge.

Martin Marra received the B.S. degree in applied math: engineering systems and computer science from Carnegie-Mellon University, Pittsburgh, PA, in 1985.

He worked for the Robotics Institute at CMU between 1983 and 1985 while pursuing his degree. In 1985 he joined the Advanced Automation Technology Section at Martin Marietta Denver Aerospace, Denver, CO, where he was involved in computer vision research and development for the Autonomous Land Vehicle Program.

Vision and Navigation for the Carnegie-Mellon Navlab

Reprinted from *IEEE Transactions on Pattern Analysis and Machine Intelligence*, Vol. 10, No. 3, May 1988. Copyright © 1988 by The Institute of Electrical and Electronics Engineers, Inc.

CHARLES THORPE, MARTIAL H. HEBERT, TAKEO KANADE, MEMBER, IEEE, AND STEVEN A. SHAFER, MEMBER, IEEE

Abstract—This paper describes results on vision and navigation for mobile robots in outdoor environments. We present two types of vision algorithms: color vision for road following, and 3-D vision for obstacle detection and avoidance. In order to evaluate these algorithms in a realistic outdoor environment, we have integrated the vision capabilities into a complete system including a self-contained mobile platform, and a software architecture for the real-time control of distributed perception and navigation modules. The resulting system is able to navigate continuously on roads while avoiding obstacles.

Index Terms—Blackboards, color vision, mobile robots, range data.

I. Introduction

MOBILE robot systems provide a unique opportunity to develop perception and navigation techniques in complex real-world environments. The tools a robot uses to bridge the chasm between the external world and its internal representation include sensors, image understanding to interpret sensed data, geometrical reasoning, and a concept of time and of the vehicle's motion over time. This paper presents a mobile robot system that includes both perception and navigation tools. The rationale for developing a whole system is that it allows us to confront our vision techniques with real-world environments in the context of actual autonomous navigation missions.

We have developed a testbed for the study of mobile robots, the Navigation Laboratory. The testbed is used for integrating perception and navigation capabilities. The sensory capabilities of the system are color vision, and 3-D vision using an active sensor. Color vision is used for finding roads in color images. 3-D vision is used for obstacle detection and terrain modeling. We have integrated the perception modules into a system that allows the vehicle to drive continuously in an actual outdoor environment. In order to integrate perception modules, navigation modules, and hardware interface, we propose a distributed architecture articulated around a knowledge database "Communication Database with Geometric

Manuscript received December 15, 1986; revised May 15, 1987. This work was supported by the Strategic Computing Initiative of the Defense Advanced Research Projects Agency, ARPA Order 5351, monitored by the U.S. Army Engineer Topographic Laboratories under Contract DACA760-85-C-0003, DARPA Contract DACA76-86-C-0019, and National Science Foundation Grant DCR-8604199.

The authors are with the Department of Computer Science and the Robotics Institute, Carnegie-Mellon University, Pittsburgh, PA 15213.

IEEE Log Number 8820104.

Reasoning'' (CODGER). The CODGER tools handle much of the modeling of time and geometry, and provide for synchronization of multiple processes. The architecture coordinates control and information flow between the high-level symbolic processes running on general purpose computers, and the lower-level control running on dedicated real-time hardware.

II. Navlab: Navigation Laboratory

The Navlab [11] is a self-contained laboratory for navigational vision system research with on-board computing facilities. Figs. 1 and 2 show the vehicle.

Navlab is based on a commercial van chassis, with hydraulic drive and electric steering. Computers can steer and drive the van by electric and hydraulic servos, or a human driver can take control to drive to a test site or to override the computer. The Navlab has room for computers on board including Sun workstations and a Warp systolic computer [1]. Navlab is currently equipped with two sensors mounted above the cab: a TV camera, and a laser range finder. The interface between the computers and the driving hardware is provided by a controller which can steer the Van along circular arcs. The vehicle has room for up to four researchers in the back. This gets the researchers close to the experiments, and eliminates the need for video and digital communications with remote computers.

III. Road Following

The first application of a mobile robot is to find and follow roads using a video camera. Two approaches could be used: line tracking and region analysis. The first approach attempts to extract the edges of the road as seen by a camera, and backprojects them on the ground plane in order to compute the appropriate steering commands for the vehicle [4], [15]. This approach assumes an accurate model of the road shape. Unfortunately, in practice the strongest edges are often the shadow edges, whereas the road edges are much weaker and do not necessarily fit the straight line model. The line tracking approach relies heavily on inferring the road geometry from the road edges. The geometric inference assumes that the road edges are parallel [15], and therefore it is very unstable when applied to noisy data. In fact, it can be shown that the inferred road may fold over or under the ground plane [5].

Fig. 1. The Navlab.

Fig. 2. Navlab interior.

The second approach classifies the pixels of a color image as on-road or off-road pixels based on the color characteristics of the road [13]. This approach does not require a geometric model as accurate as the line tracking approach. The main problem is that the dominant color features are dark (shadows) and light (sun). As a result road and nonroad pixels cannot be separated by a single threshold in RGB space. Furthermore, characteristic RGB values for a given feature (e.g., sunny road) drift, so that algorithms based on fixed threshold fail.

Previous work shows that the road following problem is difficult; existing systems work adequately for their particular environments, e.g., even illumination or straight edges. We use the second approach to find and follow roads [12]. We use multiclass adaptive color classification to classify image points as "road" or "nonroad" on the basis of their color. Since the road is not a uniform color, color classification must have more than one road model, or class, and more than one nonroad class. Because conditions change from time to time and

from place to place over the test course, the color models must be adaptive. In addition to color classification, we added texture information to the classification algorithm. Once the image is classified, the road is identified by means of an area-based voting technique that finds the most likely location for the road in the image.

A. Overview of the Road Following Algorithm

Fig. 3 shows a simple scene which we will use to explain our algorithm. As shown in Fig. 4, the algorithm involves three stages:

1) Classify each pixel.

2) Use the results of classification to vote for the best-fit road position.

3) Collect new color statistics based on the detected road and nonroad regions.

Pixel classification is done by a standard pattern classification method [6]. Each class i is represented by the means m_i and covariance matrix, Σ_i, of red, green, and blue values, and by its *a priori* likelihood, f_i, based on expected fraction of pixels in that class. Assuming Gaussian distribution, the confidence that a pixel of color X belongs to a class is computed based on the distance ($X - m_i)^t \Sigma^{-1} (X - m_i)$. Each pixel is classified with the class of highest probability. Figs. 5 and 6 show how each pixel is classified and its confidence.

Once each point has been classified, we must find the most likely location of the road. We assume the road is locally flat, straight, and has parallel sides. The road geometry can be described by two parameters as shown in Fig. 7:

1) The intercept P, which is the image column of the road's *vanishing point*. This is where the road centerline intercepts the vanishing line of the locally flat plane of the road. In other words, the intercept gives the road's direction relative to the vehicle. Since the camera is fixed to the vehicle this vanishing line is constant assuming that the vehicle's pitch and roll relative to the road are small. In the following, $r_{horizon}$ is the row position of the vanishing line in the image.

2) The orientation θ of the road in the image, which tells how far the vehicle is to the right or left of the centerline.

We set up a two-dimensional parameter space, with intercept as one dimension and orientation as the other. Each point classified as road votes for all road (P, θ) combinations to which it could belong, while nonroad points cast negative votes, as shown in Fig. 9 and Fig. 8. For a given pixel (*row, col*), the corresponding set of pairs (P, θ) is the curve:

$$(col + (row - r_{horizon}) \times tg\ \theta, \theta)$$

in parameter space. The (P, θ) pair that receives the most votes is the one that contains the most road points, and it is reported as the road. For the case of Fig. 3, the votes in (P, θ) space look like Fig. 10. Fig. 11 shows the detected position and orientation of the road. It is worth noting that since this method does not rely on the exact local

Fig. 3. Original image.

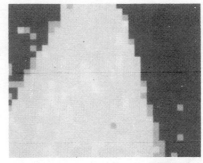

Fig. 6. Road probability image. The pixels that best match typical road colors are displayed brightest.

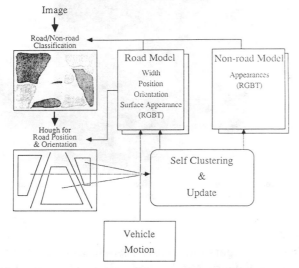

Fig. 4. Color vision for road following, including color classification, Hough transform for road region detection, and updating multiple road and nonroad models.

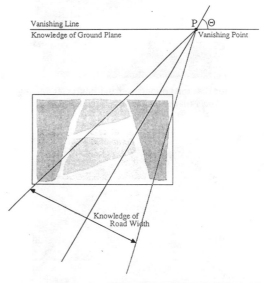

Find a good combination of (P,Θ)

Fig. 7. Road transform that considers the geometry of road position and orientation. Geometry of locally flat, straight, and parallel road regions can be described by only P and θ. Point A classified as road could be a part of the road with the shown combination of (P, θ), and thus casts a positive vote for it. Point B classified as off-road, however, will cast a negative vote for that (P, θ) combination.

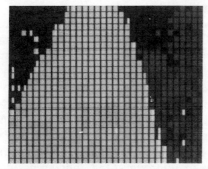

Fig. 5. Segmented image. Color and texture cues are used to label points below the horizon into two road and two off-road classes.

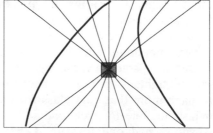

Fig. 8. A road point could be a part of roads with different orientations and vanishing points.

geometry of the road, it is very robust. The road may actually curve or not have parallel edges, or the segmentation may not be completely correct, but still the method outputs approximate road position and orientation. The main limitation is the field of view of the camera which limits the size of the parameter space. Currently, the intercept lies within the parameter space as long as the road has a curvature radius greater than 60 meters assuming a 60 degree field of view. If the road curve is sharper, the algorithm returns the closest interpretation in the parameter space.

Once the road has been found in an image, the color statistics of the road and off-road models are modified for each class by resampling the detected regions (Fig. 12) and updating the color models. The updated color statistics will gradually change as the vehicle moves into a different road color, as lighting conditions change, or as the colors of the surrounding grass, dirt, and trees vary. As

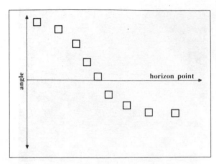

Fig. 9. The point from Fig. 8 would vote for these orientation/intercept values.

Fig. 10. Votes for best road orientation and intercept, and point with most votes (dark square), for road in Fig. 3.

Fig. 11. Detected road, from the point with the most votes shown in Fig. 10.

long as the processing time per image is low enough to provide a large overlap between images, the statistics adapt as the vehicle moves. The road is picked out by hand in the first image. Thereafter, the process is automatic, using the segmentation from each image to calculate color statistics for the next.

Actually several important additional processing steps are possible. One is to smooth the images first. This removes outliers and tightens the road and nonroad clusters. Another is to have more than one class for road and for nonroad, for instance one for wet road and one for dry, or one for shadows and one for sun. Other variations change the voting for best road. Besides adding votes for road pixels, we subtract votes for nonroad points. Votes are weighted according to how well each point matches road or nonroad classes. Finally, an image contains clues other than color, such as visual texture. Roads tend to be smooth, with less high-frequency variation than grass or leaves, as shown in Fig. 13. We calculate a normalized texture measure, and use that in addition to color in the road classification.

Fig. 12. Updating road and nonroad model colors, leaving a safety zone around the detected road region.

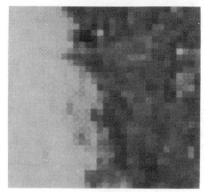

Fig. 13. Zoomed picture of road–nonroad boundary. The road (at left) is much less textured than the grass (at right).

B. Implementation

The implementation of the road following algorithm presented above runs in a loop of six steps: image reduction, color classification, texture classification, combining color and texture results, voting for road position, and color update. These steps are shown in Fig. 14, and are explained in detail below.

Image Reduction: We create a pyramid of reduced resolution R, G, and B images. Each smaller image is produced by simple 2×2 averaging of the next larger image. We found that other reduction methods, such as median filtering, are more expensive and produce no noticeable improvement in the system. We start with 480×512 pixel images, and typically use the images reduced to 30×32 for color classification. We use less reduced versions of the images for texture classification. Image reduction is used mainly to improve speed, but as a side effect the resulting smoothing reduces the effect of scene anomalies such as cracks in the pavement.

Color Classification: Each pixel (in the 30×32 reduced image) is labeled as belonging to one of the road or nonroad classes by standard maximum likelihood classification. We usually have four road and four nonroad classes. The number of classes is a compromise between having only two classes [13] (one for road and one for nonroad), which leads to poor performance in a changing environment, and having too many classes, which leads to an unreliable classification. Each class is represented by the mean R, G, and B values of its pixels, by a 3×3 covariance matrix, and by the fraction of pixels expected

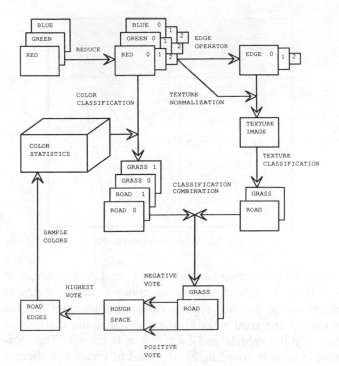

Fig. 14. Processing cycle for color vision.

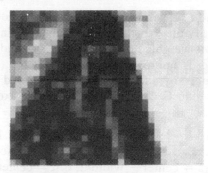

Fig. 15. Low resolution texture image, showing textures from Fig. 3. The brighter blocks are image areas with more visual texture.

a priori to be in that class. The classification procedure calculates the probability, P_i^C, that a pixel belongs to each of the classes.

Texture Calculation: The regions of paved roads tend to appear smoother in the image than nonroad regions of grass, soil, or tree trunks. The texture calculation combines a low-resolution (low-frequency) gradient image and a high-resolution (high-frequency) gradient image into a texture image. The algorithm is composed of six substeps:

• Calculate gradient at high resolution by running a Roberts operator over the 240×256 image.

• Calculate a low resolution gradient by applying a Roberts operator to the 60×64 image.

• Normalize the gradient by dividing the high resolution gradient by a combination of the average pixel value for that area and the low resolution gradient. Dividing by the average pixel value handles shadow interiors, while dividing by the low resolution gradient removes the shadow boundary.

• Normalized gradient

$$= \frac{\text{high-freq gradient}}{\alpha \times \text{low-freq gradient} + \beta \times \text{mean pixel value}}$$

Typical values for the coefficients are $\alpha = 0.2$ and $\beta = 0.8$.

• Threshold. Produce a 240×256 binary image of "microedges" by thresholding the normalized gradient. A fairly low threshold, such as 1, is usually adequate.

• Count Edges. Count the number of edges in each pixel block of 8×8 pixels. This gives a 30×32 pixel texture magnitude image. Fig. 15 shows the texture image derived from Fig. 3. Each texture pixel has a value be-

tween 0 and 255, which is the number of microedge pixels in the corresponding block of the full-resolution image.

• Texture Classification. Classify each pixel in the 30×32 image as road or nonroad on the basis of texture, and calculate a confidence P_i^T for each label. We found experimentally that a fixed mean and standard deviation for road and nonroad textures were better than adaptive texture parameters. Our best results were with road mean and standard deviation of 0 and 25, and nonroad values of 175 and 100. Effectively, any pixel block of the image with more than 35 microedges above threshold is considered textured, and is therefore classified as nonroad.

The weights and thresholds used in the texture calculation have been initially determined from a training set of 50 images.

Combination of Color and Texture Results: For each road or nonroad class i, confidence measures from color and texture are combined into one confidence P_i using the formula:

$$P_i = (1 - \alpha) P_i^T + \alpha P_i^C.$$

The weight α takes into account the fact that color is more reliable, so that the color probabilities should be weighted more than the texture probabilities. The final result is a classification of the pixels into road and nonroad, and a confidence calculated by:

$$C = Max \{ P_i, i \text{ road class} \}$$
$$- Max \{ P_i, i \text{ nonroad class} \}.$$

The final confidence C is negative for nonroad pixels, and positive for road pixels.

Vote for Best Road Position: This step uses a 2-D parameter space similar to a Hough transform. Parameter P is the column of the road's vanishing point, quantized into 32 buckets because the image on which the classification and voting are based has 32 columns. Parameter θ is the road's angle from vertical in the image, ranging from -1 to 1 radian in 0.1 radian steps. A given road point votes for all possible roads that would contain that point. The locus of possible roads whose centerlines go through that point is an arctangent curve in the parameter space. Because the road has a finite width, the arctan curve has to be widened by the width of the road at that pixel's image row. Road width for a given row is not a constant over all

405

possible road angles but is nearly constant enough that it does not justify the expense of the exact calculation. Each pixel's vote is weighted by its calculated confidence. Pixels classified as nonroad cast negative votes (with their weights reduced by a factor of 0.2) while road pixels add votes. In pseudo C code, the voting for a pixel at (row, col) is

```
for (theta = -1; theta < = 1; theta + = 0.1) {
    center = col + tan (theta) * (r - r_horizon);
    for (c = center - width/2; c < = center + width/
      2; c++) {
          parameter_space[theta] [c] + = confidence;
    }
}
```

At the end of voting, one road pair (P, θ) will have the most votes. That intercept and angle describe the best road shape in the scene.

Color Update: The parameters of the road and nonroad classes need to be recalculated to reflect changing colors. We divide the image into four regions plus a "safety zone": left off-road, right off-road, upper road, and lower road. We leave a 64-pixel wide "safety zone" along the road boundary, which allows for small errors in locating the road, or for limited road curvature. For each of the four regions, we calculate the means of red, green, and blue. We use the calculated parameters to form four classes, and reclassify the image using a limited classification scheme. Limited reclassification is based on distance from a pixel's values to the mean values of a class, rather than the full maximum likelihood scheme used in classifying a new image. This tends to give classes based on tight clusters of pixel values, rather than lumping all pixels into classes with such wide variance that any pixel value is considered likely. The limited reclassification allows road pixels to be classified as either of the two road classes, but not as nonroad, and allows nonroad pixels to be reclassified only as one of the nonroad classes. The reclassified pixels are used as masks to recalculate class statistics. Since the limited reclassification is based on the distances between pixels and mean values, the loop of classify pixels/recalculate statistics is guaranteed to converge to a classification in which no pixels can switch classes. In practice, the loop is repeated three times. The final reclassified pixels are used to calculate the means, variances, and covariances of R, G, and B for each of the classes, to be used to classify the next image.

The color update technique cannot handle sudden changes in color statistics, it is therefore important to ensure enough overlap between images. In the current implementation, the speed of the vehicle is adjusted so that there is at least 75 percent overlap between consecutive images.

Calculation of Road Position in Vehicle Coordinates: The pair (P, θ) gives the position of the road in image coordinates. The last step of the algorithm is to convert the image position into vehicle coordinates. In order to perform the conversion, the system is first cali-

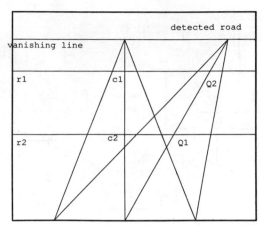

Fig. 16. Calibration procedure.

brated by computing for two rows r_1 and r_2 the number of pixels per meters, ppm_i and the column corresponding to the center of the road c_i. The calibration assumes that the width of the road w is known, and that the distance d_i between the vehicle and each row r_i is known. The location in vehicle coordinates of a road of image coordinates (P, θ), is given by the vehicle coordinates of the intersections Q_i, between the line of parameters (P, θ) and the rows r_i (Fig. 16). If Q_i is at column C_i in the image, then the distance between Q_i and the center of the vehicle is given by $x_i = (C_i - c_i)/ppm_i$.

C. Performance

We have run this algorithm on the Navlab in an outdoor environment. The failure rate is close to 0. The occasional remaining problems come from one of three causes:

• The road is covered with leaves or snow, so one road color class and one nonroad color class are indistinguishable.

• Drastic changes in illumination occur between successive pictures (e.g., the sun suddenly emerges from behind a cloud) so all the colors change dramatically from one image to the next.

• The sunlight is so bright and shadows are so dark in the same scene that we hit the hardware limits of the camera. It is possible to have pixels so bright that all color is washed out, and other pixels in the same image so dark that all color is lost in the noise.

Not every image is classified perfectly, but all are good enough to result in the detection of correct road position and orientation for navigation. We sometimes find the road rotated in the image from its correct location, so we report an intercept off to one side and an angle off to the other side. But since the path planner looks ahead about the same distance as the center of the image, the steering target is still in approximately the correct location, and the vehicle stays on the road. This algorithm runs in about 10 s per image on a dedicated Sun 3/160, using 480 × 512 pixel images reduced to 30 rows by 32 columns. We currently process a new image every 4 m, which gives about three fourths of an image overlap between images. Ten seconds is fast enough to balance the rest of the sys-

Fig. 17. Road following on a sequence of images.

tem but is slow enough that clouds can come and go and lighting conditions change between images. Fig. 17 shows the detected road on a typical sequence of nine images. This example is a typical case in which the presence of large shadow areas would preclude the use of both road edge tracking and classification based on two classes, road and nonroad.

D. Evaluation

In the course of our study of the road following problem, we have identified the following principles:

Assume Variation and Change: The appearance of a road varies from place to place and from time to time. For example, the road may be locally covered with leaves, or the lighting conditions may change in time due to cloudiness. We therefore need more than one road color model at any one time. The color models must adapt to changing conditions. In addition, we need to process images frequently so that the change from one image to the next will be moderate.

Use Few Geometric Parameters: A complete description of the road's shape in an image can be complex. The road can bend gently or turn abruptly, can vary in width, and can go up- or downhill. However, the more parameters there are, the greater the chance of error in finding those parameters. Small misclassifications in an image could give rise to fairly large errors in perceived road geometry. We found that two free parameters, orientation and distance from the vehicle, are sufficient to locally describe the road. Using only those two parameters implies that road width is fixed, the world is flat, and that the road is straight. While none of these assumptions is true over a long stretch of the road, they are nearly true within any one image; and the errors in road position that originate in oversimplifications are balanced by the smaller chance of bad interpretations.

Work in the Image: The road can be found either by projecting the road shape into the image and searching in image coordinates, or by back projecting the image onto the ground and searching in world coordinates [14]. The problem with the latter approach comes in projecting the image onto an evenly spaced grid in the world. Unless one uses a complex weighting scheme, some image pixels (those at the top that project to distant world points) will have more weight than other (lower) points. On the other hand, working directly in the image makes it much easier to weight all pixels evenly. Moreover, projecting a road shape is much more efficient than back projecting all the image pixels.

Calibrate Directly: A complete description of a camera must include its position and orientation in space, its focal length and aspect ratio, lens effects such as fisheye distortion, and nonlinearities in the optics or sensor. The general calibration problem of trying to measure each of these variables is difficult. It is much easier, and more accurate, to calibrate the whole system than to tease apart the individual parameters. The easiest method is to take a picture of a known object and build a lookup table that relates each world point to an image pixel and vice versa. Projecting road predictions into the image and back projecting detected road shapes onto the world are done by means of table lookup (or table lookup for nearby values with simple interpolations). Such a table is straightforward to build and provides good accuracy, and there are no instabilities in the calculations.

IV. PERCEPTION IN 3-D

Color vision is not sufficient for the navigation of a mobile robot. Information such as the location of obstacles requires the availability and processing of 3-D data. 3-D vision has two objectives: obstacle detection, and terrain analysis. Obstacle detection allows the system to locally steer the vehicle on a safe path. Terrain analysis provides a more detailed description of the environment which can be used for more accurate path planning or for object recognition [7].

In order to study 3-D vision for a mobile robot, we used an ERIM scanning laser range finder. The scanner produces, every half second, an image containing 64 rows by 256 columns of range values; an example is shown in Fig. 18. The scanner measures the phase difference between an amplitude-modulated laser and its reflection from a target object, which in turn provides the distance between the target object and the scanner. The scanner produces a dense range image by using two deflecting mirrors, one for the horizontal scan lines and one for vertical motion between scans. The volume scanned is 80 degrees wide and 30 degrees high. The range at each pixel is discretized over 256 levels from 0 to 64 feet.

A. Preprocessing

Our range processing begins by smoothing the data and undoing the peculiarities of the ranging geometry. The *ambiguity intervals*, where range values wrap around from 255 to 0, are detected and unfolded. Two other undesirable effects are removed by the same algorithm. The first is the presence of mixed points at the edge of an object. The second is the meaninglessness of a measurement from a surface such as water, glass, or glossy pigments. In both cases, the resulting points are in regions limited by con-

Fig. 18. Range image of two trees on flat terrain. Gray levels encode distance; nearer points are painted darker.

siderable jumps in range. We then transform the values from angle-angle-range, in scanner coordinates, to x-y-z locations. These 3-D points are the basis for all further processing.

B. Obstacle Detection and Terrain Analysis

We have two main processing modes: obstacle detection and terrain analysis. Obstacle detection starts by calculating surface normals from the x-y-z points. Flat, traversable surfaces will have vertical surface normals. Obstacles will have surface patches with normals pointed in other directions. This analysis is relatively fast, running in about 5 s on a Sun 3/75, and is adequate for smooth terrain with discrete obstacles.

Simple obstacle maps are not sufficient for detailed analysis. For greater accuracy we do more careful terrain analysis and combine sequences of images corresponding to overlapping parts of the environment into an *extended obstacle map*. The terrain analysis algorithm first attempts to find groups of points that belong to the same surface and then uses these groups as seeds for the region growing phase. Each group is expanded into a smooth connected surface patch. The smoothness of a patch is evaluated by fitting a surface (plane or quadric). In addition, surface discontinuities are used to limit the region growing phase. The complete algorithm is:

1) Edges: Extract surface discontinuities, pixels with high jumps in x-y-z.

2) Clustering: Find clusters in the space of surface normals and identify the corresponding regions in the original image.

3) Region growing: Expand each region until the fitting error is larger than a given threshold. The expansion proceeds by iteratively adding the point of the region boundary that adds the minimum fitting error.

The clustering step is designed so that other attributes such as color or curvature can also be used to find potential regions on the object. The primitive surface used to compute the fitting error can be either a plane or a quadric. The decision is based on the size of the region. Fig. 19 shows the resultant description of 3-D terrain and obstacles for the image of Fig. 18. The flat, smooth, navigable region is the meshed area, and the detected 3-D objects (the two trees) are shown as polyhedra.

Obstacle detection works at longer range than terrain analysis. When the scanner is looking at distant objects, it has a very shallow depression angle. Adjacent scanlines, separated by 0.5 degrees in the range image, can strike the ground at widely different points. Because the

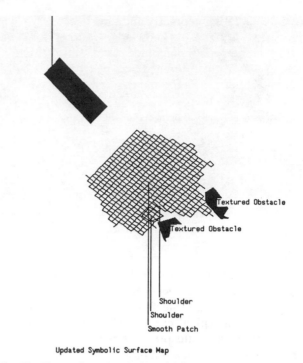

Fig. 19. The resultant description of 3-D terrain and obstacles from the image in Fig. 18. The navigable area is shown as a mesh, and the two trees are detected as "textured obstacles" and shown as black polygons.

grazing angle is shallow, little of the emitted laser energy returns to the sensor, producing noisy pixels. Noisy range values, widely spaced, make it difficult to do detailed analysis of flat terrain. A vertical obstacle, such as a tree, shows up much better in the range data. Pixels from neighboring scanlines fall more closely together, and with a more nearly perpendicular surface the returned signal is stronger and the data cleaner. It is thus much easier for obstacle detection to find obstacles than for terrain analysis to certify a patch of ground as smooth and level.

V. SYSTEM BUILDING

When computer vision is to be used as part of a larger system, the architecture of the system influences the design of the vision components. A variety of design approaches for mobile robot architectures have been discussed [2], [3], [8], [9], [15]. The main feature our approach has in common with them is the need for a distributed architecture integrating perception and navigation. We propose a system architecture called CODGER that meets the needs of the NAVLAB and provides a framework within which our vision programs are executed. The CODGER system has been designed to integrate the algorithms of Sections III and IV, and to demonstrate them on the testbed of Section II.

A. Architecture Principles for Real World Robots

Artificial Intelligence systems, including intelligent mobile robots, are symbol manipulators. In most AI systems, the symbol manipulation is based on inference, either by the logic of predicate calculus or by probabilities. In contrast, the bulk of the work of a mobile robot is based on modeling geometry and time because these are the properties utilized by knowledge sources for object modeling, motion, path planning, navigation, vehicle dynamics, and so forth. Inference may be a part of a mobile robot system, but geometry and time are pervasive. Consequently, intelligent robots need a new kind of expert system shell that provides tools for handling 3-D locations and motion.

Based on this observation, we have developed and followed the following tenets of mobile robot system design in building our system:

Modularity: Much of the deep knowledge can be limited to particular specialist modules. The effects of lighting conditions and viewing angle on the appearance of an object, for instance, are important data for color vision but are not needed by path planning. So one principle of mobile robot system design is to break the system into modules and minimize the overlap of knowledge between modules. Furthermore, modularity allows the system to have different components written in different languages or executing on specialized hardware as appropriate for each module.

Virtual Vehicle: As many as possible of the details of the vehicle should be hidden. We therefore hide the details of sensing and motion in a "virtual vehicle" interface, so a single "move" command, for instance, will use the different mechanisms of two vehicles but will produce identical behavior.

Synchronization: A system that has separate modules communicating at a fairly coarse grain will be loosely coupled. Lock-step interactions are neither necessary nor appropriate. However, there are times when one module needs to wait for another to finish, or when a demon module needs to fire whenever certain data appear. The system should provide tools for several different kinds of interaction and for modules to synchronize themselves as needed.

Real-Time Versus Symbolic Interface: A mobile robot requires a blend of high-level reasoning processes and low-level real-time processes within a single system. The high-level processes are typically event-driven and require highly variable amounts of time depending on the vehicle's state and the environment. The low-level processes, however, typically control vehicle functions such as locomotion and must run in real time. The system should provide for both real-time and asynchronous symbolic processes, and for communications between them.

Geometry and Time: Much of the knowledge that needs to be shared between modules has to do with geometry, time, and motion. An object may be predicted by one module (the lookout), seen separately by two others (color vision and 3-D perception), and used by two more (path planner and position update). During the predictions, sensing, and reasoning, the vehicle will be moving, new position updates may come in, and the geometrical relationship between the vehicle and the object will be constantly changing. Moreover, there may be many different frames of reference: one for each sensor, one for the ve-

hicle, one for the world map, and others for individual objects. Each module should be able to work in the coordinate frame that is most natural; for instance, a vision module should work in camera coordinates and should not have to worry about conversion to the vehicle reference frame. The system should provide tools that handle as many as possible of the details of keeping track of coordinate frames, motion, and changing geometry.

Distributed Control: In some systems (notably early blackboards) a single master process "knows" everything. The master process may not know the internal working of each module, but it knows what each module is capable of doing. The master controls who gets to run when. The master itself becomes a major AI module and can be a system bottleneck. In contrast, the individual modules in a mobile robot system should be autonomous, and the system tools should be slaves to the modules. The module writers should be free to decide when and how to communicate and when to execute.

We have followed these tenets in building the Navlab system. At the bottom level, we have built the CODGER "whiteboard" to provide system tools and services [10]. On top of CODGER we have built an architecture that sets conventions for control and data flow. CODGER and our architecture are explained below.

B. Blackboards and Whiteboards

The program organization of the NAVLAB software is shown in Fig. 20. Each of the major boxes represents a separately running program. The central database, called the *Local Map*, is managed by a program known as the Local Map Builder (*LMB*). Each module stores and retrieves information in the database through a set of subroutines called the *LMB Interface* which handle all communication and synchronization with the LMB. If a module resides on a different processor than the LMB, the LMB and LMB Interface will transparently handle the network communication. The Local Map, LMB, and LMB Interface together comprise the CODGER (COmmunications Database with GEometric Reasoning) system.

The overall system structure—a central database, a pool of knowledge-intensive modules, and a database manager that synchronizes the modules—is characteristic of a traditional blackboard system. Such a system is called "heterarchical" because the knowledge is scattered among a set of modules that have access to data at all levels of the database (i.e., low-level perceptual processing ranging up to high-level mission plans) and may post their findings on any level of the database. We call CODGER a *whiteboard* because it implements a heterarchical system structure, but it differs from a blackboard in several key respects. In CODGER, each module is a separate, continuously running program; the modules communicate by storing and retrieving data in the central database. Synchronization is achieved by primitives in the data retrieval facilities that allow, for example, for a module to request data and suspend execution until the specified data appears. When some other module stores the desired data,

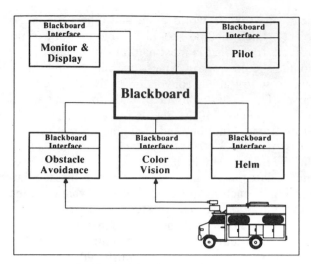

Fig. 20. Navlab software architecture.

the first module will be reactivated and the data will be sent to it. With CODGER a module programmer thus has control over the flow of execution within his module and may implement real-time loops, demons, data flows among cooperating modules, etc. Like other recent distributed AI architectures, whiteboards are suited to execution on multiple processors.

C. Data Storage and Retrieval

Data in the CODGER database (Local Map) is represented in *tokens* consisting of classical *attribute-value pairs*. A module can store a token by calling a subroutine to send it to the LMB. Tokens can be retrieved by constructing a pattern called a *specification* and calling a routine to request that the LMB send back tokens matching that specification. The specification is simply a Boolean expression in which the attributes of each token may be substituted; if a token's attributes satisfy the Boolean expression, then the token is sent to the module that made the request. For example, a module may specify:

> *tokens with* **type** *equal to "intersection" and* **traffic-control** *equal to "stop-sign"*

This would retrieve all tokens whose **type** and **traffic-control** attributes satisfy the above conditions. The specification may include computations such as mathematical expressions, finding the minimum value within an array attribute, comparisons among attributes, etc. CODGER thus implements a general database. The module programmer constructs a specification with a set of subroutines in the CODGER system.

One of the key features of CODGER is the ability to manipulate geometric information. One of the attribute types provided by CODGER is the *location*, which is a 2-D or 3-D polygon and a reference to a *coordinate frame* in which that polygon is described. Every token has a specific attribute that tells the location of that object in the Local Map, if applicable, and a specification can include geometric calculations and expressions. For example, a

specification might be

> *tokens with* **location** *within 5 units of (45, 32) [in world coordinates]*

or

> *tokens with* **location** *overlapping X*

where X is a description of a rectangle on the ground in front of the vehicle. The geometric primitives currently provided by CODGER include calculation of centroid, area, diameter, convex hull, orientation, and minimum bounding rectangle of a location, and distance and intersection calculations between a pair of locations.

CODGER also provides for automatic coordinate system maintenance and transformation for these geometric operations. In the Local Map, all coordinates of location attributes are defined relative to **WORLD** or **VEHICLE** coordinates; **VEHICLE** coordinates are parameterized by time, and the LMB maintains a time-varying transformation between **WORLD** and **VEHICLE** coordinates. Whenever new information (i.e., a new **VEHICLE**-to-**WORLD** transform) becomes available, it is added to the "history" maintained in the LMB; the LMB will interpolate to provide intermediate transformations as needed. In addition to these basic coordinate systems, the LMB Interface allows a module programmer to define *local coordinates* relative to the basic coordinates or relative to some other local coordinates. Location attributes defined in a local coordinate system are automatically converted to the appropriate basic coordinate system when a token is stored in the database. CODGER provides the module programmer with a conversion routine to convert any location to any specified coordinate system.

D. Module Architecture

Several modules use the CODGER tools and fit into a higher level architecture. The modules are:

- *Pilot:* Looks at the map and at current vehicle position to predict road location for Vision. Plans paths.
- *Map Navigator:* Maintains a world map, does global path planning, provides long-term direction to the Pilot. The world map may start out empty, or may include any level of detail up to exact locations and shapes of objects.
- *Color Vision:* Waits for a prediction from the Pilot, waits until the vehicle is in the best position to take an image of that section of the road, returns road location.
- *Obstacle Detection:* Gets a request from the Pilot to check a part of the road for obstacles. Returns a list of obstacles on or near that chunk of the road.
- *Helm:* Gets planned path from Pilot, converts polyline path into smooth arcs, steers vehicle.
- *Graphics and Monitor:* Draws or prints position of vehicle, obstacles, predicted and perceived road.

These modules use CODGER to pass information about *driving units*. A driving unit is a short chunk of the road or terrain (in our case 4 m long) treated as a unit for perception and path planning. The need for driving units comes from the limited field of view of the sensors. The environment must be divided into chunks small enough to be viewed in one frame of each sensor. The sizes and positions of the driving units is computed based on the field of view of the sensors and the speed of the vehicle. The Pilot gives driving unit predictions to Color Vision, which returns an updated driving unit location. Obstacle Detection then sweeps a driving unit for obstacles. The Pilot takes the driving unit and obstacles, plans a path, and hands the path off to the Helm. The whole process is set up as a pipeline, in which Color Vision is looking ahead 3 driving units, Obstacle Detection is looking 2 driving units ahead, and path planning at the next unit. If for any reason some stage slows down, all following stages of the pipeline must wait. So, for instance, if Color Vision is waiting for the vehicle to come around a bend so it can see down the road, Obstacle Detection will finish its current unit and will then have to wait for Color Vision to proceed. In an extreme case, the vehicle may have to come to a halt until everything clears up. All planned paths include a deceleration to a stop at the end, so if no new path comes along to overwrite the current path the vehicle will stop before driving into an area that has not been seen or cleared of obstacles.

VI. Conclusions and Future Work

The system described has successfully driven the Navlab many tens of times, processing thousands of color and range images without running off the road or hitting any obstacles. CODGER has proved to be a useful tool, handling many of the details of communications and geometry. Module developers have been able to build and test their routines in isolation, with relatively little integration overhead. Yet there are several areas that need much more work.

We drive the Navlab at 10 cm/s, a slow shuffle. Our slow speed is because our test road is narrow and winding, and because we deliberately concentrate on competence rather than on speed. But faster motion is always more interesting, so we are pursuing several ways of increasing speed. One bottleneck is the computing hardware. We have mounted a Warp, Carnegie Mellon's experimental high-speed processor, on the Navlab. We have implemented the vision and range algorithms of the Warp machine, allowing us to drive the vehicle at 50 cm/s. At the same time, we are looking at improvements in the software architecture. We need a more sophisticated path planner, and we need to process images that are more closely spaced than the length of a driving unit. Also, as the vehicle moves more quickly, our simplifying assumption that steering is instantaneous and that the vehicle moves along circular arcs becomes more seriously flawed. We are looking at other kinds of smooth arcs, such as clothoids. More important, the controller is evolving to handle more of the low-level path smoothing and following.

One reason for the slow speed is that the Pilot assumes straight roads. We need to have a description that allows for curved roads, with some constraints on maximum curvature. The next steps will include building maps as we

go, so that subsequent runs over the same course can be faster and easier. Travel on roads is only half the challenge. The Navlab should be able to leave roads and venture cross-county. Our plans call for a fully integrated on-road/off-road capability. Current vision routines have a built-in assumption that there is one road in the scene. When the Navlab comes to a fork in the road, vision will report one or the other of the forks as the true road depending on which looks bigger. It will be important to extend the vision geometry to handle intersections as well as straight roads. We already have this ability on our sidewalk system and will bring that over to the Navlab. Vision must also be able to find the road from off-road. Especially as we venture off roads, it will become increasingly important to be able to update our position based on sighting landmarks. This involves map and perception enhancements, plus understanding how to share limited resources, such as the camera, between path finding and landmark searches.

Acknowledgment

The Navlab was built by W. Whittaker's group in the Field Robotics Center, and the Warp group is led by H. T. Kung and J. Webb. The real work gets done by an army of eight staff, nine graduate students, five visitors, and three part time programmers.

References

[1] M. Annaratone, F. Bitz, J. Deutch, L. Hamey, H. T. Kung, P. C. Maulik, P. S. Tseng, and J. A. Webb, "Applications experience on Warp," in *Proc. AFIPS Nat. Comput. Conf.*, 1987.
[2] R. Bhatt, D. Gaw, and A. Meystel, "A real-time guidance system for an autonomous vehicle," in *Proc. IEEE Int. Conf. Robotics and Automation*, 1987.
[3] R. A. Brooks, "A robust layered control system for a mobile robot," *IEEE J. Robotics and Automation*, vol. RA-2, no. 1, 1986.
[4] L. S. Davis, T. R. Kushner, J. Le Moigne, and A. M. Waxman, "Road boundary detection for autonomous vehicle navigation," *Opt. Eng.*, vol. 25, no. 3, Mar. 1986.
[5] D. DeMenthon, "Inverse perspective of a road from a single image," Center for Automation Research, Univ. Maryland, College Park, Tech. Rep. 210, July 1986.
[6] R. O. Duda and P. E. Hart, *Pattern Classification and Scene Analysis.* New York: Wiley, 1973.
[7] M. Hebert and T. Kanade, "Outdoor scene analysis using range data," in *Proc. IEEE Int. Conf. Robotics and Automation*, 1986.
[8] D. Kuan and U. K. Shorna, "Model-based geometric reasoning for autonomous road following," in *Proc. IEEE Int. Conf. Robotics and Automation*, 1987.
[9] M. Parodi, J. J. Nitao, and L. S. McTamaney, "An intelligent system for an autonomous vehicle," in *Proc. IEEE Int. Conf. Robotics and Automation*, 1986.
[10] S. Shafer, A. Stentz, and C. Thorpe, "An architecture for sensor fusion in a mobile robot," in *Proc. IEEE Int. Conf. Robotics and Automation*, 1986.
[11] J. Singh *et al.*, "NavLab: An autonomous vehicle," Carnegie Mellon Robotics Inst., Pittsburgh, PA, Tech. Rep., 1986.
[12] C. Thorpe, "Vision and navigation for the CMU Navlab," in *Proc. SPIE*, Oct. 1986.
[13] M. A. Turk, D. G. Morgenthaler, K. D. Gremban, and M. Marra, "Video road-following for the autonomous land vehicle," in *Proc. IEEE Int. Conf. Robotics and Automation*, 1987.
[14] R. Wallace, K. Matsuzaki, Y. Goto, J. Crisman, J. Webb, and T. Kanade, "Progress in robot road-following," in *Proc. IEEE Int. Conf. Robotics and Automation*, 1986.
[15] A. M. Waxman, J. J. LeMoigne, L. S. Davis, B. Srinivasan, T. R. Kushner, E. Liang, and T. Siddalingaiah, "A visual navigation system for autonomous land vehicles," *IEEE J. Robotics and Automation*, vol. RA-3, no. 2, 1987.

Charles Thorpe received the B.A. degree in natural science from North Park College, Chicago, IL, in 1979, and the Ph.D. degree from Carnegie-Mellon University, Pittsburgh, PA, in 1984. His doctoral dissertation described FIDO, a stereo vision and navigation system for a robot rover.

He is a Research Scientist in the Robotics Institute of Carnegie-Mellon University. He is one of the Principal Investigators on the Navlab project. His research focuses on perception, and on the interactions between perception and planning in mobile robots.

Dr. Thorpe is a member of the Association for Computing Machinery and the American Institute of Aeronautics and Astronautics.

Martial H. Hebert received the Master's degree and Doctorate degree in computer science from the University of Orsay, France, in 1981 and 1984, respectively.

He is a Research Scientist at the Robotics Institute at Carnegie-Mellon University, Pittsburgh, PA. His past experience includes working as a Research Scientist at the Institut National de Recherche en Informatique et Automatique from 1982 to 1984. His current research includes 3-D vision for an autonomous land vehicle.

Takeo Kanade (M'80) received the Ph.D. degree in information science from Kyoto University, Japan, in 1974.

He is Professor of Computer Science and Robotics at Carnegie-Mellon University, Pittsburgh, PA. Currently he is Acting Director of the Robotics Institute. Before his 1980 appointment with Carnegie-Mellon, he was Associate Professor of Information Science at Kyoto University. He has worked on several areas in robotics and artificial intelligence: theoretical and practical aspects of computer vision, 3-D range sensing and analysis, development and control of new direct-drive arms which were first conceived and prototyped at CMU (DD Arm I and DD Arm II), and mobile robot systems. He has authored and edited three books and over sixty papers and technical reports in these areas. His current research includes: the Navlab (a van with on-board sensors and computers) vision system, the Mars Rover development, and Reconfigurable Modular Manipulator System.

Dr. Kanade served as a General Chairman of IEEE International Conference on Computer Vision and Pattern Recognition in 1983, and a Vice Chairman of IEEE International Conference on Robotics and Automation in 1986. He is the Editor of the *International Journal of Computer Vision*.

Steven A. Shafer (M'84) received the B.A. degree in computer science from the University of Florida in 1977 and the Ph.D. degree in computer science from Carnegie-Mellon University, Pittsburgh, PA, in 1983.

He is a Research Scientist studying computer vision and mobile robots in the Department of Computer Science of Carnegie-Mellon University, where he has been working since 1983. He is primarily interested in the analysis of images by computer, using optical properties of illumination, objects, and cameras. By analyzing properties such as color, gloss, and shadows, he is developing methods to base computer vision on an understanding of physics rather than the current ad hoc statistical and pattern classification techniques. His view of computer vision as a problem in measurement of physical quantities is also reflected in his work on geometric, photometric, and spectral calibration of cameras. This work is carried out in the Calibrated Imaging Laboratory at CMU, and is primarily oriented toward robotics tasks in navigation and object manipulation. He is also studying architectures for mobile robot perception, planning, and control. He is one of the authors of CODGER, the distributed blackboard system used by the NAVLAB autonomous robot van at Carnegie-Mellon University, and has been active in designing the software for control of the NAVLAB. He is also coauthor of the PHAROS "microscopic" traffic simulator, which will be used to develop a computer program to control a robot vehicle driving through traffic. He is conducting the effort at CMU to integrate into the NAVLAB system modules that are being developed at other universities and research labs. In addition to his research work, he has been active in the enhancement of the UNIX operating system at Carnegie-Mellon. He is the author and maintainer of over one hundred system programs and library subroutines, including a relational database manager (DAB) and a software upgrade program (SUP). He is the author of several software reference manuals and instructional documents, and has taught the use of UNIX at CMU and elsewhere.

Dr. Shafer has been active in professional activities in the fields of computer vision and optics (appearance measurement). He has produced two programs to perform digital image segmentation, one each at Carnegie-Mellon and at the University of Hamburg in Germany; the former program (PHOENIX) is a component of the Darpa Image Understanding Testbed compiled by SRI International. He is the author of one book and numerous papers, appearing in both the computer science and optics literature, and served as a consultant for the *Handbook of Artificial Intelligence*, the *Encyclopedia of Artificial Intelligence*, and the Time-Life book series *Understanding Computers*. He has taught several courses in computer science, including computer vision, and organized and instructed in the CMU Robotics Institute's Tutorial on Computer Vision. He is a member of professional societies for computer science, optics, and color science; he is chairman of Committee 42 of the Inter-Society Color Council, "Terminology for the Optics of Materials."

Road Boundary Detection in Range Imagery for an Autonomous Robot

UMA KANT SHARMA, MEMBER, IEEE, AND LARRY S. DAVIS, MEMBER, IEEE

Abstract—This paper describes a road following system for an autonomous land vehicle, based on range image analysis. The system is divided into two parts: low-level data-driven analysis, followed by high-level model-directed search. The sequence of steps performed in order to detect three-dimensional (3-D) road boundaries is as follows: Range data are first converted from spherical into Cartesian coordinates. A quadric (or planar) surface is then fitted to the neighborhood of each range pixel, using a least square fit method. Based on this fit, minimum and maximum principal surface curvatures are computed at each point to detect edges. Next, using Hough transform techniques, 3-D local line segments are extracted. Finally, model-directed reasoning is applied to detect the road boundaries.

I. INTRODUCTION

IN SUPPORT of the Autonomous Land Vehicle (ALV) project, which is a part of DARPA's Strategic Computing Program [1], the Computer Vision Laboratory of the University of Maryland has designed a general framework for visual navigation over terrain and roadways [2]–[4]. The objective is to endow a mobile robot with the intelligence required to sense and perceive that part of its surroundings necessary to support navigational tasks. A modular system has been designed and implemented which is currently able to drive a camera over a scale model road network strictly completely under the control of visual analysis of conventional video television imaging. No *a priori* map is provided. The vehicle maintains continuous motion, alternatively "looking ahead" and then "driving blind" for a short distance before taking another view. This paper describes a road following system based on range data analysis which is organized along very similar lines.

The navigation system described in [4] distinguishes between two different modes of processing. Generally, that system begins a task in the bootstrap mode, which requires processing an entire scene and picking out the objects of interest such as roads or landmarks. Once objects of interest are identified, the system switches to a prediction/verification mode called feed-forward, in which the location of an object as seen from a new vehicle position is estimated, thus focusing

Manuscript received October 21, 1986; revised January 29, 1988. Part of the material in this paper was presented at the SPIE Mobile Robot Conference, Cambridge, MA, October 1986. This work was supported by the Defense Advanced Research Projects Agency and the U.S. Army Night Vision and Electro-Optics Laboratory under Contract DAAK70-83-K-0018 (DARPA Order 3206).

U. K. Sharma was with the Center for Automation Research, University of Maryland, College Park, MD 20742. He is now with the Artificial Intelligence Center, FMC Corporation, Santa Clara, CA 95052.

L. S. Davis is with the Center for Automation Research, University of Maryland, College Park, MD 20742.

IEEE Log Number 8822826.

attention on a small part of the visual field. The structure of this system is shown in Fig. 1 from [4]. Unlike the system described in [4] range data are used here as the sensor input instead of intensity data. Fig. 2 shows the architecture of the range data analysis system described in this paper.

Range data provide an alternative to passive video imaging for the road detection task in the case when the road can, in principle, be geometrically separated from the background, e.g., when the road is bounded by ditches, sidewalks, etc. However, for many roads there is no such geometric separation (e.g., a road adjacent to a parking lot); in such cases, photometric segmentation techniques must be relied upon. The most important advantage of active sensors for ranging, as opposed to passive techniques such as stereo or motion, is that range acquisition is much less dependent on illumination and reflectivity, and the intensity image problems associated with shadows and surface markings do not occur. Given that high-quality range data should be widely available in the near future, it is prudent to explore the use of depth maps as sensor input to an object recognition system. A significant amount of research has already been performed on range data analysis [5]–[9].

The three-dimensional (3-D) boundary recognition process described in this paper is divided into two parts: low-level data-driven analysis followed by high-level model-directed search. The steps involved in low-level processing (surface fitting, primitive feature extraction, and 3-D linear feature extraction) are described in Section II. Section III, on high-level processing, describes reasoning and verification parts. Section IV contains the conclusions reached from this research and discusses future directions.

II. LOW-LEVEL VISION

The sequence of steps used to detect 3-D segments is as follows. Range data are first converted into the standard Cartesian coordinates based on the sensor model. The second step is to fit a surface (quadric or planar) to the neighborhood of each range pixel in the image. Based on that surface model we compute, at each range pixel, the magnitudes and directions of the minimum and maximum principal surface curvatures. Points having a sufficiently high principal surface curvature are regarded as edges and the direction of this principal curvature yields edge orientation. A Hough transform application is first used to detect two-dimensional (2-D) global lines that, after smoothing and editing, are divided into 2-D line segments. Further processing in 3-space restricts these 2-D segments to 3-D segments that are passed to the

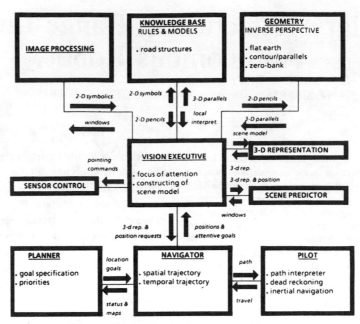

Fig. 1.   Maryland navigation system architecture.

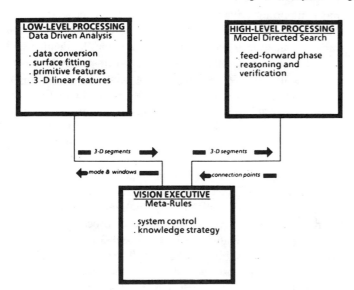

Fig. 2.   System architecture for road recognition in range data.

high-level vision algorithms. Related ideas on feature extraction relevant to the navigation problem have been described in [10], [11], [20].

A. Surface Fitting

The problem of fitting a functional model to a set of data points is a common one. Model fitting methods try to find the "best" fit, that is, they minimize some error. We fit a general quadratic model to sets of data points using a "best" fit method [12], [13].

The general quadratic equation at point (x_i, y_i) can be written in the following form:

$$z_i = a + bx_i + cy_i + dx_i^2 + ex_i y_i + fy_i^2. \qquad (1)$$

We want to fit this quadratic surface to a neighborhood in the image. This cannot be done exactly; instead, we find the surface for which the squared error in fit is minimum. The squared error of the approximation at (x_i, y_i) is defined to be

$$E^2 = (a + bx_i + cy_i + dx_i^2 + ex_i y_i + fy_i^2 - z_i)^2. \qquad (2)$$

If we have a neighborhood of k points, then the total squared error for the region is

$$E^2 = \sum_{i=1}^{k} (a + bx_i + cy_i + dx_i^2 + ex_i y_i + fy_i^2 - z_i)^2. \qquad (3)$$

We are interested in calculating the values of parameters a, b, c, d, e, and f for which the total squared error is minimum. We find these parameters by setting the derivatives of squared error with respect to the parameters to zero. This gives six equations in six unknowns. These equations can easily be solved for these parameters.

B. Edge Detection Using Surface Curvature

When detecting edges in grey-level images the scene is usually assumed to consist of regions of more or less constant intensity separated by discontinuities [12], [13], although more complex models have, of course, also been used. Gray-level edges can therefore be detected by measuring the gradient. In range data, on the other hand, the image should be modeled in space at least as a piecewise-planar surface. Because the planes can have any orientation, the gradient is not a particularly useful measure for detecting edges. Therefore, we first decide what kind of mathematical features characterize the underlying depth map function in ways that are *invariant* to changes in viewpoint which preserve the overall visibility of the surface [5].

The surface characteristic which we have chosen is the surface curvature. Surface curvature has several invariant properties; the most important of these are the following:

1) If we define a surface as the locus of a point whose coordinates are functions of two independent variables (or

parameters) u and v, then surface curvature is invariant under u, v parameterization.

2) It is also invariant under 3-D translations and rotations.

We first describe how to estimate the principal curvatures at each point in a range image, and then how these curvature values are used to detect edges. The explicit parametric form of a surface S with respect to some reference coordinate system (u, v) can be written as [14], [15]

$$S = \{(x, y, z) : x = h(u, v), y = g(u, v), z = f(u, v)\}. \quad (4)$$

We refer to this general parametric representation as $x(u, v)$ where the x-component of the x function is $h(u, v)$, the y-component of x is $g(u, v)$, and the z-component is $f(u, v)$.

There are two fundamental forms that are usually considered in the classical analysis of smooth surfaces which are referred to as the first and second fundamental forms of a surface. Let the coefficients of the equation of the first fundamental form be E, F, and G, and those for the second fundamental form be L, M, and N. The surface can be uniquely described by E, F, G, L, M, and N which uniquely determine surface shape and intrinsic surface geometry. At each point on the surface there is a direction of minimum curvature and a direction of maximum curvature for all space curves which lie on the surface and pass through that point. In classical differential geometry, a number κ is a principal curvature at each point P in the direction du/dv if and only if κ, du, and dv satisfy

$$(L - \kappa E) du + (M - \kappa F) dv = 0$$

$$(M - \kappa F) du + (N - \kappa G) dv = 0 \quad (5)$$

where

$$du^2 + dv^2 \neq 0.$$

The above is a homogeneous system of equations and will have a nontrivial solution du, dv if and only if

$$\det \begin{pmatrix} L - \kappa E & M - \kappa F \\ M - \kappa F & N - \kappa G \end{pmatrix} = 0 \quad (6)$$

or, expanding,

$$(EG - F^2)\kappa^2 - (EN + GL - 2FM)\kappa + (LN - M^2) = 0. \quad (7)$$

It can easily be shown that the discriminant of (7) is greater than or equal to zero. Thus the equation has either two distinct real roots κ_1, κ_2, the principal curvatures at a nonumbilical point, or a single real root κ with multiplicity two, the curvature at an umbilical point. The direction dv/du of the principal curvature can be computed from (5) and can be written as

$$\frac{dv}{du} = -\frac{(L - \kappa E)}{(M - \kappa F)}. \quad (8)$$

The values of E, F, G, L, M, and N can be calculated using the parameters of the quadratic equation (i.e., b, c, d, e, and f) as described in the previous section and from these we can obtain κ_1 and κ_2.

An alternative to fitting a quadratic model to each neighborhood is to fit a planar model and then estimate the surface curvatures by partially differentiating the coefficients of the planar model. This method of curvature approximation requires less computation than the quadratic model and works well at least for simple objects such as roads. The equation for a planar surface can be written in the form

$$z_i = a + bx_i + cy_i. \quad (9)$$

To calculate the principal surface curvatures at point x_i, y_i we need to compute six fundamental form coefficient functions. However, the second-order terms do not appear in (9). We can, however, calculate a, b, and c for a planar surface by the least square fit method, and the values of d, e, and f can then be estimated from the values of b and c.

For detecting edges, we determine the value of the larger of the two absolute values of κ_1 and κ_2 and its direction. The larger curvature magnitude is compared with a preset threshold value. If the curvature value exceeds this threshold, the range pixel is marked as an edge pixel and the direction of this curvature yields the edge orientation. We have selected a threshold value equal to one twentieth of the maximum curvature value in the image.

Fig. 3 contains gray-scale images of two different depth maps of roads. Fig. 3(a) is a range image obtained using the ERIM range scanner [19] at the Martin Marietta test site and Fig. 3(b) is a synthetic range image. Both of these images are 128×128 8-bit images with field of view 60° square and range resolution 1.7 in. The ERIM scanner determines the range from the scanner to the scene point for each pixel by measuring the transmit time of a modulated laser beam. The transmit time is derived by measuring the phase difference between the reference and reflected signals which corresponds to the range from the scanner to the target. Since only the phase shift is measured, the resulting values are relative instead of absolute measurements. That is, two points separated by a length equal to a complete phase shift have the same range value. The critical length is called the ambiguity interval and was equal to 37.3 ft. The frame rate was 2 per second. The results of applying both the planar and quadratic model-based edge detector to these images are shown in Fig. 4. The least square method is applied using a 5×5 neighborhood. The magnitude of curvature, encoded in gray, is shown for both the real and synthetic range images. Fig. 4(a) shows the result for the real image using quadratic approximation, while Fig. 4(d) shows the quadratic approximation for a synthetic image. The result of using the planar approximation for real and synthetic images are shown in Fig. 4(b) and (c), respectively. Figs. 5 and 6 show the result of applying edge detection using the planar approximation model on neighborhoods of different sizes. The magnitude of the curvature of edge points, encoded in gray, is shown.

C. Linear Feature Extraction

The image processing algorithms that extract the dominant linear features are developed under the assumption that roadways generate such features in a scene. We are not interested in all linear features in a scene, only in the dominant

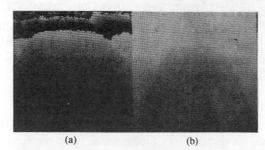

(a)   (b)

Fig. 3.   Range images of two different roads. (a) Real road. (b) Synthetic road.

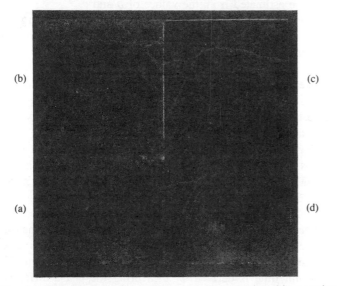

(b)   (c)

(a)   (d)

Fig. 4.   Edge detection using surface curvature. (a) Real road image using quadratic approximation. (b) Real road image using planar approximation. (c) Synthetic road image using planar approximation. (d) Synthetic road image using quadratic approximation.

(b)   (c)

(a)   (d)

Fig. 5.   Edge detection applied to the real image with surface fitting within different sizes of windows. Clockwise from the upper left 5 × 5, 9 × 9, 13 × 13, and 17 × 17 windows.

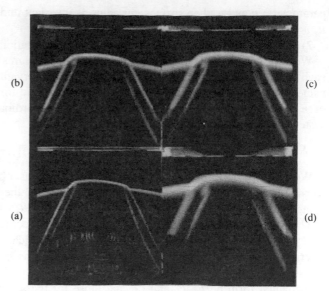

(b)   (c)

(a)   (d)

Fig. 6.   Edge detection applied to the synthetic image with surface fitting within different sizes of windows. Clockwise from the lower left 5 × 5, 9 × 9, 13 × 13, and 17 × 17 windows.

ones. These features are usually correlated with road boundaries and shoulders, all of which are important for navigating on roads. It would be possible, of course, to directly extract the 3-D line segments using Hough transform techniques applied to the 3-D edges computed in Section II-B. However, this would involve estimating more parameters and hence would require more computation than first extracting 2-D global lines. Since all 3-D lines will project into 2-D lines none of the global 3-D lines will be missed using this approach. The 2-D global lines will be segmented, based on edge density, into 2-D local line segments. These 2-D local line segments are further processed in 3-D space to extract 3-D local line segments. The following sequence of steps detects 2-D global, 2-D local, and 3-D local line segments.

1) The Hough transform [16] is used to find 2-D global lines. Each range edge element in the picture votes for all of the lines that could have passed through it. The voting takes place in a 2-D parameter plane, where each line is represented as a point. The implementation involves a quantization of the parameter plane into a rectangular grid. The grid size is determined by the acceptable errors in the parameter values and the quantization is confined to a specific region of the parameter plane determined by the range of parameter values. A 2-D array (the accumulator array or the Hough array) is then used to represent the parameter plane grid, where each array entry corresponds to a grid cell. For each range edge element, the algorithm increments the votes in all accumulator array entries that correspond to lines passing through that edge element. The parameters selected to represent lines in Hough space are r and θ, where lines are represented as

$$r = x \cos \theta + y \sin \theta. \qquad (10)$$

Here, θ is the angle of the line's normal and r is the algebraic distance of the line from the origin. It is quite simple to calculate r, because x and y are given at each point and θ is the direction of the principal curvature at this point. Fig. 7(a) corresponds to the edge detection algorithm applied to the real

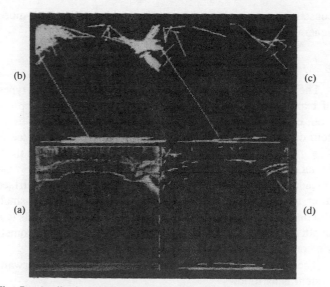

Fig. 7. Application of Hough transform to detect 2-D local line segments. (a) Edge detection applied to the real range image with surface fitting within a 13 × 13 window. (b) All the lines represented by nonempty cells of the Hough accumulator. (c) Selected peak lines. (d) 2-D local line segments obtained after editing the 2-D global lines.

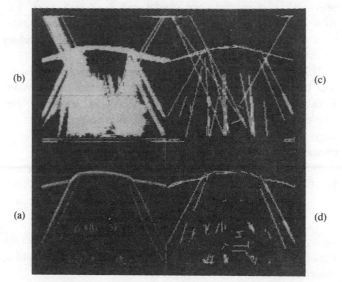

Fig. 8. Same as Fig. 7 but surface fitting within a 5 × 5 window in the synthetic image.

range image and Fig. 8(a) to the edge detection algorithm applied to the synthetic range image. Figs. 7(b) and 8(b) show all the lines corresponding to nonempty cells of the Hough accumulator. To extract the lines in an image, the Hough parameter space is searched for peaks which lie above a given threshold. These are assumed to correspond to lines in the image and participate in further processing. Figs. 7(c) and 8(c) show only the peak lines corresponding to Figs. 7(a) and 8(a), respectively.

2) The peak lines are segmented and "edited" based on the continuity of the edge points along the line. To do this, starting from the first endpoint of a peak line, we track the subsequent points on this line until we encounter a non-segment (NS) point. All other points are called segment-continuing (SC) points. A point is an SC point if any of the following

conditions holds:

a) The point has above-threshold curvature magnitude, and direction close to the global line.
b) Some point in its 3 × 3 neighborhood is an edge point (i.e., it has sufficiently high curvature magnitude) with direction, θ sufficiently close to the direction of the global line.
c) The direction θ at the point is close to the direction of the global line and at least one point in its 3 × 3 neighborhood has an above-threshold curvature magnitude.
d) The curvature magnitude at the point is above threshold and at least one point in its 3 × 3 neighborhood has an edge direction θ sufficiently close to the direction of the global line.

Any point not satisfying any of a)–d) is an NS point. The 2-D local lines are shown in Figs. 7(d) and 8(d).

3) Each 2-D local line is segmented into one or more 3-D local line segments. The algorithm employed is essentially the one described by Ramer [17] for piecewise-linear approximation of the plane curves.

a) A line is fit in 3-D space to all of the range pixels in the 2-D local line. Notice that although we could incorporate the edge information (θ and curvature magnitude) in this fitting process, for computational simplicity we do not.
b) Let d be the distance of the point P_1 on the 2-D local line that is furthest from the fitted 3-D line (cf. Fig. 9). If d is greater than a given threshold for a fitted 3-D line, the 2-D local line is split into two pieces at P_1 and the procedure is applied recursively to each piece. If d is less than a given threshold then the procedure terminates for the current segment.
c) Finally, each segment computed above is partitioned into convex and concave sets. The convexity/concavity of a range edge pixel is determined by the signs of its principal curvatures.

III. HIGH-LEVEL VISION

Our road following algorithms are based on the assumption of road continuity in the world. In order to navigate roadways, processing is focussed on the road boundaries and therefore only dominant linear features, as described in the preceding section, are extracted. We represent a road model as consisting of four attributes: i) the geometry of the left boundary, ii) the geometry of the right boundary, iii) the distance between the left and right boundaries, and iv) the shape of the road (orientations of the boundaries). All of these attributes can be verified by determining the individual and combined properties of the 3-D linear segments. The geometry of a boundary of a road can be modeled as a depression, elevation, ditch, or cliff. The prior knowledge concerning road shape may specify that a road is straight which can be verified by the orientations of the detected road boundary segments.

In the next subsection we describe the control structure for a "feed-forward" road tracking algorithm that utilizes such a geometrical model of road structure. A less restricted geomet-

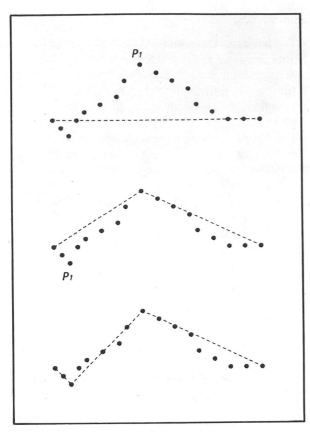

Fig. 9. Recursive line fitting.

rical model for road boundary recognition in color imagery is described in [21]. Section III-B describes the model-directed road detection process used by the feed-forward control structure.

A. Feed-Forward Processing

As mentioned previously, feed-forward implies a predictive capability in which visual processing of a prior road segment, taken together with a dead reckoning capability, is used to predict the approximate locations of important road features in a subsequent image; that is, a focus of visual attention is provided. In other words, if the vehicle knows approximately where it is in relation to the world model, it can predict (i.e., focus visual attention) on those parts of the image predicted to contain important road features. The natural continuity of roads in the world allows road tracking through the image and model updating at much higher rates than would otherwise be possible. The predictions in the feed-forward system are represented as windows in the image to be analyzed, along with a symbolic representation of the expected properties of road features that should be contained in these windows.

The task of window placement is decomposed into two cases: i) predicting the locations of the first windows near the bottom of an image, using the previously constructed world model and dead-reckoning data, and ii) predicting the locations of subsequent windows in the same image based upon the positions and orientations of the road features detected in the previous window. The size of the windows should be based on the accuracy of the dead reckoning system, the range sensor, and range data analysis. We have, in our experiments, used a constant size 32 × 32 window on both sides of the road; more generally, a variable window size should be used.

B. Model-Directed Road Detection

The system should have available to it models for the types of expected road boundary geometry, for example, depression, elevation, ditch, cliff, or flat. The cross section of these four different possibilities for a left boundary are illustrated in Fig. 10. A depression on any boundary may occur when there is flat surface after a downward slope that starts from the road boundary. An elevation occurs when there is a flat surface after an upward slope. If a surface continues after a small discontinuity it corresponds to a ditch. A cliff corresponds to a depth discontinuity for a large region. Other possible geometric configurations are, of course, possible.

The flow of control used to detect roads in the feed-forward system is as follows:

1) Square windows of size 32 × 32 are placed on the image where the important road features are expected. The windows are placed so that the road enters the window at the midpoint of the bottom edge. For the first pair of windows, the coordinates of the midpoint are determined on the basis of the previously constructed road model and dead reckoning. For subsequent windows, these coordinates will be the coordinates of the second endpoints of the detected road boundary segments in the previous window.

2) Early processing, within a given window, computes the 3-D linear segments on the left and right sides of the road. Recall that each segment has been classified as being either convex or concave by the low-level processing, and that the edge points belonging to a segment are either all convex or all concave.

3) Each 3-D linear segment is assigned a weight. The merit function adopted is

$$c_1 \cdot L - (\text{for all but the first window}) \ [c_2 \cdot \Delta C].$$

Here, L is the segment's 3-D length, ΔC its continuity with the road boundary segment in the previous window, and c_1 and c_2 are constants.

4) Concave and convex segments are sorted separately based on this merit function in the left and the right window. This yields four sorted arrays of segments.

5) Based on prior information (either map-based or the result of analyzing previous windows), a most likely geometrical model for the road boundary in both the left and the right windows is chosen. Our current implementation only supports depression detection.

Depressions are detected by first selecting the best convex linear segment as the potential road boundary. Next, a concave segment within a tolerable distance is chosen to complete the depression. This concave segment would lie to the left of the convex segment when processing the left window, and to the right when processing the right window. If a depression is detected successfully, control is transferred to step 8). If the system fails to detect a depression based on the best convex segment, the next best convex segment is chosen in the same window, and step 5) is repeated until a) four successive

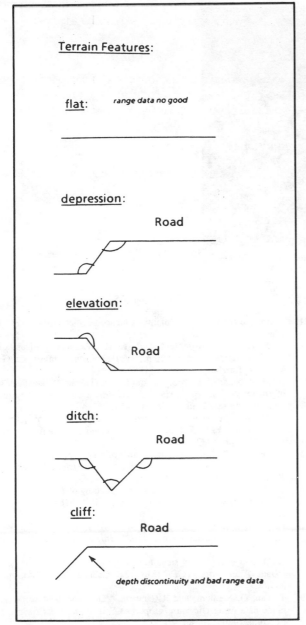

Terrain Features:

flat: *range data no good*

_____

depression:

　　　　　Road

elevation:

　　　　Road

ditch:

　　　　Road

cliff:

　　　Road

depth discontinuity and bad range data

Fig. 10. A cross section of different geometrical features on the left boundary.

failures are recorded or b) the sorted array of convex segments becomes empty.

6) In either of these (a or b) situations, we consider two possible cases for the failure:

a) First, the lower level processing might have failed to detect the segment that corresponds to the road boundary in the current, say ith, window. In this case, this window should probably be skipped, and an $(i + 1)$st window should be placed and processed.

b) Second, the high-level model-directed search might have chosen a wrong combination of boundaries in the previous, or $(i - 1)$st window, thus leading to the failure in analyzing the current window.

We have assumed, for simplicity, that the second possibility is the more likely so it is considered before case a). Furthermore,

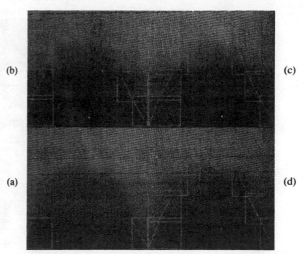

Fig. 11. The results of successive steps of feed-forward processing on a synthetic range image.

we limit backtracking to only one window. If we had not backtracked, we backtrack to stage $(i - 1)$ and a next best combination of the road boundaries in the $(i - 1)$st windows is selected.

7) To detect the next best combination, the weights of the two segments in the previous combination are compared. The segment with lesser weight is replaced by the next best convex segment on that side. Control is then transferred to step 5), in order to detect the depression based on the new segment.

8) Once potential road boundaries are detected in both the left and right windows, the next task is to compare the combined properties of these potential road boundaries against expectations concerning the width and shape (orientation of the boundaries) of the road. The calculated width is compared with the expected width. If the difference is too great, control is transferred to step 7) and the next best combination of segments is selected. Otherwise, the orientation between the two boundaries is compared with the expected orientation between the road boundaries. If the difference is too great, control is transferred to step 7) and the next best combination of segments is selected.

9) If the segments pass both verification tests, the termination of the processing of the current image is checked. If more than 20 percent of the area of the image remains to be processed, control is transferred to step 1). Otherwise, the algorithm halts.

The successive steps (clockwise from the lower left) of applying the feed-forward processing to different range images are shown in Figs. 11 and 12. The lower windows, in Figs. 11(a) and 12(a), are placed using the previously constructed world model and dead-reckoning data. The subsequent windows, in Figs. 11(b) and 12(b), Figs. 11(c) and 12(c), and Figs. 11(d) and 12(d), are placed using the positions and orientations of the road boundaries in the previous windows.

IV. CONCLUSION

We have presented low-level and high-level image processing algorithms for a road following system that supports the visual navigation of roadways by an autonomous land vehicle.

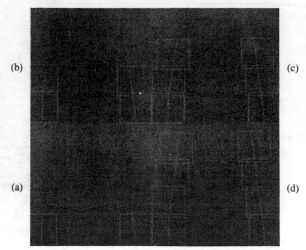

(b)   (c)

(a)   (d)

Fig. 12. Another example of feed-forward processing.

Both real and synthetic range images were used as the sensor input. *Primitive features* were extracted from the imagery using surface curvature features. *Dominant linear features*, under the assumption that the roadways generate such features in a scene, were extracted from the imagery in a feed-forward mode in which results of processing prior road segments combined with dead reckoning are used to predict the approximate locations of the important road features in subsequent images. The following remarks can be made concerning possible extensions to the work described in this paper.

1) Windows of constant size (32 × 32) were used in the feed-forward mode. The feed-forward mode should not only compute the position of the next window to be placed, but also vary its size in a sensible manner. The proper use of rectangular windows can reduce the processing time since the number of such windows required to cover linear features is small.

2) Many representations can be used in the image domain. Two such independent representations are edge-based and region-based. A representation using edge-based linear features was described in this paper. Representations using segmentation are region-based. A region-based representation for range images is described in [20]. These two representations for visual navigation of roadways, using intensity data, have been described in [18]. More research is required to determine how these representations can be used in an effective manner for range images.

3) Processing of 3-D range data is inherently expensive from a computational standpoint. For example, the planar surface fitting algorithm, which is the most time-consuming step in the system described in this paper, takes about 34 s to run on a SUN-3 workstation with floating-point coprocessor, for a 128 × 128 image using a 5 × 5 neighborhood size. The feed-forward mode of processing processes only about one half of the image and has the benefit of decreasing computational time by 50 percent. To achieve the processing speed required for real-time autonomous navigation [22], feed-forward processing should be implemented on a parallel architecture.

REFERENCES

[1] "Strategic computing," Defense Advanced Projects Agency, Oct. 28, 1983.

[2] L. S. Davis, A. Rosenfeld, and A. M. Waxman, "Visual ground vehicle navigation," Final Rep., Workshop on Autonomous Ground Vehicles, Leesburg, VA, Oct. 1983.

[3] A. M. Waxman, J. Le Moigne, and B. Srninivasan, "Visual navigation of roadways," in *Proc. 1985 IEEE Int. Conf. on Robotics and Automation* (St. Louis, MO, Mar. 1985), pp. 862–867.

[4] A. M. Waxman *et al.*, "A visual navigation system for autonomous land vehicles," *IEEE J. Robotics Automat.*, vol. RA-3, no. 2, pp. 124–141, Apr. 1987.

[5] P. Besl and R. Jain, "Invariant surface characteristics for 3D object recognition in range images," *Comput. Vision Graph. Image Process.*, vol. 33, pp. 33–80, 1986.

[6] B. Bhanu, "Representation and shape matching of 3-D objects," *IEEE Trans. Pattern Anal. Machine Intell.*, vol. PAMI-6, no. 3, pp. 340–350, 1984.

[7] R. C. Bolles, P. Horaud, and M. J. Hannah, "3DPO: A three-dimensional part orientation system," *Int. J. Robotics Res.*, vol. 5, no. 2, 1986.

[8] O. D. Faugeras, "New steps toward a flexible 3-D vision system for robotics," in *Proc. 7ICPR* (Montreal, Canada, July 1984), pp. 796–805.

[9] T. Fan, G. Medioni, and R. Nevatia, "Description of surfaces from range data using curvature properties," in *Proc. CVPR-86* (Miami, FL, June 1986), pp. 86–91.

[10] M. Hebert and T. Kanade, "First result on out-door scene analysis using range data," in *Proc. Image Understanding Workshop* (Miami, FL, Dec. 1985).

[11] A. Bergman and C. Cowan, "Noise-tolerant range analysis for autonomous navigation," in *Proc. AAAI-86* (Philadelphia, PA, Aug. 1986), pp. 1122–1126.

[12] T. Pavlidis, *Structural Pattern Recognition*. Berlin, Germany: Springer-Verlag, 1977.

[13] A. Rosenfeld and A. C. Kak, *Digital Picture Processing*. New York, NY: Academic Press, 1976.

[14] M. M. Lipschutz, *Differential Geometry*. New York, NY: McGraw-Hill, 1969.

[15] B. O'Neill, *Elementary Differential Geometry*. New York, NY: Academic Press, 1966.

[16] R. O. Duda and P. E. Hart, "Use of the Hough transform to detect lines and curves in pictures," *Commun. ACM*, vol. 15 no. 1, pp. 11–15, 1972.

[17] U. Ramer, "An iterative procedure for polygonal approximation of plane curves," *Comput. Graph. Image Process.*, vol. 1 no. 3, pp. 244–256, 1972.

[18] J. Le Moigne, A. M. Waxman, B. Srinivasan, and M. Pietikainen, "Image processing for visual navigation of roadways," University of Maryland, Computer Science TR-1536, July, 1985.

[19] D. M. Zuk and M. L. Dell'eva, "Three-dimensional vision system for the adaptive suspension vehicle," ERIM Rep. 170400-3-f, Jan. 1983.

[20] U. K. Sharma and D. T. Kuan, "Obstacle detection and intelligent navigation using range data," to be published in *Proc. IEEE Int. Workshop on Intelligent Robots and Systems* (Tokyo, Japan, Oct. 31–Nov. 2, 1988).

[21] D. T. Kuan and U. K. Sharma, "Model based geometric reasoning for autonomous road following," in *Proc. IEEE Int. Conf. on Robotics and Automation* (Raleigh, NC, Mar. 1987), pp. 416–423.

[22] U. K. Sharma and L. S. McTamaney, "Real time system architecture for a mobile robot," in *Mobile Robots II*, W. J. Wolfe and W. H. Chun, Eds., Proc. SPIE 852, pp. 25–31, 1988.

Uma Kant Sharma (M'85) was born in Meerut, India, on April 11, 1961. He received the M.S. degree in physics from Meerut University, India, in 1981, and the M.S. degree in computer science from the University of Maryland, College Park, in 1985.

During 1984–1985 he was a Research Assistant at the Computer Vision Laboratory of the University of Maryland. In 1986, he joined FMC Corporation, Central Engineering Laboratories, Santa Clara, CA, as a Computer Scientist in the Computer Vision Systems group. He has developed software related to vision-guided navigation for an autonomous mobile robot and software related to parallel processing architectures for the BBN Butterfly multiprocessor, transputer mutiprocessor, and the Pipelined Image Processing Engine (PIPE) machine. His current research interests are in computer vision and parallel algorithms for robot navigation and for commercial applications.

Larry S. Davis (S'74–M'77) was born in New York on February 26, 1949. He received the B.A. degree in mathematics from Colgate University, Hamilton, NY, in 1970, and the M.S. and Ph.D. degrees in computer science from the University of Maryland, College Park, in 1972 and 1976, respectively.

From 1977 to 1981 he was an Assistant Professor in the Department of Computer Science, University of Texas, Austin. He is currently a Professor in the Department of Computer Science, University of Maryland. He is also the Director of the University of Maryland's Institute for Advanced Computer Studies and Head of the University of Maryland's Computer Vision Laboratory. His current research interests are focused on problems of robot navigation using vision.

The Ground Surveillance Robot (GSR): An Autonomous Vehicle Designed to Transit Unknown Terrain

SCOTT Y. HARMON, MEMBER, IEEE

Abstract—The Ground Surveillance Robot (GSR) project has proceeded continuously since the Fall of 1980, and in that time an autonomous vehicle design and some degree of implementation has been achieved. The vehicle design has been partitioned into sensor, control, and planning subsystems. A distributed blackboard scheme has been developed which provides the mechanism by which these subsystems are coordinated. Vehicle position and orientation are supplied by vehicle attitude and navigation sensor subsystems. Obstacle avoidance capability has been implemented by fusing information from vision and acoustic ranging sensors into local goals and avoidance points. The influence of these points is combined through potential field techniques to accomplish obstacle avoidance control. Distant terrain characteristics are identified using information from a gray-level vision system, a color vision system, and a computer-controlled laser ranging sensor. These characteristics are used by a general planning engine to develop the desired path to a visible goal in the direction of the final goal. Progress to the final goal consists of a succession of movements from one distant but visible intermediate goal to another. The experience from implementing this autonomous vehicle has indicated the need for an integrated set of debugging tools which make the faults in subsystem hardware and software more distinguishable.

I. INTRODUCTION

THE EXTREMELY hostile conditions imposed by modern combat, outer space, and the deep ocean environments have generated the need for practical autonomous vehicles for military applications and space and ocean exploration. These relatively near-term applications will drive the sophistication and cost of autonomous vehicle technology into the realm where more mundane but more widespread applications such as automated public transportation will be possible. However, significant technology advances will be necessary before even the simplest and most crucial applications can be practically addressed. These advances will only be gained by implementing autonomous vehicle testbeds and gaining experience with the developing technology.

Several previous efforts have prepared the foundation for the autonomous vehicle development including Shaky [1], JASON [2], the RPI Rover [3], the JPL Rover [4], and the Stanford Cart [5], among others. These first generation autonomous vehicles were used to explore basic issues in

Manuscript received July 29, 1986; revised December 23, 1986. This work was supported in part by the US Marine Corps and the Naval Ocean Systems Center. This paper was presented at the SPIE Mobile Robots Conference, Boston, MA, October 1986.

The author is with Robot Intelligence International, PO Box 7890, San Diego, CA 92107, USA.

IEEE Log Number 8715001.

Reprinted from the *IEEE Journal of Robotics and Automation*, Vol. RA-3, No. 3, June 1987. Copyright © 1987 by The Institute of Electrical and Electronics Engineers, Inc. All rights reserved.

vision [1], [4], [5], planning [1], [2], [4], [5], and robot control [3], [4]. However, they were all seriously hampered by primitive sensing and computing hardware. More recent efforts have overcome many of these limitations, and very sophisticated second generation autonomous vehicle testbeds have evolved. Some of these efforts include the developments of HILARE [6], the FMC Autonomous Vehicle [7], the Autonomous Land Vehicle (ALV) [8], the various CMU mobile robots [5], and the Ground Surveillance Robot (GSR). This paper will focus on the design and implementation of only one of these recent efforts, the GSR. A more general and complete discussion of autonomous vehicle history and technical issues can be found in other sources [9].

II. PROBLEM

The design of any autonomous system must begin by defining its task very specifically. A robot's task can be described in terms of its environment and its goals (i.e., those conditions which represent the success, failure, and termination of the task). A task description contains all the information required to specify the sensing, processing, and control components of a robot necessary to accomplish that task. However, the task description only represents a statement of one part of the problem of actually building such a robot. The sensing, processing, and control components require energy handling and mechanical support to maintain their proper function. These components represent part of the reality of implementing an autonomous vehicle testbed and must be addressed with the same seriousness as applied to the more task related components.

A. Task

The GSR is designed to transit from one known geographic location (given in some absolute map coordinates) to another known geographic location over completely unknown natural terrain. The terrain can be any type over which a manned vehicle of comparable capability can traverse. Even though the locations of the starting and finishing points of the journey are well known, the specifics of the intervening terrain are completely unknown. The GSR must therefore develop a terrain map of the territory in the direction of the goal and plan its route from this information alone. However, if a computerized terrain map of the appropriate territory is available, then the GSR should be able to take advantage of this additional information to improve its performance.

Natural terrain can be described in terms of its topography,

EH0342-6/91/0000/0422$01.00 © 1987 IEEE

its surface composition, the variability of the surface, and the geometric character of the insurmountable obstacles which populate the surface (e.g., trees and rocks). The topography can include planes, continuously varying surfaces (e.g., rolling hills), and discontinuous surfaces (e.g., cliffs, ravines, and precipices). The surface composition can be described in terms of its base composition (e.g., rock, sand, and water) and the biological ground cover (e.g., grass and scrub brush). This terrain description provides sufficient information to identify impassable areas and to estimate the average speed with which a vehicle can transit trafficable surfaces.

In traversing natural terrain the GSR can transit only those areas which support the normal and shear stresses exerted by the vehicle while moving and changing direction. Unfortunately, the Earth's surface is not isotropic in either of those physical properties. Among other hazards awaiting unwary vehicles, the GSR may find ground which will not support its weight, inclined surfaces which it cannot climb or off which it may slide due to insufficient surface friction and flat surfaces with inadequate shear strength to support propulsion and steering actions. Local terrain variations can also change the vehicle's ability to propel and steer itself. Therefore, the GSR must identify traversable areas through its sensors. To complicate this task further, biological ground cover may change base material's ability to bear weight or shear stress.

In reality, other limitations are imposed upon an autonomous vehicle. Vehicles for practical applications must make the desired journey in a limited time and with a limited amount of fuel or other forms of stored energy, and several terrain characteristics can affect the transit time and fuel consumption. For instance, ground cover can impose significant drag upon the moving vehicle, thereby dramatically increasing fuel consumption to an unacceptable level. Furthermore, a field of numerous obstacles can force the vehicle to take such a tortuous path that it cannot meet the limited time criterion. The finite size of the vehicle footprint imposes additional constraints on the terrain which can be efficiently negotiated and must be considered during path identification activities. All these constraints inherent to the task of autonomous transit imply that the vehicle must not only be able to move over a surface but must also maintain continuous knowledge of its position, orientation, and velocity; avoid obstacles and other hazards; and be able to preview enough of the surroundings to identify a suitable long-range path which will take it to its goal in a finite time and with a finite amount of fuel.

B. Vehicle

The task of autonomous transit imposes direct design requirements on the vehicle. It must have sensors to perceive the environment, effectors to move from point to point, and processing which identifies actions from the sensor perceptions which will achieve the desired goals. To be self-contained, the vehicle must also supply the energy to support the processing, sensing, and effecting activities. It must also have structural features to support this equipment and protect it from unfavorable environmental conditions. Additional effectors may also be necessary to control the perspective of sensors with limited fields of view.

The vehicle sensors must perceive the vehicle's instantaneous position, orientation, local obstacles (or, more importantly, free space), and enough of the distant surroundings to plan a route toward the final goal location. The most important vehicle effectors provide its mobility. There are many different forms of mobility which involve different modes of propulsion, steering, and suspension. The choice of an existing manned vehicle which is converted for autonomous operation entirely solves the mobility design problem. However, this decision also imposes all the limitations of an existing vehicle and the difficulties of converting it to automatic control. Nevertheless, the vehicle actually meets the original project goals because its successful performance is compared to the performance of similar manned vehicles performing the same task. In addition, this solution to the mobility problem was within the project budget and has the advantage of demonstrating the concept of retrofitting an existing vehicle thus making this work more attractive to the military sponsors who own a lot of manned vehicles.

The vehicle computing presents a special challenge since it must now reside on a moving vehicle with all the harsh environment conditions that imposes (e.g., shock, vibration, contamination, elevated temperature). The vehicle computing must have sufficient capacity to interpret and integrate sensor data, plan vehicle motions based upon the sensor picture, and control the actuator responses with enough speed to effect stably the desired motion. The computing equipment must still fit within the confines of the vehicle, must not exceed the vehicle's payload, and must not consume all of the vehicle's limited fuel supply.

Furthermore, the vehicle must have an energy-handling system which converts its stored energy into forms to power computers, sensors, actuators, and vehicle locomotion, which transports the energy to where it is needed and regulates it to those conditions demanded by the consumer subsystems and which dissipates any heat generated as a result of the conversion, transportation, regulation, and consumption processes. The vehicle must also provide the structure to protect as much as possible the delicate components from the harsher aspects of environment including moisture, temperature, dust, salt mist, shock, and vibration. This implies that this structure is more than a simple enclosure on some form of mobility. These design considerations represent the reality of implementing a real robot as opposed to simply implementing a simulation of the proper components. All of the design considerations described must be sufficiently developed to implement a successful autonomous vehicle for any application. Even so, this discussion has not addressed the issues of fault tolerance and multiple degrees of freedom.

III. GSR Design

The problem of designing an autonomous vehicle to transit unknown natural terrain is discussed in terms of the basic problems outlined earlier. The GSR sensor capabilities are divided into the horizons. The closest horizon is just under the vehicle. The sensors within this horizon provide vehicle position and orientation. The next horizon identifies the limit of the vehicle's capability to identify and locate nearby

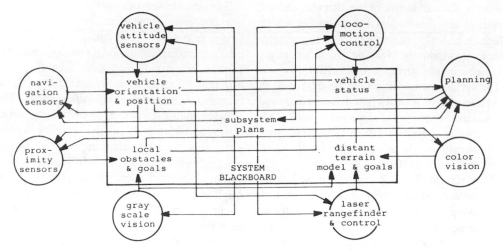

Fig. 1. Blackboard model of subsystem interaction.

obstacles. The final horizon defines the limit of the vehicle's ability to characterize distant terrain. Naturally, one would like to see all the way to the goal but that in practical situations is seldom possible because terrain features and atmospheric conditions dictate limited sensor range.

The discussion of the details of the GSR design begins with consideration of the system architecture and then proceeds to the problems discussed earlier. Each of the sensor subsystems must be tied together in some way, and the system architecture provides a way to do that.

A. System Architecture

The GSR project was started with the knowledge that we did not know precisely how to build a working autonomous system. In addition, we also knew that the project would proceed for several years and that the major component technologies would be evolving greater capability throughout this time. This knowledge significantly affected many of the design decisions. For instance, at each decision point the choice was made which maximized the opportunities for change later on. This means that an architecture was chosen which would enable the change of one part of the system without significantly changing the rest of the system. Furthermore, a choice was made to use as many commercially available resources (both hardware and software) as possible. Fortunately, this effort was undertaken after local area networks, high-speed parallel bus standards, single-board computers, microprocessor multitasking operating systems, and block structured microprocessor programming languages were reasonably well established. These resources enabled layered and modular implementation of both hardware and software.

All software except for a few hardware drivers was implemented in a reasonably high-level programming language, PLM. However, saying all this does not provide a framework within which the components can be integrated. This framework must provide well-defined physical and functional interfaces so the individual components can be implemented and tested independently and in small groups before full-scale integration is attempted. It must also be

recognized that many people are going to be involved in the design and implementation of such a complex device, and a well-defined architecture provides a critical basis for communication and coordination. These conditions dictated that the architecture provide for modular implementation. The following discussion of the GSR system architecture is partitioned into the following concepts: subsystem partitioning, the intelligent communications interfaces, and the blackboard representation.

1) Subsystem Partitioning: The first element of an architecture is the model for how subsystems are partitioned. For the GSR, tightly coupled functions must be coupled through a high bandwidth communications path. On the other hand, loosely coupled functions can be partitioned into very high-level modules with considerable shared slowly varying information. The major subsystems were grouped into high-level functions which communicate symbolically by exchanging information about the perceived world. The model of interaction between all subsystems was a blackboard model. This model of interaction is illustrated in Fig. 1. The choice of the major subsystems was made with the awareness of the capabilities provided by existing technology. However, the model supports any likely repartitioning of function. The blackboard is partitioned into the sensor data needed to make the autonomous transit discussed earlier. Vehicle status and system plans are also represented in the blackboard making communication and partitioning of control information easier and more flexible.

We endeavored to distribute as much intelligence to each subsystem as possible to loosen the coupling between them. This minimized intercomponent communications and decreased the real-time processing burden. Intelligent sensor and control components derive information directly from the real world and can exchange information directly between them. The planning mechanism sits above the sensor and control components and monitors the sensor traffic to build a picture of the world and to extrapolate expectations of the future from this current picture. Plans are derived from these expectations and sent to the sensor and control components to coordinate their future interactions. This strategy keeps the planning

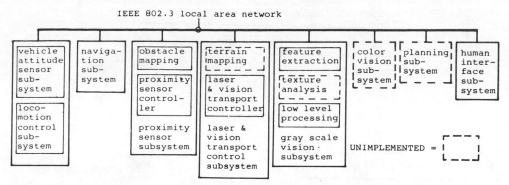

IEEE 802.3 local area network

UNIMPLEMENTED =

Fig. 2. GSR system architecture.

mechanism out of the real-time loop demanded by the vehicle control subsystems. The planning system, even with the increased throughput provided by advanced computing hardware, is always expected to be unable to deal with the real-time situations.

2) Intelligent Communications Interfaces: While the blackboard interaction paradigm provides many benefits in terms of programming the system, it is not a realistic model for implementation. Thus major subsystems were implemented in tightly coupled multiprocessor groups which communicate through a local area network as illustrated in Fig. 2. The components within each major module communicate through shared memory thus enabling the tight time sensitive coupling required by such processes as vehicle control.

Each major module is coupled to the local area network through a module called the Intelligent Communications Interface (ICI). This device makes each module think that it is effectively sharing an intelligent blackboard memory with all the other modules whether they really share common memory hardware or whether they really are communicating through a local area network. The ICI components are illustrated in Fig. 3, and these concepts for robot subsystem coordination are discussed in detail in [10], [11]. However, a brief discussion of these concepts will be presented below for the completeness of this article.

The ICI's communicate with one another through the ICI Protocol, a taxonomy of which is illustrated in Fig. 4. This taxonomy shows that communication is actually message-based, relying upon the transport layer provided by the local area network specifications (IEEE 802.3 was used). All traffic between modules is divided into world state information handled by reports and control information conveyed by plans. Reports are simply statements of world state in terms of the object attribute value representation chosen for the blackboard. These reports use ASCII coded symbolic descriptions so they could be interpreted directly by the human developers. Plans are effectively production rules which have been extended to accommodate real-time and continuous circumstances. Through plans it is possible to initiate both reporting and controlling actions as well as to control the state of existing plans. The conditions in plans represent either directly or indirectly some situation in the blackboard. Since plans are represented as objects in the blackboard the state of plans can be represented in other plan conditions, therefore making it

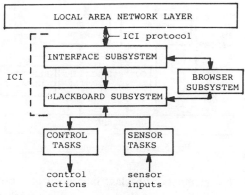

Fig. 3. Intelligent communications interface components.

possible to make plans dependent upon the state of other plans in the blackboard thus providing a very flexible control mechanism.

Fig. 5 illustrates the mechanism of the ICI. This mechanism is very similar to a production system in having components to perform plan parsing and pattern matching. This mechanism also has a component for taking reports broadcast over the network and inserting them into the local copy of the blackboard. Details of this mechanism are discussed elsewhere [10].

3) Blackboard Representation: The blackboard is a conceptual device which allows exchange of information between the GSR's subsystems. The blackboard provides a clear and consistent representation of this information which can be used by individual subsystem component developers. This means that no one person need completely understand the total system. This is certainly a realistic situation for a system as complex as the GSR.

Physically, the blackboard consists of several pieces of shared memory distributed throughout the subsystems. The ICI's provide distributed access to this shared memory as well as consistency control. Subsystems within the same module access the blackboard memory through standard blackboard interface procedures. These procedures provide well-defined mechanisms by which to read and write elements to and from the blackboard.

The blackboard is itself structured as a class tree. Each element in this tree has a list of properties to which are assigned values. This representation paradigm is commonly used in expert systems and other knowledge-based systems.

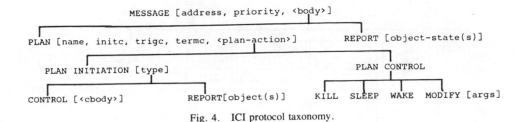

MESSAGE [address, priority, <body>]

PLAN [name, initc, trigc, termc, <plan-action>]   REPORT [object-state(s)]

PLAN INITIATION [type]   PLAN CONTROL

CONTROL [<cbody>]   REPORT[object(s)]   KILL   SLEEP   WAKE   MODIFY [args]

Fig. 4.   ICI protocol taxonomy.

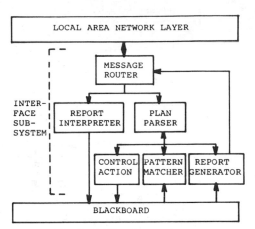

Fig. 5.   ICI interface mechanism.

Fig. 6.   Photograph of GSR.

The children elements of a class inherit all of the properties associated with the parent class and are distinguished only by the property values unique to that element. Thus the inheritance mechanism of the class tree provides an economical representation method.

Obviously, this blackboard provides the opportunity for much more than simply communicating various data values between subsystems. In fact, the blackboard provides a very powerful mechanism through which to fuse various types of sensor data [11]. Fusion of sensor data requires a consistent representation and various methods for combining the data. The representation discussed earlier must be extended to facilitate sensor data fusion. To this end, the object attribute value tuples of the class tree are expanded to include measures of accuracy, confidence, and timestamp. Most obviously, the computation of a resultant value can be derived by combining two or more sensor values through a functional dependency. This is the simplest technique for fusing data from different sensor sources since the function is predefined by the situation. The blackboard mechanism described here provides a very easy technique for this type of fusion through active functions. Active functions are devices for implementing data driven programming. Each time a value is changed to which an active function is attached, all of the values in the blackboard which are dependent upon that value are also changed to values which reflect the new change in sensor state.

Unfortunately, there are cases when two values which reflect the same value of a property in the blackboard must be fused. This situation of data fusion is much more complex. Several methods can be used to combine overlapping data including filtering, deciding, and guiding. Fusing using filtering can be illustrated by Kalman filtering or linear interpolation. Deciding is the case when some decision is

made (based upon some criteria) between which of several competing values should be used. In some situations, the newest information or the data from the most reliable source is chosen. In guiding, the value from one sensor is used to guide the processing of the data from another sensor which can provide more reliable or more accurate information about the prevailing situation. An example of guiding implemented on the GSR is when low-resolution obstacle location data are used to segment the highly complex visual sensor data to provide a more accurate picture of the location of nearby obstacles. In some cases, combinations of these techniques may be chosen.

B. Mobility and Vehicle Control

1) Mobility: The vehicle subsystem is the major effector of any mobile robot. The design of the GSR uses an existing testbed to minimize costly hardware development. Fig. 6 is a photograph of the GSR vehicle. It is an M114 armored personnel carrier which has been converted for automatic control. As can be seen from this photograph, this is a tracked vehicle which is powered by a gasoline engine. Fig. 7 shows a top view drawing of the vehicle and the placement of the various externally mounted sensors.

Several features have been added to this vehicle to permit the installation of on-board sensing and computing. These include a 10-kW auxiliary power generator set, a large air conditioner, and shock mounted equipment racks.

2) Locomotion Control: The locomotion control subsystem controls all of the equipment in the GSR that is associated with its mobility. This includes vehicle throttle, brakes, steering, and transmission shift as well as the steering mode

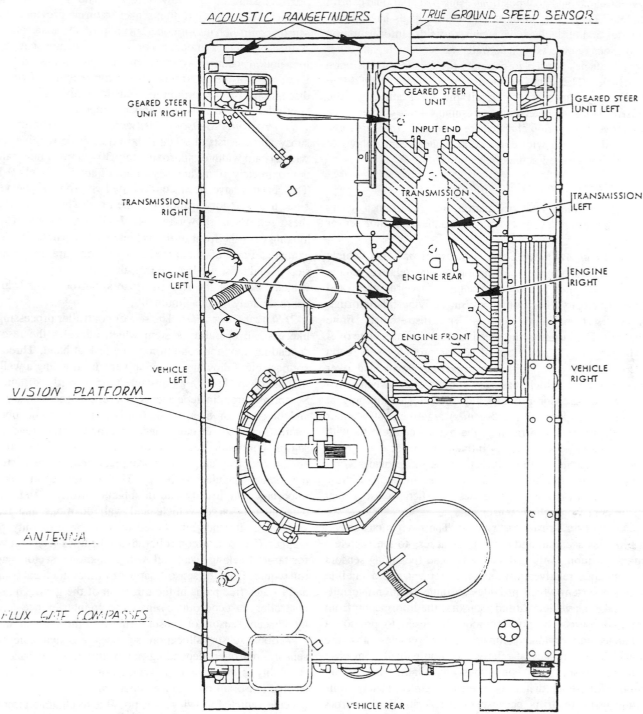

VEHICLE FRONT

ACOUSTIC RANGEFINDERS   TRUE GROUND SPEED SENSOR

GEARED STEER
UNIT RIGHT

GEARED STEER
UNIT

INPUT END

GEARED STEER
UNIT LEFT

TRANSMISSION
RIGHT

TRANSMISSION

TRANSMISSION
LEFT

ENGINE
LEFT

ENGINE REAR

ENGINE
RIGHT

ENGINE FRONT

VEHICLE
LEFT

VEHICLE
RIGHT

VISION PLATFORM

ANTENNA

FLUX GATE COMPASSES

VEHICLE REAR

Fig. 7.   Distribution of external GSR sensors.

shift, engine starting, and auxiliary power generator starting. The major control axes are actuated using electric torque motors controlled through servo amplifiers. This permits a higher level of computer control and applies well-developed technology to off-load the control computing. Steering, transmission, and throttle servo loops are closed on the actuator position so commands from the computer are given in terms of desired actuator position. The brake servo loop is closed around the applied force measured at the brake level

with a load cell. This arrangement makes the brake actuator self-adjusting as well as reduces the possibility of overstressing the mechanical components of the braking mechanism.

The locomotion control computer provides access to the locomotion actuators and, therefore, to the ultimate control of the vehicle's path. The locomotion control computer maps the positions of wall contacts, a target vehicle, local goals, and idealized obstacles (in the form of avoidance points). It also generates short-range vehicle path adjustments and computes the desired actuator positions to accomplish these changes. The desired vehicle path changes are computed from the wall, target, obstacle, and goal positions using potential field control techniques [12], [13]. In these techniques, goals have attractive fields and obstacles have repulsive fields. Information on the positions of objects such as walls, target vehicle, and other nearby obstacles is used to compute the existing forces upon the vehicle. The changes in the total potential field are computed by superpositioning all the local attractive and repulsive fields to determine a resultant force vector. The resultant vehicle motion changes are computed from this force vector and are used with conventional linear control laws to determine the desired actuator positions.

C. Vehicle Position and Orientation

Vehicle position and orientation are determined using information from the vehicle attitude and the navigation sensor subsystems.

1) Vehicle Attitude Sensor Subsystem: The vehicle attitude sensors provide information on the relative vehicle position, roll, pitch and heading angles, forward speed, and rotational velocity. The complexity in deriving this information comes from using sensors with distressingly finite limitations. For instance, the speed sensors are inaccurate over certain ranges. The track speed sensor provides very good information when the vehicle is traveling slowly and over terrain where the vehicle has very good traction, whereas the Doppler speed sensor provides reliable information only at relatively high speeds (e.g., >2 m/s) regardless of whether the tracks slip or not. Similarly, the gyrocompass drifts with time but magnetic compass is influenced by local magnetic anomalies. Neither of these direction sensors provides completely accurate information all the time. In both of these cases, an algorithm is used to decide when to use which measurement from which sensor.

2) Navigation Sensor Subsystem: The navigation subsystem provides accurate and continuous access to the vehicle's absolute position. This is done using dead reckoning sensors for continuous relative position measurement and a satellite navigation system which provides absolute position intermittently. Like the vehicle attitude sensors, the information from these two sources of position must be fused to provide a continuous and reliable estimate of the vehicle's absolute position. The estimates of relative position gained from dead reckoning sensors are continuously available and highly accurate for short distances traveled. Unfortunately, this position estimate drifts because of the accumulating errors inherent to relative position measurement. Satellite navigation provides quite accurate absolute position measurements (e.g., within 100 m) which do not drift as do the absolute estimates derived from integrating relative position information, but they are not continuously available.

D. Obstacle Avoidance

Obstacle avoidance activity can be divided into perception of obstacles and the vehicle control necessary to avoid them. The burden of local obstacle mapping is handled by the proximity sensor subsystem with some help from the vision

subsystem. The proximity sensor controller controls the positions and firing of the active acoustic sensors as well as the preprocessing of the raw returns. The proximity sensor mapper assimilates all the sensor returns provided by the sensor controller into a coherent map of the local obstacles. Actual obstacle avoidance control is accomplished by the locomotion control system which was discussed earlier. Discussion of the proximity sensor subsystem is divided into discussions of the proximity sensor hardware, the sensor controller, and the sensor mapper components.

1) Proximity Sensor Hardware: The proximity sensor subsystem consists of seven Polaroid acoustic ranging sensors with a beam width of approximately 30°, a maximum range of approximately 10 m, and resolution of approximately 0.17 m. The sensors have been concentrated in front of the vehicle since it travels forward most of the time. The arrangement of these sensors is shown in Fig. 7. Three sensors are fixed looking over the front of the vehicle each separated by 30° of azimuth. Four of the proximity sensors are mechanically steered and these are located near the left and right sides of the vehicle. Each of the steered sensors can be rotated through approximately 180° azimuthally.

2) Sensor Controller: The sensor controller processing acts like a small operating system which allocates the resources depending upon the condition of the task at hand. Three tasks are possible: following a moving vehicle, tracking a wall, and finding a way through an obstacle field toward a distant goal. Sensor resources include access to a sensor, access to a region, and blocking of another sensor because of some potential interaction (e.g., through electromagnetic interference). The sensor controller receives the raw echo, filters returns, computes object range and approximate bearing, and estimates the object angular velocity. The sensor controller puts important data directly into the blackboard (i.e., time critical data such as target vehicle and wall positions and relative velocities). It sends all filtered data to the proximity sensor mapper. The sensor controller also schedules sensor coverage for target tracking, wall following, unknown sector mapping (on request from the sensor mapper or other nonlocal sources), and locates free paths in the direction of the goal. The sensor controller must coordinate sensors so no interference is caused by adjacent sensors or sensor firing into the same area.

Various sensor allocation schemes are available to the sensor controller depending upon the task at hand. This flexibility is the primary advantage of having both fixed and steered sensors. For instance, when tracking a target vehicle to execute vehicle following, it is possible to obtain an improved target vehicle position estimate by using opposing steered sensors to triangulate upon the vehicle. The sensor controller can also use multiple returns from the same target to make crude estimates of the relative target speed. These estimates can also be used together with range and angle gates to help differentiate a target vehicle from the surroundings.

The acoustic ranging sensors are notorious for being sensitive to several forms of interference. These include reflected returns from other sensors, electromagnetic interference from other nearby sensors, and electrical noise produced by the electric motors moving the sensors as well as other sources. All of these forms of interference must be considered when determining the moving and/or firing order of the various sensors. The movement of the vehicle in an obstacle field has been simplified somewhat by using a sensor allocation algorithm which looks only for free space (as opposed to obstacles) in the direction of a distant goal. Thus, if the vehicle is traveling over a relatively uncluttered space, it can move fairly rapidly since its movement is not limited by the need to search the entire space near the vehicle for obstacles. On the other hand, if the space near the vehicle is quite cluttered then the speed must be appropriately reduced to gain sufficient information to avoid nearby obstacles in the direction of the goal.

3) Sensor Field Mapper: The proximity sensor mapper transforms all range and bearing information received from the proximity sensor controller into the absolute reference frame (using position information from the navigation sensor subsystem) and constructs a local obstacle map from that data. The information conveyed by the raw sensor data is fused into a single coherent estimate of nearby obstacle positions by representing each return by a distribution representing the probability that a reflecting obstacle was actually detected and by superpositioning the distributions associated with each sensor measurement. These distributions are formulated from *a priori* knowledge of the sensor behavior. The technique adopted is very similar to that discussed in Moravec and Elfes [14] except that they represented their local map as a grid of cells each representing a particular probability of being occupied, and our technique represents returns as line segments similar to those used by Crowley [15] and Miller [16]. This technique thus combines multiple returns from different vantage points (since the vehicle is constantly moving) to improve the accuracy of the local obstacle map. This improves the map's robustness despite the fact that it is constructed from data generated by sensors which are so susceptible to various types of noise that the measurements are individually unreliable.

The proximity sensor mapper also derives vehicle path suggestions in the form of approach and avoidance points from the existing local obstacle map. Approach points are local goals chosen from knowledge of the direction of the distant goal (formulated by the path planning subsystem) and of the availability of obstacle free corridors large enough to permit vehicle passage. Avoidance points are chosen from the points of obstacles closest to the straight line drawn between the vehicle's current position and the nearest approach point. In this way the vehicle can travel toward the closest approach point and still avoid the most important obstacle positions.

This technique also refines the detailed and constantly changing local obstacle map represented by the proximity sensor mapper to a few points which can be economically represented in the system blackboard. In this way, only a few points need to be updated when new information is received instead of the entire obstacle map which greatly decreases relatively expensive accesses to the blackboard (these are made expensive by the need for operating system context changes and reporting over the local area network). Reporting only new or unexpected deviations from the map or past

suggestions greatly improves the performance of the mapper as well as the entire system by significantly limiting communication between subsystems to maintain blackboard consistency. Approach and avoidance points are represented in the blackboard by their absolute positions computed from information on the vehicle's absolute position from the vehicle attitude and navigation sensor subsystems. This choice enables the information from multiple vehicle positions to be easily combined and is thought to be the most economical choice because there is usually only one vehicle and multiple obstacles.

The proximity sensor mapper can also request additional sensor coverage from the proximity sensor controller in the direction of unknown areas. This loose coupling between the mapper and controller limits communication between these elements to a minimum and yet permits the mapper to get further data where existing information is missing or ambiguous. Vision data can also be used to coordinate the proximity sensor resources through the sensor controller to improve the local obstacle map and to enable tracking of major terrain features from the distant terrain mapping information.

E. Terrain Modeling

1) Vision and Rangefinder Sensors: The sensors on the GSR which provide the information for terrain modeling consist of a laser range finder which can make random access range measurements within the field of view of the cameras, a high-resolution (e.g., 512×490) gray-scale solid-state camera and a low-resolution color camera. All of these sensors are mounted on a three-degrees-of-freedom vision transport platform.

The problem of developing a terrain model from the data from these sensors is divided into the subproblems of low-level gray-scale processing, low-level color processing, gray-scale texture analysis, ground cover analysis, low-level laser range finder data processing, terrain segment construction, and terrain description integration. In addition to the processing required for terrain modeling a significant amount of processing is required to control the range finding subsystem. This processing consists of a supervisor which integrates information from the vehicle attitude sensors, maps the terrain, and plans the moving mirror paths. The range finder controller controls the data point reference frame resolution, the mirror path generator, and the laser sequencer.

2) Vision Transport Platform: The vision transport platform provides for the best utilization of the terrain modeling sensors which all have relatively narrow fields of view. Fig. 8 shows the construction of the vision transport platform. This platform enables the vision sensors to be elevated 1.5 m and rotated in azimuth approximately $360°$. It also permits the camera's elevation angle to be controlled between $30°$ above and $15°$ below the horizontal. An additional axis of camera roll control is being contemplated but has yet to be implemented.

3) Representation: A proper representation must be chosen to facilitate integration of the data from the multiple sensors used for terrain modeling. Ideally, all of this different sensor data can be represented in a consistent fashion. A consistent

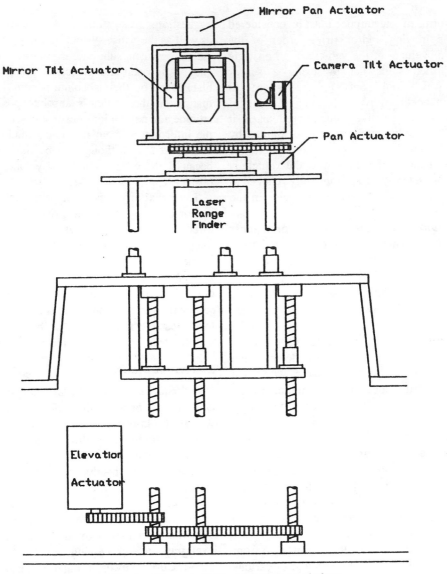

Fig. 8. Diagram of GSR vision transport platform.

representation aids in the distribution and consistency control necessary for a distributed implementation. The blackboard mechanism is used to provide a general format for representation. Within the blackboard model, the world model organizes data into a class tree with inheritance properties. Specifically, terrain data are represented as triangular segments with the properties of absolute position, orientation, adjacency, a terrain modification function, segment variability spectrum, ground cover type, and obstacle population statistics. This representation provides sufficient information to facilitate adequate route planning. This representation also avoids the problems associated with finding convex polygons while providing a very good terrain segment primitive (a triangular segment is already a convex polygon).

F. Path Planning

The planning subsystem just deals with long-range planning. Local obstacle avoidance planning is left to the control system. This decreases the direct coupling between the relatively slow high-level planning activity and the demanding

vehicle controller. Actually, the experience in constructing and testing planners for outdoor situations is very limited [3], [7], [8]. Path planning in the GSR is constructed using a generic planning engine to support the specific task of vehicle route planning over unknown terrain. This approach enables the planning problems to be separated from the specifics of the route-planning problem.

1) Route Planning: Route planning in unknown situations has the two interacting subgoals of collecting sufficient information to enable fruitful planning and of actually making progress toward the final goal. The distribution of robot activity between these competing goals must be optimized to arrive at the goal as quickly as possible.

In general, the route planner minimizes a cost function based upon the energy requirements, the estimated unknown hazards, the estimated vehicle transit speed, and the distance to the goal. This function is very similar conceptually to many other approaches to a similar route-planning problem. The GSR's planner performs this function from the present vehicle location to the final goal location if sufficient map information is available internally. However, if the terrain between the present vehicle location and the final goal is unknown, then sensor plans are generated to collect the terrain information within the sensor horizons. If the final goal is not within the sensors' ranges, then an intermediate goal is chosen which provides the best trade-off between approaching the final goal, gaining new information from the best possible vantage point, and minimizing energy consumption and travel time. In this way, the process of transiting unknown terrain consists of transiting from one intermediate goal within the sensor horizons to another intermediate goal toward the final goal location. The process of finding a path to a goal location outside of immediate sensor range over unknown terrain effectively describes the body of human knowledge called orienteering which provides various heuristics for position finding, route planning, and route following.

2) Planning Engine: Development of a planning engine for an autonomous vehicle which is appropriate for route planning in unknown natural terrain presents several challenges [17]. Ideally, orienteering knowledge can simply be encoded into the representation of some generic planning engine while the actual planning activity occurs as a natural function of the engine. This approach maximizes the flexibility of the system and readily enables the programming of competing goals. The planning mechanism should plan for the future using the present sensor information about the surroundings so that it does not interfere with the function of tight looped control systems. Since terrain information is supplied incrementally by the vehicle sensors, it should also plan from evidential, incremental and, thus, uncertain information. To make the best use of limited computational resources, the planning mechanism should be able to focus its processing demands to concentrate on the most important issues in a rapidly evolving situation. Unlike expert systems' reasoning mechanisms, this planning mechanism need not have an explanation function since once it is debugged no one will likely ask how it works, and maintaining the reasoning trace needlessly burdens the efficiency of the mechanism. In addition, the planning engine must reason in time, plan in widely different time scales (i.e., seconds to hours), and still be able to keep pace with the evolution of events in real time. It must simultaneously model several different environmental situations, approach planning as a process of continual replanning to be able to handle continuously failing plans, plan using both goals and constraints, and plan in spite of uncertainty, inaccuracy, and the unknown in the sensor data. Ideally, the planning mechanism should be completely independent of the domain, thus permitting easy modification and debugging. Finally, the planning mechanism should be as simple and uniform a mechanism as possible while still operating efficiently on a large data base. This implies that any search of the complete data base should be minimized during normal planning operation.

The requirements described present a considerable challenge to any existing planning concept. Many concepts address a small subset of these requirements but fail to approach the performance described by all of these criteria together. For this reason a completely new planning mechanism was designed for the GSR, although it could be applied to any

planning problems with similar requirements (e.g., almost any robotics situation). The first generation of this engine has been implemented, but it has not been installed on the vehicle. However, detailed discussion of the design of this system is out of the scope of this article.

IV. Implementation Status and Experience

Like many projects of this grand a nature, a significant portion of the aforementioned design remains only partially implemented. Fig. 2 illustrates with broken lines the subsystems described above which have not been implemented. However, this project is very incremental in nature, and all of the present subsystems could use various degrees of improvements. In spite of the lack of completion of this project, much useful experience has been gained. This experience is described in terms of resolved and unresolved issues.

A. Project Status

The vehicle presently has the ability to follow a target vehicle and to avoid those obstacle conditions which it can sense with the proximity sensor subsystem. The locomotion control system presently uses the potential field obstacle avoidance approach. However, its performance could be improved through various tuning actions. All proximity, locomotion, vehicle attitude, and navigation sensors have been installed aboard the vehicle, and a significant portion of the processing for these sensors has been implemented and tested. The terrain analysis and planning capabilities have not been implemented on the vehicle, although various degrees of development and prototype implementation have occurred. All of the existing modules have been integrated through the ICI's and the function of the blackboard paradigm has been demonstrated.

B. Resolved Problems

Solutions to several major technical problems inherent to autonomous systems implementation were derived and tested during the GSR's development. In particular, the solutions addressed overcoming sensor processing bandwidth limitations, coordinating planning and control processes, fusing sensor data, representing system knowledge, and coordinating the interactions between the various sensor and control subsystems.

1) Sensor Processing Bandwidth Limitations: The major sensors on the GSR provide a high volume of data flow which must be processed to extract critical information. The processing, if approached in a straightforward way, would in most cases exceed the time available to make the necessary control corrections. As a result, several steps were taken to shortcut this processing: 1) using high-resolution sensors to concentrate on the aspects of the environment most distant from the robot, 2) using low-resolution sensors closer to the robot where the quickest responses are necessary, and 3) using the information from low-resolution sensors to guide the processing of high-resolution data.

Processing in the vision sensors was structured to concentrate on the distant scenes. Concentrating on the distant both increases the time available before a particular situation is encountered and decreases the environmental detail available to the sensor which must be processed. This is not to say that the near field is ignored. Low-resolution sensors such as the acoustic proximity sensors provide the primary source for this information. This allocation proved ideal since the major necessary near-field information was used primarily for obstacle avoidance and required only coarse spatial resolution. The processing of the proximity sensor information was also broken into two parts: very nearby and more distant. Of course, the very nearby information was processed first and the more distant and, in most cases, the more processing intensive data were processed later. This processing was allocated to separate processors which were running asynchronously. The architectural approach described greatly facilitated this partitioning.

One additional way to cope with the sensor bandwidth limitations was to use the low-resolution sensors to guide the processing of the high-bandwidth sensors (such as vision). In particular, the information from the acoustic sensors was designed to aid in the segmentation of the gray level vision data.

2) Control and Planning System Coordination: Control and planning systems have historically and probably will continue to function with widely different situational constraints. Control systems must typically operate very closely to the environmental circumstances and, therefore, must have the tightest coupling to the real-time situation. Their responses must occur with very little delay. Also, in general, specific types of control processing work well within very limited circumstances and their performance degenerates when the circumstances deviate from the ideal. At this point, different types of control approaches are needed. Planning systems can usually deal with a wide variety of situations, but they impose very large time delays especially when large context changes are needed. In addition, planning systems can or should be able to function when uncertainty or inadequate sensor information is present. Control systems need fairly well-defined situations with adequate sensor information for stable control. Some difficulty arises because complex robots need the abilities of both of these systems, but they must somehow be coordinated.

The approach taken in this effort, which has proved successful, is to decouple the control and planning systems. Each is fed with the same sensor information but uses it for entirely different purposes. The control system uses the blackboard information for tight loop control of the vehicle functions while the planning system uses the prevailing sensor picture (as represented on the blackboard) to make predictions of the situational trends. It then develops future plans (for both control and sensor allocation) to deal with these expected situations. Once the initial planning has been completed and the robot has actually started executing the task the planning system continually assesses the validity and progress of the prevailing plans. If the situation seems to be deviating from one in which those plans are appropriate or effective, then replanning is done to generate an expected situation more consistent with the overall system goals. In other words, the

planner is continuously replanning throughout the task. This strategy keeps the planning mechanism out of the real-time loop demanded by the control systems. The planning system even with the increased throughput provided by advanced computing hardware is always expected to be unable to cope with the demands of real-time situations so some kind of decoupling will always be necessary.

This activity and system functional decomposition is facilitated by the distributed blackboard mechanism where the best assessment of the existing circumstances is continuously available to all robot subsystems and where control information to the subsystems can be provided through the ICI plan structure. The control system therefore operates using the desired control law until the situation changes to make that control strategy ineffective. At this point a new plan is activated, and the old plan is flushed. The ICI plan mechanism permits a whole series of interconnected contingency plans to be placed in direct access to the target subsystem. This organization enables the real-time constraints of the prevailing situation to be eased from the shoulders of the planner and placed into the domain of the coordination system.

3) Knowledge Representation: The blackboard is a conceptual device which enables exchange of information between the GSR's subsystems. The blackboard provides a clear and consistent representation of this information which can be used by individual subsystem component developers. This means that no one person participating in system implementation need completely understand the total system. This is certainly a realistic situation for a system as complex as the GSR.

Physically, the blackboard consists of several pieces of shared memory distributed throughout the subsystems. The ICI's provide distributed access to this shared memory as well as consistency control. Subsystems within the same module access the blackboard memory through standard blackboard interface procedures. These procedures provide well-defined mechanisms by which to read and write elements to and from the blackboard. The blackboard is itself structured as a class tree. Each element in this tree has a list of properties to which are assigned values. This representation paradigm is commonly used in expert systems. The children elements of a parent class inherit all of the properties associated with the parent class and are distinguished only by the property values unique to that element. Thus the inheritance mechanism of the class tree provides an economical representation method.

4) Sensor Data Fusion: A significant amount of data is fused in the GSR. This ability provides a critical flexibility enabling repeated reorganization of the sensor processing functions.

The blackboard scheme provides the opportunity for much more than simply communicating various data values between subsystems. In fact, the blackboard provides a very powerful mechanism through which to fuse various types of sensor data [12]. Fusion of sensor data requires a consistent representation and various methods for combining the data. To this end, the object attribute value tuples of the class tree must be expanded to include measures of accuracy, confidence, and timestamp. The computation of a resultant value can be derived by combining two or more sensor values through a functional dependency. This is the simplest technique for fusing data from different sensor sources, since the function is predefined by the situation. The blackboard mechanism described here provides a very easy technique for this type of fusion through active functions. Active functions are devices for implementing data-driven programming. Each time a value is changed to which an active function is attached, all of the values in the blackboard which are dependent upon that value are also changed to values which reflect the new change in sensor state.

Unfortunately, there are cases when two values which reflect the same value of a property in the blackboard must be fused. This situation of data fusion is much more complex. Several methods can be used to combine overlapping data including filtering, deciding, and guiding. Fusing using filtering can be illustrated by Kalman filtering or linear interpolation. Deciding is the case when some decision is made (based upon some criteria) between which of several competing values should be used. In some situations, the newest information or the data from the most reliable source is chosen. In guiding, the value from one sensor is used to guide the processing of the data from another sensor which can provide more reliable or more accurate information about the prevailing situation. An example of guiding implemented on the GSR is when low-resolution obstacle location data are used to segment the highly complex visual sensor data to provide a more accurate picture of the location of nearby obstacles. In some cases, combinations of these techniques may be chosen.

5) Subsystem Coordination: The blackboard mechanism provides a very powerful tool for flexible subsystem coordination. As much intelligence was distributed to each subsystem as possible to minimize intercomponent communications and decrease the real-time processing burden. Intelligent sensor and control subsystems used blackboard sensor information directly. The system planner sat above the sensor and control components and monitored the sensor traffic to build a picture of the world and to extrapolate expectations of the future from this current picture. Plans were derived from these expectations and sent to the sensor and control components to coordinate their future interactions.

Since the optimal subsystem partitioning of the GSR was not known at the time the initial design was conceived advance identification of the best interaction paths between subsystems was not possible. This situation eliminated the possibility of using a strictly hierarchical design and coordination strategy. Certainly some form of system hierarchy evolved during the implementation, but many of the details of the final result changed en route. The blackboard mechanism enabled this evolution to occur unfettered by uninformed decisions made early in the design process.

The ability to distribute plans and thus intelligence to the subsystems dynamically enabled the subsystem organization and interactions to be reconfigured at system runtime. As a result, the subsystem coordination strategy could be changed in response to varying environmental circumstances. Distributed plans also provided an additional and powerful debugging aid since system developers could make minor changes in the subsystem organization without expensive downloading each time the change was desired.

C. Unresolved Problems

Unfortunately but realistically, satisfactory solutions were not discovered for all the problems encountered during the GSR development. Interestingly, these were, for the most part, not strictly technical problems although they had technical components. These major unresolved issues include managing the complexity inherent to an autonomous system development, monitoring and debugging the system during development, dealing with the situation of exploratory system implementation and addressing the coupling of system testing with safety.

1) Complexity Management: The implementation of the GSR was a very complex process, and it produced a very complex device by all present standards. Managing this complexity proved to be very difficult. The complexity of the device came, in part, from the very flexible implementation strategy because almost any connection between any of the subsystems was possible. This provided a staggering number of possible interaction combinations. Further complexity was introduced by having a system with a large number of sensor, processing, and control components. This generated a considerable number of possible failure modes and made debugging the system a very complex and time-consuming process. Needless to say, significant complexity was introduced by the amount of software in the vehicle. When the project was initiated a considerable amount of thought was devoted to keeping track of this complexity. However, as the implementation proceeded many of these procedures were abandoned because they imposed unacceptable burdens upon the developers working in an exploratory situation where design changes were daily occurrences.

Additional complexity arose from the need to have several separate people implement the system. The work started with only three full-time people and evolved to a situation where the efforts of 12 people were dedicated to the project. This transition as well as the use of part-time personnel (such as students) produced many loose ends; many of which were never tied. Much of the system remains undocumented, and ancient bugs still remain hidden in the system despite considerable testing.

Some systematic approach to the management of this complexity is needed before sophisticated autonomous systems will evolve to practical utility. All the existing complexity management tools and techniques were either used consistently or tried and discarded in the course of this effort. Despite these measures, this problem remained a significant stumbling block throughout the effort.

2) System Monitoring and Debugging: The complexity of this system made system monitoring and debugging very difficult. The ICI mechanism proved to be the most valuable system monitoring and debugging tool available, but it was still inadequate. An integrated set of debugging tools is sorely needed which makes the dissociation of hardware and software problems easier and less time consuming. Unfortunately, the ICI mechanism did not provide adequate capability as it was implemented to enable little more than coarse-grained system performance analysis. Further tools for this purpose would greatly assist the implementation process. These tools were not developed during this effort because of the limited time and financial resources available. Thus much potential research remains in the development of system debugging and performance monitoring tools for the implementation of complex autonomous robots.

3) Exploratory Implementation: Projects which have contributed the most experience to the collected knowledge of complex hardware and software systems development usually began with well-defined design goals and plans. These are projects in which the developers knew beforehand which design elements would work and which would not. Unfortunately, autonomous systems do not have very much collected implementation experience at all. At this time, it is not clear which design elements are the most useful. Thus developers of today's autonomous systems must be prepared to alter their design ideas throughout the entire development of the autonomous system. This situation is similar to that encountered by developers of artificial intelligence systems. For these systems a new type of programming has evolved called exploratory programming for which such languages as LISP are ideally suited. The exploratory programming model is one where the developer builds and tests the system a little at a time until a final system implementation is achieved which exhibits the desired behavior.

Very sophisticated knowledge-based programming tools have evolved to support exploratory programming (e.g., expert systems shells, hot editors, and incremental compilers). However, autonomous systems developers do not have the luxury of having a set of tools already available to them to make their jobs easier. Existing complex system design tools assume that the designer knows at design time what works and what does not and very little of the entire autonomous system needs the very sophisticated tools provided by artificial intelligence programming environments. Existing concepts for system configuration management do not apply to this situation largely because the system design is sometimes changing on a daily basis. Cumbersome but very useful system documentation procedures quickly degenerate to disuse because of the rapid changes and the sizable investment of developer time required to follow them. Rapid hardware and software changes make any attempts to keep consistent system documentation very expensive to impossible. Clearly, some techniques are required to help autonomous systems designers over the hump until they can rely on sufficient past experience to have some confidence in their designs and thus to take advantage of existing proven design techniques and tools.

The lack of available design tools and theory for the development of autonomous systems also means that there is no way to design an autonomous system to achieve some complex task reliably. This means that the system must be implemented and tested incrementally throughout the design and implementation process. However, at no point along the way can the developer have any well-founded confidence that the system he is developing will meet the design goals. Once the system has been implemented, there is no way to test that system reliably to be sure that it meets the original design goals. This situation does not inspire confidence in the customer community and must be resolved before practical autonomous systems will be widely accepted.

4) Safety and System Testing: Preserving safety during autonomous system testing presents a particularly difficult situation because these are systems which are made to operate without human assistance. This is not to say that they must be tested without human supervision. However, provisions must be made in the design of the system to permit such supervised testing. This complicates the system design and implementation often requiring radio links or other forms of command links. The supervising humans must watch what is going on in the system and be able to identify any failure before it causes the system to damage itself or its surroundings. They must also be able to take sufficient control of the system to rescue it in the event of misbehavior. This situation is further complicated by the existence of additional dangerous effectors that the system has (e.g., large vehicle, dangerous lasers). Very careful testing procedures must be instituted and controlled. During the several years of GSR development we were both very careful and very lucky. Not a single accident occurred in several hundred hours of testing and demonstrations, but the potential of serious injury was omnipresent throughout this time.

V. Conclusion

This project has involved several years' work, and many lessons have been learned in that time. However, most of these lessons have involved implementation techniques as opposed to system design. Overall, the chosen design has proven quite successful, and very few deviations from the original design ideas have been taken. The most important contribution of this effort has been the development of a very flexible integration scheme which is applicable to many different robot system applications. The choice of an integration technique which forces most of the specifics of the subsystem interfaces into a database has been instrumental in the success of the implementation of the GSR to date. This has enabled the quick changes and the incremental development which are critical to any effort dealing with hitherto unknown development areas.

One area which has continued to hamper this development has been the nearly total lack of a comprehensive set of debugging tools which deal with both system hardware and software. Often one is reduced to pursuing parallel paths of software and hardware debugging techniques just to identify where the offending problem hides. A significant portion of this development effort has been devoted to pursuing system bugs. When this problem has been revealed to other robot system developers, a profusion of advice has been received. However, the most common failure mode is for all subsystems to work independently, and when they are integrated, the system performance just goes away with no indication as to the area of fault. This failure mode involves all of the subsystem developers who all simultaneously believe that their efforts are flawless. Again, this situation was the most common situation experienced during this development. An integrated set of debugging tools designed specifically for robot systems would be a truly valuable contribution to the development of future autonomous systems.

Acknowledgment

This work was performed while the author was employed by the Naval Ocean Systems Center, Code S442, San Diego, CA 92152. The author is extremely grateful to the many people who have contributed to this program for the past five years.

References

[1] N. Nilsson, "A mobile automaton: An application of artificial intelligence techniques," in *Proc. 1969 Int. Joint Conf. Artificial Intelligence,* Washington, DC; May 1969, pp. 509–520.

[2] M. Smith *et al.*, "The system design of JASON, A computer controlled mobile robot," in *Proc. IEEE Conf. Cybernetics and Society,* San Francisco, CA, Sept. 1975, pp. 72–75.

[3] S. Yerazunis, D. Frederick, and J. Krajewski, "Guidance and control of an autonomous rover for planetary exploration," in *Proc. IEEE Milwaukee Symp. Automatic Computation and Control,* Milwaukee, WI, Apr. 1976, pp. 7–16.

[4] J. Miller, "A discrete adaptive guidance system for a roving vehicle," in *Proc. IEEE Conf. Decision and Control,* New Orleans, LA, Dec. 1977, pp. 566–575.

[5] H. Moravec, "The Stanford cart and the CMU Rover," *Proc. IEEE,* vol. 71, pp. 872–884, July 1983.

[6] G. Giralt, R. Chatila, and M. Vaisset, "An integrated navigation and motion control system for autonomous multisensory mobile robots," in *Proc. 1st Int. Conf. Robotics Research,* Bretton Woods, NH, Aug.–Sept. 1983.

[7] J. Nitao and A. Parodi, "A real-time reflexive pilot for an autonomous land vehicle," *IEEE Contr. Syst. Mag.,* vol. 6, pp. 14–23, Feb. 1986.

[8] J. Lowrie *et al.*, "Autonomous land vehicle," Annual Report, MCR-85-627, ETL-0413, Martin Marietta, Denver Aerospace, Denver, CO, Dec. 1985.

[9] S. Harmon, "Autonomous vehicles," in *Encyclopedia of Artificial Intelligence.* New York: Wiley, 1986.

[10] S. Harmon *et al.*, "Coordination of complex robot subsystems," in *Proc. IEEE Conf. Artificial Intelligence Applications,* Denver, CO, Dec. 1985, pp. 64–69.

[11] S. Harmon, G. Bianchini, and B. Pinz, "Sensor data fusion through a distributed blackboard," in *Proc. IEEE Conf. Robotics and Automation,* San Francisco, CA, Mar. 1986, pp. 1449–1454.

[12] O. Khatib, "Real-time obstacle avoidance for manipulators and mobile robots," in *Proc. IEEE Conf. Robotics and Automation,* St. Louis, MO, Mar. 1985, pp. 500–505.

[13] B. Krogh, "A generalized potential field approach to obstacle avoidance control," in *Proc. SME/RI Robotics Research Conf.,* Bethlehem, PA, Aug. 1984.

[14] H. Moravec and A. Elfes, "High resolution maps from wide angle sonar," in *Proc. IEEE Conf. Robotics and Automation,* St. Louis, MO, Mar. 1985, pp. 116–121.

[15] J. Crowley, "Navigation of an intelligent mobile robot," *IEEE J. Robot. Automation,* vol. RA-1, pp. 31–41, Mar. 1985.

[16] D. Miller, "A spatial representation system for mobile robots," in *Proc. IEEE Conf. Robotics and Automation,* St. Louis, MO, pp. 122–127, Mar. 1985.

[17] S. Harmon, "Comments on route planning in unknown natural terrain," in *IEEE Conf. Robotics and Automation,* Atlanta, GA, Mar. 1984, pp. 571–574.

Scott Y. Harmon (M'78) is presently the owner of a small robotics research and development firm, Robot Intelligence International. Prior to that he pursued robotics research with the Naval Ocean Systems Center, both in San Diego, CA. He has over 25 publications in the areas of robot communications, robot computing architecture, autonomous vehicles, and mobile robot planning.

He is a member of ASME, APS, SPIE, AAAI, SME/RI, and Sigma Xi.

Guidance and Control Architecture for the EAVE Vehicle

D. RICHARD BLIDBERG, MEMBER, IEEE, AND STEVEN G. CHAPPELL, STUDENT MEMBER, IEEE

Reprinted from the *IEEE Journal of Oceanic Engineering*, Vol. OE-11, No. 4, October 1986. Copyright © 1986 by The Institute of Electrical and Electronics Engineers, Inc. All rights reserved.

(*Invited Paper*)

Abstract—For the past several years the Marine Systems Engineering Laboratory (MSEL) has directed its efforts towards the development of the technologies required for unmanned untethered submersible vehicles. The current focus of those efforts is to develop a system architecture that will allow the implementation of a knowledge-based guidance and control system. The goal of this effort is to implement a simple system which has addressed the basic problems and will allow for expansion as insight is gained from field testing the concepts using the Experimental Autonomous Vehicle (EAVE) system at MSEL. This paper considers those factors that have driven the development of an architecture which is being implemented in the EAVE vehicle system. Its intent is to focus on those issues that have guided the application of artificial intelligence (AI) techniques to meet the requirements of the system and its mission. The architecture being implemented is outlined and some of its features detailed.

I. INTRODUCTION

RECENT advances in many areas of technology are bringing the use of autonomous systems closer to reality. Many missions previously considered too dangerous for manned systems can be attempted if truly autonomous systems can be developed. The acceptance of these systems will come about only if their feasibility is proven by demonstrating that it is indeed possible to develop systems that have sufficient on-board decision-making capability to control their function in unexpected situations. This does not imply that an autonomous system must deal with all and every imaginable situation, but rather that technology exists which allows the development of a system able to perform basic tasks. Such simple systems will evolve into far more competent devices.

II. BACKGROUND

Initial autonomous vehicle development was directed at well-defined missions where the complexity of the system was relatively low and the demands placed on the hardware and software quite minimal. The undersea arctic research vehicle (UARS) at the University of Washington Applied Physics Lab was developed without microcomputer technology [1]. It was able to perform its under-ice survey mission quite well. As the potential for untethered vehicles is realized, the missions expected of them increase in complexity. This forces more and more complexity in the hardware and software and also forces consideration of new methods to fulfill the design requirements. The operation of the vehicle system changes during various phases of a mission. More and more sensors are

Manuscript received April 23, 1986.
The authors are with the Marine Systems Engineering Laboratory, University of New Hampshire, Durham, NH 03824.
IEEE Log Number 8612422.

required, navigation demands increase, and the guidance functions become more and more complex. As the thought of long-duration missions is considered, the problem of reliability increases dramatically. All of these changes result in a system with many variables, component parameters, and an overwhelming amount of data that are noisy, unreliable, yet critical to the functioning of the overall system.

What is required, then, is a design methodology which allows for more complexity by forcing modularization of the system hardware and (perhaps more importantly) the system software. This modularity would ease the inevitable design changes which occur during the system development process. Also necessary are bookkeeping techniques built into the system for handling large amounts of data, and for accessing the information content of the data. Additionally, the system must employ methods that prevent the combinatorial explosion which can result when logical solutions are driven by all possible combinations of mission, system, and control variables.

Upon examination of these goals it is realized that many of the techniques used by artificial intelligence (AI) researchers are appropriate for the design of an autonomous submersible system [2]–[6]. The problem, however, is that those techniques are used in an environment where time is really not a constraint. Their programs run in time domains of minutes or even in hours, whereas the demands placed on an autonomous submersible vehicle require decisions in the order of seconds. The problem then is to implement a methodology that allows the use of AI techniques within a definite time domain.

Without communications, remote control over system operation is impossible; all system control must be accomplished on board. This implies that not only the operator's intelligence, but also the knowledge and data base used by the operator, must be on board. As a result, we must develop methods by which to adjust operational parameters for a given situation. That situation will be uniquely defined by the environment, mission, system status, and control state. First, these methods must provide a way to *assess* the appropriate data and parameters; hence situation. Second, they must provide a strategy that interprets the operational goal, derived by the data assessment process, into *action* which determines the operation of the system.

III. ASSESSMENT

Before considering the mission planning/supervision function, first consider the functions of data assessment. The output from this assessment module is required in one form or another by nearly all functional levels in a system hierarchy. In the untethered vehicle, the situation is described uniquely by the information that is abstracted from the various sensors on board (both system monitoring sensors and environmental monitoring sensors). Those signals, developed from on-board sensors, are the only means by which the system can understand the world within which it exists. Mission requirements are forcing the inclusion of more and more sensors in vehicle systems, which are then required to process a far greater amount of information from those sensors and their signals.

Two problems must be solved. First, a method must be developed for adding, modifying the performance of, or deleting a specific sensor without major impact on the software system. The method must allow higher level functions to have an impact on sensor function; hence performance. Second, a method must be defined which would allow abstraction of the output signals from pure data to signal, to signs (representations of the physical world of being), to symbolic descriptors (representations of the personal world of meaning). This process of abstracting high-level information from lower level data must include the capacity to combine information from multiple sensors. The information should also be available in the format required by a particular module (i.e., the same data may be represented in feet, time, mean with variance of zero, etc.). As the demands placed on the sensor system increase, however, and the information describing the functioning of the system becomes more complex, an overall system awareness of the various sensors, their capabilities, limitations, and peculiarities must be maintained in an easily modified knowledge base.

Some characteristics of the assessment process that must be considered are the following.

1) The assessment process is asynchronous in nature.

2) Some system functions require access to information in a time-critical manner.

3) Many functions require information derived from data acquired over sampling intervals which are greater than the time domains within which those functions exist.

4) The assessment process is a function of contextual information resulting from various mission, system, environmental, and control parameters.

5) A quality factor must be associated with any information abstracted from data either by corroboration of multiple data sources or appropriate processing of a single data source.

IV. MISSION PLANNING SUPERVISION

An autonomous submersible can be considered as a system functionally situated in the spectrum bounded by a spacecraft and a manned submarine. On one hand, the spacecraft mission requirements have been planned in enormous quantitative detail and exhibit a deterministic structure. On the other hand, the submarine pilot, concerned with the evolving situation and the complexities of potential threats, develops a qualitative assessment of a situation and determines an operational decision based on his experience and judgment.

In current unmanned systems, operational parameters are preset before the start of the mission. It is also well understood what the implication of any instruction is on the subsystems. Assuming communications to and from the system, it is possible to understand the overall system status, in near real time, and make modifications to the system parameters that will modify the operational state of that system in the desired manner. The mission goals and subgoals can be defined quite explicitly since there exists, on a regular basis, a window when communication is possible. During this window it is possible to assess how well a subgoal is being met and modify, at a very high functional level, the subgoal parameters. This partial communication capability then drives the design philosophy used to establish the level of abstraction represented in the mission subgoal definition. Taking pictures of a planet for a spacecraft, or maintaining station at a specific location for a submarine, are conceptually similar goals. Both may have subgoals of following a vector or modifying course to adjust to some anomaly or monitoring on board resources. The difference results when one considers that, with the submarine, there are no communications. In an autonomous submersible, tasked to complete a complex mission, the subgoals must undoubtedly be modified, on board, during the course of the mission. The problem then is to develop methods by which this decision-making progress can be accomplished, and to determine at what level, between complete generality and meticulous specification, we are able to implement automated decision making in the near future.

An autonomous submersible must have the capability to plan the manner in which it is to complete its assigned mission as well as operate or control the lower level functions of the system. This implies that it contains an understanding of the complex interrelationships between various subsystems and their impact on each other. It also implies an understanding of mission goals and the impact of environmental parameters on those goals and the system. This understanding must be sufficient to develop alternative operational configurations for a specific situation. The submersible must have the ability to adapt a mission plan to anomalies in system performance and environmental situation and to reason about vehicle system resources and their impact on the attainment of, subgoals. It must be able to adapt a mission plan or mission subgoal in response to contextual parameters resulting from the mission goals, environment, system, or control state.

An important aspect of its mission planning function is the ability of the system to use on-board maps to assist in the attainment of mission goals. It seems impossible to consider an autonomous system that is tasked to perform a long or complex mission without having, on board, some limited understanding of the environment. This implies, considering the relatively limited detailed understanding of the ocean, that a method must be implemented which allows the system not only to use the knowledge in an on-board model, but also to modify that model from information derived by its own sensor suite. The on-board model would then provide some context that would bound the planning and replanning process.

Most importantly, an autonomous system must be able to cope with an unexpected event and continue to function. Many possible problems, however, can be anticipated in generic terms prior to the mission. The impact of this on the system is probably only to require rather large storage capability and fast access to the data. A truly unanticipated event presents a much more substantial problem. Considering the variability of the ocean environment, and the limited success in learning systems, this may well be the "achilles heel" of an autonomous submersible. It is therefore very important to develop fail-safe methods that understand the vehicle system's limitations.

Current unmanned untethered submersibles have not progressed very far in addressing these issues. The great majority of these systems are designed to follow a preprogrammed series of paths in a defined environment. Some have implemented obstacle avoidance strategies. Recognition of an obstacle, however, has very little impact on the predefined mission other than to avoid the object in some generic manner and return to following the defined mission path. Algorithms have been implemented in some systems, which monitor the system integrity and can fall back to some predefined fail-safe operational mode when thresholds are exceeded.

In summary, some of the key planning issues are:

1) to establish, prior to the mission, a set of parameters, data, and procedures that allow control of the operational state of the vehicle sufficient to meet the mission goals;
2) to understand, during the mission, the impact of system status, control state, and environmental parameters on the operation of the system; and
3) to automatically modify, on board, during the mission, the operational state of the vehicle as dictated by the assessed situation.

V. KNOWLEDGE-BASED SYSTEM DEVELOPMENT PROGRAM

For the past several years the Marine Systems Engineering Laboratory (MSEL) has directed its efforts to a scenario in which there is no communication other than premission configuration of the system software. This implies that any decision making required during the mission is made by the on-board computers. In an effort to implement this capability on the Experimental Autonomous Vehicle (EAVE) an architecture has evolved which is hierarchical in nature and allows for structuring decision making into time-ordered hierarchical levels. The goal of this program is to perform a feasibility demonstration of a knowledge-based guidance and control system in an unmanned untethered submersible vehicle.

Four steps have been defined to reach that goal and are currently being undertaken: 1) define a software/hardware architecture that allows implementation of AI decision-making techniques in the EAVE vehicle; 2) develop that architecture in a simulated real-time environment that models, as appropriate, the EAVE vehicle system; 3) evaluate the function of the architecture by simulating a simple mission using the EAVE system model; and 4) verify the results of that simulation by utilizing the EAVE system to perform the defined mission in the water.

To accomplish the goal, a knowledge-based guidance and control system will be implemented on the EAVE vehicle and a demonstration test performed. The demonstration will require the EAVE vehicle to perform a survey over a predefined area. The basic parameters of that survey, such as the total square area and the desired coverage, will be defined. The details such as line spacing, line direction, etc., are left to be developed by the vehicle system. A simple obstacle-avoidance sensor will be added and interfaced to the guidance system. The vehicle system will then transit through the search area performing a simple mapping survey. Objects (unknown to the vehicle system) will be placed in the search area. The vehicle will have to detect, avoid, and map those objects and still achieve its objective of completing the mapping mission. The EAVE system will be required to start from some initial point, perform the survey, and return to the starting point.

VI. THE ARCHITECTURE

Before discussing the architecture it is helpful to understand the process by which an operator controls a remotely operated underwater vehicle (ROV) since the difference between an ROV and an untethered ROV is primarily that the communication channel has been eliminated or, at the very least, dramatically limited.

If one considers the data and instruction flow in a tethered ROV, it can be seen that there is a relationship between the bandwidth limits of the communication channel and the level of information being passed through that channel. Fig. 1 helps to understand this. If we assume that current ROV's have bandwidths in the order of a 1 Mbit/s, then the requirements for decision making, on board the vehicle, remain minimal. The assessment and planning decisions are made by the human operator since he has available to him a great deal of data and can, conversely, pass many pieces of detailed control information back to the underwater vehicle.

As the communication becomes more restricted the information content of each word passed through the channel must increase. This places a greater burden on the on-board computer systems to interpret or assess raw data and transmit only the information contained in that data. Also, the on-board system must compile, from high-level instructions, the necessary control variables which determine the system response.

The architecture being implemented on the EAVE vehicle is an attempt to define a structure that mimics, on board, many of the functions that an ROV operator would perform.

Fig. 2 is a block diagram which describes a hierarchical time-ordered architecture which has been defined for the EAVE-East vehicle system. This architecture evolved from considering that an autonomous vehicle performs two major functions. First, it *assesses* the data being acquired by a suite of sensors to determine the current situation and system status. Second, the vehicle *plans* the various control sequences that are required to accomplish its mission goals.

When considering Fig. 2, five generic functions can be seen, each falling within the boundaries of a module of the hierarchy. At the lowest level are the common *real-time* systems, including sensors and effectors. At the next level is the *Data Assessment Blackboard*. This module abstracts from sensor data objective information (signs) which offers a physical description of the environment or system status (bottom is rising, power is low). The signs developed by the blackboard are input to the third module: situation assessment.

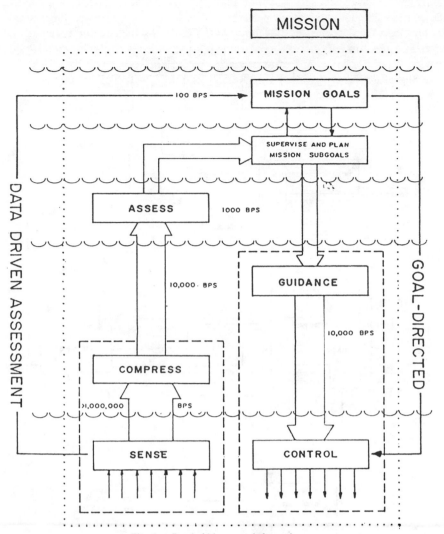

Fig. 1. Bandwidth versus information.

Situation Assessment abstracts from those signs a subjective understanding of the system and its environment (bottom is rising *too* fast or power is *too* low). These "symbols" or symbolic descriptors are used by the fourth module, *Supervisor*, to accomplish two goals: one of assessment and one of planning.

The Supervisor Module acts as the interface between the data assessment functions and the action-oriented instructions. It is driven by two sets of goals. The first set is derived from the actual mission goals, those that are determined by on-board data, maps, and mission parameters. The second set of goals are those that result when the system is required to react to problems, i.e., those that are driven by acquired data/ symbols.

An initial plan is developed by considering the on-board data (maps, mission parameters, and current status data). When a problem is recognized (as determined by the existence of a symbol), the current plan must be replanned by considering the on-board data as well as those data that have been acquired by the sensors.

The Supervisor Module is comprised of two major components: 1) a problem-solving function; and 2) a path-planning function. The problem solver is triggered by the firing of a symbol (or state-based/context-based event). Once a symbol is recognized, a local expert, developed to consider the implications of that symbol, becomes active. It considers the status of the other symbols to evaluate a set of goals that would solve the recognized problem (as defined by the symbol). It also decides the importance of the problem relative to the overall status of the system and environment. This decision then establishes some scheduling criteria (importance of completing the defined goals). The set of goals and a set of scheduling criteria is passed to a conflict-resolving expert. The conflict resolving expert adds that information to the information it has obtained from other local experts (which have been activated by other symbols). The responsibility of the conflict resolver is then to develop a set of prioritized goals that represent the best guess as to how to resolve all of the problems represented by all of the symbols that have been activated. This set of goals is then added to the goals derived from the mission definition.

The path-planning function of the Supervisor then uses the mission goals and the goals determined from the problem solver to govern the overall planning process [7].

The planning process involves developing five types of information.

1) Plan a desired set of path segments.

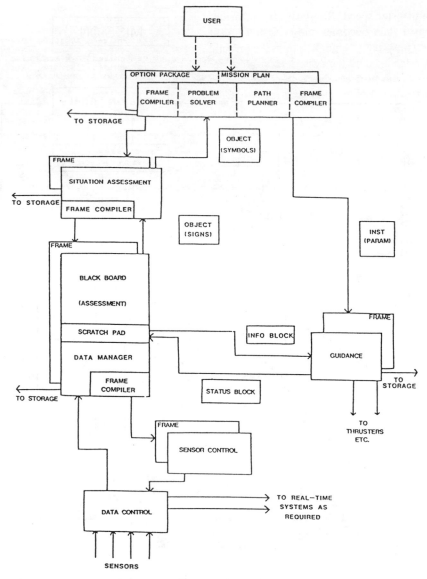

Fig. 2. Hierarchical guidance and control structure.

2) Plan a set of specific instructions (for each segment).

3) Determine the operational envelope (for each segment).

4) Determine avoidance and emergency scripts.

5) Compile the control frames[1] for the system.

In each of the five planning steps, specific information must be developed while meeting the demands placed on the process by the various goals and subgoals. The result of this process is the mission plan which determines the action taken by the fifth module, *Guidance*.

The Guidance Module's function is only to act as a complex autopilot system. Once given a specific instruction, e.g., follow a path segment, it directs the real-time control of the effectors (thrusters) such that the path following of the vehicle stays within predefined limits.

This quick overview of the function of the modules in the

hierarchy described in Fig. 2 is expanded in papers by B. Otis [8] and S. Chappell [9]. The important point is that an assessment process provides a context from which specific actions are determined and implemented. The overall process can be grouped into modules that require output with differing time constraints; i.e., a decision concerning what path should be followed might be required only once in a multiple-hour time period yet thruster speed may need to be changed every second.

VII. KNOWLEDGE REPRESENTATION

Key to the development of the architecture of Fig. 2 is the representation of the knowledge and instructions passed between various modules. When the functions required by an unmanned untethered submersible are analyzed, it is natural to organize those functions in a predominantly hierarchical structure. Representations of mission environment, and data can be considered hierarchical as well. A particular mission goal, for example, could be described in high-level terms as performing a search while at the lower level the same goal

---

[1] A control frame is a data structure containing parameters that control the behavior of a given module. Any module controls the functioning of the module(s) below it by modifying parameters in the lower module's control frame.

would require setting a thruster speed. Similarly, the information obtained from sensor data becomes more abstract as it ascends the hierarchy (i.e., data, signals, signs, symbolic descriptor) [10]. Knowledge used by the system is obtained by abstracting from basic data the situation descriptors that have a very high level of information content; data that represent the distance to the bottom can, over a period of time, be abstracted to determine that the slope of the bottom is such that a collision with the bottom is imminent if something is not changed.

In an effort to distinguish different types of system behavior and to modularize the system into separate hierarchical modules which can be dealt with individually, the following conventions have been adopted. Knowledge in the system is of two types. The first is that knowledge or information that evolves from the data *assessment* process. The second results from the planning/problem-solving *action* of the system.

A. An Assessment Information Hierarchy

Data are defined as the raw data obtained from the sensor with all of their noise and other artifacts.

Signals are the low-level data produced by system sensors. Signals represent the raw sensor data, with noise removed, and are introduced to the assessment system by the data controller. These signals are the basic element from which all higher levels of information must be extracted.

Signs are developed within the Data Assessment Blackboard. The Blackboard uses signals from the sensor manager as input and produces, from them, a three-dimensional physical world representing both the internal status and the environmental surroundings of the vehicle. Sign is the term used for the pieces that make up this higher level description. Signs are derived through the process of examining signals over time and combining information from more than one sensor at a time. Taken together, signs provide a detailed three-dimensional physical picture of both the system's surroundings and its internal status. The purpose of the Blackboard Module is to produce an adequate set of signs from the available signals.

Symbols are produced with the Situation Assessment Module. Signs from the Blackboard become the input for the processing to take place in the Situation Assessment Module. The goal of Situation Assessment is to extract from the world model, provided by the signs, a qualitative representation of the current situation. Symbols are derived by analyzing the set of signs with both mission- and system-specific knowledge. This knowledge is provided by the Supervisor Module and allows Situation Assessment to operate differently depending on the current goals of the Supervisor module. Symbols represent the two fundamental aspects of data assessment: problem recognition and goal attainment. The result of the entire assessment process should be the derivation of a set of symbols that the Supervisor Module must take into account for successful mission completion.

B. An Action Information Hierarchy

Command-based action is defined by the level of action equivalent to setting a thruster speed. It results from a direct instruction coming from a higher level module.

Skilled-based action is best represented by the transfer

TABLE I

• recognized	AVOIDING_OBJECT
• recognized	SEGMENT_COMPLETE
• recognized	HIT_SOMETHING
• recognized	INTERNAL_LEAK
• recognized	MAP_CHANGE
• determined	CURRENTS_TOO_HIGH [dir of currents]
• determined	BOTTOM_RISING_TOO_FAST
• determined	ALTITUDE_VARIATIONS_TOO_GREAT
• determined	TOO_SHALLOW
• determined	TOO_DEEP
• determined	HI_POWER_BAD
• determined	MID_POWER_BAD
• determined	LOW_POWER_BAD
• determined	USAGE_PREDICTIONS_EXCEEDED
• determined	CURRENT_USAGE_TOO_GREAT
• determined	FAULTY_sensor [NAV]
• determined	DATA_NOT_VALID [NAV]
• determined	SYSTEM_CAUSING_PROBLEM [NAV]
• determined	FAULTY_SENSOR [A/D]
• determined	DATA_NOT_VALID [A/D]
• determined	SYSTEM_CAUSING_PROBLEM [A/D]
• determined	FAULTY_SENSOR [COMP]
• determined	DATA_NOT_VALID [COMP]
• determined	SYSTEM_CAUSING_PROBLEM [COMP]
• determined	CANT_HOLD_HORIZ_POSIT
• determined	CANT_HOLD_BERING
• determined	CANT_HOLD_DEPTH
• determined	NOT_MOVING
• determined	MOVING_TOO_SLOW
• determined	MOVING_TOO_FAST
• modified map	CURRENTS_MAP
• modified map	OBJECTS_MAP
• modified map	BOTTOM_MAP

functions that exist in the Guidance Module currently implemented in the EAVE system, the algorithms that exist to implement the control of the thrusters, etc. Their function is controlled by the parameters, which are set up in the control frame, which is given to the Guidance Module by the Supervisor Module.

Rule-based action is best described by the "canned scripts" that will be developed for emergency situations such as those resulting from the "emergency triggers," such as "internal leak" or "hitting something." They consist of a sequence of instructions that will make the system respond in a very defined manner. The scripts will be parameterized such that the specific parameters will be set by the Supervisor and passed to the Guidance Module in its control frame.

Knowledge-based action will consist of a series of path segments or instructions that have been developed by the Supervisor Module to meet the proposed plan. The action resulting from that series of instructions will be in direct consequence of the assessment of the control state, system status, environmental conditions, and mission plan.

C. An Information Structure

The development of this architecture depends on one module's being completely separated from another module and relying only on input and output definitions. How the information passed between modules is developed by other modules is of little importance; only the information itself must be of interest. To meet this need, a rigid structure for representing the knowledge has been defined. As an example, Table I lists some of the currently defined symbols developed by Situation Assessment and passed to the Supervisor Module.

From those symbols, the Supervisor Module must determine the system's status and establish the constraints that will bound the planning process as it develops appropriate action commands. As defined in the architecture, each one of those symbols will be transferred to the Supervisor Module with an object attached that describes, in a rigidly structured manner, what information was used to develop the specific symbol. In this manner, the higher level module has access to a great deal of information, but does not have to use it except on occasions when it deems that necessary. The higher module may or may not choose to use the information; however, it does know what to expect. It is this rigid structure that allows the development of each module to proceed independently of any other. As an example, all of those symbols are being developed using the following structure:

```
recognize:
determine:
   (these are "symbols")

   reason being:
   caused by:
      (these are "symbols")

      factor considered:
      map modified:
         (these are "signs")

         :type
            (this is a value for a "sign")

            :value
               (determined by calculation)

            :parameter
               (determined by user)

            :variable
               (determined from data)

            - method ------
               (subroutine).
```

Table II is an example of the underlying structure which results in the symbol BAD_NAV_SNSR. The supervisor module may or may not choose to use the information; however, it does know what to expect.

VIII. SIMULATION TOOL

In order to test the proposed hierarchy architecture (Fig. 2), a "skeleton" or model of it has been implemented on the laboratory's development system (a dual 68000 based machine which runs a UNIX look-alike operating system). This skeleton is composed of six independent processes, which together form a simulation of the hierarchy. Each process represents a module or node in the hierarchy architecture. The processes communicate via the development system operating the system's queuing mechanism, which allows data exchange in packets called messages. All code is currently implemented in "C." Fig. 3 diagrams the skeleton as a six-process family, with communications lines connecting the members.

Several key requirements have influenced the design and evolution of the hierarchy simulation. These requirements reflect MSEL's experience with autonomous vehicle software development.

The first requirement was that the simulation architecture be faithful to the hierarchy architecture. The simulation must be able to exhibit the several time domains that exist in the hierarchy as well as its independent nodes. This led to the multiprocess family, which can be used to test the hierarchy communications, time domain partitions, and other global concerns.

The second requirement is that each node be developable in isolation from the rest of the family. This allows several programmers to work on different nodes, debugging them to a certain degree (for local processing correctness), without each programmer calling up the entire simulation family. In order to assure that separately developed processes communicate correctly, the interface protocol must be carefully designed and adhered to.

The third requirement is that the conversion from simulation code to actual vehicle module code be as "painless" as possible. Every possible effort must be expended to prevent a development system-specific code from getting into the vehicle module code. In cases where such preventative measures are impossible (for example, input and output routines), that specific code must be isolated from the rest of the process/module's code. This is very important since the actual vehicle hardware is still in the design stage. In order to maintain awareness of this requirement, the nodes in the simulation are often refered to as "process/modules."

The fourth requirement is that each node must "log" its operation in some manner. The log should be a text file so that it may be saved and viewed after the node has run. Such a logging capability allows for tracing a node's performance after a simulation has run in order to identify conditions that lead to problems or errors. After transporting the code to an actual autonomous vehicle, the concept of performance logging is still important (perhaps more so) for postmission analysis.

A. Mapping of Hierarchy Nodes to Simulation Nodes

The Supervisor, Situation Assessment, Guidance, and Blackboard hierarchy nodes of Fig. 2 are modeled by the processes named "super," "sitass," "guide," and "blkbrd," respectively. The fifth process, called "environ," is a simple simulation of the physics of the environment in which the vehicle will be operating, coupled with a very rudimentary Sensor Manager. These five processes are represented in Fig. 3 by appropriately named circles. Environ generates numerical data that simulate a point moving in space and time as perceived through a defined sensor suite. The environ process also simulates the consumption of vehicle power over time and contains a simple model of bottom topography. Environ is therefore the data-supplying process in the hierarchy simulation. While data flow up the left side of Fig. 3, control signals flow down both sides. Control commands, in the form of orders to do something, are issued by the Supervisor. These commands are relayed down to ever lower process/modules just as the environmental data are relayed up. In order to bring the simulation family up in an orderly manner for a simulation run, a sixth mother process (hence the name "mom") was created. This process is found in the upper left portion of Fig. 3.

TABLE II

```
determined BAD_NAV_SNSR

   reason being - DATA_NOT_VALID

      caused by:  LOSING_SIGNAL

            evaluate factor:  type_of_loss
               :type  (burst)
               :type  (random)
               :type  (cyclical)
               :type  (total)
                     :value  (% loss/unit time)
                     :value  (randomness factor)
                           :variable  (time)
                           :variable  (range reading)
                                 :parameter  (burst threshold)
                                 :parameter  (-- thresholds --)
                                      --- method ---
                                      . . . . . . .
                                      . . . . . . .

      caused by : RANGE_READING_VARIABLE

            evaluate factor:  type_of_change
               :type  (unexpected)
               :type  (spikes)
               :type  (variable)
                     :value  (predicted range)
                     :value  (time)
                     :value  (average range)
                     :value  (variance)
                           :variable  (velocity)
                           :variable  (time)
                           :variable  (range reading)
                                 :parameter  (---------)
                                      --- method ---
                                      . . . . . . . .
                                      . . . . . . . .

      caused by:  ERROR_EXISTS

            evaluate factor:  type_of_error
                  :type  (random)
                  :type  (system)
                  :type  (bias)

   reason being - SENSOR_HARDWARE

      caused by:

   reason being - VEHICLE_SYSTEM
```

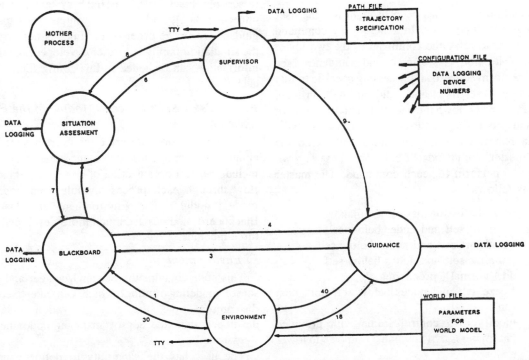

Fig. 3.   Hierarchy skeleton.

B. Mapping of Hierarchy Interfaces to Development System Queues and Files

The nodes of the hierarchy communicate with each other through uniquely defined "interfaces." In the simulation, these interfaces are mapped to unique message queues which are maintained by the development system with a special set of interprocess communication (ipc) routines. These interface lines are shown in Fig. 3 with identification numbers. All messages are in ASCII text to allow easy human viewing and manipulation of interface contents. In an effort to prevent process/module code from becoming dependent on the development system ipc package (or any other input/output (I/O) scheme) and, therefore, nonportable to other systems, an I/O "adapter" has been implemented. This adapter consists of a generic input routine and a generic output routine (which are included in each process/module) through which all nodes perform their interprocess I/O. All development system ipc specific code is isolated in this adapter. When the actual vehicle hardware is finalized and the modules' interconnects built, only the I/O adapter routines will need modification to convert the simulation code (which relies on the development system ipc routines) to actual vehicle code (which will rely on yet to be specified I/O routines).

The I/O adapter provides another service. In order to answer the question of how to provide for independent process/module development (requirement number two, above) the adapter has been designed to be able to access the development system file system as well as the ipc mechanism. This means that any process's I/O can involve other running processes or static files. This feature allows a user to work on one node without having to invoke the entire family. He need only supply the proper files to take on the form of interfaces, and the I/O adapter handles the rest.

C. Control Commands of the Hierarchy Simulation

Once the simulation family has been brought up, the user controls it through the Supervisor. In general, any command causes the Supervisor to relay the command to the two lower process/modules (Situation Assessment and Guidance, see Fig. 3) and then perform some local processing specific to that command. Situation Assessment and Guidance will, in turn, relay the command further down the net, and then do their own specific command processing. Lower nodes follow the same algorithm. As a result, all process/modules understand the same set of "master commands," but each has its own particular details to perform for each command. The master commands are as follows:

help	print the list of master commands
cstart	coldstart self and nodes below self
wstart	warmstart self and nodes below self
init	initialize self and nodes below self
run	do "normal" processing
pause	freeze self and nodes below self until next run
framod	modify own control frame with provided parameters; may involve "framod" to lower nodes

tscope	time scope; "expand" time to slow cycle rates relative to real time
shutdown	shut self and nodes below self down.

D. Viewing Individual Process/Module Output

Given that there are five members in the simulation family, the user will find it very useful (indeed mandatory) to view the output of more than one process at a time. The optimal configuration would be one terminal per process/module, especially for demonstrations. In most cases, however, one or two terminals would suffice. Additionally, it is not feasible to tie up the entire development system (use all of its ports) whenever the hierarchy simulation is in operation. To answer these concerns, each process/module's log output has been designed such that it can be directed to any arbitrary terminal or file. The target terminal or file is specified in a configuration file, which each process/module opens and reads when it comes up. The config file is shown in Fig. 3, upper right. By editing the config file prior to a simulation run, the user can specify where each process/module's log data will go. The present format of the config file is shown in the following example:

super	/dev/tty0
sitass	/dev/null
guide	guidance.log
blkbrd	/dev/tty1
environ	/dev/tty3.

Each process has a line of configuration information in the config file. The lines are indexed by process/module name, which is the first field in each line. The second field contains the log destination in UNIX path name format. In the above example, the Supervisor will send its log output to terminal 0, the Blackboard to terminal 1, and the Environment to terminal 3. Situation Assessment sends log output to the "null device" (a sort of "black hole" in UNIX systems, which can receive unneeded data), and Guidance output goes to the file name guidance log. The order of the lines is not important, and as the simulation family is further developed, other configuration information can be added to this config file as new fields for each line.

E. Details of Each Process/Module in the Hierarchy Skeleton

The next sections describe the current state of each of the simulation's process/modules in some detail. Such details include general organization of code and data structures, data flow through each process/module, and operation of each process/module. The Environment and Data Assessment Blackboard modules are much more developed than the other three modules.

F. Environment

This node contains the Sensor Manager and a collection of sensor modeling routines which generate "sensor data." At the present time, the data generation code is far more developed than the sensor manager code, hence the name "environ."

The user has the capability to define constants such as

voltage and current start levels and drop rates, maximum cruise speed and dimensions of the "world." During the simulation, problems can be imposed on the vehicle such as a leak in one of the tubes or collision with obstacles.

The Environment is somewhat of a special case when compared to the other process/modules in the simulation. Since it is the source of all environmental data for the simulation, it is also the "problem generator" for the hierarchy simulation. In order to test the other modules' reactions to problem situations such as collision, faulty power levels, and pressure tube leaks, the Environment must be able to deliver the sensor data that represent those problems. Since problems are transitory and utterly random in nature, it has been decided that they will be instigated at the user's whim during the running of a simulation. This means that the user must be able to communicate with the Environment (to instigate problems) as well as the Supervisor (to issue vehicle commands) during a simulation run. Therefore, the Environment has a bidirectional log port (for input and output) unlike the other four modules, which all have unidirectional (output) log ports.

The log output generated by the Environment is so voluminous (20 values each cycle) that the output is formatted in a screen-oriented manner. This introduces special screen control codes into a log output stream. While this log output can be captured in a file (as with the other process/modules) for postmission analysis, the file can only be played back on a video terminal.

G. Simulated Data

1. Time: The Environment is not currently "locked" into real time. It generates the passage of time by incrementing a counter on each cycle. Each increment of the "clock" represents one second.

2. Position: The x, y, z, bearing, altitude, and headroom values are generated by a simple motion simulator which moves a single point (representing the vehicle) through space and time in an ideal world (no surface ice, flat bottom). The z value is depth according to a pressure transducer, headroom is depth according to an upward-looking sonar. The motion simulator accelerates a point to a maximum speed, cruises at that speed, and then decelerates to the target position. The bearing is calculated similarly: the "vehicle" accelerates to a maximum angular velocity, "cruises" at that rotation rate, and decelerates to the target bearing. The motion simulator is driven by path commands which specify which x, y, z, and bearing to go to. The simulator is quite general, since it can generate movement in all four variables simultaneously. It also makes no effort to simulate the hydrodynamics of any particular vehicle.

3. Power: The current EAVE vehicle has three power systems: high (24 V), mid (16 V), and low (8 V). A voltage and current level are calculated for each system on each cycle. The levels decrease linearly over time from an initial value with each tick of the environmental "clock."

4. Impact and Leak Sensors: The impacts and leaks are currently simple on/off values. There is an impact sensor for each axis. The value is set "on" for one cycle if the vehicle meets with a restriction in motion along that axis. For example, if the "vehicle" is moving in both x and y and attempts to pass through an obstacle, the x and y impact sensors will be triggered for one cycle. The user can specify (in the world file) the locations and sizes (always cubes) of two obstacles prior to running the Environment. A third "instantaneous" obstacle can be caused to appear directly in front of the "vehicle" at any time by selecting the collision problem.

There are four leak sensors, one for each electronics pressure tube on the current EAVE vehicle. A leak is turned on by the user during simulation by selecting the leak problem and it remains on until selected again (that is they toggle on or off with each selection).

H. External Data Files

The Environment accesses two external data files, "world" and "path" when it comes up. These files may be found in Fig. 3 on the right. The world file contains all constants which define the world (e.g., depth of the bottom) and vehicle-specific constants (e.g., maximum cruise speed). There are currently 29 parameters in the world file.

The path file contains the motion commands that drive the motion simulator. It is of the following format.

start	x	y	z	b
go	x	y	z	b
go	x	y	z	b
•				
•				
•				
end.				

The "start" command gives the motion simulator its initial position in the world. A stream of "go" commands gives the motion simulator target positions to attain. The "end" command stops motion simulations.

The implementation and operation of the Environment Process/Module is documented in [11].

I. Data Assessment Blackboard

This node represents the first step in the data assessment process. It incorporates some (but not all) of the "blackboard" concepts pioneered by various systems and goes by that name. There are two basic parts to the procedural component of the Blackboard: the data base manager and the set of knowledge sources.

The data base manager performs the mundane data base management chores: receive data from the Environment and Guidance, store the values in the data base, and maintain the ring buffers. It has the added duty of activating knowledge sources according to the data it receives. It is responsible for the activation of the "always-on" (or ground-level) knowledge sources.

The concept of knowledge sources is the most important idea borrowed from the literature on Blackboards [12]–[14]. A knowledge source is a unit of encapsulated knowledge pertaining to one restricted item: in this case, a particular sensor reading. A knowledge source is named after the sensor whose readings it processes. Consequently, each sensor has at least one knowledge source associated with it. A scheduling scheme has been implemented which allows the knowledge

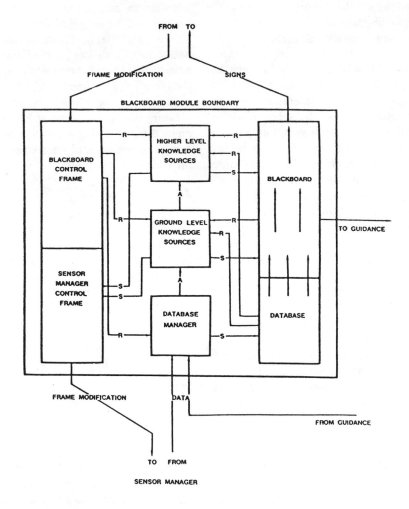

S = STORE OPERATION
R = RETRIEVE OPERATION
A = ACTIVATE OPERATION

Fig. 4.   Block diagram of Blackboard Module.

sources to be activated in a LISP-like manner. The knowledge sources are currently *C* functions.

The concept of higher level knowledge sources has been implemented in the form of a pair of knowledge sources which are activated only when it is determined by the navigation knowledge source that the vehicle is moving in the x-y plane (laterally). One checks for bottom variations, the other for surface objects and/or topography. The scheduler is able to handle such "sometimes-active" knowledge sources. The scheduler itself is also isolated in one file and may be considered as a knowledge source in its own right.

The basic flow of data through the Blackboard Process/ Module is diagrammed in Fig. 4. The central column of three boxes in the diagram represents the code space of the Blackboard Module. The data handled by the Blackboard are divided into four varieties, two (data base and blackboard) on the right of the diagram and two (control frames) on the left of the diagram. The arrows within the Blackboard Module boundary refer to operations (functions) that occur during Blackboard execution.

When data arrives at this module (either from the Environ-

ment or Guidance), it is stored in the data base by the data base manager which also activates the proper ground-level knowledge source (GLKS). That GLKS will then retrieve the data it needs from the data base and post any results in the blackboard data space. The GLKS may also activate a higher level knowledge source (HLKS). As data is stored in the data base and the various knowledge sources are fired (which results in new data being posted on the Blackboard), there is an evolution of data into signs at the top of the Blackboard (see right side of Fig. 4). At all times, the data manager and knowledge sources retrieve their controlling parameters from the Blackboard Control Frame. Any decisions that require altering the behavior of a sensor are stored in the Sensor Manager Control Frame. Control of Blackboard Module behavior is effected by Situation Assessment sending down modifications to the Blackboard Control Frame. In a similar manner, the Sensor Manager (which exists in the Environment) is controlled by the Blackboard transmitting the Sensor Manager Control Frame.

When running, the Blackboard consumes raw data from the Environment and produces signs. Currently, the signs are reported out the Blackboard's log port for human viewing, and not to Situation Assessment. The only signs sent up to Situation Assessment are those counting received Environment and Guidance data sets.

J. Supervisor

The current version of the Supervisor serves mainly as a placeholder in the hierarchy simulation. It handles commands issued by the user and services the interfaces that connect it to Situation Assessment and Guidance. It simulates the function of path planning by accessing a user-supplied path file (Fig. 3).

This module will presumably be the most "AI intensive" module in the entire hierarchy. Consequently, implementing it in C will be difficult. The most productive route to follow seems to be to develop the Supervisor in a symbolic environment up to a certain degree of usefulness. The measure of usefulness is yet to be determined, however, it will certainly involve the Supervisor's ability to plan its own path, monitor the execution of that path, and replan in the face of unexpected problems. Transporting such a program to the actual vehicle environment will pose difficult problems above and beyond the standard "downloading" problems. Most important is the question of how to translate, compile, or otherwise transport a symbolic language and its environment into an embedded computer system. The recent acquisition by MSEL of a computer system which supports both LISP and C and high-speed communication between the two environments has provided some of the tools needed for experimentation with this problem.

K. Guidance

This node in the hierarchy is unique in that it is fully implemented in the current EAVE vehicle. However, it has been deemed inappropriate to transport the vehicle implementation to the development system at this time. Consequently, only the most important guidance function has been implemented in the hierarchy simulation. This function is the determination of a path segment made good. To accomplish this, Guidance checks each position reported by the Environment for a match with the target position as specified by the Supervisor. When a match is found, Guidance reports "path made good" to the Blackboard.

L. Situation Assessment

The current version of Situation Assessment is similar to the Supervisor in that it is presently a placeholder. This node is essentially the continuation of the data assessment process started by the Blackboard node. The difference is in the level of abstraction. The output of the Blackboard (signs) have meaning in the physical world, while the Situation Assessment output (symbols) have meaning relative to the vehicle's "personal point of view." Assessing situations is the process of understanding the current state of things (situation) and producing symbols which represent that situation in the planner's current context. These symbols serve as triggers for the Supervisor.

While the assessment process in this node is structurally

very similar to that of the Blackboard, the level of abstraction is much higher. More concrete work on this node must therefore wait for other work concerning the Blackboard's signs, and the symbols expected by the Supervisor. The problem is that situations define a context, and contexts are difficult concepts to describe. The definition of recognizable situations must follow the development of the description language.

IX. SUMMARY

Unmanned untethered submersibles are becoming a viable tool to compliment the capabilities of manned and unmanned tethered submersibles. They offer the ability to perform complex missions in a very cost-effective manner.

The Marine Systems Engineering Laboratory is focusing its efforts on developing and implementing a knowledge-based guidance and control system for the EAVE vehicle. A promising approach to this goal begins with a system architecture that is organized into hierarchical levels, where each level represents a different time domain and class of problem. With this approach it appears possible to implement decision making within the overall time requirements.

Knowledge-based system engineering is expanding dramatically, but one must still question the technology and how much judgment may be programmed into an untethered vehicle. One must also question whether sufficient time is available for real-time systems to develop necessary decisions. These are difficult questions. To develop a knowledge-based adaptive guidance and control system for an untethered vehicle is an ambitious task. A successful effort, however, would provide insights applicable to a wide range of real-time system problems.

REFERENCES

[1] *Undersea Vehicles Directory: 1985*, Busby Associates, Inc., Arlington, VA.
[2] M. Stefik, *An Examination of a Frame-Structured Representation System*, Comput. Sci. Dep., Stanford Univ., Stanford, CA.
[3] M. Stefik, "Planning with constraints," Ph.D. dissertation no. STAN-CS-80-784, Comput. Sci. Dep., Stanford Univ., Stanford, CA, 1980.
[4] K. DeJong, "Intelligent control: Integrating AI and control theory," Navy Center for Applied Research in AI, NRL, Washington, DC.
[5] L. Fagan, "VM: Representing time-dependent relations in a medical setting," Ph.D. dissertation, Comput. Sci. Dep., Stanford Univ., Stanford, CA, 1980.
[6] H. P. Nii and E. A. Feigenbaum, "Rule-based understanding of signals," in *Pattern Directed Inference Systems*. New York: Academic, 1978.
[7] R. Wilensky, *Planning and Understanding—A Computational Approach to Human Reasoning*. Reading, MA: Addison-Wesley, 1983.
[8] B. Otis, "Knowledge-based data assessment for an autonomous underwater vehicle," in *Proc. 4th Int. Symp. Unmanned Untethered Submersible Technology*, June 1985, p. 324.
[9] S. Chappell, "A Planner/Supervisor Module for an autonomous vehicle," in *Proc. 4th Int. Symp. Unmanned Untethered Submersible Technology*, June 1985, p. 309.
[10] J. Rasmussen, "Skills, rules, and knowledge; Signals, signs, and symbols, and other distinctions in human performance models," *IEEE Trans. Syst. Man Cybern.*, vol. SMC-13, no. 3, May/June 1983.
[11] M. Momenee, "The Environment Module," Univ. of New Hampshire, Durham, NH, Marine Syst. Eng. Lab. Rep., May 22, 1986.
[12] L. D. Erman *et al.*, "The Hearsay-II speech-understanding system: Integrating knowledge to resolve uncertainty," *Computing Surveys*, vol. 12, pp. 213–253, 1980.

[13] B. Hayes-Roth, "BB1: An architecture for blackboard systems that control, explain, and learn about their own behavior," Stanford Univ. Comput. Sci. Dep., Stanford, CA, Tech. Rep. STAN-CS-84-1034, 1984.

[14] B. Hayes-Roth, "A blackboard architecture for control," *Artificial Intelligence*, vol. 26, pp. 251–321, 1985.

D. Richard Blidberg (M'83) received the B.S. degree in electrical engineering from the University of New Hampshire, Durham, in 1967.

In 1967 he worked at Woods Hole Oceanographic Institution and the UNH Engineering Design and Analysis Laboratory. He then became Manager of Seabed Survey Operations in the Arctic and Eastern Canada for Ocean Research Equipment, Inc. Since 1976 he has been Associate Director of the Marine Systems Engineering Laboratory, University of New Hampshire, Durham. He has published numerous technical articles and serves on several scientific committees. His current research interests include knowledge-based guidance and control, applied underwater acoustics, acoustic navigation, underwater robotics, and the application of microcomputer technology to instrumentation for ocean science and engineering.

Mr. Blidberg is a member of AUVS and ACM. He was Cochair of the Technical Session on Autonomous ROV Developments at the ROV '85 Conference. He also organized a series of symposia on unmanned untethered submersible technology; the fourth in the series was held at the University of New Hampshire in June 1985. He serves as project leader for autonomous submersibles in the U.S./French Cooperative Program in Oceanography.

★

Steven G. Chappell (S'78) received the B.S. degree in botany from the University of New Hampshire, Durham, in 1978. He is presently enrolled in the M.S. program for computer science at UNH with all course work completed.

He joined the Marine Systems Engineering Laboratory (MSEL) at the University of New Hampshire in 1981. Since then, he has been designing, implementing, and maintaining software for MSEL's experimental autonomous vehicle (EAVE). His interests include real-time operating systems, autonomous systems, parallel processing, and artificial intelligence programming techniques applied to real world problems. He is currently working on his Master's thesis which will cover the implementation of a prototype mission planner for an autonomous vehicle.

A vision system for a Mars rover

Brian H. Wilcox, Donald B. Gennery, Andrew H. Mishkin, Brian K. Cooper,
Teri B. Lawton, N. Keith Lay, and Steven P. Katzmann

Jet Propulsion Laboratory
California Institute of Technology
4800 Oak Grove Drive
Pasadena, California 91109

ABSTRACT

A Mars rover must be able to sense its local environment with sufficient resolution and accuracy to avoid local obstacles and hazards while moving a significant distance each day. Power efficiency and reliability are extremely important considerations, making stereo correlation an attractive method of range sensing compared to laser scanning, if the computational load and correspondence errors can be handled. Techniques for treatment of these problems, including the use of more than two cameras to reduce correspondence errors and possibly to limit the computational burden of stereo processing, have been tested at JPL. Once a reliable range map is obtained, it must be transformed to a plan view and compared to a stored terrain database, in order to refine the estimated position of the rover and to improve the database. The slope and roughness of each terrain region are computed, which form the basis for a traversability map allowing local path planning. Ongoing research and field testing of such a system is described.

1. INTRODUCTION

Recently, interest in planetary rovers has revived. Current NASA planning is for a possible launch of a Mars rover and sample return mission in 1998, if funding is forthcoming. A Jet Propulsion Laboratory (JPL) study[1] and a workshop[2] have helped to define the characteristics that a Mars rover launched in the 1990's should have. A small-scale internal research project at JPL is developing a few of the needed techniques for the Semiautonomous Mobility (SAM) concept for local navigation by a Mars rover. (Our current experimental rover testbed is shown in Figure 1.) This paper discusses several aspects of the vision system being developed and tested to demonstrate the necessary capability for local navigation in the SAM concept.

2. OVERVIEW: SEMIAUTONOMOUS MOBILITY CONCEPT FOR A MARS ROVER

Because of the long signal time to Mars (anywhere from 6 minutes to 45 minutes for a round trip at the speed of light), it is impractical to have a Mars rover that is teleoperated from Earth (that is, one in which every individual movement would be controlled by a human being). Therefore, some autonomy on the rover is needed. On the other hand, a highly autonomous rover (which could travel safely over long distances for many days in unfamiliar territory without guidance from Earth and obtain samples on its own) is beyond the present state of the art of artificial intelligence, and thus can be ruled out for a rover launched before the year 2000.

The Semiautonomous Mobility (SAM) concept (Figure 2) is an approach incorporating a level of autonomy intermediate between the two extremes just mentioned. The vision system work discussed in this paper is consistent with the SAM concept, which is described below.

In the SAM method, local routes are planned autonomously using images obtained on the vehicle, but they are guided by global routes planned less frequently by human beings using a topographic map, which is obtained from images produced by a vehicle orbiting Mars. The orbiter could be a precursor mission which would map a large area of Mars in advance, or it could be part of the same mission and map areas only as they are needed.

The sequence of operations in the portion of SAM involving Earth is as follows. As commanded from Earth, the orbiter takes a stereo pair of pictures (by taking the two pictures at different points in the orbit) of an area to be traversed (if this area is not already mapped). These pictures might have a resolution of about one meter, although poorer resolution could be used. The pictures are sent to Earth, where they are used by a human operator (perhaps

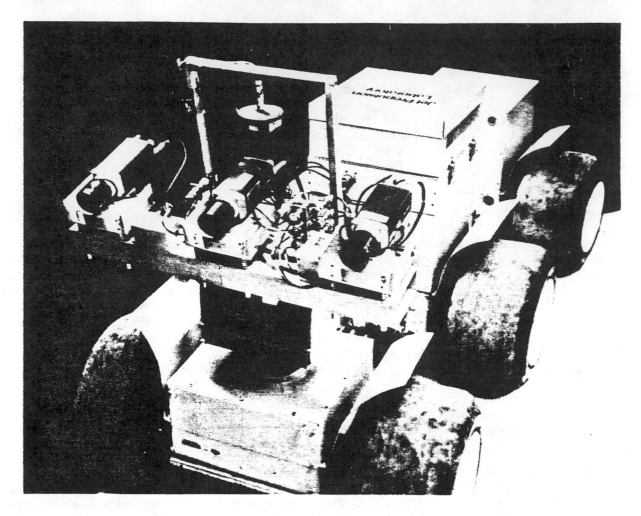

Figure 1. JPL Planetary Rover Testbed

with computer assistance) to designate an approximate path for the vehicle to follow, designed to avoid large obstacles, dangerous areas, and dead-end paths. This path and a topographic map for the surrounding area are sent from Earth to the rover. This process repeats as needed, perhaps once for each traverse between sites where experiments are to be done, or perhaps once per day or so on long traverses.

The sequence of operations in the portion of SAM taking place on Mars is as follows. The rover views the local scene and, by means discussed below, computes a local topographic map. This map is matched to the local portion of the global map sent from Earth, as constrained by knowledge of the rover's current position from other navigation devices or previous positions, in order to determine the accurate rover position and to register the local map to the global map. The local map (from the rover's sensors) and the global map (from the Earth) are then combined to form a revised map that has high resolution in the vicinity of the rover. This map is analyzed by computation on the rover to determine the safe areas over which to drive. A new path then is computed, revising the approximate path sent from Earth, since with the local high resolution map small obstacles can be seen which might have been missed in the low-resolution pictures used on Earth. Using the revised path, the rover then drives ahead a short distance (perhaps ten meters), and the process repeats.

With the computing power that it will be practical to put on a Mars rover in the 1990's, the computations needed to process a stereo pair of images and perform the other calculations needed may require roughly 60 seconds. If these are needed every 10 meters and it takes the rover 30 seconds to drive 10 meters, the resulting average rate of travel is 10 meters every 90 seconds, which is 11 cm/sec or 10 km/day. If a ten-kilometer path is designated from Earth each time, only one communication per day is needed, and the rover could continue to drive all night, using strobe lights for illumination. The method is more reliable than autonomous operation, because of the human guidance and the overview that the orbital data provides. However, for backup in case of communication failure, the ability for more

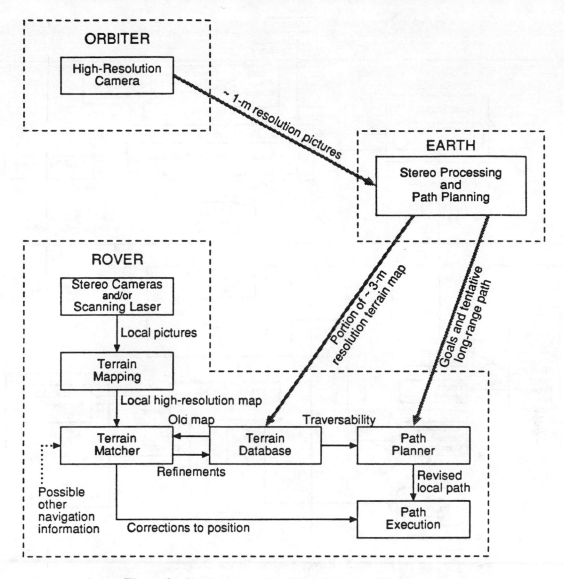

Figure 2. Semiautonomous Operation of a Mars Rover

autonomous operation is present, since the rover can plan its own path based on the area that it can see, provided that some global navigation system is available. Missions further in the future, after more sophisticated algorithms have been developed and more confidence in their abilities has been gained through experience, may be allowed to operate autonomously most of the time.

Additional discussion of the SAM concept and other technical issues related to a Mars rover in the 1990's can be found elsewhere.[3]

3. RESEARCH IMPLEMENTATION OF ROVER VISION SYSTEM

An experimental system with the principal elements of the SAM concept is currently being implemented at the JPL Robotics Laboratory. The system block diagram for the entire rover system is shown in Figure 3. The major subsystems are perception (vision), route planning, vehicle control, the contol executive, and the rover vehicle itself. Route planning functions are performed by a Symbolics 3670 LISP machine; an LSI 11/73 handles raw sensor data collection and low level vehicle control; all other system functions operate on a VAX 11/750. The test course is an approximately 100- by 400-meter area located in the Arroyo Seco immediately adjacent to JPL.

Figure 4 details the elements of the vision system. The major operations of the implemented vision system are briefly described below. (More detailed descriptions will be provided in forthcoming papers.)

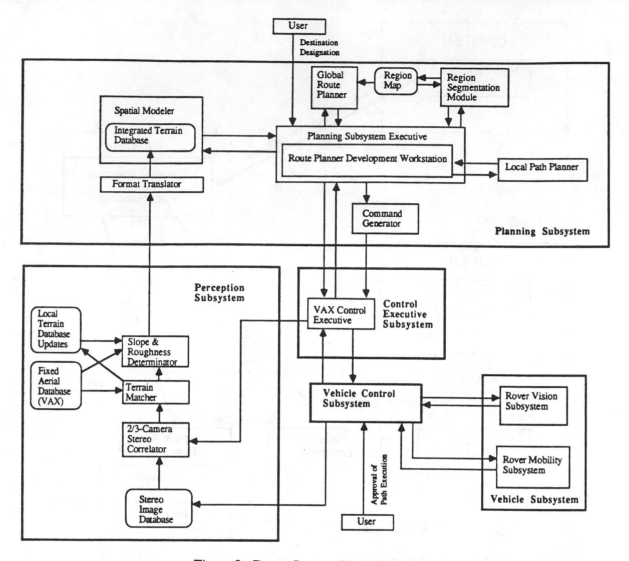

Figure 3. Rover System Block Diagram

3.1 Stereo Correlation Approaches

Two stereo correlation approaches have been implemented. Experimentation will continue with both to confirm their relative advantages and disadvantages.

3.1.1 Two camera stereo. The two camera stereo correlation approach (based on work by Gennery [4,5]) operates on a pair of images from the left and right cameras on the rover Area correlation is used, with same-size windows into the two images being moved in one-pixel increments relative to each other. Area correlation was chosen, rather than edge correlation, because a Mars rover (as well as the current experimental rover) will operate in a natural terrain environment. Scenes in natural environments tend to contain significant texture, but few well-defined edges.

3.1.2 Three camera stereo. Simple algorithms for two-camera stereo are plagued with correspondence errors, and more sophisticated algorithms, such as that described above, require such extensive computations that considerable computation time may be needed by a radiation-hardened, space-qualified, and low power on-board computer to perform them. An alternative is to use more than two images to resolve correspondence problems by back-triangulating into the redundant images for confirmation of a correct match. Since the probability of a match in another image is much lower than a match between additional adjacent pixels (because adjacent pixels are highly correlated), the total number of pixels that need to be examined in order to get any desired level of correctness is reduced. Thus, in principal, one can reduce the size of the "window" being matched between images (whether by area- or edge-based techniques), use simpler techniques for testing for correspondence, and still get acceptably low levels of

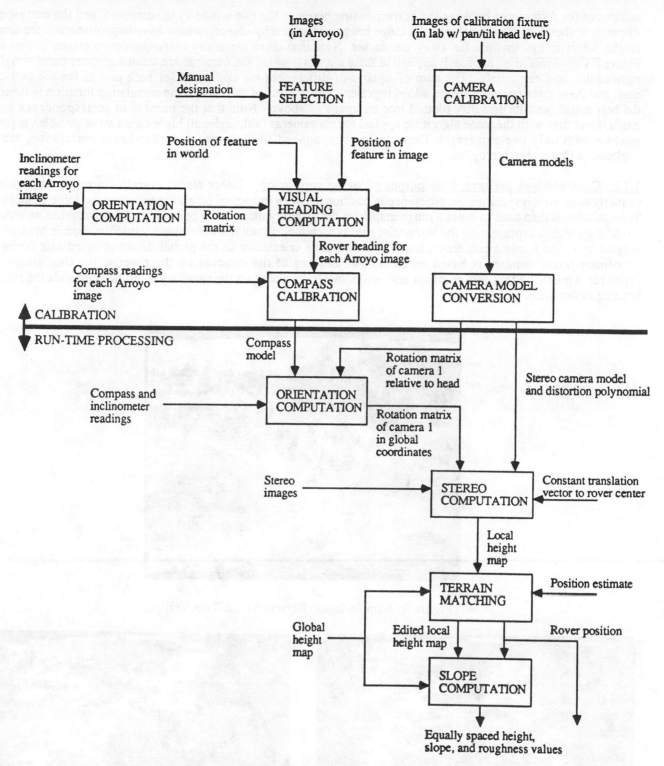

Figure 4. Vision And Attitude Processing For the Rover System

errors. This would greatly reduce the numbers of computations needed for stereo vision, if it could be demonstrated experimentally. This is the focus of the current effort in three-camera stereo.

To illustrate these concepts, Figure 5 shows a typical image returned from our test vehicle. Figure 6 shows what a simple two-camera correlation algorithm does with a stereo pair of this scene. This algorithm correlates a 3-pixel-high by 9-pixel-wide window in one image with 128 possible matching windows in the other image. (The cameras were mounted parallel, so that the search is strictly along the scan lines of the image.) The best value for the sum of the squares of the differences between the corresponding pixels in the two windows is computed, and the corresponding disparity is shown as the intensity of the range image. Thus nearby objects, which have large disparity, are shown as bright, while things that are far away are darker. Note that there are many correspondence errors in this image. Figure 7 shows this same approach applied to three cameras, where the cameras are treated as three pairs -- right/left, right/center, and center/left. The sum-of-squares-of-differences are computed for each pair as for the two camera case, and these three correlations are added together. The minima of this composite correlation function is taken to be the best match, and its disparity plotted into an image as above. Note that the number of correspondence errors is much lower than with the same algorithm applied to two cameras (although still higher than more complex algorithms produce with only two cameras). These techniques, applied to both area and edge-based correlation, are being explored in this research program.

3.1.3 <u>Computations performed on output of stereo correlation.</u> Either stereo correlation approach results in a disparity map, which specifies the difference in position between images of pixels representing the same point in space. Triangulation is then used to build a range map from the disparity map. Each point in the range map has an associated covariance matrix representing the uncertainty in its location. Finally, a coordinate transformation is applied to the range map to put it into a coordinate frame with the same orientation as the global database coordinate frame. The coordinate transformation is based on the known heading of the cameras on the rover at the time images were captured; a magnetic flux gate compass and inclinometers mounted on the camera pan/tilt head provide the necessary heading information.

Figure 5. Sample Image Returned from Test Vehicle

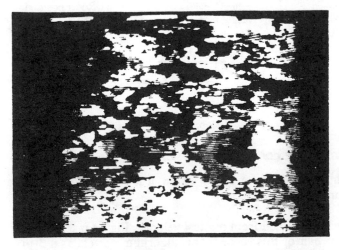

Figure 6.
Range Map from Two-Camera Stereo Processing

Figure 7.
Range Map from Three-Camera Stereo Processing

3.2 Local/global terrain matching

The terrain matcher merges the elevation map generated by stereo correlation with the elevation map in the database by a process of correlation and averaging, which also produces the best estimate of vehicle position as that which produces the best correlation. However, this computation is more complicated than ordinary correlation because of the following: the points are not equally spaced, there may be significant uncertainties in their horizontal positions, and there may be occasional mistakes in the stereo data.

The terrain matcher performs a variable resolution search, beginning at the lowest resolution to produce a rough match, then refining the match as the resolution is increased. The uncertainty estimates associated with each point in the height maps are incorporated into the representations used for matching. In the matching process, points in the local database that disagree significantly with the global database are eliminated. Note that a feature in the local database must be large enough to appear in the lower resolution global database to be thrown out. Features in the local database which are too small to show up in the global database are retained. The process for rejecting points is similar to that in the ground surface finder approach previously described by Gennery.[4,5]

The output of the terrain matcher is: 1) the position of the rover in the global coordinate frame, together with a goodness of fit measurefor the determined location, and 2) an edited local height map with apparently erroneous points removed.

3.3 Slope and roughness determination

Traversability can be determined by analyzing the database to determine the slope and roughness of the ground at each horizontal position. This can be done by local least-square fits of planar or other surfaces and analysis of the residuals.

The slope and roughness determination module accepts as input the current rover position, the corresponding local height map, the global database, and any existing local height maps generated previously from rover data collected at earlier stopping points along the route. The rover position provides the offsets necessary to properly position the local database in the global database. Only a rectangular subset of the global database in the vicinity of the rover location is read in for integration with the local database. A Gaussian function is defined around each (equally spaced) output point, and a least squares fit is performed of all local database points that fall close enough to the output point, with weights derived from the Gaussian function and the covariance matrices of the input points. A plane is fit in three dimensions to the data; the z value of the plane at the (x, y) coodinate of the output point becomes the output height. The scatter of the points used for the least squares fit is used to determine the roughness measure.

The output of the slope and roughness computation is a height map of equally spaced points. Associated with each point is the following data: the slope at the location, a 3 x 3 covariance matrix indicating the uncertainty in the height and slope, a roughness value, and an uncertainty measure associated with the roughness value.

All of this information is sent to the planning subsystem resident on the Symbolics machine; a local route plan in generated, and rover motion commands are sent back to the VAX to be implemented. The vehicle traverses a local path, a set of stereo images is acquired, and the cycle repeats. The details of the route and path planning approach are discussed in other papers.[6,7]

3.4 Calibration

Since the terrain matching process assumes that the local and global data have the same orientation and scale factor, it is important that the instruments be calibrated. Accurate stereo camera calibration ensures that the depth map will not have distortion or scale errors. Accurate calibration of the compass and inclinometers, together with the camera calibration, ensures that the orientation is correct.

The cameras are calibrated independently in the laboratory by taking pictures of a calibration fixture and performing a least-squares adjustment for the calibration parameters, as described in detail elsewhere.[8] The inclinometers have an accurate factory calibration. The analog circuitry and A/D converter that processes their output is calibrated by observing the digital result for various measured voltages from the inclinometers at various tilts.

A real Mars rover will probably have an inertial heading reference. However, our experimental vehicle uses a magnetic compass consisting of a two-axis flux-gate magnetometer, which is pendulously suspended so that it is level (when the rover is stationary). A linear model is used for the compass, consisting of three terms for each axis, which represent a bias and the amplitudes of the two quadrature components. (This model subsumes internal errors, misalignment, and constant local magnetic variation.) To solve for these six parameters, a least-squares adjustment is done, using measurements taken at several known headings. Each of these heading measurements is produced by manually designating a feature, in an image from one of the cameras, that is at a known geographical position. The image position of the feature, together with the known geographical positions of the feature and the rover when calibrating and the known camera model, enables the heading of the rover to be computed.

4. CURRENT TESTING STATUS

Laboratory simulations are currently being performed of Arroyo Seco route traversals by the experimental rover system. These simulations rely on actual images, inclinometer and compass readings collected during teleoperated runs of the vehicle through a 50 meter arroyo test course. Terrain matching has not been successful with real data to date; this is probably due to the features in the local database (for the fairly flat areas used so far) being too small to be matched against the features present in the 3-meter resolution Arroyo Seco global database available. A new 1-meter resolution global database is now being prepared from digitized aerial photographs of the arroyo. After the 1-meter database is installed, further simulations should demonstrate the functionality of the entire rover system, including stereo correlation, terrain matching, slope and roughness computation, automated route planning, and vehicle control. System runs in the arroyo test course should occur during November of 1987.

5. CONCLUSIONS

The safe guidance and control of the rover on Mars at speeds of several km/day will require substantial on-board sensing and computing resources. A reliable vision system is critical to the successful operation of such a rover. Ongoing research at JPL is demonstrating the feasibility of using stereo correlation approaches, terrain matching, roughness and slope computation, together with automated route planning, to guide a rover along a route and around obstacles in a natural terrain environment.

6. ACKNOWLEDGMENT

The research described in this publication was carried out by the Jet Propulsion Laboratory, California Institute of Technology, funded by the Director's Discretionary Fund.

7. REFERENCES

1. J. R. Randolph (ed.), "Mars Rover 1996 Mission Concept," JPL D-3922, Jet Propulsion Laboratory, Pasadena, California (1986).
2. J. C. Mankins (ed.), "Proc. Technology Planning Workshop for the Mars Rover," Jet Propulsion Laboratory, Pasadena, California (1987).
3. B. Wilcox and D. Gennery, "A Mars Rover for the 1990's," J. of the British Interplanetary Society, Vol. 40, 484-488 (1987).
4. D. B. Gennery, "Modelling the Environment of an Exploring Vehicle by Means of Stereo Vision," AIM-339 (Computer Science Dept. Report STAN-CS-80-805), Stanford University, (Ph.D. dissertation) (1980).
5. D. B. Gennery, "A Stereo Vision System for an Autonomous Vehicle," Fifth International Joint Conference on Artificial Intelligence, MIT, 576-582 (1977).
6. R. S. Doshi, D. J. Atkinson, and J. White, " Reasoning with Inaccurate Spatial Knowledge," SPIE Robotics (1987).
7. J. M. Cameron, R. S. Doshi, and K. G. Holmes, "Route Planner Development Workstation, Volume 3A - Operational Extensions," JPL Technical Report, JPL D-2733 Vol. 3A (1987).
8. D. B. Gennery, T. Litwin, B. Wilcox, and B. Bon, "Sensing and Perception Research for Space Telerobotics at JPL," IEEE International Conference on Robotics and Automation, 311-317 (1987).

Ambler

An Autonomous Rover for Planetary Exploration

John Bares, Martial Hebert, Takeo Kanade, Eric Krotkov,
Tom Mitchell, Reid Simmons, and William Whittaker

Carnegie Mellon University

For centuries people have been fascinated by Earth's planetary neighbors. There has been much speculation, in science and science fiction, about what lies on and under their surfaces. Despite considerable study, our knowledge remains very limited. Orbiting vehicles cannot examine internal features, and stationary vehicles, like the three Soviet and two US landers on Mars, miss what is over the horizon, atop mountains, and in ravines.

Active exploration of other planets could answer many questions about the nature and origins of our solar system. Sending astronauts or remotely controlled vehicles is a possibility, but a manned expedition is highly unlikely in the near future, and conventional teleoperation is impractical because of the long signal times (for example, up to 45 minutes for a round trip to Mars at the speed of light). A more promising approach is to launch an unmanned prospector and a vehicle to return collected samples to Earth—for instance, NASA's proposed Mars Rover and Sample Return mission.[1] The broad objectives of such a mission would be to observe and gather materials representative of the planet's geophysical, meteoro-

Extremely self-reliant, this six-legged robot will prioritize its goals and decide on its course of action as it explores the rugged terrain of a place like Mars.

logical, and biological conditions and to return a varied selection of samples. Since the payload of the return vehicle is limited, the mission requires a sophisticated on-site system that can explore, assay, and select.

Recently we initiated a research program that addresses the central robotics challenges of designing a roving explorer capable of operating with minimal external guidance. The purposes of this research are to confront issues not faced

by laboratory robots, to identify and formulate the difficult problems in autonomous exploration, and to generate insights, principles, and techniques for their solutions. We are not attempting to satisfy all the requirements of a system that would be flown to another planet (for example, space-qualified processors). Instead, we are building a prototype legged rover, called the Ambler (loosely an acronym for Autonomous MoBiLe Exploration Robot), and testing it on full-scale, rugged terrain of the sort that might be encountered on the Martian suface.

To undertake an extraterrestrial prospecting mission, we must extend existing robotic technology. Because the rover will be beyond the reach of timely aid from Earth, it must exhibit extreme self-reliance. It must be able to navigate, explore, and sample, and to know, moreover, what tasks do and do not lie within its capabilities. Particular issues critical to autonomous planetary exploration include locomotion, rough-terrain navigation, sample acquisition, perception, self-awareness, task autonomy, safeguarding, and system integration. Although semiautonomous and remotely assisted systems may be practical for some tasks,[2] our

EH0342-6/91/0000/0457$01.00 © 1989 IEEE

research strategy is to strive for full autonomy wherever possible, with the rover deciding when to ask for missing information.

In this article we present an overview of our research program, focusing on locomotion, perception, planning, and control. We summarize some of the most important goals and requirements of a rover design and describe how locomotion, perception, and planning systems can satisfy these requirements. Since the program is relatively young (one year old at the time of writing), this article aims to identify issues and approaches and to describe work in progress rather than to report results. Although our discussion concentrates on a Mars mission, we expect many of the technologies developed in our work to be applicable to other planetary bodies and to terrestrial concerns such as hazardous waste assessment and remediation, ocean floor exploration, and mining.

Locomotion

Our data on the Martian landscape indicates that an explorer would encounter a wide variety of terrain features, including a canyon 4,800 kilometers long by 7 kilometers deep, a mountain 27 kilometers high, and numerous sand dunes, rock fields, and craters. Figure 1 illustrates the barren, rugged terrain viewed by the Viking 2 lander.

Since the locomotion system must safely transport the vehicle over vast expanses of irregular terrain, perhaps the most important design criterion for the locomotion system is traversability: it must be able to navigate over extremely rugged terrain. Specifically, the rover should be capable of traversing a one-meter step, negotiating a 60 percent slope, and maintaining an average velocity of approximately one kilometer per day (these specifications, although somewhat arbitrary, reflect plausible assumptions[1] about the scale of objects on the Martian surface and about potential missions). Energy efficiency poses an additional design constraint because total on-board power generation is expected to be less than one kilowatt. As the dominant energy consumer, the locomotion mechanism must be extremely efficient. Another design consideration is that the locomotion mechanism must provide a stable platform for sensors and sample acquisition tools.

These design criteria admit a wide variety of possible locomotion candidates,

Figure 1. Martian terrain viewed by the Viking 2 lander.

including mechanisms that roll, walk, combine rolling and walking, or perform so-called hybrid locomotion.[3] Rolling machines have wheels or tracks in continuous contact with the terrain; they propel themselves by generating traction parallel to the terrain surface. Walkers suspend themselves over the terrain on discrete contact points and maintain principally vertical contact forces throughout propulsion; this allows more tractable models of terrain interaction than are possible for wheels. In addition, walking mechanisms isolate the robot's body from the underlying terrain and propel the body along a smooth trajectory independent of surface irregularities.

After comparing these candidates (see article by Bares and Whittaker[4] for a trade-off analysis of locomotion mechanisms with respect to these constraints), we selected legged locomotion because of its superior rough-terrain traversability characteristics, its theoretical efficiency, and its ability to keep sensors and sampling equipment steady and stable.

Our initial Ambler configuration consists of six legs stacked coaxially at their shoulder joints (Figure 2). Each leg is

mounted at a different elevation on the central axis of the body and can rotate fully around the body. Each leg (Figure 3) consists of two revolute joints (shoulder and elbow) that move in a horizontal plane to position the leg, and a prismatic joint at the end of the elbow link that effects a vertical telescoping motion to extend or retract the foot. Thus, the locomotor has 18 degrees of freedom. The planar reach (combined length of shoulder and elbow links) of a leg is 2.5 meters and the vertical stroke (telescopic distance) is 1 to 2 meters, depending on the position of the leg on the stack. The average overall height of the Ambler is approximately 3.5 meters, and its nominal width is approximately 3 meters. With these dimensions, the Ambler can step over obstacles 1 meter high while maintaining a level body trajectory.

The Ambler body, a cylinder one meter in diameter situated below the leg stack, will contain equipment for power generation, computing, sample analysis, and scientific instrumentation. Sample acquisition tools can be mounted on legs or on the underside of the body. Communication equipment can be mounted either

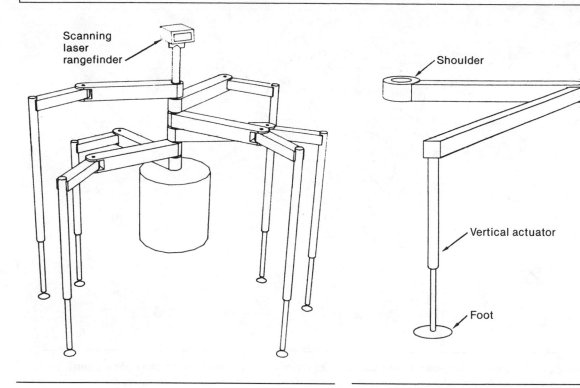

Figure 2. Sketch of the Ambler.

Figure 3. Side view of an Ambler leg.

above the leg stack or in the body. Perception sensors will be mounted above the leg stack, where they will have large fields of view; other high-resolution sensors can be placed under the body or directly on legs.

This configuration has a number of benefits. First, decoupling the vertical and horizontal joints simplifies walk planning and motion control by reducing complex, six-dimensional problems to smaller ones. Second, the sampling tools under the body have a clear view of and close proximity to the terrain that they must access; body movements position and orient the tools, reducing the number of degrees of freedom they require. Third, during locomotion the legs isolate the body and sensors from terrain roughness.

In operation the Ambler will walk over rugged terrain much as one poles a raft floating over a rough lake bottom. The six vertical actuators in the Ambler's legs level the body over terrain, while the planar joints propel the body. As the body advances, one leg at a time moves ahead of the walker, much as the pole is placed ahead of the raft. A unique result of the stacked leg configuration is that an over-

lapping gait is possible—that is, a gait in which a rear leg moves ahead, or "recovers," past forward supporting legs. Figure 4 illustrates this gait. An overlapping gait requires fewer foot placements, saving energy (energy is expended whenever a foot interacts with terrain) and reducing demands on perception and planning. While one leg is recovering, the five other legs support the body. The stability of the stance can be maximized by maintaining the center of gravity inside a conservative support polygon. Inside this region, the vehicle remains stable even if one (and possibly more) of the legs ceases to support it, either due to failure or slippage.

Experience with existing walking mechanisms (see article by McGhee[5] or Raibert[6] for an overview) suggests that they are difficult to coordinate due to their complexity, suffer large energy losses due to actuator conflict, and can be unreliable upon failure of one or more legs. We designed the Ambler to overcome these problems.

Unlike those of other walkers, the Ambler's actuator groups for body sup-

port and propulsion are orthogonal; a subset of the planar joints propels the body, while the vertical actuators support and level the body. The Ambler can level itself without propelling and propel without leveling, having no power coupling between the two motions. This should make it both easier to control and more efficient than other walking mechanisms.

The Ambler locomotor configuration is a dramatic improvement in reliability over conventional walking mechanisms. Because the legs are stacked above the body and can rotate by 2π about their shoulder joints, any leg can operate in any body sector. Thus, any functional leg can reposition itself to substitute for any failed leg, and three legs would have to fail to cause immobilization.

Perception

The Ambler needs timely and detailed perception to plan effective locomotive and sampling strategies and to monitor their execution. The perception system's task is to build and maintain represen-

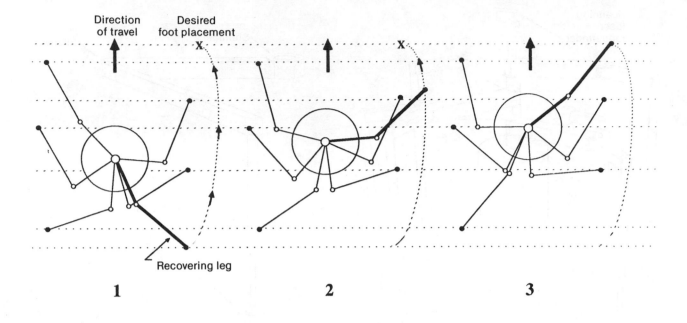

Figure 4. Overlapping gait. To advance the walker, the recovering leg (drawn thicker) overlaps the two right supporting legs. Depending on a leg's location on the central stack, some weaving around supporting legs may be necessary for it to move past forward legs.

tations—which we call terrain maps—of terrain (geometry, soil type) and discrete objects (size, shape).

Perceiving and mapping rugged, outdoor terrain are significant challenges. Current machine perception techniques can be applied with some success to man-made, structured, indoor scenes. Unlike industrial systems, however, the rover will have little need to recognize or describe regular geometric shapes and will not be able to capitalize on the powerful features (such as symmetry, smoothness, constant illumination) utilized to perceive worlds consisting of blocks. We must develop new techniques for constructing maps of natural, unstructured, outdoor environments.

The problem of building and maintaining those maps raises several issues: (1) representation of data at different levels of resolution, (2) construction of maps and descriptions from different sensors, and (3) effective use of the maps.

Representation. The perception system must provide an environmental representation that is appropriate for a wide vari-

ety of tasks, each with different requirements. For example, locomotion and sampling require detailed, local representations, while navigation and mission planning demand broad, global descriptions. To uniformly accommodate these diverse needs, we have selected a hierarchical representation scheme that describes terrain and objects at varying levels of resolution. At each level of resolution we describe the environment in two ways: on a geometric grid and as object descriptions. Together these comprise a terrain map at one scale.

We define an elevation map on a regular grid. Each grid square records information about the terrain in that area—for instance, its elevation above a ground plane. Other terrain attributes include the following: the uncertainty of the estimated elevation; roughness; slope; labels indicating whether the terrain is unknown (has never been observed) or occluded (currently not observed because it lies in a shadow cast by another object); mineralogical composition; and a measure of traversability derived from slope, roughness, and other properties.

Object descriptions include the size, shape, and location of particular objects such as a boulder; symbolic terrain descriptions, such as hill, valley, saddle, and ridge, which may be useful for identifying promising routes or sample sites; paths the vehicle has followed; locations that have been sampled; and viewpoints from which observations have been made.

Constructing terrain maps. Constructing terrain maps requires sensing and interpretation, ranging from low-level data collection to high-level scene modeling. This section first focuses on the lowest level of abstraction (sensors) and then describes an intermediate level of abstraction (local surface geometry). It does not address the highest level of object identification and semantic interpretation.

We will equip the Ambler with a battery of different sensors to collect multispectral data. Our primary sensor is a scanning laser rangefinder that measures both reflectance and range.[7] The scanner directly recovers the environment's three-dimensional structure, operating more rapidly and reliably than other vision tech-

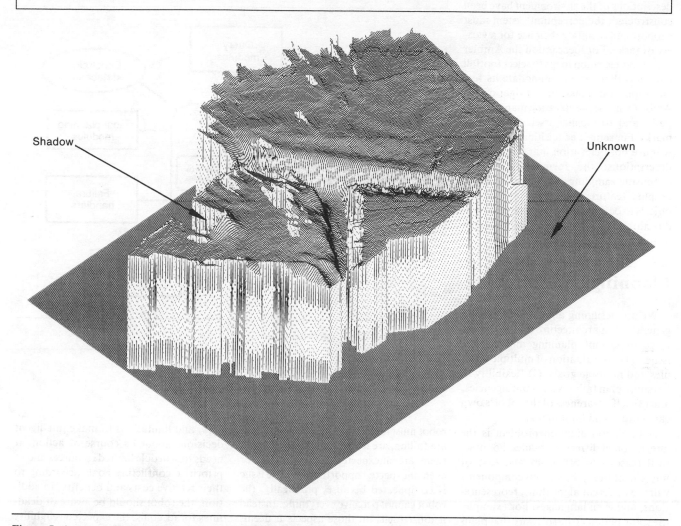

Shadow

Unknown

Figure 5. A composite elevation map constructed by merging four rangefinder views of rugged terrain at a construction site. The grid size is 10 centimeters.

niques such as binocular stereo and motion. In the near future we will also use a pair of color cameras for determining material properties from color and texture, for long-range viewing, and for stereo viewing to back up the rangefinder. We also plan to incorporate an inertial reference sensor, inclinometers, tactile sensors on sampling tools, and other imaging devices. However, the perception system need not be limited to passively interpreting data; it can actively use the Ambler vehicle itself as a sensor to determine soil cohesion and friction parameters either directly, by measuring leg joint torques while walking, or indirectly, by comparing the soil in its footprints to nearby soil.

As an example of how the perception system operates, we will consider the interpretation of rangefinder signals in terms of local surface geometry (interested readers can find details in a technical report by Hebert, Kanade, and Kweon[7]). This involves creating an elevation map from a range image, matching it to another elevation map, and merging the two maps to form a composite map.

To construct elevation maps from range images, we have developed an algorithm that operates at arbitrary resolutions. It computes the locations where rays emitted by the sensor strike the terrain, and then it refers the intersection points and an estimate of their uncertainty to a reference grid, thus creating an elevation map.

To merge elevation maps from successive viewpoints, we have developed a two-stage algorithm to determine the correspondence between two elevation maps. The first stage matches a sparse set of geometric features extracted from the two maps, whose output is the estimated rigid transformation T relating the two sets of features. The second stage takes T as an initial estimate and refines it by gradient descent, iteratively minimizing an error functional defined over all the data points in the two maps. Once we know T, we can apply it to merge the maps; Figure 5 illustrates a composite elevation map constructed by merging four rangefinder views of the rugged terrain at a construction site.

461

Using terrain maps. Once maps and descriptions of the environment have been constructed, the perception system must support and facilitate their use for a variety of tasks. For locomotion the Ambler will access elevation maps to select footfall locations that can accommodate its feet and support its mass. For navigation the Ambler will use the elevation maps to plan paths and to localize itself from landmarks. For sample acquisition the Ambler will use both elevation maps and object descriptions: the former to identify promising sampling sites based on topographic features; the latter to identify objects to be sampled, determine approach directions, and select control regimes (such as force and position).

Planning and control

We are designing and implementing a general robot architecture that addresses three important planning and control issues: (1) coordination of multiple modules and multiple goals, (2) flexibility in handling plan failures and contingencies, and (3) self-awareness of the robot's own capabilities and limitations.

A basic coordination problem is the integration of different modules. For practical reasons of efficiency and ease of implementation, the various components will use different algorithms, representations, and even languages. For example, the route planner (the software that plans routes) may use geometric algorithms, while the mission planner might use symbolic techniques. The architecture must facilitate the exchange of data between modules and flexible yet efficient control flow management.

Another important coordination issue is the handling of multiple goals. The Ambler will have multiple, often conflicting, goals, such as navigation, science, and health goals. The control architecture must prioritize and schedule goal achievement on the basis of a cost-benefit analysis that accounts for factors such as the environment, the robot's capabilities, and its past actions. For example, if the Ambler detects a potential sedimentary rock while traveling, it might decide to detour to obtain a sample if the rock is close, if the robot has no pressing deadlines, and if the robot has not already acquired sufficient sedimentary samples.

The issue of flexibility concerns how well the robot reacts to changes in its uncertain, dynamic environment. The robot must notice indications that its plans are failing, are no longer applicable, that there are unexpected contingencies, or even unexpected opportunities. The issue is complicated because, practically, the robot cannot perceive everything. Instead it must focus on those aspects it deems most important, deploying monitors to check on specified conditions.

The robot also needs flexibility in deciding how to handle problems indicated by the changes it detects. Instead of merely passing an error up the goal tree, for instance, the robot might try re-achieving the goal (for example, chipping a rock again), adding a new subgoal (detouring around an obstacle), or even attending to a different goal altogether. Once the robot decides how to handle a problem, however, the recovery processes can utilize the same algorithms that created the robot's initial plans. For example, once deciding to add a detour, the robot can use its path planner, treating the detour as if it had been planned from the start.

The issue of self-awareness is particularly important for planetary exploration robots, since they are remote from human assistance and must be largely responsible for monitoring their environment and choosing acceptable actions. The robot needs knowledge of its resources, capabilities, and limitations to make intelligent decisions about its course of action. It needs to schedule limited resources and to prioritize conflicting goals according to their relative costs and benefits. In addition, the robot should be aware of deadlines for its goals, the expected reliability of its planners, and the time necessary to execute its plans.[8]

Although our robot control architecture is being designed to provide many tools for constructing a planetary explorer, it will be open enough for easy experimentation and extension. The architecture is a distributed system with centralized control. It supports modules running on separate machines in different languages (currently, C and Lisp), communicating via coarse-grained message passing. Building on ideas from the Navlab project,[9] communication and data transfer are totally transparent to the individual modules.

Whereas planning, sensing, and actuation are distributed, control of when and how to attend to goals is centralized (Figure 6). The central control receives messages from other modules and routes them to the appropriate module to be handled. The control module also schedules the available computational and physical resources and maintains goal trees to help in error recovery and decision making.

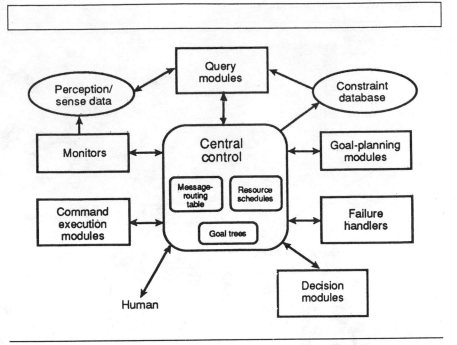

Figure 6. Task control architecture.

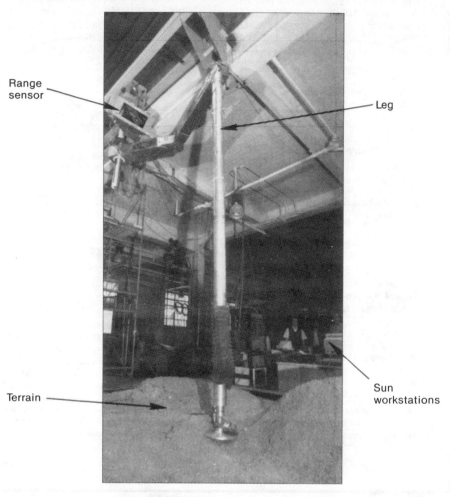

Range sensor

Leg

Terrain

Sun workstations

Figure 7. Single-leg testbed.

Although centralized control has the potential of being a bottleneck, we believe that global control is crucial for robots that have to contend with multiple goals and limited resources. At the same time, we are examining how to combine centralized control architectures with architectures that provide fast, reflexive actions (see Brooks,[10] for example).

The architecture directly supports different classes of messages for constructing robot systems. Query messages, which are used to access perceptual and internal sense data, return information and control to the requesting module. Goal messages, which plan actions by issuing other goal or command messages, are nonblocking, so that control returns before the subgoals are actually achieved. This nonblocking feature facilitates error recovery and the coordination of multiple tasks, since the central control can decide when to suspend or achieve goals.

Other message classes include (1) command messages, nonblocking messages used to control actuators, (2) constraint messages, used to add information to a database, (3) monitor messages, which check on the status of the robot or its environment, (4) decision messages, used to prioritize goals, and (5) failure messages, which indicate that either a plan-time or execution-time error has been detected.

Although the Ambler is designed to be largely autonomous, it is important that humans can intervene and operate it remotely when the situation warrants. The architecture meets this need by allowing humans to send or receive messages, at any level of abstraction, just as any other module can. This feature helps in development of the Ambler because it lets us substitute human input for as-yet-unwritten modules.

Experimental testbeds

Although simulations are often useful abstractions of the problems an autonomous robot will face, they are never as revealing as operating the actual mechanism. Our philosophy is to embed our ideas in working mechanisms that operate in natural environments. While we build the six-legged vehicle described here, we are testing our locomotion, perception, planning, and control ideas on two operational testbeds.

The first testbed is a one-legged version of the Ambler (Figure 7). It enables us to begin integrating the component technologies into a complete but simplified system that can demonstrate single-leg walking, using a few frames of range data, simple walk planning, and simple error recovery. A full-scale leg has been built and mounted on a carriage that travels along rails on the ceiling to simulate body motion. A scanning laser rangefinder mounted above the leg provides data for building terrain maps. A large sandbox under the testbed contains different soil types and obstacles. A rudimentary version of the planning and control architecture, running on Sun workstations, enables the leg to lift, recover, and land at a chosen position. We are also using this testbed to study foot-terrain interactions such as foot slippage and sinkage, and power consumption during footlift and footfall.

The other testbed is a Heathkit Hero 2000, a commercially available robot, which we are using to explore ideas for combining navigation and sample collection in an indoor environment. The Hero mechanism is wheeled and has an arm and gripper. For perception it has a base sonar and a rotating head sonar, and we have added a sonar on its wrist. In addition, a camera mounted in the ceiling of our lab gives the Hero a global overhead view. Using its vision system, the Hero plans paths to objects, and, once in the vicinity of an object, uses its sonars to locate and grasp the object. At present the Hero picks up and deposits plastic cups and cans, and soon we expect it to be able to retrieve printer output and to schedule the achievement of multiple, conflicting goals.

onstraints inherent in the task of autonomously exploring another planet have driven our design of a robotic rover. In particular, the task demands a system that efficiently and reliably navigates over rough terrain, reliably perceives rugged terrain and irregularly shaped objects, and exhibits extreme self-reliance in performing a multitude of tasks. We have incorporated these constraints into our design for the Ambler. By pursuing the issues of locomotion, perception, planning, and control in the design of a robot that operates on another planet, we hope to find solutions for the problems involved in sending intelligent machines to Mars and beyond. □

Acknowledgments

We would like to acknowledge the contributions of the entire Ambler research team at Carnegie Mellon University, whose collective efforts have produced the ideas reported here.

This research was sponsored by NASA under contract NAGW-1175. The views and conclusions contained in this article are those of the authors and should not be interpreted as representing the official policies, either expressed or implied, of NASA or the United States government.

References

1. J.R. Randolph, ed., *Mars Rover 1996 Mission Concept*, Technical Report D-3922, Jet Propulsion Laboratory, Pasadena, Calif., 1986.

2. B. Wilcox and D. Gennery, "A Mars Rover for the 1990's," *J. British Interplanetary Society*, Vol. 40, 1987, pp. 484-488.

3. D.D. Wright and R.E. Watson, "Comparison of Mobility System Concepts for a Mars Rover," *Proc. SPIE Conf. Mobile Robots II*, Vol. 852, Society of Photo-Optical Instrumentation Engineers, Cambridge, Mass., Nov. 1987, pp. 180-187.

4. J. Bares and W. Whittaker, "Configuration of an Autonomous Robot for Mars Exploration," *Proc. World Robotics Conf.*, Society of Mechanical Engineers, Gaithersburg, Md., May 1989.

5. R. McGhee, "Vehicular Legged Locomotion," in *Advances in Automation and Robotics*, G.N. Saridis, ed., Jai Press, Greenwich, Conn., 1985.

6. M. Raibert, Introduction to Special Issue on Legged Locomotion, *Int'l J. Robotics Research*, Vol. 3, No. 2, Summer 1984, pp. 2-3.

7. M. Hebert, T. Kanade, and I. Kweon, *3-D Vision Techniques for Autonomous Vehicles*, Technical Report CMU-RI-TR-88-12, Robotics Institute, Carnegie Mellon University, 1988.

8. T. Dean and M. Boddy, "An Analysis of Time-Dependent Planning," *Proc. Seventh Nat'l Conf. Artificial Intelligence*, St. Paul, Minn., 1988, pp. 49-54.

9. C. Thorpe et al., "Vision and Navigation for the Carnegie Mellon NAVLAB," *Annual Review of Computer Science*, Vol. 2, 1987, pp. 521-556.

10. R.A. Brooks, "A Robust Layered Control System for a Mobile Robot," *IEEE J. Robotics and Automation*, Vol. RA-2, No. 1, Mar. 1986, pp. 14-23.

John Bares is pursuing a PhD degree at Carnegie Mellon, researching configuration techniques for autonomous, rugged-terrain walking machines, including the Ambler. His research interests include configuration, simulation, design, and implementation of teleoperated and autonomous field robots.

Bares received the BS and MS degrees in civil engineering from Carnegie Mellon University in 1985 and 1987, respectively.

Martial Hebert is a research scientist at the Robotics Institute at Carnegie Mellon University. His research focuses on the range data in computer vision for object recognition and autonomous vehicle navigation. He is involved in the development of a 3-D vision system for the Navlab, an autonomous vehicle for road and cross-country navigation, as well as in the Ambler project.

Hebert received the doctorate in computer science from the University of Orsay, France, in 1984. He is a member of the IEEE Computer Society.

Takeo Kanade is a professor of computer science and codirector of the Robotics Institute at Carnegie Mellon University. Before joining CMU in 1980, he was an associate professor of information science at Kyoto University, Japan.

Kanade has worked on various problems in vision, sensors, manipulators, and mobile robots. He has written and edited three books and written more than 70 papers in these areas. He is the principal investigator in three robotics research programs at CMU. He also chairs CMU's newly established robotics PhD program.

Kanade received his PhD in information science from Kyoto University in 1974. He is a member of the IEEE Computer Society. He served as general chair of the 1983 IEEE International Conference on Computer Vision and Pattern Recognition and vice chair of the 1986 IEEE International Conference on Robotics and Automation. He is the editor of the *International Journal of Computer Vision*.

Eric Krotkov is a research scientist studying machine perception and mobile robots at the Robotics Institute of Carnegie Mellon University. Before joining CMU in 1988, he worked as a postdoctoral fellow in the robotics group of the Laboratorie d'Automatique et d'Analyse des Systemes, Toulouse, France. Krotkov has worked in several areas of robotics and artificial intelligence: three-dimensional computer vision, mobile robot systems, and philosophy of perception.

Krotkov received his BA in philosophy from Haverford College in 1982 and his PhD in computer and information science from the University of Pennsylvania in 1987. He is a member of the ACM and the IEEE Computer Society.

Tom Mitchell is a professor of computer science at Carnegie Mellon University and an affiliated faculty member of the Robotics Institute. He taught in the Computer Science Department at Rutgers University from 1978 until moving to Carnegie Mellon in 1986. His current research focuses on robots that learn and general architectures for problem solving and learning.

Mitchell earned his BS degree in 1973 from MIT and his MS and PhD degrees from Stanford University in 1975 and 1978, respectively. In 1983 he received the IJCAI Computers and Thought Award in recognition of his research in machine learning, and in 1984 an NSF Presidential Young Investigator Award.

Reid Simmons is a research associate in the School of Computer Science at Carnegie Mellon University. His doctoral thesis developed and analyzed techniques for combining associational and causal reasoning for planning and interpretation tasks. His current research is in robot control architectures that can handle multiple, conflicting tasks in uncertain and changing environments.

Simmons earned his BA degree in 1979 from SUNY at Buffalo and his MS and PhD degrees in artificial intelligence from MIT in 1983 and 1988, respectively. He is a member of ACM, AAAI, and Sigma Xi.

William Whittaker is a senior research scientist with the Robotics Institute at Carnegie Mellon and a senior lecturer in the university's Department of Civil Engineering. He is also director of the Field Robotics Center. His research interests center on mobile robots in unpredictable environments and include computer architectures to control mobile robots, modeling and planning for nonrepetitive tasks, complex problems of objective sensing in random or dynamic environments, and integrations of complete field robot systems.

Whittaker received his BS from Princeton in 1973 and his MS and PhD from Carnegie Mellon in 1975 and 1979, respectively. He received Carnegie Mellon's Teare Award for Teaching Excellence. *Science Digest* named him one of the top 100 US innovators in 1985 for his work in robotics.

Readers may write the authors at the Robotics Institute, Carnegie Mellon University, Pittsburgh, PA 15213.

Chapter 5: An Introduction to Legged Locomotion

Mobile robots based on wheeled platforms share many research issues and approaches with robots using legs for locomotion. Many problems that must be addressed in the development of autonomous robots — including robotic perception and world modeling, planning, and decision making — are not necessarily influenced by the mode of locomotion employed by the robot. On the other hand, many issues are specific to walking, hopping, or running machines; for example, gait selection, foot placement, and dynamic balance control.

While it is beyond this tutorial's scope to provide a detailed introduction to legged locomotion, we have included two review papers on legged robotic locomotion. Raibert provides a historical survey of work on legged robots, discusses work in dynamically stable locomotion, and outlines some basic problem areas that remain to be solved. Kumar and Waldron also provide a historical review, and discuss gaits, dynamic models, control, and motion planning — with particular reference to statically stable walking machines.

Written by leading authorities in the field, these papers and references cited therein provide an excellent starting point for researchers interested in walking machines. Additional research can be found in the references cited below.[1,2]

References

1. M. Raibert, *Legged Robots That Balance,* MIT Press, Cambridge, Mass., 1986.
2. S.M. Song and K.J. Waldron, *Machines That Walk,* MIT Press, Cambridge, Mass., 1989.

16 Legged Robots

Marc H. Raibert

16.1   Why Study Legged Machines?

Aside from the sheer thrill of creating machines that actually run, there are two serious reasons for exploring legged machines. One reason is mobility: There is a need for vehicles that can travel in difficult terrain, where existing vehicles cannot go. Wheels excel on prepared surfaces such as rails and roads, but perform poorly where the terrain is soft or uneven. Because of these limitations only about half the earth's landmass is accessible to existing wheeled and tracked vehicles, whereas a much greater area can be reached by animals on foot. It should be possible to build legged vehicles that can go to the places that animals can now reach.

One reason legs provide better mobility in rough terrain is that they can use isolated footholds that optimize support and traction, whereas a wheel requires a continuous path of suport. As a consequence, a legged system is free to choose among the best footholds in the reachable terrain whereas a wheel is forced to negotiate the worst terrain. A ladder illustrates this point: Rungs provide footholds that enable systems to climb, but the spaces between the rungs prevent the wheeled system from making progress.

Another advantage of legs is that they provide an active suspension that decouples the path of the body from the paths of the feet. The payload is free to travel smoothly despite pronounced variations in the terrain. A legged system can also step over obstacles. The performance of legged vehicles can, to a great extent, be independent of the detailed roughness of the ground.

The construction of useful legged vehicles depends on progress in several areas of engineering and science. Legged vehicles will need systems that control joint motions, sequence the use of legs, monitor and manipulate balance, generate motions to use known footholds, sense the terrain to find good footholds, and calculate negotiable foothold sequences. Most of these tasks are not well understood yet, but research is under way. If this research is successful, it will lead to the development of legged vehicles that travel efficiently and quickly in terrain where softness, grade, or obstacles make existing vehicles ineffective. Such vehicles may be useful in industrial, agricultural, and military applications.

A second reason for exploring legged machines is to understand how humans and animals use their legs for locomotion. A few instant replays

Reprinted from *Robotics Science*, 1989, pages 563-594, "Legged Robots" by M.H. Raibert, by permission of The MIT Press, Cambridge, Massachusetts.

on television will reveal the large variety and complexity of ways athletes can carry, swing, toss, glide, and otherwise propel their bodies through space, maintaining orientation, balance, and speed as they go. Such performance is not limited to professional athletes; behavior at the local playground is equally impressive from a mechanical engineering, sensory-motor integration, or computational point of view. Animals also demonstrate great mobility and agility. They use their legs to move quickly and reliably through forest, swamp, marsh, and jungle, and from tree to tree. They move with great speed and efficiency.

Despite the skill we apply in using our own legs for locomotion, we are still at a primitive stage in understanding the principles that underlie walking and running. What control mechanisms do animals use? The development of legged machines will lead to new ideas about animal locomotion. To the extent that an animal and a machine perform similar locomotion tasks, their control systems and mechanical structures must solve similar problems. Of course, results in biology will also help us to make progress with legged robots. This sort of interdisciplinary appoach has already become effective in other areas where biology and robotics have a common ground, such as vision, speech, and manipulation.

16.2   Research on Legged Machines

The scientific study of legged locomotion began just over a century ago when Leland Stanford, then governor of California, commissioned Eadweard Muybridge to find out whether or not a trotting horse left the ground with all four feet at the same time. (For milestones in the development of legged robots, see table 16.1.) Stanford had wagered that it never did. After Muybridge proved him wrong with a set of stop-motion photographs that appeared in *Scientific American* in 1878, Muybridge went on to document the walking and running behavior of over forty mammals, including humans (Muybridge 1955, 1957). Even after 100 years, his photographic data are of considerable value and beauty, and survive as a landmark in locomotion research.

The study of machines that walk also had its origin in Muybridge's time. An early walking model appeared in about 1870 (Lucas 1894). It used a linkage to move the body along a straight horizontal path while the feet moved up and down to exchange support during stepping (see figure 16.1). The linkage was originally designed by the famous Russian mathematician

470

Table 16.1
Milestones in the development of legged robots

1850	Chebyshev	Designs linkage used in early walking mechanism (Lucas 1894).
1872	Muybridge	Uses stop-motion photography to document running animals.
1893	Rygg	Patents human-powered mechanical horse.
1945	Wallace	Patents hopping tank with reaction wheels that provide stability.
1961	Space General	Eight-legged kinematic machine walks in outdoor terrain (Morrison 1968).
1963	Cannon, Higdon and Schaefer	Control system balances single, double, and limber inverted pendulums.
1968	Frank and McGhee	Simple digital logic controls walking of Phony Pony.
1968	Mosher	GE quadruped truck climbs railroad ties under control of human driver.
1969	Bucyrus-Erie Co.	Big Muskie, a 15,000-ton walking dragline is used for strip mining. It moves in soft terrain at a speed of 900 ft/hr (Sitek 1976).
1977	McGhee	Digital computer coordinates leg motions of hexapod walking machine.
1977	Gurfinkel	Hybrid computer controls hexapod walker in USSR.
1977	McMahon and Greene	Human runners set new speed records on *tuned track* at Harvard. Its compliance is adjusted to mechanics of human leg.
1980	Hirose and Umetani	Quadruped machine climbs stairs and climbs over obstacles using simple sensors. The leg mechanism simplifies control.
1980	Kato	Hydraulic biped walks with quasidynamic gait.
1980	Matsuoka	Mechanism balances in the plane while hopping on one leg.
1981	Miura and Shimoyama	Walking biped balances actively in three dimensional space.
1983	Sutherland	Hexapod carries human rider. Computer, hydraulics, and human share computing task.
1983	Odetics	Self-contained hexapod lifts and moves back end of pickup truck (Russell 1983).
1987	OSU	Three-ton self-contained hexapod carrying human driver travels at 5 mph and climbs over obstacle.

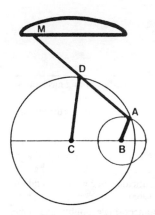

Figure 16.1
Linkage used in an early walking machine. When the input crank AB rotates, the output point M moves along a straight path during part of the cycle and an arched path during the other part of the cycle. Two identical linkages are arranged to operate out of phase so at least one provides a straight motion at all times. The body is always supported by the feet connected to the straight-moving linkage. $AB = 1$; $CD = AD = DM = (3 + \sqrt{7})/2$; $BC = (4 + \sqrt{7})/3$. After Lucas (1894).

Chebyshev some years earlier. During the eighty or ninety years that followed, workers viewed the task of building walking machines as the task of designing linkages that would generate suitable motions when driven by a source of power. Many designs were proposed (e.g., Rygg 1893; Nilson 1926; Ehrlich 1928; Kinch 1928; Snell 1947; Urschel 1949; Shigley 1957; Corson 1958; Bair 1959; Morrison 1968) (see figure 16.2), but the performance of such machines was limited by their fixed patterns of motion, since they could not adjust to variations in the terrain by placing the feet on the best footholds. By the late 1950s it had become clear that linkages providing fixed motion would not do the trick and that useful walking machines would need *control* (Liston 1970).

One approach to control was to harness a human. Ralph Mosher used this approach in building a four-legged walking truck at General Electric in the mid-1960s (Liston and Mosher 1968). The project was part of a decade-long campaign to build advanced teleoperators, capable of providing better dexterity through high-fidelity force feedback. The walking

Figure 16.2   ▶
Mechanical horse patented by Lewis A. Rygg in 1893. The stirrups double as pedals so the rider can power the stepping motions. The reins move the head and forelegs from side to side for steering. Apparently the machine was never built.

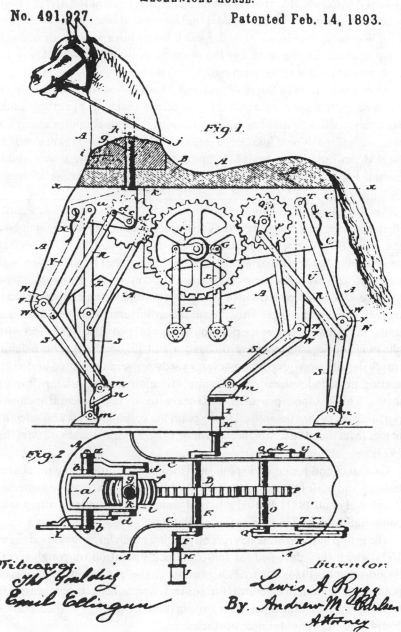

L. A. RYGG.
MECHANICAL HORSE.

No. 491,927.   Patented Feb. 14, 1893.

Fig. 1.

Fig. 2.

Witnesses
Thos Spalding
Emil Ellingun

Inventor
Lewis A. Rygg
By. Andrew M. Carlsen
Attorney

machine Mosher built stood 11 feet tall, weighed 3,000 pounds, and was powered hydraulically. It is shown in figure 16.3. Each of the driver's limbs was connected to a handle or pedal that controlled one of the truck's four legs. Whenever the driver caused a truck leg to push against an obstacle, force feedback let the drive feel the obstacle as though it were his or her own arm or leg doing the pushing.

After about twenty hours of training Mosher was able to handle the machine with surprising agility. Films of the machine operating under his control show it ambling along at about 5 mph, climbing a stack of railroad ties, pushing a foundered jeep out of the mud, and maneuvering a large drum onto some hooks. Despite its dependence on a well-trained human for control, the GE Walking Truck was a milestone in legged technology.

An alternative to human control became feasible in the 1970s: use of a digital computer. Robert McGhee's group at the Ohio State University was the first to use this approach successfully (McGhee 1983). In 1977 they built an insectlike hexapod that would walk with a number of gaits, turn, walk sideways, and negotiate simple obstacles. The computer's primary task was to solve kinematic equations in order to coordinate the eighteen electric motors driving the legs. This coordination ensured that the machine's center of mass stayed over the polygon of support provided by the feet while allowing the legs to sequence through a gait (figure 16.4). The machine traveled quite slowly, covering several yards per minute. Force and visual sensing provided a measure of terrain accommodation in later developments. The hexapod provided McGhee with an experimental means of pursuing his earlier theoretical findings on the combinatorics and selection of gait (McGhee 1968; McGhee and Jain 1972; Koozekanani and McGhee 1973).

Gurfinkel and his coworkers in the USSR built a machine with characteristics and performance quite similar to McGhee's at about the same time (Gurfinkel et al. 1981). It used a hybrid computer for control, with analog computation aiding in kinematic calculations.

The group at Ohio State has recently built a much larger hexapod (figure 16.5), which they designed for self-contained operation on rough terrain (Waldron et al. 1984). It carries a gasoline engine for power, several computers and a human operator for control, and a laser range sensor for terrain preview. At the time of this writing this machine has walked at about 5 mph and negotiated simple obstacles.

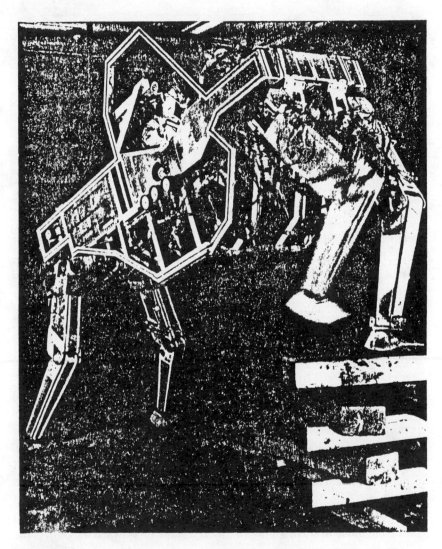

Figure 16.3
Walking truck developed by Ralph Mosher at General Electric in about 1968. The human driver controlled the machine with four handles and pedals that were connected to the four legs hydraulically. Photograph courtesy of General Electric Research and Development Center.

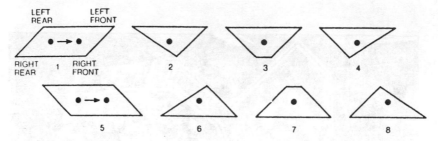

Figure 16.4
Statically stable gait. The diagrams shows the sequence of support patterns provided by the feet of a quadruped walking with a crawling gait. The body and legs move to keep the projection of the center of mass within the polygon defined by the feet. A supporting foot is located at each vertex. The dot indicates the projection of the center of mass. Adapted from McGhee and Frank (1968).

Hirose realized that linkage design and computer control were not mutually exclusive. His experience with clever and unusual mechanisms—he had built seven kinds of mechanical snake—led to his design of a special leg that simplified the control of locomotion and could improve efficiency (Hirose and Umetani 1980; Hirose et al. 1984). The leg was a three-dimensional pantograph that translated the motion of each actuator into a pure Cartesian translation of the foot. With the ability to generate x, y, and z translations of each foot by merely choosing an actuator, the control computer was freed from the arduous task of performing kinematic solutions. The mechanical linkage was helping to perform the calculations needed for locomotion. The linkage was efficient because the actuators performed only positive work in moving the body forward.

Hirose used this leg design to build a small quadruped, about 1 yard long. It was equipped with touch sensors on each foot and an oil-damped pendulum attached to the body. Simple algorithms used the sensors to control the actions of the feet. For instance, if a touch sensor indicated contact while the foot was moving forward, the leg would move backward a little bit, move upward a little bit, then resume its forward motion. If the foot had not yet cleared the obstacle, the cycle would repeat. The use of several simple algorithms like this one permitted Hirose's machine to climb up and down stairs and to negotiate other obstacles without human intervention (Hirose 1984).

These three walking machines, McGhee's Gurfinkel's, and Hirose's, represent a class called *static crawlers*. Each differs in the details of construction and in the computing technology used for control, but shares a common

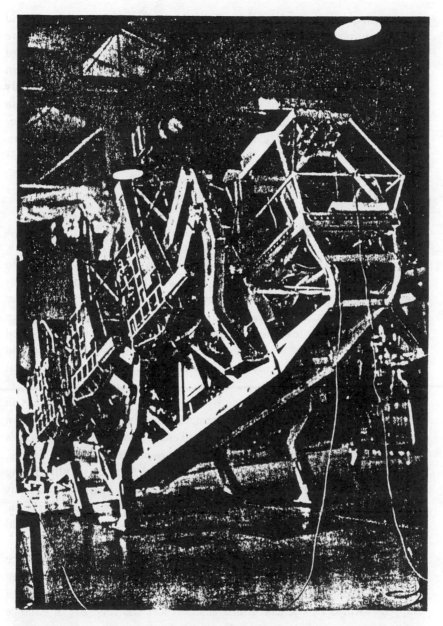

Figure 16.5
The hexapod walking machine developed at Ohio State University. It stands about 10 feet
tall, 15 feet long, and weighs 3 tons. A 90-horsepower motorcycle engine provides power to
18 variable displacement hydraulic pumps that drive the joints. The legs use pantographs
linkages to improve energy efficiency. The operator normally provides steering and speed
commands while computers control the stepping motions of the legs.

approach to balance and stability. They all keep enough feet on the ground to guarantee a broad base of support at all times, and the body legs move to keep the center of mass over this broad support base. The forward velocity is kept low enough so that stored kinetic energy can be ignored in the stability calculation. Each of these machines has been used to study rough terrain locomotion in the laboratory through experiments on terrain sensing, gait selection, and selection of foothold sequences. Several other machines that fall into this class have been studied in the intervening years, for example see (Russell 1983; Sutherland and Ullner 1984; Ooka et al. 1985).

16.3 Dynamics and Balance Improve Mobility

We now consider the study of dynamic legged machines that balance actively. These systems operate in a regime where the velocities and kinetic energies of the masses are important determinants of behavior. In order to predict and influence the behavior of a dynamic system, we must consider the energy stored in each mass and spring as well as the geometric structure and configuration of the mechanism. Geometry and configuration taken alone are not adequate to model a system that moves with substantial speed or has large mass. Consider, for example, a fast-moving vehicle that would tip over if it stopped suddenly with its center of mass too close to the front feet.

The exchange of energy among its various forms is also important in dynamic legged locomotion. For example, there is a cycle of activity in running that changes the form of the stored energy several times: the body's potential energy of elevation changes into kinetic energy during falling, then into strain energy when parts of the leg deform elastically during rebound with the ground, then into kinetic energy again as the body accelerates upward, and finally back into potential energy of elevation. This sort of dynamic exchange is central to an understanding of legged locomotion.

Dynamics also plays a role in giving legged systems the ability to balance actively. A statically balanced system avoids tipping and the ensuring horizontal accelerations by keeping its center of mass over the polygon of support formed by the feet. Animals sometimes use this sort of balance when they move slowly, but usually they balance actively.

A legged system that balances actively can tolerate departures from static equilibrium, tipping and accelerating for short periods of time. The control system manipulates body and leg motions to ensure that each tipping interval is brief and that each tipping motion in one direction is compensated by a tipping motion in the opposite direction. An effective base of support is thus maintained over time. A system that balances actively can also tolerate vertical acceleration, such as the ballistic flight and bouncing that occur during running.

The ability of an actively balanced system to depart from static equilibrium relaxes the rules govering how legs can be used for support, which in turn leads to improved mobility. For example, if a legged system can tolerate tipping, then it can position its feet far from the center of mass in order to use footholds that are widely separated or erratically placed. If it can remain upright with a small base of support, then it can travel where there are closely spaced obstructions or where is a narrow path of firm support. The ability to tolerate intermittent support also contributes to mobility. Intermittent support allows a system to move all its legs to new footholds at one time, to jump onto or over obstacles, and to use short periods of ballistic flight for increased speed. These abilities to use narrow base and intermittent support generally increase the types of terrain a legged system can negotiate. Animals routinely exploit active balance to travel quickly on difficult terrain; legged vehicles will have to balance actively, too, if they are to move with animal-like mobility and speed.

16.4 Research on Active Balance

The first machines that balanced actively were automatically controlled inverted pendulums. Everyone knows that a human can balance a broom on his finger with relative ease. Why not use automatic control to build a broom that can balance itself? Claude Shannon was probably the first to do so. In 1951 he used the parts from an erector set to build a machine that balanced an inverted pendulum atop a small powered truck. The truck drove back and forth in response to the tipping movements of the pendulum, as sensed by a pair of switches at its base. In order to move from one place to another, the truck first had to drive away from the destination to unbalance the pendulum, then proceed toward the destination. In order to balance again at the destination, the truck moved past the destination until the pendulum was again upright with no forward veocity, then moved back to the destination.

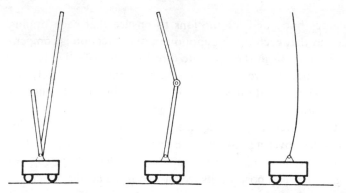

Figure 16.6
Cannon and his students built machines that balanced inverted pendulums on a moving
cart. They balanced two pendulums side by side, one pendulum on top of another, and a
long limber inverted pendulum. Only one input, the force driving the cart horizontally, was
available for control. Adapted from Schaefer and Cannon (1966).

At Shannon's urging Robert Cannon and two of his students at Stanford
University set about demonstrating controllers that balanced two pendulums at once. In one case the pendulums were mounted side by side on
the cart, and in the other they were mounted one on top of the other (figure
16.6). Cannon's group was interested in the single-input multiple-output
problem and in the limitations of achievable balance: how could they use
the single force that drove the cart's motion to control the angles of two
pendulums as well as the position of the cast? How far from balance could
the system deviate before it was impossible to return to equilibrium, given
such parameters of the mechanical system as the cart motor's strength and
the pendular lengths?

Using analysis based on normal coordinates and optimal switching
curves, Cannon's group expressed regions of controllability as explicit
functions of the physical parameters of the system. Once these regions were
found, their boundaries were used to find switching functions that provided control (Higdon and Cannon 1963; Higdon 1963). Later, they extended these techniques to provide balance for a flexible inverted pendulum
(Schaefer 1965; Schaefer and Cannon 1966). These studies of balance for
inverted pendulums were important precursors to later work on locomotion and the inverted pendulum model for walking would become the
primary tool for studying balance in legged systems (e.g., Hemami and
Weimer 1974; Vukobratovic and Stepaneko 1972; Vukobratovic 1973;

Hemami and Golliday 1977; Kato et al. 1983; Miura and Shimoyama 1984). It is unfortunate that no one has yet extended Cannon's elegant analytical results to the more complicated legged case.

The importance of active balance in legged locomotion had been recognized for some years (e.g., Manter 1938; McGhee and Kuhner 1969; Frank 1970; Vukobratovic 1973; Gubina, Hemami, and McGhee 1974; Beletskii 1975a), but progress in building physical legged systems that employ such principles was retarded by the perceived difficulty of the task. It was not until the late 1970s that experimental work on balance in legged systems got underway.

Kato and his coworkers built a biped that walked with a *quasi-dynamic* gait (Ogo et al. 1980; Kato et al. 1983). The machine had ten hydraulically powered degrees of freedom and two large feet. This machine was usually a static crawler, moving along a preplanned trajectory to keep the center of mass over the base of support provided by the large supporting foot. Once during each step, however, the machine temporarily destabilized itself to tip forward so that support would be transferred quickly from one foot to the other. Before the transfer took place on each step, the *catching* foot was positioned to return the machine to equilibrium passively. No active response was required. A modified inverted pendulum model was used to plan the tipping motion. In 1984 this machine walked with a quasi-dynamic gait, taking about a dozen 0.5-m steps per minute. The use of a dynamic transfer phase makes an important point: A legged system can exhibit complicated dynamic behavior without requiring a very complicated control system.

Miura and Shimoyama (1980, 1984) built the first walking machine that balanced itself actively. Their *stilt biped* was patterned after a human walking on stilts. Each foot provided only a point of support, and the machine had three actuators: one for each leg that moved the leg sideways and a third that separated the legs fore and aft. Because the legs did not change length, the hips were used to pick up the feet. This gave the machine a pronounced shuffling gait reminiscent of Charlie Chaplin's stiff-kneed walk.

Control for the stilt biped relied, once again, on the inverted pendulum model of its behavior. Each time a foot was placed on the floor, its position was chosen according to the tipping behavior expected from an inverted pendulum. Actually, the problem was broken down as though there were two planar pendulums, one in the pitching plane and one in the rolling plane. The choice of foot position along each axis took the current and desired state of the system into account. The control system used tabulated descriptions of planned leg motions together with linear feedback to perform the necessary calculations. Unlike Kato's machine, which came to static equilibrium before and after each dynamic transfer, the stilt biped tipped all the time.

Dynamic bipeds that balance are now being studied in several laboratories around the world. Miura (1986) has edited a videotape that reports work, including new machines by Kato, Arimoto, Masubuchi, and Furusho.

Matsuoka (Matsuoka 1979) was the first to build a machine that ran, where running is defined by a period when all feet are off the ground at one time. Matsuoka's goal was to model repetitive hopping in humans. He formulated a model with a body and one massless leg, and he simplified the problem by assuming that the duration of the support phase was short compared with the ballistic flight phase. This extreme form of running, in which nearly the entire cycle is spent in flight, minimizes the influence of tipping during support. This model permitted Matsuoka to derive a time-optimal state feedback controller that provided stability for hopping in place and for low speed translations.

To test his method for control, Matsuoka built a planar one-legged hopping machine. The machine operated at low gravity by rolling on ball bearings on a table that was inclined 10° from the horizontal in an effective gravity field of 0.17 g. An electric solenoid provided a rapid thrust at the foot, so the support period was short. The machine hopped in place at about 1 hop per second and traveled back and forth on the table.

16.5 Running Machines

Running is a form of legged locomotion that uses ballistic flight phases to obtain high speed. To study running, my coworkers and I have explored a variety of legged systems and implemented some of them in the form of physical machines. To study running in its simplest form, we built a machine that ran on just one leg. The machine hopped like a kangaroo, using a series of leaps. A machine with only one leg allowed us to concentrate on active balance and dynamics while avoiding the difficult task of coordinating many legs. We wanted to know if there were algorithms for walking and running that are independent of gait and that work correctly

Figure 16.7
Planar hopping machine traveling at about 0.8 m/sec (1.75 mph) from right to left. Lines
made by light sources attached to the machine indicate paths of the foot and the hip.

for any number of legs. Perhaps a machine with just one gait could suggest
answers to this question.

The first machine we built to study these problems had two main parts:
a body and a leg. The body carried the actuators and instrumentation
needed for the machine's operation. The leg could telescope to change
length and could pivot with respect to the body at a simple hip. The leg
was springy along the telescoping axis. Sensors measured the pitch angle
of the body, the angle of the hip, the length of the leg, the tension in the leg
spring, and contact with the ground. This first machine was constrained to
operate in a plane, so it could move only up and down and fore and aft and
rotate in the plane. An umbilical cable connected the machine to power and
a control computer.

The only way this one-legged machine can run is to hop. The running
cycle has two phases. During one phase, called *stance* or *support*, the leg
supports the weight of the body and the foot stays in a fixed location on
the ground. During stance, the system tips like an inverted pendulum.
During the other phase, called *flight*, the center of mass moves ballistically,
with the leg unloaded and free to move. (See figures 16.7 and 16.8.)

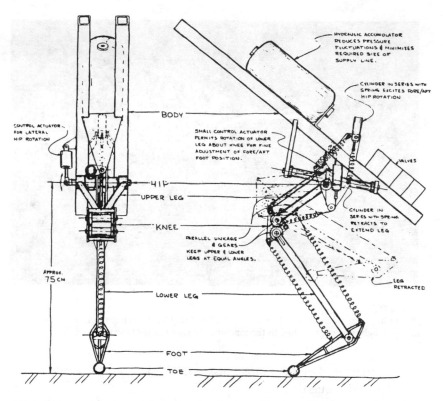

Figure 16.8
Ben Brown and I had this early concept for a one-legged hopping machine that was to operate in three dimensions. This version never left the drawing board.

16.5.1 Control of Running Was Decomposed into Three Parts

We found that a simple set of algorithms can control the planar one-legged hopping machine, allowing it run without tipping over. Our approach was to consider the hopping motion, forward travel, and posture of the body separately. This decomposition lead to a control system with three parts:

• *Hopping.* One part of the control system excites the cyclic hopping motion that underlies running, and regulates the height to which the machine hops. The hopping motion is an oscillation governed by the mass of the body, the springiness of the leg, and gravity. During support, the body bounces on the springy leg, and during flight, the system travels on a ballistic trajectory. The control system delivers a vertical thrust with the leg during each support period to sustain the oscillation and to regulate its amplitude. Some of the energy needed for each hop is recovered by the leg spring from the previous hop.

• *Forward Speed.* A second part of the control system regulates the forward running speed. This is done by moving the leg to a specified forward position with respect to the body during the flight portion of each cycle. The position of the foot with respect to the body when landing has a strong influence on tipping and acceleration behavior during the support period that follows. The body will either continue to travel with the same forward speed, accelerate to go faster, or slow down, depending on where the control system places the foot. To calculate a suitable forward position for the foot, the control system takes account of the actual forward speed, the desired speed, and a simple model of the system's dynamics. The algorithm works correctly when the machine is hopping in place, accelerating to a run, running at a constant speed, and slowing to a stationary hop.

The rule for placing and positioning the foot is based on a kind of symmetry found in running. In order to run at constant forward speed, the instantaneous forward accelerations that occur during a stride must integrate to zero. One way to satisfy this requirement is to shape the running motion so that forward acceleration has an odd symmetry throughout each stride—functions with odd symmetry integrate to zero over symmetric limits. If $x(t)$ is an odd function of time, then $x(t) = -x(-t)$. If $x(t)$ is even, then $x(t) = x(-t)$. A symmetric motion is produced by choosing an appropriate forward position for the foot on each step. In principle, symmetry of this sort can be used to simplify locomotion in systems with any number of legs and for a wide range of gaits.

• *Posture.* The third part of the control system stabilizes the pitch angle of the body to keep the body upright. Torques exerted between the body and leg about the hip accelerate the body about its pitch axis, provided that there is good traction between the foot and the ground. During the support period there is traction because the leg supports the load of the body. Linear feedback control operates on the hip actuator during each support period to restore the body to an upright posture.

Breaking running down into the control of these three functions simplifies locomotion. Each part of the control system acts as though it influences just one component of the behavior, and the interactions that results from imperfect decoupling are treated as disturbances. The algorithms implemented to perform each part of the control task are themselves quite simple, although the details of the individual control algorithms are probably not so important as the framework provided by the decomposition.

Using the three-part control system, the planar one-legged machine hops in place, travels at a specified rate, maintains balance when disturbed, and jumps over small obstacles. Top running speed is about 2.6 mph.

16.5.2 Locomotion in Three Dimensions

The one-legged machine just described was mechanically constrained to operate in the plane, but useful legged systems must balance themselves in three-dimensional space. Can the control algorithms used for hopping in the plane be generalized for hopping in three dimensions? A key to answering this question was the recognition that animal locomotion is primarily a planar activity, even though animals are three-dimensional systems. Films of a kangaroo hopping on a treadmill first suggested this point. The legs sweep fore and aft through large angles, the tail sweeps in counter-oscillation to the legs, and the body bounces up and down. These motions all occur in the sagittal plane, with little or no motion normal to the plane.

Sesh Murthy realized that the plane in which all this activity occurs can generally be defined by the forward vector and the gravity vector. He called this the *plane of motion* (Murthy 1983). For a legged system without a preferred direction of travel, the plane of motion might vary from stride to stride, but it would be defined in the same way. We found that the three-part decomposition could be used to control activity within the plane of motion.

We also found that the mechanisms needed to control the remaining *extraplanar* degrees of freedom could be cast in a form that fit into the original three-part framework. For instance, the algorithm for placing the foot to control forward speed became a vector calculation. One component of foot placement determined forward speed in the plane of motion, whereas the other component caused the plane to rotate about a vertical axis, permitting the control system to steer. A similar extension applied to body posture. The result was a three-dimensional three-part control system that was derived directly from the one used for the planar case.

To explore these ideas, we built a second hopping machine, which is shown in figure 16.9. This machine has an additional joint at the hip to permit the leg to move sideways as well as fore and aft. This machine travels on an open floor without mechanical support. It balances itself as it hops along simple paths in the laboratory, traveling at a top speed of 4.8 mph.

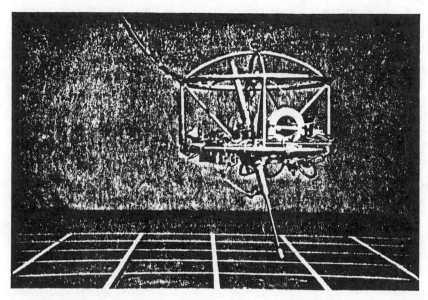

Figure 16.9
Three-dimensional hopping machine used for experiments. The control system operates to regulate hopping height, forward velocity, and body posture. Top recorded running speed was about 2.2m/sec (4.9 mph).

16.5.3 Running on Several Legs

Experiments on machines with one leg were not motivated by an interest in one-legged vehicles. Although such vehicles might very well turn out to have merit, our interest was in getting at the basics of active balance and dynamics in the context of a simplified locomotion problem. In principle, results from machines with one leg could have value for understanding all sorts of legged systems, perhaps with any number of legs. Wallace and Seifert saw merit in vehicles with one leg. Wallace (1942) patented a one-legged hopping tank that was supposed to be hard to hit because of its erratic movements. Seifert (1967) proposed the *Lunar Pogo* as a means of efficient travel on the moon.

Our study of locomotion on several legs has progressed in two stages. For a biped that runs like a human, with strictly alternating periods of support and flight, the one-leg control algorithms apply directly. Because the legs are used in alternation, only one leg is active at a time. One leg is placed on the ground at a time, one leg thrusts on the ground at a time, and one leg exerts a torque on the body at a time. We call this sort of running a *one-foot gait*. Assuming that the behavior of the other leg does not interfere, the one-leg algorithms for hopping, forward travel, and posture can each be used to control the active leg. Of course, to make this workable, some bookkeeping is required to keep track of which leg is active, which leg is idle, and it is required to keep the idle leg "out of the way."

Jessica Hodgins and Jeff Koechling demonstrated the effectiveness of this approach by using the one-leg algorithms to control each leg of a planar biped. The machine can run with an alternating gait, run by hopping on one leg, it can switch back and forth between gaits, and it can run fast. Top recorded speed is 11.5 mph. The biped has also done forward flips and aerials (Hodgins and Raibert, 1987), as shown in figure 16.10.

In principle, this approach could be used to control running on any number of legs, so long as just one touches the ground at a time. Unfortunately, when there are several legs this approach runs into difficulties. There is a conflict between the need to provide balance and the need to move the legs without collisions. For the legs to provide balance the feet

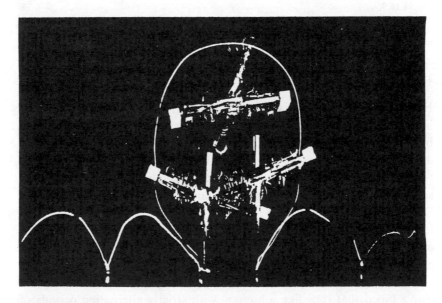

Figure 16.10
Planar biped doing a flip. The three images were made at the touchdown before the flip, the halfway point of the flip, and, the lift-off after the flip.

must be positioned so as to sweep under the center of mass during support. This argues for attaching all the legs to the body directly at or below the center of mass. However, the legs must be able to swing without colliding with one another, suggesting separation between the hips (shoulders). In principle, it is not impossible for a quadruped to run using this approach, however they rarely to (Hildebrand, private communication).

An alternative is to use the legs in pairs. Suppose that we introduce a new control mechanism that coordinates the legs of a pair to act like a single equivalent leg—what Ivan Sutherland (1983) called a *virtual leg*. Such coordination requires that the two legs of a pair exert equal forces on the ground, that they exert equal torques on the body, and that the position of each leg's foot with respect to its hip be the same. To control locomotion, the three-part control algorithms described previously specify the behavior of each virtual leg, while the control system ensures that the physical legs move so as to obey the rules required for virtual leg behavior.

Using this approach there are three quadruped gaits to consider: the *trot*, which uses diagonal legs in a pair; the *pace*, which uses lateral legs in a pair; and the *bound*, which uses the front legs as a pair and the rear legs as a pair.

We argue that the quadruped is like a virtual biped, that a biped is like a one-legged machine, and we already know how to control one-legged machines. A control system for quadruped running consists of a controller that coordinates each pair of legs to act like one virtual leg, a three-part control system that acts on the virtual legs, and a bookkeeping mechanism that keeps track. Figure 16.11 shows a four-legged machine that trots, paces, and bounds with this sort of control system. Also see tables 16.2 and 16.3.

16.5.4 Computer Programs for Running

The behavior of the running machines just described was controlled by a set of computer programs that ran on our laboratory computer. These computer programs performed several functions, including

- sampling and filtering data from the sensors,
- transforming kinematic data between coordinate systems,
- executing the three-part locomotion algorithms for hopping, forward speed, and body attitude,
- controlling the actuators,
- reading operator instructions from the console, and
- recording running behavior.

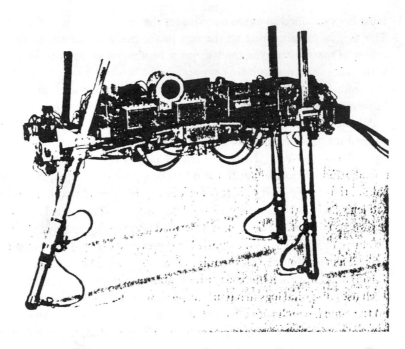

Figure 16.11
Quadruped machine that runs by trotting. *Virtual legs* are used to make trotting like biped running, which in turn is controlled with one-leg algorithms.

The control computer was a VAX-11/780 running the UNIX operating system. In order to provide real-time service with short latency and high-bandwidth feedback, the real-time control programs were implemented as a device that resided within the UNIX kernel. The device driver responded with short latency to a hardware clock that interrupted through the UNIBUS every 8 msec. All sensors and actuators were also accessed through interfaces that connected to the UNIBUS. Each time the clock ticked the running machine driver programs sampled and scaled data from the sensors, estimated joint and body velocities to determine the state of the running machine, executed the three-part locomotion algorithms, and calculated a new output for each actuator.

The programs performing these tasks were written in a mixture of C and assembly language. Assembly language was used where speed was of primary concern, such as in performing the kinematic transformations used to convert between coordinate systems. In some cases tabulated data were

Table 16.2
Details of state machine for biped and quadruped[a]

State[b]	Trigger event	Action
1 LOADING A	A touches ground	Zero hip torque A Shorten B Don't move hip B
2 COMPRESSION A	A air spring shortened	Erect body with hip A Shorten B Position B for landing
3 THRUST A	A air springs lengthening	Extend A Erect body with hip A Keep B short Position B for landing
4 UNLOADING A	A air spring near full length	Shorten A Zero hip torque A Keep B short Position B for landing
5 FLIGHT A	A not touching ground	Shorten A Don't move hip A Lengthen B for landing Position B for landing

a. The state shown in the left column is entered when the event listed in the center column occurs. During normal running states advance sequentially. During states 1–5, leg A is in support and leg B is in recovery. During states 6–10, these roles are reversed. For biped running A refers to leg 1 and B refers to leg 2. For quadruped trotting each letter designates a pair of physical legs.

b. States 6–10 repeat states 1–5, with A and B interchanged.

Table 16.3
Summary of leg laboratory research

1982	Planar one-legged machine hops in place, travels at a specified rate, keeps its balance when disturbed, and jumps over small obstacles.
1983	Three-dimensional one-legged machine runs and balances on an open floor.
1983	Murphy finds passive stability in bounding gait of simulated model.
1984	Data from cat and human runners exhibit symmetries like those used to control running machines.
1984	Quadruped running machine runs with trotting gait. The one-leg algorithms are extended to control this machine.
1985	Planar biped runs with one- and two-legged gaits and changes gait while running. Top speed is currently 11.5 mph.
1986	Monopod runs with new leg design. It uses an articulated structure and a leaf-spring for a foot.
1986	Planar biped does forward flip and aerial.

used to further increase the speed of a transformation. For instance, trigonometric functions, square roots, and higher order kinematic operations were evaluated with linear interpolation among precomputed tabulated data. One such evaluation takes about 25 μsec. The kinematic relationship between the quadruped's fore/aft hip actuator length and the forward position of the foot in body coordinates in an example of a function that was evaluated with a table.

The control programs were synchronized to the behavior of the running machine by a software finite state machine. The state machine made transitions from one state to another when sensory data from the running machine satisfied the specified conditions. For instance, the state machine made a transition from COMPRESSION to THRUST when the derivative of the support leg's length changed from negative to positive. Figure 16.12 and table 16.2 give some detail for the state machine that was used for the biped and quadruped running machines. We found that using a state machine along with properly designed transition conditions aided the interpretation of sensory data by providing noise immunity and hysteresis.

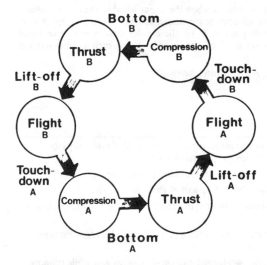

Figure 16.12
Simplified diagram of state machine that synchronizes the control programs to the behavior of the running machine. This state machine is for the biped and quadruped machines, but the one-legged state machines are similar. State transitions are determined by sensory events related to the hopping motion. A different set of control actions are put into effect in each state, as indicated in table 16.2.

Whereas sensory data determine when the state machine makes transitions, the resulting states determine which control algorithms operate to provide control. For instance, when the biped is in the THRUST A state, the control programs extend leg A, exert torque on hip A, shorten leg B, and position foot B.

In addition to the real-time programs that control the running machines, a top-level program was used to control the real-time programs. The top-level program permitted the user to initiate a running experiment, select among control modes, examine or modify the variables and parameters used by the control programs, specify sensor calibration data, mark variables for recording, and save recorded real-time data for later analysis and debugging. Each of these functions was accomplished by one or more system calls to the driver. The top-level program had no particular time constraints, so it was implemented as a time-sharing job that was scheduled by the normal UNIX scheduler.

16.6 Experiments in Animal Locomotion

Earlier in this chapter I said that studying legged machines could help us to understand more about locomotion in animals. Detailed knowledge of concrete locomotion algorithms with well-understood behavior might help us formulate experiments that elucidate the mechanisms used by humans and animals for locomotion. For instance, one might ask if animals control their forward speed as each of the running machines do, by choosing a forward position for the leg during each flight phase, with no adjustments during stance. Is there any decomposition of control? There are several questions like these that I am studying in collaboration with Thomas McMahon of Harvard University. The questions we would like to answer are summarized in the following (each is based on observations or ideas that arose in the course of studying legged machines).

16.6.1 Animal Experiments Motivated by Robot Experiments

Algorithms for Balance. The legged machines and computer simulations we have studied all use a specific algorithm for determining the landing position of the foot with respect to the body's center of mass. The foot is advanced a distance $x_f = T_s \dot{x}/2 + k_{\dot{x}}(\dot{x} - \dot{x}_d)$. This calculation is based on the expected symmetric tipping behavior of an inverted pendulum.

Do animals use this algorithm to position their feet? To find out, one must measure small changes in the forward running speed, in the angular momentum of the body during the flight phase, and in the placement of the feet. Rather than look at average behavior across many steps, as is done in studies of energetics and neural control, we must look at error terms within each step.

Symmetry in Balance. Symmetry plays a central role in simplifying the control of the dynamic legged robots described earlier. The symmetry of interest specifies that motion of the body in space, and of the feet with respect to the body, are even and odd functions of time during each support period (see figure 16.13). These are interesting motions because they leave the forward and angular motion of the body unaccelerated over a stride, and therefore lead to steady state travel. Do animals run with this sort of symmetry? To find out we are examining film data for several quadrupedal animals. Preliminary measurements show that the galloping and trotting cat sometimes runs with symmetric motions (Raibert 1986b; Raibert 1986c). We would like to know

- How universal is the use of symmetry running by animals, both in terms of different gaits and different animals?
- What is the precision of the observed symmetry?
- Does the symmetry extend to angular motion of the body, as the theory predicts?
- How do asymmetries in the mechanical structure of the body and legs influence motion symmetry?

Virtual Legs. To control the quadruped running machine the control system synchronizes the behavior of pairs of legs that provides support during stance. The synchronization has three parts, requiring that the legs of a pair exert equal vertical forces on the ground, exert equal hip torque on the body, and displace their feet equal distances from the hip or shoulder. The result of such synchronization is called a *virtual leg* (Sutherland and Ullner 1984) because it makes two legs act like one equivalent leg located halfway between the pair. The virtue of virtual legs is that they simplify the control algorithms for balance and dynamic control.

Do animals use virtual legs? To answer this question one must examine the horizontal and vertical forces exerted on the ground by both legs during double support in trotting or pacing. From these measurements one can find the differential force or impulse. One would expect the differential

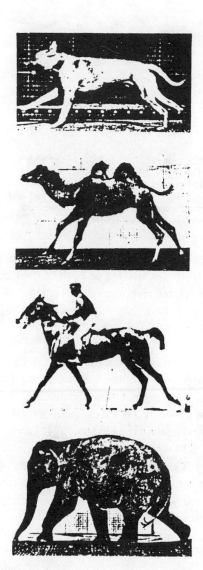

Figure 16.13
Symmetry in animal locomotion. Animals shown in symmetric configuration halfway through the stance phase for several gaits: rotary gallop (top), transverse gallop (second), canter (third), and amble (bottom). In each case the body is at minimum altitude, the center of support is located below the center of mass, the rearmost leg was recently lifted, and the frontmost leg is about to be placed. Photographs from Muybridge (1957).

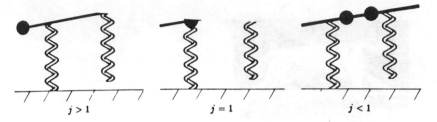

$$j > 1 \qquad\qquad j = 1 \qquad\qquad j < 1$$

Figure 16.14
The dimensionless moment of inertia, $j = J/(md^2)$, predicts passive stability of the body's pitching motion for a simple simulated model. (Left) For $j > 1$, a vertical force on the left foot causes the right hip to accelerate upward. The model has no passive pitch stability. (Center) For $j = 1$, the system acts as two separate oscillators, with neutral stability. (Right) When $j < 1$, an upward force on the left leg causes the right hip to accelerate downward. The model has passive pitch stability. From Murphy and Raibert (1985).

force to be zero if the system were using a control strategy based on virtual legs. An important manipulation will be to disturb one or both of the feet during stance, and to measure the active force response of the legs. Asymmetries in the distribution of mass in the system may require a somewhat generalized version of the virtual leg, in which a fixed ratio of forces, torques and displacements prevails.

Distribution of Body Mass.　The distribution of mass in the body can have a fundamental influence on the behavior of a running system. In earlier work we defined the dimensionless group representing the normalized moment of inertia of the body, $j = J/(md^2)$ where J is the moment of inertia of the body, m is the mass of the body, and d is half the hip spacing (Murphy and Raibert 1985). As shown in figure 16.14 when $j = 1$ the hips are located at the centers of percussion of the body. Through computer simulations of a simplified model we found that when $j < 1$ the attitude of the body in the sagittal plane can be passively stabilized when running with a bounding gait. However, when $j > 1$ stabilization was not passively obtained.

This finding has implications for the interaction between the mechanical structure of an animal and the control provided by the nervous system. Could it be that bounding animals do not actively control pitching of their bodies, but control only forward running speed and direction? A first step toward answering this question is to measure values of j for a variety of quadrupedal animals, and to relate the measurements to their preferred modes of running, trotting or pacing (no pitching) vs. bounding or galloping (pitching). Does the value of j for a quadruped vary with its trot-to-gallop transition speed? Anatomical measurements like those of Fedak, Heglund, and Taylor (1982) will provide much of the data needed to answer this question.

Yaw Control. How do human runners keep themselves from rotating about the yaw axis? Control of this degree of freedom in the one-legged hopping machine is difficult because it does not have a foot that can exert a torsional torque on the ground. But humans have long feet that might be used to develop substantial torsional traction on the ground about the yaw axis. A first step in exploring this question would be to measure the torsional torque humans exert on the ground during running and to relate the measurements to yaw motions and yaw disturbances of the body.

16.6.2 Philosophy of Interaction between Robotics and Biology

It is interesting to combine the study of animals with the study of machines. Biological systems provide both great motivation by virtue of their striking performance, and guidance with the details of their actions. They are existence proofs that give us a lower bound on what is possible. Unfortunately, biological systems are often too complicated to study—there are many variables, precise measurement is difficult, there are limitations on the experimenter's ability to manipulate the preparation, and perhaps, an inherent difficulty in focusing on the information-level of a problem.

On the other hand, laboratory robots are relatively easy to build. Precisely controlled experiments are possible, as are careful measurements and manipulations, and the "subject" can be redesigned when necessary. However, the behavior of these experimental robot systems is impoverished when compared with the biological counterpart. They are easy to study, but they do not perform nearly so well as biological systems.

Analysis of living systems and synthesis of laboratory systems are complementary activities, each with strengths and weaknesses. Together, these activities can strengthen one another, leading to fundamental principles that elucidate the domain of both problems, independent of the particular implementation. Because machines face the same physical laws and environmental constraints that biological systems face when they perform similar tasks, the solutions they use may embrace similar principles. In solving the problem for the machine, we generate a set of plausible algorithms for the biological system. In observing the biological behavior, we explore plausible behaviors for the machine. In its grandest form, this approach lets the study of robotics contribute to both robotics and biology and lets the study of biology contribute to both biology and robotics.

16.7 The Development of Useful Legged Robots

The running machines described in the previous section are not useful vehicles or even prototypes for such vehicles. They are experimental apparatus used in the laboratory to explore ideas about legged locomotion. Each was designed to isolate and examine a specific locomotion problem, while postponing or ignoring many other problems. Let us now step back and ask what problems remain to be solved before legged robots are transformed into practical machines that do useful work.

16.7.1 Terrain Sensing

Perhaps the deepest problem limiting current walking machines, as well as othe forms of autonomous vehicle, is their inability to perceive the shape and structure of their immediate surroundings. Humans and animals use their eyes to locate good footholds, to avoid obstacles, to measure their own rate and direction of progress, and to navigate with respect to visible landmarks. The problem of giving machines the ability to see has received intensive and consistent attention for the past twenty-five or thirty years. There has been steady progress during that period. Current machines can see well enough to operate in well-structured and partially structured environments, but it is difficult to predict when machines will be able to see well enough to operate autonomously in rough outdoor terrain. I do not expect to see such autonomous machine behavior for at least ten years.

Sensors simpler than vision may be able to provide solutions to certain parts of the problem under certain circumstances. For instance, sonar and laser range data may be used to detect and avoid nearby obstacles. Motion data may be used for measuring speed and direction of travel with techniques that are substantially simpler than those needed to perceive shape in three dimensions.

16.7.2 Travel on Rough Terrain

Complete knowledge of the geometry of the terrain, as might be supplied by vision or these other senses, would not in itself solve the problem of walking or running on rough terrain. A system travelling over rough terrain needs to know or figure out what terrain shapes provide good footholds, which sequence of footholds would permit traversal of the terrain, and how to move so as to place the feet on the available footholds. It will be necessary to coordinate the dynamics of the vehicle with the dynamics of the terrain.

There are several ways that terrain becomes rough and therefore difficult to negotiate:

- not level;
- limited traction (slippery);
- areas of poor or nonexistent support (holes);
- vertical variations:
    - minor vertical variations in available footholds (less than about one-half the leg length);
    - major vertical variations in available footholds (footholds separated vertically by distances comparable to the dimensions of the whole leg);
- large obstacles between footholds (poles);
- intricate footholds (e.g., rungs of a ladder).

The techniques that will allow legged systems to operate in these sorts of terrain will involve the mechanics of locomotion, kinematics, dynamics, geometric representation, spatial reasoning, and planning. Although coarse- and medium-grain knowledge of the terrain will be important, I expect techniques that make legged systems inherently insensitive to fine-grain terrain variations to play an important role too. Ignoring the hard sensing issues mentioned earlier, I believe the perception and control mechanisms required for legged systems to travel on rough terrain will require a substantial research effort, but the important problems can be solved within the next ten years if they are pursued vigorously.

16.7.3 Mechanical Design and System Integration

When these sensing and control problems are solved, it will remain to develop mechanical designs that function with efficiency and reliability. Useful vehicles must carry their own power, control computers, and a payload. A host of interesting problems present themselves, including such matters as energy efficiency, structural design, strength and weight of materials, and efficient control. For instance, the development of materials and structures for efficient storage and recovery of elastic energy will be particularly important for legged vehicles. I expect that early useful legged vehicles can be built with existing mechanical and aerospace technology, but performance will improve rapidly as designs are refined, embellished, and improved.

Acknowledgments

This research was supported by a grant from the System Development Foundation and a contract from the Defense Advanced Research Projects Agency.

References

Ehrlich, A. 1928. Vehicle Propelled by Steppers. Patent Number 1,691,233.

Frank, A. A. 1970. An approach to the dynamic analysis and synthesis of biped locomotion machines. *Medical and Biological Engineering* 8:465–476.

Gurfinkel, V. S., Gurfinkel, E. V., Shneider, A. Yu., Devjanin, E. A., Lensky, A. V., and Shitilman, L. G. 1981. Walking robot with supervisory control. *Mechanism and Machine Theory* 16:31–36.

Hemami, H., and N. Golliday, C. L., Jr. 1977. The inverted pendulum and biped stability. *Mathematical Biosciences* 34:95–110.

Higdon, D. T., and Cannon, R. H., Jr. 1963. On the control of unstable multiple-output mechanical systems. In *ASME Winter Annual Meeting*.

Hirose, S. 1984. A study of design and control of a quadruped walking vehicle. *International J. Robotics Research* 3:113–133.

Hirose, S., and Umetani, Y. 1980. The basic motion regulation system for a quadruped walking vehicle. *ASME Conference on Mechanisms*.

Hodgins, J., and Raibert, M. H., 1987. Planar biped goes head over heels. *ASME Winter Annual Meeting*, Boston.

Hodgins, J., Koechling, J., and Raibert, M. H. 1985. Running experiments with a planar biped. *Third International Symposium on Robotics Research*, Cambridge: MIT Press.

Kato, T., Takanishi, A., Jishikawa, H., and Kato, I. 1983. The realization of the quasi-dynamic walking by the biped walking machine. In *Fourth Symposium on Theory and Practice of Robots and Manipulators*, A. Morecki, G. Bianchi, and K. Kedzior (eds.). Warsaw: Polish Scientific Publishers. 341–351.

Koozekanani, S. H., and McGhee, R. B. 1973. Occupancy problems with pairwise exclusion constraints—an aspect of gait enumeration. *J. Cybernetics* 2:14–26.

Liston, R. A. 1970. Increasing vehicle agility by legs: the quadruped transporter. Presented at *38th National Meeting of the Operations Research Society of America*.

Liston, R. A., and Mosher, R. S. 1968. A versatile walking truck. In *Proceedings of the Transportation Engineering Conference*. Institution of Civil Engineers, London.

Lucas, E. 1894. Huitieme recreation—la machine a marcher. *Recreations Mathematiques* 4:198–204.

Matsuoka, K. 1980. A mechanical model of repetitive hopping movements. *Biomechanisms* 5:251–258.

McGhee, R. B. 1968. Some finite state aspects of legged locomotion. *Mathematical Biosciences* 2:67–84.

McGhee, R. B. 1983. Vehicular legged locomotion. In *Advances in Automation and Robotics*, G. N. Saridis (ed.). JAI Press.

McGhee, R. B., and Frank, A. A. 1968. On the stability properties of quadruped creeping gaits. *Mathematical Biosciences* 3:331–351.

McGhee, R. B., and Jain, A. K. 1972. Some properties of regularly realizable gait matrices. *Mathematical Biosciences* 13:179–193.

McGhee, R. B., and Kuhner, M. B. 1969. On the dynamic stability of legged locomotion systems. In *Advances in External Control of Human Extremities*, M. M. Gavrilovic and A. B. Wilson, Jr. (eds.) Jugoslav Committee for Electronics and Automation, Belgrade, 431–442.

Miura, H. 1986. Biped locomotion robots. Unpublished videotape.

Miura, H., and Shimoyama, I. 1984. Dynamic walk of a biped. *International J. Robotics Research* 3:60–74.

Morrison, R. A. 1968. Iron mule train. In *Proceedings of Off-Road Mobility Research Symposim*. International Society for Terrain Vehicle Systems, Washington, 381–400.

Murphy, K. N., and Raibert, M. H. 1985. Trotting and bounding in a planar two-legged model. In *Fifth Symposium on Theory and Practice of Robots and Manipulators*, A. Morecki, G. Bianchi, and K. Kedzior (eds.). Cambridge: MIT Press, 411–420.

Murthy, S. S., and Raibert, M. H. 1983, 3D balance in legged locomotion: modeling and simulation for the one-legged case. In *Inter-Disciplinary Workshop on Motion: Representation and Perception*, ACM.

Muybridge, E. 1955. *The Human Figure in Motion*. New York: Dover Publications. First edition, 1901 by Chapman and Hall, Ltd., London.

Muybridge, E. 1957. *Animals in Motion*. New York: Dover Publications, First edition, 1899 by Chapman and Hall, Ltd., London.

Raibert, M. H. 1986. *Legged Robots That Balance* Cambridge: MIT Press.

Raibert, M. H. 1986. Symmetry in running. *Science*, 231:1292–1294.

Raibert, M. H., and Brown, H. B., Jr. 1984. Experiments in balance with a 2D one-legged hopping machine. *ASME J. Dynamic Systems, Measurement, and Control* 106:75–81.

Raibert, M. H., Brown, H. B., Jr., and Chepponis, M. 1984. Experiments in balance with a 3D one-legged hopping machine. *International J. Robotics Research* 3:75–92.

Raibert, M. H., Chepponis, M., and Brown H. B. Jr. 1986. Running on four legs as though they were one. *IEEE J. Robotics and Automation*, 2.

Russell, M., Jr. 1983. Odex I: the first functionoid. *Robotics Age* 5:12–18.

Schaefer, J. F., and Cannon, R. H., Jr. 1966. On the control of unstable mechanical systems. *International Federation of Automatic Control*. London, 6c.1–6c.13.

Seifert, H. S. 1967. The lunar pogo stick. *J. Spacecraft and Rockets* 4:941–943.

Shigley, R. 1957. *The Mechanics of Walking Vehicles*. Land Locomotion Laboratory, Report 7, Detroit, Michigan.

Sitek, G. 1976. Big Muskie, *Heavy Duty Equipment Maintenance* 4:16–23.

Snell, E. 1947. Reciprocating Load Carrier. Patent Number 2,430,537.

Sutherland, I. E., and Ullner, M. K. 1984. Footprints in the asphalt. *International J. Robotics Research* 3:29–36.

Urschel, W. E. 1949. Walking Tractor. Patent Number 2,491,064.

Vukobratovic, M., and Stepaneko, Y. 1972. On the stability of anthropomorphic systems. *Mathematical Biosciences* 14:1–38.

Waldron, K. J., Vohnout, V. J., Pery, A., and McGhee, R. B. 1984. Configuration design of the adaptive suspension vehicle. *International J. Robotics Research* 3:37–48.

Wallace, H. W. 1942. Jumping Tank Vehicle. Patent Number 2,371,368.

A REVIEW OF RESEARCH ON WALKING VEHICLES

Vijay R. Kumar
Department of Mechanical Engineering and Applied Mechanics
University of Pennsylvania
Philadelphia, PA 19104

Kenneth J. Waldron
Department of Mechanical Engineering
The Ohio State University
Columbus, OH 43210

ABSTRACT

Research into artificial legged locomotion systems with particular reference to statically stable walking vehicles is reviewed. The primary technical problem areas are briefly discussed and the principal references in the literature are cited.

1. INTRODUCTION

Mobility in Robots

Many robotic applications today require or would benefit from mobility. However, most present day robots are stationary or have, at most, limited motion along guideways in one or two directions and operate only on smooth level floors. This is because, until recently, it was not economical or practical to have mobile robots, particularly in situations involving uneven terrain. With recent advances in robotics and allied sciences (especially the development of relatively inexpensive digital computers), more generalized mobile robots have become feasible and very attractive prospects for appropriate applications. Researchers have realized the enormous potential of having mobility in robots performing repetitive tasks in industrial environments and for transportation in hostile surroundings and mobile robotics has become a very active area of research.

Legged Locomotion

Legged locomotion systems offer advantages over wheeled or tracked systems in appropriate situations. Legged systems can provide mobility in environments which are cluttered with obstacles where wheeled or tracked vehicles could not. They can usually provide superior mobility in difficult terrain or soil conditions. Further, such systems are also inherently omni-directional, whereas wheeled and tracked systems are directional, two degree of freedom systems. Finally, a passive suspension, as used in conventional vehicles is subject to unpredictable dynamics which compromise its performance as an instrument platform. Fully terrain adaptive legged vehicles (McGhee 1984) can completely isolate the vehicle

Reprinted from *Robotics Review 1*, 1989, pages 243-266, "A Review of Research on Walking Vehicles" by V.R. Kumar and K.J. Waldron, by permission of The MIT Press, Cambridge, Massachusetts.

body from small wavelength terrain irregularities, thereby providing a stable and predictable instrument platform.

As will be seen in the paper, there are significant research issues which are important, or unique to the design and operation of legged locomotion systems. These issues have been the focus of a very active research effort in recent years (Hirose 1984; Mosher 1969;Okhotsimski et al 1977; Orin 1982; Raibert and Sutherland 1982; Russell 1983; Waldron and McGhee 1986).

Objective and Scope

In this paper, some of the recent advances in legged locomotion are described, and a critical assessment of the state of the art in walking vehicles is presented. Component design and scaling, prime-movers, actuators and transmissions, and computer control have been discussed exhaustively elsewhere (McGhee 1984; Waldron *et al* 1984). The main thrust of this paper is on motion planning, control and coordination. Also, for the most part, we limit our discussion to statically stable vehicles. An excellent analysis and review of dynamically stable vehicles is presented by Raibert (1985). Biped locomotion which is largely dynamically stable is also outside the scope of this paper. The interested reader can refer to (Kato *et al* 1983) and (Muira and Shiyoma 1985) as a starting point.

In this paper, we present a review of those walking machines that have been built so far, which are of most interest scientically, and the key research contributions that led to the design and fabrication of these systems in Section 2. The present state of the art is briefly discussed. Section 3 focusses on statically stable gaits. A brief history of research in this area is presented. The gait selection problem, practical issues in implementation, and automatically adapting the gait to the environment are of interest here. In Section 4, we discuss the kinematic and dynamic models that have been used for walking vehicles. From a kinematic point of view, the actuation scheme is a parallel one. Further, typically, such systems are redundant in actuation. The problem of allocating the load on the vehicle between the actuators (and feet) is analyzed and progress in this area is reviewed. In Section 5, we elaborate on the problem of motion planning for vehicles in unstructured environments, especially, uneven terrain. Finally, the research problems that need to be investigated and future challenges are described.

2. REVIEW OF WALKING ROBOTS

Walking machines possess an immense potential for rough-terrain locomotion, and several articulated legged vehicles have been built in the past two decades to demonstrate this potential (Hirose 1984; Mosher 1969; Okhotsimski *et al* 1977; Orin 1982; Raibert and Sutherland 1983; Russell 1983; Waldron and McGhee 1986). This area of science is relatively new and until now, most of the research has been geared to understanding the mechanics of locomotion and control and coordination of articulated limbs. Some of the significant projects in the area of legged locomotion, both past and current, are described in this section.

The first legged machine completely controlled by computers was built at the University of South California (McGhee and Frank 1968). The

Phoney Pony was a four legged machine which had two degrees of freedom in each leg which were coordinated by a computer. It was inspired by a machine built by Mosher (1969) at General Electric Corporation. The GE quadruped was a 3000 lb. machine in which the operator controlled the twelve degrees of freedom in the four legs by his hands and feet through a master-slave type valve controlled hydraulic servo system. No computer coordination was used. This proved to be extremely cumbersome and it demonstrated the necessity of automation in the coordination of the legs in a walking machine. In 1969, the largest off-road vehicle, *Big-Muskie*, a 15000 ton walking dragline was built for strip shining by Bucyrus-Erie Company (Sitek 1976). Work on analog computer controlled bipeds was pioneered in Yugoslavia (Vukobratovic, Frank and Juricic 1970) and in Japan (Kato *et al* 1983). At this stage, walking machines required a lot of operator participation but several proofs of the concept now existed.

In the USSR, significant theoretical work on locomotion and gaits was done by Bessonov and Umnov (1973). A few years later, Okhotsimski *et al* (1977) developed a six-legged walking vehicle. Unfortunately, very few details on this machine are available to us. It was thought that the number of legs reflected a tradeoff between stability and complexity and six was judged to be an optimal number. Although the machine was powered externally it was under complete computer control and was also equipped with a scanning range-finder. This machine was the first to demonstrate application of artificial intelligence concepts to legged locomotion systems. The legs were similar to those of insects and had a total of 18 degrees of freedom. The intelligence structure was hierarchical and the decision-making process was organized into a situation-action dictionary. Perception and motion planning algorithms were developed for locomotion in unstructured terrains. It seems to have been successful in adapting its gait to the surroundings to some extent. The problem of force allocation between the legs of the machine was also studied.

The *OSU Hexapod* was a 300 lb. six legged machine built at the Ohio State University (McGhee 1977; Orin 1982). It had "insect type" legs similar to its Russian counterpart and was driven by electric motors controlled by SCRs. It was powered externally and controlled through a PDP-11/70 computer. The Hexapod was equipped with force sensors at the feet for force control and gyroscopes for attitude control. Work on the design of walking machines was accompanied by research on intelligent control schemes and terrain-adaptive locomotion. Among several gait experiments, a *follow-the-leader* gait was implemented on the machine (Özgüner *et al* 1984). The operator used a hand-held laser to designate candidate footholds, which were used for the front legs if the computer found them acceptable. Two CID television cameras were used to detect the laser beam. The four legs behind the two fore legs stepped on the footprints selected for the fore legs.

The *PVII* was a small quadruped walking machine constructed at the Tokyo Institute of Technology (Hirose and Umetani 1980). It weighed only 10 kg. and had a 10 watt power consumption. It was powered externally and connected via an umblical cord to a minicomputer. A hierarchical control system was implemented whose functions included navigation, planning, gait control, posture regulation and the generation of commands for the servomechanisms. The machine demonstrated the ability to climb stairs and to avoid small obstacles through tactile sensing at the feet. More recently, another quadruped, *TITAN III* has been developed (Hirose *et al* 1985). It is a 320 watt machine which is also equipped with wheels as alternative locomotion elements. The emphasis in design has been on improving the efficiency of locomotion.

In 1983, Ivan Sutherland built the first six legged machine which was entirely self-contained and had an on-board computer and power supply at Carnegie Mellon University (Sutherland and Ullner 1984) . This also had a total of 18 degrees of freedom which were hydraulically actuated. The hydraulic circuits were designed to make the legs move "usefully" without being digitally controlled by a microprocessor. The operator controlled the speed, direction and attitude of the body. The questions of optimal parameters for gaits or foothold selection for navigation were not addressed. About the same time, also at CMU, Raibert built a one-legged hopping machine (Raibert and Sutherland 1983). It was the first successful statically unstable but dynamically stable machine. The project was a study in dynamic balancing and height and attitude control through hopping. Similar efforts later led to the design and fabrication of a four-legged machine based on the same principle (Raibert 1986; Raibert *et al* 1986).

The ODEX is a six-legged walking machine with an axisymmetric leg configuration built by Odetics Inc. in 1983 (Russell 1983). It had an unprecedented strength to weight ratio (5.6 when stationary and 2.3 while walking) and reasonable agility. It had eighteen degrees of freedom too but proved to be very easy to control. The mode of control was teleoperator-like and the vehicle received commands through a joystick via radio telemetry. The actions of the legs were coordinated by on-board computers but the operator used the camera system to view the surroundings and appropriately direct ODEX. ODEX uses a tripod gait used by many six-legged arthropods. A newer version of the machine is being evaluated for maintenance tasks in nuclear power plants.

Gradually, with an improvement in the understanding of the mechanics of locomotion, research on planning and coordination of the legs began to gain momentum. Kessis *et al* (1985) reported a four-level control architecture for an autonomous six-legged hexapod built at the University of Paris in 1980. The lowest level (leg level) involved the control of individual legs using a variable leg compliance to adapt to uneven terrain. The second level (gait level) generated gaits according to the commands from level three (plan interpreter). Level four (planner) coped with the perception and modeling of the universe and planned actions according to the situation encountered. The terrain model was built by a rule based production system and path planning used the A^* algorithm in a 3-dimensional environment. In Japan, adaptive gait control strategies to negotiate forbidden footholds have become the subject of active research (Hirose 1984). Similar efforts in the USA have lead to the development of algorithms for foothold selection and gait control for walking on unstructured terrain (McGhee and Iswandhi 1979; Kwak 1986; Kumar 1987; Qui and Song 1988).

Recently, a research effort at the Ohio State University has resulted in the successful design, fabrication and testing of a six-legged vehicle called the Adaptive Suspension Vehicle (ASV) (Waldron *et al* 1984a; Waldron *et al*

1986; Song and Waldron 1988). The ASV is designed to carry an operator but it may eventually operate as a completely autonomous unmanned system. It is a proof-of-concept prototype of a legged vehicle designed to operate in rough terrain that is not navigable by conventional vehicles (Waldron and McGhee 1986). It is 3.3 meters (10.9 feet) high and weighs about 3200 kg (7000 lb.). It presently operates in a supervisory control mode. To this end, it possesses over 100 sensors, 17 onboard single board computers and a 900 c.c. motorcycle engine rated at 50 kW (70 hp) continuous output. It has three actuators on each of the six legs thus providing a total of 18 degrees of freedom. The 18 degrees of freedom are hydraulically actuated through a hydrostatic configuration. The engine is coupled through a clutch to a 0.25 kW/hr flywheel which smooths out the fluctuating power requirements of the pumps.

The ASV senses over 100 control variables. The most important sensor for motion planning is an optical scanning rangefinder which is a phase modulated, continuous wave ranging system with a range of approximately 30 feet and a resolution of 6 inches at maximum range (Zuk and Dell'Eva 1983; Klein *et al* 1987). It has a field of view of 40 degrees on either side of the body longitudinal axis and from 15 to 75 degrees below the horizontal. A inertial sensor package consisting of a vertical gyroscope, rate gyroscopes for the pitch, roll and yaw axes, and three linear accelerometers provide information to determine body velocity and position. Leg position feedback is used from the legs in the support phase for the purpose of correcting for gyro and integration drift in the inertial reference system. Thus, there is considerable scope for sensor cross checking and error detection. The leg control system is based on a force control scheme, when the leg is on the ground, and on a position control scheme, when the leg is in transfer phase. Thus the position, velocity and pressure difference across each of the eighteen hydraulic actuators are monitored during operation.

The ASV, unlike its predecessors is completely computer controlled and mechanically autonomous. The operator performs the function of path selection and specifies the linear velocities of the vehicle in the fore-aft and lateral directions, and the yaw velocity. The roll and pitch rates and the velocity in the vertical direction are automatically regulated by the guidance system. In addition to path selection, the operator may also suppy important gait parameters such as stride length and duty factor, and set the mode of control (cruise/dash, terrain-following or large obstacle mode).

In conclusion, the walking machines that have been built so far, have been bulky, heavy, slow and awkward. Consequently, they have not been useful as vehicles or even prototypes. However, as laboratory vehicles, they have served as proofs of concept. In particular, they have demonstrated that active coordination of mechanical articulated limbs in order to drive a vehicle is a very practical idea. The need for intelligent control to minimize human participation in coordinating the multiple degrees of freedom has been underscored. These efforts have spurred research on biological systems to better understand the neural control of locomotion and has bridged the gap between work on animal walking and robot walking.

3. GAITS

Introduction

A vast variety of examples of legged locomotion can be found in nature. Several studies of animal gaits have been undertaken with a view to discovering strategies for control and coordination in locomotion. These studies have lead to the development of a mathematical theory for gait analysis. Very briefly, the history of research in this area, the mathematical models, and the application of the theory to gaits for walking robots is discussed here.

Gait Stability

Broadly speaking, animal locomotion can be classified into two categories. The first type is the one exhibited by insects. Insects are arthropods and have a hard exoskeletal system with jointed limbs. They use their legs as struts and levers and the legs must always support the body during walking, in addition to providing propulsion. In other words, the metachronal or sequential pattern of steps must ensure static stability. The vertical projection of the center of gravity must therefore always be within the support pattern (the two dimensional convex polygon formed by the contact points). This kind of locomotion has been described as *crawling* and the legs have to provide at least a tripod of support at all times.

Another kind of locomotion may be observed in humans, horses, dogs, cheetahs and kangaroos which have a more flexible structure. These animals require dynamic balance, which is a less stringent restriction on the posture and gait of the animal. The animal may not be in static equilibrium; to the contrary, there may be periods of time when none of the support legs are on the ground, as is observed in trotting horses, running humans and, of course, hopping kangaroos.

Until now, most efforts to build dynamically balanced robots have been confined to bipeds (Kato *et al* 1983; Muira and Shimoyama1985) and hopping machines (Raibert 1985). This is because the complexities of the locomotion system in biological creatures have prevented proper understanding of the involved mechanics and controls. Also, the present state of the art in digital computers has allowed the implementation of only simplified dynamic models for walking machines. Specifically, in these systems, pertubations in body attitude and altitude and the corresponding rates are corrected by small changes in the stride to produce a limit cycle stability. Coordination is much more complex as compared to statically stable machines. However, grouping of legs permits extension of control strategies developed for one or two legs to four legged dynamic locomotion (Raibert *et al* 1986). As mentioned earlier, a detailed critique of such machines is outside the scope of this paper. Further work in this direction is detailed in (Glower and Özgüner 1986; Goldberg and Raibert 1987; Agrawal and Waldron 1988).

The present generation of walking machines almost exclusively emulates the mechanism of walking in insects. Control in such machines is obviously simpler, and is well within the reach of modern technology. Also, the statically stable crawl typified by insects is better suited to heavy machines with rigid structures like the present day machines. A brief account of the evolution of a comprehensive theory for gait analysis for statically stable systems follows.

Biological Systems

A vast variety of examples of legged locomotion can be found in nature and many quantitative studies of animal gaits have been undertaken with a view to discovering strategies for control and coordination in locomotion. See for example, (Alexander 1984; Delcomyn 1981; Gambaryan 1974; Gray 1968; Hildebrand 1960; Pandy *et al* 1988; Pearson and Franklin 1984). The earliest studies of gaits date back to 1872 when Muybridge used successive photographs to study the locomotion of animals (Muybridge 1957) and later human locomotion (Muybridge 1955). However, Hildebrand (1965) was probably the first researcher to analyze gaits quantitatively. Since then, many reports have been published on this subject (Alexander 1984; Delcomyn 1981; Gambaryan 1974; Gray 1968) motivated in some cases by an interest in legged robots (Pandy *et al* 1988; Pearson and Franklin 1984).

Wilson's report (1966) on insect gaits is particularly informative. He developed a model for insect gaits by proposing a few simple rules. A wave of protraction runs from posterior to anterior, and *contralateral* (on opposite sides of the body) legs of the same segment alternate in phase. The protraction time is constant and the retraction time is varied to control the frequency. Further, the intervals between the steps of the hind and middle *ipsilateral* (on the same side of the body) legs, and those between the middle and the fore ipsilateral legs are constant. Though these rules are not observed in all species (Wilson 1966, Wilson 1976), most of these rules have been validated qualitatively with a variety of insects when walking on smooth horizontal surfaces. The classical alternating *tripod* gait exhibited by fast moving locusts and cockroaches is a good example (Delcomyn 1981). More recently, a cinematic analysis of locusts undertaken to study locomotion characteristics on uneven terrain has been described (Pearson and Franklin 1984). Studies by Cruse (1976) on stick insects have shown that the posterior legs are placed close to the support sites of the immediately anterior ipsilateral legs. This *follow-the-leader* behavior was observed in horses by Hildebrand and in domestic (Nubian) goats (Pandy *et al* 1988) and may be even a general strategy for walking on uneven terrain.

Walking Vehicles

Tomovic (1961) and Hildebrand (1965) were the first to study gaits quantitatively. Hildebrand developed the concepts of a *gait formula* and a *gait diagram* to describe symmetric gaits of horses (see (Song and Waldron 1988) for definitions of new terms). McGhee (1968) started the development of a general mathematical theory of locomotion based on a finite state concept. A leg was defined as a sequential machine with an output state 1 representing the support phase, where the leg is in contact with the supporting surface, and an output state 0 representing the transfer phase. In physiology, the support phase is the period of retraction and the transfer phase is the period of protraction. McGhee and his coworkers developed several mathematical tools such as *event sequences, gait formulae and gait matrices* to study stepping patterns (see McGhee and Frank (1968), McGhee and Jain (1972), and McGhee(1984)).

The main emphasis of investigations on statically stable gaits was the relationship between the stepping patterns and the static stability engendered by the gait. To this end, McGhee and Frank (1968) defined the *static stability margin* for gaits as a measure of the static stability of gaits (see Figure 1). A computationally more tractable measure called the *longitudinal stability margin*, as shown in Figure 1 , was also proposed by them. They identified the statically stable gaits for a quadruped as creeping gaits, of which the regular crawl gait (crawl gaits were defined by Hildebrand (1965)) was shown to be the most optimal in terms of the longitudinal stability margin. McGhee and Jain (1972) attempted to explain the bias shown by animals towards certain gaits through a characteristic called *regular realizability*.

The study of gaits has led to the definition and classification of different gaits. In particular, *periodic* and *regular* gaits were found to possess interesting properties. In a periodic gait, it is sufficient to describe one locomotion cycle in order to specify the gait. In a regular gait, the fraction of time in a locomotion cycle that a leg spends in support phase, or the *duty factor*, is the same for all legs. Other important quantities are the *time period* of a cycle, and the *stroke*, or the distance through which a foot translates relative to the vehicle body through the locomotion cycle. Bessenov and Umnov (1973) used numerical experimentation for hexapods to demonstrate that the optimal gaits are regular and *symmetric*. A gait is symmetric if the motion of contralateral adjacent legs are exactly half a cycle out of phase. These optimal gaits were described as "wavy" gaits. This agreed with Wilson's observation of a metachronal pattern of steps in insect walking. The *wave gait* is a symmetric, periodic and regular gait in which any adjacent ipsilateral pair of legs differ in phase by the same amount. The regular crawl gait, which was shown to be optimal for quadrupeds (McGhee and Frank 1968) could be classified as a wave gait. It was discovered through numerical experimentation that for a given duty factor, wave gaits were optimally stable gaits for hexapods as well as quadrupeds. More recently, Song and Waldron (1987) carried out a detailed analysis and classification of gaits, in which the wave gait was analytically proven to be the optimal gait for hexapods. The variation in load carying capacity with load and speed for walking vehicles was investigated by Huang and Waldron (1987) for wave gaits.A more extensive treatment of statically stable gaits can be found in the work reported in references (Kumar and Waldron 1988; Song and Waldron 1988).

Statically stable gaits and stepping patterns on even terrain are now extensively researched and are well-understood. The focus of research on gaits has shifted from optimization based on static stability to more practical considerations such as foothold selection and motion planning (Lee and Orin 1988). In particular, terrain - adaptive walking has become an extremely important area of research. If an accurate description of the terrain geometry is available, a pre-defined sequence of steps can be identified for traversing a path (Qui and Song 1988). One such example is the *large obstacle gait* (Song and Waldron 1988). However this is clearly an ideal situation. In a practical case, an intelligent system is required for terrain - adaptive locomotion (Kumar and Waldron 1989).

The follow-the-leader gait was studied by Özgüner *et al* (1984) as a method of decreasing the burden of foothold selection on the operator. It was demonstrated on the OSU Hexapod. The need for a gait which could be automatically adapted to varying terrain conditions lead to the development of aperiodic gaits, for which the large obstacle gait is a good example. An aperiodic gait called the *free gait*, was proposed by Kugushev and Jaroshevskij (1975) to address the problem of automatic foothold selection in a real world situation. It was later modified by McGhee and Iswandhi (1979) for the OSU Hexapod. The free gait algorithm sought to maximize the stability as well as the "availability" of legs by appropriately specifying the footholds in space as well as time for each leg. The utility of the free gait was subsequently demonstrated through simulations of the ASV (Kwak 1986) and on the actual machine (Patterson *et al* 1983; Klein *et al* 1987). However, the free gait required the terrain to be discretized into cells each of which had to be classified as GO or NO-GO. Further, it did not address the problem of locomotion on rough terrain. Another problem was that the algorithm involved a heavy penalty in terms of computational load. An adaptive gait controller that was designed to overcome these shortcomings was proposed by Kumar and Waldron (1989) for motion planning on uneven terrain. A periodic, optimally stable gait (that is, the wave gait) was selected for locomotion, but transitory aperiodicity and asymmetry in the support patterns were allowed to enable the system to circumvent variants in the terrain relief. Once more static stability was an important criterion, but the measures of static stability allowed for deviations from ideal conditions of motion along the axis of symmetry and a two-dimensional terrain (Kumar and Waldron 1988a).

Though work on terrain-adaptive walking has been reported, all of the past work has involved numerical descriptions of the terrain-vehicle systems accompanied by research on quantitative analysis of walking and numerical measures of kinematic optimality of gaits. The possibility of symbolic descriptions of the task, vehicle-terrain interaction and control of the task must also be investigated. Such techniques are perhaps better suited to terrain description and adaptive gait control and might prove to more powerful.

4. DYNAMICS AND CONTROL

Introduction

In this section, we describe the dynamics of the vehicle-terrain system and the problems associated with the control of walking robots. With few

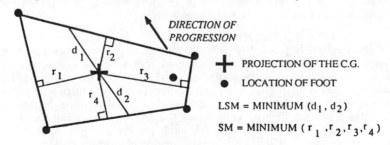

Figure 1   Measures of Stability for Statically Stable Gaits

exceptions (Glower and Özguner 1986; Shih and Frank 1987), quasi-static models of the system have been used for analysis (McGhee and Orin 1976, Kumar and Waldron 1988b). In other words, the legs have been assumed to massless compared to the vehicle body. This was a reasonable assumption for statically stable vehicles of the type being currently built. Further, the legs as well as the body were assumed to be rigid.

We first discuss the kinematics of walking vehicles. Next, a discussion of the equations of motion and the nature of static indeterminacy in the problem is presented. A critique of past and current research in the areas of trajectory planning and coordination follows. A parallel is drawn between walking vehicles and other robotic systems with parallelism in actuation, and the common features in the problem of coordination of such systems are described.

Kinematics of Legged Systems

Consider an earth-fixed reference frame E, and a body-fixed reference frame B with the origin at the center of gravity, O. Let the transformation from E to B be denoted by $^{B}\Gamma_{E}$ and its inverse by $^{E}\Gamma_{B}$. v_O is the velocity of the center of gravity of the vehicle and ω_B is the angular velocity of the vehicle. The position of a point P on a foot is denoted by r_P and the velocity v_P. Quantities measured relative to B possess a leading superscipt, B, and position or velocity vectors relative to O are denoted by a trailing subscript P/O.

The position of a point P on a foot, at the center of contact between the foot and the ground is given by

$$r_P \quad = \quad r_O + r_{P/O}$$
$$= \quad r_O + {}^{B}r_{P/O}$$

The velocity of P, v_P can be found by differentiating this equation:

$$v_P \quad = \quad v_O + v_{P/O}$$
$$= \quad v_O + {}^{B}v_{P/O} + \omega_B \times {}^{B}r_{P/O}$$

Therefore,

$$^{B}v_{P/O} \quad = \quad v_P - v_O - \omega_B \times {}^{B}r_{P/O}$$

If the leg were in transfer phase, its desired absolute velocity, v_P, would be computed independent of the kinematic constraints induced by the vehicle-terrain interaction, by a trajectory planning algorithm as detailed by Orin (1982). The above equation would then yield $^{B}v_{P/O}$. On the other hand, if P belongs to a leg in support phase, v_P must equal zero, since P must be at rest with respect to the ground. In such a situation:

$$^{B}v_{P/O} \quad = \quad -v_O - \omega_B \times {}^{B}r_{P/O}$$

Thus the desired leg velocity with respect to the vehicle body can be computed in either case. Once the desired foot velocity is known, the desired joint velocities can be computed quite simply by inverting the "leg Jacobian". This is analogous to the resolved motion rate control algorithms which involves the inversion of the "arm Jacobian" to determine the desired joint rates from the desired velocity of the hand. Decoupling the three translational degrees of freedom for each leg by using a pantograph mechanism results in efficient rate decomposition algorithms (Waldron *et al* 1984; Hirose 1984).

If the angular velocity of the vehicle is zero, the foot contacts of all the legs in support phase have the same velocity with respect to the body. This velocity is equal and opposite to the vehicle velocity. In this simple case, all the foot velocities are related to the gait parameters, β (duty factor), R (stroke), and T (cycle time period):

$$v_{P/O} = \frac{R}{\beta T}$$

The legs stroke in a direction parallel to the relative velocity vector which is, in general, arbitrarily directed . If ω_B is not zero, $v_{P/O}$ is constantly changing in direction and magnitude and is different for different legs. Notice also that even if the vehicle body linear and angular velocity are constant, the relative leg velocity varies with position. In fact, no leg velocity may equal $v_{P/O}$ which makes the above equation invalid. However, an instantaneous stroke may be defined based on the instantaneous relative velocity ($v_{P_i/O}$, for a point P_i on the ith foot) for each leg to assist gait planning (Kumar and Waldron 1989).

Dynamics of Legged Vehicles

Based on the assumptions outlined earlier, the equations of motion assume a very simple form. The desired accelerations and current velocities are known in a typical case. Let the resultant forces and moments of the foot forces (reaction forces exerted by the ground on the feet)be denoted by R and M respectively in the reference frame B. If ω and v are the angular velocity and the velocity of the center of gravity of the vehicle and H is the angular momentum of the body in the body fixed frame, and g is the acceleration due to gravity,

$$\mathbf{R} = \frac{d}{dt}(m\mathbf{v}) + \omega \times m\mathbf{v} - {}^B\Gamma_E \, mg$$

$$\mathbf{M} = \frac{d}{dt}(\mathbf{H}) + \omega \times \mathbf{H}$$

It is convenient to model the interaction of the feet with the ground by frictional point contact which was used by Salisbury and Roth (1983) for their analysis of multifingered grippers. According to the point contact model, the contact interaction can be represented by a pure force through an appropriate point, which may be called the center of contact. Note this center of contact may not be the geometric center of contact of the feet. However, this model does not take into account the frictional characteristics of the foot-terrain interface. A foot force is valid only if the force does not tend to pull the terrain (the component of force normal to the terrain can only be along one of two directions). In addition, the foot force must satisfy the appropriate laws of friction. If Coulomb's model of friction is accepted, the angle formed by the components of the foot force normal and tangential to the ground must be within a certain limit so that the foot does not slip (McGhee and Orin 1976; Orin and Oh 1981).

If $\mathbf{F}_i$ denotes the contact forces at the ith contact, whose center is at r_i (x_i, y_i, z_i) in the reference frame B, the following equations must be satisfied:

$$\sum_{i=1}^{i} \mathbf{F}_i = \mathbf{R} \qquad \sum_{i=1}^{n} (\mathbf{r}_i \times \mathbf{F}_i) = \mathbf{M}$$

where n is the number of feet on the ground. The above equations represent a system of linear equations which must be solved for the F_i. This is because, to produce a desired acceleration at the current velocity the foot forces given by the expressions above must be exerted on the body. The actual implementation can use a variety of schemes to control the support legs such as force control or active compliance control schemes (Klein *et al* 1983). Nevertheless, force set-points do need to be supplied to such control systems. Since the foot forces are not completely determined by the above equation, the problem is statically indeterminate. The determination of these foot forces subject to the frictional constraints is an important problem in coordination and is discussed next.

Coordination

Algorithms which decompose the external (including inertial) force system into foot forces are called coordination algorithms. The word coordination is used to mean the level of control at which the positions, rates, forces, or torques to be commanded from the actuator servos are computed, based on a relatively small number of command inputs from a supervisory control system. The corresponding analog in serial chain manipulators can be found in algorithms used to decompose desired end-effector rates into joint rates (Paul 1981).

The kinematics of a walking vehicle involves simple closed chains and multiple frictional contacts between the actively coordinated articulated legs and the passive terrain. Multifingered grippers and multiple cooperating arms belong to the same class of systems. Walking machines, along with multifingered grippers and multiarm systems can also be treated as parallel manipulation systems. Redundancy in manipulation systems with parallelism has been studied with reference to multiple arms (Zheng and Luh 1988), multifingered grippers (Abel *et al* 1985; Holzmann and McCarthy 1985; Hollerbach and Narasimhan 1986; Salisbury and Roth 1983; Kerr and Roth 1986; Kumar and Waldron 1988c) and walking vehicles (Orin and Oh 1981; Klein and Chung 1987; Gardner, Srinivasan and Waldron 1988; Kumar and Waldron 1988b). *Kinematic redundancy* in serial manipulators, in which the number of actuators is greater than the dimension of the task space, has been shown to be mathematically isomorphic to *static indeterminacy* (redundancy) in parallel systems such as walking robots (Waldron and Hunt 1987).

Optimization is a logical choice for analysis of under-determined systems and there are several mathematical techniques that can be employed for this purpose. McGhee and Orin (1976) combated the problem of static indeterminacy by optimizing the foot forces to obtain a minimum energy consumption. Linear programming was also used by Orin and Oh (1981). However, simulations showed that the associated computational load was unacceptable for real-time performance. In addition, the solution may exhibit unacceptable chatter behavior (Klein and Chung 1987). Methods based on the pseudo inverse have been described by Klein, Olson and Pugh (1983). Experiments on the OSU Hexapod demonstrated improved stability and better performance on even terrain. However, these methods fail to yield satisfactory solutions which satisfy the frictional constraints unless an iterative procedure is employed (Waldron 1986; Song and Waldron 1988). This problem becomes worse in uneven terrain. A comparison of different methods for walking vehicles is presented by Kumar and Waldron (1988b). Further, they also suggest using a compliant model for the legs instead of assuming that they are rigid. The displacement equations necessitated by geometric compatibility automatically resolve the indeterminacy in the problem. The basic idea here is that the leg compliances can be selected so that legs with poorer footholds are made softer, while legs with good footholds are rigid. Clearly the desired compliance can be electronically simulated using feedback. Though the resulting force distribution does not completely satisfy the frictional constraints, the legs in which these constraints are violated support a very small fraction of the load and therefore do not jeopardize the stability of the vehicle.

Another problem arises at the planning level of control. In order to predict whether or not a given (planned) configuration is stable, it is essential to determine if a valid set of foot (finger) forces (that does not violate the frictional constraints) can be commanded to maintain equilibrium. However, in this case, it is sufficient to be able to determine if *a* valid solution exists. In other words, it is not necessary to find an optimal solution to the problem. Kumar and Waldron (1988b) used the first phase of the Simplex Method for linear programming to this end. Recall that Phase I of the method merely seeks a feasible solution while Phase II optimizes the feasible solution (Gass 1985). The number of computations is considerably reduced. Further, the planning algorithms are not required to be implemented in real-time.Thus, this approach would be adequate in terms of computing time as well as efficacy.

For real-time control, an efficient technique for obtaining a solution to the equations of motion while satisfying the friction angle constraints still remains to be found. Perhaps, the answer lies in a parallel computing scheme for linear programming. Further, it is necessary to perform a dynamic analysis for faster vehicles. This is especially true for heavy vehicles - even if the accelerations are small, the inertial forces would be large. The idea of varying leg compliances could still be implemented in the dynamic case. However, now the leg impedance or admittance must be specified and controlled.

5. MOTION PLANNING AND CONTROL

Machine Intelligence in Mobile Robots

In most of the mobile robotic systems developed so far, a common feature has been the recognition for the need of a hierarchical structure of intelligence and hierarchical decomposition of the task (of locomotion) and the model of the perceived world. The higher levels define tasks or subgoals for the lower levels and monitor their status. As we go up the hierarchy, on a time scale there is a decrease in the frequency of updating sensory information, and on the length scale, the world taken into consideration is larger but with fewer details. At each successive level down the hierarchy, there is a decrease in the generality and scope of the search and greater resolution. Four levels may be identified in the hierarchy - the *planner* or the route layout module, the *navigator* or the path selection module, the *pilot* or the guidance module and the *controller* (Isik and Meystel 1988). The *cartographer* is concerned with maintaining maps for the path selection and guidance modules at the required resolution and range and with appropriate detail.

Planner

The route layout is done at the highest level. It involves planning a route for the vehicle taking into account the general characteristics of the robot's ability to adapt to different terrains and surmount various obstacles. At this level, the basic element of locomotion (leg or wheel) is not of concern. In a completely autonomous system, it may be the only level which interacts with a human operator. This interaction may be limited to off-line specification of the task. The planner prescribes a set of subgoals for the next lower level. It works on a model of the world which extends typically to a hundred body lengths. Presently, there is little or no research in evidence in this regime of machine intelligence.

Navigator

The navigator is primarily concerned with path selection. It uses the terrain preview data and information from other sensors to chart a "best" course for several "vehicle body lengths" to realize the subgoal command from the upper level. Obstacle avoidance is an integral part of such a process. It in turn prescribes subgoals for the pilot level to meet the requirements of the selected path. It maintains a terrain map and a model of the world confined to a few (typically ten) body lengths which possesses a higher level of resolution than the planner's model.

Most of the work on autonomous navigation has been at the path selection level. It is specific to wheeled or walking vehicles only to the extent that the characteristics of the vehicle have to be known (dimensions of the body, size of the tire or foot, maximum stride length and so on). However most of the work reported in this field has been with reference to wheeled robots. This problem is similar to the problem of planning manipulator transfer movements without explicit programming of the motion (Lozano-Peréz 1983). Given the initial and final location (starting point and subgoal) of the mobile robot, the optimal path that circumvents obstacles has to be found. The concept of shrinking the robot to a single reference point while expanding the obstacle regions (Udupa 1977) has proved to be useful. Lozano-Peréz (1981) has formalized this approach with the concept of a configuration space. This involves approximating the obstacles by polyhedra. In two dimensions the shortest collision free path is composed of straight lines joining the origin to the destination through a set of vertices of obstacle polygons. A *V-graph* or visibility graph in which each link represents a straight line between two points which can "see" each other, can be built and a search routine is used to obtain the optimal path (Davis and Camacho 1984; Nillson 1980). A similar method was employed to navigate SHAKEY, an integrated mobile wheeled system built at the Stanford Research Institute. In 3-dimensions however, this method gets a little more complicated. Lozano-Peréz (1983) uses *slices* (a projection of any space into a lower dimension space) to overcome this problem.

Thompson (1977) describes a path planning module for the JPL rover. This approach avoids the construction of a V-graph for the whole space but builds (expands) the graph as and when needed. Similarly, Koch *et al* (1985) propose the concept of *sectors* (sectors of a circle containing the current position of the vehicle and the goal with the vehicle at the center) which yields a subset of the visibility graph. Thus they overcome the problem of combinatorial explosion.

Methods of representing the robot world have received considerable attention, since the actual search for a suitable path can be easily performed by standard algorithms like the A^* algorithm (Nilsson 1980). Brooks (1983a; 1983b) modeled the free space (space outside the obstacles) as a union of generalized cones. This eventually leads to a more efficient utilization of space. He used generalized cones to build a connectivity graph which was then the input to the path search routine. HILARE was a three wheeled robot built in France (Giralt *et al* 1979) whose model of the world was also represented by a connectivity graph. Typically, the nodes of a connectivity graph would be places or rooms and links, traversible boundaries. The rooms could be further decomposed into polygonal cells representing free space.

In legged locomotion, modeling of the environment typically involves the interpretation of terrain elevation data (Klein *et al* 1987). Two general approaches can be identified. The first method involves object identification or feature extraction (Poulos 1986). The environment is partitioned into regions corresponding to physical features. This is followed by a symbolic description of the regions and matching the descriptions with a preprogrammed set of object descriptions. This is a complex problem which has been studied with reference to structured environments with some success. However, the complexity greatly increases in an unstructured terrain and the required computations are prohibitively time consuming.

The lesser used region classification technique is based on inferring local topographic features for each *voxel* (VOLume piXEL) from the discrete elevation data, which are then used to build a symbolic model of the terrain. This alleviates problems of symbolic description and eliminates problems of matching descriptions to known patterns. Olivier and Özguner (1986) propose a similar method in which they characterize voxels by fitting local quadric surfaces through each voxel employing the least squares technique with the elevation data. The Hessian matrix and its eigenvalues are then used to classify each voxel as a peak, pit, ridge, ravine, flat area, safe hill, or an unsafe hill. Poulos (1986) pursued this method and demonstrated its feasiblity for real-time operations on a Symbolics® machine.

Richbourg *et al* (1987) and Ross and McGhee (1987) used a cost function for each voxel in conjunction with a *wavefront* method. The basic idea here is if the space is divided into regions each of which can be associated with a cost of traversal, the locally optimal path can be found by a combination of refraction and defraction optics (Snell's law). This analogy to optics is immediately seen if the cost rate is equated to the refractive index in optics.

Vegetative cover and varying soil properties still pose problems as elevation data only yields geometric information. Further, though a good optical terrain scanner can provide a geometric description of the immediate terrain, better sensors are required for global path selection. Another problem arises due to self-occlusion, in which the terrain is obscured from the sensors by local peaks and high-rises leading to uncertainty. Clearly, more work on reasoning and planning in such environments is needed.

The path selection algorithms must also be used at a lower level of guidance to enable fine tuning of the planned path. This is because the lower level would typically have a more accurate and detailed description of the terrain, and obstacles which were not evident at higher levels would become apparent at the lower levels. However, the same basic algorithms that are used for path selection can be used for such a local path modification strategy as long as higher resolution maps and models are available .

Pilot

The pilot plans a sequence of elementary acts of motion in space and time to generate a path between the goals prescribed by the path selection level. Minor deviations from the prescribed path may be tolerated to circumvent small obstacles. The pilot has a short term memory and its perception of the world is confined to one or two body lengths. This level may be virtually absent in wheeled locomotion where the motion is almost fully constrained or specified by the path selection level. However, a legged locomotion system is free to select footholds in space and time. That is, when and where it "samples" the terrain. Thus optimization of gait parameters, optimal foothold selection, dynamic balancing and attitude control of the vehicle body are tasks which must be performed at this level. The choice of legs as elements of locomotion introduces a new level of intelligence and a new degree of complexity into motion planning for the body. This level has also been described as the guidance module by some researchers. The pilot treats the leg as a finite state machine (McGhee 1968) and delegates the lower level task of planning trajectories for the legs and the associated problem of leg collision avoidance to the controller.

Thomson (1977) has identified a guidance module in wheeled vehicles which translates the planned path into a set of commands for the actuators and uses feedback to control the vehicle limiting the deviation from the prescribed path to a few evasive movements. However, this description would seem to encompass the function of the next lowest level (controller) too. In general, the pilot's function could be stated as being limited to walking vehicles.

The operation of the pilot has been more actively pursued for legged locomotion. In the four-level hierarchy of the University of Paris hexapod (Kessis *et al* 1985), the gait level and the plan interpreter would appear to constitute the guidance module. Hirose (1984) describes a gait control level partitioned into a local motion-trace generator (which responds to the commands of the navigation planning module) and an adaptive gait controller. For the ASV, the free gait has been used as an automatic heuristic technique to optimize the vehicle walk over unstructured terrain (McGhee and Iswandhi 1979) and the guidance algorithm is based on this free gait (Patterson, Reidy and Brownstein 1983). This enables automatic selection of footholds and gait parameters but the process is quite inefficient and slow. Current efforts are directed towards *supervisory* control of the ASV which is aimed at implementing the "horse-intelligence" in the "rider-horse-system" analogy. This will at least enable supervisory control in rough terrain with man providing the "rider-intelligence".

Kumar and Waldron (1989) classify the functions at this level into 6 categories. These are local path modification or local navigation, foothold selection, gait optimization, adaptive gait control, automatic body posture regulation and checking vehicle stability. They use a finite state approach to planning and use a numerical representation of the terrain vehicle system. In this study, algorithms based on heuristics are developed for foothold selection, posture regulation and gait control. The efficacy of these algorithms has been demonstrated through simulations (Kumar 1987).

Controller

The lowest level, the controller, is the only level which interacts directly with the actuators. It represents the "spinal" level associated with control of individual joints in natural systems and involves real-time servo control loops and sensory feedback at the actuator level. This is the only area in this hierarchy which is reasonably well researched and documented and is the foundation for all robotic technology. In legged locomotion systems, this also entails leg trajectory planning using proximity sensors, and actuation, and also incorporates "cerebellar" intelligence to a certain extent. The corresponding level in wheeled systems controls the actuation of the wheels. This level involves very little intelligence and it is also possible to identify the controller as a "plan execution" module and exclude it from this model of the intelligence structure. Since the technology involved is not unique to walking vehicles it has not been discussed in detail here. However, information about specific systems can be found elsewhere (Waldron *et al* 1984; Sutherland and Ullner 1984; Orin 1982; Hirose and Umetani 1980).

6. CONCLUDING REMARKS

In the foregoing, we have reviewed the principal problem areas associated with legged locomotion systems with particular reference to statically stable walking machines, and have cited many of the principal sources of information in the literature. The emphasis has, of course, been on scientific information. For this reason, references to some machines which have not been scientifically described have not been included.

Some of the problem areas cited above are unique to legged locomotion systems. More usually, problem areas have features in common with other types of robotic systems, although there may be differences in the emphasis. For this reason, research in legged locomotion has drawn heavily on robotics research in general. There is, correspondingly, great potential for transfer of results and techniques generated for legged systems, to other types of robotic systems.

REFERENCES

Abel, J.M., Holzmann, W., and McCarthy, J.M. 1985. On grasping objects with two articulated fingers. *IEEE J. Robotics and Automation*. Vol. RA-1. No. 4 : 211-214.

Agarwal, S.K. and Waldron, K.J. 1988 (Dec.). Impulse model of an actively controlled running machine. *ASME Winter Annual Meeting*. Chicago (to be presented).

Alexander, R.M. 1984. The gaits of bipedal and quadrupedal animals. *Int. J. Robotics Research*, Vol. 3, No. 2, 1984, pp. 49-59.

Bessonov, A.P. and Umnov, N.V. 1973. The analysis of gaits in six-legged vehicles according to their static stability. *First Symp. Theory and Practice of Robots and Manipulators*. Amsterdam: Elsevier Scientific Publishing Company.

Brooks, R.A. 1983a. Solving the find-path problem by good representation of free space. *IEEE Trans. Systems, Man, Cybernetics*. SMC-13 (3) pp. 191-197.

Brooks, R.A. 1983b. Planning collision-free motions for pick-and-place operations. *Int. J. Robotics Research*. Vol. 2. No. 4 : 19-44.

Cruse, H. 1976. The control of body position in the stick insect (*carausius morosus*), when walking over uneven terrain. *Biological Cybernetics*. No. 24: 25-33.

Davis, R.H. and Camacho, M. 1984. The application of logic programming to the generation of paths for robots. *Robotica*. Vol. 2: 93-103.

Delcomyn, F. 1981. Insect locomotion on land. *Locomotion and Energetics in Arthropods*. Eds. C.F. Herried and C.R. Fourtner. Plenum Press. New York.

Gambaryan, P.P. 1974. *How Mammals Run*. John Wiley and Sons. New York.

Gardner, J.F., Srinivasan, K., and Waldron, K.J. 1988. A new method for controlling force distribution in redundantly actuated closed kinematic chains. *1988 ASME Winter Annual Meeting*. Chicago.

Gass, S.I. 1985. *Linear Programming*. McGraw-Hill. New York.

Giralt, G., Sobek, R., and Chatila, R. 1979. A multi-level planning and navigation system for a mobile robot. *Proc. Sixth IJCAI*. Vol. 1. Tokyo. Japan.

Glower, J.S., and Özguner, G. 1986. Control of a quadruped trot. *IEEE Conference on Robotics and Automation*. San Fransisco: 1496-1501.

Goldenberg, K.Y. and Raibert, M.H. 1987. Conditions for symmetric running in single and double support. *IEEE Conference on Robotics and Automation*. Raleigh. North Carolina: 1890-1895.

Gray, J.1968. *Animal Locomotion*. London:WiedenField andNicolson.

Hildebrand, M. 1960. How animals run. *Scientific American..* 148-157.

Hildebrand, M. 1965. Symmetrical gaits of horses. *Science.* Vol. 150.

Hildebrand, M. Analysis of tetrapod gaits. 1976. *Neural Control of Locomotion.* Eds. R.M. Herman, *et al.* Plenum Press. New York.

Hirose, S. and Umetani, Y. Sept 1980. The basic motion regulation system for a quadruped walking machine. *ASME Paper 80-DET-34.*

Hirose, S. 1984. A study of design and control of a quadrupedal walking vehicle. *Int. J. Robotics Research.* Vol. 3. No. 2: 113-133.

Hirose, S., Masui, T., Hidekazu, K., Fukuda, Y. and Umetani, Y. 1985. TITAN III: a quadruped walking vehicle. *Robotics Research* 2. ed. H. Hanafusa and H. Inoue. MIT Press: 325-332.

Hollerbach, J. and Suh, K.C. 1985. Redundancy resolution of manipulators through torque optimization. *IEEE Conf. on Robotics and Automation.* St. Louis, Missouri: 1016-1021.

Hollerbach, J. and Narasimhan, S. 1986. Finger force computation without the grip Jacobian. *IEEE Conf. on Robotics and Automation.* San Fransisco: 871-875.

Holzmann, W., and McCarthy, J.M. 1985. Computing the friction forces associated with a three fingered grasp. *IEEE J. Robotics and Automation.* Vol. RA-1. No. 4.

Huang, M. and Waldron, K.J. 1987. Relationship between payload and speed in legged locomotion. *IEEE Conf. on Robotics and Automation.* Raleigh. North Carolina: 533-536.

Isik, C. and Meystel, A.M. 1988. Pilot level of a hierarchical controller for an unmanned Robot. *IEEE J. Robotics and Automation.* 4(3): 241-255.

Kato, T., Takanishi, A., Jishikawa, H., and Kato, I. 1983. The realization of quasi-dynamic walking by the biped walking machine. *Fourth Symp. on Theory and Practice of Robots and Manipulators.* A. Morecki, G. Bianchi, and K. Kedzior. eds. Warsaw: Polish Scientific Publishers: 341-351.

Kerr, J. and Roth, B. 1986. Analysis of multifingered hands. *Int. J. Robotics Research.* Vol. 4. No. 4: 3-17.

Kessis, J.J., Rambaut, J.P., Penné, J., Wood, R. and Mattar, N. 1985. Hexapod walking robots with artificial intelligence capabilities. *Theory and Practice of Robots and Manipulators.* A. Morecki, G. Bianchi, and K. Kedzior. eds. Kogan Page, London: 395-402.

Klein, C.A., Olson, K.W., and Pugh, D.R. 1983. Use of force and attitude sensors for locomotion of a legged vehicle over irregular terrain. *Int. J. Robotics Research*. Vol. 2. No. 2: 3-17.

Klein, C.A. and Chung, T.S. 1987. Force interaction and allocation for the legs of a walking vehicle. *IEEE J. Robotics and Automation*. RA-3. No. 6.

Klein, C.A., Kau, C.C., Ribble, E.A., and Patterson, M.R. 1987. Vision processing and foothold selection for the ASV walking machine. *SPIE Conf. - Advances in Intelligent Robotics Systems*. Cambrige. MA.

Koch, E., Yeh, C., Hillel, G., Meystel, A. and Isik, C. 1985. Simulation of path planning for a system with vision and map updating. *IEEE Conf. on Robotics and Automation*. St. Louis, Missouri: 146-160.

Kugushev, E.I., and Jaroshevskij, V.S. 1975. Problems of selecting a gait for an integrated locomotion robot. *Proc. Fourth IJCAI*, Tilisi. Georgian SSR. USSR: 789-793.

Kumar, V. 1987. Motion planning for legged locomotion systems on uneven terrain. *Ph.d. Dissertation*. The Ohio State University, Columbus, Ohio.

Kumar, V., and Waldron, K.J. 1988a. Analysis of omnidirectional gaits for walking vehicles on uneven terrain. *Seventh Symp. on Theory and Practice of Robots and Manipulators*. A. Morecki, G. Bianchi, and K. Kedzior. eds.: 37-62.

Kumar, V., and Waldron, K.J. 1988b. Force distribution in walking vehicles.*Trends and Developments in Mechanisms, Machines and Robotics*. ASME. DE-15. Vol. 3. A. Midha. ed.: 473-480.

Kumar, V. and Waldron, K.J. 1988c. Force distribution in closed kinematic chains. *IEEE J. Robotics and Automation*. in press.

Kumar, V., and Waldron, K.J., 1989. Adaptive Gait Control for a Walking Robot. *J. Robotic Systems* (in press).

Kwak, S. A Computer Simulation Study of a free gait motion coordination algorithm for rough-terrain locomotion by a hexapod walking machine. *Ph.d. Dissertation*. The Ohio State University, Columbus.

Lee, W. J. and Orin, D.E. 1988. The kinematics of motion planning for multilegged vehicles over uneven terrain. *IEEE J. Robotics and Automation*. 4(2):204-212.

Lozano-Peréz, T. 1983. Spatial Planning: A configuration space approach. *IEEE Trans. on Computers*. Vol. C-32. No. 2.

Lozano-Peréz, T. 1981. Automatic planning of manipulator transfer movements. *IEEE Trans. on Systems, Man, Cybernetics*. SMC-11 (10): 681-689.

McGhee, R.B. 1968. Finite state control of quadruped locomotion. *Mathematical Biosciences*. No. 2: 57-66.

McGhee, R.B. and Frank, A.A. 1968. On the stability properties of quadruped creeping gaits. *Mathematical Biosciences*. No. 3: 331-351.

McGhee, R.B., and Jain, A.K. 1972. Some properties of regularly realizable gait matrices. *Mathematical BioSciences*. No. 13: 179-193.

McGhee, R.B. 1977. Control of legged locomotion systems. *Proc.Joint Automatic Control Conference*. San Fransisco: 205-215.

McGhee, R.B., and Iswandhi, G. 1979. Adaptive locomotion of a multilegged robot over rough terrain.*IEEE Transactions on Systems, Man and Cybernetics*. SMC-9(4): 176-182.

McGhee, R.B. 1984. Vehicular legged locomotion. *Advances in Automation and Robotics*. ed. G.N. Saridis. Greenwich. Connecticut: Jai Press.

Mosher, R.S. Exploring the potential of a quadruped. 1969. *Int. Automotive Engineering Congress*. Paper no. 690191. NewYork: SAE.

Muira, H., and Shimoyama, I. 1984. Dynamic walk of a biped. *Int. J. Robotics Research*. Vol. 3. No. 2: 60-74.

Muybridge, E.1955. *The Human Figure in Motion*, Dover, New York, 1955.

Muybridge, E. 1957. *Animals in Motion*, Dover, New York, 1957.

Nilsson, N. 1980. *Principles of Artificial Intelligence*. Tioga Publishing. California.

Okhotsimski, D.E., Gurfinkel, V.S., Devyanin, E.A., and Platonov, A.K. 1977. Integrated walking robot development. *Machine Intelligence*. Vol. 9. Eds. J.E. Hayes, D. Michie and L.J. Mikulich.

Olivier, J.F. and Özguner, F. 1986. A navigation algorithm for an intelligent vehicle with a Laser Range Finder. *IEEE Conf. on Robotics and Automation.*. San-Fransisco.

Orin, D.E. 1982. Supervisory control of a multilegged robot. *Int. J. Robotics Research*. Vol. 1. No. 1: 79-81.

Orin, D.E. and Oh, S.Y. 1981 (June). Control of force distribution in robotic mechanisms containing closed kinematic chains," *J. Dynamic Systems, Measurements, and Control*, Vol. 102: 134-141.

Özguner, F., Tsai, S.J., and McGhee, R.B. 1984. An approach to the use of terrain-preview information in rough-Terrain locomotion by a hexapod walking machine. *Int. J. Robotics Research*. 3(2): 134-146.

Pandy, M.G., Kumar, V., Berme, N., and Waldron, K.J. 1988. The dynamics of quadrupedal locomotion. *J. Biomechanical Engineering*. 110(3): 230-237.

Patterson, M.R., Reidy, J.J., and Brownstein, B.B. 1983. Guidance and actuation techniques for an adaptively controlled vehicle. *Final Report, Contract MDA 903-82-c-0149*. Battelle Columbus Division, Ohio.

Paul, R.P. 1981. *Robot Manipulators, Mathematics, Programming and Control*. The MIT Press, Cambridge.

Pearson, K.G., and Franklin, R. 1984. Characteristics of leg movements and patterns of coordination in locusts walking on rough terrain. *Int. J. Robotics Research*. Vol. 3.No. 2: 101-112.

Poulos, D.D. 1986. Range image processing for local navigation of an autonomous land vehicle. *M.Sc. Thesis*. Naval Postgraduate School. Monterey, California.

Qui, X. and Song, S.M. 1988. A strategy of wave gait for a walking machine traversing a rough planar terrain. *Trends and Developments in Mechanisms, Machines and Robotics*. ASME. DE-15. Vol. 3. A. Midha. ed.: 487-496.

Raibert, M.H., and Sutherland, I.E. 1983. Machines that walk. *Scientific American*. 248 (2): 44-53.

Raibert, M.H.1985. *Legged Robots that Balance*. MIT Press. Cambridge. Massachusetts.

Raibert, M.H., Chepponis, M., Brown, H.B. Jr. 1986. Running on four legs as though they were one. *IEEE J. Robotics and Automation*. RA-2 (2): 70-82.

Raibert, M.H. 1986. Running with symmetry. *Int. J. Robotics Research*. 5(4): 3-19.

Richbourg, R.F., Rowe, N.C., Zyda, M.J. and McGhee, R.B. 1987. Solving global two-dimensional routing problems using Snell's law and A* search. *IEEE Conf. on Robotics and Automation*. Raleigh. North Carolina: 1631-1636.

Ross, R.S., Rowe, N.C., and McGhee, R.B. 1987. Dynamic multivariate terrain cost maps for automatic route planning. *AAAI Workshop on Planning Systems for Autonomous Mobile Robots*. Seattle. Washington.

Russell, M. 1983. Odex 1: The first Functionoid. *Robotics Age*. 5(5): 12-18.

Salisbury, J.K., and Roth, B. 1983. Kinematic and force analysis of articulated mechanical hands. *J. Mechanisms, Transmissions, and Automation in Design*. Vol. 105: 35-41.

Shih, L. and Frank, A.A. 1987. A study of gait and flywheel torque Effect on legged machines using a dynamic compliant joint model. *IEEE Conference on Robotics and Automation*, Raleigh, N.Carolina: 527-532.

Sitek. G. 1976. Big Muskie. *Heavy Duty Equipment Maintenance*. Vol. 4:16-23.

Song, S.M, and Waldron, K.J. 1987. Geometric design of a walking machine for optimal mobility. *J. Mechanisms, Transmissions, Automation in Design*. Vol. 109. No. 1: 21-28.

Song, S.M, and Waldron, K.J. 1987. An analytical approach for gait study and its application on wave gaits. *Int. J. Robotics Research*. Vol. 6. No. 2.

Song, S.M, and Waldron, K.J. 1988. *Machines That Walk: The Adaptive Suspension Vehicle*. MIT Press. Cambridge. Massachusetts.

Sutherland, I. and Ullner, M.K. 1984. Footprints in the asphalt. *Int. J. Robotics Research*. Vol. 3. No. 2: 29-36.

Tomovic, R. 1961. A general theoretical model of creeping displacement. *Cybernetica IV* : 98-107 (English Translation).

Thompson, A.M. 1977. The navigation system of the JPL robot. *Proc. Fifth IJCAI*. Cambridge. MA: 749-757.

Udupa, S.M. 1977. Collision detection and avoidance in computer Controlled Manipulators. *Proc. Fifth IJCAI*. M.I.T.: 737-748.

Vukobratovic, M., Frank, A.A., and Juricic, D. 1970. On the stability of biped locomotion. *IEEE Trans. on Biomedical Engineering*. Vol. 17. No. 1: 25-36.

Waldron, K.J., Song, S., Wang, S., and Vohnout, V.J. 1984. Mechanical and geometric design of the Adaptive Suspension Vehicle. *Theory and Practice of Robots and Manipulators*. Proc. Romansy '84: Fifth CISM-IFToMM Symp., Ed. A. Morecki, G. Bianchi, and K.Kedzior: 295-306.

Waldron, K.J., Vohnout, V.J., Pery, A., and McGhee, R.B. 1984. Configuration design of the Adaptive Suspension Vehicle. *Int. J. Robotics Research.* Vol. 3. No. 2: 37-48.

Waldron, K.J. and McGhee, R.B. 1985. The Adaptive Suspension Vehicle Project. *Unmanned Systems.* Summer. 1985.

Waldron, K.J. 1986. Force and motion management in legged locomotion. *IEEE J. on Robotics and Automation.* Vol. RA-2. No. 4.

Waldron, K.J. and McGhee, R.B. 1986. The Adaptive Suspension Vehicle. *IEEE Control Systems Magazine.* 6(6): 7-12.

Waldron, K.J. and Hunt, K.H. 1988. Series-parallel dualities in actively coordinated mechanisms. *Robotics Research 4.* R.Bolles and B. Roth. eds.: 175-182.

Wilson, D.M. 1966. Insect walking. *Annual Review Entomology.* Vol. 11.

Wilson, D.M. 1976. Stepping patterns in tarantula spiders.*J. Experimental Biology,* No. 47: 133-151.

Zheng, Y.F. and Luh, J.Y.S. 1988. Optimal load distribution for two industrial robots handling a single object. *IEEE Conf. on Robotics and Automation.* Philadelphia. PA: 344-349.

Zuk, D.M., and Dell'Eva, M.L. 1983. 3-D vision system for the Adaptive Suspension Vehicle. *Final Report.* Ann Arbor, Michigan: Environment Research Institute of Michigan.

About the authors

S. Sitharama Iyengar — chairman of the Computer Science Department and professor of computer science at Louisiana State University — has directed LSU's Robotics Research Laboratory since its inception in 1986. He has been actively involved with research in high-performance algorithms and data structures since receiving his PhD in 1974, and has directed over 10 PhD dissertations at LSU. He has served as principal investigator on research projects supported by the Office of Naval Research, the National Aeronautics and Space Administration, the National Science Foundation/Laser Program, the California Institute of Technology's Jet Propulsion Laboratory, the Department of Navy-NORDA, the Department of Energy (through Oak Ridge National Laboratory, Tennessee), the LEQFS-Board of Regents, and Apple Computers.

In addition to this two-volume tutorial, he has edited two other books and over 150 publications — including 85 archival journal papers in areas of high-performance parallel and distributed algorithms and data structure for image processing and pattern recognition, autonomous navigation, and distributed sensor networks. Iyengar was a visiting professor (fellow) at JPL, the Oak Ridge National Laboratory, and the Indian Institute of Science. He is also an Association for Computing Machinery national lecturer, a series editor for *Neuro Computing of Complex Systems*, and area editor for the *Journal of Computer Science and Information*. He has served as guest editor for the *IEEE Transactions on Software Engineering* (1988); *Computer* magazine (1989); the *IEEE Transactions On System, Man, and Cybernetics;* the *IEEE Transactions on Knowledge and Data Engineering;* and the *Journal of Computers and Electrical Engineering*. He was awarded the Phi Delta Kappa Research Award of Distinction at LSU in 1989, won the Best Teacher Award in 1978, and received the Williams Evans Fellowship from the University of Otago, New Zealand, in 1991.

Alberto Elfes — a robotics scientist at the Department of Computer Sciences, IBM T.J. Watson Research Center, Yorktown Heights, New York — is one of the principal investigators at that institution's Autonomous Robotics Laboratory. He obtained his E.Eng. degree in electronics engineering and the M.Sc. in computer science from the Instituto Tecnólogico de Aeronáutica, Brazil. From 1976 to 1982 he was on the faculty of ITA's Computer Science Department, where he performed research and directed dissertations in computer graphics, image processing for remote sensing applications, pattern recognition and computer-based medical diagnosis, and where he was awarded CNPq (National Council for Scientific and Technological Development) and FAPESP (Foundation for Research Support of the State of São Paulo) research awards and fellowships. He obtained his PhD in electrical and computer engineering at Carnegie Mellon University in 1989. His doctoral research was performed at the Mobile Robot Lab, CMU, where he concentrated on distributed architectures for mobile robot control, and stochastic approaches to robotic perception and navigation. From 1987 to 1989 he was a researcher with the Engineering Design Research Center and the Robotics Institute, CMU, where he worked on integrated environments for computer-aided design and on large-scale architectures for autonomous mobile robots. Since 1989 he has been with IBM's Autonomous Robotics Laboratory, where he is conducting research on the dynamic planning and control of robotic perception, and on real-time architectures for robotic control. He has published over 35 research papers in refereed journals, conference proceedings, and books in the areas of computer-based medical diagnosis, pattern recognition, computer-aided design, autonomous mobile robots, and computer vision.

IEEE Computer Society

IEEE Computer Society Press Publications

Monographs: A monograph is an authored book consisting of 100-percent original material.

Tutorials: A tutorial is a collection of original materials prepared by the editors, and reprints of the best articles published in a subject area. Tutorials must contain at least five percent of original material (although we recommend 15 to 20 percent of original material).

Reprint collections: A reprint collection contains reprints (divided into sections) with a preface, table of contents, and section introductions discussing the reprints and why they were selected. Collections contain less than five percent of original material.

Technology series: Each technology series is a brief reprint collection — approximately 126-136 pages and containing 12 to 13 papers, each paper focusing on a subset of a specific discipline, such as networks, architecture, software, or robotics.

Submission of proposals: For guidelines on preparing CS Press books, write the Editorial Director, IEEE Computer Society Press, PO Box 3014, 10662 Los Vaqueros Circle, Los Alamitos, CA 90720-1264, or telephone (714) 821-8380.

Purpose

The IEEE Computer Society advances the theory and practice of computer science and engineering, promotes the exchange of technical information among 100,000 members worldwide, and provides a wide range of services to members and nonmembers.

Membership

All members receive the acclaimed monthly magazine *Computer*, discounts, and opportunities to serve (all activities are led by volunteer members). Membership is open to all IEEE members, affiliate society members, and others seriously interested in the computer field.

Publications and Activities

Computer magazine: An authoritative, easy-to-read magazine containing tutorials and in-depth articles on topics across the computer field, plus news, conference reports, book reviews, calendars, calls for papers, interviews, and new products.

Periodicals: The society publishes six magazines and five research transactions. For more details, refer to our membership application or request information as noted above.

Conference proceedings, tutorial texts, and standards documents: The IEEE Computer Society Press publishes more than 100 titles every year.

Standards working groups: Over 100 of these groups produce IEEE standards used throughout the industrial world.

Technical committees: Over 30 TCs publish newsletters, provide interaction with peers in specialty areas, and directly influence standards, conferences, and education.

Conferences/Education: The society holds about 100 conferences each year and sponsors many educational activities, including computing science accreditation.

Chapters: Regular and student chapters worldwide provide the opportunity to interact with colleagues, hear technical experts, and serve the local professional community.